Lead and Lead Alloys

Properties and Technology

Wilhelm Hofmann †

English Translation of the
Second Revised German Edition
by Lead Development Association, London

Springer-Verlag Berlin Heidelberg GmbH 1970

Dr.-Ing- habil. WILHELM HOFMANN †
Professor an der Technischen Hochschule
Braunschweig

Title of the Original Edition
Blei und Bleilegierungen, 2. Auflage
(Reine und angewandte Metallkunde in Einzeldarstellungen Bd. 6)

The editorship of the book after the death
of Professor HOFMANN was taken over by

Dr.-Ing. GERWIG VIBRANS
Abteilungsvorsteher und Professor
im Institut für Werkstoffkunde und Herstellungsverfahren
der Technischen Universität Braunschweig

ISBN 978-3-662-27030-1 ISBN 978-3-662-28508-4 (eBook)
DOI 10.1007/978-3-662-28508-4

Originally published by Springer-Verlag Berlin Heidelberg New York in 1970.
Softcover reprint of the hardcover 1st edition 1970

Library of Congress Catalog Card Number 69-15931

Title-No. 1550

Lead and Lead Alloys

Properties and Technology

Wilhelm Hofmann †

English Translation of the
Second Revised German Edition
by Lead Development Association, London

Springer-Verlag Berlin Heidelberg GmbH 1970

Dr.-Ing- habil. WILHELM HOFMANN †
Professor an der Technischen Hochschule
Braunschweig

Title of the Original Edition
Blei und Bleilegierungen, 2. Auflage
(Reine und angewandte Metallkunde in Einzeldarstellungen Bd. 6)

The editorship of the book after the death
of Professor HOFMANN was taken over by

Dr.-Ing. GERWIG VIBRANS
Abteilungsvorsteher und Professor
im Institut für Werkstoffkunde und Herstellungsverfahren
der Technischen Universität Braunschweig

ISBN 978-3-662-27030-1 ISBN 978-3-662-28508-4 (eBook)
DOI 10.1007/978-3-662-28508-4

Originally published by Springer-Verlag Berlin Heidelberg New York in 1970.
Softcover reprint of the hardcover 1st edition 1970

Library of Congress Catalog Card Number 69-15931

Title-No. 1550

Foreword

The first edition of this monograph (1941) arose from the research work which I conducted in the Bleiforschungsstelle (Lead Research Centre), Berlin, together with Professor Hanemann and Dr. Angelica Schrader.

After World War II literature on lead had expanded considerably and preparatory work of almost ten years' duration was necessary before the 2nd edition of the book "Blei und Bleilegierungen" could be published in 1962. I welcomed proposals made by the Lead Development Association and the Springer-Verlag to have an English translation published and at the same time to bring it up to date. While the English edition is in preparation, international research on lead is being actively pursued. This is due to the initiative of the International Lead Zinc Research Organization under Dr. S. F. Radtke, to the collaboration of American, European and Australian research centres and to the International Lead Conferences.

A book on lead and lead alloys would have been impossible without the cooperation of the lead industry and I should like to thank sincerely all those firms who have freely supplied data and information for it as well as the many co-workers in the Institut für Werkstoffkunde of the Technical University of Brunswick.

Special thanks are due to the Lead Development Association of London, the Lead Industries Association Inc. and the International Lead Zinc Research Organization of New York which drew up the original plans for an English edition and provided the necessary financial support for the translation. I should also like to acknowledge the work of Mr. B. Fullman who has translated the volume with expert thoroughness and that of Messrs. Hughes and Wesson of the LDA who have been responsible for the organisational planning and detailed supervision of all aspects of the translation.

With the translation of this volume I hope to have brought the report on my systematic work on lead to a certain conclusion. I would be grateful if the book could be read with understanding, for our knowledge of lead and its alloys has grown so extensively that in a work of moderate proportions as this, many details must necessarily be excluded; the decision on subject matter was not always easy and many items could be improved if more time had been available. If the book on lead were to be rewritten thermochemists, electrochemists and theoreticians on elasticity would be required, besides technologists, to support the work of the metallurgist.

In recent years the pace of research and development has increased enormously, resulting in the evolution of completely new technologies. In spite of the demand thereby created for new and exotic materials with special properties, industry will continue to depend in large measure on classical materials and conventional processes.

The older metals, including lead, will therefore always maintain their importance alongside the new materials, even in a world of rapidly changing technologies.

Brunswick (Braunschweig),
July 1965

Wilhelm Hofmann

Editor's Foreword

Prof. Dr.-Ing. habil. WILHELM HOFMANN, who died suddenly on 16th November, 1965, had already completed the manuscript of the English edition before his death, together with the main Foreword reproduced here. I have made every attempt to ensure that publication should be carried out as Prof. HOFMANN would himself have wished. This has been done with the cooperation of Frau Dr. MARGARETE BLUTH and the support of Frau AURELIE HOFMANN.

Brunswick (Braunschweig),
July 1969

Gerwig Vibrans

Contents

A. General

I. History, Geochemistry, Production 1
II. Smelter Production 5
Types of Lead 5
III. Physical Properties 9
1. Atomic Weight 9
2. Crystal Structure, Density, Thermal Expansion 10
3. Elastic Properties 12
4. Melting Point, Boiling Point, Vapour Pressure 13
5. Specific Heat, Latent Heat of Fusion and of Evaporation, Self-Diffusion 15
6. Surface Tension, Grain Boundary Energy and Internal Friction . . . 16
7. Thermal and Electrical Conductivity, Heat Transfer 19
8. Magnetic, Acoustic and Optical Properties 20
9. Use of Lead for Radiation Protection 21

B. Lead and Lead Alloys

I. Preparation of Lead Microsection for the Examination of Structure . . 25
II. Binary Alloys 29
1. General Survey. Solidification Phenomena 29
2. Pb–Ag. Lead-Silver 32
3. Pb–Al. Lead-Aluminium 34
4. Pb–As. Lead Arsenic 35
5. Pb–Au. Lead-Gold 37
6. Pb–Ba. Lead-Barium 38
7. Pb–Bi. Lead-Bismuth 40
8. Pb–Ca. Lead Calcium 44
a) Structure and General Properties of the Alloys 44
b) Mechanical Properties and Age-Hardening 47
9. Pb–Cd. Lead-Cadmium 50
10. Pb–Co. Lead-Cobalt 53
11. Pb–Cr. Lead-Chromium 53
12. Pb–Cu. Lead-Copper 54
13. Pb–Fe. Lead-Iron 56
14. Pb–Ga. Lead-Gallium 58
15. Pb–H. Lead-Hydrogen 58
16. Pb–Hg. Lead-Mercury 58
17. Pb–In. Lead-Indium 59
18. Pb–K. Lead-Potassium 61
19. Pb–Li. Lead-Lithium 61
20. Pb–Mg. Lead-Magnesium 63

21. Pb–Mn. Lead-Manganese 66
22. Pb–Na. Lead-Sodium 67
23. Pb–Ni. Lead-Nickel 69
24. Pb–O. Lead-Oxygen 70
25. Pb–S. Lead-Sulphur 74
26. Pb–Sb. Lead-Antimony 75
a) Production and Structure of Alloys 76
b) Casting Properties. Physical Properties 79
c) Mechanical Properties 82
d) Age-Hardening 86
(α) Relationship to Antimony Content and Additions p. 86. — (β) Pretreatment. Age-Hardening at Elevated Temperature p. 89. — (γ) Effect of Deformation on Age-Hardening p. 90. — (δ) Addendum p. 93.
27. Pb–Se. Lead-Selenium 94
28. Pb–Si. Lead-Silicon 94
29. Pb–Sn. Lead-Tin 95
a) Structure of the Alloys 97
b) Technological Properties 97
30. Pb–Sr. Lead-Strontium 100
31. Pb–Te. Lead-Tellurium 101
32. Pb–Ti. Lead-Titanium 105
33. Pb–Tl. Lead-Thallium 105
34. Pb–U. Lead-Uranium 106
35. Pb–Zn. Lead-Zinc 107

III. Ternary Alloys 109

1. Pb–Ag–Al. Lead-Silver-Aluminium 109
2. Pb–Ag–Cu. Lead-Silver-Copper 110
3. Pb–Ag–S. Lead-Silver-Sulphur 112
4. Pb–Ag–Sb. Lead-Silver-Antimony 113
5. Pb–Ag–Sn. Lead-Silver-Tin 114
6. Pb–Ag–Zn. Lead-Silver-Zinc 116
7. Pb–Al–Bi. Lead-Aluminium-Bismuth 120
8. Pb–Al–Cu. Lead-Aluminium-Copper 121
9. Pb–Al–Mg. Lead-Aluminium-Magnesium 121
10. Pb–Al–Sn. Lead-Aluminium-Tin 122
11. Pb–As–Cu; Pb–As–Fe; Pb–As–Ni. Lead-Arsenic-Copper, Lead-Arsenic-Iron, Lead-Arsenic-Nickel 123
12. Pb–As–Sb. Lead-Arsenic-Antimony 126
13. Pb–Ba–Ca. Lead-Barium-Calcium 128
14. Pb–Bi–Cd. Lead-Bismuth-Cadmium 129
15. Pb–Bi–Sb. Lead-Bismuth-Antimony 131
16. Pb–Bi–Sn. Lead-Bismuth-Tin 132
17. Pb–Bi–Te. Lead-Bismuth-Tellurium 134
18. Pb–Bi–Zn. Lead-Bismuth-Zinc 135
19. Pb–Ca–Na. Lead-Calcium-Sodium 136
20. Pb–Cd–Mg. Lead-Cadmium-Magnesium 137
21. Pb–Cd–Sb. Lead-Cadmium-Antimony 137
22. Pb–Cd–Sn. Lead-Cadmium-Tin 139
23. Pb–Cd–Tl. Lead-Cadmium-Thallium 141
24. Pb–Cd–Zn. Lead-Cadmium-Zinc 142

25. Pb–Cu–Ni. Lead-Copper-Nickel 143
26. Pb–Cu–O. Lead-Copper-Oxygen 144
27. Pb–Cu–S. Lead-Copper-Sulphur 147
28. Pb–Cu–Sb. Lead-Copper-Antimony 149
29. Pb–Cu–Sn. Lead-Copper-Tin 151
30. Pb–Cu–Te. Lead-Copper-Tellurium 153
31. Pb–Cu–Zn. Lead-Copper-Zinc 154
32. Pb–Hg–Na. Lead-Mercury-Sodium 155
33. Pb–In–Sn. Lead-Indium-Tin 156
34. Pb–Mg–Sb. Lead-Magnesium-Antimony 158
35. Pb–Mg–Sn. Lead-Magnesium-Tin 159
36. Pb–Na–Sn. Lead-Sodium-Tin 161
37. Pb–S–Sn. Lead-Sulphur-Tin 161
38. Pb–Sb–Sn. Lead-Antimony-Tin 163
a) Structure of Alloys 163
b) Hardness and Strength 167
c) Static and Dynamic Compression Tests 170
39. Pb–Sb–Zn. Lead-Antimony-Zinc 171
40. Pb–Sn–Zn. Lead-Tin-Zinc 173
41. Comparative Investigation of the Mechanical Properties of 37 Ternary Alloys 174
42. Concluding Remarks 175

IV. Specific Properties of Lead and its Alloys 177
1. Straining and Recrystallisation 177
a) Deformation of Lead: General 177
b) Textures 182
c) The Recovery and Recrystallisation of Lead: General 184
d) Movement of Large Angle Grain Boundaries in Recrystallisation . 186
e) Observations on Strain Hardening and Softening 189
f) The Recrystallisation Threshold. Additions which Inhibit Recrystallisation 190
g) The Course of Grain Growth with Time. Recrystallisation Diagrams. Grain Coarsening 194
2. Mechanical Properties (in the-Short-Time Test) 197
a) Measured Values from the Tensile Test 197
b) Types of Stress-Strain Diagrams 201
c) Hardness 202
(α) Brinell Hardness p. 202. — (β) Relationship between Hardness and Tensil Strength p. 204. — (γ) Diamond Cone Hardness p. 204. — (δ) Dynamic Hardness p. 205.
d) Compression. Impact and Notch Tests 205
3. Creep Strength 207
a) Basic Phenomena 207
b) Time Laws of Transient Creep 208
c) Processes in Steady State Creep 209
d) Dependence of Steady State Creep on Temperature and Stress . . 212
e) The Tertiary Stage of Creep and the Related Embrittlement Phenomena 216
f) Performance and Evaluation of Creep Tests 217
g) Effect of Prestrain on the Creep Rate 220

h) Creep under Multi-Axial Tension and Pressure and under Bending Stress . . . 222
i) Tests on Single Crystals . . . 227
j) Is There True Creep Limit? . . . 229
k) Creep-to-Rupture . . . 230
l) Dynamic Creep Strength . . . 231
m) Alloy Structure and Creep Strength . . . 233
n) Results for Individual Lead Grades and Alloys . . . 236
4. Endurance (Fatigue Strength) . . . 244
5. Hardening and Heat Treatment of Lead. Various Conditions of a Lead Alloy . . . 262

V. Lead as a Corrosion-Resistant Material . . . 268
1. Fundamental Properties . . . 268
2. Resistance to Sulphuric Acid . . . 272
a) Fundamentals . . . 272
b) Effect of Small Additions to Lead . . . 273
c) Corrosion-Resistance of Antimonial Lead . . . 282
d) Stress-Corrosion and Corrosion-Fatigue in Sulphuric Acid . . . 284
e) Corrosion in Nitrosylsulphuric Acid . . . 287
f) Further Phenomena, Summary . . . 289
3. Resistance to Inorganic and Organic Acids and other Chemicals . . . 290
a) Halogen Acids . . . 290
b) Nitric Acid and Mixed Acids . . . 291
c) Phosphoric Acid . . . 292
d) Sulphurous Acid . . . 292
e) Organic Acids . . . 292
f) Alkalis and Ammonia . . . 294
g) Salt Solutions . . . 295
h) Gases . . . 297
i) General . . . 297
4. Lead Anodes . . . 298
5. The Behaviour Lead towards Water . . . 301
a) Distilled Water without Access of Air . . . 301
b) Distilled Water with Access of Air . . . 302
c) Tap Water, Sea Water . . . 304
6. The Behaviour of Lead in the Ground . . . 308
a) Soil Corrosion . . . 308
b) The Behaviour of Ducted Cables . . . 313
c) Stray Current Corrosion and Electrical Methods of Protection . . . 314
7. Damage by Animals . . . 317
8. Indications of Further Causes of Corrosion . . . 319

C. The Technical Processing of Lead

I. Foundry Alloys . . . 321
1. Melting and Casting . . . 321
a) General Properties of Lead Melts . . . 321
b) Melting . . . 321
c) Attack of Lead Melts on the Crucible Materials . . . 322
d) Drossing (Oxidation) of Lead Melts . . . 326
e) Casting . . . 334

2. Lead Storage Batteries 341
a) General Development 341
b) Production and Properties of Storage Battery Plates 344
c) The Durability of Storage Battery Plates. Special Phenomena . . 347
3. Pressure Die Casting 356
4. Lead Alloys in the Printing Industry 357
a) Working Methods in General 357
b) Composition, Structure and Properties of Alloys 358
c) Casting Properties and the Occurrence of Defects 361
5. Bearing Metals 365
a) Mode of Operation of Plain Bearings 365
b) Properties of Bearing Metals and Manufacture of Bearings 369
c) Running Behaviour of the Bearing Material 372
d) Composition, Structure and Properties of White Metals and Leaded Bronzes 380
6. Fusible Alloys 387

II. Plastic Working (Technology) 390
1. General Discussion of the Extrusion of Lead 390
2. Cable Sheathing 400
a) General Design of Cables 400
b) Production of Cable Sheathing 404
c) Types of Lead, and Lead Alloys for Cable Sheathing 410
d) Special Phenomena and Defects 420
3. Pipes, Wires, Bends (Traps) 428
a) Methods of Productions 428
b) Structure, Special Phenomena and Defects 431
c) Wall-Thickness and Creep Resistance of Commercial Lead and Antimonial Lead Pipe 439
4. The Rolling of Lead 441
5. Impact Extrusion and Deep Drawing 447
6. Lead in Powder Metallurgy 449

III. Metallic Coatings. Use of Sheet Lead. Jointing Methods 453
1. Coating with Lead 453
a) General 453
b) Hot Lead Coating (Dip Coating) 453
c) Homogeneous Lead Lining 454
d) Spray-coating with Lead 461
e) Lead Electroplating 463
f) Lead Cladding 465
g) Further Methods 465
h) Red Lead Priming 465
2. Use of Sheet Lead in Chemical Technology and in Building 465
3. Welding and Soldering of Lead 468
4. Lead in Soft Solders 470
a) Solder Alloys and the Soldering Process 470
b) Properties of Soft Solders and Soldered Joints 473

Bibliography (also List of Names) 480

Subject Index 531

A. General

I. History, Geochemistry, Production

Lead, together with gold, silver, copper, iron, tin and mercury, was used technically before the birth of Christ. The Egyptians probably used lead, as well as copper, silver and gold, as early as 5,000 years B. C; the Phoenicians developed the lead deposits at Rio Tinto in Spain in 2,300 B. C. and the Chinese probably made lead coins in 2,000 B. C. Lead deposits in the Mediterranean and in Great Britain were worked in antiquity. Roman water pipes are particularly well known; they were produced by rolling sheet lead (which had been made by hammering) into tubes, and filling the joint with molten lead, thereby forming a weld (Hofmann [552]). In the early Middle Ages, lead production began in Bohemia (Příbram about 750) and in the Unterharz (Rammelsberg since 968) and, in the twelfth and thirteenth centuries, the lead deposits in Saxony (Freiberg), in the Rhineland and in the Oberharz were discovered.

Statistics of world production and use of lead are available from 1801 onwards. Between 1826 and 1850, world production of lead had already risen above 100,000 tons per year, and in the following quarter of a century reached 225,000 tons. Great Britain, Spain, Germany and the United States of America were the most important producers. The London Metal Exchange was founded in 1862. With the beginning of the present century, the volume of production in Europe had already been overtaken by that from overseas (U.S.A., Canada, Mexico, Australia). After the first World War, world production (if the falls of 1931/32 and 1945 are disregarded) rose above 1.5 million metric tons per year and has now almost reached the 3 million mark. The Paley Committee, set up by President Truman in 1951, made certain assumptions and calculated that the lead requirements of the Western World (disregarding scrap) would amount to 2,700,000 tons in 1975 as compared with 1,628,000 tons in 1950. The consumption of refined lead in the Western World for 1963 is reported as 2.3 million tons (metric) [844]. Even though the shortage of lead (which was forecast) has not occurred, much of this report still deserves attention today (Ketzer [658]). The reasons for the temporary over-supply of lead are discussed by Friedensburg [342].

The position of lead has changed relative to the other non-ferrous metals, as shown by the diagram in Fig. 1 [842]. Aluminium and magnesium, whose sources of raw materials far exceed those of the heavy metals, have moved markedly into the foreground. In some applications lead, like other metals, has been replaced by synthetic materials.

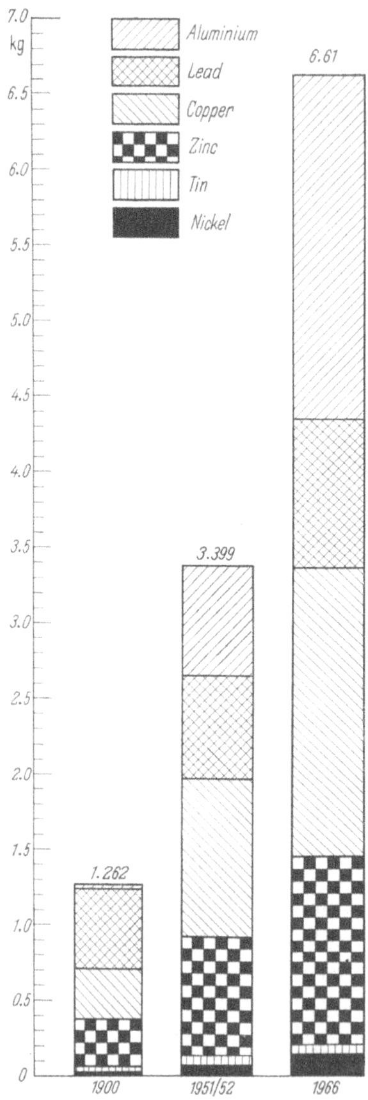

Fig. 1. Annual world consumption of non-ferrous metals in kg per head of population. (According to statistics compiled by Metallgesellschaft).

The ores of lead, of which the most important is galena, PbS, were formed in conjunction with the solidification of the magmatic rocks where the crystallisation of silicates was the main feature. The rare elements and volatile constituents became enriched in the residual melt, which, with falling temperature, acquired a progressively aqueous character (hydrothermal stage). The residual solution penetrated into cracks in the rock and led to the deposition of metallic ores among other minerals. In certain deposits, ores were also formed by chemical decomposition of the neighbouring rock by leaching (Metasomatosis). Only rarely did such an enrichment take place that the term "deposit" can be justified. Lead ores are frequently accompanied by zinc ores, and galena usually contains a certain amount of silver and traces of other noble metals. A post war estimate gave the available lead resources as 28 million tons (Dunham [262]) and recently Ketzer mentioned 35 million tons [659]. Separate estimates have been made of the reserves of the five accepted types of lead-zinc deposits, and distribution of these types on the earth's surface is shown in Fig. 2 [262]. Dunham pointed out that the contribution to world production of the ore deposits in the pre-Cambrian shield areas of

Canada, the Far East, Scandinavia, Australia and Africa (hypothermal) only became significant in the present century and that this type of deposit perhaps offered the best prospects for the future. While lead only occurs in the earth's crust in the proportion of 0.0015% (MASON [803a]), and is thus among the less abundant metals, nevertheless its trace occurrence down to a depth of one kilometre (similarly to other metals) exceeds that in ore deposits by more than 10^5 (NODDACK [903], GOLDSCHMIDT [405]).

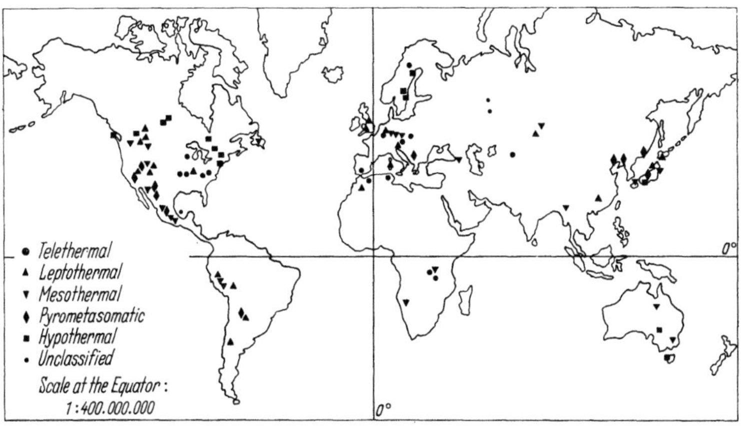

Fig. 2. Distribution of the principal lead and zinc mines of the world. (According to DUNHAM [262]).

By improved ore-dressing methods, it is possible to exploit deposits of low concentration. For lead, contents down to a lower limit of about 2%, in exceptional cases even lower, are regarded as economically workable.

In addition to the lead available from ore, treatable scrap lead should also be taken into account. The world consumption of lead for the period 1,800 to 1958 is estimated at 100,000,000 tons of which about 50,000,000 tons should be available for recovery [842].

The following data (Tables 1 and 2) on mine and smelter production of lead are taken from the statistics of Metallgesellschaft A. G. [842, 844]:

The data of the U. S. Bureau of Mines in Table 2 illustrate some of the different applications of lead.

In Great Britain in 1963 the lead consumption of 393,300 tons of lead was somewhat differently distributed. 28% were used for cable sheathing, 19.7% for accumulators (grids and lead oxide), 7.9% for lead tetraethyl, 6.1% for other chemical compounds. 2.0% for white lead, 17.3% for sheet and pipe, 1.4% for shot, 1.3% for foil and collapsible tubes, 1.5% for other semi-finished products, 4.1% for solder, 5.9% for other alloys and 4.9% otherwise.

Table 1. *World Production of Lead Ores and Metallic Lead*

	Mine Production (Lead Content in 1000 metric tons)			Production of Refined Lead (excluding Remelted Lead) in 1000 metric tons		
	1938	1966	1968	1938	1966	1968
Austria	—	4.9	5.9	9.3	10.7	11.3
Belgium	—	—	—	88.0	86.5	95.3
Bulgaria	—	100.0	—	—	90.3	—
Czechoslovakia	3.0	15.0	—	4.6	20.0	—
Denmark	—	—	—	—	9.9	9.7
Finland	0.3	4.6	4.3	—	—	—
France	4.2	26.7	26.4	39.0	141.9	145.4
Germany	89.5	66.4	—	175.9	272.9	—
Greece	4.2	8.9	11.2	4.1	7.2	7.9
Italy	41.1	37.5	36.0	43.3	64.6	76.1
Netherlands	—	—	—	—	14.9	17.5
Poland	5.3	51.3	—	13.1	43.7	—
Portugal	—	1.7	1.3	—	—	1.2
Rumania	6.4	20.0	—	6.4	20.0	—
Spain	38.0	62.4	71.2	31.8	56.2	62.2
Sweden	8.6	69.2	70.1	—	59.5	60.9
United Kingdom	30.2	2.0	0.0	11.0[1]	174.6	234.2
Yugoslavia	77.4	102.6	111.8	8.6	97.5	94.8
Rest of Europe	0.2	48.4	—	0.1	2.5	—
Europe (excluding European USSR)	*308.4*	*621.6*	—	*435.2*	*1172.9*	—
Burma	84.9	18.0	18.0	81.4	18.0	12.0
China	2.0	100.0	—	0.9	100.0	—
India	—	4.1	3.8	—	2.5	2.0
Iran	—	17.0	—	—	—	—
Japan	12.2	63.1	62.9	10.6	118.6	161.4
Korea[2]	5.7	66.7	—	5.7	50.0	—
Turkey	5.6	3.6	6.8	0.9	2.0	—
USSR (including European USSR)	69.0	425.0	—	69.0	475.0	—
Rest of Asia	5.8	6.3	—	0.1	0.5	—
Asia (including European USSR)	*185.2*	*703.8*	—	*168.6*	*766.6*	—
Algeria	5.4	2.9	—	—	—	—
Belgian Congo	4.9	—	—	—	—	—
French Equatorial Africa	2.8	—	—	—	—	—
Morocco	20.3[3]	79.6	—	—	18.7	23.0
Northern Rhodesia	0.3	19.9	—	0.3	19.9	22.8
South-Africa	0.1	—	—	—	—	—
South-West Africa	18.0	102.4	60.9	—	75.3	55.5
Tunisia	18.1	16.4	22.0	23.8	14.6	17.5
Rest of Africa	0.6	2.2	—	3.2	—	—
Africa	*70.5*	*223.4*	—	*27.3*	*128.5*	—

[1] Production from Ore. — [2] Only North Korea. — [3] French Morocco

Table 1: *World Production of Lead Ores and Metallic Lead (continued)*

	Mine Production (Lead Content in 1000 metric tons)			Production of Refined Lead (excluding Remelted Lead) in 1000 metric tons		
	1938	1966	1968	1938	1966	1968
Argentina	23.7	31.5	28.0	9.9	22.0	37.8
Bolivia	13.2	21.2	20.4	—	—	—
Brazil	—	23.2	18.8	—	17.4	15.8
Canada	221.8	275.6	329.7	181.7	167.7	183.3
Chile	0.9	1.3	—	—	—	—
Guatemala	—	0.7	—	—	—	—
Mexico	282.4	182.1	164.2	222.3	181.1	171.3
Peru	58.0	144.8	164.9	29.3	88.8	86.4
USA	335.4	289.7	334.8	340.2	532.2	796.9
Rest of America	0.7	12.0	—	0.1	1.1	—
America	*936.1*	*982.1*	1081.8	*783.5*	*1010.3*	1291.5
Australia	*278.8*	*367.6*	372.8	*227.3*	*215.8*	—
World Total	**1779.0**	**2898.4**	—	**1641.9**	**3294.1**	—

Table 2. *Use of Primary and of Remelted Lead in USA 1966*[1]

Use	t	%	Use	t	%
Ammunition	70,216	5.9	Lead Oxide for		
Bearing Metals	18,628	1.6	Accumulators	205,442	17.4
Bronze and Brass	22,206	1.9	Type Metal	28,506	2.4
Cable Sheathing	59,892	5.1	White Lead	7,240	0.6
Lead for Packing	53,226	4.5	Red Lead and		
Foundries	3,309	0.3	Litharge	81,193	6.9
Collapsible Tubes	10,456	0.9	Other Paints[2]	18,823	1.6
Foil	4,284	0.4	Lead Tetraethyl	223,965	19.0
Pipe	16,516	1.4	(Anti-knock)		
Sheet	24,502	2.1	Other Uses	56,018	4.7
Solder	65,897	5.6		1179,520	100.0
Accumulator Plates					
(Antimonial Lead)	209,201	17.7			

II. Smelter Production

Types of Lead

The production of metallic lead from ore and scrap will be summarised only, since the subject is outside the scope of this book. For a thorough study reference should be made, for example, to the publications of Feiser [309], Tafel [1158] and Thews [1179].

[1] Including remelted lead, lead in alloys, and lead-content of Ores directly converted to paints and salts.

[2] Including lead-content of zinc oxide containing lead.

Table. 3. *Composition of some Grades of Soft Lead*

	Ag	As	Bi	Cd	Cu	Fe
Electrolytic Trail (Tadanac)	0.0003	Trace	0.0005		0.0005	Trace
Harris soft Smelter Lead (Norddeutsche Affinerie)	0.0006		0.040		0.0003	0.0001
Feinblei Harz 99.99 (Lautenthal)	0.0007	0.0003	0.003 to 0.005	0.0001	0.0002	0.0005
Smelter Lead Hüttenblei Oker	0.001	0.0003	0.008 to 0.05		0.0003	0.0005
Broken Hill	0.00023	0.0001	0.0030	0.0001	0.0004	0.0004
Monteponi (Italy)	0.0007	0.0005	0.0060	0.0002	0.0002	0.0009
ASTM Designation: B 29—49 Max. Contents Corroding Lead[2]	0.0015	0.0015	0.05		0.0015	0.002
Chemical Lead[3]	0.002 min, to 0.020		0.005		0.040 min to 0.080	0.002
Acid Lead[4]	0.002		0.025		0.040 min to 0.080	0.002
Copper Lead[5]	0.020		0.10		0.040 min to 0.080	0.002
Common Desilverised Lead A[6]	0.002		0.15		0.0025	0.002
Common Desilverised Lead B[6]	0.002		0·25		0.0025	0.002
Soft Undesilverised Lead[7]	0.002		0.005		0.04	0.002
According to DIN 1719 of 1963						
Feinblei 99.99	0.001	0.001	0.005		0.001	0.001
Feinblei 99.985	0.001	0.001	0.01		0.001	0.001
Kupferfeinblei	0.0025	0.001	0.01		0.04-0.08	0.001
Smelter Lead 99.94	0.001	0.001	0.05		0.001	0.001
Smelter Lead 99.90	0.003	0.001	0.09		0.005	0.001
Remelted Lead 99.75[8]					0.1	
Remelted Lead 98.5[8]					0.5	

[1] Lead (by difference) min. per cent.
[2] Lead which has been refined to a high degree of purity.
[3] Undesilverised lead produced from Southeastern Missouri ores.
[4,5] Acid Lead and Copper lead are made by adding copper to fully refined lead.

(Weight Per Cent). Data in part from Feiser [309]

S	Sb	Sn	Zn	Pb[1]	
	0.0005	Trace	Trace	99.997	
	0.0003		0.0002	99.95	
	0.0003	0.0003	0.0002	>99.99	
	0.0003	0.0003	0.0002	99.94 to 99.985	
0.0004	0.0037		0.0003	99.9915	
	0.0010	0.0005	0.0006	99.985	
	Sb + Sn 0.0095		0.0015	99.94	Ag + Cu max. 0.0025
			0.001	99.90	As + Sb + Sn max. 0.002
			0.001	99.90	As + Sb + Sn max. 0.002
			0.002	99.85	As + Sb + Sn max. 0.015
			0.002	99.85	As + Sb + Sn max. 0.015
			0.002	99.73	As + Sb + Sn max. 0.015
			0.002	99.93	As + Sb + Sn max. 0.015
	0.001	0.001	0.001	99.99	
	0.002	0.001	0.001	99.985	
	0.002	0.001	0.001	99.9	
	0.002	0.001	0.001	99.94	
	0.002	0.001	0.001	99.90	
			0.05	99.75	
			0.05	98.5	

[6] Fully refined desilverised lead.

[7] Soft undesilverised lead is used in the trade to describe the type of lead produced from ores of the Joplin, Mo. District.

[8] Is not always well suited for welding.

In smelting lead, the first product, formed by a roasting process, is lead bullion, of purity 95 to 99%. In addition to the noble metals, it contains, as impurities, antimony, arsenic, tin, copper and bismuth. Most of the copper segregates in the kettle at low temperatures because the solubility of copper in lead, just above the latter's melting point, is very low. After the dross (which has a high copper content) has been skimmed off, the rest of the copper is removed by stirring sulphur into the bath. The work lead which has been decoppered down to 0.01%, now goes to the next process. The melt is brought to a temperature of 700 to 750 °C in a reverberatory furnace and air or steam blown in, or over, the surface. The elements tin, arsenic and antimony in that order are oxidised before lead, and carried into the scum. The operation can be achieved in single charges or continuously and in both cases efforts are made to obtain a low concentration of these impurities in the furnace, in order to accomplish a high rate of oxidation (p. 331). For example, in the continuous operation at Port Pirie, Australia, a concentration of 0.03% Sb and 0.001% As is constantly maintained in the furnace. The removal of tin, arsenic and antimony can also be carried out by means of oxidising melts (sodium hydroxide, chloride and nitrate) as in the Harris process.

The lead, after the refining treatment described above, is now desilverised. This is effected, usually, by the Parkes zinc-desilverising process, the basis of which is discussed under the ternary alloy system lead-silver-zinc. Here also, the last residues of copper are removed. The desilverised lead is essentially a eutectic melt of lead and zinc and the latter is extracted now usually by vacuum dezincing, in which the main part of the zinc is evaporated and recovered in metallic form. Only in a few cases is zinc nowadays removed by oxidising with air or steam (Poland). Another process is chlorine dezincing according to Betterton.

Grades of lead with high bismuth content are unsuitable for some purposes, and removal of the bismuth is economic for contents above 0.1%. Nowadays, it is generally carried out by the Kroll-Betterton process which involves treatment of the melt with calcium and magnesium Bismuth forms compounds with these metals, which rise to the surface and are removed as dross. The lower attainable limit of bismuth is 0.007 to 0.009%. According to Davey [235] end-contents of only 0.004%. Bi are obtained by treatment with potassium-magnesium (Jollivet process). In the so-called Dittmer process the bismuth content could be reduced to 0.001% by precipitation with sodium. Removal of the bismuth is also possible by electrolytic refining of lead. The process, however, is economic only if the cost of current is low or the bismuth content exceeds 0.5%.

Lead produced by the processes mentioned or others has a purity of 99.90 to 99.99%. Within this range the grades of lead differ only by their bismuth content.

By present smelting techniques, it is possible to produce commercial lead of any desired degree of purity, independent of the nature of the raw material. If the user desires certain admixtures, for example small copper contents for lead cable sheathing, these are added subsequently. The following Table 3 shows compositions of some known types of lead, German and other, and the maximum contents of impurities according to the German, American and British standard specifications (DIN 1719, ASTM Designation: B 29—49 and BS 334 respectively). It is impossible to deal with methods of analysis of lead within the framework of this book, and attention is drawn to special books on analysis of metals [193].

Grades of lead of especially high purity have been produced for scientific purposes. They are of great value for the investigation of fundamental properties, such as corrosion, plastic deformation, recovery and recrystallisation. Special reference must be made to lead of purity up to 99.9995% produced by RUSSELL [1038] by double electrolysis, and reproduced by STAHL [496].

Lead of purity 99,9995% is nowadays supplied by some smelters and offered among others by Bergmetall Goslar in Germany and by the Trail works in Canada. The highest degree of purity was formerly attributed to lead made in the Kaiser-Wilhelm-Institut für Physikalische Chemie, Berlin-Dahlem, where the last traces of noble metals particularly had been eliminated. The gold content still amounted to 10^{-10}%, the silver content to 10^{-8}% [1078]. Zone melting after PFANN [955] offers a possibility of producing an even purer lead; BOLLING and WINEGARD [104] mention an estimated degree of purity of 99.9999% (TILLER [1198], further, IGAKI [602a] and GREENWOOD [426a]).

III. Physical Properties

1. Atomic Weight

Lead is in the Fourth Group of the Periodic Table with atomic number 82. In its compounds it is di- and tetra-valent. In the 1963 International Table, the atomic weight is given as 207,21. Ordinary lead is a mixture of isotopes of mass numbers 204, 206, 207, 208, and the atomic weights of lead of different origins vary by 0.03 to 0.04 units, owing to different ratios of isotopes (BAXTER [67]).

2. Crystal Structure, Density, Thermal Expansion

That lead is a member of the cubic crystal system is clear, even without investigation by X-ray diffraction, from the shape of the crystals as grown from vapour, and from the appearance of etched specimens (HUMFREY [601], YAMAMOTO [1294]). The crystal structure is face-centred cubic (A 1-type) (EWALD [300]). At 20 °C the edge length of the unit cell is $a = 4.9400 \pm 0.0002$ kX units (99.99% lead) or according to a more recent determination at 20 °C, 4.9401 ± 0.0001 kX units (99.999% lead) (KLUG [678]). 1 kX unit = 1.00202 Å. An atomic radius of $\frac{a\sqrt{2}}{4} = 1.75$ Å is derived from the crystal structure; the co-ordination number is 12 (cubic close-packed). As with other metals, a fairly close packing of atoms can be assumed in the molten state also. STEEB [1143a] by Fourier analysis of X-ray interference in molten lead, derived a co-ordination number of about 9.4. (DANILOVA, on the other hand, found hardly any difference of co-ordination number between solid and molten lead [229]). An alloy of lead with about 27 to 37 At.-% Bi crystallises as a close packed hexagonal. KLEMENT [670a] extrapolated from this alloy to pure lead and estimated for lead that the same structure (c. p. h.) would exist at pressures above 0.16 megabar. According to HUME-ROTHERY [600a] lead in the metal crystal is to be regarded as divalent.

As with all other face-centred cubic metals, microsections of recrystallised lead show twinning lamellae. These are cases of coherent twins according to the Spinel Law, with the octahedral surface as a plane of symmetry and interface. That aluminium is an exception in the occurrence of recrystallisation twins, is attributed by SEEGER [1098] to the fact that in trivalent aluminium a particularly high stacking fault energy is required for twin formation. THORNTON and HIRSCH [1194] give data on the stacking fault energy of lead and a series of other metals. Lead does not exhibit allotropy. Earlier contrary assumptions are based on intercrystalline breakdown in lead acetate-nitric acid solution (HELLER [506], THIEL [1181]) or on grain growth at higher temperatures (EDA [269], HARGREAVES [490]), and are therefore unsound.

The average value of six density determinations of rolled Kahlbaum (99.996) and electrolytic lead at 25 °C was 11.3307 g/cm³ (TIMMERHOFF [1199]). From this and by means of the co-efficient of the volume expansions of $87.3 \cdot 10^{-6}$/°C, a value of 11.336 g/cm³ can be calculated for 20 °C. The average of five determinations on electrolytic, remelted and hammered lead at 19.94 °C was 11.337 g/cm³. Radium lead from uranium ores (atomic weight 206.3) has a density of 11.288 g/cm³ and thus the same atomic volume as ordinary lead (RICHARDS [1008]). A density of 11.345 g/cm³ at 20 °C can be calculated from the lattice constant

$a = 4.9401$ kX units (KLUG [678]), the atomic weight and the Avogadro's number $N_L = 6.02308 \cdot 10^{23}$. According to current views, the greater value of the calculated density can be attributed to the presence of unoccupied lattice-positions in the actual crystal. FEDER and NOWICK [307] conclude from measurements of thermal expansion up to the melting point that the concentration of vacancies at that temperature is $\leq 1.5 \cdot 10^{-4}$. An energy of ≥ 0.53 eV is necessary for the formation of a vacancy. D'HEURLE [248a] gives, for lead at the melting point, a concentration of vacancies of $2 \cdot 10^{-4}$ At. %

GERTSRIKEN and SLYUSAR [375] determined the activation energy of the formation of vacancies from measurements of electrical resistance and of thermal expansion at $E_d = 10.2$ kcal/mole. This value is about one-third of the activation energy of self-diffusion. The number of vacancies just below the melting point, when compared to the number of occupied positions in the lattice, was estimated at $10 \cdot 10^{-4}$, a value which differs considerably from those given above.

A slight increase of hardness of quenched unalloyed 99.99% lead on storage was explained by HOOKER as due to freezing of excess vacancies on quenching and to their subsequent aggregation [587].

The density of molten lead is (at temperatures shown).

327 °C 10.686	350 °C 10.658	400 °C 10.597	450 °C 10.536	500 °C 10.477	550 °C 10.418
600 °C 10.359	650 °C 10.302	700 °C 10.245	750 °C 10.188	800 °C 10.132	850 °C [237] 10.078

A later determination by MATSUYAMA [811] gave values which were not very different. KNAPPWOST and RESTLE [680] gave the density of molten lead at the melting point as 10.785 ± 0.0173 g/cm³.

The co-efficient of linear expansion of lead was investigated from −253 °C almost up to the melting point. The results of various authors show satisfactory agreement, see LANDOLT-BÖRNSTEIN [729]. From the work of HIDNERT and SWEENEY [522], which covers the widest range of temperature, the following may be given as average co-efficients of linear expansion α from −250 up to 20 °C, $25.1 \cdot 10^{-6}$; 20 to 100 °C, $29.1 \cdot 10^{-6}$; 20 to 300 °C, $31.3 \cdot 10^{-6}$. The true co-efficient of linear expansion between 20 and 295 °C, according to EUCKEN and DANNÖHL [294] may be calculated from

$$\alpha = 28.77 \cdot 10^{-6} - 20.95 \cdot 10^{-11}\, t + 60.30 \cdot 10^{-12} \cdot t^2$$

An X-ray determination by STOKES and WILSON [1150] gave the function $\alpha = 28.15 \cdot 10^{-6} + 23.6 \cdot 10^{-9} \cdot t$ with a total error for α of

$\pm$ 0.3 · 10^{-6}. For the increase of volume on melting, values have been given between 3.44 (Endo [279]) and 3.61% (Sauerwald [1050]). In the theory of the liquid state it is assumed that on melting a crystal, spherical holes of the size of the atoms are formed; their number can therefore be calculated approximately from the increase of volume on melting (Fürth [351], Scheil [1056]). The coefficient of volume expansion of molten lead, a little above the melting point, is 12.9 · 10^{-5}/°C (Landolt-Börnstein [729]).

3. Elastic Properties

In the literature there are values for the elastic modulus of lead at room temperature between 1,493 kg/mm² (Schaefer [1054]) and 2,040 kg/mm² (Koch and Dieterle [686]). The most probable value for the quasi-isotropic material can be assumed as 1,700 kg/mm² (Kohlrausch [695]). Goens [399] determined the following principal elastic constants, at 20 and −273 °C, by frequency measurements on vibrating single-crystal rods of lead:

	S_{11}	S_{12}	S_{44}	$S_{11}-S_{12}-{}^1/_2\,S_{44}$	C_{11}	C_{12}	C_{44}
°C	S_{ik} in 10^{-13} cm²/dyn				C_{ik} in 10^{11} dyn/cm²		
20	93.0	−42.6	69.4	101	4.83	4.09	1.44
−273	67.5	−31.0	52.8	72.2	6.7	5.7	1.89

The results obtained by Goens (adiabatic conditions) are in good agreement with those of Swift and Tyndall [1157] who used a static method (isothermal conditions). Prasad and Wooster [976] carried out exact determinations of the elastic constants by means of diffuse scattering of X-rays. The change in the value of C_{11} to C_{12} (determined directly by this method) was apparent in considerable changes in the quantities S_{ik}.

A calculation of the elastic constants of the polycrystal from the values for the single crystal is not available. In a compilation by Birch [83] the value 4,333 kg/mm² is given for the compressive modulus at tmosphe ricpressure and 20 °C (cf. [1082a]). The values for the torsional or shear modulus vary from 550 (Schaefer [1054]) to 724 (Kikuta [665]) to 780 kg/mm² (Koch and Dannecker [685]). Kohlrausch [695] gives the Poisson ratio as 0.44. The Poisson ratio can be calculated from the above values of the elastic modulus and the compression modulus as 0.434, and the shear modulus as 593 kg/mm². These values should be regarded as the most reliable ones. In careful measurements on unstretched specimens, CHALMERS [188] determined an elastic limit in the order 0.09 kg/mm², and on stretched specimens 0.30 kg/mm², and offered information on the elastic after-effect. The elastic and shear moduli were measured from room temperature up to the neighbourhood of the melting point (KOCH and DIETERLE [686], KIKUTA [665], KOCH and DANNECKER [685]), the usual fall with rising temperature being found for both. KÖSTER [691] measured the variation of the elastic modulus with temperature by a dynamic method, in the course of a comparative investigation of the metals of the Periodic System. For each degree Centigrade rise of temperature, in the temperature ranges -180 up to 0°C, 0 to 100 °C, 100 to 200 °C, 200 to 300 °C, the elastic modulus falls by 1.8; 1.9; 2.2; 2.8 kg/mm². At the transition point for superconductivity (7.22 °K) it amounts to 3,150 kg/mm² (WELBER and QUIMBY [1255]). The percentage fall of the shear modulus with increasing temperature is somewhat greater than that of the elastic modulus (KOHLRAUSCH), so that the ratio elastic modulus/shear modulus and the Poisson ratio approach their upper limiting values 3/1 and 0.5 respectively (SCHAEFER [1054]). Reference should also be made to measurements on molten lead by GORDON [408] who determined the velocity of propagation in longitudinal sound waves over a temperature interval from 328 °C to 370 °C and therefrom evaluated the parameters in a simple hole theory. For the temperature of melting point he gives the value

$$v = 1{,}776 \pm 4 \text{ m/s}.$$

Lead is distinguished by a high damping power towards mechanical vibrations (see also p. 467). FÖRSTER and KÖSTER [329] gave the value of $4.57 \cdot 10^{-3}$ for the natural logarithmic damping decrement of cast lead. WEERTMAN and SALKOVITZ [1252] determined the damping (internal friction) of very pure lead (expressed as the loss of energy per half-vibration-period divided by the potential energy of the system) as about 10^{-2}. This corresponds to a logarithmic decrement of $5 \cdot 10^{-3}$. The main purpose of the work was to study the effect of additions of between 0.01 and 1 At.-% of Bi, Cd or Sn on the values of damping and elastic modulus. WELBER and QUIMBY [1255] investigated the damping of lead in the superconductivity range of temperature. The internal friction (damping) increases exponentially with the temperature (HIKI [526a]).

4. Melting Point, Boiling Point, Vapour Pressure

According to an evaluation of the literature by KOHLRAUSCH [695], a melting point of 327.3 °C can be accepted. In the pressure range 150 to 2,000 atmopheres, the melting point is increassed by 0.00803 °C for each increase of pressure of one atmosphere (JOHNSTON and ADAMS [627]). A Russian investigation by BUTUZOV and GONIKBERG [165] found in the region of 8,000 to 12,000 atmospheres a rise of melting point of 6.8° per 1,000 atmosphere increase of pressure, and in the region of 20,000 to 30,000 atmospheres an increase of 5.4°.

Disregarding older measurements, the most exact directly-observed value of the boiling point of lead is (1740 ± 10) °C (LEITGEBEL [740], FISCHER [323]). A similar value can be derived from the vapour pressure curve (Fig. 3). Curves showing the free energy of evaporation have recently been compiled for a large number of metals by EVANS [296].

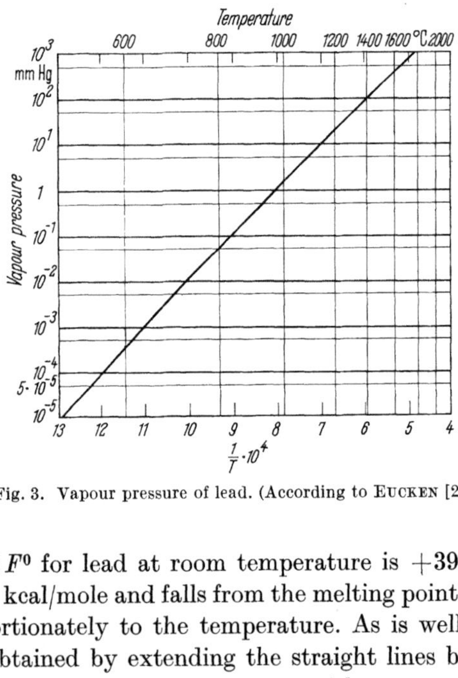

Fig. 3. Vapour pressure of lead. (According to EUCKEN [239, 729]).

The value ΔF^0 for lead at room temperature is +39.4, at the melting point +30.5 kcal/mole and falls from the melting point to 0 at the boiling point, proportionately to the temperature. As is well known, negative values are obtained by extending the straight lines beyond the boiling point. The change of vapour pressure with temperature is of special importance with regard to the toxicity of lead vapour (p. 340). PALATNIK and KOMNIK carried out experiments on the condensation of lead on substrates, maintained at a temperature between 50 and 300 °C [929a], which indicated that there are two critical condensation temperatures. Above the critical temperature T_{K2} no condensation of the metal can take place. Between a lower critical temperature T_{K1} and the upper temperature T_{K2} crystal nuclei of solid lead are formed directly from the vapour phase, whereas below T_{K1} the condensation occurs through an intermediate liquid phase. The metal films deposited below and above T_{K1} have different properties.

5. Specific Heat, Latent Heat of Fusion and of Evaporation, Self-Diffusion

Data on the specific heat of lead between 1 °K and room temperature are given by HOROWITZ and co-workers [595] and by MEADS and co-workers [830]. KUBASCHEWSKI and EVANS [715] give in their book, for the region between room temperature and melting point, the approximate formula $C_p = a + bT + cT^{-2}$ cal/degree mole. With their values for the constants $a = 5.63$ and $b = 2.33 \cdot 10^{-3}$, the specific heat C_p at 25 °C is found to be 6.32 cal/(degree gram-atom) and $c_p = 0.0305$ cal/g degree respectively. The specific heat of solid at the melting point is calculated as $C_p = 7.03$ cal/(degree gram-atom) and $c_p = 0.034$ cal/g degree respectively. Similar values are obtained by using the approximation for the solid state, given by DOUGLAS and DEVER [256]. For the region up to 900 °C these authors give the approximation $c_p = 1.5272 \cdot 10^{-1} - 1.493 \cdot 10^{-5}$ T Joule/g degree, 1 Joule being 0.239 cal. For the melt at 327 °C the values obtained are $c_p = 0.0353$ cal/g degree and $C_p = 7.31$ cal/(degree gram-atom) and at 600 °C $c_p = 0.0343$ cal/g degree and $C_p = 7.11$ cal/(degree gram-atom). The minimum in the specific heat at about 450 °C, observed by FÖRSTER and TSCHENTKE [330] is not expressed by the formula.

Values of reliable recent determinations of the latent heat of fusion may be cited as follows: 1.15 ± 0.03 kcal/gram-atom and 5.55 cal/g by KUBASCHEWSKI [715]; and 1.158 ± 0.006 kcal/gram atom and 5.59 cal/g by OELSEN and co-workers [916]. DOUGLAS and DEVER [256] give the latest heat of fusion as 5.50 cal/g.

According to EUCKEN [293], the latent heat of vaporisation at the boiling point amounts to 42.70 cal/gram atom and 206 cal/g. KUBASCHEWSKI [715] finds practically the same values.

NACHTRIEB and HANDLER [888] determined the self-diffusion of lead single crystals of 99.999% purity (after evaporation of radium D) by means of measurements of β-radiation.

The following values to the diffusion constant were determined:

Temp. (°C)	173.8	206.3	225.9	253.6
D (cm²sec⁻¹)	$4.81 \cdot 10^{-13}$	$2.21 \cdot 10^{-12}$	$6.41 \cdot 10^{-12}$	$2.41 \cdot 10^{-11}$

Temp. (°C)	285.1	304.4	322.4
D (cm²sec⁻¹)	$8.79 \cdot 10^{-11}$	$1.94 \cdot 10^{-10}$	$4.09 \cdot 10^{-10}$

The values follow from the equation $D = 0.281 \cdot \exp(-24{,}210/RT)$ cm²/sec. Preference should be given to the value of the activation energy as found by this equation, over the figure of 27,900 cal/gram atom as determined by v. HEVESY, SEITH and KEIL [517]. An application of hydrostatic pressure inhibits the self-diffusion, i.e. the activation energy is slightly elevated and the frequency factor D_0 is considerably lowered [598a] HUDSON and others. By diffusion measurements on polycrystalline lead OKKERSE [919] found values of the diffusion coefficients for lattice self-diffusion ($D = 1.17 \cdot \exp(-25{,}700/RT)$ cm²/sec), and for the grain boundary self-diffusion ($D = 0.81 \cdot \exp(-15{,}700/RT)$ cm²/sec) assuming the thickness of the grain boundaries to be 10 Å. In a further investigation [920] on bicrystals, he showed that the grain boundary diffusion depends on the angle between the dislocation lines and the direction of diffusion. When the two directions are parallel, diffusion takes place far more quickly than if the direction of the diffusion is perpendicular to the dislocation lines.

According to measurements by ROTHMAN and HALL [1036], the diffusion coefficient for self-diffusion in molten lead between 606 and 930 °K amounts to $D = (9.15 \pm 0.30) \cdot 10^{-4} \exp [(-4{,}450 \pm 330)/RT]$ cm²/sec.

6. Surface Tension, Grain Boundary Energy and Internal Friction

Various measurements of surface tension (surface energy) of molten lead are set out in Fig. 4. The lowest and highest values, 397 and 536 dyn/cm, express, for complete wetting, rises of 8.26 and 9.98 mm respectively in a tube of 1 mm diameter. If a salt cover of chlorides of lead and potassium is used, the interfacial tension is only about half as high. Latterly, determinations of the surface energy of solids have gained in importance. From the values of the latent heat of sublimation. FRICKE [339] has calculated the free surface energies for the interface gas-solid of the cubic and octahedral faces of lead as 871 and 745 erg/cm² respectively at 25 °C. Following German [367], the cubic octahedral faces were exposed in evaporation experiments in high vacuum. The value for polycrystalline lead at the melting point was given as 700 erg/cm². As is shown by comparison with Fig. 4, the surface energy falls abruptly on melting.

In polycrystalline metals, crystals of different orientation meet at the grain boundaries. In a thin layer, disturbances in the atomic arrangement permit transitions from one crystal lattice to another. In the case of the so-called small-angle grain boundaries (difference of orientation $< 20°$) a series of edge dislocations is involved. The defects in crystal structure indicate an increase of energy depending on the difference of orientation which, when related to the unit of interface, is called grain boundary energy. The grain boundary energy beetween two lead crystals having a common cubic edge is shown in Fig. 5 as a function of the angle of distortion (AUST and CHALMERS [31a] and LÜCKE [776]). The relative free grain boundary energy of coherent twins, in terms of the

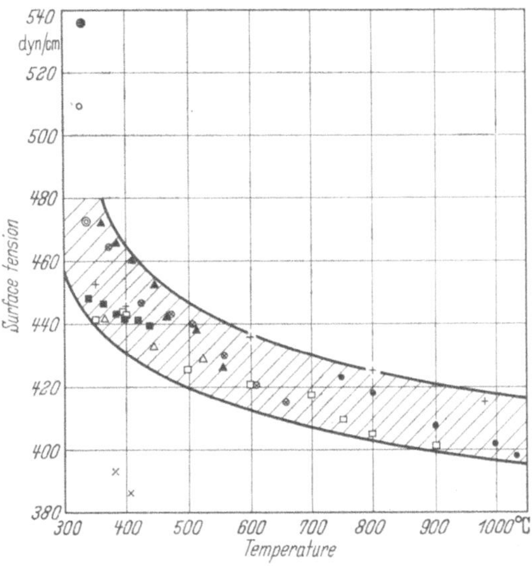

Fig. 4. Surface tension of molten lead. (According to various investigators).

● in CO_2 QUINCKE 1858 [729]
⊙ in CO_2 GRUNMACH [729]
○ in CO_2 SIEDENTOPF 1897 [729]
+ in H_2 BIRCUMSHAW 1926 [84]
△ in H_2 HOGNESS 1921 [578]
• in H_2 DRAHT und SAUERWALD 1927 [258]
▲ in H_2 MELFORD und HOAR 1956 [833]
□ in N_2 STAHL 1957 [1143, 496]
■ in N_2 GREENAWAY 1948 [420]
× in Vacuum HAGEMANN 1914 [729]
⊗ in Vacuum MATSUYAMA 1927 [809]

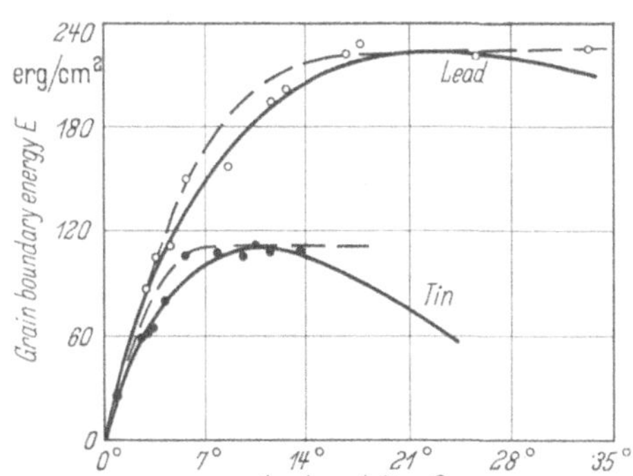

Fig. 5. Grain boundary energy between two single crystals of lead with a common cubic plane as a function of the angle of rotation Θ. Values for tin given for comparison. (According to AUST and CHALMERS [31a, 776]).

average grain boundary energy was determined in zone-melted lead as 0.050 by BOLLING and WINEGARD [106]. Addition of 0.1 At.-% Ag increased the relative free grain boundary energy to 0.077. The number of twins observed in both metals, was in inverse ratio to the relative grain boundary energy of the twin boundaries.

Measurements of the internal friction of molten lead have been carried out repeatedly. The following values from older measurements by GERING and SAUERWALD [366] can be regarded as reliable:

Temperature in °C	Viscosity η in Poises	Fluidity $\frac{1}{\eta}$
350	0.02648	37.8
400	0.02315	43.1
450	0.02057	48.6
500	0.01850	54.1
550	0.01681	59.5
600	0.01540	64.9

These values are in satisfactory agreement with the results obtained by other authors, e.g. ESSER, GREIS and BUNGARDT [290], YAO and KONDIC [1297], ROTHWELL [1036a], OFTE and WITTENBERG [918a].

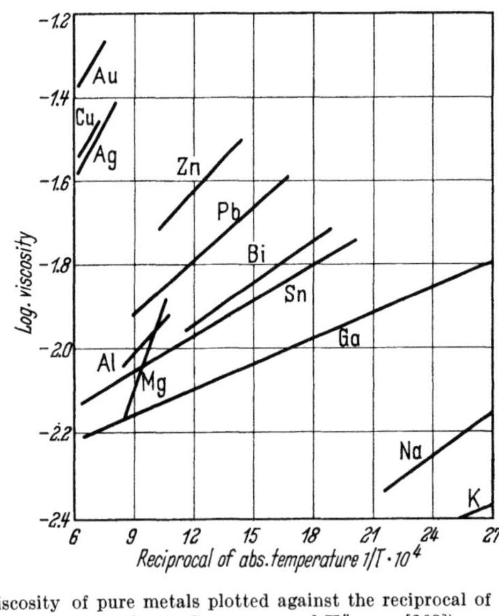

Fig. 6. Viscosity of pure metals plotted against the reciprocal of temperature. (According to GEBHARDT and KÖSTLIN [363]).

The dependence of the viscosity in (CGS units) on temperature can be expressed by the equation $\eta = A \cdot e^{B/T}$, as shown in a summary by GEBHARDT and KÖSTLIN [363] in Fig. 6. Still better agreement with the experimental values is given by Andrade's equation $\eta \cdot V^{1/3} = A\, e^{c/VT}$.

V is the specific volume, T the temperature in °K, the constants A and c have the values $2.753 \cdot 10^{-3}$ and 83.15 for lead [1036a]. BORCHERS [109c] considers that sound impulses observed at phase changes of lead alloys arise from the friction between crystallites on volume contraction.

7. Thermal and Electrical Conductivity, Heat Transfer

The thermal conductivity of lead has been repeatedly measured from the neighbourhood of absolute zero up to the region of melting. For solid lead the results obtained in two laboratories agreed well (BIDWELL [80, 81], KONNO [696]), and may be represented by a smooth curve. For the molten region, the more recent measurements deserve preference. The following values of the thermal conductivity λ in cal/sec · degree · cm were derived from the graphs:

°C	−250	−200	−100	0	100	200
λ	0.122_5	0.112_2	0.095_2	0.084_6	0.079_4	0.076_3
°C	300	327 (solid)	327 (liquid)	400	500	600
λ	0.073_8	0.073_3	0.058_4	0.058_9	0.059_3	0.059_7

As molten lead finds extended application for heat transfer (p. 41), heat transfer coefficients are important. For the combination molten lead and steel, at bath temperatures of about 450 and 750 °C, the heat transfer coefficients were determined as 1,112 and 1,075 kcal/m² · h · degree respectively (BÜHLER [140]).

Lead is superconducting slightly above absolute zero. According to the most recent measurements on zone-melted lead, the temperature of the transition point is (7.175 ± 0.005) °K (PEARSON and TEMPLETON [942]). From measurements of the magnetization on lead alloys in the temperature range of super-conductivity, conclusions may be drawn as to the amount of solved alloying element (LIVINGSTON [754b]). Above the transition point, lead is used in low-temperature physics as a material for resistance thermometers (MEISSNER and FRANZ [831]). The specific electrical resistance of vacuum-melted lead (purity $> 99.998\%$) was given by EUCKEN and SCHÜRENBERG [295] as $19.28 \cdot 10^{-6}$ ohm cm at 0 °C, and the temperature coefficient as $4.22 \cdot 10^{-3}$/°C (KOHLRAUSCH [695]). Table 4 shows some selected values of electrical resistance at temperatures between −253 and 856 °C. The data for low temperatures are taken from the Handbook of Chemistry and Physics [475]. The values up to 460 °C are taken from the work of PIETENPOL and MILEY [962]; the lead used contained small traces of gold, arsenic, antimony and copper. The melt, in wire form, was held together by the oxide skin formed on the surface. The accuracy of measurement was 1%. Above 460 °C, the values given are those of MATSUYAMA [808]. Further measurements in the molten region were carried out by FÖRSTER and TSCHENTKE [330] by an electrodeless method (cf. ROLL and co-workers [1023] for further work).

Table 4. *Specific Electrical Resistance ϱ of Lead*

Temp. (°C)	−252.9	−203	−103	0	20	200	300	320
ϱ (10^{-6} Ωcm)	0.59	4.42	11.8	19.28	20.648	36.478	47.938	54.761

Temp. (°C)	330	400	460	527	682	731	776	856
ϱ (10^{-6} Ωcm)	96.735	101.418	104.878	105	112	114	117	120

AUBREY offers a contribution on the determination of the Fermi surface of lead and of BRILLOUIN zones [29a].

8. Magnetic, Acoustic and Optical Properties

Lead is diamagnetic. The specific susceptibility per gram at 18 °C amounts to $\chi = -0.12 \cdot 10^{-6}$ (HONDA [584]).

The velocity of sound in lead, i. e., the velocity of propagation of longitudinal elastic waves in a rod, was determined as 1,560 m/sec by WOOD and SMITH [1284]. If the velocity of sound is calculated according to the formula $v = \sqrt{\frac{E}{\varrho}}$, and if the values of elastic modulus and density given on pages 12 and 10 are used, then $v = 1212.5$ m/sec.

The high damping of mechanical and acoustic vibrations in lead resulted in the development of isolating materials which effectively attenuate mechanical oscillation and noise (antivibration pad p. 467). LOTTERMOSER and MEYER [766a] reported on electro-acoustic measurements of the sound spectra of organ pipes of tin-plated lead, tin, copper, zinc and wood. From a musical point of view it is desirable that organ pipes emit the harmonic partial tones as continuously as possible in all directions and that they produce only a small amount of discord and noise. The results of measurements provide a physical argument for preferring lead and tin to zinc in the manufacture of organs. The choice of lead and tin is a consequence of the low bending stiffness and the high internal friction of these metals. It is now obvious that lead is preferred for registers of which a "full, soft" sound is expected. For "Principals" tin is preferred because of its "brightness and splendour". Copper is also suited for such voices.

For a wavelength of 5,890 Å the following optical constants were given: refractive index 2.01, absorption co-efficient 3.48, reflectivity 62% (WARTENBERG [1238]). The radiating power of lead was measured at various temperatures by SCHMIDT and FURTHMANN [1073].

9. Use of Lead for Radiation Protection

In order to clarify the use of lead for protection against X-rays and gamma-rays, consideration should be given to the extinction law governing the passage of these rays through matter $J = J_0 e^{-\mu d}$ (COMPTON [210]). Here J/J_0 signifies the fraction of radiation passing through, d the thickness of material and μ the linear absorption co-efficient. μ is the sum of the photoelectric or true absorption co-efficient τ and the scattering co-efficient σ, namely $\mu = \tau + \sigma$. For the absorption co-efficient the relation is (KOHLRAUSCH [695]):

$$\tau = \text{constant} \cdot \varrho \cdot Z^3 \cdot \lambda^3$$

ϱ = density, Z = atomic number, λ = wavelength (Å) =12.34/working voltage of the X-ray tube in kilovolts). The scattering co-efficient is in certain cases the classical value $\sigma = 0.2\,\varrho$. In scattering, the radiation suffers a change of direction and softening, and is further weakened by repetition of the scattering and absorption processes. The formula for τ shows that, owing to Z being present as the third power, and multiplied by the density ϱ, the absorption co-efficient of lead will be quite high in comparison with most of the common metals. Comparison of the formula given for τ and σ shows however that with decreasing wavelength the influence of σ in attenuating the radiation becomes progressively more important than that of τ. In practice the influence of the material in the region of the shortest wavelengths depends only on the ratio between the scattering co-efficient σ and the density ϱ. The superiority of lead in these regions is not so marked, and, consequently, concrete (with addition of heavy materials) is often used as the most economic protection against radiation.

X-rays are generally produced at operating voltages up to 300 kV; for radiation of shorter wavelengths, artificial radioactive isotopes are now available. To produce radiation of the same wavelength with X-ray tubes, the latter should be operated at voltages of at least 600 kV (Ir 192) up to 1300 kV (Co 60). Still shorter wavelengths are provided by the betatron, recently introduced for the testing of materials. It is manufactured for voltages of 15 or 31 megavolts [695]. Where low space requirements are important, lead is the most usual protective material employed against X- and gamma radiation. In addition to its high atomic number and density. lead has the further advantage that on irradiation with neutrons, it does not become radioactive, i.e., does not become itself a source of radiation (p. 24). This assumes the absence of impurities which would be activated by the radiation, that is the use of lead of defined purity [611a].

The concept of dosage is important for radiation protection. According to German Standard Specification DIN 6809 of 1963, "X-rays and Gamma Radiation in Medicine and Biology, Rules for Dosimetry", the Röntgen (R) is defined as the unit of ionic dosage, $1\ \mathrm{R} = 2.58 \cdot 10^{-4}\ \frac{\text{Coulomb}}{\text{Kilogramme}} \left(\frac{\mathrm{C}}{\mathrm{kg}}\right)$. The ionic dosage J of an ionising radiation is the limiting value of the quotient of the electrical charge ΔQ of the ions of one sign (positive or negative) which are produced directly or indirectly in an air volume element of mass $\Delta m = \varrho_L \cdot \Delta v$ (Δv volume; ϱ_L density of air) by the radiation, and of the mass Δm:

$$J = \frac{dQ}{dm} = \frac{1}{\varrho_L} \frac{dQ}{dv}$$

According to the German Standard Specification DIN 6811 of 1962, "Medical X-Ray Equipment up to 300 kV: Radiation Protection Rules

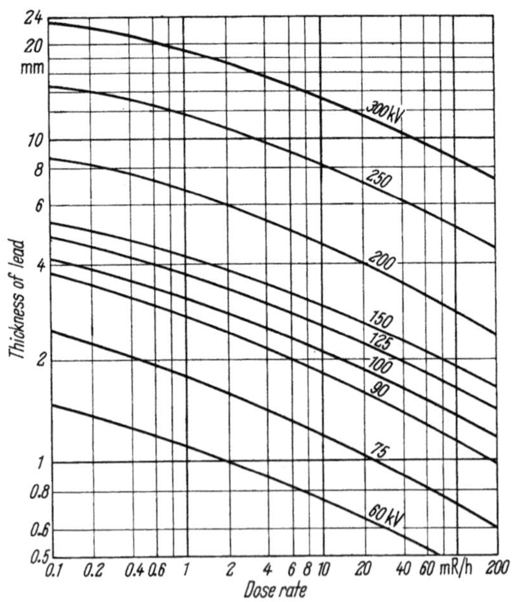

Fig. 7. Reduction of direct X-radiation (effective radiation) by lead layers of various thickness. Assumed tube current 1 mA. The distance measured from the focus is 1 m. The parameter is the operating voltage of the X-ray tube. The data are valid for the use of constant direct current.

for Manufacture", and in agreement with a decision of the International Committee on Radiation Protection of November 1956, the maximum permissible dose for operators is 5 R per year, that is an average of 0.1 R per week. Fig. 7 gives, according to DIN 6811, the dosage output in milliröntgens per hour in the effective cone of rays behind a lead shield, if the tube current amounts to 1 mA and the distance from the focus is 1 metre. If this local output is multiplied by the square of the distance (in metres) and the reciprocal value of the nominal steady current (in mA) of the tube, then it is possible to read from Fig. 7 the necessary thickness of lead for various tube voltages. In addition to the effective radiation, the scattered radiation also must be considered. The specification provides data on the thicknesses of lead necessary to reduce the scattered radiation.

Table 5 shows the thicknesses necessary for protection against radium and mesothorium radiation of various activities. While the intensity of primary radiation in an X-ray tube is proportional to the tube current, in radioactive material it is proportional to the activity of the substance, which is expressed in millicuries (mC). In the case of radium metal, one gram of radium corresponds to 1,000 mC.

Table 5. Minimum Distance of the Working Position from the Surface of the Protective Container (Lead) for Radium and Mesothorium in relation to the Thickness of the Lead Protection and the Activity of the Source. The Table is based on a Dosage Output of about 0.1 R per week at the Working Position. According to DIN 6804 of 1964

Lead Thickness in mm	10	30	50	75	100	120	150
Minimum distance in m at							
10 mC	1.40	0.90	0.55	0.35	0.20	0.20	0.20
50 mC	3.50	1.75	1.05	0.70	0.35	0.35	0.20
100 mC	—	2.60	1.60	0.90	0.55	0.35	0.20
200 mC	—	3.50	2.60	1.25	0.70	0.55	0.35
500 mC	—	—	3.50	1.75	1.05	0.70	0.35
1,000 mC	—	—	5.20	2.60	1.40	0.90	0.55
2,000 mC	—	—	—	3.50	2.60	1.25	0.70
5,000 mC	—	—	—	6.10	3.50	2.60	1.05
10,000 mC	—	—	—	7.80	4.40	2.60	1.40

In radiation protection lead is used particularly as sheet, as lead bricks, as blocks (Holmes [583a]) or in the form of homogeneous lead lining (p. 454). Furthermore, lead can be cast into the space between two steel sheets. If an intense development of heat is to be expected (as in reactor construction) a system of cooling pipes is incorporated in the lead. Mention may also be made here of the application of lead as a constituent of leaded rubber, lead glass or cement.

The absorption cross-section for thermal neutrons in lead (Cap [175]) is 0.2 barn (1 barn $= 10^{-24}$ cm^{-2}).

In shielding nuclear reactors, both, neutron radiation and gamma radiation must be absorbed. The absorption of neutrons is effected by (among others) hydrogen, lithium, boron and cadmium, and the absorption of gamma rays by concrete of appropriate composition and thickness, steel or lead (p. 21). It is advantageous that lead does not thereby itself become a source of secondary radiation. Consequently, combinations of lighter and heavier elements, among the latter lead in particular [611a, 504a], are effective for shielding reactors. This subject is in a state of considerable flux.

Lead and Lead Alloys

I. Preparation of Lead Microsections for the Examination of Structure

It must be borne in mind when preparing specimens and finishing microsections that lead undergoes deformation even under low pressure and this deformation usually gives rise immediately to recrystallisation which changes the structure of the material under investigation. Consequently it is essential to avoid all unnecessary deformation, excessive friction and heating.

Specimens are taken from the material under investigation by means of a fret-saw; cooling with alcohol is recommended. The surface of the microsection must be prepared by filing, turning or planing. The deformed surface layer thus produced must be completely removed by subsequent grinding and particularly by etch polishing. Use can be made here of an energetic macro etching solution, e.g. a solution of ammonium molybdate or a solution of 100 parts of water + 40 parts of hydrogen peroxide (30%) + 10 parts glacial acetic acid (Jones [631]). Small specimens are embedded in thermo-plastics, for example Palatal, supplied by Badische Anilin- und Sodafabrik. Similarly carnauba wax has also proved suitable. Further treatment will be described, with particular reference to the instructions given by A. Schrader [1079, 1081].

The specimen can be hand-ground with emery paper No. 120, 220, 320 (15 minute emery), 400 (60 minute emery), 700 (200 minute emery), by use of a wax, e.g. carnauba wax (the time indicates the period of elutriation of the emery). The wax prevents the smearing of the lead and the embedding of emery in the surface. It is better, however, to grind wet with waterproof polishing paper of grain numbers 220, 320, 400 and 600, as the running water removes the dust which is formed. Before grinding, the paper is covered with carnauba wax. In some circumstances grinding is followed by pre-polishing on a wheel (120 revs. per minute) covered by a felt cloth which is kept wet, while plenty of soap and fine 60-minute elutriated emery is used in the process. Then the specimen is polished by wool cloth on a wheel (600 to 800 revs. per minute) with distilled water, a few drops of 0.2% nitric acid (1.40) and No. 1 alumina as polishing medium, until the large scratches disappear and the surface of the specimen becomes bright.

As all the scratches do not disappear in polishing, fine polishing is carried out on a slow wheel with chamois leather (moistened with distilled water) in a thick paste of No. 1 alumina with a trace of ammonium tartrate, without great pressure. In the course of fine polishing, etching (after rinsing under the tap) is carried out several times with Vilella solution (VILELLA and BEREGEKOFF [1222]), and the etch immediately polished off again. All adhering alumina is removed by rinsing under a strong stream of water. Fine polishing can also be done on a wheel carrying silk velvet or rayon velvet. Under the fine velvet fabric are placed another one or two layers of velvet. The polishing medium is distilled water with No. 1 alumina and about 1 g/l of ammonium tartrate.

From time to time the development of the structure on the etched section is observed under the microscope. SCHULZ [1086] recommends pickling off the oxide skin formed on etching, with VILELLA solution (see below) in a solution of three parts glacial acetic acid and one part H_2O_2 (30%). For observation, the section is rinsed, dipped in absolute alcohol and dried in hot air. The finished surface of the section must not be touched again, not even with cotton wool, as everything leaves scratches. Photomicrographs must be made as soon as possible, as sections tarnish rather quickly in air. They can suitably be kept in dessicators. If satisfactory structural pictures are not obtained, this is usually due to the fact that the surfaces are still smeared from the grinding and polishing. In most cases this layer disappears by further etching and fine polishing.

The process described above permits the production of sections for the highest magnification and the highest demands of microscope techniques. Very good results can also be obtained by shaving off the surface of the specimen with a sharply ground microtome. The direction of the cut should be along a diagonal to minimise the shock of the initial impact of the blade (LUCAS [770]; BASSETT and SNYDER [52]; REINACHER [1000]). At first the shavings should be of 10 μm, but later of 1 μm thickness. The cut surfaces are not further ground, but polished and etched as described above. This method also gives satisfactory detail of structure at the highest magnifications.

For many purposes the processes described above are too time-consuming and the resolution obtained is, in fact, unnecessary. If merely a survey of grain size is required, then macroscopic surface etching suffices, which, in the case of rolled and extruded lead products, does not necessitate grinding and polishing. If macro etching is required to give information about the interior of the specimen, it should be carried out on carefully turned surfaces. The cutting tool must be sharply ground so that it cuts but does not squeeze, and feed and cutting speed should be low (ASTM Standards [28]). Better still is diamond turning, according to SCHULZ. Particularly clean etchings can be obtained if the specimens are first handpolished on felt, using the following solution: 100 cm³ water, 40 cm³ H_2O_2 (30%), 10 cm³ glacial acetic acid.

Deformation of the specimen surfaces is not wholly avoided in any of the processes described. However if the precautions indicated are observed, then even in the case of pure lead (99.99%), which easily recrystallises, the structural picture obtained is reliable, i.e. it corresponds to the internal condition of the material. The mechanical fine polishing

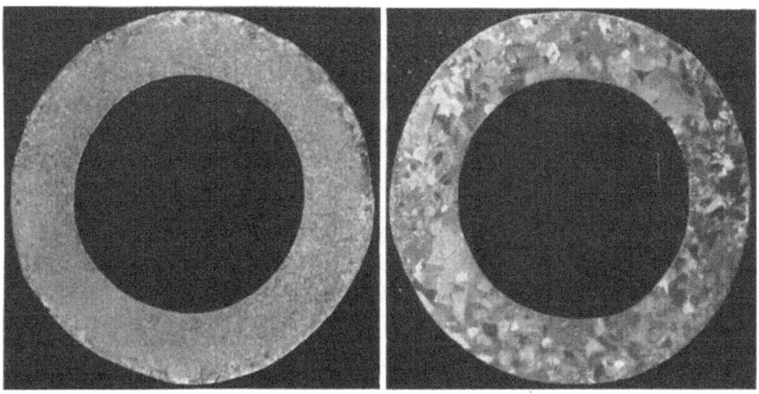

Fig. 8a and b. Macrographs of a lead pipe after etching with electrolyte A5 [683].
a) polished for a short time, b) polished 1 hour. ×2

described above can be avoided by employing Jacquet's method [610, 611]. The following process has proved sound at the author's laboratory in Brunswick. After wet grinding, polishing is carried out on wool cloth for some time (see above), then etch-polishing in electrolyte A 5, according to KNUTH-WINTERFELDT [683], several times for a few seconds at a current strength of 1 A and finally with 5 A.

Specimens for macroscopic observation are pretreated in a similar way. For etching, the section is dipped in a small metal dish which forms the cathode. Current is led to the specimen (which forms the anode) through the forceps. Fig. 8a, b, shows the effect of polishing on the structure, in the case of a lead pipe. The deformation caused by cutting the pipe and grinding the surface has resulted in a superficial fine-grained recrystallisation. This fine-grained zone was only removed completely after one hour's polishing.

The most important, proved, etching media for lead and lead alloys were compiled by SCHRADER [1080] in her "Ätzheft" (Etching Manual).

These are given below together with some other solutions taken from the literature:

Macroscopic etching of grain surfaces	Normal Vilella etching medium: 16 cm^3 nitric acid (1.40), 16 cm^3 glacial acetic acid, 68 cm^3 glycerol. Etch for a few minutes.
Development of microstructure and grain boundaries	As above. Etch for a few seconds.
Macroetching of pure lead, lead-bismuth, lead-tellurium, lead-nickel	100 cm^3 H_2O, 25 cm^3 glacial acetic acid. 20 cm^3 H_2O_2 (30%), 20 cm^3 HNO_3 (1.40). Etch for 2 to 10 seconds.
Development of microstructure for particularly fine structural phenomena, e.g. precipitates	Diluted Vilella etching medium: 8 cm^3 nitric acid (1.40), 8 cm^3 glacial acetic acid, 84 cm^3 glycerol. Etch for a few seconds.
Development of microstructure for lead-tin and lead-cadmium alloys	Vilella etching medium "extra" 8 cm^3 nitric acid (1.40), 16 cm^3 glacial acetic acid, 76 cm^3 glycerol.
Macroscopic etching	20 cm^3 nitric acid (1.40), 80 cm^3 distilled water. Etching time about 10 minutes. Heat if thick layers of deformed structure must be removed. Rinse in running water, then in alcohol and dry quickly.
Macroscopic etching	(a) 15 g ammonium molybdate, 100 cm^3 distilled water. (b) 58 cm^3 nitric acid (1.40), 42 cm^3 distilled water. Mix equal quantities of solutions (a) and (b). Etch for 10 seconds or longer with constant movement. Quickly rinse in water. Carefully dab off the deposit with cotton wool under running water, then rinse in alcohol and dry.
For lead-antimony-tin alloys (differentiation of Sb and SbSn)	(a) 10 g ammonium persulphate, 100 cm^3 distilled water. (b) 30 g tartaric acid, 100 cm^3 distilled water Mix 5 cm^3 of (a) with 2 cm^3 of (b).
For lead-antimony alloys up to 2% Sb [28]	3 parts of glacial acetic acid, 1 part of H_2O_2 (9%). Etch for 10 to 30 minutes depending on depth of deformed layer. If necessary, subsequent treatment in conc. nitric acid.
For lead-antimony alloys [28]	3 parts glacial acetic acid, 1 part H_2O_2 (30%). Etch for 6 to 15 seconds. In the present author's experience, etching is effected with H_2O_2 (30%) alone.
For unalloyed lead and lead-calcium alloys [28]	2 parts glacial acetic acid, 1 part H_2O_2 (30%). Etch for 8 to 15 seconds.
For lead-antimony above 2% Sb [28], lead-tin up to 3% Sn, unalloyed lead [52]	Electrolytic etching in 60 cm^3 pure perchloric acid and 40 cm^3 water. The specimen is the cathode, a platinum spiral the anode. Etch for 0.75 to 1.5 minutes.

GIFKINS and NICHOLLS developed an etching solution which offers an excellent contrast for the observation of lead in polarized light: 7 parts of glacial acetic acid, 1 part of H_2O_2 (30%), 1 part of ethyl alcohol, 2 parts of water [382a].

For the development of the fine structures of antimonial lead, type metal and lead-rich white metal, a mixture of 1 cm^3 nitric acid (1.40) and 100 cm^3 distilled water can also be used. After etching, the section is kept for a short time in tap water. The lead is dark-etched.

Comprehensive presentations of metallographic technique are available in publications in English by VILELLA [1221] and by WORNER [1287]. The latter publication contains details which are too numerous to be discussed here. WORNER particularly emphasises that if the best results are to be obtained from the microscope, then the etching solution must be individually selected for each alloy.

II. Binary Alloys

1. General Survey. Solidification Phenomena

As the binary alloys in this section are arranged for convenience in alphabetical order, certain analogies between the equilibrium diagrams and relationships with the Periodic System are not immediately apparent. Consequently a short compilation, Table 6, has been provided of the more closely investigated alloys, according to the type of equilibrium diagram (HANSEN [488].)

Table 6. *Types of Binary Systems of Lead. Atomic Radii in Å for Co-ordination Number 12. Value for Lead 1.75 Å*

I. Miscibility gap in the liquid state, no intermediate phases.
Pb–Al 1.43; –Co 1.26; –Cr 1.28; –Cu 1.28; –Fe 1.27; –Ga 1.39; –Mn 1.31; –Ni 1.24; –Si 1.34; –Zn 1.37.

II. Eutectic without intermediate phases.
Pb–Ag 1.44; –As 1.48; –Cd 1.52; –Sb 1.61; –Sn 1.58 (see below).

III. Systems with intermediate phases.
  a) Eutectic on the lead side:
  Pb–Au 1.44; –Ba 2.25; –K 2.36; –Li 1.57; –Mg 1.60; –Na 1.92; –Pd 1.37; –Pt 1.38.
  b) Peritectic on the lead side:
  Pb–Bi 1.82 (see below); –Ca 1.97.
  c) Liquidus curve rising steeply from the lead side:
  Pb–Ce 1.82; –Ge 1.39 (no intermediate phase); –La 1.86; –Pr 1.82; –S; –Se 1.6; –Sr 2.16; –Te 1.7; U.

IV. Large region of solid solution on the lead side.
Pb–Bi 1.82 (see above); –Hg 1.55; –In 1.57; –Sn 1.58 (see above); –Tl 1 .71

The atomic radii are indicated for the co-ordination number 12, i.e. for close packing, corresponding to the structure of the lead. The equilibrium diagram must (at least as a concept) be based on atomic percentages.

The division of the systems into groups has taken place in such a way that their order approximately corresponds to an increasing ability to alloy with lead. The compilation reveals relationships with the Periodic System and the atomic radii.

I. Systems with miscibility gaps on an average involve metals with the smallest atomic radii, thus, having the greatest differences of atomic radii with lead. They belong chiefly to the group of transition metals. Of the metals enumerated, a measurable solubility in solid lead has only been detected for zinc, whereas that for nickel must still be regarded as doubtful.

II. In the systems having a eutectic without intermediate phases, cadmium, antimony and tin have a greater solubility in solid lead than the elements with greater difference of atomic radius, silver and arsenic.

III. The alkali and alkaline earth metals (as far as this has been investigated) form throughout intermediate phases with lead. These are, in part, the so-called Zintl phases, in which the heteropolar types of linkage appears (Dehlinger [240]). Throughout there is a temperature-dependent solid solution region on the lead side, combined with age-hardenability of these alloys. The atomic radii are in the neighbourhood of the value for lead, but, in part, higher. Potassium with the largest atomic radius has only limited miscibility with lead in the liquid state. The three noble metals (gold, palladium, platinum) in this group certainly form intermediate phases with lead, but no solid solutions. The homologous elements, sulphur, selenium and tellurium, alloy with lead in the order shown. Sulphur, with the smallest atomic radius and marked non-metallic character, probably forms a miscibility gap with lead; tellurium even goes into solid solution in lead to a certain extent. The intermediate phases, PbS, PbSe, PbTe, can be described as predominantly heteropolar.

IV. In the systems with the largest solid solution region on the lead side, elements, whose atomic radius is close to that of lead, are involved throughout. Experience shows that extensive solid solution formation between metals is only present if the atomic radii differ by no more than 14 to 15%. The elements which satisfy this condition with respect to lead are in the region of 1.5 to 2.0 Å. This condition is fulfilled by the elements named here. While the approximate equality of atomic radii is a necessary condition for extensive solubility in the solid state, it is however not a sufficient condition. Tellurium has indeed an atomic radius similar to lead, but only a very small solid state solubility in lead. In the case of the combination lead-tellurium, a second condition for sufficient solubility in the solid state is not fulfilled, namely the condition that the alloying partners should have similar values of electronegativity. Electronegativity is understood to mean the tendency of an atom to attract electrons. The electronegativity of lead is given by DARKEN and GURRY [231a], according to GORDY, as 1.5; the elements with greater solubility in lead have electronegativities between 1.0 (Hg) and 1.8 (Bi), whereas of the elements with smaller solubility, tellurium as a value of 2.1 and gold of 3.1. However it is not yet possible to make any certain predictions of the amount of solubility in the solid state, even from simultaneous consideration of the atomic radius and electronegativity (WABER [1233a]). The question of the solubility of alloying elements in lead is also discussed in a comparative systematic investigation by Armour Research Foundation [24a] on fifteen binary alloys of lead. This investigation is particularly significant because the alloys were made from zone-melted 99.9999 lead. The investigation was focussed on the question of solid solution strengthening, oxygen absorption and resistance to recrystallisation. The results will be treated in later sections of this book. In this connection mention may be made of an investigation by TYZACK [1210], in which the change of the lattice constants of pure lead on addition of Sb, Bi, Sn, In, Tl, Cd and Hg was determined. All the elements named, apart from Bi, contracted the lead lattice, as is to be expected on the basis of the differences on the atomic radii.

As regards the analytical determination of alloying elements in lead, reference may be made to special publications [193]. Further, mention may be made of spot tests for the detection of alloying constituents (HIGGS and EVANS [526]), in view of the ease with which they can be carried out.

More extensive knowledge of crystal growth has become available during the last ten years. WINEGARD [1278a] gives an excellent survey. If a pure metal crystallises with a positive temperature gradient, layers of platelets grow (at their edges), giving a terraced interface, where the platelets are parallel to the closest packed planes (111). If the temperature gradient ahead of the growing crystals is negative, i.e. if the melt is thermally supercooled, a pure metal crystallises in dendrites. The dendritic stalks are parallel to the cube-edge directions. The dendrites have this orientation, because several (111) platelet systems develop simultaneously. The dendrites grow faster if supercooling increases.

In order to understand how an alloy solidifies, we consider the distribution coefficient K_0. K_0 is the ratio of the concentration of solute in the solid state to the concentration in the melt. For a simple eutectic phase diagram $K_0 < 1$.

According to TILLER and RUTTER [1198] a liquid layer with a higher concentration of solute exists at the interface if $K_0 < 1$. The concentration in the boundary layer is C_0/K_0 where C_0 is the average content of solute in the liquid. The liquidus temperature within the boundary layer is lower than the liquidus temperature of the original composition. Thus even with a positive temperature gradient, "constitutional supercooling" may exist in the boundary layer.

On account of constitutional supercooling, dendrites may grow from liquid alloys in spite of a positive temperature gradient. With less constitutioned supercooling the alloys crystallise in a cellular structure, where parallel rod-like elements grow perpendicularly out of the interface. The solute elements and the impurities accumulate between the rods.

2. Pb–Ag. Lead-Silver

The phase diagram (Fig. 9) is illustrated by photomicrographs of hypoeutectic alloys (Fig. 10, 11). The eutectic of quench-cast specimens

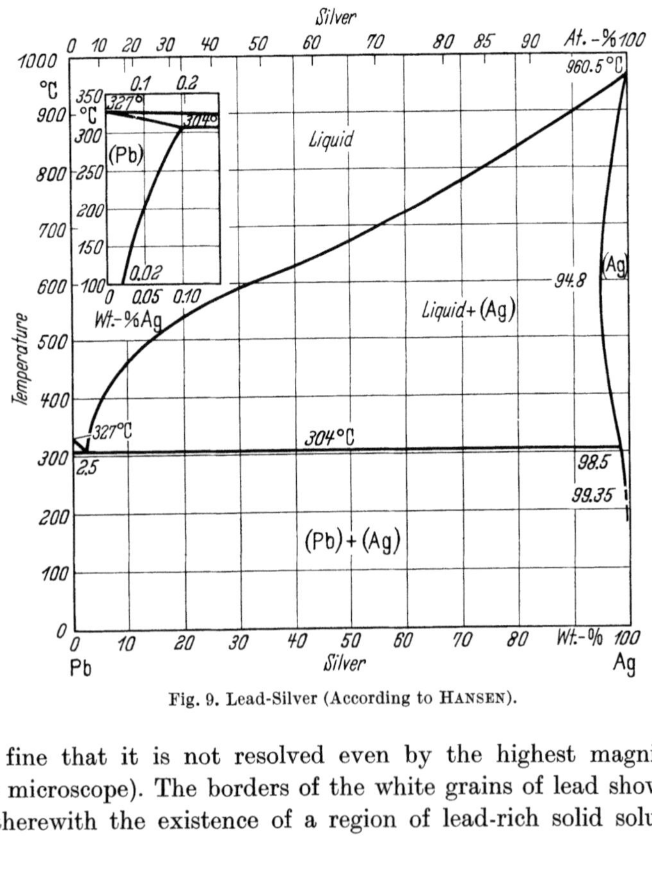

Fig. 9. Lead-Silver (According to HANSEN).

is so fine that it is not resolved even by the highest magnification (light microscope). The borders of the white grains of lead show coring and therewith the existence of a region of lead-rich solid solution. If the eutectic alloy is allowed to solidify in air, the silver crystals appear in the form of needles, which preferably meet at angles of 60 and 120° (RHINES and TIMPE [1007]). The solubility curve on the lead side was redetermined by FUKE and KONDO [353] and is the basis of Fig. 9. Highly supersaturated solid solutions with 10 Wt.-% Pb and above are obtained on the silver side by electrodeposition of alloys according to

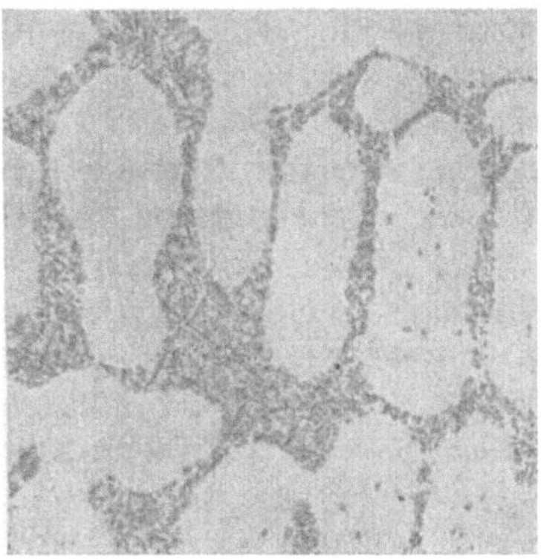

Fig. 10. Section from a cast lead-silver anode. Lead solid solution. Lead-silver eutectic. ×500

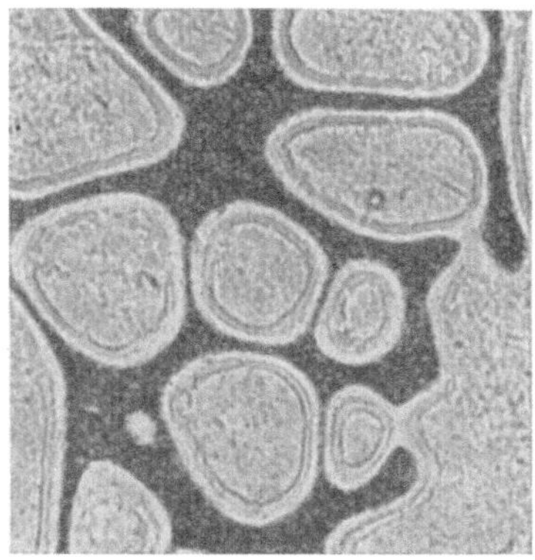

Fig. 11. 1% Ag. Chill-cast. Light: lead solid solution with coring; dark: lead-silver, eutectic. ×1,500

RAUB [993] and ENGEL [994]. HEIDENREICH [503] described peculiar phenomena of hardness increase and softening in alloys with silver content up to 0.026 Wt.-%. In comparison with unalloyed lead, the alloys have essentially higher strength values. The maximum hardness occurs in quenched specimens; storage leads to a softening. The pronounced age-hardening is attributed to the formation of clouds of dissolved atoms near and in the defects of the basic crystal lattice. These defects form the nuclei for the formation of an intermediate phase on breakdown of the solid solution. The intermediate phase has the hexagonal close-packed structure with $a = 2.92$ and $c = 4.76$ Å, and the formula Ag_4Pb. Perhaps the special properties of alloys of low silver content, described in what follows, can be brought into relationship with the processes discussed (see also p. 241).

The high resistance to recrystallisation of the alloys is especially noteworthy. In this respect, with lead-calcium, they are outstanding among the alloys investigated (p. 191) (HOFMANN and HANEMANN [561], RUSSELL [1038]). On the other hand if very slowly cooled specimens were rolled to foil, X-ray investigation showed immediate recrystallisation. This proves that the initiation and the course of recrystallisation is in practice influenced only by the dissolved part of the silver. Rolled alloys above about 0.1% Ag, after heat treatment at 160 to 180 °C, are distinguished by fineness of grain similar to that of lead-tellurium alloys (Kröner [709], Greenwood and Worner [433], p. 101). Additions of silver to lead increase its creep strength. A marked effect was already observed at contents below 0.005% in purest electrolytic lead (99.9995%) (Russell [1039]). After rolling, the specimens were annealed for 15 minutes at 100 °C in order to bring about a defined initial condition. An optimum position was found between 0.01 and 0.05% Ag (Greenwood and Worner [433, 422]). On the other hand, Phelps [958] found an increase of creep strength at a stress of 35.2 $kg/cm^2 = 500$ lb/in^2 up to silver contents of 0.01 Wt.-% only. v. Malotki [569a] confirmed the high creep strength of an alloy with 0.01 Wt.-% Ag by measurements on single crystals grown from the melt.

Among the physical properties of this series of alloys, mention may be made of the density (Matthiessen [812]), the magnetic susceptibility (Spencer and John [1140]), the diffusion coefficients (Seith [1102]) for the passage of solid silver into molten lead between 326 °C and 530 °C (Preekshot [979b]), the electrical resistance of melts (Roll and Uhl [1024]) and their thermodynamic properties (Aldred [14a]).

By and large, lead alloyed with silver shows good resistance to sulphuric acid (p. 273). In practice, hypoeutectic alloys are used as anodes in the electrolytic extraction of zinc (p. 298). The eutectic lead-silver alloy also has limited significance as a soft solder of high melting point (Campbell [173]).

3. Pb–Al. Lead-Aluminium

The lead-aluminium phase diagram is of technical interest because some age-hardened lead bearing metals, particularly Bahnmetall, contain additions of aluminium as a deoxidant (see below). On the other hand, additions of lead are made to aluminium alloys in order to improve their machinability by machine tools. The system contains a miscibility gap in the molten state, the most probable course of which, in the light of an evaluation of the literature, (Hansen [488]) is reproduced in Fig. 12. Careful investigation revealed the exact position of the eutectics on the lead side (Campbell [169], Dardel [231]) and on the aluminium side. The liquidus curve rises steeply on the lead side from the eutectic point (0.021 Wt.-% Al). The data from the literature, given by Hansen [488] however show very marked individual variations. The measurements of thermodynamic magnitudes by Wilder [1271a] at 900 °C are in good agreement with earlier data.

The significance of the aluminium-lead diagram can also be recognised in powder metallurgy. Nowadays friction materials (derived from copper) for clutches and brakes are produced by means of powder metallurgy methods. The copper contains additions of alloying elements, as well as substances which have a special effect on friction and wear, such as graphite, lead, alumina (GOETZEL [400], PIEPER [961], HOFMANN [574]). Perhaps aluminium, also, instead of copper, is not without interest as a basis for such materials. Molten aluminium can receive up to 10 Wt.-% of lead in the form of an emulsion or colloidal suspension (CAMPBELL [169], SCHEIL [1057]).

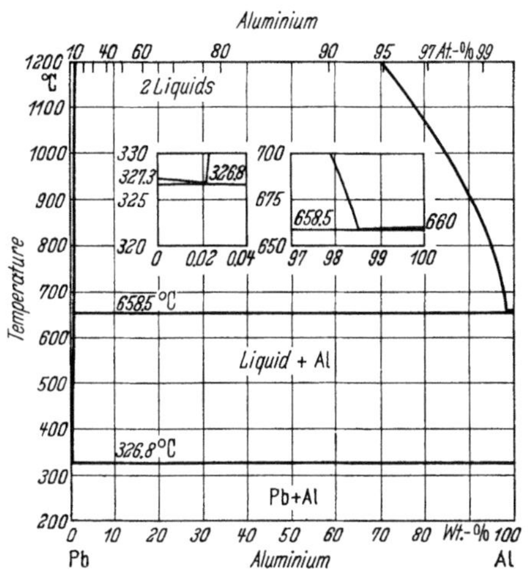

Fig. 12. Lead-Aluminium. (According to HANSEN).

4. Pb–As. Lead-Arsenic

As an illustration of the phase diagram (Fig. 13), a hypoeutectic alloy in the cast state is shown in Fig. 14. The data on the concentration of the eutectic vary somewhat between 2.5 and 4.3 Wt.-% As (HEIKE [504]).

In the Staatliches Materialprüfungsamt at Berlin-Dahlem the properties of the alloys and the solid-state solubility (Fig. 15) were very closely investigated (BAUER and TONN [63]). Particular reference will be made to this work. On melting, a loss of arsenic is to be expected, which is applied in pratice in the refining of lead (p. 8). An alloy with 1.82% As with three repetitions of casting and remelting was held for six hours at 550 °C. The arsenic content sank to 1.44%. The hypoeutectic alloys show no segregation even on slow solidification. In the hypereutectic alloys the arsenic crystallising primarily shows a strong tendency to rise upwards in the melt. At a composition of 8% As, segregation can no longer be suppressed, not even by casting in a strongly cooled chill-mould. The tendency to piping is less in the hypoeutectic and in the eutectic alloys than in pure lead. Similarly the shrinkage of the alloys is lowered as compared with refined lead, and indeed falls from the value 1.02% for refined lead to 0.82% at compositions of 0.1 to 8% As. The density of the alloys sinks with the increasing arsenic content in an almost linear way from 11.34 down to 10.775 g/cm³ at 5% As. According to measurements by HARVEY [496c₁], the surface tension of the alloys also falls with the arsenic content.

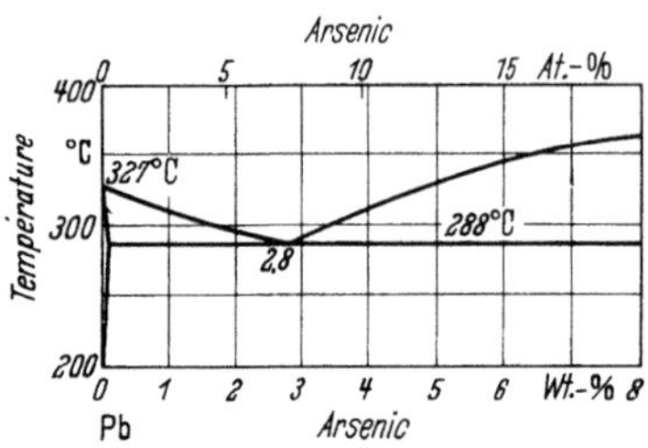

Fig. 13. Lead-Arsenic. (According to HANSEN).

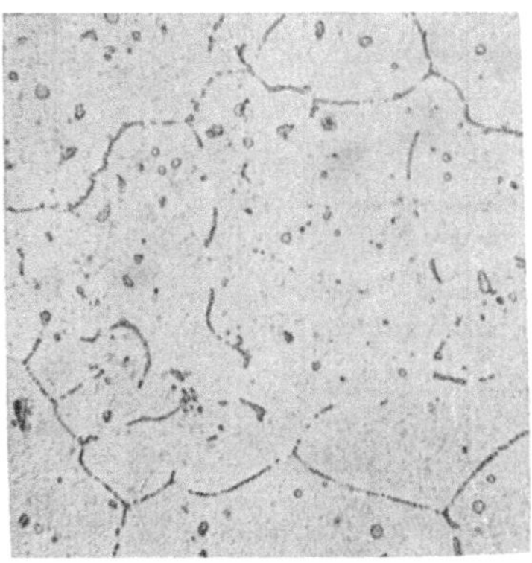
Fig. 14. 0.25% As. Cast. Eutectic particulary. ×500

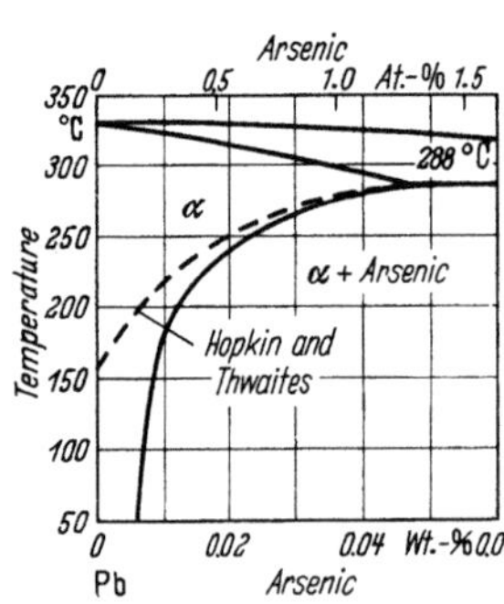

Fig. 15. Lead-Arsenic (According to BAUER and TONN as well as HOPKIN and THWAITES).

The hardness of the materials, particularly at low arsenic contents, depends in high degree on the pretreatment (Fig. 16). The hardness of the cast alloys increases steeply up to a content of about 0.5% As, then more slowly. A considerable increase of hardness is achieved by homogenising at 280 °C and quenching. It depends on the complete saturation of the lead solid solution with arsenic. The course of the curve shows a solubility of 0.05% As in lead at 280 °C. The fall in solubility with decreasing temperature (Fig. 13) could be determined by hardness measurements, as alloys which after quenching from the homogeneous region and annealed in the heterogeneous region (α + arsenic) soften, owing to breakdown. This softening also occurred on storage at room temperature (Fig. 16).

A series of alloys with an arsenic content up to 0.02% was recently prepared from purest TADANAC lead (HOPKINS and THWAITES [588]). The hardness values of the specimens quenched from 300 °C lay on a straight line with a slope of 1.5 Vickers units per 0.01% As. The alloys were therefore homogeneous at 300 °C, in accordance with the data of BAUER and TONN [63]. The hardness measurements on the alloys quenched from 160 °C showed, on the other hand, a lower solubility limit (<0.001 % As) than there indicated. The transition point for superconductivity, corresponding with the low solubility of arsenic in lead, is practically the same as for pure lead [832].

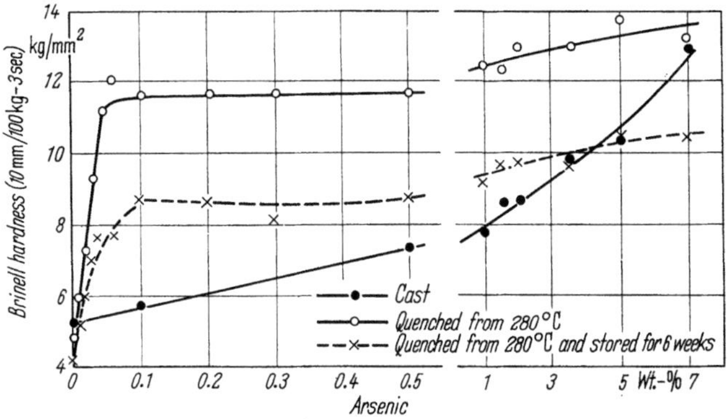

Fig. 16. Brinell hardness of lead-arsenic alloys. (According to BAUER and TONN). (Cf. Fig. 133).

The compressive strength of lead-arsenic alloys is not very high and as a result of softening at room temperature not sufficiently permanent to justify use of the binary alloys as bearing metal, type metal and so forth. A series of compression tests were carried out on alloys in which softening on storage had already set in. The tested specimens from alloys low in arsenic had a very crumpled surface, a sign of coarse grain. The roughness abated at an arsenic content of 1% and disappeared in the hypereutectic alloys.

The binary alloys, above all, play a part in the production of soft shot (p. 339). The alloy with 0.85 Wt.-% As is distinguished by a very small volume shrinkage on solidification. It has therefore been suggested as an alloy for lead blocks for radiation protection, where castings free from pores are required [1256a]. EMICKE [275] recommends the use of the binary alloys as materials for cable sheathing (p. 410). Further, arsenic additions play a part in alloys containing several metals, e.g. in antimonial lead for water pipes (p. 440) and for storage battery grids (p. 355).

5. Pb–Au. Lead-Gold

The system is of interest because unrefined smelter lead in general contains gold up to a maximum of 10 g/ton (FEISER [309]). It is also noteworthy that the classical diffusion investigations of ROBERTS-AUSTEN [1014], were carried out on this system. The equilibrium diagram contains, on the lead side, a eutectic lead-$AuPb_2$ at 85 Wt.-% Pb and

215 °C (HANSEN [488]). According to SPENGLER [1141 a], FUJIKI and co-workers [351 a] showed the phase $AuPb_3$ by metallography and X-ray diffraction. It is formed peritectically near the eutectic temperature. KLEPPA [675 a] reports the heat of solution of Ag and Au in liquid lead (at 450 °C) and the heat of formation of Au_2Pb (at 350 °C). Diffusion experiments on gold in lead indicated a solubility of 0.03 and 0,08 At.-% Au in lead at 170 and 200 °C respectively (HANSEN [488]). Recent investigations on the diffusion of gold in lead have been carried out by an optical method by SCHOPPER [1077] and by means of radioactive isotopes by ASCOLI and co-workers [26, 26 a]. The measurements made by the latter authors took place at temperatures between 190 and 300 °C and indicated an activation energy of 8.91 ± 0.18 kcal/mole at polycrystals and of 9.35 ± 0.07 kcal/mole at single crystals. BOLLING and WINEGARD [105] studied the effect of additions of gold and silver on the grain growth of lead in relation to temperature and concentration (see Fig. 203, p. 192).

By an electrodeless method ROLL and UHL [1024] determined the electrical resistance of molten lead-gold alloys, and found a marked maximum on the resistance isotherms in the region of the intermediate phase $AuPb_2$. Mention may be made of a new thermodynamic study on molten lead-gold alloys (KLEPPA [673, 675 a]). Lead-gold alloys can be electrodeposited (RAUB [993 a]).

6. Pb–Ba. Lead-Barium

The lead-barium equilibrium diagram (HANSEN [488], GRUBE and DIETRICH [444]), Fig. 17, is eutectic on the lead side but for the rest

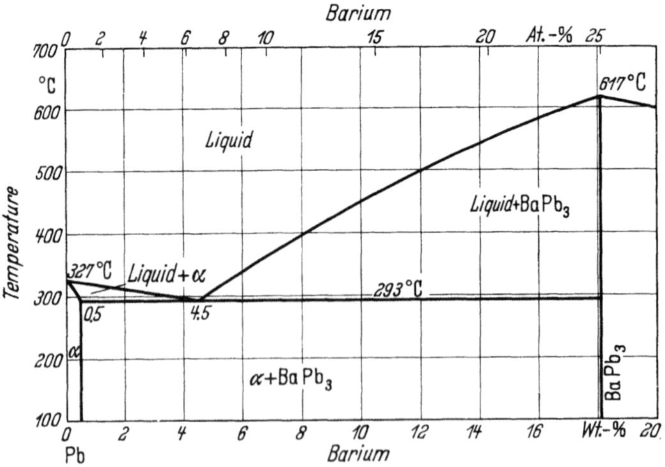

Fig. 17. Lead-Barium. (According to HANSEN).

shows similarity with the lead-calcium diagram. A hypoeutectic alloy is shown in Fig. 18. $BaPb_3$, in the form of white "needles", crystallises from hypereutectic alloys. Therefore the cubic crystal system, as in the corresponding calcium compound, should not be present.

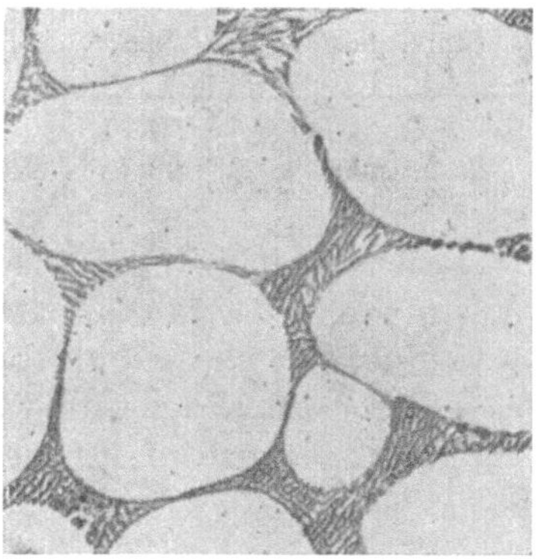

Fig. 18. 0.8% Ba, solidified in a crucible. Lead solid solution. Network of eutectic. × 500

At the eutectic temperature, lead takes up about 0.5 Wt.-% Ba in solid solution, at 130 °C 0.42 Ba. Homogenised alloys with barium contents down to 0.024% age hardened on storage after quenching (Fig. 19). Accordingly, the solubility limit of barium in lead at room temperature is to be put at about 0.02 Wt.-% (SCHMID [1069]). The percentage increase of hardness is highest in alloys with about 0.6% Ba, which are near to the solubility limit at the eutectic temperature. Table 7 gives the values of cast hardness of "normally cooled" reguli, according to GRUBE [444].

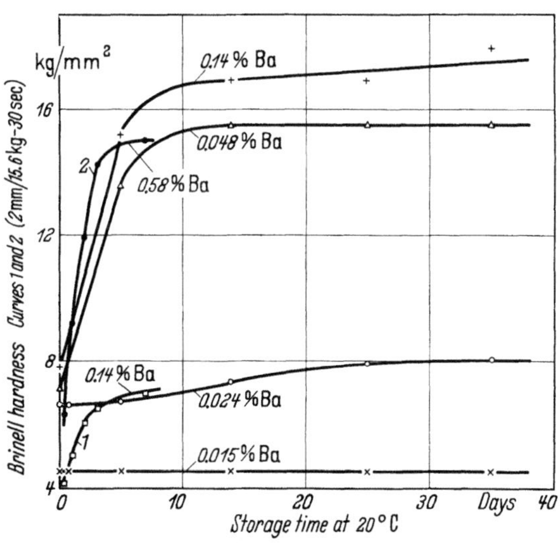

Fig. 19. Age-hardening of lead-barium alloys after quenching from 250 °C. Period of heat-treatment 7 days. (According to SCHMID). Curves 1 and 2 by GRUBE and DIETRICH.

Pure lead-barium alloys have found only slight application. As barium is very reactive, alloys of high content decompose in air; moreover the barium content falls on remelting more than is the case in the

Table 7. *Brinell Hardness of Cast Lead-Barium Alloys*
Ball diameter 2.5 mm, load 15.6 kg, 30 s

Wt.-% Ba	0	0.29	0.90	1.14	1.89	3.14	3.57
Brinell hardness	4.0	8.3	15.1	14.8	15.7	19.8	25.4
Wt.-% Ba	4.15	5.61	6.20	6.86	8.53	10.36	
Brinell hardness	28.2	28.5	31.3	35.0	37.3	45.4	

lead-calcium alloys. Again, alloys of low barium content lack the hardness necessary for bearing metal. On the other hand, barium together with other alkaline earth metals, e. g., calcium, comes into consideration as a component of age-hardenable lead alloys and brings about here an additional increase of hardness (p. 386). For possible application in radiation protection (see p. 129).

7. Pb–Bi. Lead-Bismuth

While the system lead-bismuth, according to the earlier views, was a simple eutectic system, recent investigations have all indicated an intermediate, incongruent-melting phase (β) with the hexagonal

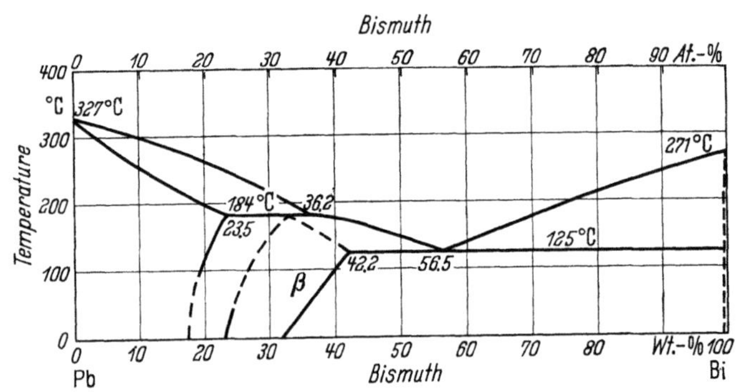

Fig. 20. Lead-Bismuth. (According to HANSEN).

close-packed structure. Latterly it has also been detected under the microscope (HO [535], MITANI [859], SOLOMON and MORRIS-JONES [1135]). The existing investigations of the equilibrium diagram (Fig. 20) were effectively supplemented by the application of thermodynamic analysis by OELSEN [914] (see also GERSHMAN [369], DOUGLAS and DEVER [256]). OELSEN also obtained information on the solid state from his measurements. Among other things it appears noteworthy that the heats of formation of the molten alloys are negative, whereas those of the solid alloys have a positive sign. This means that the solid alloys are formed with absorption of heat.

Numerous investigations have been directed to the properties of the alloys in the molten state, including vapour pressure and thermodynamic activity of lead and bismuth in their alloys (KLEPPA [675], GONSER [407]), density (MATSUYAMA [811]), diffusion (GRACE [412], NIWA [901], ROTHMAN [1036]), electrical resistance in relation to temperature and concentration (MATSUYAMA [808]). An irregularity in the curve of the electrical resistance may be related to the presence of an intermediate phase in the molten state, as frequently the special type of bond in the solid state still makes itself noticeable, even in the melt.

Bismuth and lead are among the elements of low neutron absorption; their alloys have low melting points and are of interest for heat transfer in nuclear reactors (see the section on Heat Transfer). A large number of investigations have therefore been directed to the question of heat transfer to steel, on agitated alloy melts of eutectic composition. Mention may be made for example of JOHNSON [623, 624], SEBAN [1096], LUBARSKY [769]. Further, in this connection, the attack of lead-bismuth melts on materials of construction is important (cf. AKUTAGAWA [11a]). Additions of titanium or zirconium to the melt are said to reduce the attack (KAMMERER and co-workers [643]). GANGLER [355] found that at 1,093 °C only molybdenum, purely ceramic materials and cermets (with the exception of ZrC) showed satisfactory resistance. The solubility of molybdenum in pure lead at 1200 °C is below 0.005 Wt.-% [14]. All high temperature alloys, on the other hand, were noticeably attacked at 815 °C. The resistance of molybdenum between 760 and 1093 °C was also emphasised by GRASSI [414] (cf. CATHCART and MANLY [184]. WILKENSON, HOYT and RHUDE [1273]). STACHURA [1142a] gives a survey of the properties of the eutectic alloy important in reactor technique.

As lead contracts in solidification, while bismuth expands, there must be an alloy which solidifies without change of volume. This is on the bismuth side of the system (WIEDEMANN [1268], TAKASE [1163]). In the solidification of the eutectic alloy (55.9% Bi), PRESTON and BROOMFIELD [980] were able to measure a contraction of $1.52 \pm 0.1\%$ of the volume of the solid phases. Various investigations have been made of the physical properties of lead-bismuth alloys in the solid state, to the electrical resistance (v. HOFE and HANEMANN [548], SCHULZE [1088]), the density (GOEBEL [391]), and the thermal conductivity (SCHULZE [1088] and MIKRYUKOV [852]).

Although the lead solid solution region resulted satisfactorily from X-ray back reflection photographs and from electrical resistance measurements, the photomicrographs of alloys which were far within this region exhibited structures which could first be regarded as precipitates (Fig. 21). The correct interpretation was obtained by preparing an X-ray back reflection photograph of a microsection with "precipitates". Then the X-ray photograph was repeated in the same part of the section after electrolytic deep etching: while the first picture showed interference rings containing many spots, the second one only showed few interference spots. This indicates fine grain in the polished, coarse grain in the deep-etched section surface. Thus, in polishing, recrystallisation has taken place within the outer layer of the section surface. In Fig. 20, it is noteworthy that the change in the surface only occurs within the grain. The original grain boundaries are retained, as here the deformation is prevented. In a precipitation process, the precipitates would be expected preferentially at the grain boundaries.

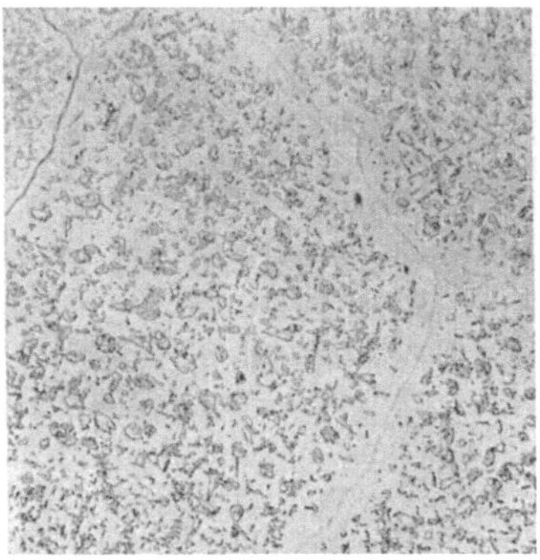

Fig. 21. 1% Bi. Cast. The dark areas inside the grains have originated by recrystallisation of worked surface layers. ×150

The phenomenon described above was never observed in the polishing of soft lead sections. Rather does it occur from about 0.1% Bi onwards, becoming more marked with increasing bismuth content. Thus we have as a characteristic of bismuth-containing lead, that recrystallisation is accelerated as compared with that of pure lead. Measurements of the self-diffusion of lead showed (in agreement with the above observations) an increase of the diffusion coefficient with rising bismuth content.

The hardness of lead is little increased by additions of bismuth (e.g. Di Capua and Arnone [179]). Thus it can be understood that even alloys with higher bismuth contents (e.g., above 12% Bi) can easily be rolled. According to the result shown in Fig. 22 (Thompson [1189]), the hardness increases slightly up to a content of 20% Bi, remains approximately constant between 20 and 45% Bi, and then increases again. No age-hardening of cast specimens was observed in the region up to 35% Bi not even after seven months. On the other hand, the hardness of alloys with above 35% Bi in the cast state and after quenching from 120 °C increased slightly on storage, in the latter case within a day. The suspicion, that we have an age-hardening phenomenon on the basis of the solubility curve of the hexagonal β-phase as against bismuth, was confirmed in later investigations (Fig. 23). Tensile tests were carried out on cast rods of 48 mm gauge length at the uniform strain rate of 12 mm/min. The curves of tensile strength and elongation at fracture showed irregularities in the region of 25 to 35% Bi which could be attributed to the occurrence of the then unknown β-phase (Fig. 22).

The creep rate of bismuth-containing lead with bismuth contents up to 0.1%, after rolling as well as after subsequent annealing, was measured for stresses of 35 and 24 kg/cm² (GREENWOOD and WORNER [432]).

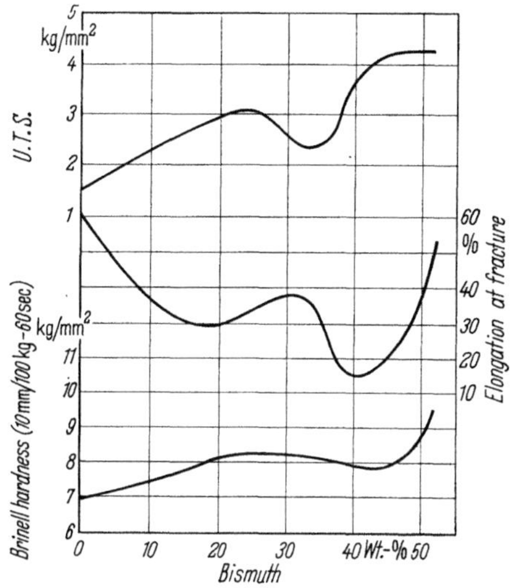

Fig. 22. Mechanical properties of cast lead-bismuth alloys. (According to THOMPSON [1189]).

Particularly after annealing, only insignificant differences as compared with soft lead (99.9915%) were found. Investigations at the Bleiforschungsstelle showed a similar result. Also for the fatigue strength of bismuth-containing lead, no noteworthy deviations from the behaviour of soft lead were found.

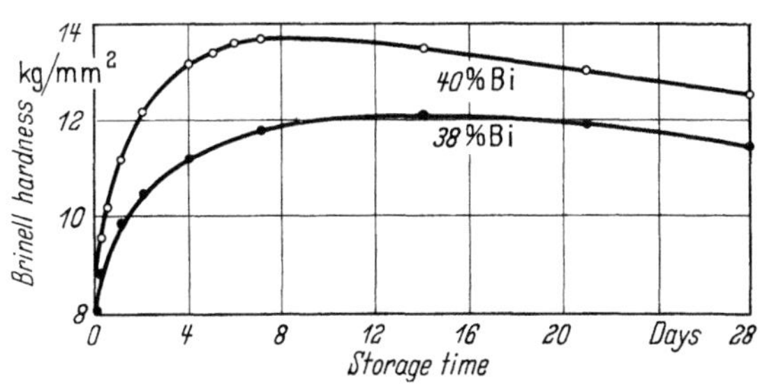

Fig. 23. Age-hardening of the hexagonal lead-bismuth phase (β). (According to Ho [535]).

While the grain size of lead in the cast and in the recrystallised condition is reduced by many alloying additions, this is not the case for additions of bismuth, as they go into solid solution (p. 196). A slight refining of grain in extruded cable sheathing at a bismuth content of 0.18% was first found (EMMERICH [278]).

Bismuth has frequently been described as "injurious" to lead. In fact, lead-bismuth alloys possess certain peculiarities. Thus their marked inclination to recrystallisation should be noted. As, however, this rarely results in any more intense grain growth than in soft lead, it cannot be regarded as a special disadvantage. The mechanical properties of bismuth-containing lead show hardly any deterioration as compared with soft lead. Bismuth appears to be without special effect on the behaviour of lead towards the corrosive effects of the atmosphere, of water, of soil, and presumably also of cold sulphuric acid (pp. 275 and 308).

According to the results of laboratory experiments, bismuth contents up to 0.05% are without effect on the durability and the self-discharge of lead-antimony grids for storage batteries. On the other hand, a content of 0.05% Bi in Planté-type plates caused a deterioration in the dimensional stability and the durability (Grosheim-Krisko [443]). A bismuth content is also disadvantageous to the resistance of lead to sulphuric acid at higher temperatures. Here, however, there can be no doubt about the compensating effect of other additions, particularly copper. Bismuth-containing lead is undesirable for the production of lead-alkaline earth alloys, as calcium, for instance, is removed by bismuth from the lead (p. 8). Finally, bismuth-containing lead is not suitable for the production of white lead and of optical glasses (George and Ensslin [365], Butcher [161]).

Lead-bismuth alloys were applied or proposed as solder for glass-to-metal joints, and for the purposes of printing or reproduction [147], since details of the surface are said to be very sharply reproduced (Cournot [220]). The molten alloys are recommended as a liquid bath for heat treatment (Smith [1129]). Above all, bismuth is an important component to polymetal alloys of lead, in the first place the low-melting solders. Small additions of bismuth are, however, also used in type metal and bearing metals on the lead-antimony-tin basis, as the casting properties are said to be improved thereby (Thompson [1189]).

8. Pb–Ca. Lead-Calcium

a) Structure and General Properties of the Alloys. In accordance with the equilibrium diagram in Fig. 24, primary $CaPb_3$ crystallises from alloys with more than 0,07% Ca on solidification. At calcium contents between 0.07 and 0.10%, in agreement with the equilibrium diagram, a re-solution of the $CaPb_3$, which has crystallised out, is observed on annealing close below the peritectic at 328.3 °C (Schumacher and Bouton [1090]). The fact that the alloy shown in Fig. 25 is already single phase in the cast state is related to the inhibition of crystallisation of $CaPb_3$ on rapid passage through the two-phase region of melt + $CaPb_3$.

As shown by FALKENHAGEN [304], it is possible to obtain alloys with calcium contents up to 0.16% as supersaturated solid solutions on sudden solidification of melt by means of an evacuated sucking-chill mould.

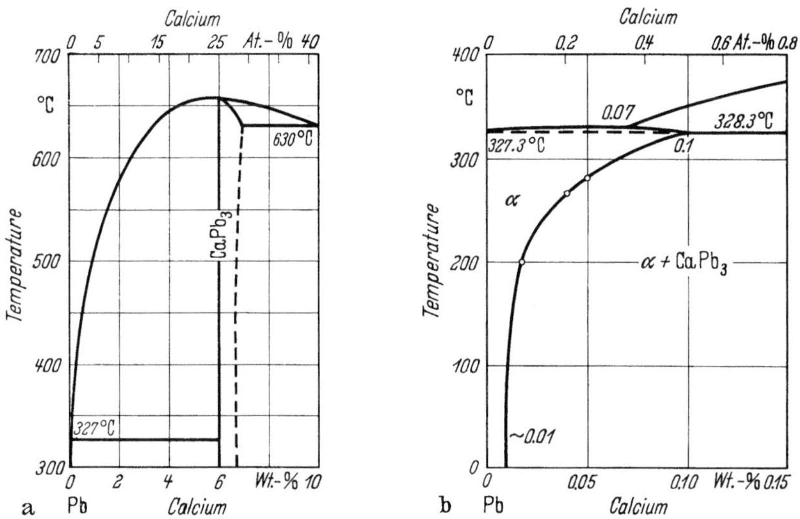

Fig. 24. Lead-Calcium. (According to HANSEN)

Further examples of similar behaviour among the lead alloys are lead-sodium and lead-tellurium (pp. 67 and 101). Characteristic of the alloy in Fig. 25, and general in alloys of low calcium content, are the remarkable serrated grain boundaries. These serrations are sometimes so marked that parts of the crystals appear constricted. An explanation of this phenomenon is still lacking.

Honeycomb-type coring is frequently observed in castings of lead-calcium alloys (Fig. 26). It has been described also in other cases of supersaturated solid solutions from the melt, e.g. in the systems Al–Mn and Al–Ti (FALKENHAGEN [304], HANEMANN [479], BÜCKLE [138]). As experiments on an alloy with 0.15% Ca showed the occurrence of honeycombs and the absence of $CaPb_3$ crystals on casting in a chill mould heated to 200 °C, and on the other hand the occurrence of $CaPb_3$ on casting in a cold mould (v. GÖLER [393]); this observation and its interpretation may need re-examination (compare [1033]).

Primary $CaPb_3$ occurs in the form of cubes or star-shaped dendrites, as shown in Fig. 27 for a cast alloy. The crystals appear white after polishing and tarnish dark after etching. The formation of $CaPb_3$ inclusions owing to precipitation in the solid state is illustrated in

Fig. 28. The cubic face-centred crystal lattice belongs to the structure type of $AuCu_3$ (ZINTL and NEUMAYR [1306]). The density is 9.40 g/cm³.

Attention may also be drawn to the difference in grain size between single phase and two-phase solidified lead-calcium alloys, as may be

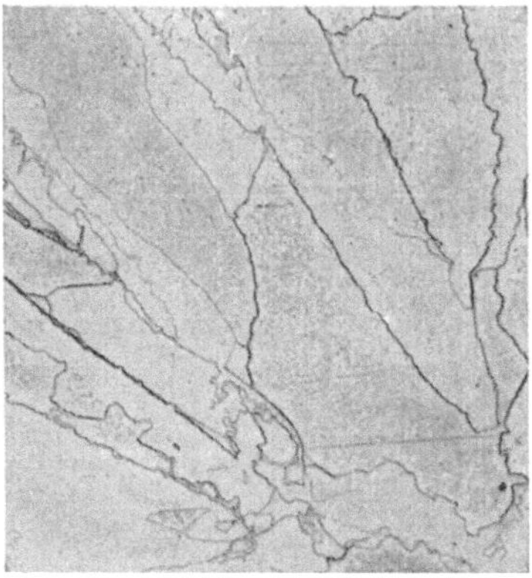

Fig. 25. 0.096% Ca. Cast. Lead solid solution. ×50

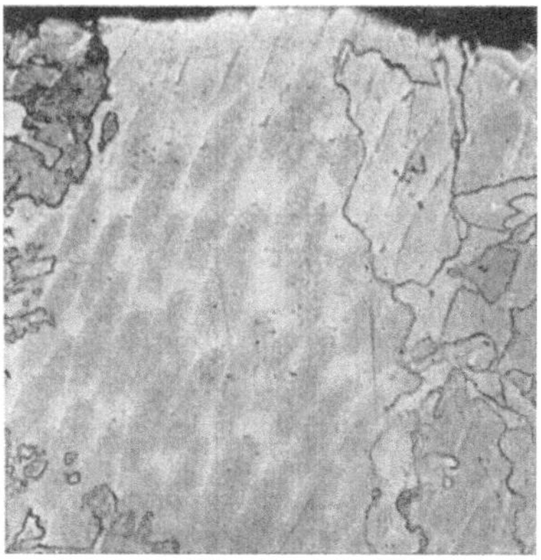

Fig. 26. 0.1% Ca. Battery grid with honeycomb type coring. Honeycomb structure not affected by later formation of smaller grains. ×100

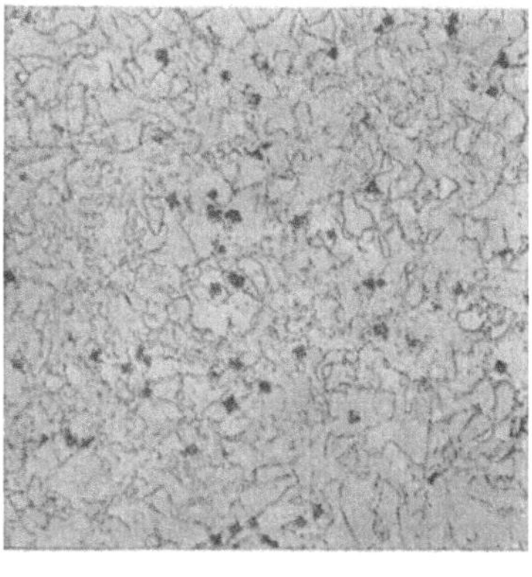

Fig. 27. 0.185% Ca. Cast. Dark spots are primary crystals of $CaPb_3$. Matrix is lead solid solution. ×150

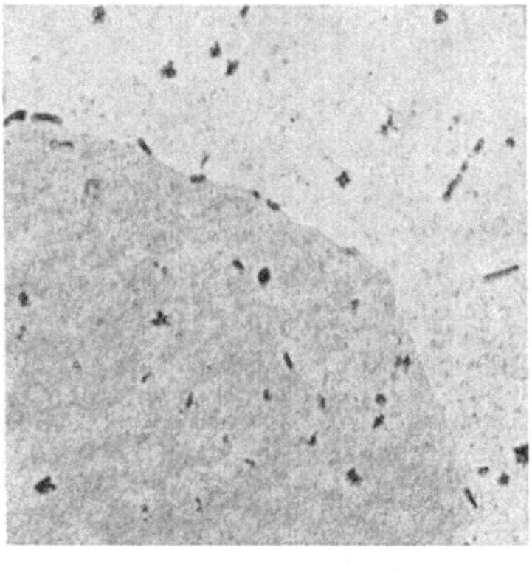

Fig. 28. 0.096% Ca. Heat-treated at 290 °C and cooled during 3 days. Dark small rods and stars: precipitates of $CaPb_3$. ×300

recognised by comparison of Figs. 25 and 27, taking note of the magnification. The fine grain of the alloys of high calcium content was held responsible for the low creep strength of these alloys in comparison with those of lower calcium content (pp. 50, 211 and 233).

Lead-calcium alloys are best prepared with the aid of a hardener alloy, which is obtained by interaction of molten lead with a slag containing calcium. Hardener alloys with 2.5 to 4% Ca are obtained, which, under some circumstances, may also show a content of alkali metal (sodium). Hardener alloys with 2 to 4% Ca can also be prepared by interaction of CaC_2 with Al and Pb at 1,150 °C (RODYAKIN [1017]).

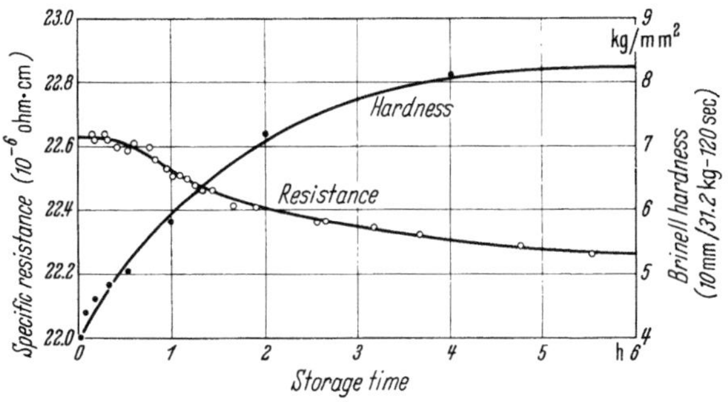

Fig. 29. 0.07% Ca. Change of hardness and of electrical resistance of an alloy quenched from 315 °C [502].

Electrolysis of molten $CaCl_2$–KCl mixtures with Pb as cathode has also been used on the laboratory scale (MASLANKA-ORMANOWA [803]). The precautions to be observed in alloying are mentioned on p. 322. The density of the alloys is reduced by 0.029 g/cm³ for each 0.1% Ca (v. GÖLER and WEBER [398]). Their casting properties are indicated in the section on storage batteries (p. 354).

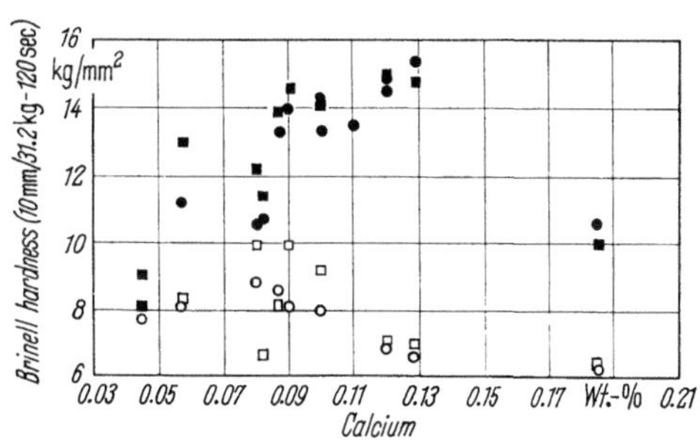

Fig. 30. Age-hardening of cast lead-calcium alloys [502].
○ 1 day after air cooling;
□ 10 days after air cooling;
● 1 day after quenching;
■ 10 days after quenching.

b) Mechanical Properties and Age-hardening. The age-hardening of lead-calcium alloys is very pronounced, above all at low calcium contents, so that an indication of mechanical properties without consideration of the age-hardening state seems insufficient. As Fig. 29 shows, the incrase of hardness takes place within a few hours. In a series of experiments the alloys were cast from 500 °C into an iron chill mould at 200 °C, removed from the mould immediately at the end of solidification and either quenched or cooled in air. The hardness attained by the air-cooled alloys was lower than that of the quenched alloys (Fig. 30). The Brinell hardness of the air-cooled and of the quenched alloys has a maximum at the calcium content of 0.085 and 0.13% respectively. An increase of the calcium content of the cast alloys above ≈ 0.10% Ca thus produces no essential increase of hardness (p. 265).

A number of investigations have been directed to the effect of heat-treatment of the alloys on the age-hardening. No increase of hardness over that in air cooled castings was obtained by a homogenising heat treatment and subsequent quenching. This can be understood in the light of the remarks made above about the suppression of crystallisation

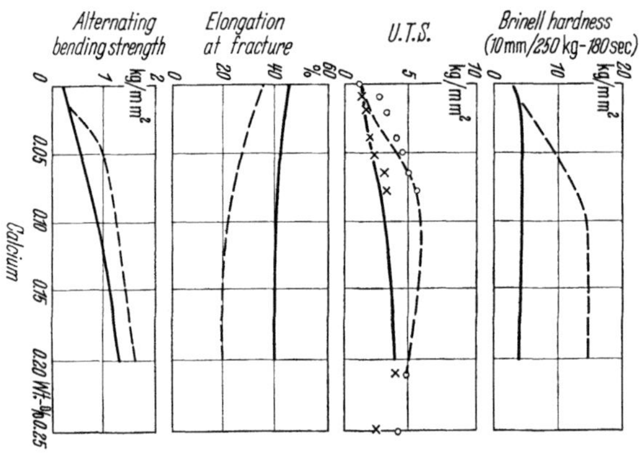

Fig. 31. Mechanical properties of extruded lead-calcium alloys. Alternating bending strength for 2 · 10⁷ alternations. Broken line: age-hardened, continuous curves: soft. Curves by v. Göler [393]. Data of [1090] shown as o and ×.

of Pb_3Ca on rapid solidification of the alloys. As the only means of increasing the hardness further, additions, for instance of lithium, barium or sodium, should be considered. Additions of antimony or bismuth have the opposite effect, as they bind calcium in the form of high melting intermetallic compounds (metallides), which segregate from the melt (Kroll [712]). The hardness and the tensile strength of heat-treated alloys, like those of cast alloys, show a maximum at about 0.1% Ca. Fig. 31 gives the result of such an investigation. The soft condition is obtained in the usual way by slow cooling of the specimens or extrusion at low temperatures. The age-hardened condition occurs after homogenising the specimens just below the melting point and quenching. The diagram also contains data on the elongation at fracture and the fatigue strength (v. Göler [393]).

The strength of age-hardened alloys depends on the quenching temperature, as Fig. 32 shows for a limited number of alloys with calcium contents between 0.02 and 0.06%. After quenching from about 300 °C, the alloy with the highest calcium content attains the highest tensile strength, whereas at the quenching temperature of 250 °C, the alloy with 0.04% Ca becomes the strongest. This regularity is in agreement with the course of the solubility curve (Fig. 24). In order that an alloy should fully age-harden after heat treatment, it must be quenched from the α-field of the equilibrium diagram (p. 265). The corresponding situation holds for non-heat-treated extrusion alloys. For full age-hardening, they must be extruded in the phase area of solid solution. If, for example,

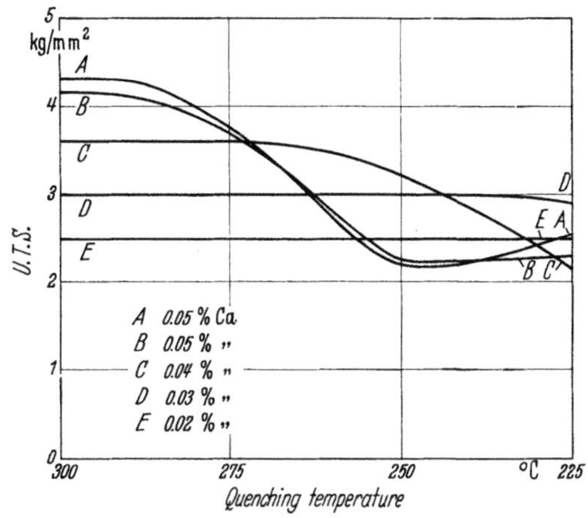

Fig. 32. Lead-Calcium. Tensile strength of alloys quenched from various temperatures after 10 days storage. (According to DEAN and RYJORD [238]).

250 °C is assumed as the extrusion temperature, the calcium content should not amount to more than 0.035%.

The age-hardening of lead-calcium-alloys is accelerated by the use of a raised storage temperature. By this means (in contrast with the lead-antimony alloys) a slight increase of strength can still be realized in some circumstances (Fig. 33).

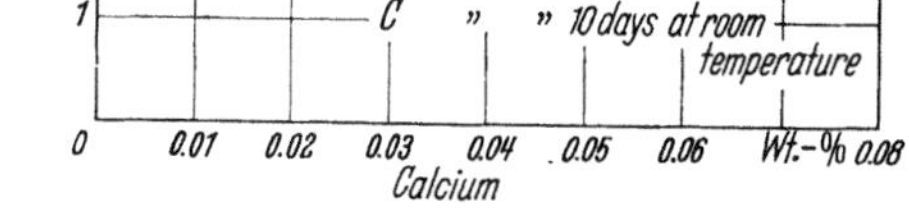

Fig. 33. Lead-Calcium. Tensile strength of alloys quenched from 300°C after storage at various temperatures. (According to DEAN and RYJORD [238]).

The age-hardening is connected with the breakdown of the supersaturated solid solution, as is shown by the course of the electrical resistance (Fig. 29). Some observations indicate the possibility of detecting the precipitates by means of the microscope. In a specimen with 0.096% Ca, after $1^1/_2$ years of storage during which the hardness did not noticeably alter, areas could be observed which were essentially more darkly etched than the rest of the grain. The areas of dark etching attack had irregular boundaries, and it seems as though they penetrated into and absorbed the light-coloured areas. Here and there little rod-shaped precipitates could also be seen. An electron microscope investigation of such specimens is urgently desired.

The high resistance of age-hardened lead-calcium alloys to recrystallisation after room temperature working is discussed on (p. 192.)

The creep strength of the alloys is considered elsewhere in connection with the behaviour of other types of lead (pp. 240 and 354). For this reason only a few special questions will be discussed here. The various cast lead-calcium alloys of Fig. 308 show considerable differences in creep behaviour. It seems peculiar that the alloys of higher calcium content sometimes creep more rapidly than those with lower calcium content. The reason for this might be found in the fact that the alloys of higher calcium content solidify with finer grain than those poorer in calcium, and further that the calcium-rich alloys have sometimes a heterogeneous structure and then are more feebly age-hardened than the solid solutions of lower calcium concentration which have solidified homogeneously. The type of cooling after casting also has an effect on the age-hardening and thus on the creep rate.

The fatigue strength of the alloys is treated elsewhere (p. 259).

Lead-calcium alloys have been suggested for various purposes in the place of lead-antimony alloys, e.g. for pipe, wire, cable sheathing and storage battery grids. The fact that these suggestions have not been fully exploited is due particularly to the difficulty of maintaining an exact calcium content on melting and casting. However the alloys are still in use today in individual cases. In the hardened lead bearing metals, calcium, among other additions, forms an important alloying constituent.

9. Pb–Cd. Lead-Cadmium

The equilibrium diagram of the system (Fig. 34) was extended by Oelsen [913, 918], compare Nagasaki [890] by consideration of the heat contents of the alloys with the aid of quantitative thermal analysis Thus a clue could be obtained to the solubility of cadmium in solid lead, which is hardly possible by ordinary thermal analysis. Disregarding the details of the solid model set up by Oelsen (for example curvature of the specific heat-content plane of the melts at about 50% Cd), it can be seen that the specific heats of the melts are formed almost additively from those of the molten components, and further that the lead-cadmium melts are almost "regular" mixtures. Schürmann [1085] further evaluated the results of the measurements by Oelsen and derived the position of the miscibility gap in the region of supercooled melts, and the vaporisation equilibria. Reference must also be made to the measurements of the heat of mixing in the region of the molten alloys and to the thermodynamic deductions of KLEPPA [676].

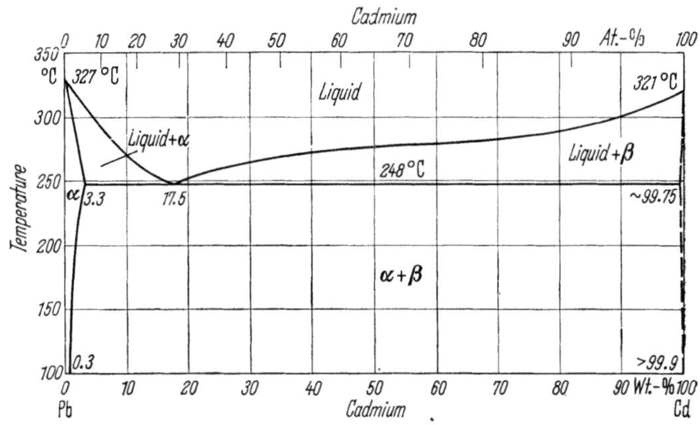

Fig. 34. Lead-Cadmium. (According to HANSEN as well as ROLLASON and HYSEL [1031]).

The solubility of cadmium in solid lead was determined at 232 °C as 2.5 Wt.-% (PASTERNAK [937]), at room temperature as about 0.3 Wt.-% (Fig. 34). In supersaturated alloys breakdown occurs on storage already (p. 267). It can clearly be seen in Figs. 35 and 36. Further, in Fig. 35 it is noteworthy that in the honeycomb segregation zones only the solid solution concentration was increased on solidification, but still no eutectic occurred. Particularly beautiful, oriented precipitates occur

Fig. 35. 1% Cd. Cast. Lead solid solution with honeycomb-type of orientated precipitates formed on storage. ×500

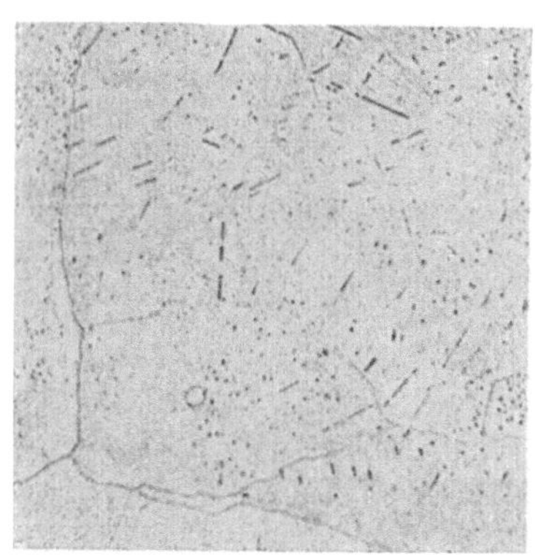

Fig. 36. 1% Cd, quenched after 4 days at 235 °C. Precipitates of cadmium. ×500

in the alloy with 1% Cd on slow cooling after homogenisation. The precipitates emanate from the grain boundaries, i. e., they develop discontinuously (LIVINGSTON [754c]).

Density and resistance measurements in the liquid state were carried out (MATSUYAMA [808, 811]). Disregarding a few small irregularities, the densities of the alloys fall in an approximately linear way from the value for lead to that for cadmium (8.64 g/cm³) (GOEBEL [391]). NIWA [901] gives values for the diffusion of cadmium in lead.

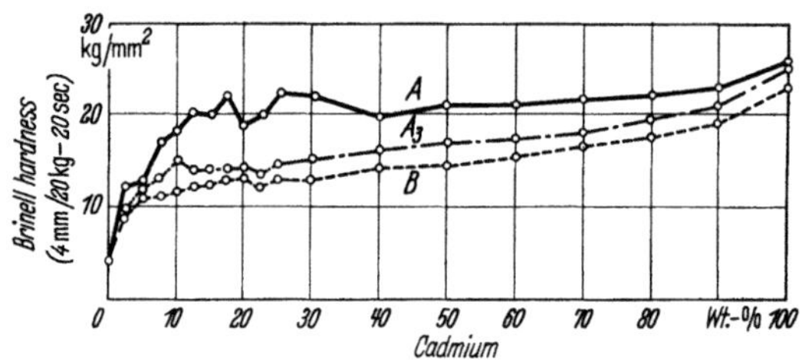

Fig. 37. Hardness of lead-cadmium alloys. *A* chill cast; A_3 the same after 3 months; *B* slowly cooled. (According to GOEBEL [391]).

Since the supersaturated lead solid solution, as mentioned already, breaks down gradually at room temperature, the mechanical properties of the alloys must depend to a high degree on the pretreatment and on the duration of storage. In fact, the curves in Fig. 37 show that the hardnesses of the quenched cast alloys fall after three months of storage

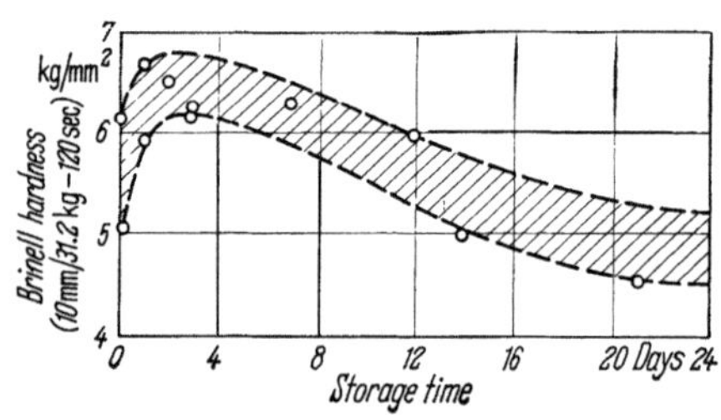

Fig. 38. Change of hardness on storage of quenched alloy with 2% Cd.

almost to the values of the slowly-cooled specimens. Noteworthy also on the curve of the cast alloys is the steep increase of hardness up to the eutectic composition, as in the lead-antimony alloys. The fall of hardness of supersaturated alloys on storage was confirmed in the author's own investigations (Fig. 38). In this connection the transient occurrence of age-hardening should be noted.

Further strength values are given in Table 8 (Cournot [219]).

Table 8. *Strength Values for Lead-Cadmium Alloys*

Wt.-% Cd		Tensile Strength kg/mm²	Elongation at Fracture %	Brinell Hardness		Simple Bending Number (Bending angle 90°)
Weighed	Analysis			at 20 °C [1]	at 175 °C [2]	
0	0	1.5	31	4.8	1.1	7
1.5	1.74	3.4	15.5	8.3	2.1	5
3.0	2.89	4.8	13.5	11.4	2.4	3

[1] ball diameter 10 mm, 250 kg [2] ball diameter 10 mm, 175 kg

The hardness measurements and tensile tests were carried out on cast specimens, namely the tensile tests on round rods of cross-section 150 mm² and gauge length 140 mm. For the alternating bend tests, rolled specimens were used. The details of the experimental method are not complete, so the values only provide a first approximation. A comparison of the hardness values with those in Fig. 38 suggests that the experiments were carried out on freshly cast specimens.

Melts of lead with higher cadmium content are prone to considerable drossing [913]. Binary lead-cadmium alloys have found little application. They have been proposed, for instance, for storage battery grids of lowest possible selfdischarge (Vinal [1224]). Small additions of cadmium, with antimony and tin, form constituents of cable sheathing and pipe alloys; larger amounts are found particularly in low melting solders (p. 387).

10. Pb–Co. Lead-Cobalt

Pelzel [944] and Stevenson [1145a] determined the solubility of cobalt in molten lead. The value found by Stevenson for the solubility at 700 °C is more than one order of magnitude below that found by Pelzel (0.12 Wt.-%). According to Schenck [1058a], the lead content of the cobalt-rich phase at 1550 °C is only 0.98 Wt.-% and the cobalt content of the lead-rich phase 0.33 Wt.-%, roughly in agreement with the results of Stevenson.

11. Pb–Cr. Lead-Chromium

To the rather incomplete diagram set up by Hansen [488], Alden and co-workers [14] gave supplementary values for the solubility of chromium in molten lead:

Dissolved amount Cr in Wt.-%	0.16	0.06	0.05	0.04	0.03	0.02	0.01
Temperature °C	1,210	1,104	1,048	1,014	1,002	992–964	908

12. Pb–Cu. Lead-Copper

Until very recently the data of the miscibility gap of the diagram (Fig. 39) were quite variable. This is because it is very difficult to suppress the stratification on cooling of alloys from the region above the miscibility gap. In addition to the values of the critical point of 65% lead and 1,000 °C (Friedrich and Waehlert [345], Briesemeister [133]), which were obtained by chemical analysis of the layers,

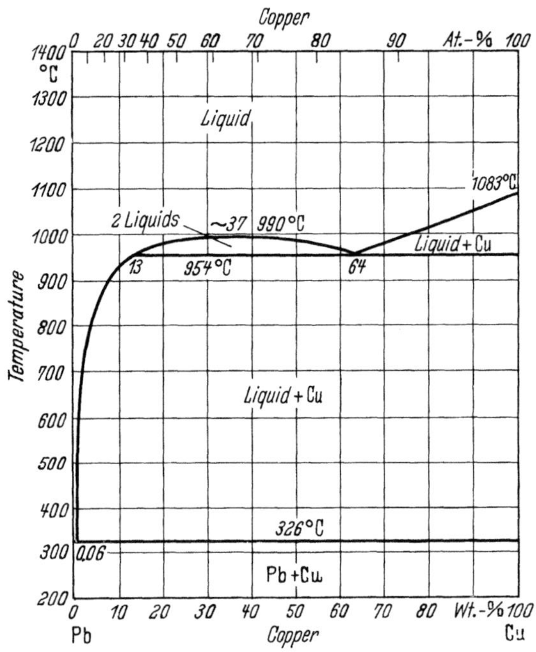

Fig. 39. Lead-Copper. (According to Hansen).

there was an indication of temperatures above 1,500 °C (Bornemann and Wagenmann [115]), which was based on conductivity measurements. The contradictions were traced back to the fact that above 1,000 °C a stratification of the finely divided emulsion should not occur and a homogeneous melt should be simulated (Claus [201]). A more recent investigation brought further elucidation for the extension of the miscibility gap (Bish [87]). The electrical resistance of a melt of 65% Pb, remainder copper, was measured at various distances from the bath surface by means of an immersed probe of insulated tungsten wires. At a temperature of 1,007 °C, no change occurred in the resistance in relation to the depth of measurement, whereas measurements at 994 °C gave a discontinuous increase of resistance at a certain depth. After the critical point had been found in a further investigation at below 1,100 °C (using thermodynamic considerations), the view of BORNEMANN [115] can be regarded as finally refuted and the diagram set up according to

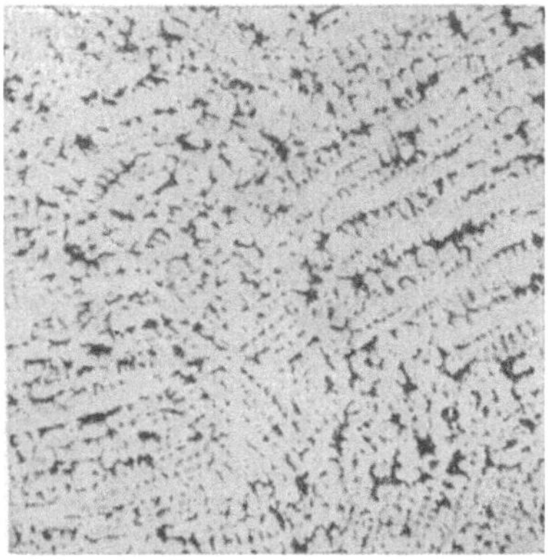

Fig. 40. 77% Cu, remainder Pb, cast in chill-mould. Primary crystals of copper (light) as dendrites, residual fields of copper-lead monotectic (dark). ×125

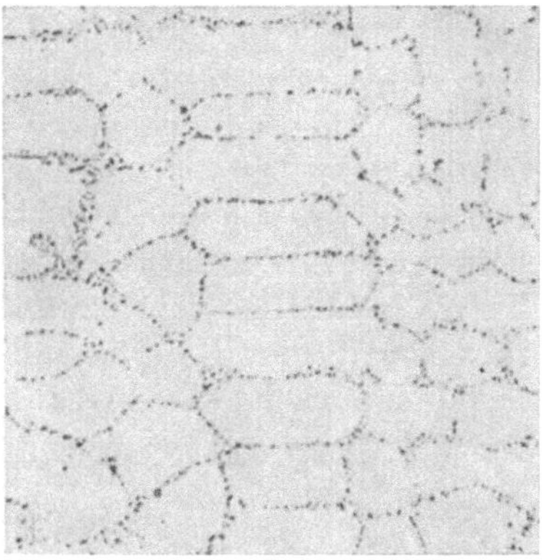

Fig. 41. 0.04% Cu, cast in chill-mould. Practically pure eutectic structure. Lead crystallised as dendrites. ×150

Fig. 39. SEITH and co-workers [1105] have confirmed it. The position of the liquidus curves on the lead side between 954 °C and the melting point of lead is the result of a thorough new determination (KLEPPA [677]). The copper-rich alloys, which are of great importance as leaded bronzes, usually have lead contents which are still outside the miscibility gap. The change in the miscibility gap by further additions was investigated principally in the Institute GUERTLER [449, 451] and by OSBORG [926]. The diffusion coefficients for the diffusion of copper in molten lead between 478 and 750 °C were determined by GORMAN [409]. Copper migrates in the melt as a divalent ion.

Examples of the structure of cast alloys on the copper side and on the lead side of the system are shown in Fig. 40 and 41. Copper can only be removed from the lead by segregation down to the eutectic content of 0.06 Wt.-%. The solubility of copper in solid lead is certainly remarkably small, since, for instance, a copper content of 0.007% can still be detected by the microscope (GREENWOOD and ORR [430]).

Disregarding the leaded bronzes and developments in powder metallurgy (p. 451), only amounts of copper below 0.1% have practical significance. These small copper contents in lead effect a considerable grain-refinement and particularly structural stability at high temperatures (p. 187). As regards creep and fatigue strength of the alloys, reference is made to the appropriate sections of this book (p. 240 onwards).

The chemical industry frequently favours copper-containing lead, owing to its mechanical properties and structural stability. In addition it shows good resistance to sulphuric acid at high temperatures (pp. 6 and 276). For cable sheathings also, additions of copper of the same amount are of great significance. In order to alloy lead with copper, a hardener alloy can be used. This is prepared by dissolving copper in lead at red heat and care must be taken to distribute the copper uniformly in the lead. In hypoeutectic alloys, uniform distribution offers no difficulties, but at higher copper contents, on the other hand, there is the risk of segregation. Complete data on the rate of solution of copper in lead and its dependence on temperature and concentration are given by WARD [1237]. STEVENSON added the flow velocity of the melt as a further parameter [1145b]. MALENBERG gives values for the density of liquid and of solid alloys [792a].

13. Pb–Fe. Lead-Iron

Up to 1,600 °C, iron and lead show only slight mutual solubility, and form no intermetallic compound (ISAAC and TAMMANN [605]). LORD and PARLEE recently determined the solubility of lead in molten iron [766]. Lead was supplied to the molten iron in the form of vapour, so that absorption in emulsion form in the molten iron could be excluded. The analysis of the iron layer gave the following values for the solubility of lead:

1,550 °C	0.22–0.26%	1,650 °C	0.34–0.40%
1,600 °C	0.27–0.33%	1,700 °C	0.37–0.43%

STEVENSON [1145a] measured the solubility of iron in lead from 700 °C to 1,300 °C. (e.g. 0.002% Fe at 800 °C; 0.01% at 1,000 °C; 0.06% at 1,300 °C).

OELSEN and co-workers [917] studied the distribution of Ca, Sn, As, Sb, Ag and Au between lead melts and carbon-saturated iron melts. The underlying technical problem was the question of the degree to which the elements mentioned can be removed from the pig iron by washing with lead. As at 1,250 °C small contents of copper are distributed in approximately equal contents between the iron and the lead melts, very large quantities of lead would have to be used to remove the copper from the pig iron. On the other hand, silver and gold are taken up by lead to a very large extent; conversely, arsenic remains almost completely in the pig iron.

The mutual solubility in the solid state is vanishingly small. According to magnetic measurements, it amounts on the lead side to

$2 \cdot 10^{-4}$ to $4 \cdot 10^{-4}$% (TAMMANN and OELSEN [1170]). The traces of iron which can frequently be detected chemically in lead must therefore be present as mechanical inclusions. As lead practically does not alloy with iron, lead can be melted in iron pots. In the coating of iron by means of molten lead, on the other hand, in order to obtain good adhesion, lead alloys must be used or, as in homogeneous lead lining, intermediate layers of tin (p. 454). A satisfactory metallic joint between iron and lead, without an intermediate layer, is obtained by means of cold pressure welding. A slice of lead is placed between two rods of soft steel and the zone of contact is deformed so energetically by application of pressure that after a part of the slice of lead has flowed out, the steel also is bulged (BURAT and HOFMANN [148]). The process can be conceived as a wetting of the boundary surfaces in the solid state. Wetting of the iron surface by molten lead was convincingly demonstrated by PELZEL [950].

In regard to leaded free-machining steels, attention is drawn to the work of HOUDREMONT [598]. Use is there made of lead contents between 0.2 and 0.5%, which are present in the form of mechanical inclusions. Leaded wrought steels with 0.15 to 0.35% Pb also find application in practice (ROSE [1032]). Data on the mechanical properties (including fatigue strength) of such steels are given by BARDGETT [41].

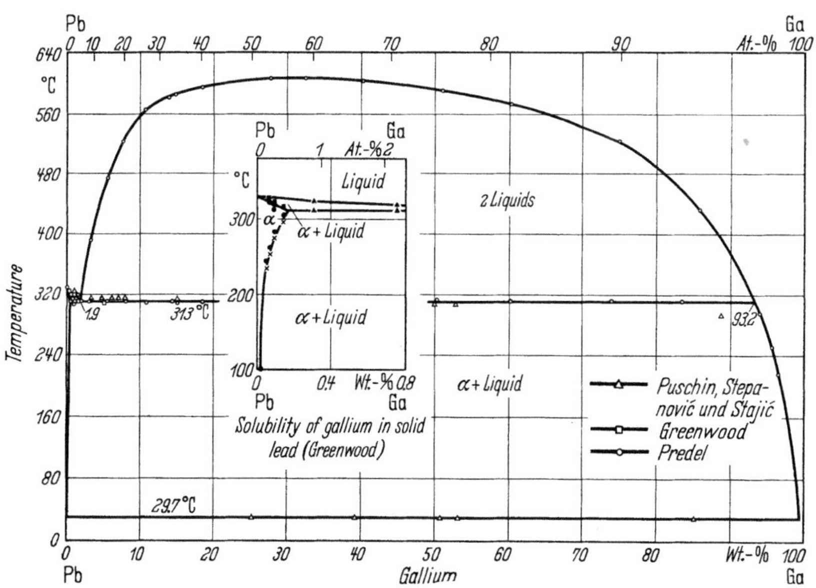

Fig. 42. Lead-Gallium. (According to PREDEL as well as GREENWOOD).
× 2 phases observed, • 1 phase observed.

14. Pb–Ga. Lead-Gallium

According to PREDEL [978], the miscibility gap in the liquid state extends at the monotectic temperature (313 °C) from 2.4 to 94.5 At.-% Pb (Fig. 42).

According to GREENWOOD [426], the region of solid solution of gallium in lead extends from 0.02 Wt.-% Ga (0.06 At.-%) at 110 °C to 0.17 Wt.-% Ga (0.5 At.-%) at the monotectic temperature. In accordance with the course of the solubility curve, the alloys exhibit age-hardening phenomena. PREDEL [979a] has now also established the complete ternary diagram of lead-gallium-cadmium.

15. Pb–H. Lead-Hydrogen

SIEVERTS and KRUMBHAAR [1121] could not determine any measurable solubility of hydrogen in either molten or solid lead. A repetition of these measurements, by a similar method, by OPIE and GRANT [922] on molten lead at 500 to 900 °C indicated that the solubility was markedly temperature-dependent. The quantity of dissolved hydrogen amounted here to 0.25 Nml per 100 g lead at 600 °C. Thorough investigations by MAATSCH [567] did not confirm this result. By means of a measuring arrangement due to LIESER and WITTE [750], a solubility of less than 0.01 Nml hydrogen per 100 g lead was found at 600 °C, within the range of experimental error. Nor could MANNCHEN and BAUMANN [793] detect any solubility of hydrogen in lead by means of the hot extraction process.

16. Pb–Hg. Lead-Mercury

The broad region of solid solution of mercury in lead is characteristic of the phase diagram (Fig. 43). While however the solubility limit was earlier (HANSEN [487]) assumed at about 35 Wt.-%, according to TYZACK [1210] there exists in this region of concentration an intermediate phase of approximate composition $HgPb_2$. Its tetragonal face-centred unit cell has the edge lengths $a = 4.97$ kX, $c = 4.50$ kX. The determinations of diffusion constants of mercury in lead (GERTSRIKEN, BUTZIK and GOLUBENKO [374]), as well as the activities of lead in amalgams, and the electrode potentials of lead amalgams in lead chlorate (HARING, HATFIELD and ZAPPONI [492]) will not be discussed in any further detail. The hardness of the alloys in the cast state increases in the following way with the mercury content, GOEBEL [390]:

Wt.-% Hg	0	1	2	3	4	5	6	7
4/20–20 Brinell hardness kg/mm²	4.1	4.8	5.6	6.7	7.0	7.5	8.3	9.2

Ingot segregation was not observed in castings (BAUER and ARNDT [58]).

Age-hardening or softening on storage is not to be expected in the alloys. Mercury has often been used as a hardening constituent in lead,

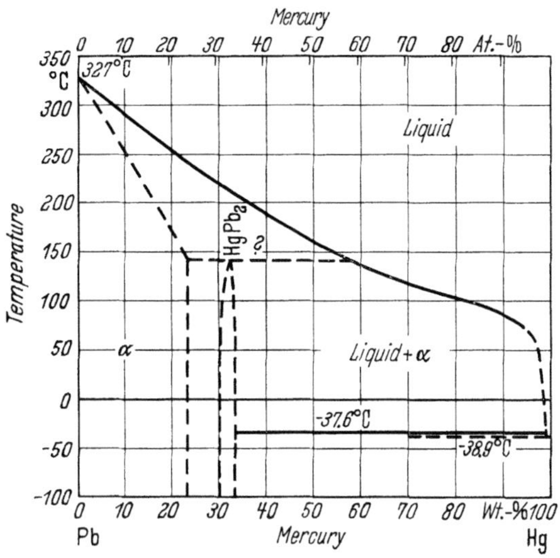

Fig. 43. Lead-Mercury. (According to HANSEN).

i.e. FRARY bearing metal (p. 128, developed in USA, contains 0.25% Hg in addition to maximum 2% Ba and 1% Ca (GRANT [413]). Low-melting solders can be developed, based on lead-mercury, but there are hygienic considerations against their use. BURGAN [152] gives data on the vapour pressure of the alloys, and the activities of mercury and of lead deduced from it.

17. Pb–In. Lead-Indium

Indium, a very soft metal (Brinell hardness of 1) with a good chemical stability, similar to tin in many ways, has generated some technical interest in recent years. Earlier it was thought that a continuous solid solution range existed between indium and lead. The fact that indium, contrary to lead, has a tetragonal body centred crystal lattice casts doubt on this opinion (HANSEN [487]). Later researches have now shown in fact that not only is there no continuous solid solution region between the two components but that, in addition, between the area of indium and that of the lead solid solution, there is a further tetragonal body centred phase (VALENTINER [1215], KLEMM [672], KOGAN [692b],

MOORE [867]). Lead can hold up to 60 At.-% indium in solid solution, resulting in some increase in hardness (Fig. 44). The heat of solution of lead-indium alloys is positive (SCHEIL [1057 a]). GRATSIANSKY and co-workers conducted surface tension measurements [414] on molten alloys as well as corrosion tests in sulphuric and critric acids and found that 50 to 75 At.-% Pb was the limit of resistance [416].

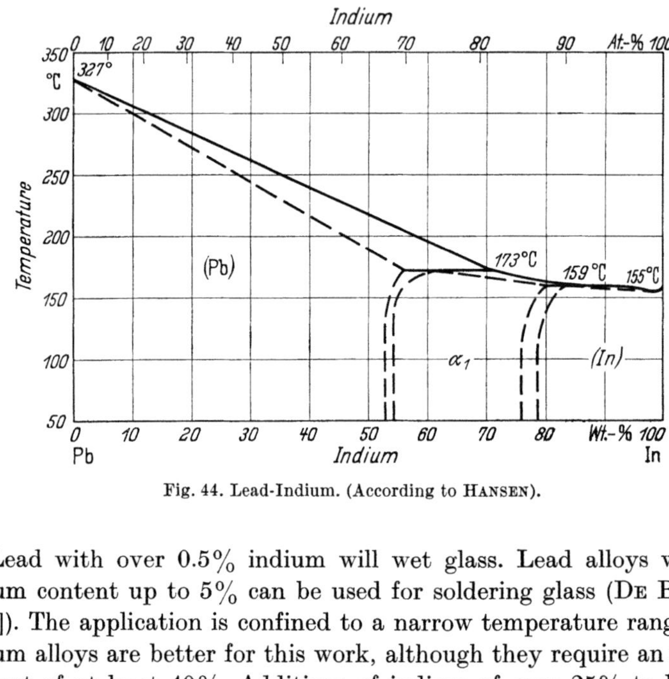

Fig. 44. Lead-Indium. (According to HANSEN).

Lead with over 0.5% indium will wet glass. Lead alloys with an indium content up to 5% can be used for soldering glass (DE BRUYNE [136]). The application is confined to a narrow temperature range. Tin-indium alloys are better for this work, although they require an indium content of at least 40%. Additions of indium of over 25% to lead-tin solders increase their alkali resistance (GRYMKO [447]). Whilst lead silver (3%) solders with an addition of 1 to 2% indium have an increased strength (JAFFEE [617]), this is not the case with lead-tin alloys. Various low melting point alloys have indium additions to lower the melting point still further (KEIL [650]). Lead bronze bearings for aircraft and road vehicles can be electroplated with lead indium coatings.

With suitable heat-treatment, indium diffuses into the lead surface and increases its corrosion-resistance towards certain lubricants ([835 a]; p. 371; compare ZAITSEV [1300 a]). Silver-lead-indium layers of thickness 20 to 60 μ on steel have been developed as bearings for the highest loads. The electrodeposited silver layer is made to size and obtains a defined layer thickness of lead and of indium (or tin) again electrodeposited (THEWS [1180]). Before fitting the bearing, the indium is diffused into the lead at 170 to 175 °C. A solid solution with about 16%

In is formed. Bearings of this type are said to have higher fatigue strength and load-carrying capacity than any other bearing materials. At high loads and speeds they are said to have 30 times the life of white metal bearings, and 10 times the life of leaded bronze bearings ([835b], SNELLING [1130]).

18. Pb–K. Lead-Potassium

The production of lead-potassium alloys by reaction of lead with carbon and potassium hydroxide or other potassium compounds is said to take place even more easily than has been indicated for the analogous case of lead-sodium alloys (PUTNAM [984]; p. 67).

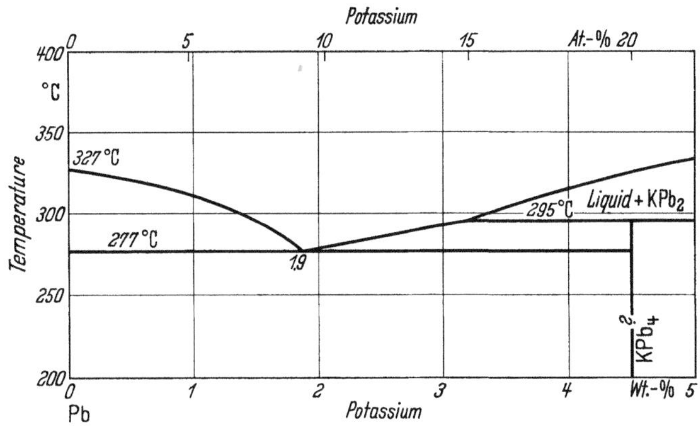

Fig. 45. Lead-Potassium. (According to HANSEN).

The equilibrium diagram is still incomplete on the potassium side (HANSEN [488]). The miscibility gap in the molten state at middle concentrations above 568 °C can hardly be of technical significance. According to Fig. 45 there is a eutectic lead KPb_4 on the lead side. While in earlier work no age-hardening of the lead-rich alloys was found (TAMMANN and RÜDIGER [1171]), JENCKEL and HAMMES [618] found a hardness of 15 Brinell units in a cast alloy with 0.1% K. This indicates a high age-hardenability of the alloys. Potassium in quantities up to 0.06% is a constituent of SATCO bearing metal (v. GÖLER and WEBER [398]), in which it obviously acts as a hardening constituent.

19. Pb–Li. Lead-Lithium

On alloying lead with lithium, an intense heat effect occurs (CZOCHRALSKI and RASSOW [228]), an indication of heteropolar linkage forces (p. 30). The equilibrium diagram was set up on the basis of thermal analyses and resistance measurements (Fig. 46). According to this the compound LiPb, richest in lead, shows a transformation in the solid state, which is reminiscent of the transition between ordered and disordered atom distribution in known alloy systems (e.g. β-brass). Investigations by means of resistance measurements at variable temperatures and pressures gave possibilities of interpretation (WILSON [1278]), but X-ray work by various authors (WILSON [1278], NOWOTNY [906]) still produced marked contradictions in the determination of the

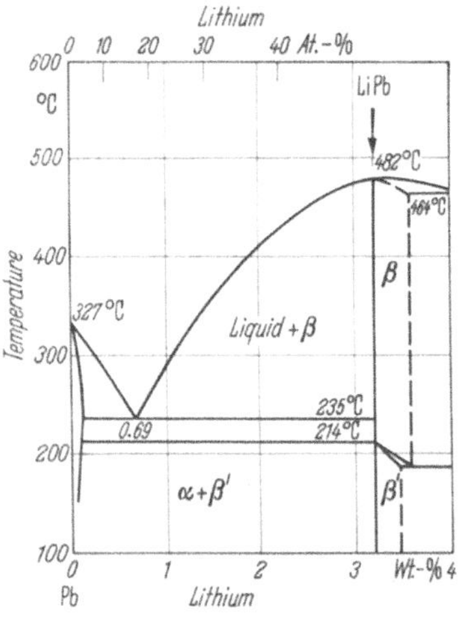

Fig. 46. Lead-Lithium. (According to GRUBE and KLAIBER [445]).

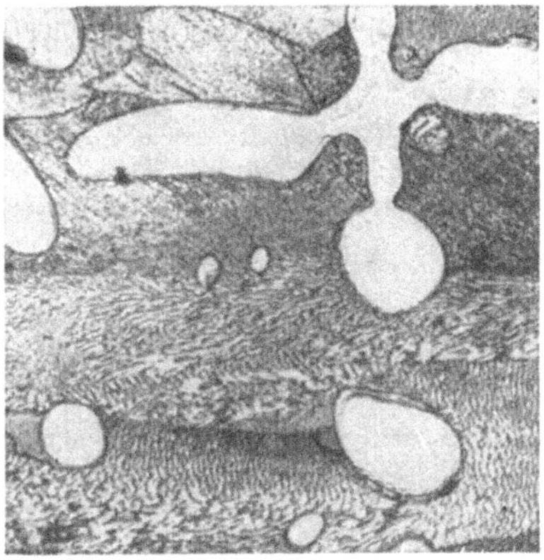

Fig. 47. 0.58% Li. Solidified in crucible. Light area: lead solid solution. Eutectic lead-LiPb. ×200

MILLER indices and the calculation of density. The equilibrium diagram (Fig. 46) is illustrated (Fig. 47) by the structure of a hypoeutectic alloy. According to GRUBE and KLAIBER [445], the solubility on the lead side falls from 0.11% Li at 235 °C to 0.03% at 120 °C. POGODIN [968] found somewhat lower values of solubility, namely 0.09, 0.07, 0.06, 0.04 and 0.01% Li at temperatures 235, 200, 170, 120 and 20 °C. The age-hardenability of the lead-rich alloys down to 0.02% Li can thus be understood.

Age-hardening curves are illustrated in Fig. 48. Age-hardening takes place vigorously after chill-casting, as well as after extrusion at high temperatures and subsequent quenching. The tensile strength and the elongation at fracture of extruded lead-lithium wires were determined after storage for one month and longer (BURKHARDT [155]). The course of age-hardening was not followed up. On addition of 0.01% Li, the tensile strength rose above 2 kg/mm^2. No marked inhibition of recrystallisation was found after working age-hardened alloys (HOFMANN and HANEMANN [561]), as for example, in the case of lead-calcium alloys (p. 193).

Up to a lithium content of 2.15% the alloys are stable in air even on long standing. On the other hand, specimens with higher concentrations showed lower stability (CZOCHRALSKI and RASSOW [228]). Alloys with Li and Sn can be strengthened to 15,000 psi by quenching from the melt (REICHEL [998a]).

Binary lead-lithium alloys have not attained any significance. The alloys up to 0.10% Li, possibly with additions of up to 0.18% Cd or 0.5% Sb were recommended for cable sheathing; here a heat treatment for

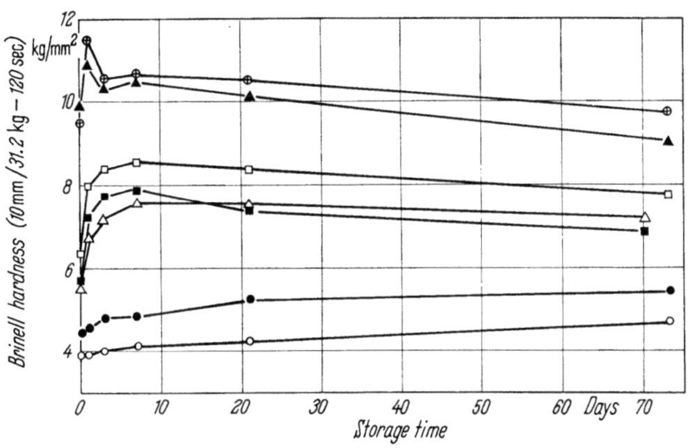

Fig. 48. Age-hardening of lead-lithium alloys. (According to v. HANFFSTENGEL and HANEMANN [484]). ○ 0.01% Li, cast; ● 0.02% Li, cast; □ 0.04% Li, cast; ■ 0.04% Li, cast, heat-treated, quenched; △ 0.04% Li, extruded, quenched from 235 °C; ▲ 0.07% Li, cast; ⊕ 0.10% Li, cast.

age-hardening should follow extrusion [841]. Similar proposals also relate to lead-lithium-bismuth alloys (OSBORG [927]). On the other hand, lithium has achieved a certain significance as a constituent of Bahnmetall (a bearing metal). At a content of 0.04%, it increases the hardness without promoting metal loss on melting or by corrosion, as might perhaps be the case for larger additions of sodium (GRANT [413], v. GÖLER [393]). In a varied alloy of this type the lithium content was lowered to 0.02%, that is, about the solubility limit at room temperature, in order to oppose softening on heating the bearing metal.

20. Pb–Mg. Lead-Magnesium

The diagram for the lead-rich alloys is characterised by a eutectic of lead solid solution with Mg_2Pb (Fig. 49). The position of the eutectic and the course of the liquidus curve up to magnesium contents of 3 Wt.-% were recently redetermined by HORSLEY [596]. Viscosity measurements by GEBHARDT [361] gave a very marked maximum at the constant temperature of 550 °C in the region of the compound Mg_2Pb. From this a reciprocal action was deduced between the lead and the magnesium atoms in the melt. In agreement with this, SCHEIL [1058] found from

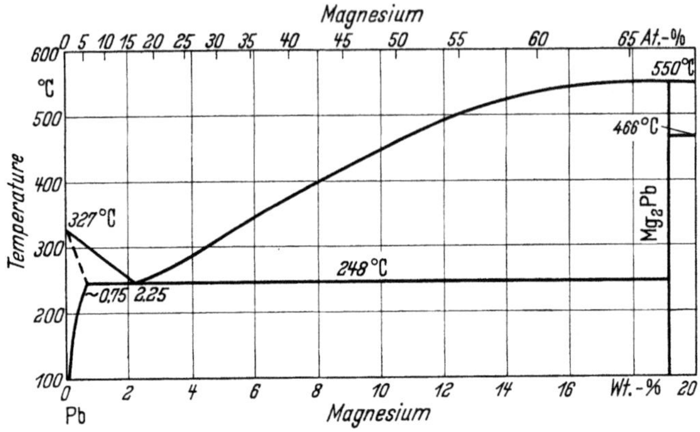

Fig. 49. Lead-Magnesium. (According to HORSLEY and HANSEN).

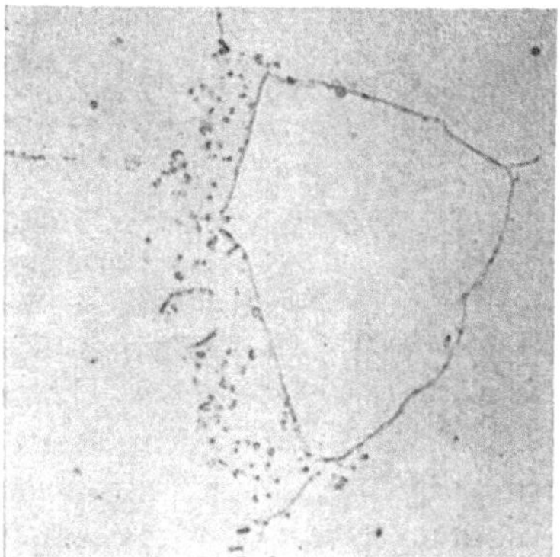

Fig. 50. 0.2% (weighed addition) Mg. Cast. A little lead-Mg_2Pb eutectic grain boundaries. ×500

vapour pressure measurements that the activities of magnesium in the alloys were lower than was to be expected in an ideal solution. The limit of solid solution at high temperatures, hitherto accepted, remained practically unaltered by a new determination by KURNAKOV [722]. On the other hand, extrapolation gave a solubility of only 0.12% Mg in lead at 100 °C. This fits in well with the value of 0.06% Mg at room temperature, derived from resistance measurements (EUCKEN [295]). An alloy with 0.2% Mg already contains (Fig. 50) some eutectic as a result of coring. The fact that the eutectic part of the alloy with 0.5% Mg (Fig. 51) depends only on coring, is shown by homogenising this alloy. The eutectic then disappears (Fig. 52).

Table 9 offers some clues for the age-hardening of the alloys. The age-hardening is connected with a microscopic precipitation of Mg_2Pb, therefore possibly giving way to a resoftening on further coarsening of the precipitate.

Table 9. *Age-Hardening of Lead-Magnesium Alloys*

Magnesium (weighed) in Wt.-%	In Cast Condition				Quenched from 220 °C (6 days)			
	0.2	0.5	2.5	3.5	0.2	0.5	2.5	3.5
Brinell hardness kg/mm²								
at once	5.8	9.75	10.7	14.1	5.0	8.8	12.4	16.6
,, ,, after 1 day	5.9	9.5	15.3	19.5				
,, ,, ,, 6 mos.					5.8	9.8	12.3	16.1

As favourable additions, 0.5 to 0.7% Mg are indicated (KURNAKOV and VIDUSOVA [721]). Having regard to the corrosion behaviour, these values are certainly far too high. The stability is very small, particularly in the hypereutectic alloys (GOEBEL [391]), which is due to the properties of the structural constituent Mg_2Pb. This compound has a fluorspar

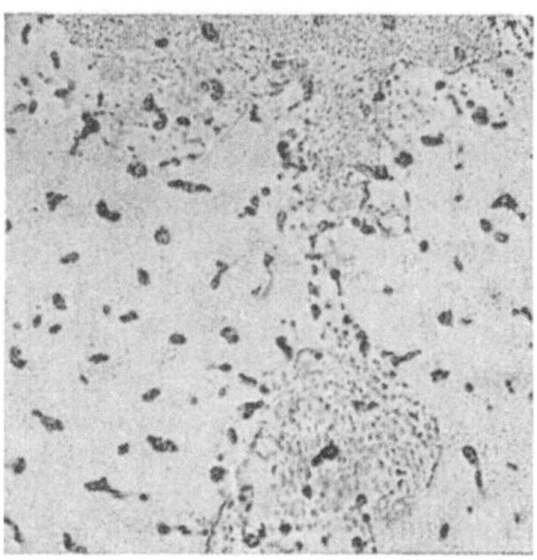

Fig. 51. 0.5% (weighed addition) Mg. cast. Lead solid solution interspersed with eutectic (dark). Two areas with spot-precipitates of Mg_2Pb. ×500

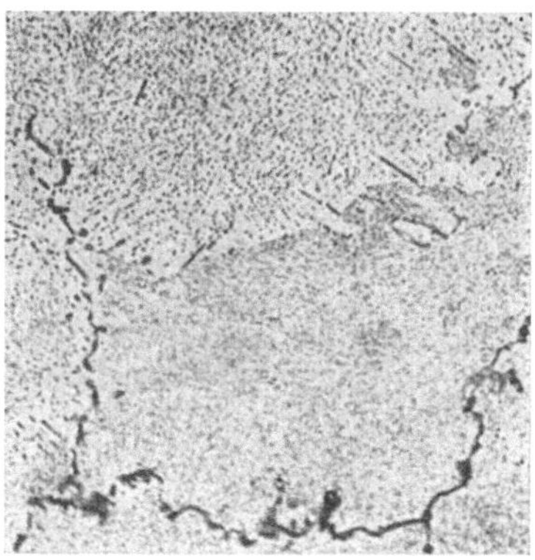

Fig. 52. Previous alloy homogenised at 220 °C, quenched in oil and stored for 20 days. Orientated precipitates. Intercrystalline corrosion. ×500

structure (SACKLOWSKI [1044]), and so no typically metallic crystal lattice. On melting the compound, contraction occurs (KUBASCHEWSKI [716]). Its electrochemical potential is scarcely more noble than that of magnesium (KREMANN and GMACHL-PAMMER [704], JENGE [620]). The pure compound is violently decomposed by water in the cold. Mg_2Pb preferentially forms precipitates on the grain boundaries even in alloys of the lowest concentration, and thus leads to intercrystalline corrosion (Figs. 50, 52). Intercrystalline corrosion occurred in rolled strips with only 0.05% Mg (GREENWOOD [431]). The alloy became embrittled after a storage period of only three months. On this account, magnesium additions in technical alloys should be less than the concentration indicated (SCHMID [1069]).

On account of their good mechanical properties, the alloys were recommended as a substitute for lead-tin alloys in sheathing for telephone cables (STENQUIST [1145]), but owing to their insufficient resistance to corrosion, were not successful. The same is true for use as bearing metal, which has been protected by many patents (SCHMIDT [1074]). The bearing materials were temporarily on the market, but failed to some extent. This application required careful treatment and regard to the following aspects (KROLL [711]): the alloy must not be made with scrap lead, as in some circumstances the impurities form compounds with magnesium which rise to the surface of the melt and decompose in moist air with formation of poisonous metal hydrides (for example arsine). The alloys cannot be stored, and the castings must therefore be used immediately. Turnings decompose very rapidly and have no scrap value, but only ore value. In addition they are inflammable. KURTEPOV [724] gives data on the corrosion behaviour of the alloys over the total range of concentrations.

Finally, attention may be drawn to a peculiar application of lead-magnesium alloys (JORDAN [637]). As wild ducks may take up stray lead shot in mud as food, and are said to get poisoned in some circumstances, shot made of lead-magnesium has been suggested for wild duck shooting. The shot disintegrates in the moisture within a few hours.

21. Pb–Mn. Lead-Manganese

PELZEL [945] investigated the solubility of manganese in molten lead in the temperature range 475 to 1,000 °C. Fig. 53 illustrates the result. According to WILLIAMS [1274c], there is a miscibility gap at

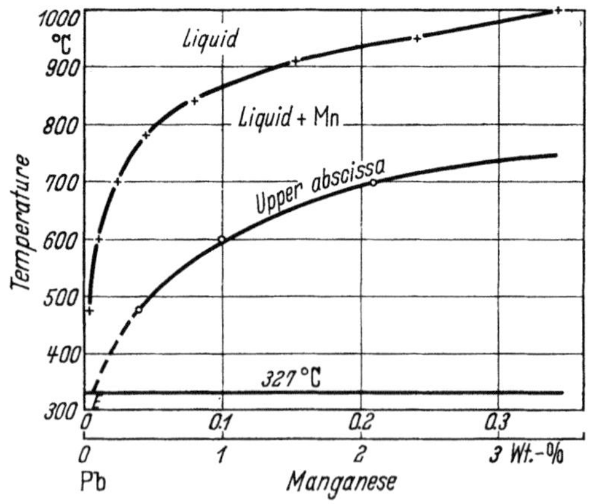

Fig. 53. Solubility of manganese in molten lead. (According to PELZEL). Eutectic E at <0.01% Mn.

1,200 °C above a manganese content of 12 Wt.-% in lead. From the lowering of the temperature of transition to superconductivity by manganese content, BARTH [49a] fixes a solubility of the order of 0.02 At.-% in lead at room temperature.

22. Pb–Na. Lead-Sodium

Sodium forms with lead a fairly large region of solid solution, very dependent on temperature. Although sodium has a larger atomic radius than lead, the lattice constant of the solid solution falls with the sodium content (AGEEW [10]). On this account the density of lead sinks less than calculated on alloying with sodium. The density falls by 0.079

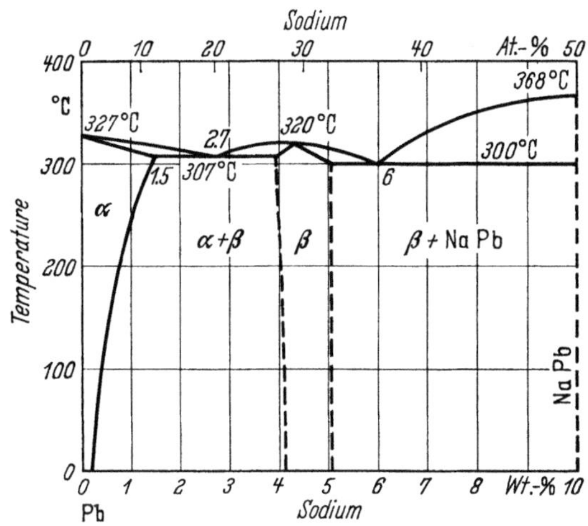

Fig. 54. Lead-Sodium. (According to HANSEN).

g/cm³ per atomic % of sodium (v. GÖLER and WEBER [398]). The β-phase involved in the eutectic (Fig. 54) is a solid solution of the compound Pb_3Na (not existing as such) with sodium. It has the same lattice as lead but with partially ordered atomic distribution. The high age-hardening of the alloys makes it possible to use hardness measurements on alloys homogenised and quenched from various temperatures, for re-examination of the solubility curve. As is shown by the results of such experiments (Fig. 55) the measurements must be carried through over very long periods of storage since only then does the age-hardening of the sodium-poor alloys take place. From these observations, which were in agreement with the results of microscopy (SCHULZ [1086]), the limit of the solid solution region at room temperature was fixed at about 0.2%.

The age-hardening of lead-sodium alloys is connected with a precipitation of excess dissolved sodium, which can be partly detected by the microscope. For the practical application of the alloys it is important that age-hardening also takes place on cast and non-heat-treated specimens. The hardness maximum has here been determined at a content of 0.8% Na. Obviously a higher supersaturation of the solid

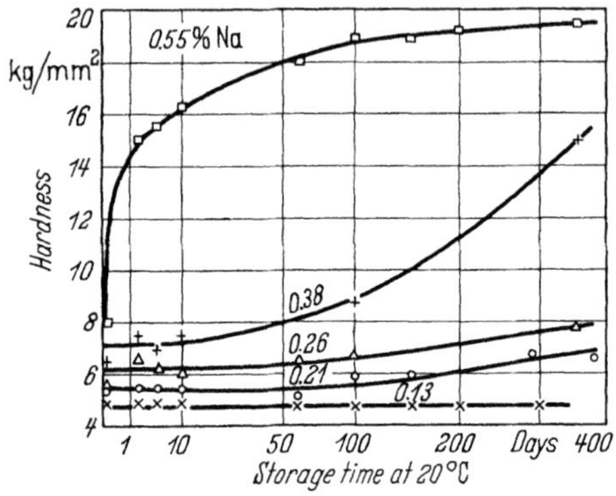

Fig. 55. Age-hardening of lead-sodium alloys. Specimens homogenised by heating 6 hours at 290 °C, and quenched. (According to SCHMID [1069]).

solution cannot be attained in the cooling conditions of normal casting. Under extreme conditions, such as are given in the abrupt solidification of the melt in a suction chill mould, practically homogeneous alloys up to a sodium content of 3% could be obtained (p. 45). The age-hardening of cast, slowly cooled alloys is somewhat less than that of alloys obtained by quench-casting (GOEBEL [391]). The final hardness is, however, not much lower, as the initial value after slow cooling is higher than after quench-casting, which is perhaps due to an age-hardening during the cooling. BURKHARDT [155] gives results of tensile tests for alloys up to 0.3% Na.

GOEBEL [391] has given 0.75 to 0.85% as the highest limit of the sodium content for the applicability of binary alloys. Such alloys however decompose in air and therefore can be considered only for special purposes. An alloy with 10% Na (corresponding to the formula NaPb) has great technical significance as an intermediate product in the synthesis of the anti-knock compound lead tetraethyl (MORGENTHALER [870]). The alloy is prepared by melting refined smelter lead and metallic sodium in closed vessels. Owing to its brittleness it can be pulverised in a hammer mill. On reacting NaPb in stirred autoclaves with ethyl chloride, lead tetraethyl is formed with the separation of sodium chloride and lead powder. The production of the high-sodium alloy is said to be cheapened if lead is reacted at high temperatures with petroleum coke and caustic soda (according to an idea by ROSSITER), instead of beginning with sodium. Use is made of the reaction (PUTNAM [984]) $6\ NaOH + 2\ C = 2\ Na_2CO_3 + 2\ Na + 3\ H_2$. A sodium content of 1.30% was present in the bearing metal "Noheet" suggested in 1905. The stability was so low that for storage the bearing metal had to be sealed withoil (GRANT [413]). On this account the sodium content was later on reduced by addition of calcium and lithium, Bahnmetall being thus developed. The high age-hardening of Bahnmetall depends mainly on the supersaturation with sodium at room temperature. The comparatively high sodium content gives rise to some disadvantages, particularly the high loss of alloying metals in remelting and the softening of the bearings on heating. Attempts have been made to avoid these disadvantages by variations of the Bahnmetall composition (p. 386). There is a Russian paper on the anodic corrosion of the alloys (SCHACHKELDIAN [1053]).

23. Pb–Ni. Lead-Nickel

Having regard to the miscibility gap in the molten state, and to the course of the liquidus curves (Fig. 56), the production of alloys of high nickel content requires certain precautions. The melts must be quench-cast from high temperature in order to avoid any breakdown on cooling. Fig. 57 gives an example of such an alloy. The solubility of nickel in molten lead below the monotectic temperature has been determined several times (PELZEL [944], DAVEY[1], ALDEN [14], CAVENAUGH [184a]); Fig. 58 is based on the results. The course of solidification near the lead side is still insufficiently precise to permit a decision whether there is here a case of a eutectic or a peritectic. The rate of solution of nickel in agitated lead melts was measured by STEVENSON [1145b]. The solubility of nickel in lead in the solid state was determined by a magnetic method (TAMMANN and OELSEN [1170]). Comparatively high solubilities were obtained, e.g., 0.195% Ni at 327 °C, 0.023% Ni at 180 °C. A re-examination by other methods seems urgently necessary, since, for instance, the considerable grain refinement in technical lead-nickel alloys, containing a few multiples of 0.01% Ni (p. 196 is evidence against a larger region of solid solution. Owing to their good resistance to corrosion in sulphuric acid, combined with improved fatigue strength at elevated temperatures, the alloys have achieved practical significance. Before the last war, "Edelblei" (Fig. 59) produced by the firm of KESSLER at BERNBURG, was known in Germany. Reference may be made to addition of nickel in some white metals (p. 382).

[1] Verbal information.

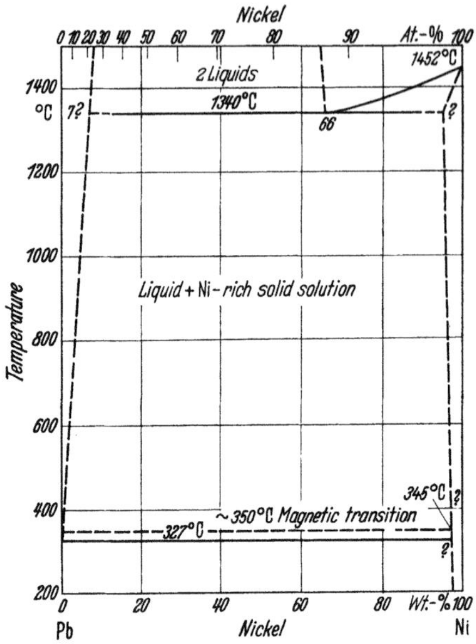

Fig. 56. Lead-Nickel. (According to HANSEN).

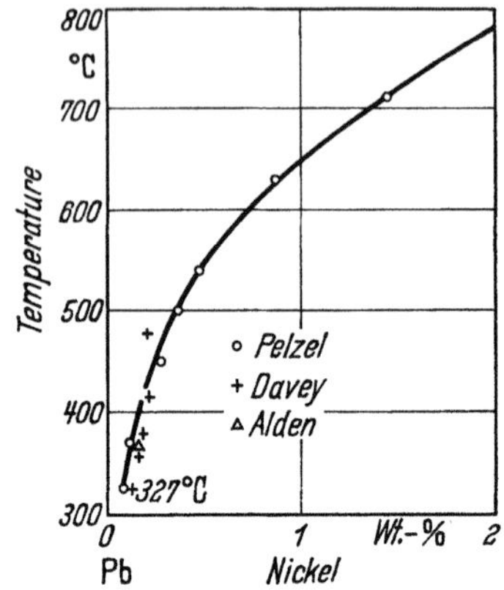

Fig. 58. Lead-Nickel. Liquidus curve on the lead side. (According to DAVEY, PELZEL and ALDEN [944, 14]).

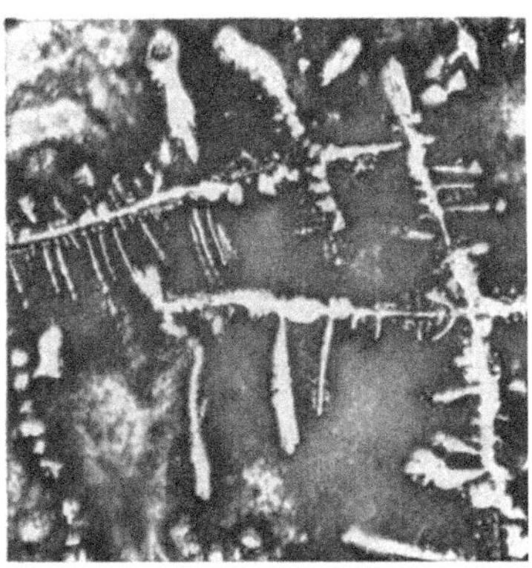

Fig. 57. 5% Ni. Cast. White primary crystals of nickel in relief against soft matrix of lead. ×300

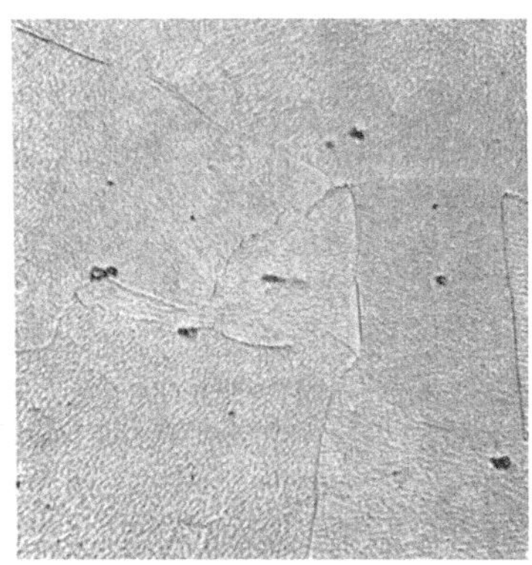

Fig. 59. "Edelblei" with 0.025% Ni. Individual nickel inclusions in lead. ×500

24. Pb–O. Lead-Oxygen

In this system, not only the solubility of oxygen in the metallic phase but also the nature of the different oxides is of interest. Together with their crystallographic data, they have recently been described by KATZ [646] on the basis of careful chemical and X-ray investigations, and the data collected in a diagram. In this way contradictions between the views of earlier authors seem to have successfully cleared up. The yellow lead monoxide β-PbO, which crystallises in the rhombic system, has long been known. The red monoxide, α-PbO stable at lower temperatures, is likewise rhombic, but simultaneously pseudotetragonal. In contrast to β-PbO, it can take up oxygen above the stoichiometric composition. In this way the divergence from the tetragonal crystal symmetry increases.

While the transition temperature of yellow lead monoxide-red lead monoxide is generally given as 489 °C, a temperature of about 585 °C seems more probable, according to an evaluation of the literature by HANSEN [488]. Molten and then solidified lead monoxide (presumably yellow lead monoxide) has, according to KRYSKO [714a] a density of 9.62 g/cm³ at room temperature, and 8.53 g/cm³ at 880 °C. Tetragonal lead monoxide has a density of 9.32 g/cm³. Minium (red lead) belongs to the tetragonal crystal system. By a small deviation from the formula Pb_3O_4 it is said to become rhombic, pseudotetragonal. Lead dioxide (peroxide) PbO_2 can become impoverished of oxygen in the homogeneous phase down to the formula $PbO_{1.87}$. This change is connected with a very feeble lattice contraction in the direction of the tetragonal a-axis, the c-axis remaining unchanged. As a further phase of a special kind, a solid solution designated PbO_n was established which crystallises in the rhombic system, as well as pseudocubic. It is homogeneous within the limits $n = 1.33$ to $n = 1.55$, and changes its colour continuously with the oxygen content from brick red ($n = 1.33$) to black ($n = 1.55$). Within these limits is the composition Pb_2O_3. PbO_n is regarded as a solid solution of Pb_2O_3 with lead or oxygen respectively. The solid solutions with contents of $n = 1.55$ to $n = 1.57$ decompose and thus lose oxygen down to the composition $PbO_{1.41}$ in the homogeneous region described. Finally, as a further phase, KATZ [646] has established an oxide of formula $PbO_{1.87}$. Details of this extensive work cannot be reproduced here.

The solubility of oxygen in molten and in solid lead has frequently been discussed (p. 289), particularly in regard to the question of corrosion-resistance. The older determinations of oxygen content of lead yielded results which were too high and can therefore be omitted. The most exact determinations, such as were carried out in the last 20 years in many countries, gave results in good agreement. Gravimetric determinations with very high weight of sample gave, in technical lead grades, oxygen contents of only 0.00013 to 0.00056% (WORNER [1286]). At the British Non-Ferrous Metals Research Association in London, a manometric method of analysis was developed by BAKER [38]. In this hydrogen acts under reduced pressure on the molten lead. The water vapour formed first freezes in a cooling vessel, and at the end of a reduction is vaporised by warming the cooling vessel. The pressure of the unsaturated water vapour gives the oxygen content with great accuracy.

BARTELD [49] used the manometric method of determination. He passed highly purified hydrogen at normal pressure over the boat containing the molten lead. At a temperature of 700 °C and with a

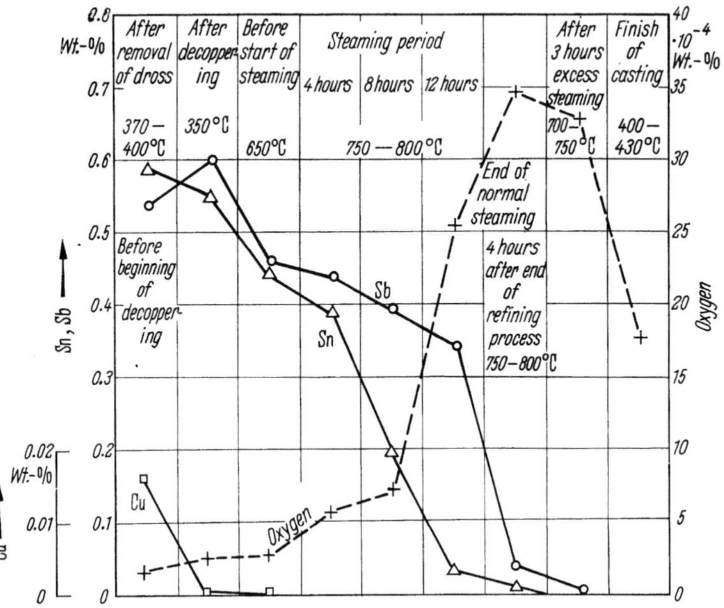

Fig. 60. Oxygen-contents of lead melts at various stages of the fire refining process. (According to BARTELD).

sample of 200 g lead, reduction is complete in half an hour. It is best to use specimens in compact form in order to reduce the content of surface oxygen to a minimum. Determinations were carried out on numerous German smelter and scrap grades of lead by this method. The values obtained were between 0.0001 and 0.0018%, mostly under 0.001% (10 g/ton = 10 ppm wt.); it was not possible to state a difference in oxygen content between smelter lead and scrap lead. Fig. 60 shows the change in oxygen content during the fire refining of scrap lead in the *Metallwerke Unterweser*. It is interesting that before the refining operation, the scrap lead has an oxygen content as low as that of the work lead of *Bleikupferhütte Oker*. The value is given in the column "after decoppering". With increasing temperature, an opportunity is given to the melt in the refining furnace to absorb oxygen. The diagram shows that oxygen dissolves only slowly in the lead; that means, it is first consumed in the oxidation of tin, arsenic and antimony. Only after the removal of these impurities from the melt does the lead take up more oxygen. After cooling for more than 12 hours, the oxygen content has decreased to half. The changes occurring in the oxygen content during refining by the Harris process at the *Norddeutsche Affinerie* were followed up in the same way. The course of the curve thus obtained was similar.

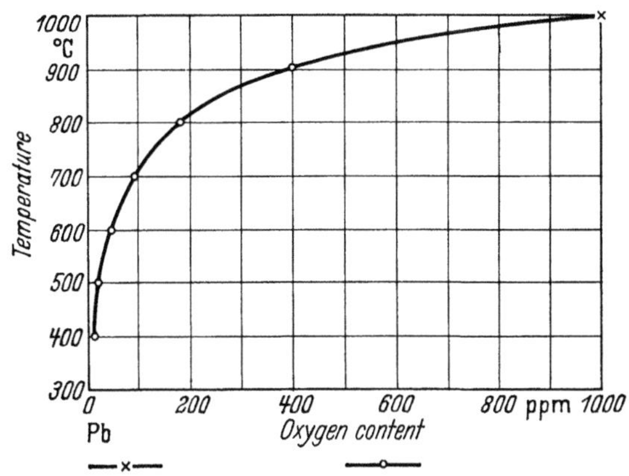

Fig. 61. Solubility of oxygen in molten lead —x— DANNATT and RICHARDSON [230]. —o— BARTELD [48].

For the determination of the solubility of oxygen in lead, BARTELD [49] held melts in open crucibles at constant temperatures between 350 and 800 °C until the equilibrium value of the oxygen solubility had set in. After skimming the melt, a ladle sample was taken and cast in a thick-walled chill mould. Turnings were taken from the rapidly solidified regulus and the oxygen content determined. As there was a risk in sampling that during the cooling and solidification of the melt a part of the oxygen content would be lost (MCMASTER [828]), further oxygen determinations were made on specimens cooled with extreme rapidity (BARTELD [48]). The suction tube (closed with an adhering lead foil) of an evacuated cooled thick-walled copper chill-mould was dipped into the melt. After penetrating the drossed surface, the lead foil melted, and the melt shot into the wedge shaped space between the copper sides and at once solidified. On such specimens there were in fact found higher values of the oxygen content which are in harmony with the results of DANNATT and RICHARDSON [230] and may come closer to the values of the true solubility (Fig. 61). On the other hand, the solubility limit at 750 °C (0.42 Wt.-% oxygen) estimated by BRADHURST [121] seems excessively high.

Rolling tests and hardness measurements were carried out in a study of the effect of different oxygen contents on the mechanical properties of lead. While an effect of oxygen on hardness could scarcely be established, a superficial flaking was observed on rolling specimens with about 0.005 Wt.-% oxygen. In the application of the lead impact test to the determination the kinetic energy of forging hammers, KIENZLE [663] had found that the resistance of lead to deformation depended on the casting temperature. Oxygen determinations on Kienzle's specimens gave the relationship shown in Fig. 62 between the oxygen

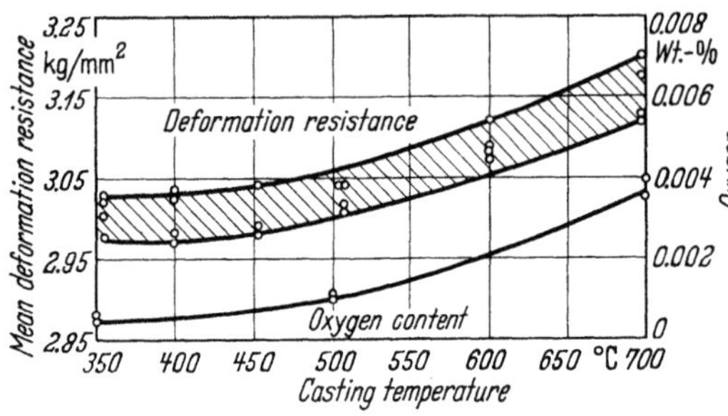

Fig. 62. Mean deformation resistance and oxygen-content of cast lead.

content of lead and its resistance to deformation. In the systematic investigation of 15 binary alloys of lead [24a] mentioned on p. 31 the question of oxygen absorption by the melts was also thoroughly examined. It was established that oxygen is taken up by the melt, if the alloying element differs only slightly from lead in its affinity for oxygen, e.g., in the case of alloys with bismuth or silver. At a high affinity for oxygen, the alloying element, on the other hand, acts as a deoxidant and hinders the absorption of oxygen. If oxygen is taken up, it causes an increase of strength, in agreement with the result shown in Fig. 62. In the corrosion of lead by hot (90 °C) concentrated sulphuric acid, BARTELD [48] found an injurious effect of oxygen content in lead above 0.002%; on the other hand, the attack of lead in nitrosyl sulphuric acid was scarcely affected, not even by oxygen contents 10 times higher.

Even below its melting point lead oxide destroys the protecting skin of chromium oxide on chromium or on stainless steel and therefore leads to a catastrophic oxidation of these materials (SAWYER [1051a]).

25. Pb–S. Lead-Sulphur

In the diagram, the occurrence of lead sulphide, the melting point of which is at 1,135 °C (KOHLMEYER [694]), should be noted. It possesses the sodium chloride structure and as a semi-conductor has been the object of various investigations. FRANKLIN and WAGNER [337a] have determined the dislocation density and the slip systems in lead sulphide (PbS). There is no miscibility gap in the molten state in this part of the diagram (LEITGEBEL and MIKSCH [741]) (Fig. 63). CHENG and ALCOCK [197a] determined the sulphur content in molten lead in equilibrium with mixtures of H_2 and H_2S and from this calculated the solubility of sulphur in molten lead in equilibrium with PbS, and the free energy of solution between 500 and 680 °C. Lead sulphide is formed by the contact of solid lead with liquid sulphur. The rate of growth is time independent and increases exponentially with rising temperature (DORIN [254a]).

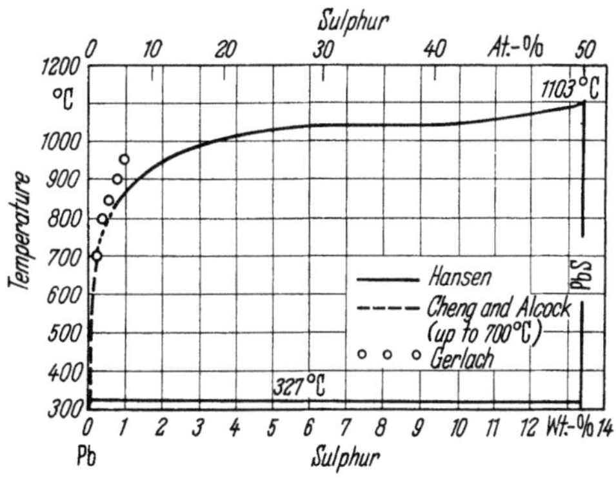

Fig. 63. Lead-Sulphur. (According to HANSEN).

According to GREENWOOD [435] the solubility of sulphur in solid lead at 300 °C is less than 0.0001 Wt.-%. GERLACH [366a] determined, at 700 to 950 °C, the molar fraction x_s of sulphur dissolved in lead to be

$$\lg x_s = -3{,}384/T + 3.535$$

The role of sulphur in lead deserves attention in various respects. Addition of about 1% sulphur prevents stratification in lead-copper alloys (GUERTLER [451]). Traces of sulphur in type metals have shown a surprising activity in regard to the nature of the crystallisation of antimony (LÖHBERG [758]).

26. Pb–Sb. Lead-Antimony

If it is desired to increase the hardness of lead for practical applications, then the readiest alloying element is, in general, antimony. The lead-antimony alloys have been known for long, simply as hard lead, and are used extensively as the most important group of alloys, i.e. for pipe and sheet, cable sheathing, collapsible tubes, storage battery grids, anodes, sulphuric acid fittings, and units for radiation protection. The alloys are grouped in the German standard specification DIN 17641 of 1962 "Lead-Antimony Alloys (Hard Lead)".

Pressure pipes (for 6 and 10 atmospheres nominal pressure) and waste pipes for water systems, of soft lead and of hard lead, are laid down in German standard specifications DIN 1261, 1262 and 1263. Antimony, further, is an essential constituent of alloys of lead (with several metals) which are used as type metal and as bearing metal. Outwardly the alloys of high antimony content are distinguished from soft lead by the fact that they retain their bright appearance in air for a longer time.

a) Production and Structure of Alloys. The alloys are produced either by stirring antimony below its melting point into molten lead or, on the other hand, by adding lead to molten antimony. For the production of alloys of low antimony content it is best to begin with smelter antimonial lead of about eutectic composition or a master alloy of lower antimony content. The master alloy can be added to lead at a lower temperature. In Germany known standard master alloys of high purity, as required for example for storage battery manufacture, are antimonial lead 9 and antimonial lead 9 X.

The composition of antimonial lead 9 is 8.7 to 9.0% Sb. Permissible impurities in per cent are Ag 0.005, As 0.01, Bi 0.02, Cu 0.01, Fe 0.01, Sn 0.01, Zn 0.001, Ni trace, free from other noble metals.

The composition of antimonial lead 9 X is 8.7 to 9.0% Sb. Permissible impurities in per cent are Ag 0.008, As 0.015, Bi 0.04, Cu 0.02, Fe 0.01, Sn 0.02, Zn 0.001, Ni trace, free from other noble metals.

The lead-antimony diagram has recently been revised by RAYNOR [996] on the basis of a critical survey of the literature (Fig. 64). According to this the eutectic temperature is to be taken as 252 ± 0.5 °C, its composition as 11.1% Sb. Deviations from these values are due primarily to the strong tendency of melts to supercooling. By addition of 0.01 Wt.-% S, the eutectic temperature is lowered by 1.9 °C, the eutectic concentration moved towards the antimony side by 0.3% (KNOLLE and LÖHBERG [682]). According to DASSOJAN [232], addition of 0.1% of ebonite to lead-antimony storage battery alloys refines the structure (owing to the sulphur content of the ebonite) and improves the castability. The direct chemical analysis of the eutectic residual melt of a slowly cooled alloy gave the value 11.6% Sb (HOFMANN and ENGEL [559]), in good agreement with the figure given by RAYNOR and with older determinations of a similar kind (BLUMENTHAL [93], and also QUADRAT and JIŘIŠTĚ [985]). According to this, the earlier antimony content of the eutectic, taken as 13% (HANSEN [487]), is to be regarded as too high. OELSEN [915] carried out a very careful thermal analysis of the alloys, by means of calorimetric and electro-chemical measurements (see also EREMENKO [287] and DILLER [250]) and, in addition to the solidification diagram, obtained further thermodynamic magnitudes, such as the affinity of formation and the entropy of mixing of the melts. The structures of a hypereutectic, a eutectic and a hypoeutectic alloy are shown in Fig. 65, Fig. 66, Fig. 304. As lead can take up a maximum of 3.45% Sb in solid solution, alloys with lower antimony contents should contain no eutectic. Actually however, as a result of coring, eutectic is present at contents of 2% Sb (Fig. 67) and far below.

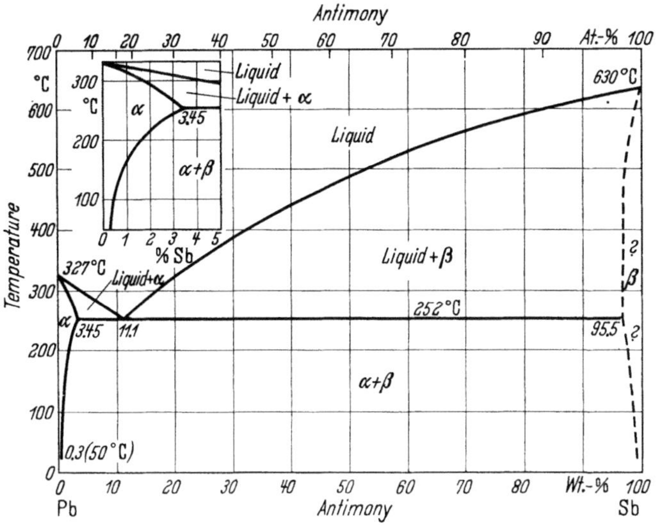

Fig. 64. Lead-Antimony. (According to RAYNOR).

Alloys with only 0.1% Sb show this coring even more markedly (Fig. 68), the antimony content being greatly increased in the dark honeycomb area. Coring was detected even at an antimony content of 0.01% (SIMON and JONES [1122a]). On annealing the alloys, the coring represented in the two last figures disappears, and homogeneous alloys are formed, corresponding to the phase diagram. In addition, the last-named authors [1124] described inverse segregation in hypoeutectic alloys. In this case eutectic residual melt is pressed to the surface. As a result of supercooling, antimony crystallises as the last component with the rhombohedral (111)-planes and the hexagonal base respectively, parallel to the surface.

In the former Bleiforschungsstelle, observations have been made [577] on formation of precipitates in the solid state as a result of solubility decreasing with falling temperature. Antimony-segregates occur in homogenised alloys with 2% Sb after cooling extending over days (Fig. 69). Alloys which were cooled at the rate of 1 °C/minute (i.e., still slowly) showed only the beginning of precipitates on the grain boundaries (Fig. 70). On air cooling, precipitates are no longer observed in the structures. Here X-ray diffraction provides reproducible values of the lattice constants, which are considerably lowered in comparison with pure lead (HOFMANN, SCHRADER and HANEMANN [577], also OBINATA and SCHMID [908)]. This must be interpreted in the sense that on the

Fig. 65. 15% Sb. Cast. Primary crystals of antimony in eutectic matrix. ×150

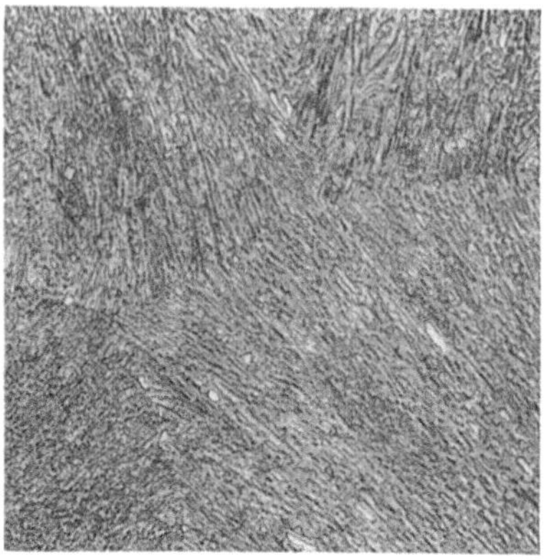

Fig. 66. 11.9% Sb. Cast. Pure eutectic structure. Lead crystallites can be distinguished by means of the various orientations of the antimony lamellae. ×500

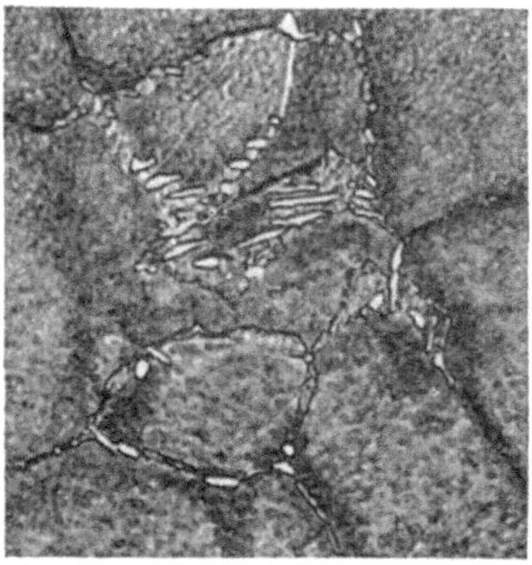

Fig. 67. 2% Sb. Cast. Lead solid solution with solid state precipitates. Eutectic at the grain boundaries and in the more darkly etched other areas. ×1,500

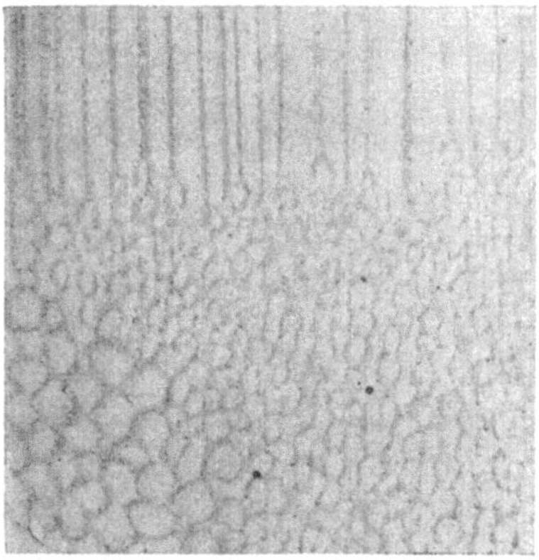

Fig. 68. 0.1% Sb. Cast. Honeycomb type of coring. ×150

air cooling of homogeneous alloys, all the antimony remains in solid solution. Measurements of electrical resistance have led to the same result. In heterogeneous alloys, that is, predominantly in alloys of high antimony content, the suppression of precipitates is more difficult. The sluggish precipitation of antimonial lead is of importance in so far as it permits age-hardening even in air-cooled parts (p. 89, further p. 263). LIVINGSTON [754b and c] examined precipitation phenomena in lead-antimony, lead-tin and lead-cadmium alloys by means of magnetisation curves in the region of superconductivity.

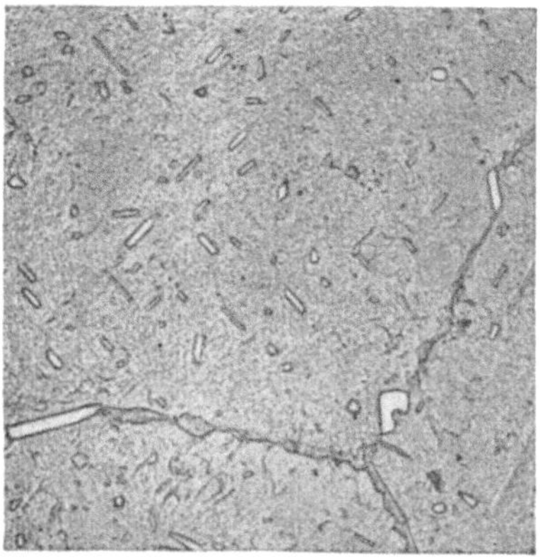

Fig. 69. 2% Sb. Heat treated at 235 °C and cooled during 7 days. Orientated segregate of antimony in form of platelets. × 600

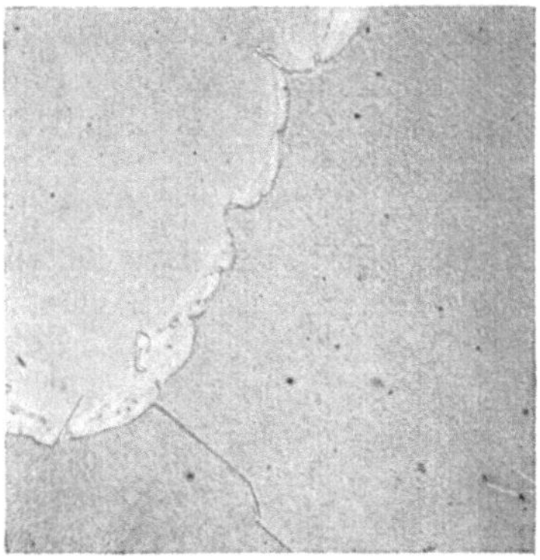

Fig. 70. 2% Sb. After heat treating at 245 °C, cooled at the rate of 1 °C/min. × 500

b) Casting Properties. Physical Properties. The shrinkage of near-eutectic alloys on solidification is, as shown in Table 10, lower than that of lead (Dean, Zickrick and Nix [239]).

Table 10. *Solidification-Shrinkage of Some Lead-Antimony Alloys*

Antimony in Wt.-%	0	10	12	16	100
Solidification-shrinkage in Vol.-%	3.85	2.31	2.47	2.06	1.45

As shown in Table 11, the longitudinal shrinkage of the alloys (p. 334) is also less than that of lead (v. Göler [393], Wüst [1292]).

Table 11. *Longitudinal Shrinkage of Lead-Antimony Alloys*

Antimony in Wt.-%	0	3	6.9	14.68	14.68	19.2	19.2
Casting temp. °C	—	—	—	450°	500°	650°	750°
Longitudinal shrinkage %	0.97[1]	0.65[1]	0.64[1]	0.56	0.56	0.54	0.54

In some basic researches, Bauer and Sieglerschmidt [62] determined the dependence of shrinkage on the dimensions and temperature of the mould, on the ratio of the weight of the melt to that of the mould, etc. The values found were below those calculated from the formula

[1] Calculated from $\alpha(t_{sol} - 20\,°C)$ which gives too high values. α = linear coefficient of expansion; t_{sol} = solidus temperature in °C (v. Göler [393]).

$\alpha(t_{sol} - 20\,°C)$. For an alloy with 16% Sb and 0.21% graphite, which was presumably regarded as a bearing metal, the shrinkage obtained was:

	Sand-cast	Chill-cast
Initial mould-temperature 20 °C	0.54	0.52
Initial mould-temperature 125 to 148 °C	0.56	0.48

The hypereutectic alloys have a strong tendency to segregate; as a result the upper parts of the castings become enriched in antimony (Fig. 71a). The lower layers, in the extreme case, have the eutectic

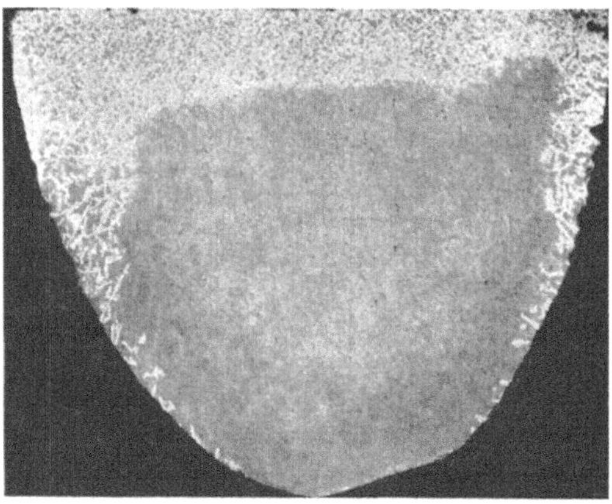

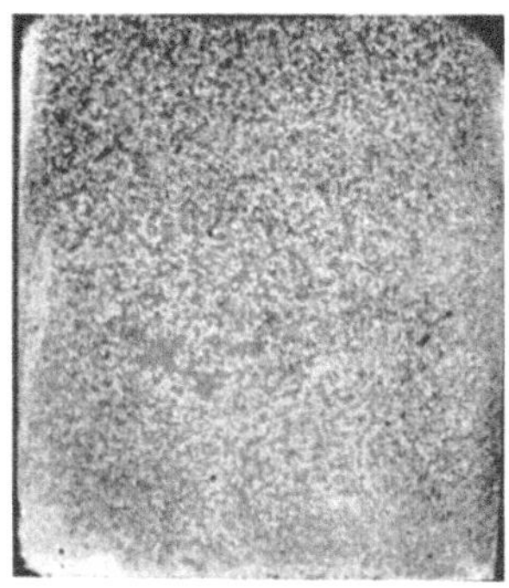

Fig. 71. a) 15% Sb. Slowly cooled; ×1.6 b) 15% Sb. Cast in cold mould. ×2

composition (p. 77). Segregation is due to the difference in density between antimony and the melt and is influenced by the rate of cooling and the viscosity of the melt. It can be almost completely avoided by sudden cooling (Fig. 71b); a low casting temperature is advantageous. In taking samples for analytical purposes, regard must be paid to the segregation [193].

Instead of older measurements of the viscosity (Plüss [966], Grosheim-Krisko [441]), reference will be made here only to some new work (Fig. 72) by Gebhardt and Köstlin [362]. At a given temperature, the viscosity of melts falls from the lead side to a minimum between 40 and 60 At.-% of antimony. The diffusion coefficient of antimony in molten lead in the temperature range 450 to 600 °C is according to Niwa and co-workers ([901], compare [252a, 214a]),

$$D = 0.0025 \exp \frac{-6{,}400}{RT} \left[\frac{\text{cm}^2}{\text{sec}}\right]$$

The densities of the alloys in the cast state have been determined frequently (DEAN and ZICKRICK [239], GOEBEL [391], HIDNERT [521], YURKOV [1298]) (Fig. 73). By means of the densities, a rapid determination of antimony in antimonial lead can be carried out (RICHTER [1011]). GREENAWAY [420] gives on the basis of his measurements, values of the surface tension and density of lead, antimony and five lead-antimony alloys in the molten state. The surface tension, measured at 140 °C above the liquidus curve, falls from the lead side to the antimony side of the system, at first rapidly and then more slowly.

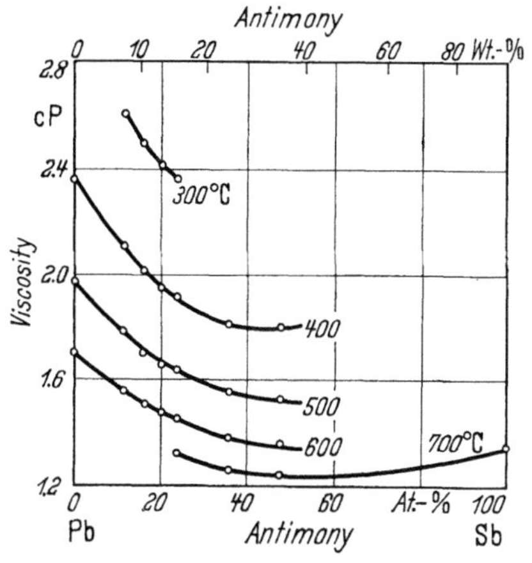

Fig. 72. Viscosity of lead-antimony alloys. (According to GEBHARDT and KÖSTLIN).

At an antimony content of 10%, the electrical (and thermal) conductivity falls by 23% of the value for lead (DEAN and ZICKRICK [239]). KONOZENKO [697] carried out extensive measurements of the electrical conductivity and the thermoelectric power between 0 and 300 °C, on neareutectic alloys and also investigated the Hall effect at room temperature. HIDNERT [521] determined the thermal expansion of cast

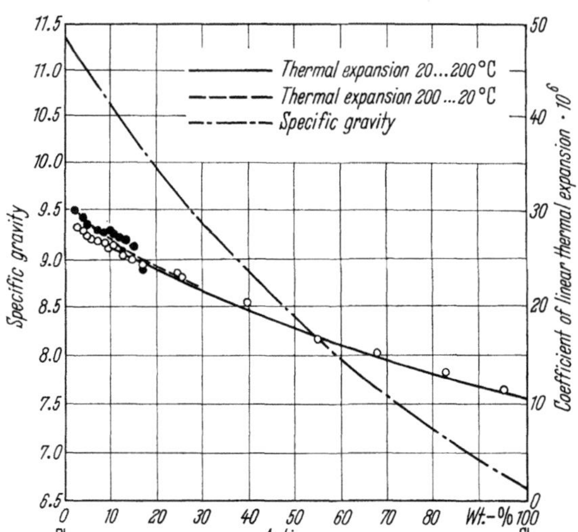

Fig. 73. Lead-Antimony. Density and average linear coefficient of thermal expansion. (According to HIDNERT [521]).

alloys with antimony contents of 2.9 to 98% between −12 and 200 °C. In this region, which may be regarded as the region of heterogeneous alloys, a linear relationship was found between the thermal coefficient of expansion and the concentration in At.-%. Instead, Fig. 73 is set up on the basis of Wt.-%. At the beginning of the investigation the alloys

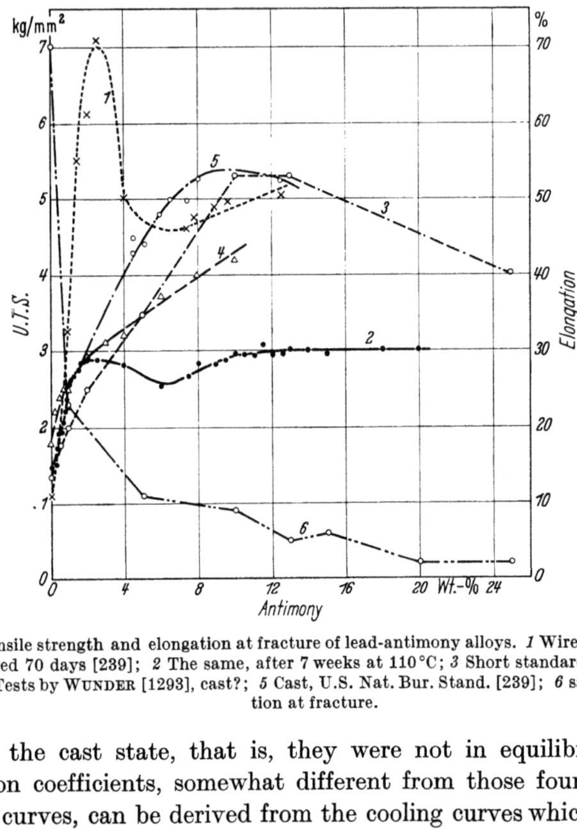

Fig. 74. Tensile strength and elongation at fracture of lead-antimony alloys. *1* Wires, quenched from 235 °C, stored 70 days [239]; *2* The same, after 7 weeks at 110 °C; *3* Short standard test bars, cast? [1261]; *4* Tests by WUNDER [1293], cast?; *5* Cast, U.S. Nat. Bur. Stand. [239]; *6* same as *3* elongation at fracture.

were in the cast state, that is, they were not in equilibrium. Other expansion coefficients, somewhat different from those found from the heating curves, can be derived from the cooling curves which are joined to the first heating curve.

c) Mechanical Properties. The strength properties of lead-antimony alloys, particularly in the region 0.5 to 3% Sb, depend in a high degree on the history of the material (cf. the following section and p. 263). The nature of the pretreatment can result in differences of over 100% in the strength values in the region of the concentration indicated. Data on mechanical values of the alloys must therefore include the history. Small amounts of impurities also have a great effect on the strength properties (p. 87). In Fig. 74, the tensile strengths of alloys after various pretreatments are compared with one another. Curves 1 and 2 correspond to the two extreme cases of maximum age-hardening and of complete softening. The similarity of the course of the curves in cases 3 to 5 leads to the conclusion that they refer to cast alloys as indicated for curve 5. The curves for cast alloys lie between those of the age-hardened alloys and those of the alloys softened by annealing at 110 °C, an indication that a certain degree of age-hardening must be expected even

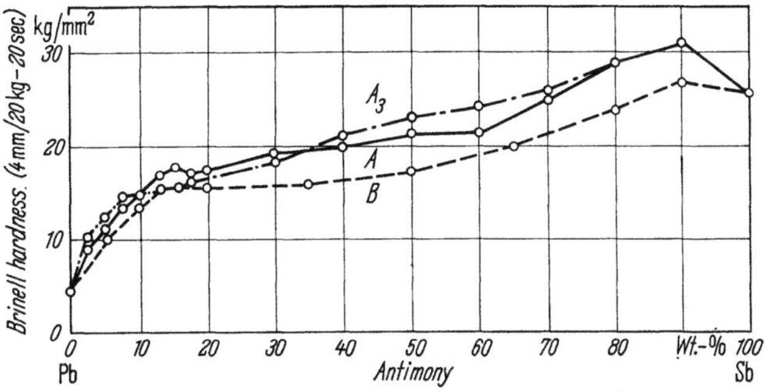

Fig. 75. Hardness of lead-antimony alloys. Purity of original lead, 99.9%. *A* chill-cast; *B* slowly cooled; A_3 cast, stored 3 months. (According to GOEBEL [391]).

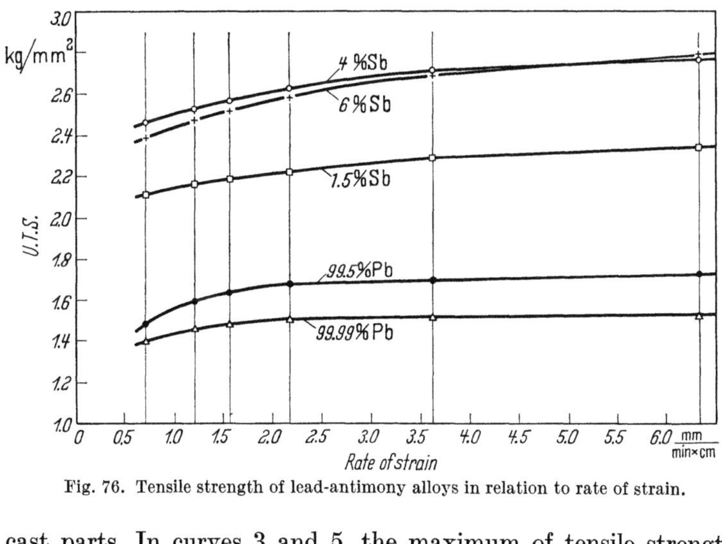

Fig. 76. Tensile strength of lead-antimony alloys in relation to rate of strain.

in cast parts. In curves 3 and 5, the maximum of tensile strength at about the eutectic composition should be noted. The fine-lamellar development of the eutectic (Fig. 66) makes this maximum intelligible. With the occurrence of primary antimony crystals in hypereutectic alloys, the tensile strength falls again, while the hardness still increases slowly (Fig. 75). The torsion modulus of the alloy series shows, according to measurements by AOYAMA [22], a continuous increase with increasing lead content and with falling temperature.

In Fig. 76 the tensile strength of hypoeutectic alloys of different antimony contents is shown in relation to the rate of test [814]. The alloys of higher purity (from 99.99% lead) were annealed for 24 hours at 100 °C, in order that the equilibrium at room temperature should be approximately reached.

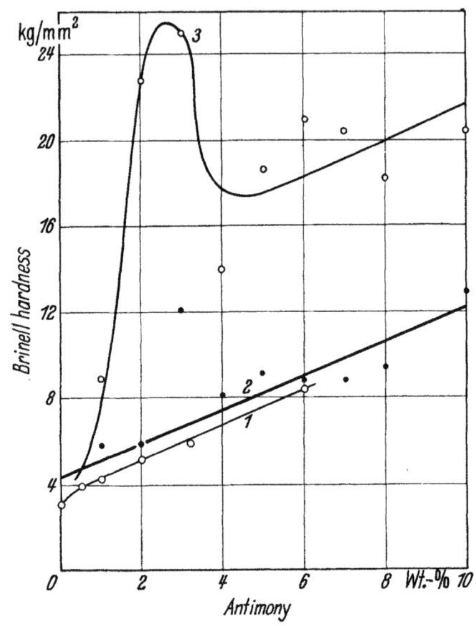

Fig. 77. Hardness of lead-antimony alloys. *1* Heat-treated and quenched [577]; *2* Quenched from 235 °C [239]; *3* Preceding specimens, stored.

The hardness, as well as the strength, depends on the pretreatment. According to Fig. 75, the hardness, in relation to the antimony content, rises steeply and regularly up to the eutectic composition, and, in the initial stages thereafter, alters only slightly. Other investigators found the end of the steep rise at lower concentrations. The hardness of slowly cooled specimens is the lowest owing to breakdown of the solid solution. According to Fig. 75, alloys between 2.5 and 10% antimony age-harden somewhat after casting in a cold chill mould and storage for three months. Fig. 77 shows the hardness for the region of lower antimony content. The values for homogenised and quenched alloys are very low and show a uniform increase with the antimony content. On storage of these alloys there is a considerable increase of hardness as a result of age-hardening (curve 3 of the figure). The representation of hot-hardness of the alloy series in Fig. 78 must be regarded as the first approximation.

The workability of lead-antimony alloys falls with increasing antimony content. Hypereutectic alloys can no longer be rolled. The elongation at fracture can be regarded as a crude criterion for the workability of alloys. It falls with increasing antimony content, as shown by the values for cast alloys in Fig. 74. For known tensile strength it is possible to get a rough idea of the magnitude of the elongation at fracture from the nomogram in Fig. 214. As the elongation at fracture

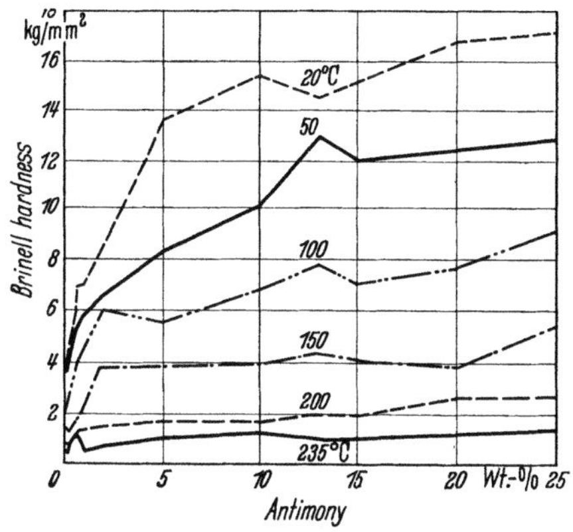

Fig. 78. Hot-hardness of lead-antimony alloys. (According to WERNER [1261]).

is related to the pretreatment of the alloys, a summarising diagram does not rigorously apply to individual cases. From the impact tests, described on p. 207, PELZEL [949] also obtained some views on the workability of lead-antimony alloys. He determined the inception of fracture in relation to the antimony content and the temperature. His account shows that, on the one hand, the alloys in the neighbourhood of the solubility limit, and on the other hand, the hypereutectic alloys below about 100 °C do not tolerate much compression.

While in the tensile test pure lead breaks after about 100% reduction in area, this is no longer true for alloys of high antimony content (e.g. 4% Sb). From a flat maximum at about 2% Sb, the bending value of alloys in the alternate bending test falls uniformly with the antimony content; for example in one experiment the bending values at 0, 1.5, 4 and 6% Sb were 23, 23, 16 and 9 respectively (MAYER-WEGELIN [814]).

Moreover the drawing ratio D/d ($D =$ highest diameter of disc which can still be drawn; $d =$ diameter of die) in the deep-drawing test reaches a maximum value at small antimony contents. Perhaps this maximum can be attributed to the grain-refining action of small antimony contents.

Further data on mechanical properties, particularly those obtained in the compression test, are given in the section lead-antimony-tin. The creep strength of the alloys is reported on p. 236, and the fatigue strength on p. 260.

d) Age-Hardening. *(a) Relationship to Antimony Content and Additions.* A rough survey of the composition of age-hardening lead-antimony alloys is given in curve 1 of Fig. 74. According to this the maximum age-hardening of homogenised alloys approximately coincides with the

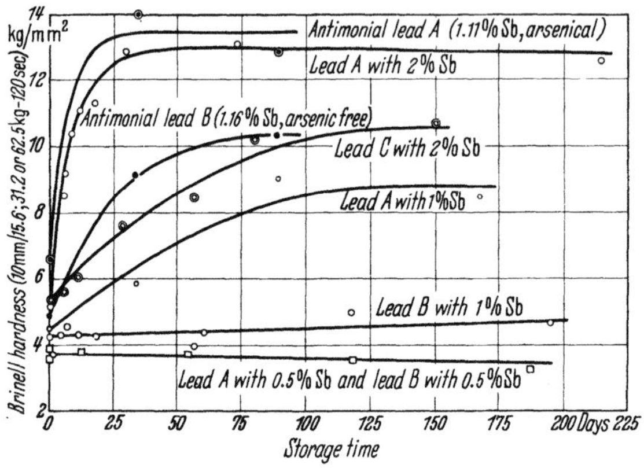

Fig. 79. Storage time of lead-antimony alloys at room temperature. Lead A: Bi 0.022%, Cu 0.014%, Pb 99.961%; Lead B: Pb 99.994%; Lead C: Pb 99.990%.

maximum solubility of antimony in lead, as is to be expected. The action of small amounts of additions on the course of age-hardening was investigated in the former Bleiforschungsstelle (Lead Research Institute) on alloys produced from the qualities of lead *A*, *B* and *C* shown in Fig. 79, with Kahlbaum antimony (purest antimony with a trace of Si).

As the figure shows, both the rate as well as the amount of age-hardening were increased by the small amounts of additions in A quality lead. The alloy of very pure lead B (99.994%) with 1% Sb scarcely age-hardens at all. The action of copper and the even more powerful action of arsenic were tested in special series of experiments (Bluth and Hanémann [94]). The antimony contents were 0.5, 1.0 and 2.0%. The alloys were homogenised and quenched from 240 °C. At an antimony content of 2%, the most effective arsenic content was only 0.05%. It accelerated the age-hardening to such an extent that it was complete after 3 days. The hardness of the age-hardened alloy amounted to 22 Brinell units, compared with 10.4 for the fully age-hardened arsenic-free alloy. Fig. 80 shows the results for an antimony content of 1%. Even in alloys with 0.5% Sb, the small arsenic addition of 0.05% increased the hardness markedly. This amounted, after 100 days, to about 7 Brinell units, compared with 3.6 for the almost non-age-hardening purest alloy. An increase in the arsenic content above 0.05% brought no further improvement.

The experimental results on the action of arsenic agreed, in essentials, with earlier work (Seljesater [1107]). These extended not only to age-hardening at room temperature, but also at 50 and 100 °C (see below). At room temperature an arsenic content of a few multiples of 0.01% was found to have a considerable effect. The effect is still present at a heat treatment temperature of 50 °C, but disappears at 100 °C. PELZEL

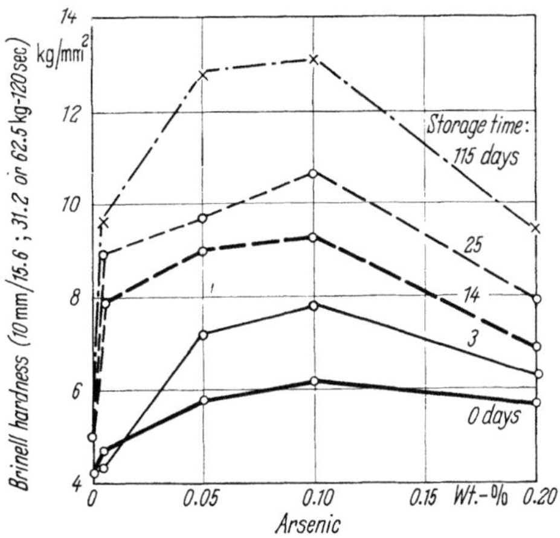

Fig. 80. Effect of arsenic on the age-hardening of an alloy with 1% Sb. Cf. previous figure. (According to BLUTH and HANEMANN [94]).

[948] recently confirmed the action of arsenic contents of the magnitude of 0.1% on the age-hardening of lead-antimony alloys and also tested the action of other additions.

In the work at the Bleiforschungsstelle (Lead Research Institute) an effect of copper on the age-hardening was also determined. Thus copper contents of 0.005 and 0.05% markedly accelerated the age-hardening of alloys with 1% Sb and 0.02% to 0.05% As in comparison with the same copper-free alloys. In Fig. 79 there are also indicated two technical lead-antimony alloys with about 1.1% Sb. Alloy A hardens more than alloy B because of its arsenic content. An effect of copper in addition to that of arsenic is suspected here.

HOPKIN and THWAITES [588] began with a still purer TADANAC lead with 0.00003% Ag: 0.0001% Cu; 0.0001% Zn; 0.0001% Fe; 0.0001% Bi; 0.0002% Tl; 0.0001% Sb; 0.0001% Cd; traces of As, Se + Te, Au and In (Ni, Co, Sn not detectable); and 99.975% antimony containing 0.006% As. In agreement with earlier work, the age-hardening ceased in an alloy with 0.85% Sb made from the pure metal. An arsenic content of 0.001% already sufficed to induce a considerable age-hardening (Fig. 81). The degree of increase of hardness, and the rate of age-hardening, did not rise when the arsenic content was increased to 0.02%, but the curves as a whole were moved to a parallel course with greater hardness (Fig. 81). Apart from arsenic, age-hardening could be induced also by addition of 0.01% Ag; on the other hand, equal contents of tin and copper were without action. This last result raises the question whether

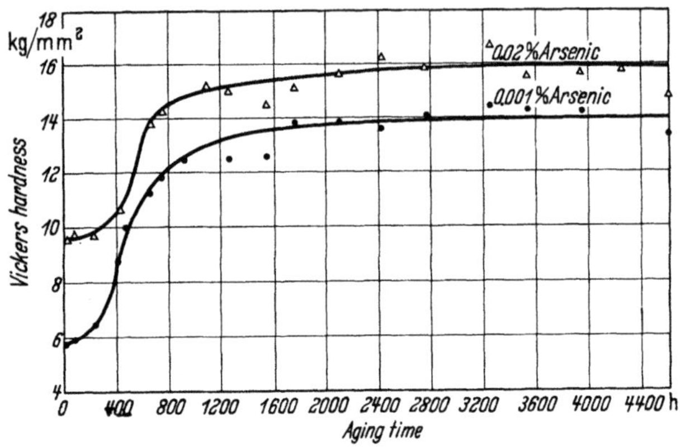

Fig. 81. Effect of arsenic contents on the age-hardening of an alloy with 0.85% Sb. (According to HOPKIN and THWAITES).

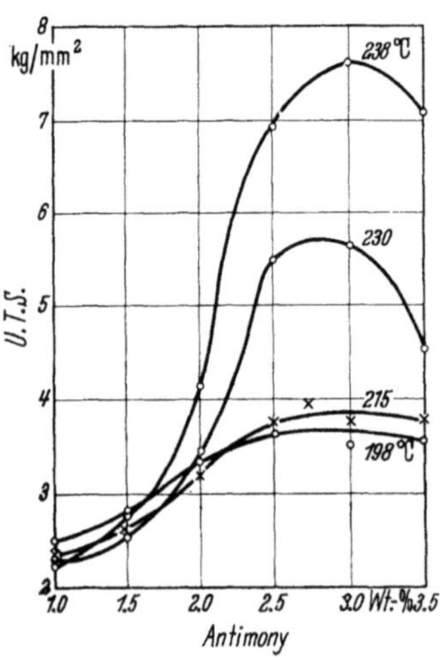

Fig. 82. Age-hardening of lead-antimony alloys in relation to quenching temperature. Storage time 1 day. (According to DEAN, ZICKRICK, NIX [239]).

the effect of copper found earlier was not due to some other impurity not detected.

The above mentioned authors also tested the effect of quenching temperature on age-hardening. Quenching the arsenical alloys from 160 °C gave only slight age-hardening in comparison with a quenching temperature of 300 °C. In explanation for this it is stated that at 160 °C the antimony is indeed in solid solution, but not the arsenic (p. 126). HUME-ROTHERY[600] offers a mode of interpretation for the role of arsenic which, while not very credible, is nevertheless interesting as a working hypothesis. According to this, arsenic could form an inter-stitial solid solution with the lead, therefore expanding the lead lattice. As the dissolved antimony atoms certainly substitute the lead atoms, therefore contracting the lead lattice, then energy considerations would favour the formation of a cloud of antimony atoms round the arsenic atoms as a preliminary to precipitation.

(β) Pretreatment. Age-hardening at Elevated Temperature (p. 265). The age-hardening of lead-antimony alloys assumes the presence of a supersaturated solid solution at room temperature. This condition is found in castings and hot extruded parts, provided that the cooling is not too slow. Consequently age-hardening occurs here without any heat-treatment. Specimens in which the condition mentioned is not satisfied

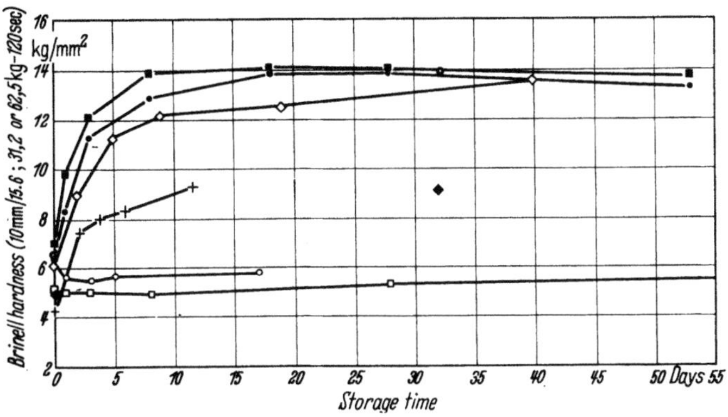

Fig. 83. Age-hardening of antimonial lead A (containing As) and B (As-free). o Antimonial lead A extruded at room temperature; ■ Antimonial lead A, the same heat-treated for 10 minutes at 250 °C then quenched; • Antimonial lead A, air cooled from 250 °C; □ Antimonial lead A, cooled from 250 °C in $1^1/_2$ hours; ◇ antimonial lead A, extruded at 250 °C air-cooled; o Antimonial lead A, cast, heat-treated, quenched; + Antimonial lead B, quenched from 250 °C; ◆ Antimonial lead B, cast, heat-treated, quenched.

(see below) harden only after a short heat-treatment. For higher antimony contents, the heat-treatment temperature should be a little below the eutectic, so that, on the one hand, as much antimony as possible is dissolved and, on the other, no melting of the eutectic occurs (Fig. 82). Ten minutes suffice for the heat-treatment time (Fig. 83). Longer homogenisation heating does not bring about any further increase of hardness. Air cooling gives lower rates of age-hardening with the resulting hardness initially constant. After furnace cooling, age-hardening is almost entirely absent because a breakdown of the supersaturated solid solution has occurred. Age-hardening is also absent if the alloys are extruded or rolled at low temperatures. In this case, during or after working, recrystallisation occurs, which is connected with precipitation of the excess dissolved antimony, corresponding to the lower solubility at the temperature of working (p. 91).

It might be thought that the age-hardening of lead-antimony alloys could be accelerated by using higher aging temperatures, as is known, for example, from the case of duralumin. In fact, an alloy with 0.85% Sb and 0.02% As requires, after quenching from 300 °C, storage for about 40 days in order to reach the maximum of hardness. For an aging temperature of 50 °C, perhaps four days suffice for this purpose. However room temperature age-hardening resulted in an end hardness of almost 16 Vickers units, but at 50 °C only 13 units were attained. Moreover, in this case, after a total of 100 days heat treatment, the hardness sank further by 1 Vickers unit (HOPKIN and THWAITES [588]). Still more marked is the softening after completed age-hardening at 100 °C [577].

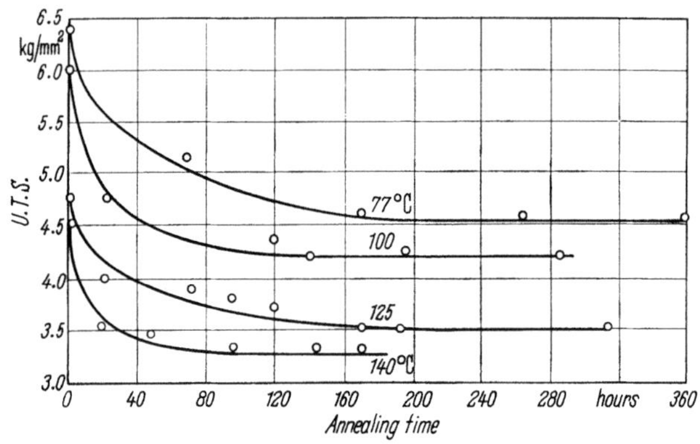

Fig. 84. Softening of an age-hardened lead-antimony alloy during annealing. Tensile strength in age-hardened condition $\sigma_B = 10.1$ kg/mm². (According to DEAN, ZICKRICK, NIX [239]).

At this temperature, combined with a fall of hardness, the formation of coarse oriented precipitates of antimony occurs in the structure.

Softening also takes place if specimens, age-hardened at room temperature, are subsequently annealed at elevated temperatures (DEAN and ZICKRICK [239]). Fig. 84 relates to an alloy with 2.5% Sb, which was extruded to wire, then (for the purpose of age-hardening) homogenised for fifteen minutes at 238 °C, quenched and stored for ten days. At an annealing temperature of 77 °C, the tensile strength falls considerably in 180 hours and then remains constant up to a heat-treatment period of 360 hours. At 140 °C, softening becomes complete, the strength falling below the value for the freshly quenched condition. The behaviour of alloys stored for a very long period is described on p. 267.

(γ) Effect of Deformation on Age-Hardening. It is necessary to distinguish between deformation before age-hardening, e.g. immediately after quenching the alloy to room temperature, or after attainment of complete age-hardening. Experiments of the first kind, among others, were carried out in the Bleiforschungsstelle (Lead Research Institute) [577]. The alloys were homogenised, quenched and immediately cold-rolled with 50% reduction. As characteristic, the behaviour of alloys with 2% Sb (Fig. 85) may be mentioned. They show, in the first days, a considerable increase in hardness, that is an acceleration of age-hard-

ening compared with unworked alloys, as is generally to be expected. After days or weeks, softening begins, which proceeds to an earlier or later end, according to the degree of purity of the alloy and the amount of deformation. Investigation of the softened specimens under the microscope showed complete disappearance of the elongated grains,

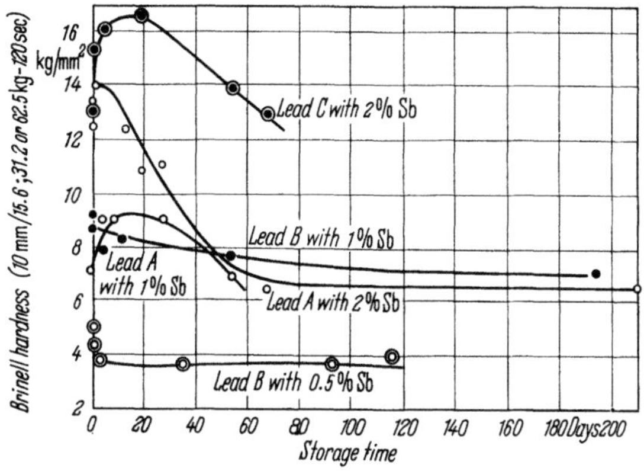

Fig. 85. Lead-Antimony. Change of hardness of alloys quenched and worked immediately. (Cf. Fig. 79).

therefore denoting recrystallisation (compare Fig. 206). In the crystals formed by recrystallisation, visible precipitation of excess dissolved antimony had occurred. It was possible to detect the precipitates even in alloys with 0.5% Sb, and thus to confirm under the microscope the low solubility of antimony in lead at room temperature (Fig. 86).

A very pure alloy with 0.85% Sb, which after quenching from the region of solid solution does not age-harden (Hopkin and Thwaites [588]), underwent, on subsequent stretching by 5 and 10% respectively, an immediate increase of hardness from 5.1 to 6.6 and 7.2 Vickers units respectively. In the course of 8 months there occurred a further slight increase of hardness to 7.5 and 8.5 Vickers units respectively.

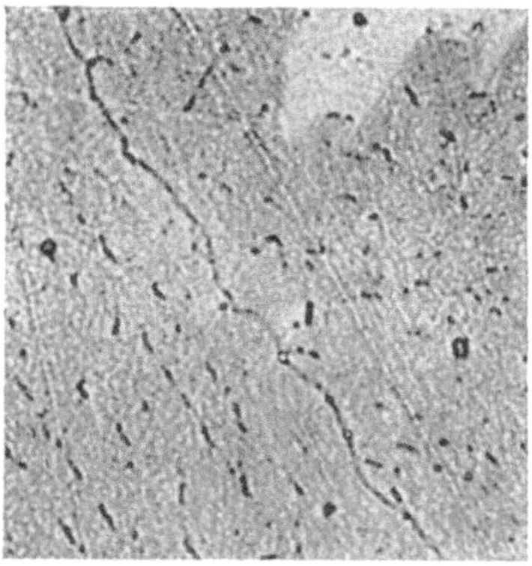

Fig. 86. 0.5% Sb. Quenched, rolled immediately with 50% reduction. After 3 days. Precipitates. ×1,500

In his thesis, Hooker [586] concerned himself exhaustively with the age-hardening of lead-antimony alloys of technical purity and in this connection considered the effect of deformations (elongations up to 10%) immediately after quenching and after a longer period of storage.

The antimony content of the alloys varied between 0 and 0.88%, the temperature of experiment between room temperature and 150 °C. Fig. 87 gives, as an example from his work, the course of the hardness of undeformed specimens. Curve *a* is typical for antimony contents within the solubility limits at room temperature. The increase of hardness after quenching is attributed to the deformations produced by the quench-

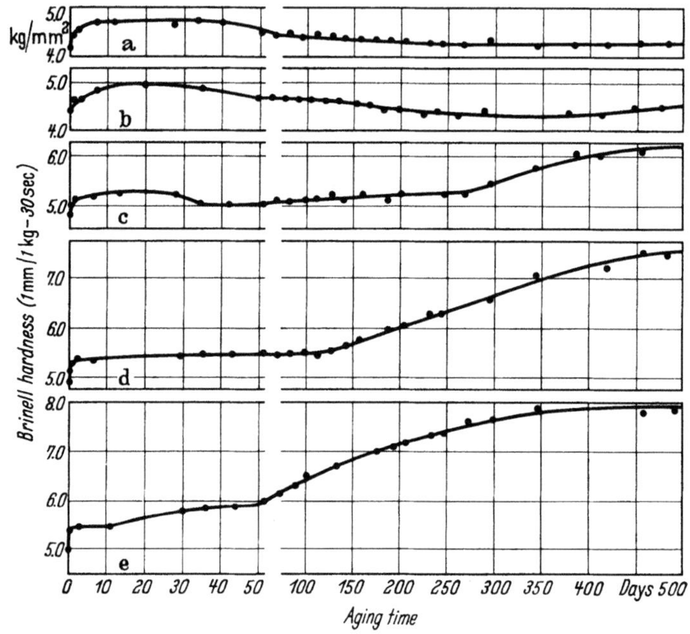

Fig. 87a—e. Changes in hardness of lead-antimony alloys on storage at room temperature. (According to HOOKER). a) 0.15% Sb; b) 0.40% Sb; c) 0.65% Sb; d) 0.75% Sb; e) 0.88% Sb.

ing stresses. All further curves are explained by a superposition of the effect of quenching and of age-hardening. In the deformed specimens there is the additional effect of deformation, from which the irregular course of the curves (not reproduced) may be understood. The photomicrographs of the specimens showed partly discontinuous precipitation, i.e., coarse antimony inclusions in a light coloured ground mass in the neighbourhood of former grain boundaries. This phenomenon is attributed to recrystallisation. The occurrence of recrystallisation in relation to the antimony content and to the temperature is shown in Fig. 88. It seems noteworthy that the beginning of recrystallisation is shifted by the dissolved part of the antimony to higher degrees of deformation. The highest values of the degree of deformation in relation to the antimony content thus coincide in the diagram with the solubility limit of antimony in lead at the different temperatures (dashed lines in Fig. 88).

The effect of deformation on age-hardened alloys is discussed comprehensively elsewhere (p. 194).

(δ) Addendum. The author recently became acquainted with a comprehensive study by BORCHERS and co-workers [109, 109b, 146] on the age-

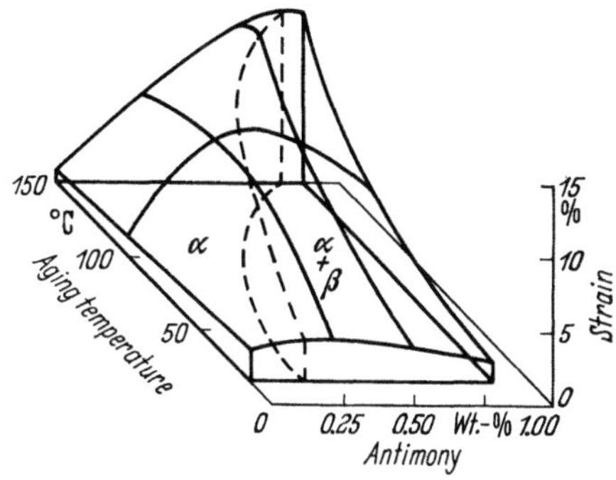

Fig. 88. Lead-Antimony. Degree of deformation necessary for initiation of recrystallisation in relation to temperature and antimony-content. (According to HOOKER).

hardening of cast lead-antimony alloys. The antimony content varied between 2 and 20%. In this study, many of the observations described above were confirmed, e.g., the effect of a temperature increase on the rate and on the amount of age-hardening as well as the decrease of age-hardening at higher antimony contents. A change in the homogenisation temperature had no effect worth mentioning on the age-hardening, as far as the alloys were in the region of solid solution. While the alloy with 8% Sb reached its maximum hardness of 18.6 kg/mm² after a homogenisation period of half an hour at 240 °C (initial hardness after quenching 13.4 kg/mm²), the cast alloy with 2% Sb required a period of 110 hours homogenisation to reach a hardness maximum of 19.8 kg/mm². On the other hand, the same alloy after heat treatment for thirty minutes showed a hardness maximum on only 16 kg/mm² after 500 hours storage at 20 °C. Complete homogenisation therefore requires longer periods of heat treatment in alloys low in antimony. The age-hardening process was repeated five times in the alloy with 2% Sb after respective homogenisation, without any change in the course of age-hardening. The effect of additions of arsenic and copper on the rate and on the amount of age-hardening appeared in a particularly striking way in a technical alloy with 8% Sb, 0.10% As, 0.05% Cu (p. 88).

Storage of the alloys at 0 °C brought about an increase of hardness above that on storage at room temperature, of course after correspondingly longer periods of storage. After pre-storage at −30 °C (during which an increase of hardness took place), all the alloys reached their highest hardness on storage at room temperature. From this it is concluded that at —30 °C, in addition to the precipitation hardening, there occurs "a single-phase segregation" (coherent segregation), which increases the subsequent precipitation at room temperature. An initial participation of single-phase segregation is also suspected in room temperature age-hardening. In agreement with this are earlier unpublished observations by H. Bückle and the present author (short-time increase of electrical resistance after quenching, before the subsequent decrease).

27. Pb–Se. Lead-Selenium

In analogy with the lead-sulphur diagram, the compound PbSe is found in this system melting at 1,076 °C and similarly possessing the sodium chloride structure. The lower part of the liquidus curve (Fig. 89)

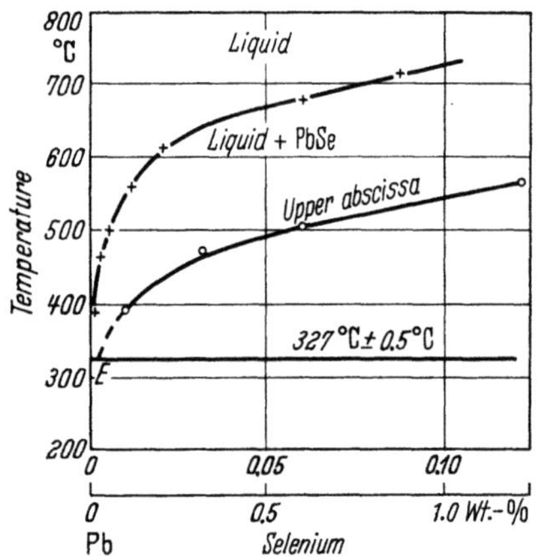

Fig. 89. Solubility of selenium in molten lead. (According to Pelzel). Eutectic E at <0.005% Se.

was determined by Pelzel [945]. He accepts the eutectic at 0.005 Wt.-% and 327.2 °C, indicated by Greenwood [435]. In contrast with earlier determinations (Hansen [488]), Nozato [907] assumed a miscibility gap in the molten state between 3.2 and 9.6 Wt.-%. This could not be confirmed by Seidman [1101a]. Greenwood [435] gives the solubility of selenium in solid lead as 0.0015 Wt.-% at 300 °C.

28. Pb–Si. Lead-Silicon

According to Hansen [488], there are only a few results of investigations on this system up to the present. Terasawa reports in several investigations [1177, 1177a] on the mechanical and technological properties of alloys and of lead-antimony-silicon alloys and on their age-hardening.

29. Pb–Sn. Lead-Tin

The phase diagram (Fig. 90) was derived from a critical evaluation of various investigations made up to now (RAYNOR [995]). According to this the solubility of tin in lead at room temperature amounts to

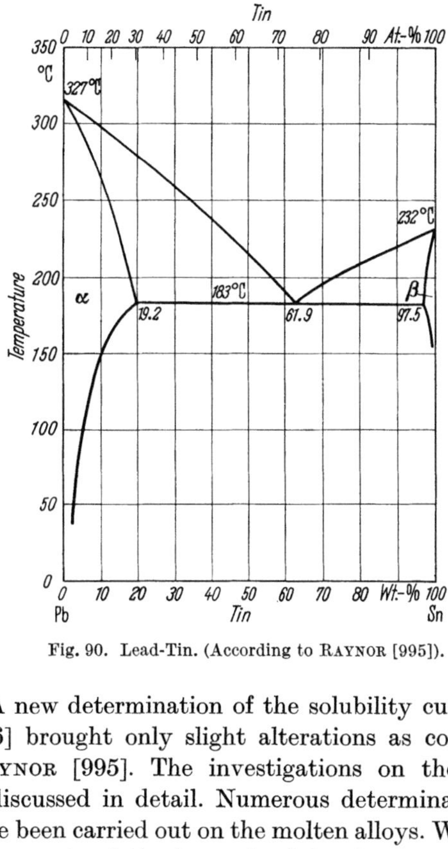

Fig. 90. Lead-Tin. (According to RAYNOR [995]).

$1.3 \pm 0.5\%$. A new determination of the solubility curve by CAHN and TREAFTIS [166] brought only slight alterations as compared with the values of RAYNOR [995]. The investigations on the phase diagram need not be discussed in detail. Numerous determinations of physical properties have been carried out on the molten alloys. Worthy of mention are the measurements of the heat of mixing in the molten and in the solid state (WEIBKE and KUBASCHEWSKI [1253], SCHÜRMANN [1085a]), the thermodynamic activity of lead (PREDEL [979]), the surface tension (BIRCUMSHAW [85], CLAUS and BLANK [202], HOAR [536]), the density and the viscosity (SATO [1048], FISHER and PHILLIPS [325, 326], GEBHARDT [362], JONES [636], BASTIEN and DARMONY [55]), the isothermal transfer from solid lead to liquid lead-tin (LOMMEL [764b]), the electrical resistance (MATSUYAMA [808], ROLL [1022a]), and the diffusion constant of tin (NIWA [901]). The specific volume alters in an approximately linear way at 400 °C with the tin content of the alloy; the density falls by about 0.6% for a temperature increase of 50 °C (FISHER and PHILLIPS

[325]). The minimum viscosity found by FISHER [325, 326] at the eutectic composition of the alloy was not confirmed (Fig. 91) by GEBHARDT [362]. HOLLOMON and TURNBULL [582] carried out fundamental investigations on the nucleation of the alloys on solidification. By subdivision of the melt into drops of 30 to 15 μ diameter they eliminated the action of impurities and attained supercoolings up to 100 °C. The dendritic crystallisation of the alloys (MORRIS, TILLER, RUTTER and WINEGARD [872]) and the solidification of the eutectic were the object of further researches. The thickness of the eutectic lamellas is proportional to $V^{-\frac{1}{2}}$, where V is the freezing velocity. High values of V result in a transition from the lamellated to the globular form of the eutectic (TILLER [1197a]). In the growth of lamellae of eutectic, the tin is the leading phase (WINEGARD, MAJKA, THALL and CHALMERS [1279], TOBER [1202]); there is no relationship of orientation between the lead- and tin-rich phases (TOBER [1202]). In the solidification of cast alloy sheets, a surface layer of 5 to 10 μ thickness becomes enriched with eutectic (DETERT [248]). The phenomena of super-cooling and of nucleation in supersaturated solid alloys, particularly in the temperature region of 150 °C, were followed up several times calorimetrically and resistometrically (STOCKDALE [1146]), BORELIUS [111], BORELIUS, LARRIS and OHLSSON [112], TURNBULL and TREAFTIS [1208], BORELIUS and LARSSON [113], DE SORBO and TURNBULL [1136], MURPHY and ORIANI [884]). The measured heat effect of precipitation was smaller by an order of magnitude of 100 cal/mole than the calculated value. The difference was traced back to the internal stresses produced by precipitation (BORELIUS and SÄFSTEN [114]). Precipitation is connected with a recrystallisation of the matrix (TIEDEMA and BURGERS [1197]). According to TURNBULL and TREAFTIS [1208], the precipitates form discontinuously, i.e., hemispherical regions of precipitate (which they call cells) grow out from the grain boundaries into the inside of the grains. In the cells, lead and tin phases are arranged as parallel lamellae. The distance between lamellae falls with decreasing temperature of precipitation and with increasing supersaturation.

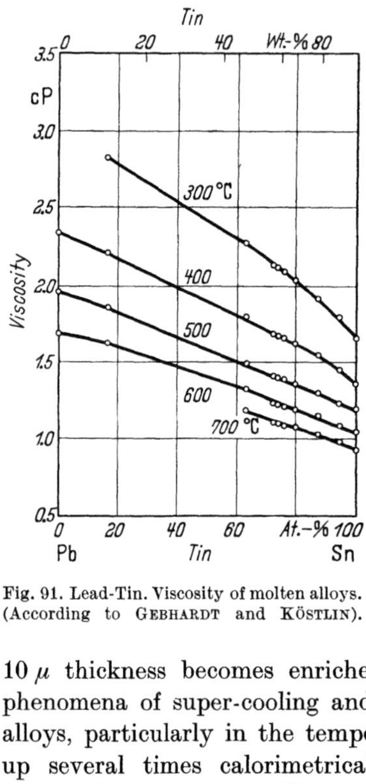

Fig. 91. Lead-Tin. Viscosity of molten alloys. (According to GEBHARDT and KÖSTLIN).

Turnbull and Treaftis conclude from the measurements that the cell growth is controlled by diffusion of tin along the cell boundaries (cf. also Sulonen [1156a] and Livingston [754c]).

a) Structure of the Alloys. The structure of an alloy with 16% Sn is shown in Fig. 92. The alloy still lies within the solid solution region

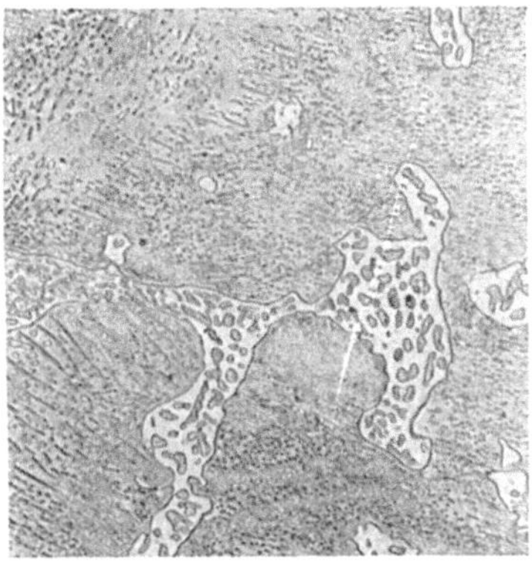

Fig. 92. 16% Sn. Cast. Lead solid solution with streaky segregate of tin. Lead (dark) – tin (white) eutectic. ×500

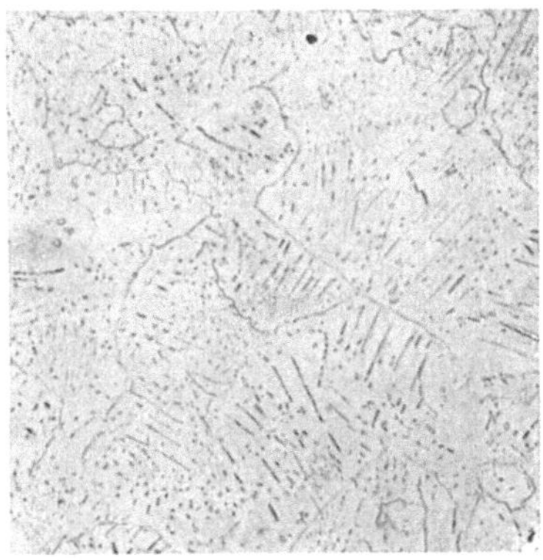

Fig. 93. 3% Sn. Extruded alloy. Lead solid solution with streaky and granular precipitates of tin. ×150

for higher temperatures, so that the occurrence of eutectic material, as usual in such cases, can be explained by coring. With regard to the softening of the alloys (discussed below), the coarse character of the decomposition occurring on cooling or storage should be noted (p. 267). This can be recognised especially well in the micrograph of the technical lead-tin pipe in Fig. 93, which was prepared three months after extrusion. The breakdown of highly supersaturated solid solutions cannot be suppressed even by quenching the alloys.

The transformation of white tin into grey tin below 18 °C was also established on alloys with 50% Pb (that is on hypoeutectic alloys) in organ pipes (Cohen [205]). The minimum content of tin at which tin pest can occur, is not known.

b) Technological Properties. Yanagihara [1295] investigated the relationship between the castability of alloys, expressed by the length of the path of flow to solidification, and the casting temperature. The shrinkage of some alloys was indicated as follows (Wüst [1292]; cf. Bochvar and Dobatkin [96]):

Casting Temperature	Tin Content	Amount of Shrinkage
650°	18.27%	0.56%
550°	70.01%	0.44%
550°	80.90%	0.55%

The smallest shrinkage was found in the eutectic alloy. The densities of the alloys in the solid state have been measured several times (REINGLASS [1002]). The values up to 30% Sn are rather dependent on the pretreatment. This can bring about changes up to 2% (GOEBEL [391]), which is related to the breakdown of the alloys on storage (see below). Fig. 94 contains only the values for quench-cast alloys.

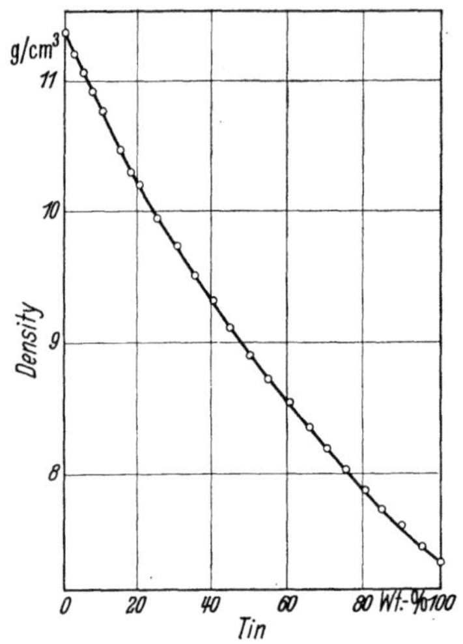

Fig. 94. Density of lead-tin alloys. (According to GOEBEL [391]).

The hardness of the alloys in the quench-cast state rises steeply up to about 15% Sn (that is, about the solubility limit in the solid state) and after that only slowly. The stored and slowly cooled alloys have lower hardnesses as the breakdown of the solid solutions mentioned above has set in (Fig. 95). In data on the strength properties of lead-tin alloys, therefore, not only should the pretreatment be considered but also the period of storage of the material. The action of these factors and the differences in testing methods explain the considerable scatter of the curves in Fig. 95. The effect of pretreatment ought to disappear to a certain degree only at tin contents within the solubility limit at room temperature. The softening of supersaturated lead-tin alloys on storage was investigated (Fig. 96) on homogenised and quenched, very pure alloys over a period of two years (PETRI [954]). Here the alloy with 6% Sn showed transient age-hardening. Softening occurs more rapidly, the higher the tin content. At 10% Sn it was practically finished after five weeks, at 6% Sn it was merely suggested after 2 years, and at 3% Sn it was not detectable; the process is accelerated by working and subsequent recrystallisation. The softening by cold working is particularly marked in the eutectic alloy (UNCKEL [1213]). In these tests the hardness fell from 25 Brinell units in the cast alloy to 2 Brinell units and less in the rolled alloy. This surprisingly great difference (among others) can be explained by the fact that the period of loading in the hardness test amounted to 10 minutes, therefore representing a hardness-creep test. Here the fine grain of the specimens, recrystallised after cold working, certainly plays an essential part (UNCKEL [1213]). The impoverishment of the solid solution in tin, connected with the softening, could be followed by the X-ray method as well as by conductivity measurements in an alloy with 19% Sn (RICHARDSON, WHITE and WEAVER [1009], LARIKOV [732]).

Like the hardness, the tensile strength of the alloys has a maximum at about the composition of the eutectic (AOYAMA and FUKUROI [22]).

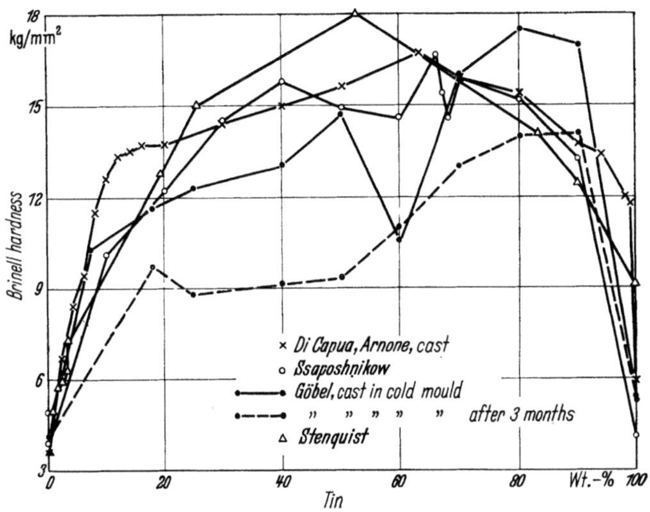

Fig. 95. Brinell hardness of lead-tin alloys. Ref. in [487].

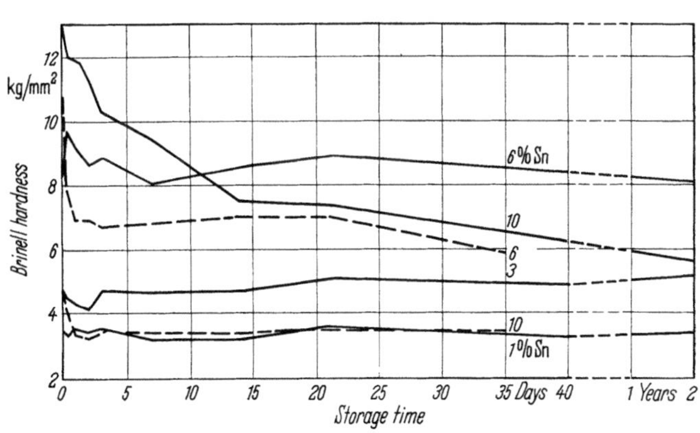

Fig. 96. Change of hardness of lead-tin alloys on storage. Continuous lines: heat-treated and quenched. Broken lines: rolled to 75% reduction after quenching.

Further data on the strength properties of cast lead-tin alloys are given in the section of soft solders. KALISH [642] carried out investigations on the strength of tin-rich alloys with 0 to 50% Pb at low temperatures.

Impact compression tests at −20, 20 and 100 °C on cylindrical specimens of lead-tin alloys, (as well as static compression tests) in no case showed surface cracks or development of fracture, not even at a thickness reduction of 50% or more (HEYN and BAUER [520]). This explains the fact that all the alloys can be rolled (SPERRY [1142]). From the compression tests the energies of deformation and the compressive stresses, necessary to achieve a specific reduction of height, could be deduced. The values found varied with the tin content in the same manner as the hardness figures.

Measurements of the torsion modulus of the alloy seiers at −195 and at 18 °C showed a continuous fall of its value (p. 12) with the tin content and an increase with falling temperature (AOYAMA and FUKUROI [22]). The damping of the alloys is above that of pure lead and of pure tin respectively (BARDUCCI [42]).

In storage, the lead-tin alloys do not assume the grey appearance of soft lead. The difference can already be clearly recognised at a tin content of a few per cent. Lead-tin shares with lead-antimony the role of being the most important alloy group of lead. Tin contents up to 3% are used for cable sheathing, higher tin contents for pressure die casting and principally for soft solders. Organ pipes are produced with tin contents of 45 and 75% respectively (p. 20). Along with other alloying elements, principally antimony, tin is an essential component of bearing and type metals, which exemplify multimetal alloys.

30. Pb–Sr. Lead-Strontium

The structure of the phase Pb_3Sr primarily crystallised on the lead side was found similar to that of Pb_3Na and Pb_3Ca respectively (ZINTL and NEUMAYR [1306]). Further the X-ray investigation showed existence of a certain solubility of strontium in solid lead. From this, an analogy of the lead side of the phase diagram with that of the lead-calcium system may be drawn. In fact, the structure of an alloy with about 0.6% Sr could be entirely interpreted in this sense (Fig. 97). The lead solid solution contains a very fine segregate of Pb_3Sr in the form of small rods. Detailed data on the properties of the alloys are not available. Presumably they are age-hardenable. Strontium was used as an alloying addition in some age-hardenable lead bearing metals.

Fig. 97. 0.6% (weighed addition) Sr. Solidified in crucible. Probably cube-shaped, peritectically transformed primary crystals of Pb_3Sr. Matrix lead solid solution. ×500

31. Pb–Te. Lead-Tellurium

According to GREENWOOD [435] there is a eutectic at 0.025 Wt.-% Te and at a temperature of 326.7 °C on the lead side of the phase diagram. Particularly at low tellurium contents, the surface tension of the molten alloys is lower than the values for pure lead. From this HARVEY [496c] deduces an enrichment of tellurium atoms in the surface. According to

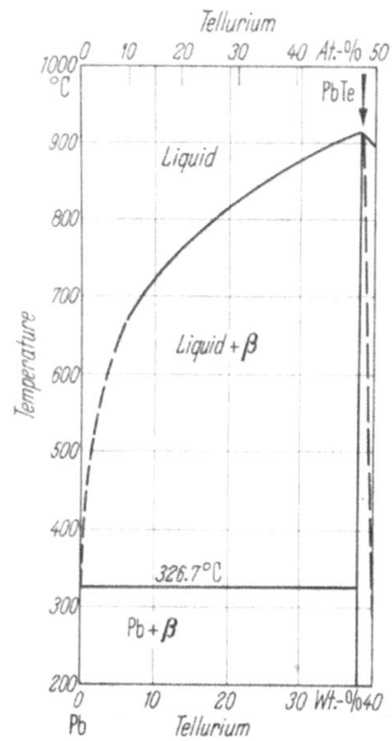

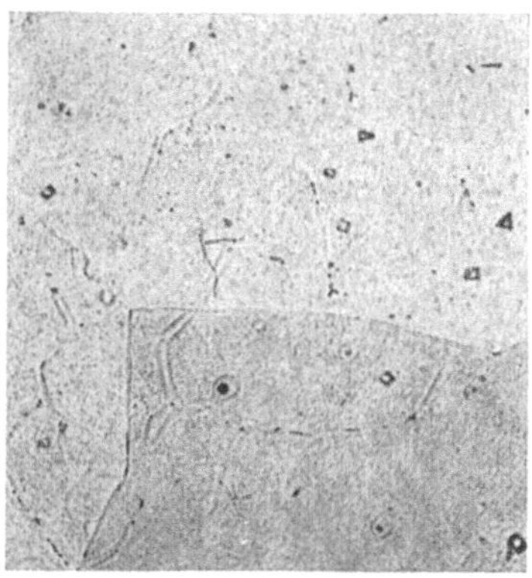

Fig. 99. 0.10% Te. Cast. Cube form primary crystals of PbTe. Lead solid solution with branched coring. ×500

Fig. 98. Lead-Tellurium. (According to HANSEN).

MILLER [855], the compound PbTe has no detectable region of solubility (Fig. 98). It is important as a semiconductor. Measurements by MACHONIS [789a] are devoted to the charge carriers in the system PbTe–SnTe. Data on the production and properties of lead telluride are given by BRADY [122]. The maximum solubility of tellurium in solid lead is given by this author as 0.004 Wt.-%. On the other hand, SINGLETON [1126] deduces a solubility of about 0.05 Wt.-% Te from the relation of the properties of the alloys to the tellurium content. A solid solution limit of only 0.004 Wt.-% Te can scarcely be reconciled with the pronounced age-hardenability of the alloys. The contradiction can perhaps be explained by the fact that the alloys on solidification tend to form supersaturated solid solutions beyond the equilibrium solubility (p. 45). By abrupt solidification of melts in a suction chill mould, FALKENHAGEN [304] produced extremely supersaturated solid solutions with tellurium contents up to 0.2 Wt.-%. (Supersaturated solid solutions from the melt were produced in a similar way in aluminium alloys.) The solubility of tellurium in solid lead can be recognised under the microscope in the coring of cast (Fig. 99) and extruded alloys.

The alloys were developed in England. Consequently, the description of properties principally arises from the investigations carried out there (SINGLETON and JONES [1126]). Lead-tellurium alloys (0.06% Te), both

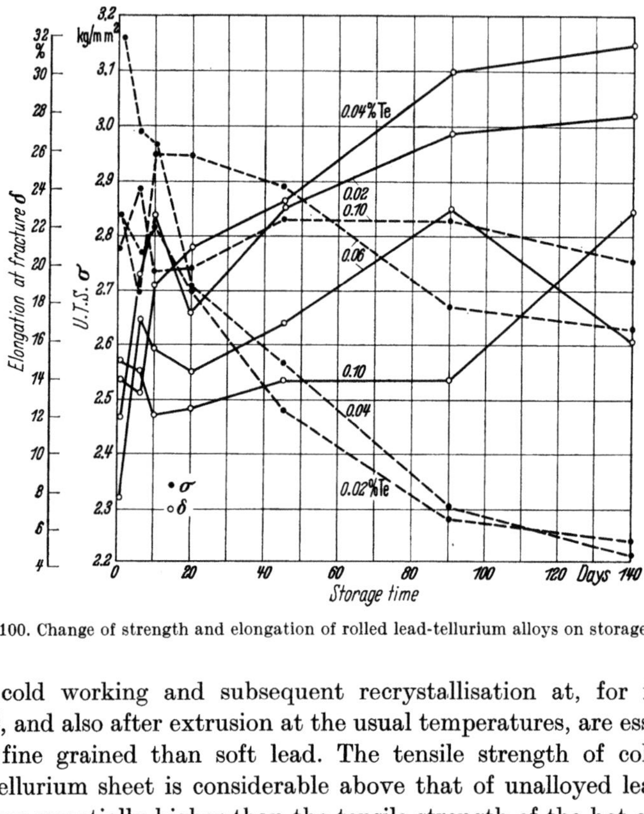

Fig. 100. Change of strength and elongation of rolled lead-tellurium alloys on storage [1126].

after cold working and subsequent recrystallisation at, for instance 250 °C, and also after extrusion at the usual temperatures, are essentially more fine grained than soft lead. The tensile strength of cold-rolled lead-tellurium sheet is considerable above that of unalloyed lead. It is however essentially higher than the tensile strength of the hot-extruded material, from which it was concluded that the alloys are hardenable. In regard to the values in Fig. 100 it may be mentioned that the reduction on rolling was 93.75% and that the temperature of rolling fell from 100 °C to room temperature. The high tensile strength of specimens of low tellurium content is not permanent. On long storage loss of strength occurs, which is combined with an increase of elongation at fracture and (according to the results of structural investigation) with recrystallisation. Only in the alloys with 0.10% Te, no perceptable loss of strength occurred in a period of 140 days, according to Fig. 100. However, that it begins in course of time (contrary to the view put forward by SINGLETON) can be concluded from the results of another series of experiments in the same work, and this was also surmised in the discussion of this work (cf. Fig. 101). The tensile strength of sheets with 0.085% Te, which were cold-rolled to 92% reduction, fell in twelve days from 3.22 kg/mm² to 2.66 kg/mm² (see below). The authors [1126] further emphasise the fact

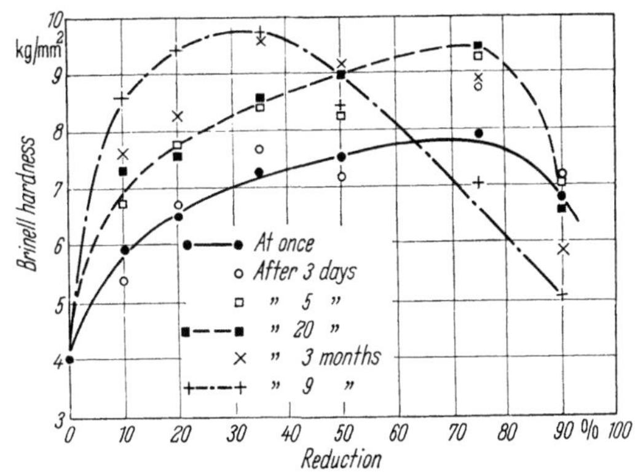

Fig. 101. Age-hardening and softening of a cold-rolled alloy with 0.10% Te [564].

that contrary to other metals the increase of strength of lead-tellurium by cold working is not combined with a loss of ductility, The rolled material can be further deformed without the occurrence of fracture. The slow loss of strength of rolled lead-tellurium alloys on storage is accelerated by annealing. For example, after rolling to 87% reduction, recrystallisation was complete at 100 °C in $2^1/_2$ hours (Beck [68]).

Investigations by the author [564] showed that resistance to recrystallisation and work-hardenability of the alloys are not present in the degree expected by Singleton. Cast alloys with 0.10% Te, prepared by the use of a hardener alloy (Kröner [640]), were rolled down to various reductions. As can be seen in Fig. 101, the age-hardening only increases slowly at higher degrees of working; indeed at the highest it falls, as a result of crystal recovery or recrystallisation. The recrystallisation of the most highly worked specimens proceeds further in the course of time, and, as can be recognised in the fall of hardness after 9 months, is then almost complete. The less highly rolled specimens also become increasingly soft after long storage (Fig. 102).

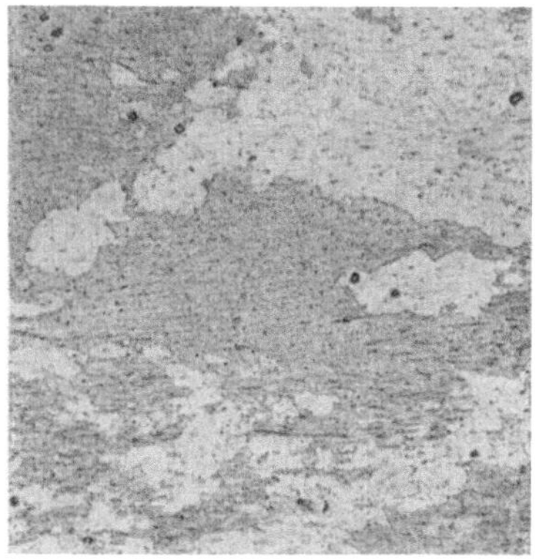

Fig. 102. 0.10% Te. Cast specimen cold-rolled to 75% reduction and stored 4 weeks. Dark: deformed structure. Light: region of recrystallisation. ×150

The age-hardening, observed at low degrees of rolling, mainly 10% and 20%, is of a new kind. It is only initiated by working the specimens, as is shown by comparison with material which has only been cast. The question remains open whether the age-hardening of slightly worked lead-tellurium alloys is not perhaps relieved by recrystallisation after very long periods, as is, for example, the case in lead-antimony alloys worked before age-hardening (p. 91). HADDOW[1] observed no essential loss of strength in a 40% cold-worked alloy with 0.065% Te after one year (p. 397).

In further experiments it was found that age-hardening is less if the alloys are homogenised before rolling. The specimens so treated also recrystallise more quickly than those rolled in the cast condition. This phenomenon could be related to the above-mentioned supersaturation of the lead-rich solid solution after solidification.

In contrast with cold-rolled lead-tellurium, the hot extruded material is in a very soft state. The strength values from a large number of measurements for two different rates of tensile testing may be quoted (SINGLETON and JONES [1126]) (Table 12).

Table 12. *Tensile Strength and Elongation at Fracture of Hot Extruded Lead-Tellurium Alloys*

Composition	Rate of Tensile Test mm/min	Tensile Strength σ_B kg/mm²	Elongation % for gauge length 203 mm
0.05% Te	51	1.86–2.04	55–67
Soft Lead	51	1.40–1.69	30–65
0.05–0.06% Te	2.38	1.62–1.72	85–100
Soft Lead	2.38	1.15–1.37	20–65

The high elongation at fracture of the hot-extruded alloys is noteworthy. Most of the elongation is uniform elongation and arises from the work-hardenability of the material (p. 201 and Fig. 215). It is most pronounced at a tellurium content of 0.05 to 0.06%. The instability of the work-hardening is an objection to the alloy's practical utilisation in "frost-resistant" water pipes. Moreover it is not possible to count on a high creep strength in these materials (GREENWOOD and WORNER [432]). The alloys have high fatigue strength and are therefore used as cable sheating alloys (KRÖNER [709], GREENWOOD [421]). Their good resistance to sulphuric acid is also utilised in various places in the construction of chemical apparatus (HIERS and STEERS [525]).

[1] According to a recent written communication to the author.

32. Pb–Ti. Lead-Titanium

According to an evaluation of the literature by HANSEN, the diagram contains the intermediate phases Ti_2Pb and Ti_4Pb. The lead side is not yet elucidated, so the diagram is not reproduced.

33. Pb–Tl. Lead-Thallium

The heats of mixing of molten lead-thallium alloys have a negative sign (SCHEIL [1057 a]). In the lead-thallium equilibrium diagram (Fig. 103) the large region of solubility on the lead side should be noted. In agreement with this, an alloy with 9% Tl exhibited only homogeneous solid

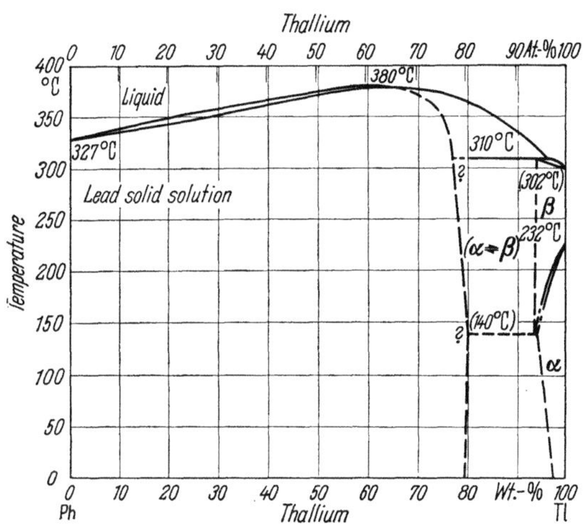

Fig. 103. Lead-Thallium. (According to HANSEN).

solutions, with coring. The lattice constant of the solid solution falls only slightly up to a content of 10% Tl, but beyond this a higher contraction of the lattice takes place (TANG and PAULING [1172]). The existence of superlattices within the solid solution region has frequently been suspected. The data about this are however very contradictory (HANSEN [488]). Alloys with thallium contents up to 54% remain bright in air for several weeks, whereas the thallium-richer alloys tarnish black within several hours (ÖLANDER [912]).

Numerous investigations on the different mechanical, physical and chemical properties of the alloys have been made. Attention may be directed to the hardness, the yield pressure, the density, the electrochemical potential, the thermoelectric power, the hydrogen overvoltage, the magnetic susceptibility, the electric conductivity, the thermal conductivity and superconductivity, and with regard to the literature, reference should be made to HANSEN [487, 488].

According to GIFKINS [380, 381], the creep rate of the alloys falls up to a content of 0.5% Tl, then increases markedly up to 8% Tl. There follows a decrease of the rate of creep up to 26% Tl; at a content of 40% Tl cracks occur in the course of creep. The decrease of grain size with the thallium content plays a part in the interpretation of these phenomena.

The alloys were considered as an anode material for the electrolytic production of zinc (cf. p. 298). Their microstructure is however of interest in another aspect, as thallium is found in traces in some kinds of lead. An increase in the fatigue strength of lead by thallium, which is mentioned in the patent literature (German Patent Application PA 44/0571, of. 19. 4. 1944), could not be confirmed in unpublished experiments by the author.

34. Pb–U. Lead-Uranium

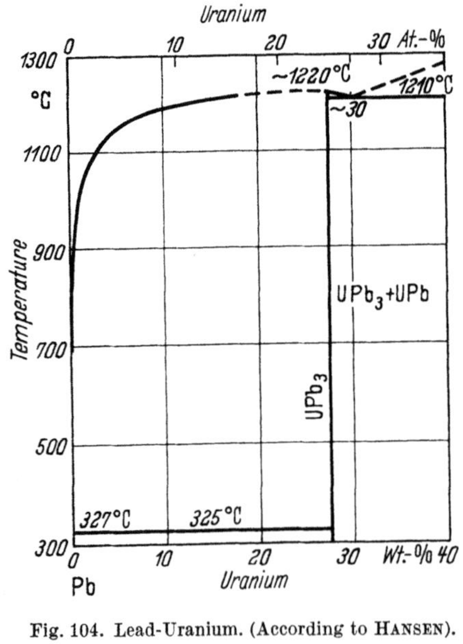

Fig. 104. Lead-Uranium. (According to HANSEN).

An evaluation of the literature of the system for the whole region of concentration of the alloys is given by HANSEN [488]; Fig. 104 shows the lead corner of the diagram.

35. Pb–Zn. Lead-Zinc

The equilibrium diagram was recently drawn up anew on the basis of a technique of differential thermal analysis (Fig. 105) (SEITH and co-workers [1104, 1105], HANSEN [488]). According to this miscibility is complete over 800 °C. ROSENTHAL [1034] dealt with the determination of the thermodynamic constants of molten alloys. He confirmed, in essentials, the phase diagram hitherto accepted as valid, but

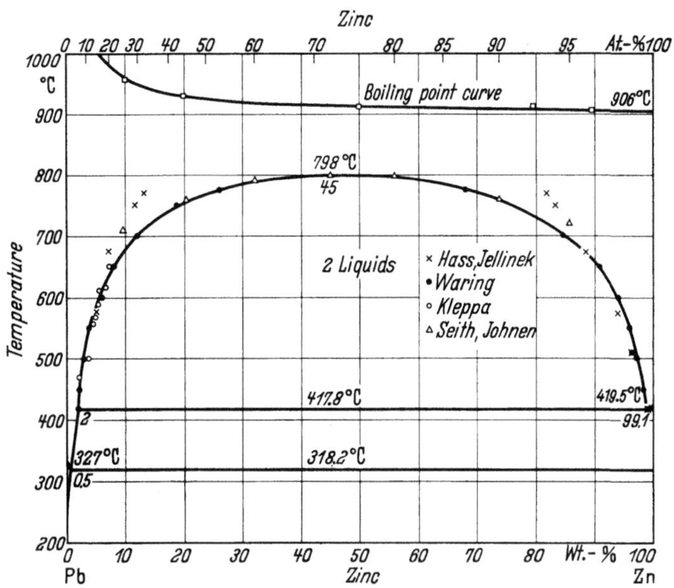

Fig. 105. Lead-Zinc. (According to HANSEN).

gives the maximum of the miscibility gap at about 900 °C. The measurements of zinc vapour pressure by BUDA [137] deserve interest, with regard to the vacuum dezincing of lead.

On the basis of thermal analysis, a solubility of 0.05 to 0.06% Zn in solid lead at 318 °C was determined (HODGE and HEYER [537]). By and large, this result has been confirmed by later investigations of structure (BRAY [127]). According to LUMSDEN [783], the solubility at the eutectic temperature amounts even to 0.10 Wt.-% Zn.

The phase diagram forms one of the fundaments of the desilverisation of lead by zinc (p. 119). Figs. 106 and 107 represent the structure of a hypoeutectic and a eutectic alloy. The eutectic can be regarded as an oriented coalescence of very thin plates of zinc with lead. The interfaces are the base of hexagonal zinc and an octahedral plane of lead, as was demonstrated by LUIS [566a] on single crystals of the eutectic alloy. Fig. 108 shows an alloy from the region of the miscibility gap. In spite of rapid cooling, the regulus is enriched with zinc droplets at the top. Fig. 109 shows the upper part of the regulus under greater magnification.

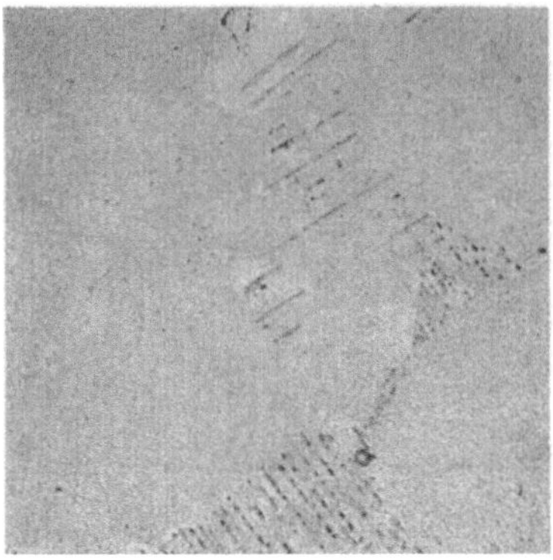

Fig. 106. 0.2% (weighed addition) Zn. Solidified in crucible. Lead solid solution. Lead-zinc eutectic. ×500

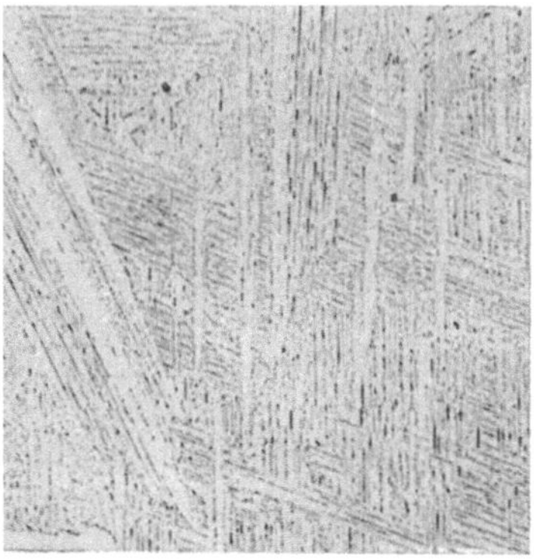

Fig. 107. 0.6% (weighed addition) Zn. Solidified in crucible. Purely eutectic structure. ×150

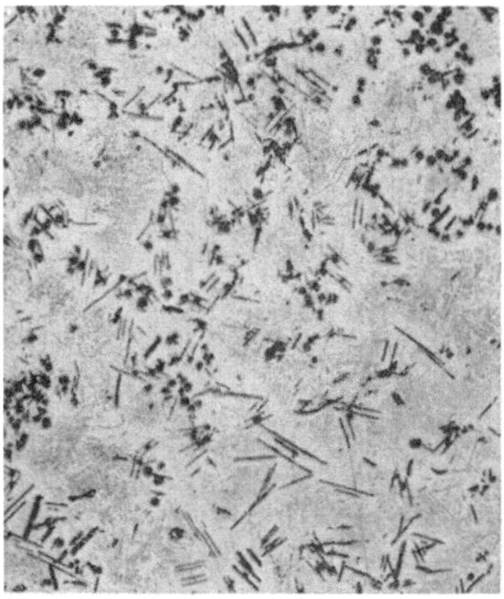

Fig. 108. 3% (weighed addition) Zn. At top solidified specimen enriched with solidified droplets of Zn. At bottom a few droplets and small rod-shaped primary crystals of zinc and lead-zinc eutectic. ×50

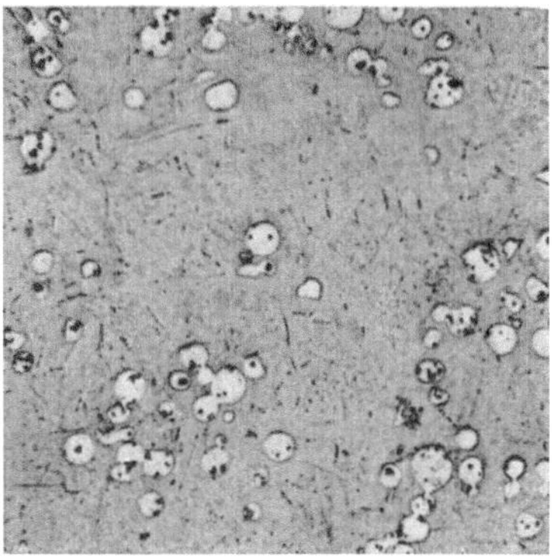

Fig. 109. Previous alloy. Upper layer some solifidied droplets of zinc, partly with straight boundary lines. Eutectic matrix. ×150

The primary crystals of zinc precipitated during passage through the hypereutectic part of the liquidus curve here seldom appear acicular, because they crystallise on to the zinc droplets solidified at 418 °C. On this account the latter show straight boundary lines in places.

In Germany, towards the end of the last war, alloys of lead with 0.5 to 1% zinc were proposed for cable sheathing as a substitute for the alloy with 0.6% Sb. According to an investigation carried out by PFENDER and SCHULZE [957] from this point of view, addition of 0.8% Zn raised the value of the Brinell hardness of lead from 4.4 to 5.5 kg/mm^2, and the tensile strength of extruded pipe from 1.5 to 1.8 kg/mm^2. The fatigue strength of the lead-zinc alloy is similar to that of lead with 0.5% Sb; moreover, the pressure necessary for the production of cable sheathing from lead-zinc is lower than if antimony is used as the hardening additive. The creep strength of zinc-containing lead also appears to be higher than that of pure lead (GREENWOOD and WORNER [434]). The rate of creep between 200 and 300 days, under a stress of 24.6 kg/cm^2 for zinc contents of 0.01%, 0.05%, and 0.1% amounted to 1.38, 0.955 and $0.76 \cdot 10^{-4}$%/h respectively. Creep tests under a stress of 35 kg/cm^2 passed off without intermediate recrystallisation of the specimens, as long as the zinc content was above 0.01%. As regards a possible application of zinc in lead alloys, attention is drawn to pp. 176 and 388.

III. Ternary Alloys

1. Pb–Ag–Al. Lead-Silver-Aluminium

To determine the miscibility gap in the molten state, thermal analyses were carried out on alloys in the middle of the gap, so that on slow cooling, approximately equal thickness layers of the two melts were formed. The solid specimen was cut up and the composition of the two layers determined by chemical analysis. The values given by CAMPBELL and co-workers [172] for the solidification temperatures of the lower density melts, may well indicate the temperatures for the start of solidification. Therefore the boundary of the miscibility gap, drawn in Fig. 110, is approximately the curve which separates it from the surfaces of primary solidification. The stratification curve passes from the aluminium corner with falling temperature to a minimum of 548.6 °C, where the eutectic trough from the binary eutectic, aluminium solid solution (α) — ξ-phase enters. Presumably there is a four-phase plane at 548.6 °C, from which there is formed, from the melt with 67.5% Ag, 30.71% Al and 1.73% Pb, a melt near the lead corner and two phases, an aluminium-rich solid solution aluminium-silver and ξ. The triangle of the assumed four-phase plane is indicated in Fig. 110 by lines with dots and dashes. The critical point of the stratification curve, according to the tie lines in Fig. 110, presumably lies near the lead-silver side at about

40% lead. The maximum temperature of the stratification curve (736.3 °C) occurs on the basis of temperature measurements at the composition 5% Al and 15% Pb. The compositions of the higher density melts were mainly very close to the lead-silver side. Therefore the fields of primary crystallisation could not be determined in the lead corner.

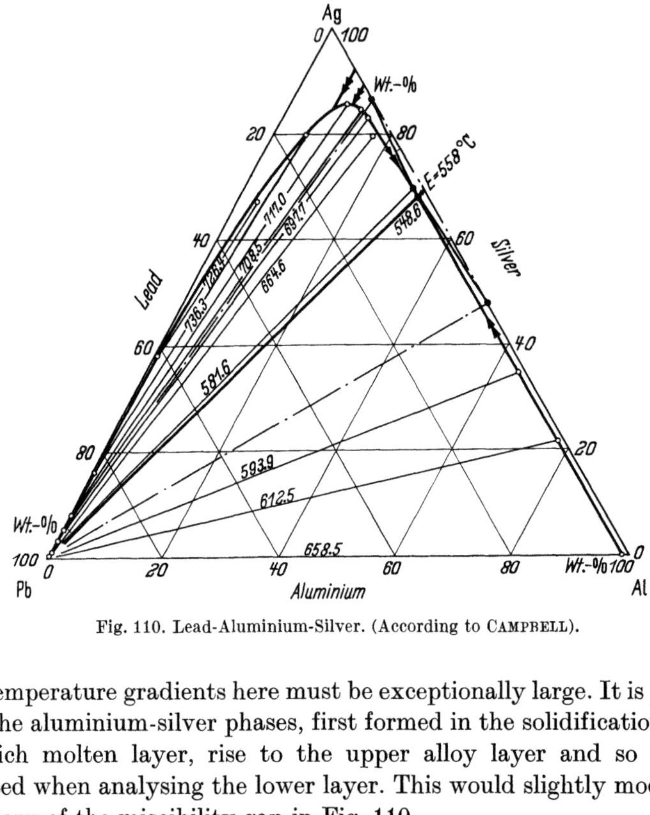

Fig. 110. Lead-Aluminium-Silver. (According to CAMPBELL).

The temperature gradients here must be exceptionally large. It is possible that the aluminium-silver phases, first formed in the solidification of the lead-rich molten layer, rise to the upper alloy layer and so are not detected when analysing the lower layer. This would slightly modify the boundary of the miscibility gap in Fig. 110.

The temperature of the lead-silver eutectic could not be lowered by making aluminium additions, and chemical analysis did not provide any indication of the presence of a ternary eutectic in the lead corner. No solubility of lead in the solidified aluminium-silver phase could be detected by thermal analysis.

2. Pb–Ag–Cu. Lead-Silver-Copper

The miscibility gap emerging from the lead-copper system closes in the ternary system, but its details were not accurately determined in the earlier work. Thermal analysis of the alloys lying between the miscibility gap and the copper corner in Fig. 111 showed only three with a non-uniformity, indicating the monotectic three-phase reaction. According to JÄNECKE [612], the extension of the miscibility gap indicated in Fig. 111 is deduced by evaluating the rectilinear part of the isotherms of the "solidification surface" due to FRIEDRICH and LEROUX [344].

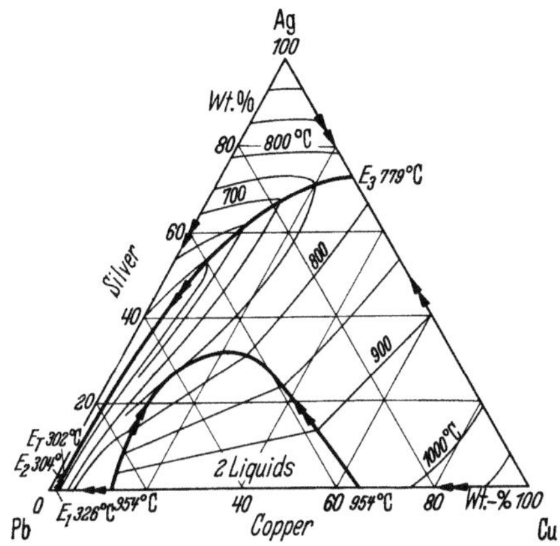

Fig. 111. Lead-Copper-Silver. (According to FRIEDRICH and LEROUX, also JÄNECKE).

In general this diagram is quite straight forward. The binary eutectic troughs emerging from the three boundary systems intersect at a ternary eutectic point E_T, 2% Ag and 0.5% Cu, i.e. near the binary lead-silver eutectic. LLEWELYN [755] has investigated possible applications of the alloys for high-stress bearings. The properties of interest for the bearing materials were determined on 26 alloys. Apart from Brinell hardness measurements, tests were made of the ability to bond to a steel backing and the tendency to segregation. The most promising alloy discovered was Cu:Ag:Pb = 30:30:40, which is just outside the miscibility gap, and Table 13 shows the results obtained in a hot tensile test with this alloy.

Table 13. *Comparison of Mechanical Properties of a Lead-Silver-Copper Alloy with a High-Tin White Metal* (LLEWELYN [*755*])

Alloy	Test Temperature °C	Tensile Strength kg/mm²	Elongation at Fracture %	Elasticity Modulus kg/mm² · 10³
30% Cu,	20	9.93	—	4.78
30% Ag, and	100	7.25	1.0	3.80
40% Pb	150	6.93	1.0	3.02
	180	6.77	1.5	2.67
3—4% Cu,	20	6.93	15.0	5.35
6.5—7.5% Sb,	100	6.93		4.25
Remainder Tin	150			2.67

The strain rate corresponded to an increase in load of about 150 kg/cm²· min. Fig. 112 compares the hot hardnesses of the alloy with other known bearing metals. In corrosion resistance the alloy behaves similarly to the usual leaded bronzes, and its friction and wear under dry running conditions in the Amsler machine were satisfactory. The ability

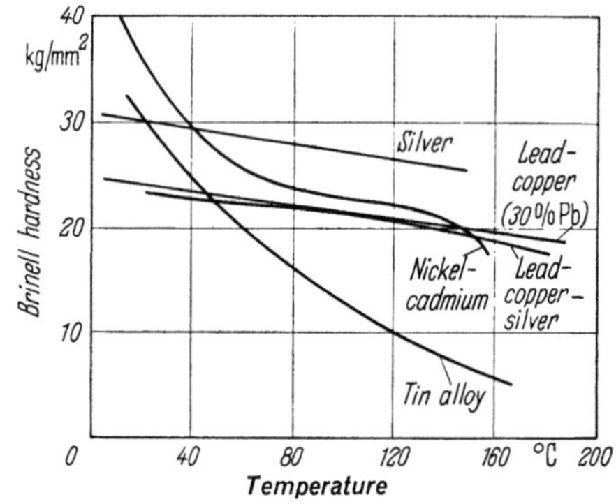

Fig. 112. Hot-hardness of various bearing metals. (According to LLEWELYN).

to embed hard constituents by deliberately contaminating the lubricant with sand was investigated by running tests of 10 hrs. duration. The amount of embedded sand was determined from the residue after dissolving the bearing metals in acid. The ability of the ternary alloy to embed hard constituents was compared with tin bearing metal and cast silver and found to be in the ratio 1.0:1.73:0.55. This ternary diagram is also of interest in the chemical industry where lead with copper and silver additions (see p. 6) is used.

3. Pb–Ag–S. Lead-Silver-Sulphur

The fundamental structure of the ternary system within the lead-lead sulphide-silver sulphide-silver quadrilateral was derived by JÄNECKE [612], and further work done by VOGEL [1229]. The partial system (Fig. 113) was drawn up by thermal analysis of the silver-lead sulphide section, and using previous work from the literature on the boundary systems. The diagram contains the four solidification fields of the phases of the Pb, PbS, Ag_2S, Ag boundary systems. The miscibility gap from the silver-silver sulphide side reaches far into the solidification field of lead sulphide. Eutectic troughs run from the curve bounding the miscibility gap to the binary eutectics e_3 and e_1 of the boundary systems. F_1e_3 represent the binary lead sulphide-silver eutectic, and likewise F_2e_1, the lead sulphide-silver eutectic. Three invariant equilibria are found. In the four-phase plane, F_1-silver-F_2-lead sulphide, the reaction Melt F_1 + Melt $F_2 \leftrightharpoons$ Silver + Lead sulphide occurs. e_3 represents the ternary eutectic of lead sulphide with silver sulphide and metallic silver, and its composition practically coincides with that of the binary lead sulphide-silver sulphide eutectic. Similarly e_1 signifies a ternary eutectic of lead with silver and lead-sulphide where the composition is

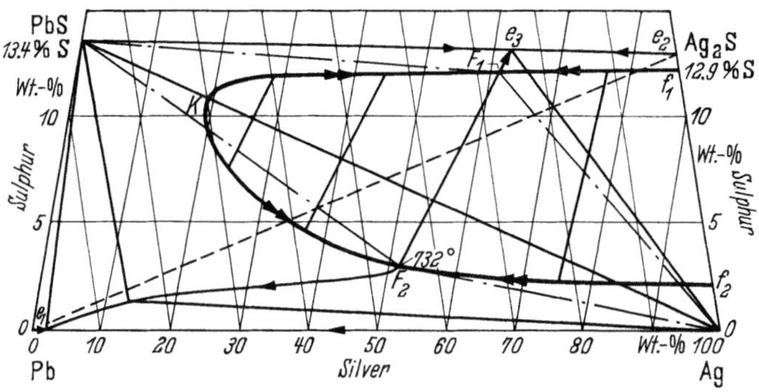

Fig. 113. Partial system Lead-Silver-Sulphur. (According to VOGEL).

practically identical with that of the lead-silver eutectic. The boundary of the miscibility gap near the lead sulphide corner was deduced by observing its disappearance from the microstructure of the alloys. The critical point K ist at about 20% Ag, 11% S and 69% Pb.

4. Pb–Ag–Sb. Lead-Silver-Antimony

BLUMENTHAL [92] determined the basic features of this ternary system by thermal analysis while reinvestigating the lead-antimony

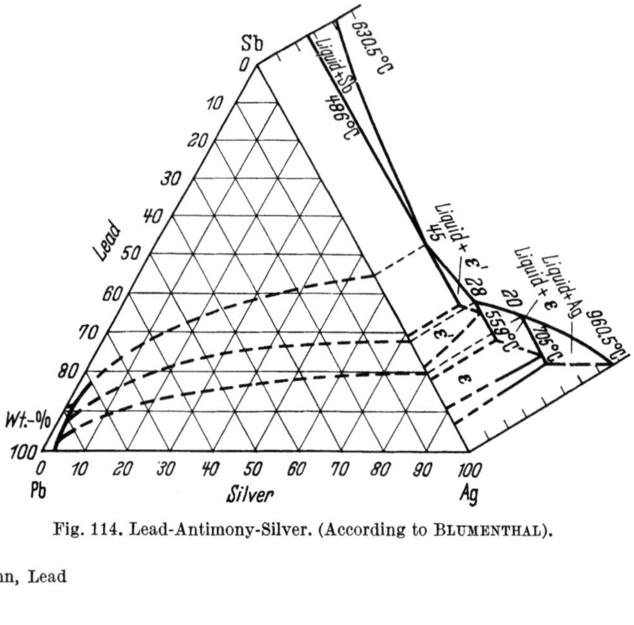

Fig. 114. Lead-Antimony-Silver. (According to BLUMENTHAL).

system. Fifty melts were investigated in the sections parallel to the lead-antimony side, having 0.4, 0.8, 2 and 3% Ag and also in sections with 5 and 10% Sb, parallel to the lead-silver side of the system. The assumption that the troughs from the three binary eutectics pass to a ternary eutectic point was confirmed. The ternary eutectic is close to the lead-antimony eutectic at 11.4% Sb, 1.4% Ag, 87.2% Pb and 244.7 ± 0.5 °C. Further, the two troughs from the peritectics of the silver-antimony system extend to two peritectic four-phase planes. They correspond to the reactions, Melt $T' + \varepsilon \leftrightharpoons \varepsilon'$ + Pb and Melt T + Ag $\leftrightharpoons \varepsilon$ + Pb. The melt T' has 92.3% Pb, 6.0% Sb, 1.7% Ag at 272 °C, and T (estimated) 96.4% Pb, 1.5% Sb, 2.1% Ag at 297 °C. Fig. 114 shows a more schematic division of the ternary system into crystallisation fields and Fig. 115 the precise position of the troughs near the lead corner. Investigations on the effect of small silver additions on the properties of lead-antimony alloys in storage batteries are mentioned in the appropriate section (p. 355).

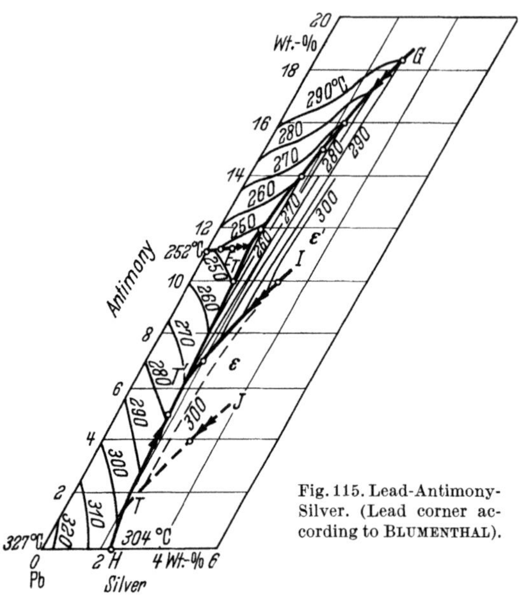

Fig. 115. Lead-Antimony-Silver. (Lead corner according to BLUMENTHAL).

5. Pb–Ag–Sn. Lead-Silver-Tin

PARRAVANO [933] compiled this ternary system (Fig. 116) based on a binary silver-tin system having only one incongruently melting intermetallic compound Ag_3Sn. The peritectic curve, running from point P_2 of the boundary system, unites with the eutectic trough emerging from the boundary lead-silver system at point S at 300 °C. Present knowledge of the silver-tin system (HANSEN [488]) indicates that the ζ-phase must be involved in the related four-phase equilibrium, and that the four-phase plane is bounded by the corner points S, $\varepsilon(Ag_3Sn)$, ζ and, Pb solid solution. The course of the transition curve emerging from P_1 is not known. An accurate determination by EARLE [266] gives the ternary eutectic of Pb with Sn and Ag_3Sn at 62.5% Sn, 36.15% Pb, 1.35% Ag and 178 °C. In addition he has determined the isotherms of the liquidus surfaces in the lead corner more accurately (Fig. 117). The eutectic trough passing from the lead-silver eutectic to the ternary eutectic has a temperature maximum at point *E* (composition 1.75% Ag, 0.7% Sn) between 309 °C and 310 °C. As one would expect point *E* lies on the line connecting the lead corner and the compound Ag_3Sn. It forms the eutectic between Pb and Ag_3Sn. EARLE gave no further details concerning the solidification fields in the region to the right of the eutectic trough in Fig. 117, so the diagram requires further investigation.

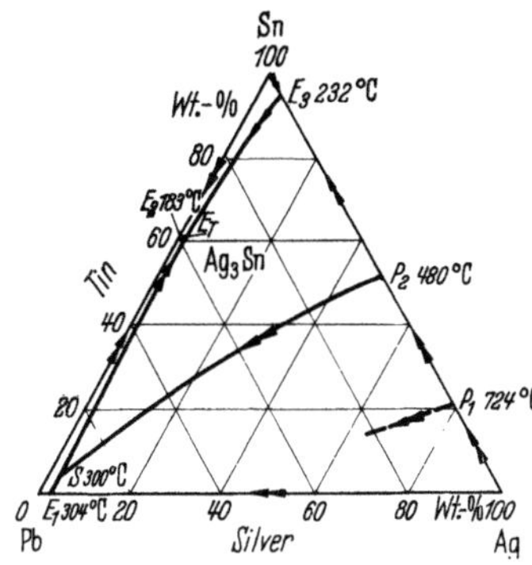

Fig. 116. Lead-Silver-Tin. (According to PARRAVANO).

GARRE and VOLLMERT [358] and also EARLE [266], provided qualitative data on the solid solution region in the lead corner, showing that it extends from the lead corner in the direction of Ag_3Sn. The solubility decreases with falling temperature, since the slowly cooled specimens are softer than those quenched from 250 °C. However, there is no age-hardening. The alloys have achieved a certain importance as soft solders. EARLE cites silver contents of 1.5 to 1.75% and tin contents of 0.7 to 1% as the optimum composition and describes the use of

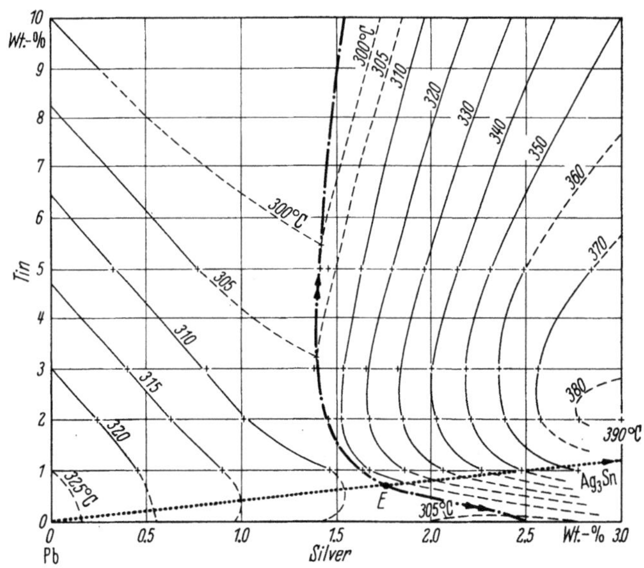

Fig. 117. Isotherms of liquidus surfaces in the lead corner of the Lead-Silver-Tin system. (According to EARLE).

similar alloys for soldering food cans with soldering machines (body-makers). Solders of near ternary eutectic composition also have practical application. Their silver content should be below 1.5%, otherwise there is a risk of silver loss through precipitation of Ag_3Sn. The silver addition to lead-tin solders has a certain advantage in reducing the solution of silver when making small soldered joints in metallic silver (Keil [651], Weigert [1254]).

6. Pb–Ag–Zn. Lead-Silver-Zinc

Because of the importance of this ternary system in understanding the Parkes zinc desilverising process, it was investigated at the beginning of the last century. The extent of the miscibility gap found in the old investigations — by Wright [1290, 1291], Mathewson and Scott [806]

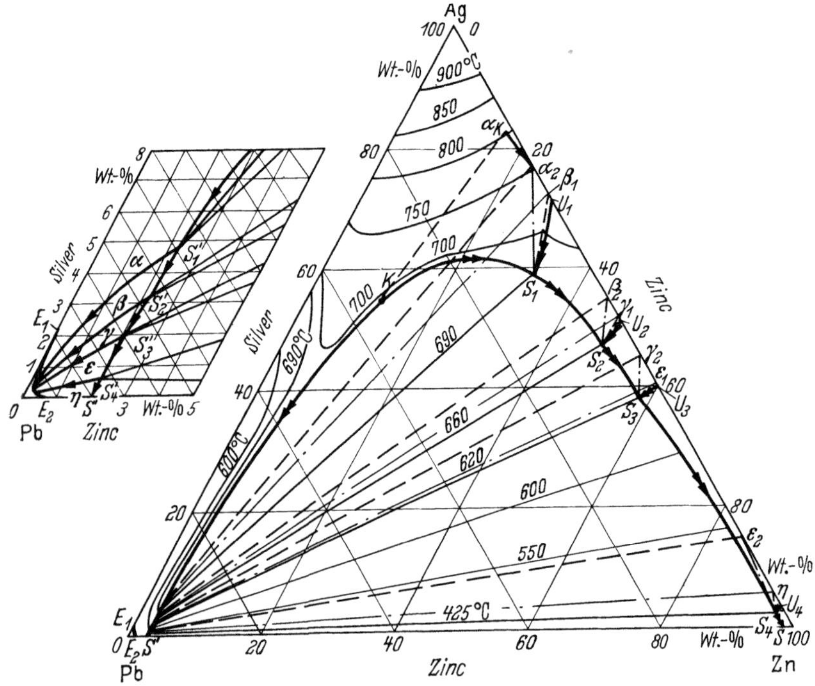

Fig. 118. Lead-Silver-Zinc. (According to Henglein and Köster).

Bogitsch [102] — was described in the first edition of this book and so will not be repeated here. More precise results were obtained in a work by Kremann [705] using thermal analysis. The melts were obviously not stirred so that stratification occurred, but this had hardly any influence on the temperature for the beginning of solidification. The boundary of the miscibility gap towards the silver-zinc boundary system at the temperature at which solidification began, was determined from the maximum of the arrest point. The approximate direction of the tie-lines can also be determined from the arrest points given in tables. The thermal data however do not give the critical point sufficiently accurately, as the boundary curve of the miscibility gap is very flat in its neighbourhood. HENGLEIN and KÖSTER [509], by extensively using

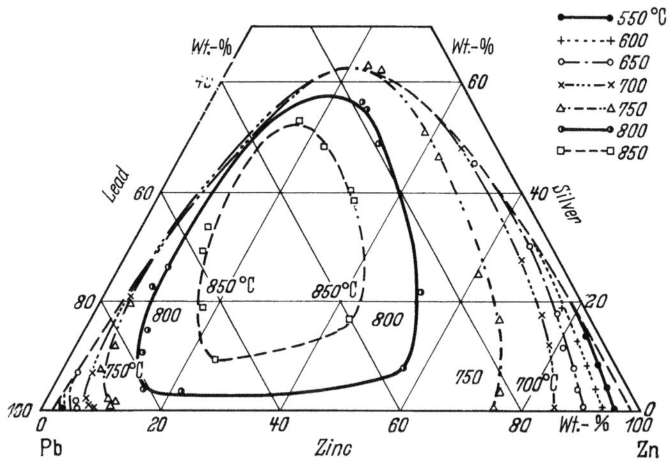

Fig. 119. Isothermal sections through the miscibility gap in the Lead-Silver-Zinc system. Boundary of the miscibility gap towards the spaces with solid phases. (According to SEITH and HELMOLD).

the fundamental data provided by KREMANN [705], sketched a complete ternary system shown in Fig. 118. The boundary curve of the miscibility gap was again determined experimentally, and agreement with the results found earlier was good. The miscibility gap touches the solidification surfaces of all the phases of the silver-zinc boundary system. A solid solution α_K is in equilibrium with the critical melt of composition about 55% Ag, 34% Pb, 11% Zn. The curve $\alpha_K - \alpha_2$ represents the silver-rich solid solution, which is in contact with the melts on the segregation curve to the left and right of K. Troughs run from the four peritectic points of the silver-zinc system in the direction of the segregation curve. Four intersection points S_1 to S_4 with the related invariant equilibria are produced:

$$S_1 + \alpha_2 \leftrightharpoons S_1' + \beta_1 \text{ at } 690\,^\circ\text{C} \qquad S_3 + \gamma_2 \leftrightharpoons S_3' + \varepsilon_1 \text{ at } 620\,^\circ\text{C}$$
$$S_2 + \beta_2 \leftrightharpoons S_2' + \gamma_1 \text{ at } 660\,^\circ\text{C} \qquad S_4 + \varepsilon_2 \leftrightharpoons S_4' + \eta \text{ at } 425\,^\circ\text{C}$$

Here S_i', $(i = 1, \dots 4)$, signifies the limit in the lead corner of the tie-lines emanating from S_i.

SEITH and HELMOLD [1103] investigated the shape and boundary of the miscibility gap accurately. After complete separation of the two melts in a crucible, part of the lower melt was removed through a hole in the bottom. Specimens of the upper layer could then be pipetted off. The boundary curve of the miscibility gap determined in this way (Fig. 119) agrees well with earlier work. The tie-lines (which for clarity have been omitted from Fig. 119) spread fanwise from the lead corner,

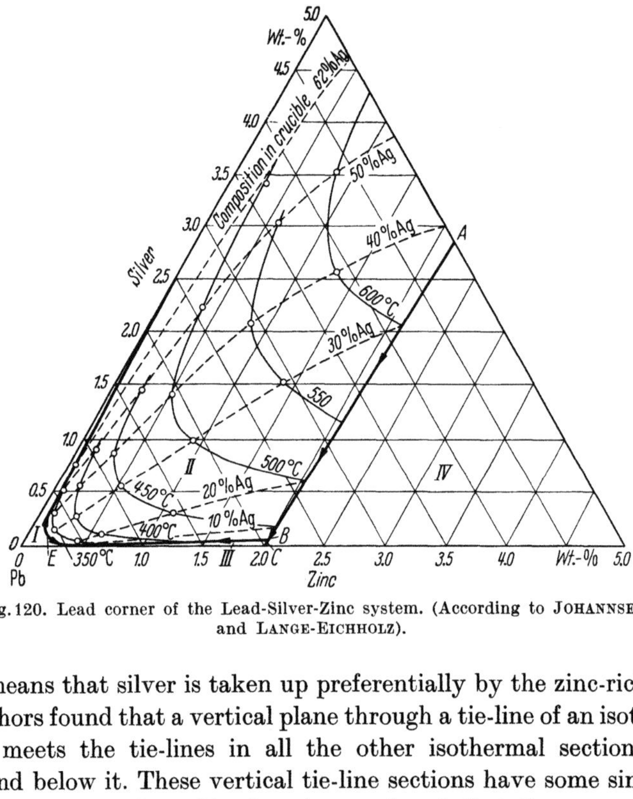

Fig. 120. Lead corner of the Lead-Silver-Zinc system. (According to JOHANNSEN and LANGE-EICHHOLZ).

which means that silver is taken up preferentially by the zinc-rich melt. The authors found that a vertical plane through a tie-line of an isothermal section meets the tie-lines in all the other isothermal sections lying above and below it. These vertical tie-line sections have some similarity to quasi-binary sections. Further, it was found that the centres of all the tie-lines of the two-phase area were in one plane. The miscibility gap closes at an absolute maximum of above 850 °C, and the boundary surface of the miscibility gap towards the silver corner falls vertically from 750 °C to 700 °C.

The basic structure of the lead corner investigated by HENGLEIN and KÖSTER [509] is shown in an enlarged drawing beside Fig. 118. A eutectic trough $E_2 - E_1$ runs from the lead-zinc eutectic to the lead-silver eutectic. The solidification surfaces of the phases of the silver-zinc system are close together between the miscibility gap and the eutectic trough. The reactions of the phases with one another and with the melt in this part of the diagram were not investigated further. The interactions on the eutectic trough, indicated by the intersections with the curves of the threephase equilibria, are of no practical consequence in zinc desilverisation, which ends at about 330 °C, i.e. even before the melt reaches the eutectic trough.

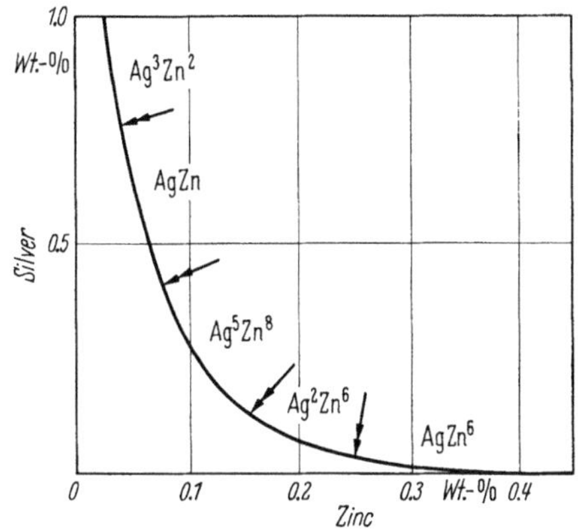

Fig. 121. Eutectic channel in the Lead-Silver-Zinc system. (According to JOLLIVET).

In zinc desilverisation, silver contents of the order of 0.2% are reduced to about 5 g silver per ton of lead by two or three additions of zinc, added as metal or as a poor grade zinc-rich dross. The zinc addition hardly exceeds 1% of the weight of the charge, that is to say zinc desilverisation takes place between the lead corner and the boundary curve of the miscibility gap in the molten state. It is important to know the exact position of the eutectic trough and the isotherms to understand this process in detail. Fig. 120 gives the eutectic curve derived by JOHANNSEN and LANGE-EICHHOLZ [621]. The dashed curves indicate the composition paths of the melts with falling temperature. In order to obtain these curves, lead melts were held in crucibles, made of lead-zinc alloys, corresponding to the compositions indicated on the dashed curves, until equilibrium set in, and the changes in composition of the lead melts were determined. Similar curves could be obtained by noting the change of composition of the melts with falling temperature. The points of intersection of tangents to the curves with the silver-zinc side, give the compositions of the silver-zinc solid phases in equilibrium with the corresponding points on the curves. Desilverisation must be done so that after the last zinc addition on cooling, the composition of the melt enters at a point on the 330 °C isotherm lying at 0.58% Zn or somewhat higher. At 0.58% Zn, the 330 °C isotherm approaches the boundary lead-zinc system so closely that the melt only contains the desired low silver content. The presence of a ternary eutectic, postulated by JOHANNSEN and LANGE-EICHHOLZ [621], does not appear to be relevant in understanding zinc desilverisation. Although the projection of the course of the eutectic trough was very accurately evaluated by JOLLIVET [628], see Fig. 121, it is still uncertain whether the trough falls from the zinc-lead side to the lead-silver side, or whether, as assumed by KREMANN [705], it passes over a flat temperature maximum. JOHANNSEN and LANGE-EICHHOLZ [621] give precise results from experience of desilverisation in a works and on the economical operation of the process.

7. Pb–Al–Bi. Lead-Aluminium-Bismuth

The structure of the binary aluminium-bismuth phase diagram is analogous to that of the aluminium-lead. The miscibility gaps of the binary systems merge continuously into the ternary system. Therefore there is only a small region of single phase melts in the aluminium corner WILDER [1271 a].

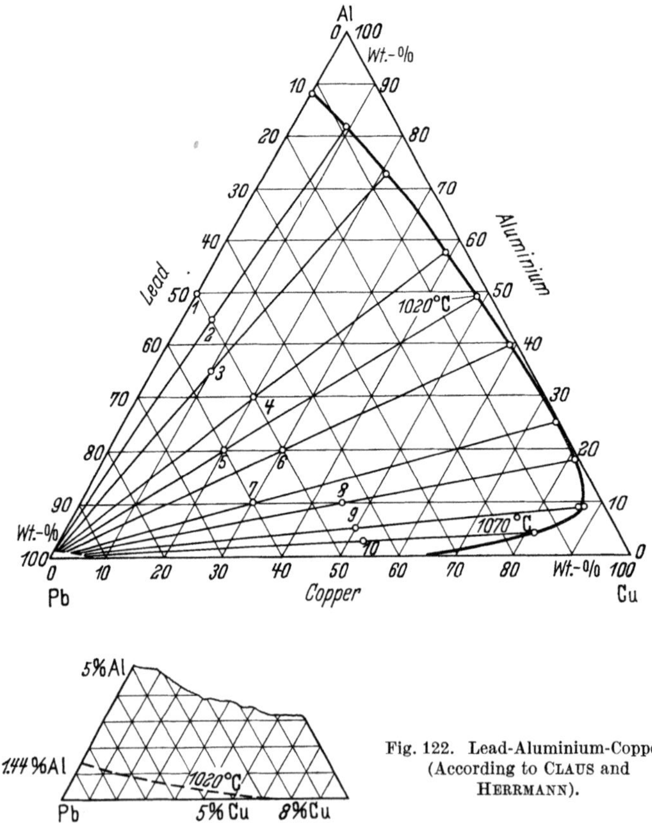

Fig. 122. Lead-Aluminium-Copper. (According to CLAUS and HERRMANN).

8. Pb–Al–Cu. Lead-Aluminium-Copper

Claus and Herrmann [203] determined the extent of the miscibility gap in the ternary system by means of ladle samples, taken at 1,020 °C and 1,070 °C. From analyses of ladle samples from the upper part of the melt, having regard to the binary diagrams Al–Pb and Cu–Pb, the approximate extent of the miscibility gap in the molten state was found (Fig. 122). The boundary of the miscibility gap in the aluminium corner agrees with the determination by Kempf and van Horn [654] that the solubility of lead in molten aluminium is lowered by a copper content of 5%.

9. Pb–Al–Mg. Lead-Aluminium-Magnesium

The application of aluminium-magnesium alloys with lead additions for free cutting alloys gave rise to the first study of the ternary system (Bauer [56]). A number of alloys were investigated by thermal analysis and microscopy, and some also by X-ray diffraction. The only phases present are the binary ones of the boundary systems. As illustrated in Fig. 123, the miscibility gap (p. 34) of the aluminium-lead system closes in the ternary system at about 28% Mg. Beyond the miscibility gap there extends up to about 50% Mg a region of alloys which because of their content of Mg_2Pb, disintegrate in air. The quasi-binary section Al_2Mg_3 (δ) — Mg_2Pb has a eutectic at about 45% Al_2Mg_3 and 439 °C. In the partial region Mg–Al_2Mg_3–Mg_2Pb troughs extend from the three binary eutectics to the ternary eutectic point at 50% Mg, 16% Al, 34% Pb and 405 °C. As shown by the ternary eutectic four-phase plane drawn in the diagram, there is, at the eutectic temperature, considerable solubility of lead and aluminium in magnesium and of lead in Al_2Mg_3. The decrease of solubility in the aluminium corner and in the Al_2Mg_3 corner with falling temperature is shown by the arrowed curves in Fig. 123. The region of technically usable alloys in the aluminium corner is limited by the occurrence of lead segregates. It extends up to 2.5% Pb with a maximum 3% Pb. An aluminium-magnesium alloy with 7% Mg and 2.5% Pb showed good technical properties and gave rise to the short chips desired in machining. Hanemann and Schrader [480] have provided some supplementary observations to the work of Bauer [56] regarding the region bounded by the points Al–Al_8Mg_5(β)–Mg_2Pb–Pb. Two four-phase planes were found by thermal analysis. An alloy with 34.4% Mg and 1.57% Pb gave an arrest-point at 448.7 °C. Presumably involved in this equilibrium is a melt near to the aluminium-magnesium system, α(Al-solid solution), (β(Al_8Mg_5) and a lead-magnesium-rich melt, from which, on further cooling, Mg_2Pb crystallises.

There is a ternary eutectic $E_T \leftrightharpoons \alpha + Pb + Mg_2Pb$ in the lead corner at 0.5% Al, 2.5% Mg at 252.3 °C. The binary eutectic Pb–Mg_2Pb is found at 252.4 °C. Beneath these two four-phase equilibria, the aluminium corner is divided into the two- and three-phase spaces α + Pb,

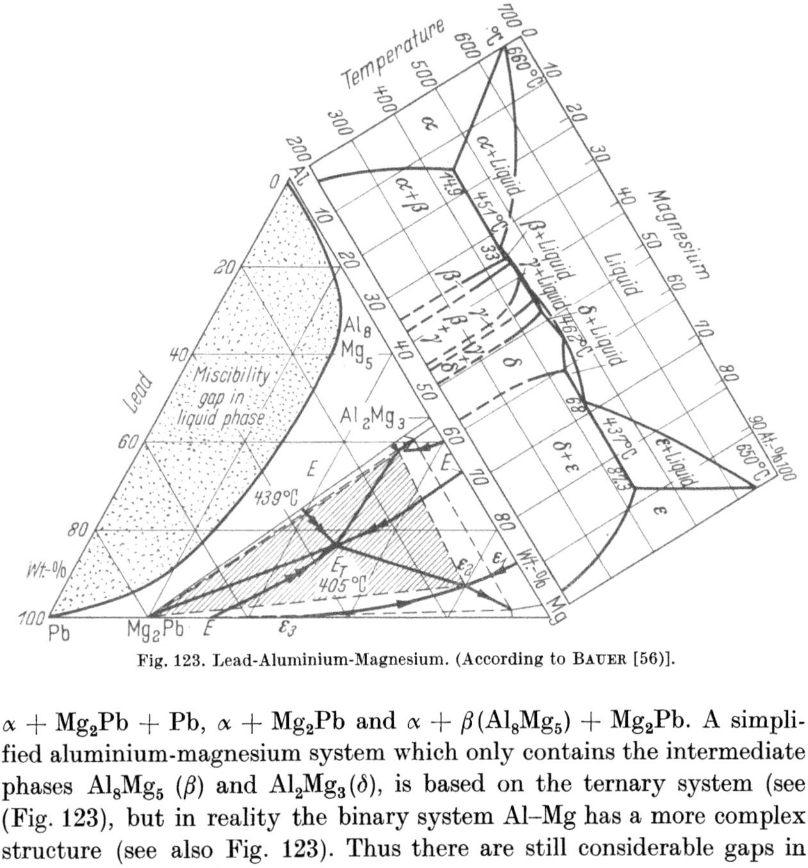

Fig. 123. Lead-Aluminium-Magnesium. (According to BAUER [56)].

$\alpha + Mg_2Pb + Pb$, $\alpha + Mg_2Pb$ and $\alpha + \beta(Al_8Mg_5) + Mg_2Pb$. A simplified aluminium-magnesium system which only contains the intermediate phases Al_8Mg_5 (β) and $Al_2Mg_3(\delta)$, is based on the ternary system (see (Fig. 123), but in reality the binary system Al–Mg has a more complex structure (see also Fig. 123). Thus there are still considerable gaps in our knowledge of this ternary system.

10. Pb–Al–Sn. Lead-Aluminium-Tin

As aluminium-tin alloys have a certain interest as bearing metals, the question arose of how much lead could be added to these alloys without getting into the region of the miscibility gap which comes from the aluminium-lead boundary system. DAVIES [236], at the Fulmer Research Institute, has looked into this problem. He sucked molten samples of these alloys (cleanly separated and in equilibrium) into small quartz tubes and, by chemical analysis, found the boundary of the miscibility gap at three temperatures (Fig. 124). For clarity, the tie-lines are only shown for 800 °C.

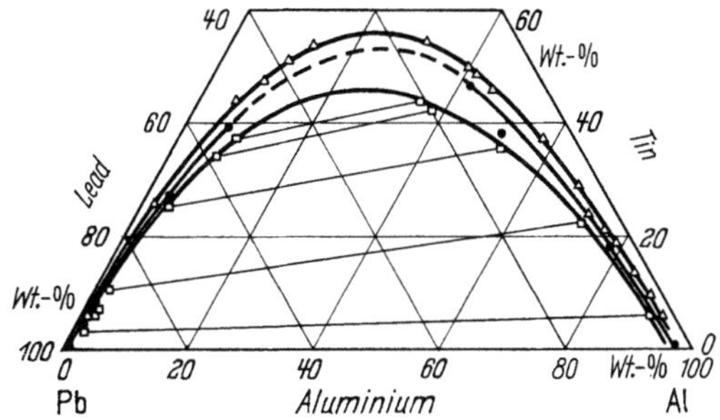

Fig. 124. Lead-Aluminium-Tin. Expansion of miscibility gap at 650 °C (△), 730 °C (●) and 800 °C □). (According to DAVIES).

CAMPBELL [170] found, in the immediate neighbourhood of the binary lead-tin eutectic, a ternary eutectic with 38.1 Wt.-% Pb, 61.7 Wt.-% Sn, 0.08 Wt.-% Al at 183.0 °C. According to his report the ternary eutectic alloy is relatively very hard and, as a solder, has notably good shear strength.

11. Pb–As–Cu; Pb–As–Fe; Pb–As–Ni. Lead-Arsenic-Copper, Lead-Arsenic-Iron, Lead-Arsenic-Nickel

In smelting sulphide ores of lead or copper-lead, containing other metals, speiss often occurs as a fourth phase in addition to the usual three

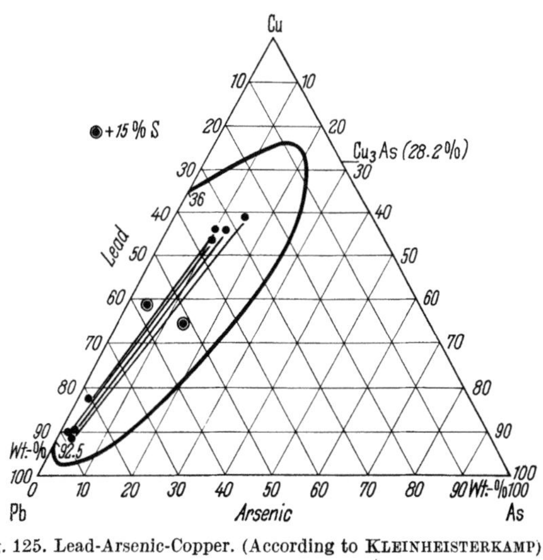

Fig. 125. Lead-Arsenic-Copper. (According to KLEINHEISTERKAMP).

layers: slag (oxide), matte (sulphide) and crude lead. It lies between the matte and crude lead and contains the arsenides and antimonides of

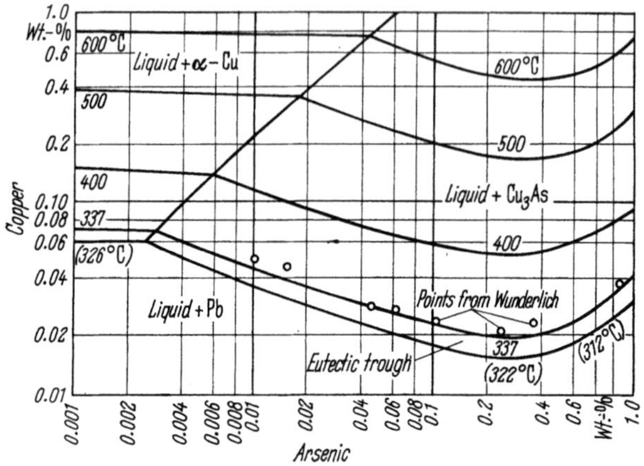

Fig. 126. Lead-Arsenic-Copper. Lead corner. (According to DAVEY [235a]).

iron, nickel and cobalt together with considerable quantities of copper, iron and lead. KLEINHEISTERKAMP [670] put forward some fundamental ideas on equilibria during the formation of metallurgical speiss. In the

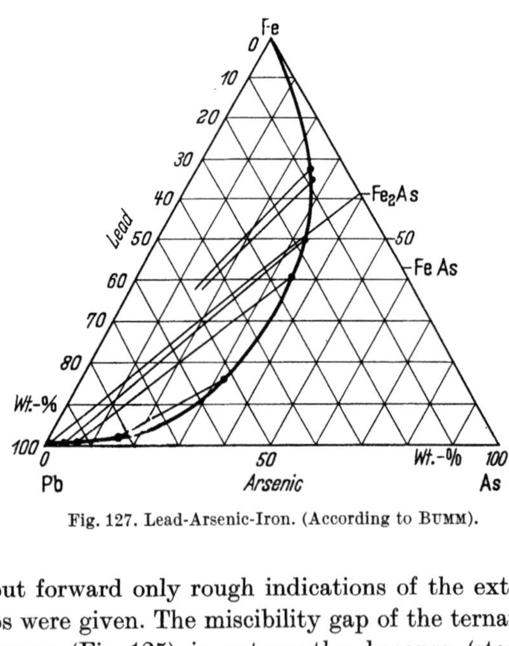

Fig. 127. Lead-Arsenic-Iron. (According to BUMM).

diagrams put forward only rough indications of the extent of the miscibility gaps were given. The miscibility gap of the ternary lead-arsenic-copper diagram (Fig. 125) is noteworthy because (starting from the boundary lead-copper system) additions of arsenic first cause widening and only with high contents does it close. The solubilities of copper, arsenic, and copper plus arsenic in lead formed the basis of the determination of the lead corner of the ternary system by DAVEY [235a] (Fig. 126). The free energy of formation of Cu_3As_{solid} was estimated as $G = -2{,}000 - 11.7$ T cal/mole. The lead corner contained solidification fields for copper, lead and Cu_3As. Because of their contour,

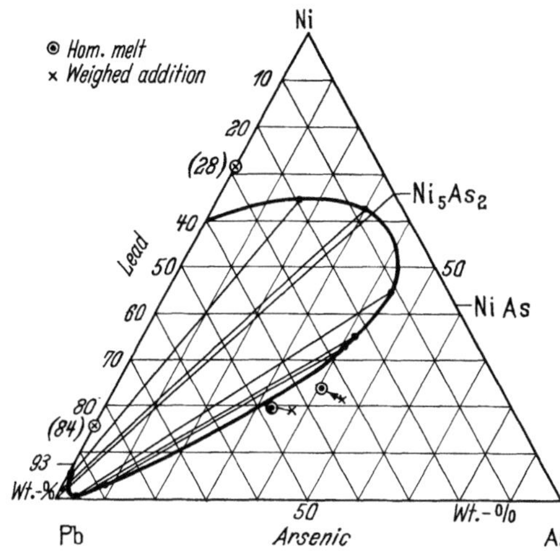

Fig. 128. Lead-Arsenic-Nickel. (According to KLEINHEISTERKAMP).

all the troughs in Fig. 127 have a eutectic character. It must be concluded from DAVEY's data, that the Pb–Cu_3As–Cu section is quasi-binary and has a quasi-binary eutectic with a temperature maximum near the lead corner. In that case, the partial triangle Pb–Cu_3As–Cu contains a ternary eutectic E_T near the lead-copper binary eutectic. Even with the addition of some 0.1% arsenic, the solubility of copper in molten lead falls from 7.3% to about 2.5%.

The ternary system lead-arsenic-iron (Fig. 127) was studied for a thesis by BUMM [143]. He investigated the possibility of alloying lead-based bearing metals with iron in addition to small arsenic and antimony contents. For this reason the extent of the miscibility gap was only determined for the moderately high temperature regions.

The miscibility gap in the ternary lead-arsenic-nickel diagram (Fig. 128) fundamentally has a shape similar to the lead-arsenic-copper. Investigations by PORTEVIN [972] and KLEINHEISTERKAMP produced identical results. With 3.3% arsenic present, lead only dissolves 0.6% nickel.

12. Pb–As–Sb. Lead-Arsenic-Antimony

According to an investigation by ABEL and REDLICH [1] only two solid phases, a lead solid solution and a continuous antimony-arsenic solid solution series, exist in the ternary lead-antimony-arsenic system. A binary eutectic trough runs from the lead-arsenic system with falling

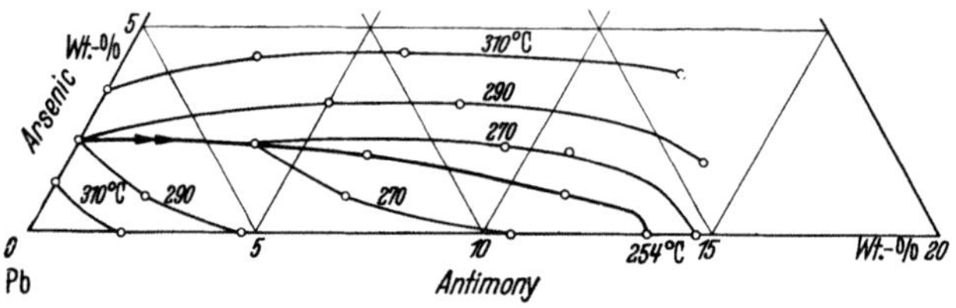

Fig. 129. Lead-Antimony-Arsenic [1]. Lead corner on the basis of thermal analysis, redrawn in Wt.-%. o Recalculated points. The channel perhaps has a minimum.

temperature to the lead-antimony system (Fig. 129). Fig. 130 shows the structure of a near eutectic alloy, and Fig. 131 that of a hypereutectic alloy. The primary antimony-arsenic solid solution here has an inclination towards acicular growth, whereas primary antimony in binary alloys shows more cubic development. In an alloy with higher

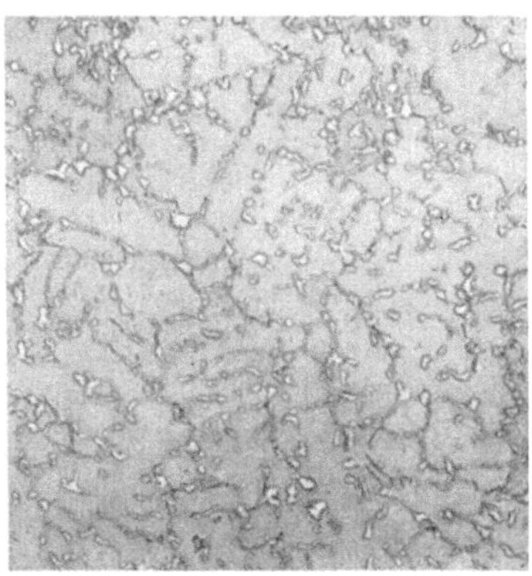

Fig. 130. 2.5% As, 4% Sb. Cast. Ground mass: lead solid solution White: antimony-arsenic solid solution as a component of the eutectic. ×300.

antimony content two kinds of pre-eutectic segregates (Fig. 132) can clearly be observed, but it is doubtful whether the differently developed crystals belong to an uninterrupted solid solution series. Also, observations by VÄTH [1214] on an alloy with 15% Sb and 4% As, in which two pre-eutectic phases (namely grey neddles and irregular white crystals) obviously occur, can hardly be reconciled with the diagram in Fig. 129.

The area of solubility of antimony and arsenic in solid lead has not been investigated, but it must be quite close to the lead-antimony boundary system, as the solubility of arsenic in lead is low.

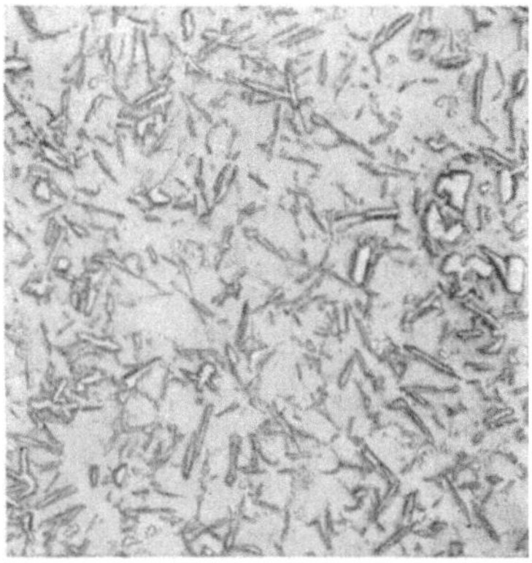

Fig. 131. 4.3% As, 4,0% Sb. Cast. Larger white primary crystals, antimony-arsenic in high relief. Eutectic ground mass. ×300

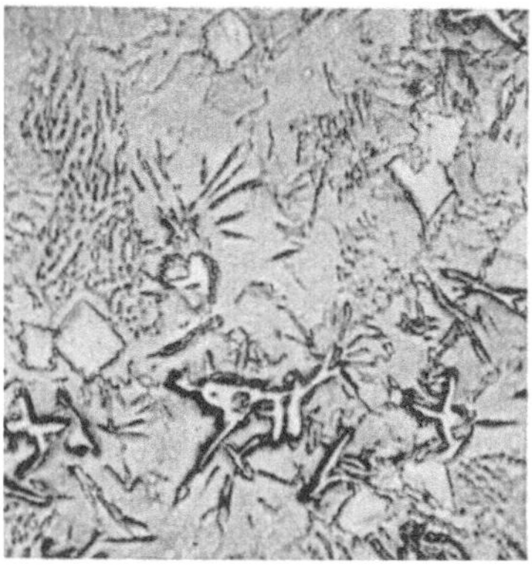

Fig. 132. 2.4% As, 8.0% Sb. Cast. Grey cubic crystals in weak relief, and white branched crystals in high relief. Eutectic ground mass. ×500

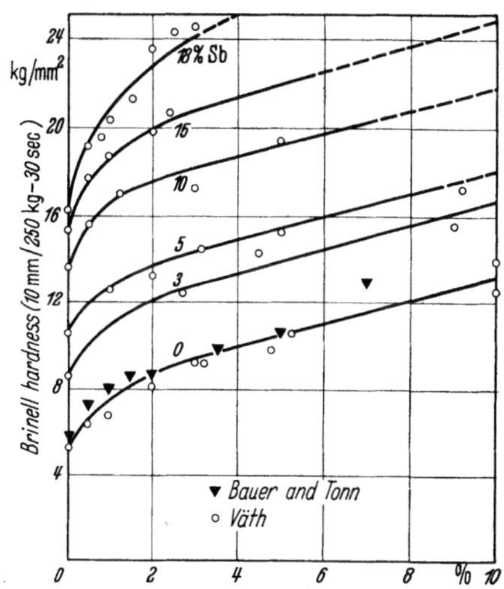

Fig. 133. Hardness of cast lead-arsenic and lead-antimony-arsenic alloys. (According to Väth).

While small additions of arsenic to low antimonial lead alloys accelerate age-hardening (p. 87), hardly any noticeable age-hardening is to be expected in alloys with a higher arsenic content, because, on the one hand, it is far outside the area of solid solution, and on the other, binary lead-arsenic alloys become softer rather than harder on storage (p. 36).

Väth [1214] carried out density, hardness and compressive strength measurements on the ternary alloys. According to his work on an alloy series with constant antimony content, the Brinell hardness increases steeply as the arsenic content is increased up to the eutectic trough, but after this it increases uniformly but less steeply. In the alloy series outside the eutectic trough (15% and 18% Sb) the Brinell hardness curves seem to spread fanwise from one another (Fig. 133). While the compressive strength of hypereutectic alloys, containing arsenic alone, or arsenic and a small amount of antimony, is low, the compressive strength (and compressive yield point) of hypereutectic lead-antimony alloys increases considerably with additions of a few per cent of arsenic; although a certain embrittlement must indeed be taken into consideration. A marked tendency to segregation in hypereutectic lead-antimony alloys, is reduced by additions of arsenic. This is attributed to grain refinement and "acicular" development of the primary crystals (Ackermann [5], v. Göler and Scheuer [396], Väth [1214]). The friction properties of alloys with antimony and arsenic contents of up to 5% have been examined (v. Schwarz [1092]). Certain compositions (1% to 7% As, 5% to 15% Sb, with other additions) seemed promising for use as bearing metal (Metals Handbook [835b], v. Schwarz [1093]). The alloys are also used in the production of hard shot (p. 339), and see also storage batteries. (p. 355).

13. Pb–Ba–Ca. Lead-Barium-Calcium

This ternary diagram has not been investigated in detail, but some indications are given through the work of Cowa, Simpkins and Hiers [221] on Frary or Ulco metals, which are essentially alloys of lead with barium (up to 2%) and calcium (up to 1%). Although the solidification range of a binary alloy with 1.2% Ba is between 317 °C and 291 °C, and of one with 0.8% Ca between 440 °C and 327°C, it is between 446 °C and 284 °C for a ternary alloy with the same barium and calcium contents — a further break in the cooling curve obviously occurring at 316 °C. Owing to a lowering of the solidus temperature compared with the binary alloys, a ternary eutectic is to be suspected. Further details are lacking, since only a few compositions of alloys were investigated. Some observations led to an assumption that in Frary metal up to 0.4% Ba and 0.2% Ca, went into solid solution in lead. As in the two binary systems, the solubility falls with decreasing temperature, as shown by the age-hardenability of the alloys indicated in Table 14.

Age-hardening properties were improved by a heat-treatment (obviously homogenising), particularly in the case of extruded alloys. Obviously a breakdown of the solid solution occurred on extruding and

Table 14. *Age-Hardening of Lead-Barium-Calcium Alloys*

Ba (Wt.-%)	Ca (Wt.-%)	Brinell Hardness BH kg/mm²		
		After 1 day	After 7 days	After 4 weeks
2.00	0.77	25.7	29.2	31.5
1.20	0.77	17.6	26.1	28.4
0.40	0.77	13.8	21.8	24.5
1.00	0.50	18.7	22.2	23.5

subsequent cooling. The age-hardening of alloys with lower barium and calcium contents might be even more marked, but such alloys were not tested in the work [221]. In spite of their high barium content, the alloys could be easily remelted, which was attributed to certain impurities. In the U.S.A., Frary metal had a wide range of applications, for example as bearings in tramcars, but it was not wholly satisfactory as it may have been too brittle; it was therefore replaced by other alloys (Grant [413]). It also has a possible application for radiation protection [504a].

14. Pb–Bi–Cd. Lead-Bismuth-Cadmium

The binary lead-bismuth system was previously regarded as a simple eutectic system, and with this assumption, the ternary systems of lead and bismuth plus another element were devised. However, an intermediate hexagonal close-packed phase (ε) was discovered in the lead-bismuth system, so the diagram was correspondingly altered (Fig. 19). Thus the ternary diagrams of lead, and bismuth with another element also had to be redrawn. The reinvestigation by Ho [535] and the present author was done by thermal analysis, microscopy and X-ray diffraction (Fig. 134). An etchant with 4 cm³ nitric acid (1.40), 20 cm³ glacial acetic acid and 76 cm³ glycerin was particularly valuable for distinguishing the α and β phases. The α phase showed with a light brown colour, and the β solid solution was dark brown. The other phases could also be distinguished easily under the microscope.

A more accurate investigation of one single vertical section at 5% Cd sufficed to redetermine this system, and the results of Barlow [45] were used as far as possible for the rest. In part these were modified in view of recent changes to the boundary systems and as a result of thermal analysis investigations. The ternary system shown in Fig. 129 has two four-phase planes. The peritectic four-phase plane at 127.6 °C represents the reactions $S + \alpha \leftrightharpoons \beta_1 + \text{Cd}$, and the concentrations at the corner points of this four-phase plane are:

S:	50.0% Pb	8.0% Cd	42.0% Bi
β_1:	67.2% Pb	1.6% Cd	31.2% Bi
α:	77.7% Pb	2.3% Cd	20.0% Bi

The corners of the eutectic four-phase plane (93.3 °C) are at:

β_2:	61.2% Pb	0.8% Cd	38.0% Bi
γ:	1.0% Pb	0.6% Cd	98.4% Bi
Cd:	no solubility assumed		

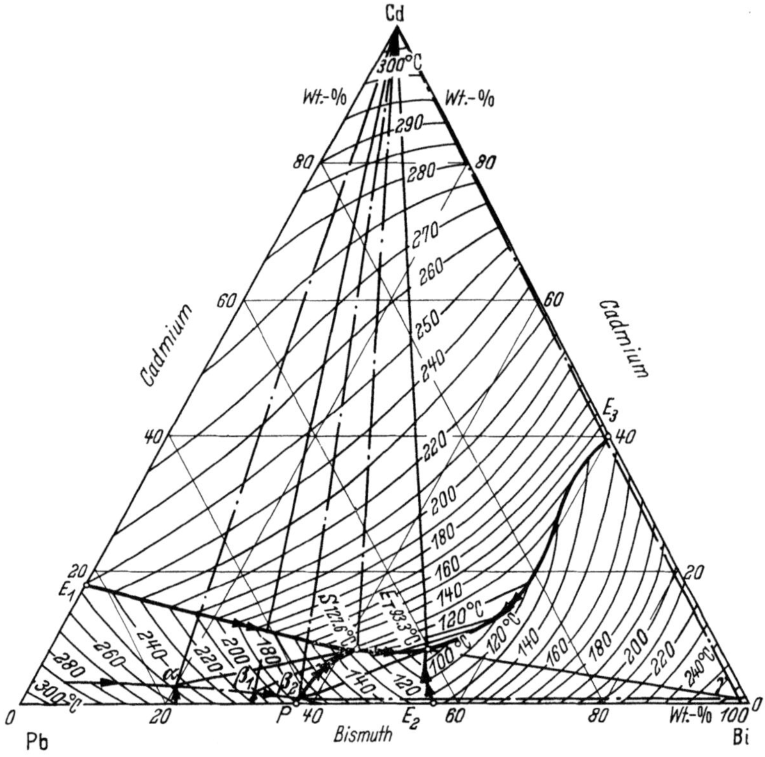

Fig. 134. Lead-Cadmium-Bismuth. (According to Barlow, also Ho and co-workers).

According to Barlow [45], the ternary eutectic point is at 40.2% Pb, 8.15% Cd, 51.65% Bi and 91.5 °C, and an alloy with this composition showed the expected structure (Fig. 135). Kremann and Langbauer [706] carried out measurements of potential against cadmium in n-$CdSO_4$

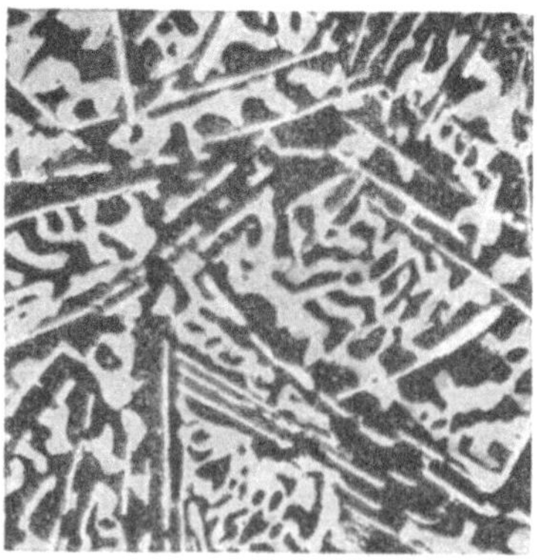

Fig. 135.
8.15% Cd. 51.65% Pb. Solidified in crucible. Dark: β-phase lead-bismuth. White: cadmium. Light grey: bismuth. Ternary eutectic alloy. ×500

over the whole range of the ternary system. Further, the results of extensive hardness measurements made by DI CAPUA [176] on the cast alloys and alloys annealed at 85 °C are available. It is interesting that the ternary eutectic of the cast alloys, similar to the lead-bismuth-tin system, corresponds to a hardness minimum. The minimum disappeared on annealing, in fact all the specimens became softer. These alloys, particularly when further additions (e.g. tin) are made, are used as low-melting solders.

15. Pb–Bi–Sb. Lead-Bismuth-Antimony

As bismuth is a common impurity in lead, this ternary system has significance although the ternary alloys have not found any special application. The antimony-bismuth system has a continuous series of

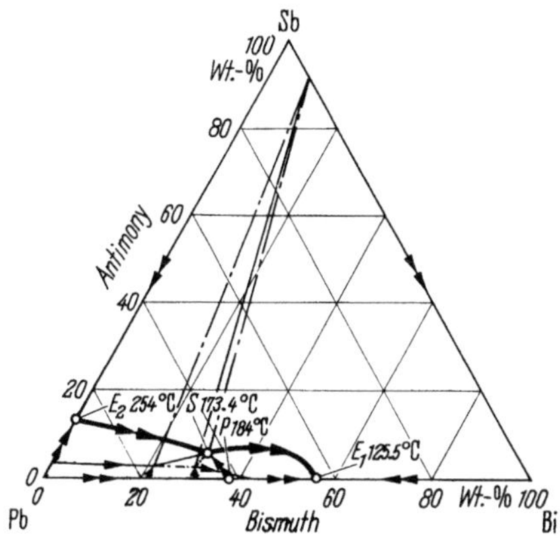

Fig. 136. Lead-Antimony-Bismuth. (According to v. HOFE and HANEMANN).

solid solutions. An early investigation of the ternary system (MORGEN, SWENSON, NIX and ROBERTS [869]) showed a binary eutectic trough passing from the boundary lead-antimony system to the lead-bismuth system. After a further incongruently melting phase (β) had been found in this system, the ternary system was reinvestigated by v. HOFE [548], using thermal analysis and X-ray diffraction. The result, Fig. 136 shows a peritectic trough coming from the lead-bismuth system and entering the above-mentioned eutectic trough at S. This gives rise to a peritectic four-phase plane which corresponds to the reaction, Melt S + Lead solid solution $\leftrightharpoons \beta$ + Bi–Sb solid solution, taking place at 173.4 °C.

MORGEN and co-workers [869] provided some information on the solid solution region in the lead corner by electrical resistance measurements and age-hardening experiments. It was shown that 1% Bi hardly affects the solubility of antimony in lead or the age-hardening of the alloys. On the other hand, higher bismuth contents lower the age-hardenability, as they reduce the melting points of alloys and the solubility of antimony in solid lead.

16. Pb–Bi–Sn. Lead-Bismuth-Tin

The equilibrium diagram evaluated by CHARPY [189] required revision because the β-phase, unknown to him, had since been found. Based on CHARPY's work, but amplified by thermal analysis, microscopy and X-ray diffraction examination of the three vertical sections at 5%, 15% and 65% Sn, Ho [535] devised the ternary diagram reproduced in Fig. 137. It contains a peritectic four-phase plane with the corner points:

α:	70.6% Pb	20.0% Bi	9.4% Sn
β_1:	63.9% Pb	27.9% Bi	8.2% Sn
S:	40.8% Pb	32.6% Bi	26.6% Sn
δ_1:	2.9% Pb	5.4% Bi	91.7% Sn

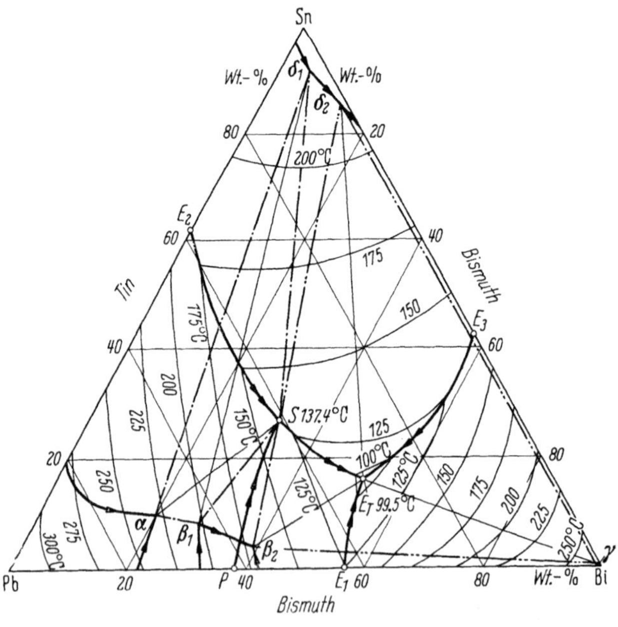

Fig. 137. Lead-Bismuth-Tin. (According to CHARPY, also Ho and co-workers).

The composition of the ternary eutectic point E_T was taken from CHARPY [189] to be at 32% Pb, 52% Bi and 16% Sn. The corner points of the ternary eutectic four-phase plane are at:

β_2:	57.0% Pb	39.5% Bi	3.5% Sn
γ:	0.5% Pb (assumed)	99.0% Bi	0.5% Sn
δ_2:	0.8% Pb	13.5% Bi	85.7% Sn

The temperature of the ternary eutectic was found to be 99.5 °C. GERSHMAN [370] re-examined the ternary diagram by thermodynamic analysis and confirmed the existing diagram.

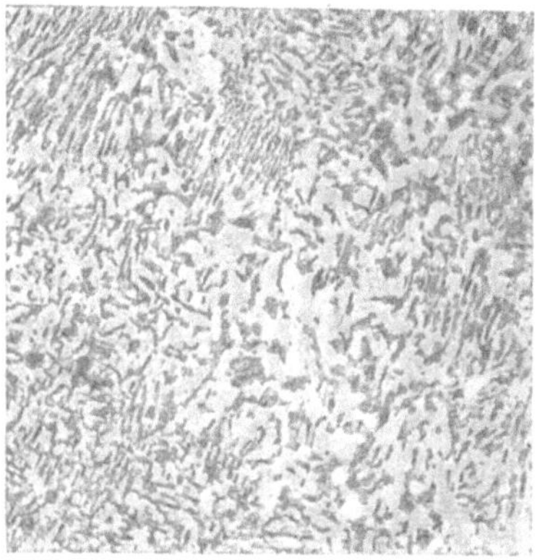

Fig. 138. 52% Bi, 16% Sn. Dark: β-phase lead-bismuth. White: tin. Light grey: bismuth. Ternary eutectic alloy. ×500

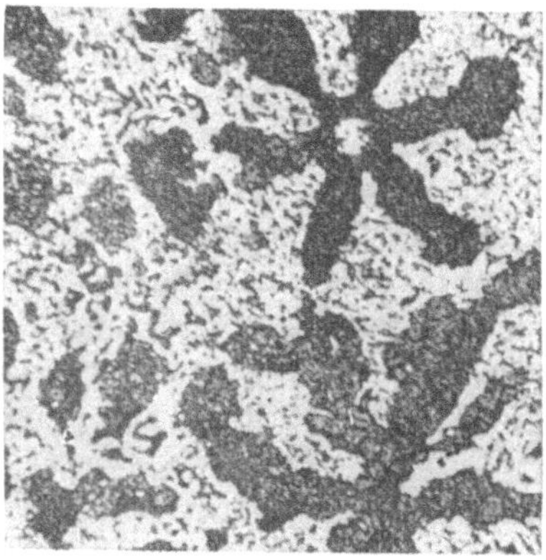

Fig. 139. 44% Bi, 15% Sn. Remainder lead. Cast. Hexagonal primary crystal of β (dark) in eutectic ground mass of $\beta + \delta$ and $\beta + \gamma + \delta$. (According to HO, HOFMANN and HANEMANN). ×500

Fig. 138 illustrates the structure of the ternary eutectic alloy and Fig. 139 represents an alloy from the primary solidification field of β. In the alloys near the ternary eutectic, solid state transformations have been observed several times by various methods. Particular reference may be made to a work concerning Roses metal having 27.545% Pb, 48.902% Bi and 23.553% Sn (FLEISCHMANN [328]). Here the so-called σ-modification solidifies from the melt, and on cooling to 60 °C it changes to the α, which, in turn, transforms to β on heating above 76 °C. This β is not identical however with the definite β phase in the diagram (Fig. 137). The $\alpha \leftrightarrows \beta$ transformation is reversible and can be repeated as often as desired, superheating and supercooling both being possible within a restricted temperature range. The $\alpha \rightarrow \beta$ transition takes place with a 0.61% reduction in volume and an increase in resistance. The volume change was previously observed by ERMANN and SPRING [288] in 1827.

Unlike the $\alpha \leftrightarrows \beta$ transformation, σ could only be obtained by remelting, which leads one to suspect that "σ" arises through non-equilibrium conditions during solidification, or that with certain supercooling (similar to that described in the lead-tin system) sudden segregation of a supersaturated solid solution occurs.

The $\alpha \leftrightarrows \beta$ transformation was investigated further by dilatometry in some comprehensive work on 43 different alloys by ISIHARA [606]; and the above result was confirmed. However, no satisfactory interpretation as to the nature of the transformation was made.

KEIL [650] communicated further observations which are related to the transformation described above. Following quenching a small casting rose in temperature by about 10 °C above room temperature. The phenomenon lasts about 20 minutes, while the hardness, measured with a low load, falls from about 15 to 5 kg/mm², the originally brittle material having become plastic.

CLARA DI CAPUA [178] carried out an extensive research programme on the Brinell hardnesses of the ternary alloys both in the cast state and after fifteen days' annealing at 80 °C. The conclusion was that the hardnesses of all the alloys (except perhaps the ternary eutectic alloy) were reduced by the heat treatment, which may well arise through segregation of supersaturated solid solutions in the cast state. Of the cast alloys, the ternary eutectic had the minimum hardness.

These alloys are important as low-melting alloys and solders. Owing to their rapid volume increase on solidification, they are particularly suitable as embedding material and for anchoring (KEIL [650]). WILKENSON [1272] studied the attack of melts having 32 Wt.-% lead, 16 Wt.-% tin, ~ 52 Wt.-% bismuth, on various metals such as Al, Armco iron, cast iron, Be, Ti, Zr and on a number of steels. Molten alloys of similar compositions might possibly be of interest for heat transfer media in reactor technology. Of the many results it is worth noting that beryllium was still resistant at 500 °C and molybdenum even at 800 °C. Oxygen in the melt retarded attack on stainless steel (oxide film!). but conversely accelerated the solution of carbon steel and cast iron.

Finally, mention should be made of an investigation by NIWA [900], where the effect of bismuth on the diffusion co-efficient of tin in lead-tin melts is reported.

17. Pb–Bi–Te. Lead-Bismuth-Tellurium

The PbTe–Bi section is quasi-binary and has a eutectic at 98.5 Wt.-% PbTe, 1.5 Wt.-% Bi and 266 °C. The mutual solubility of lead telluride and bismuth is exceedingly small (O'SHEA [927a]).

18. Pb–Bi–Zn. Lead-Bismuth-Zinc

Information on the boundary lead-bismuth and lead-zinc systems should be obtained from the section on the binary systems. Like the lead-zinc system, the bismuth-zinc has a miscibility gap in the molten state, extending at the monotectic temperature of 416 °C from 98.1%

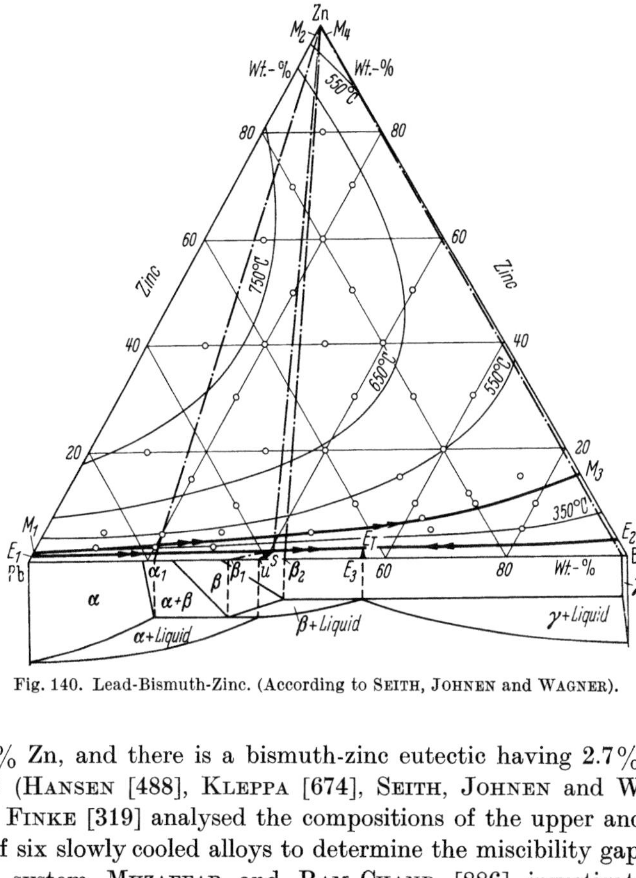

Fig. 140. Lead-Bismuth-Zinc. (According to SEITH, JOHNEN and WAGNER).

to 15.5% Zn, and there is a bismuth-zinc eutectic having 2.7% Zn at 254.5 °C (HANSEN [488], KLEPPA [674], SEITH, JOHNEN and WAGNER [1105]). FINKE [319] analysed the compositions of the upper and lower layers of six slowly cooled alloys to determine the miscibility gap in the ternary system. MUZAFFAR and RAM CHAND [886] investigated the boundary of the miscibility gap more accurately and the solidification field lying near the boundary lead-bismuth system, so the ternary eutectic suspected by JÄNECKE [615] could be confirmed. SEITH and co-workers [1105] have recently developed a differential thermal analysis method so that the beginning of segregation can be accurately measured. They also investigated the stratification region in this system. At the same time some corrections were made to the early determinations of the system, so this ternary diagram is now fixed very precisely (Fig. 140).

The miscibility gap of the ternary system thus forms a continuous transition between the two lead-zinc and bismuth-zinc boundary systems. The temperature fall on the stratification curve from the monotectic M_1 on the lead-zinc side to the monotectic M_3 on the bismuth-zinc side amounts to only a few degrees. The solidification field for zinc joins the two boundary curves of the miscibility gap towards the zinc corner and the boundary lead-bismuth system. The binary eutectic troughs, $E_1 - E_T$, $E_2 - E_T$ and $E_3 - E_T$, from the three boundary systems, unite at the ternary eutectic E_T with the composition 55% Bi, 43% Pb, 2% Zn at 124 °C (the boundary of the eutectic four-phase plane through β_2 — Zn — γ — β_2). A peritectic trough runs from point U of the boundary lead-bismuth system into the ternary system, which represents the reaction Melt $+ \alpha \leftrightharpoons \beta$. It meets the eutectic trough $E_1 - E_T$ at the point S at 160 °C. The peritectic four-phase plane α_1 — Zn — S — β, represents the reaction α + melt $\leftrightharpoons \beta$ + Zn. From E_1, along the eutectic trough, the binary eutectic lead + zinc solidifies to the point S, and from S the eutectic β + zinc. Thus, point E_T represents a ternary eutectic of β with bismuth and zinc. The solubility of zinc in the solid phases of the lead-bismuth boundary system is so small that it can be ignored within the ternary diagram determination. Division of the ternary diagram into one-, two- and three-phase areas in the solid state can easily be done due to Fig. 140.

19. Pb–Ca–Na. Lead-Calcium-Sodium

The following progress of the isotherm of the liquidus surface could be obtained from a sketch of the lead corner of the ternary system derived by Ageew [9], which extends up to about 5% Ca and 4% Na:

	350 °C			400 °C			450 °C			500 °C		
% Ca	0.26	0.21	0.92	0.39	0.61	1.45	0.96	1.0	2.43	1.69	3.0	4.87
% Na	0	2.0	3.8	0.	2.38	3.39	0	1.78	2.64	0	0.82	0.91

From the peritectic of the lead-calcium side, having 0.13% Ca, a trough runs almost parallel to the other boundary system. There is a break in the trough at the end of the diagram at the composition 0.65% Ca and 4.0% Na. On the curve there is an invariant point at 0.1% Ca and 2.26% Na where another equilibrium curve enters from the adjacent eutectic point of the lead-sodium side at 2.26% Na. According to A. Lange and B. Beyer [729a] the invariant point represents a ternary eutectic at 306 °C. It is believed that the peritectic line mentioned assumes a eutectic character as the sodium content of the alloy is increased. From age-hardening experiments the authors draw conclusions on the extent of the area of solid solubility. The solubility of sodium in lead at 300 °C (1.4 Wt.-%) is reduced to about half this value with some 0.1 Wt.-% Ca. According to LANGE, the temperature of the liquidus surface given by AGEEW is too low.

20. Pb–Cd–Mg. Lead-Cadmium-Magnesium

JÄNECKE [616] carefully investigated the structure of this ternary system, in order to throw some light on the binary magnesium-cadmium system. Only the lead corner of the ternary system is presented here, the

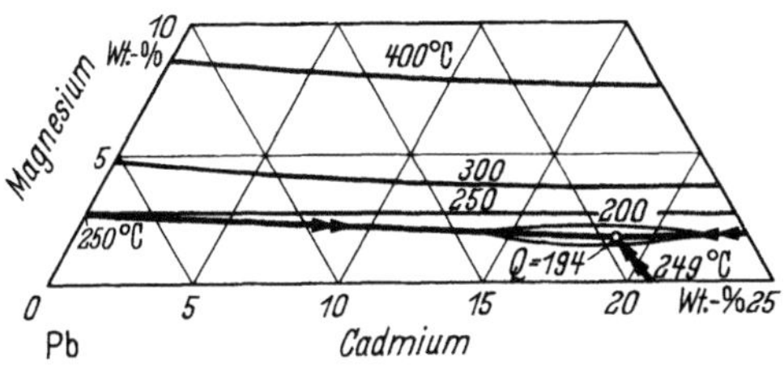

Fig. 141. Lead-Cadmium-Magnesium. (According to JÄNECKE).

atomic percentages given by JÄNECKE being converted into weight percentages. The lead corner has a ternary eutectic at the composition 18.9 Wt.-% Cd, 1.8 Wt.-% Mg, 79.3 Wt.-% Pb, and temperature 194 °C (Fig. 141). It corresponds to the reaction: $E_T \leftrightarrows$ Lead solid solution + Cadmium solid solution + Mg_2Pb. This diagram still needs the details revised, particularly in the regions of the solid solutions. Because the compound Mg_2Pb is readily decomposed, practical application of the alloys in the lead corner is unlikely to occur (see p. 65).

21. Pb–Cd–Sb. Lead-Cadmium-Antimony

The binary antimony-cadmium system is divided into two eutectic systems by the congruently melting compound CdSb. According to ABEL and ADLER [2] a quasi-binary section exists from the compound CdSb to the lead corner of the ternary system (Fig. 142). Through this section, which contains a eutectic (E_3) Pb–CdSb with 4.2% Cd, 4.5% Sb, at 275 °C, the ternary diagram breaks down into two partial systems, each of which has a ternary eutectic. The eutectic Pb–CdSb–Cd has 17.7% Cd, 2.6% Sb at 236 °C and the eutectic Pb–CdSb–Sb 1.3% Cd, 11.3% Sb at 242 °C. Figs. 143 and 144 illustrate two alloys in the cadmium-rich partial system. The first of these contains primary CdSb but the second is almost completely ternary eutectic. Figs. 145 and 146 are taken from the antimony-rich partial system. The primary CdSb has (Fig. 145) a different appearance from that in the previous partial system (Fig. 143), and the alloy (Fig. 146) is almost wholly ternary eutectic. All the alloys solidified in air in a crucible.

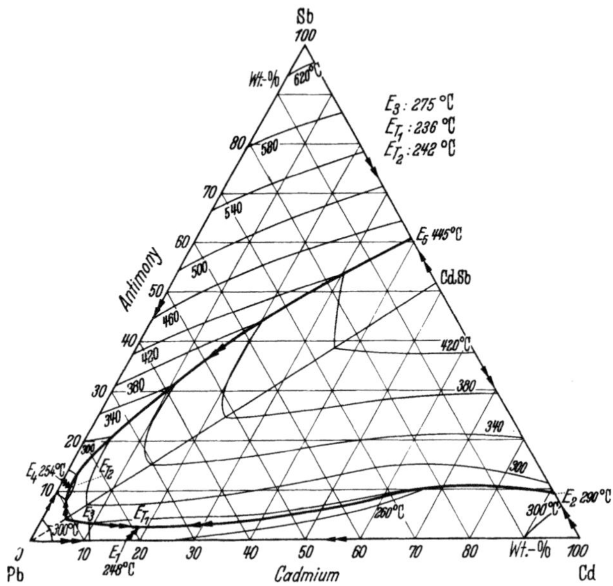

Fig. 142. Lead-Antimony-Cadmium, redrawn in Wt.-%.

Work by GARRE and MÜLLER [357] showed the eutectic in the quasi-binary section to be somewhat different from that given above with 10% CdSb, i.e. 5% Cd, 5% Sb, at 260 °C. The solubility of CdSb in lead was shown to drop from 5.5% CdSb at 260 °C to about 0.4% at room temperature. Alloys with more than 0.4% CdSb age-hardened, particularly after quenching from 200 °C and the electrical resistance fell as usual. The maximum hardness alloy of over 15 BHN had the composition 2.5% CdSb, the remainder lead. Age-hardening of this alloy was inhibited by the presence of zinc and tin. WATERHOUSE [1244] likewise detected considerable age-hardening of the alloys. Weight losses of the quasi-binary section alloys in sulphuric acid have also been determined.

Owing to their superior hardness and fatigue strength, lead-rich alloys were used instead of commercial lead for water pipes and cable sheathing, e.g. in England, the alloy with 0.5% Sb and 0.25% Cd (p. 414). The alloys with higher antimony and cadmium contents, and the quaternary alloys, obtained with additions of tin, have been tested for suitability as bearing materials (ACKERMANN [4]). [see also p. 355.]

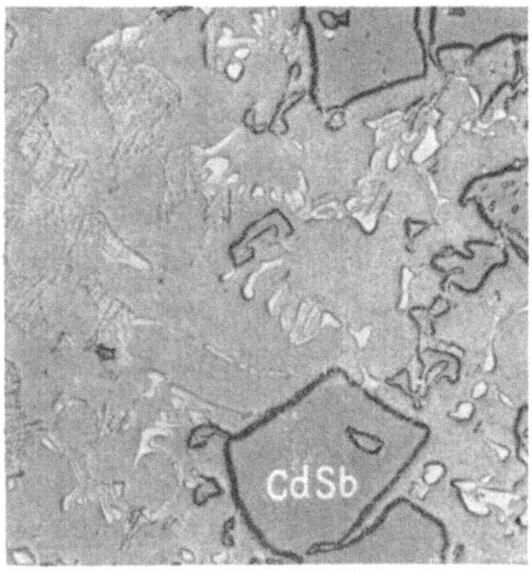

Fig. 143. 15% Cd, 5% Sb. Unetched. Dark grey: CdSb. White: cadmium. Light grey: lead solid solution. Primary crystals CdSb, binary (lead + CdSb) and ternary eutectic. ×150

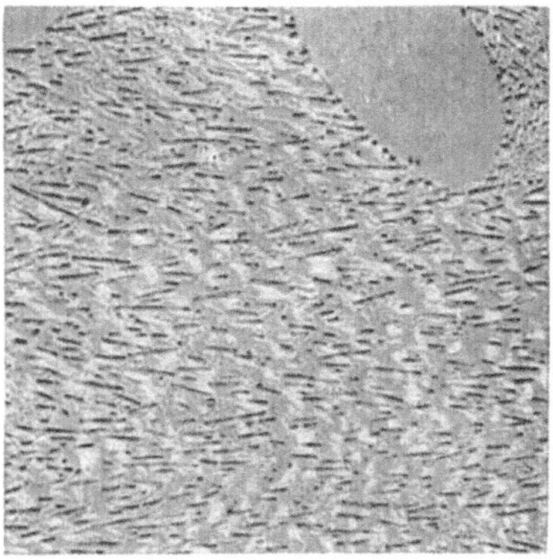

Fig. 144. 17.5% Cd, 2.6% Sb. Preponderant ternary eutectic of lead (grey) with CdSb ("needles") and cadmium (white). ×500

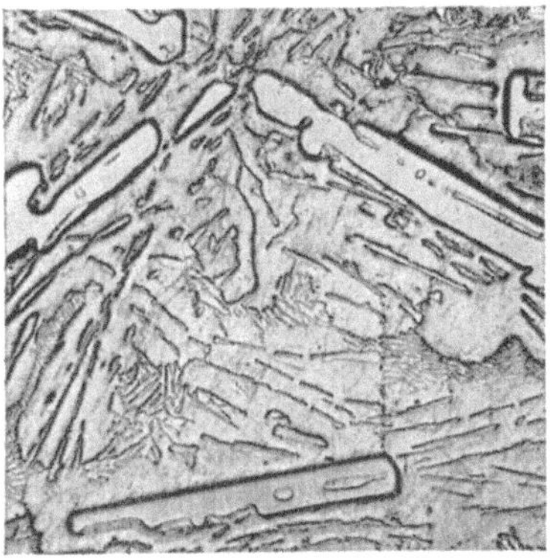

Fig. 145. 5% Cd, 10% Sb. Rod-shaped primary crystals of CdSb. Ground mass predominantly ternary eutectic lead + antimony + CdSb. ×150

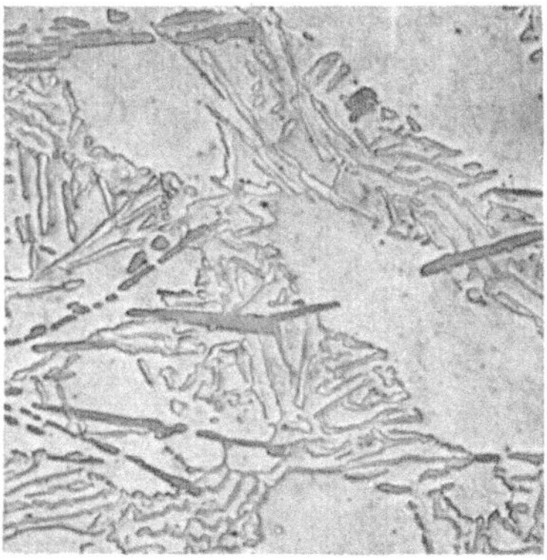

Fig. 146. 1.3% Cd, 11.3% Sb. Ternary eutectic of lead, antimony (white) and CdSb (dark grey). ×500

22. Pb–Cd–Sn. Lead-Cadmium-Tin

This ternary system (Fig. 147) has, according to STOFFEL's work [1149], a ternary eutectic at 32% Pb, 18% Cd, 50% Sn and 145 °C. The eutectic corresponds to the simultaneous solidification of the lead-, cadmium- and tin-rich solid solutions from the melt. It is built up in a layered structure with the scheme ABCBA, in which *A* indicates the tin-lamellae, *B* the lead-lamellae and *C* the cadmium lamellae (KERR [657 a]). In the meantime a new phase (β) was found in the cadmium-tin diagram (cf. HANSEN [488]), which forms peritectically at 223 °C and decomposes eutectoidally at 133 °C. β, however, only arises with very slow solidification of the alloys. The ternary diagram therefore had to be amended and a peritectic trough emerging from the boundary cadmium-tin system added. If the peritectic trough meets the eutectic trough S $\leftrightharpoons$ Sn + Pb at the ternary eutectic, phase β would be involved (in addition to the lead- and cadmium-rich solid solution) rather than

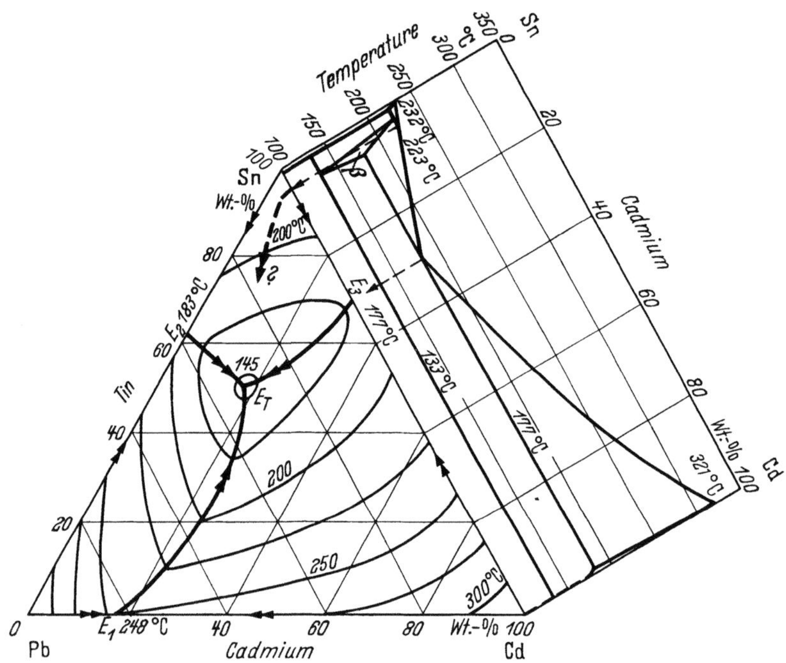

Fig. 147. Lead-Cadmium-Tin. (According to STOFFEL).

(the tin-rich solid solution. In fact STOFFEL [1149] has noted that the solidified ternary alloys undergo a transformation between 118 °C and 112 °C, which is less marked towards the lead corner. This transformation is similar in character to the eutectoidal breakdown of β at 133 °C in the cadmium-tin system. BOTSCHWAR and GOREW [117], as well as TAMMANN and MORITZ [1169], have made observations on the distribution of the constituents in the ternary eutectic.

Lead alloys with high percentages of tin and cadmium form low-melting solders, as also those with further additions (e.g. bismuth). Those with low tin and cadmium contents are important as cable sheathing alloys in view of their high fatigue strength, e.g. the alloy containing 1.5% Sn and 0.25% Cd used in England. According to WATERHOUSE and

Willows [1244] the alloys in the lead corner age-harden, though in general only temporarily. However, all alloy compositions are quite plastic. A wide variety of these alloys have become standard for soft soldering.

23. Pb–Cd–Tl. Lead-Cadmium-Thallium

Clara di Capua [176] made a careful investigation of this ternary system using thermal analysis; it was revised by Jänecke [613]. He found that in the boundary lead-thallium system, the fields of $PbTl_2$ solid solution and of lead solid solution, which are separated by a two-phase area, are distinguishable. A peritectic three-phase equilibrium

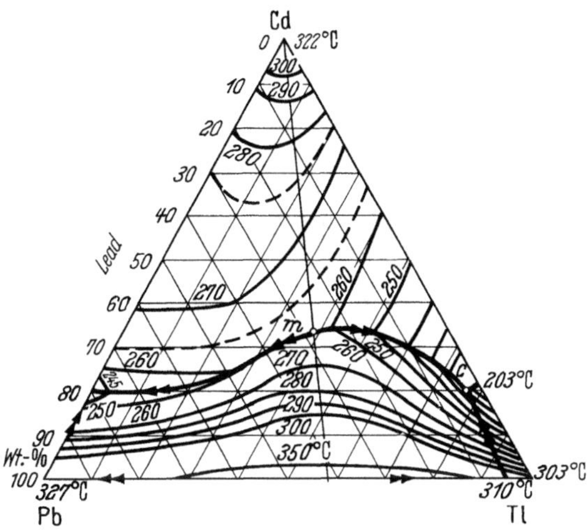

Fig. 148. Lead-Cadmium-Thallium. (According to di Capua and Jänecke).

should be present at 379 °C, corresponding to the transition of the two phases into one another. The ternary system shown in Fig. 148 is di Capua's [176] original conception without the peritectic assumed by Jänecke [613]. The other peritectic trough which he inserted in the diagram is also omitted, since it has not been verified. Some small modifications by Jänecke [613] to the shape of the liquidus surfaces are used in the Fig. 148. The diagram is traversed by a eutectic trough between the lead-cadmium and thallium-cadmium boundary systems. The trough has a temperature maximum at the point m. A melt of this composition is in equilibrium with two solids — a cadmium-rich solid solution near the cadmium corner of the diagram and a lead-thallium solid solution, whose position is given by the extrapolation of the line joining Cd and m to the lead-thallium side of the system. The peritectic trough from the lead-thallium system enters the eutectic trough at point c, forming one corner of a peritectic four-phase plane at 211 °C. This reaction is Melt c + Solid solution $PbTl_2 \leftrightharpoons$ Thallium-rich solid solution + Cadmium-rich solid solution. No more precise details of the solid solution regions are available.

24. Pb–Cd–Zn. Lead-Cadmium-Zinc

The lead-cadmium and cadmium-zinc systems are both simple binary eutectics. The miscibility gap of the boundary lead-zinc system stretches well into the ternary system, and its boundary was determined

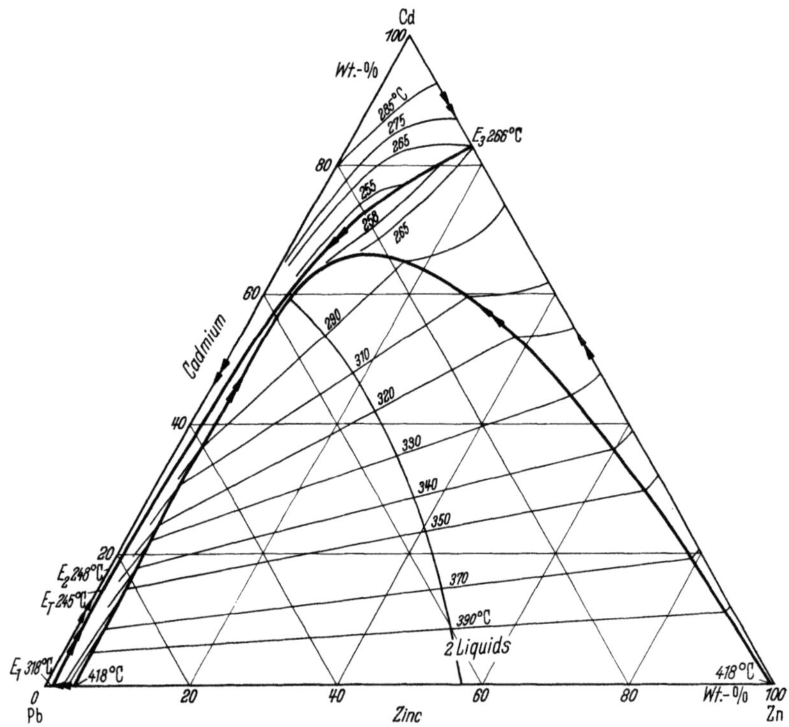

Fig. 149. Lead-Cadmium-Zinc. (According to Cook [214]).

by investigating the stratification of slowly cooled alloys microscopically. Further, isotherms at the start of crystallisation were determined by thermal analysis. The straight line part of the isotherms, which also represent tie-lines of the appropriate melts, indicated the boundary curve of the miscibility gap, which was in good agreement with the microscopic investigation. The stratification curve is represented approximately by the contour shown in Fig. 149. The miscibility gap only rests on the crystallisation field of zinc, and does not touch any trough of the three-phase equilibrium. The part of the liquidus surface of zinc, lying to the left of the miscibility gap, falls steeply towards the lead-cadmium system. The three binary eutectic troughs emerging from the boundary systems unite at 17.3% Cd, 81.7% Pb, 1% Zn and 245 °C, the ternary eutectic. No data are available on the solid solution region in the lead corner.

25. Pb–Cu–Ni. Lead-Copper-Nickel

Lead and copper, like lead and nickel, have only limited miscibility in the molten state, whereas copper and nickel have unlimited mutual solubility in both the molten and the solid states. According to an in-

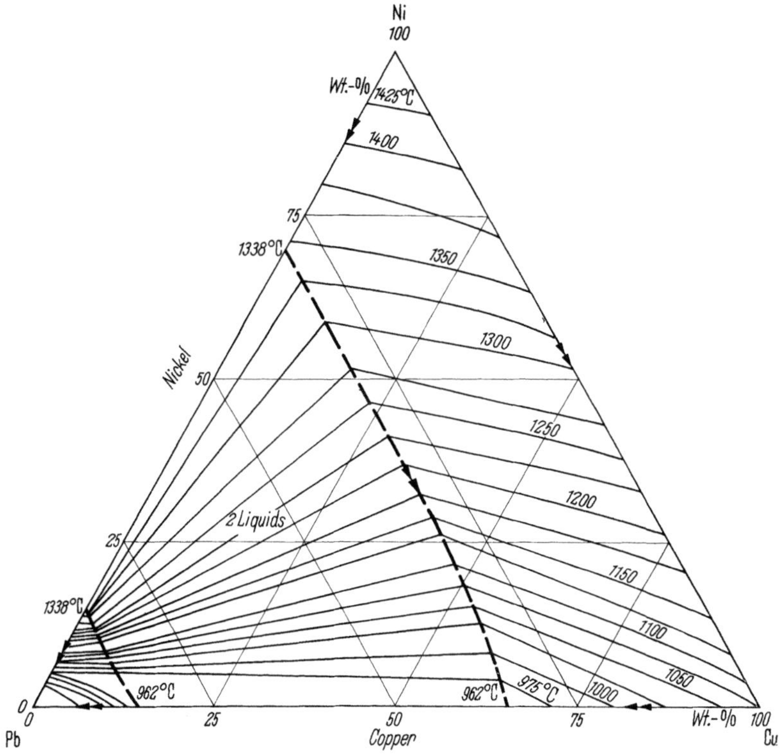

Fig. 150. Lead-Copper-Nickel on the basis of thermal analysis. (According to PARRAVANO).

vestigation of the ternary system by PARRÁVANO [934], using thermal analysis, the miscibility gap extends through the whole ternary system. Fig. 150 gives: the two bounding curves of the miscibility gap, the position and temperature of the tie-lines (which limit the miscibility gap towards the monotectic three-phase area) and the isotherms for the inception of solidification in the region outside the miscibility gap. Quite near the lead corner a trough runs from the lead-nickel system to the lead-copper eutectic, which is, however, not indicated on the diagram.

GUERTLER [453] studied the tendency towards stratification in these ternary alloys and from this also obtained information about the miscibility gap in the molten state. While segregation occurs in molten

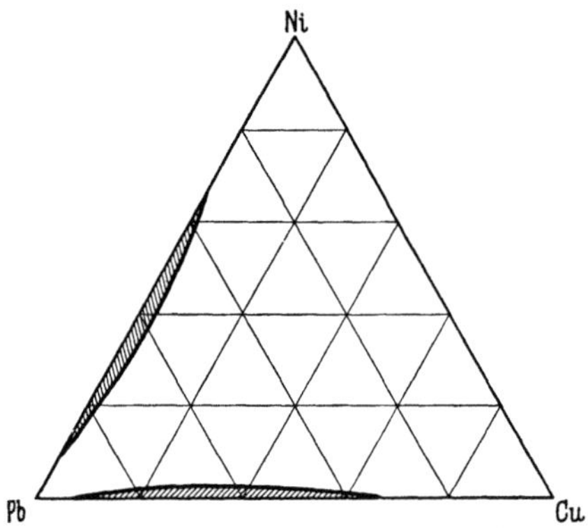

Fig. 151. Stratification of lead-copper-nickel alloys. (According to GUERTLER, also CLAUS).

lead-copper and lead-nickel alloys, it was surprisingly absent for the large part of the ternary alloys. It was concluded from this that the miscibility gap of the lead-copper system closes on the addition of 2.5% Ni, and that of the lead-nickel system with 6% Cu, as shown in Fig. 151.

NEMILOW [893] confirmed the broad outlines of GUERTLER's work and by thermal analysis produced a rough diagram of the isotherm at the inception of solidification. PELZEL [947] also confirmed the closing of the miscibility gap in the lead-copper side of the ternary system, and found the critical point, 60% Pb, 2% Ni, remainder Cu at 965 °C. A redetermination of the ternary diagram would now be desirable, using all the latest facilities. The solubility rate of a copper-nickel alloy of 55 Wt.-% Cu in agitated lead melts was measured by STEVENSON [1145 b].

26. Pb–Cu–O. Lead-Copper-Oxygen

This system was investigated a few years ago, independently, but simultaneously, by KOHLMEYER and the present author [565] and by GEBHARDT and OBROWSKI [364]. The two investigations essentially gave identical results. From the boundary copper-cuprous oxide system, the miscibility gap extends without interruption to the lead-oxygen system.

In the equilibria of the copper-rich melt with slag (Fig. 152), it is worth noting that the tie-lines are in about the direction from the copper corner to the lead oxide corner. It could be concluded from this that the reactions between metal melts and oxide melt essentially occur according to the reaction, $Cu_2O + Pb \leftrightharpoons 2\,Cu + PbO$. Volatilisation experiments by SCHNEIDER [1076] show that the oxygen in the ternary system alloys combines preferentially with the lead. This reaction is accompanied by

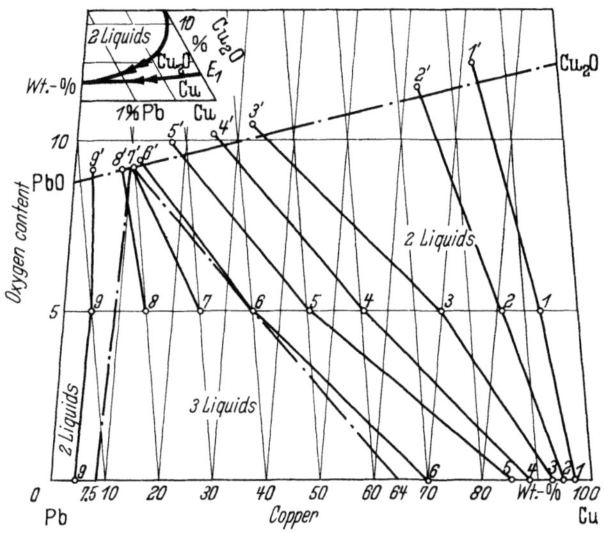

Fig. 152. Metal-Slag equilibria from settling tests at 1,150 °C. (According to HOFMANN and KOHLMEYER). (Miscibility gap Cu–Pb according to HANSEN [487]).

an intense evolution of heat, the mixtures of lead and cuprous oxide being heated to above the melting point of lead. At about 600 °C the reaction becomes so violent, that temperature rises of 200 to 300 °C occur within half a minute. With prolonged heating the reaction also takes place at a little above the melting point of lead. Both investigations agree that the solidification field for cuprous oxide in the copper corner disappears with the addition of a few per cent of lead. The monotectic trough of the three-phase equilibrium "Oxide-rich melt $\leftrightharpoons$ Copper-rich melt + Cuprous Oxide", emerging from the boundary copper-cuprous oxide system, enters, with a few per cent of lead, the eutectic trough Melt $\leftrightharpoons$ Copper + Cuprous Oxide emerging from the boundary system. From this follows that the miscibility gap in the molten state first meets the solidification field for cuprous oxide at the copper corner, on the other hand from the invariant point, already mentioned, it meets a narrow solidification field for copper. It must be concluded therefore that the oxide-rich melts on the other side of the miscibility gap also border on the narrow solidification field for copper. In addition, the diagram contains a region in which three melts are in equilibrium with each other. As it is already known that the miscibility gap in the copper-lead diagram closes near to 1,000 °C (p. 54), the three melt region can only be present below that temperature. The cuprous oxide-lead monoxide section is quasi-binary and, according to GEBHARDT [364], has a eutectic

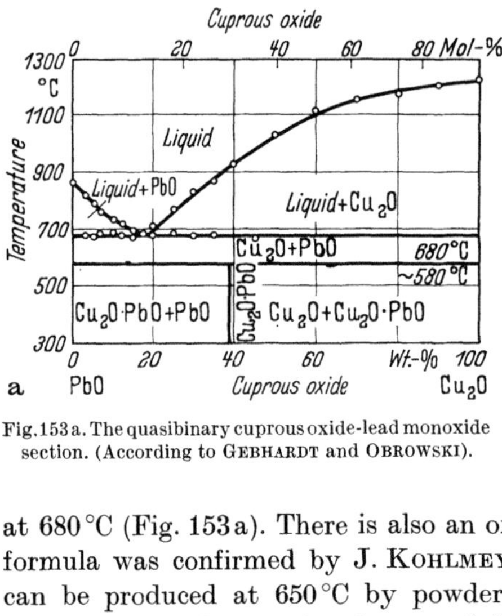

a

Fig. 153a. The quasibinary cuprous oxide-lead monoxide section. (According to GEBHARDT and OBROWSKI).

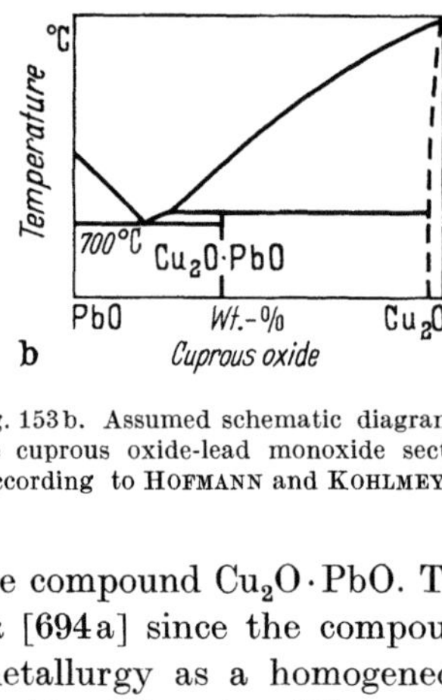

b

Fig. 153b. Assumed schematic diagram of the cuprous oxide-lead monoxide section. (According to HOFMANN and KOHLMEYER).

at 680 °C (Fig. 153a). There is also an oxide compound $Cu_2O \cdot PbO$. This formula was confirmed by J. KOHLMEYER [694a] since the compound can be produced at 650 °C by powder metallurgy as a homogeneous phase from stoichiometric mixtures of the oxides. There is a difference in views insofar as GEBHARDT [364] states that the eutectic of the quasi-binary section consists of cuprous oxide and lead monoxide, and that the oxide-compound forms at about 580 °C in the solid state; the other authors [565] assumed an oxide compound-PbO eutectic and close above this a peritectic "Melt $+ Cu_2O \leftrightharpoons Cu_2O \cdot PbO$" (Fig. 153b). Observations by PAWLEK (by verbal communication) and [694b], but contrary to [364], find that the compound $Cu_2O \cdot PbO$ is stable up to the region of the melt. GEBHARDT [364] has put forward a schematic ternary system (Fig. 154) and gives in tabular form the following summary of the course of the equilibrium in the part of the system he investigated. Only a few important points of the ternary system can be considered here. Separation into three melts only occurs below the temperature of the two critical points K_1 and K_2, which can be assumed to be 1100 °C. The oxide-rich melt moves with falling temperature along the curve $K_2 U_2'''$, the lead-rich melt moves along $K_1 U_2''$ and the copper-rich melt along $K_1 U_2'$. The area of the three melts lies on a four-phase plane, which includes a crystal region in the copper corner as the additional corner point. At the temperature of this four-phase plane, the melt U_2' decomposes into droplets of the U_2'' and U_2''' melts with simultaneous solidification of copper. The boundary curve of the miscibility gap on the side of the cuprous oxide-lead monoxide section must touch the solidification fields of $Cu_2O \cdot PbO$ and of Cu. The ternary eutectic point U_4 assumed in this region is worth noting. From a melt of this (which is of almost identical composition to the eutectic E_4 of the quasi-binary

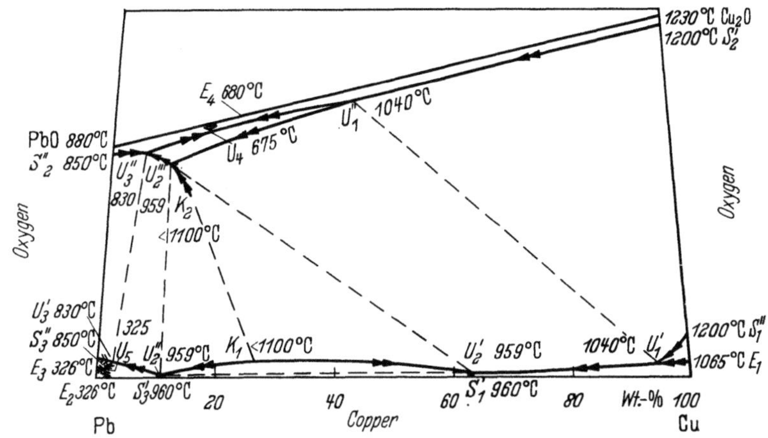

Fig. 154. Melting equilibria in the system Cu–Pb–PbO–Cu_2O (schematic in part). (According to GEBHARDT and OBROWSKI).

section) Cu_2O, PbO and Cu solidified simultaneously. Finally it is assumed that close to the lead corner there is a further ternary eutectic melt, $E_3 \leftrightarrows Cu + Pb + PbO$. While the investigations mentioned provide a first impression of the ternary diagram, what has been said indicates that further investigations will be necessary for a complete interpretation. This is already clear from the fact that lead monoxide occurs in two modifications and apart from them numerous other lead oxides exist, as indicated in the chapter on binary systems. This ternary system is important because leaded bronzes are melted under oxidising conditions, in order to reduce their hydrogen content (SCHNEIDER [1076, 575, 576]). Oxygen determinations on some arbitrarily chosen leaded bronzes gave contents of 0.0104 Wt.-% and 0.0045 Wt.-% oxygen.

27. Pb–Cu–S. Lead-Copper-Sulphur

The general arrangement of this ternary system was suggested by JÄNECKE [614] based on previous work by GUERTLER and LANDAU [452], and is shown schematically in Fig. 155. In the quadrilateral Pb–PbS(F)–Cu_2S(E)–Cu, there are only the phases of the four corner points. The molten state miscibility gap emerging from the boundary lead-copper system extends through the solidification surfaces for Cu, Cu_2S, PbS up to the neighbourhood of the boundary lead-lead sulphide system. The miscibility gap emerging from the copper-sulphur system

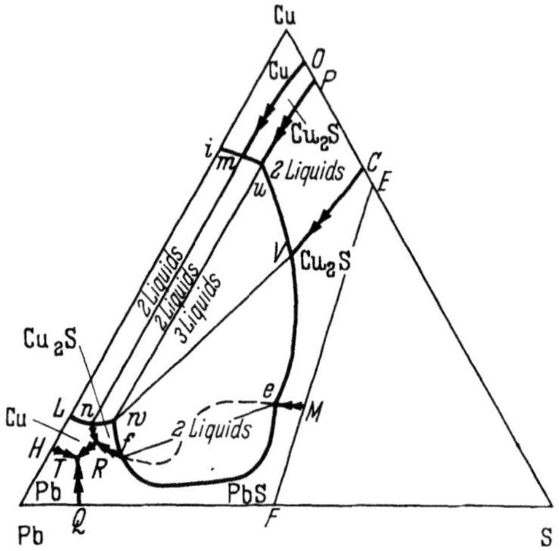

Fig. 155. Lead-Copper-Sulphur. (Sketch in At.-% according to JÄNECKE).
E: Cu_2S; *F*: PbS.

enters at the points *u* and *v*. This gives an invariant equilibrium, formed by the four-phase plane, Cu_2S-melt *v* — melt *w* — melt *u*, and at the temperature of this four-phase plane the $u + v \leftrightarrows w + Cu_2S$ reaction occurs. A three-phase area of three melts lies on the triangle *u*–*v*–*w*. Binary eutectic troughs, emerging from the boundary Cu–Cu_2S and Cu_2S–PbS systems, meet the miscibility gap at the points *m* and *e*. This gives rise to two further four-phase planes: Melt $m \leftrightarrows Cu + Cu_2S +$ Melt *n* and Melt $e \leftrightarrows Cu_2S + PbS +$ Melt *f*. Outside the miscibility gap the four-phase reaction, $Cu_2S +$ Melt $R \leftrightarrows Cu + PbS$, takes place with falling temperature. The end of solidification is given by the ternary eutectic, Melt $T \leftrightarrows Pb + Cu + PbS$. The ternary eutectic point nearly coincides with the lead corner.

DAVEY [235a] determined the lead corner of the ternary diagram from accurate figures for: the solubility of copper and of sulphur in lead, the simultaneous solubility of copper and sulphur in lead, and the free energy of formation of Cu_2S and PbS. He assumed in his thermodynamic calculations that there is a high degree of association between the copper and the sulphur atoms in the melt. The distribution of the phases (Fig. 156) varies somewhat from JÄNECKE's scheme (Fig. 155) in so much as the solidification fields for copper and PbS touch in the latter's work whereas DAVEY's diagram shows that the solidification fields for Cu_2S and Pb join. The trough between the Cu/Cu_2S fields has a eutectic

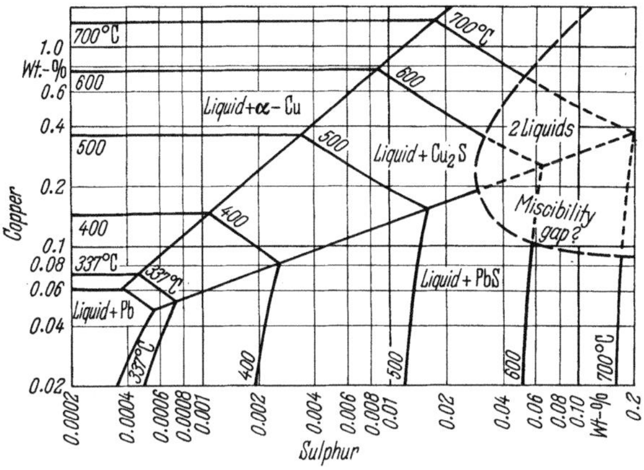

Fig. 156. Lead-Copper-Sulphur. Lead corner, second estimate. (According to DAVEY [235a]).

character; and the Cu_2S/PbS trough should also have a eutectic character at high temperatures, but on the other hand a peritectic character at lower temperatures. The point of change is 465 °C at 0.009% S; 0.127% Cu.

28. Pb–Cu–Sb. Lead-Copper-Antimony

SCHRADER [1082] carried out a careful determination of the ternary system by microscopy, thermal analysis and X-ray diffraction. According to her work there are, with the antimony, lead and copper solid solutions, three intermediate antimony-copper phases occurring as primary crystals (Figs. 157, 158).

SCHRADER's views on the lead corner of the diagram are no longer valid since the binary antimony-copper system, on which it is based, has since been proved to be more complicated below 500 °C. To the three intermediate phases known, thirty years ago, Cu_2Sb, β and ε (below 469 °C), HANSEN has added two new phases η and $\varkappa$, or even three new phases (SCHUBERT [448a]). If the lead corner is ignored, however, the diagram compiled by SCHRADER could still be valid (Fig. 157). The solidification surface for Cu_2Sb and β extends close to the lead corner, and the miscibility gap extends well into the β solidification field. The bounding curve of the miscibility gap runs through a minimum at S_1—S_2 and a maximum at M_2–M_3 up to the critical point K. Extrapolation of the tie-line M_3M_2 gives the direction of the saddle M_2M_1 of the liquidus surface. An investigation by the author [558] shows that β crystallises as a superlattice of the body-centered cubic ($A2$) structure type — its primary crystals appear in the form of cubes and octahedra.

The phase fields in the lead corner have in the meantime been redetermined (Fig. 158) by DAVEY [235a] by the methods indicated on

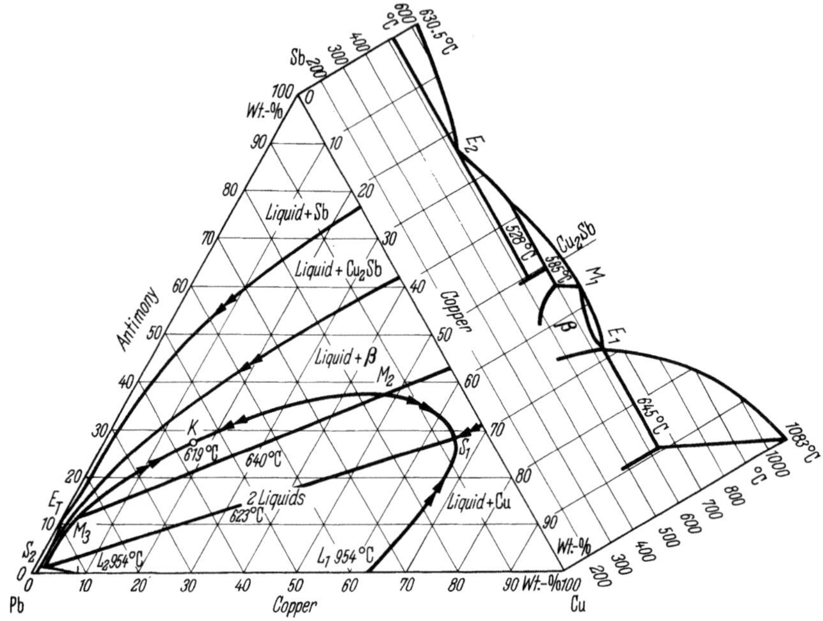

Fig. 157. Lead-Antimony-Copper. (According to SCHRADER and HANEMANN).

page 148. Here the main interest was with regard to the decoppering of lead. The solidification fields in the copper-antimony system for the recently found phases η and $\varkappa$ are indicated in Fig. 158 by the hatching. Presumably the boundary curves of the ε/α–Cu and ε/Cu_2Sb solidification fields are peritectic. The ternary point at 325 °C would express the equilibrium $S + Cu = Pb + \varepsilon$; the point at 324 °C the equilibrium $S + \varepsilon = Pb + Cu_2Sb$. The reactions on the phase boundaries are however yet to be established in detail.

Within this ternary diagram (if small additions of arsenic up to 1.5% and nickel up to 0.3% are disregarded) is the bearing metal, standardised in Germany as Lagerhartblei 12, having 10.5 to 13% Sb and 0.3 to 1.5% Cu (p. 380). It contains lamellar primary crystals ("needles") of Cu_2Sb in a predominantly eutectic ground mass. Lead cable sheaths with 0.5 to 1% Sb generally contain additions of some 0.01% Cu. This serves not only for grain refinement, but to allow a reduction in the pressure to be made when manufacturing the cable sheath. F. and W.

GLANDER [387] found in laboratory tests on alloys containing 0.5 to 1% Sb, that the pressure could be reduced by about 7% with the addition of 0.06% Cu, and this was confirmed on a work's press. In order to explain this phenomenon, it may be assumed that the copper is present in the

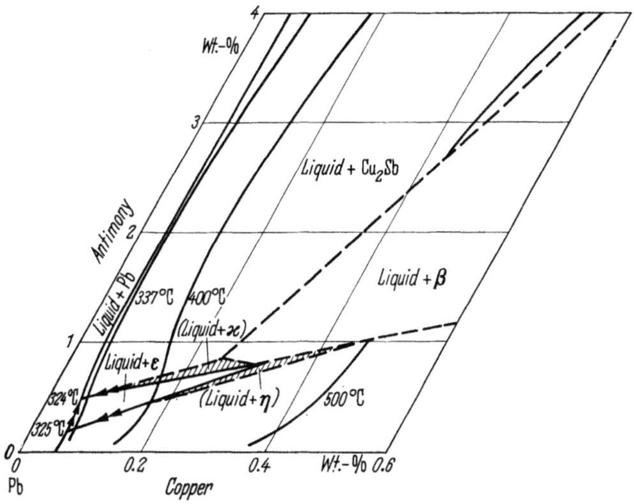

Fig. 158. Lead-Antimony-Copper. (Lead corner according to DAVEY).

alloy as Cu_2Sb. Thus on adding 0.06% Cu the lead-antimony solid solution is impoverished in antimony by about 0.1%. The inclusions of Cu_2Sb obviously have less influence on the pressure needed than the antimony dissolved in the lead.

29. Pb–Cu–Sn. Lead-Copper-Tin

The extent of the miscibility gap in the molten state was determined by BRIESEMEISTER [133] from ladle samples. As the samples were only taken from the upper part of the melt of lower specific gravity, it was only possible to determine the portion of the isotherms at some distance from the lead corner (Fig. 159). The boundary curve of the miscibility gap was approximately determined by investigating the stratification under the microscope. This, in turn, demonstrated that in spite of the comparatively high rate of cooling, the melts were still almost entirely segregated. The part of the stratification curve leading towards the copper corner was determined by GIOLITTI [385] and BLUMENTHAL [91] who evaluated the breaks in the cooling curves following primary solidification. This agreed satisfactorily with the investigations. Further study was made (VESZELKA [1220] and JÄNECKE

[612]) of this part of the boundary curve, and the two four-phase reactions which arise through contact with the solidification field for copper and that for α-, β- and γ-bronze. This part of the diagram is of importance in the solidification of leaded bronzes.

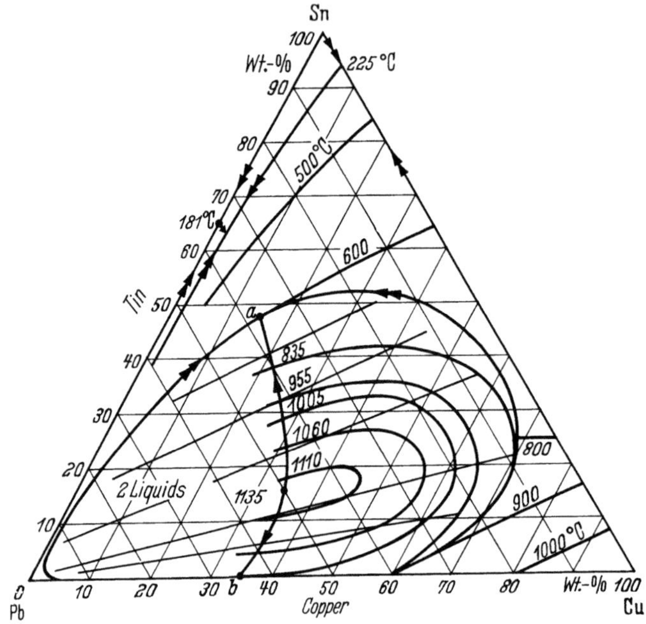

Fig. 159. Lead-Copper-Tin. (According to BRIESEMEISTER and JÄNECKE).

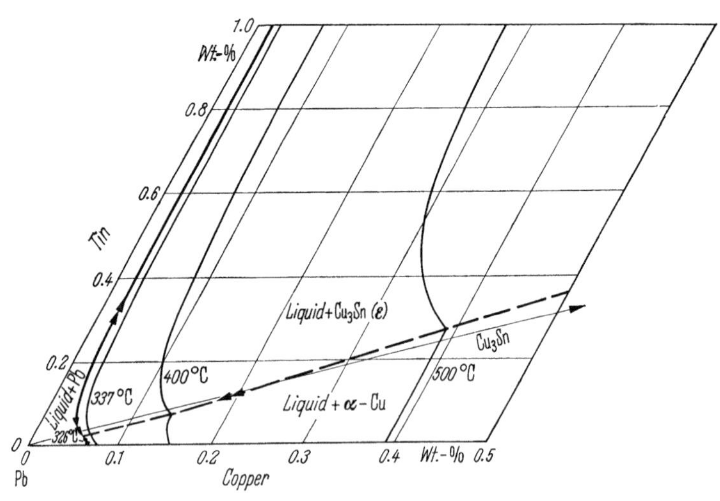

Fig. 160. Lead-Copper-Tin. (According to DAVEY).

The miscibility gap of the ternary system is wider than that in the binary and it extends to higher temperatures. Thus, the critical curve *a–b* shown in Fig. 159 has a temperature maximum of 1135 °C and composition of about 35% Cu and 15% Sn. JÄNECKE [612] considers that there is a ternary eutectic point outside the miscibility gap close to the binary Pb–Sn eutectic. This ternary eutectic consists of the lead and tin phases and the η-phase of the copper-tin alloys. DAVEY [235a] investigated the lead corner of the ternary system (Fig. 160) in a similar way to that described for lead-arsenic-copper and lead-copper-sulphur. A eutectic trough extends from the binary lead-copper eutectic parallel to the boundary lead-tin system and over a temperature maximum, towards the tin corner. The ternary eutectic has the composition 0.02% Sn and 0.057% Cu at 326 °C. This system is not only of importance in the structure of leaded tin bronzes, but also in understanding white metals.

30. Pb–Cu–Te. Lead-Copper-Tellurium

GRAVEMANN and WALLBAUM [418] determined the partial region Cu–Cu_2Te–PbTe–Pb of the ternary system by thermal analysis and microscopy and the results, together with theoretical inferences, led to

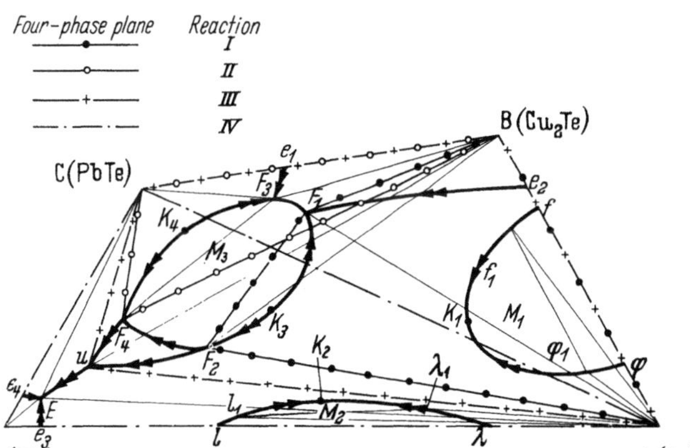

Fig. 161. Ideal diagram of the system Cu–Cu_2Te–PbTe–Pb. (According to GRAVEMANN and WALLBAUM).

Fig. 161. Here the Cu_2Te–PbTe section is quasi-binary and has a eutectic e_1. The miscibility gaps in the Pb–Cu and Cu–Cu_2Te boundary systems terminate in the ternary. The ternary system contains an additional miscibility gap bounded by the curve K_3, F_2, F_4, K_4, F_3, F_1. The intersection points of the troughs forming the three-phase equilibria lead to three four-phase planes having the corner points F_2, F_1, Cu_2Te, Cu; F_4, PbTe, Cu_2Te; u. PbTe, Cu_2Te, Cu and finally to the eutectic four-phase plane Pb, PbTe, Cu. The ternary eutectic occurs at 325.5 °C, very close to the lead corner of the diagram, and the temperatures of the three four-phase reactions were determined as 792 °C, 625 °C and 595 °C.

31. Pb–Cu–Zn. Lead-Copper-Zinc

In 1914 PARRAVANO, MARETTI and MISETTI [935] investigated the details of the entire ternary system by thermal analysis. BAUER and HANSEN [59], in their work on the effect of adding a third metal to brass, discussed the Italians' investigations. Further research was done by NICLASSEN [896] but was based on a more limited amount of experimental work. More attention will therefore be paid to the results of PARRAVANO and his co-workers. They were precise about the boundary of the miscibility gap in the molten state and the direction of the tie-lines and isotherms. The miscibility gap extends from the boundary lead-copper system through the entire ternary system to the boundary lead-zinc system. In the lead corner there is a small hollowed-out area where the liquidus surfaces of all the solidified phases of the ternary system in equilibrium are closely crowded together. Similar surfaces of primary solidification (apart from that for lead) are situated in the wide region between the boundary copper-zinc system and the curve bounding its miscibility gap. HENGLEIN and KÖSTER [509] have studied this part of the diagram and the lead corner further on the basis of PARRAVANO's work [935] and have compiled the ternary diagram shown in Fig. 162. In connection with the copper-zinc system with falling temperature, the solidification surfaces for α-, β-, γ-brass, the δ-, ε- and η-phase (zinc) follow successively from the copper corner. The peritectic troughs running from the boundary Cu–Zn system in the direction of the miscibility gap, cut the boundary curve of the latter at the invariant points S_1 to S_5. A point on the boundary curve in the lead corner is related to each of the invariant points. Five four-phase planes are produced, all based on the same pattern of reaction. The consecutive reactions are:

$S_1 + \alpha_2 \rightleftharpoons S_1' + \beta_1$ at 886 °C $S_4 + \delta_2 \rightleftharpoons S'_4 + \varepsilon_1$ at 590 °C
$S_2 + \beta_2 \rightleftharpoons S_2' + \gamma_1$ at 825 °C $S_5 + \varepsilon_2 \rightleftharpoons S'_5 + \eta_1$ at 422 °C
($S_3 + \gamma_2 \rightleftharpoons S_3' + \delta_1$ at 690 °C)

The third four-phase reaction is bracketed, to indicate that the δ-phase breaks down eutectoidally at low temperatures. For clarity on the diagram, only the two first four-phase planes are indicated, δ_1 and δ_2 being omitted. A eutectic trough runs from the lead-copper eutectic containing 0.06% Cu at 326 °C to the boundary lead-zinc system at 0.5% Zn and 318 °C. The trough is not straight, however, as assumed for clarity in the diagram, but, from an investigation by JOLLIVET [629], very curved, so that it approaches the lead corner closely at a point corresponding to 0.004 Wt.-% Cu and the same amount of zinc. The isotherms of the liquidus surfaces in the lead corner have equally curved

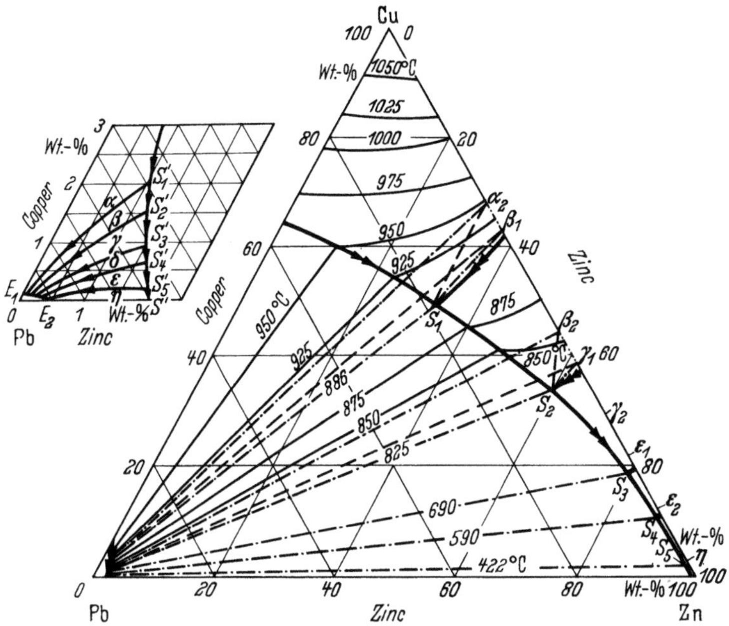

Fig. 162. Lead-Copper-Zinc. (According to HENGLEIN and KÖSTER).

contours, but there is no metallographic examination of this part of the ternary system available. There is practically no solubility of lead in the copper-zinc phases. This ternary diagram helps one to understand the solidification processes in leaded brasses. However it also has a metallurgical interest in that in desilverising lead with zinc, there is a simultaneous decoppering, and the lead corner investigated by JOLLIVET [629] is particularly important for understanding this phenomenon. There is no indication from the investigations whether the eutectic trough enters the binary lead-zinc eutectic or terminates at a ternary eutectic.

32. Pb–Hg–Na. Lead-Mercury-Sodium

GOEBEL [389, 390] investigated the properties of alloys in this equilibrium diagram up to the composition 7% Hg and 4% Na. A eutectic trough was shown to extend from the lead-sodium system parallel to the lead-mercury boundary system. In Fig. 163 a dashed line is shown since in view of the different structures of the two eutectic phases, they remain separated, so the curve is continuous. The solubility limit in the lead-sodium system is assumed here to be 0.8% Na, but according to Fig. 54 it amounts to about 1.5 Wt.-%. With the addition of mercury,

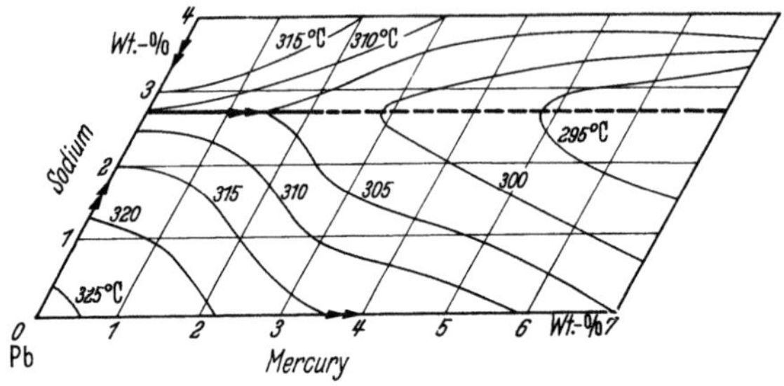

Fig. 163. Lead-Sodium-Mercury. (According to GOEBEL).

the solubility of sodium appears to be extended. For example, with 3% Hg, 4.6% Na can be dissolved. To verify this, an alloy with weighed amounts of 3% Hg and 3% Na was produced and it, in fact, proved to be one phase if solid granular precipitates were ignored.

The ability of lead-sodium alloys to age-harden was reduced by mercury additions, so in order to achieve the same hardness with mercury as the binary alloys with under 1% Na (which normally show very marked age hardening) an increase in the sodium content was required. Bending and compression tests show that the ductility of lead-sodium alloys is increased by mercury additions. The maximum hardness and, at the same time, ductility were shown in the alloys having 4 to 5% Hg and 2 to 3% Na. For practical purposes, particularly in bearing metals, no more than 2% sodium was recommended. Corrosion resistance was also improved by the mercury addition. HALLA, HERDY [473] measured the actitivity of molten alloys in the system. On melting these alloys, mercury vapour is evolved and so they have not found much practical application.

33. Pb–In–Sn. Lead-Indium-Tin

This system has recently been investigated by CAMPBELL and coworkers [171] using thermal analysis. Of the boundary systems, lead-tin and indium-tin, both contain a eutectic. It was shown by careful temperature measurement that no ternary eutectic is present in the ternary system. A eutectic trough runs from the lead-tin system with a steady fall in temperature to the indium-tin system (Fig. 164). As the binary lead-indium and indium-tin systems contain peritectic three-phase equilibria, peritectic reactions must also be present in the ternary system. However, they could not be detected on the cooling curves, so it was impossible to complete the ternary system by dividing the liquidus

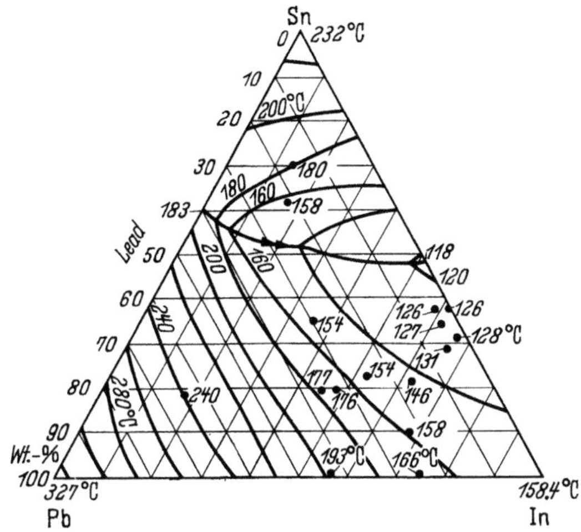

Fig. 164. Lead-Indium-Tin. (According to CAMPBELL).

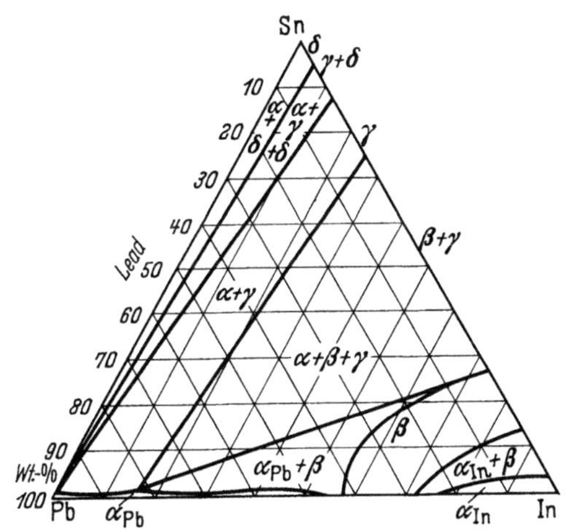

Fig. 165. Lead-Indium-Tin. Phase areas room temperature. (According to CAMPBELL).

surfaces into solidification fields of the individual phases. It was possible, though, by *X*-ray diffraction and microscopy, to determine the regions of the solid phases in a horizontal section. For this investigation the alloys were cooled in 24 hours from a temperature just above the beginning of solidification down to room temperature. Thus the horizontal section shown in Fig. 165 approximately represents the equilibria at room temperature. The region of homogeneity of the β-phase running from the lead-indium to indium-tin side is worth noting. This phase has, according to KLEMM and co-workers [671], a tetragonal body-centred

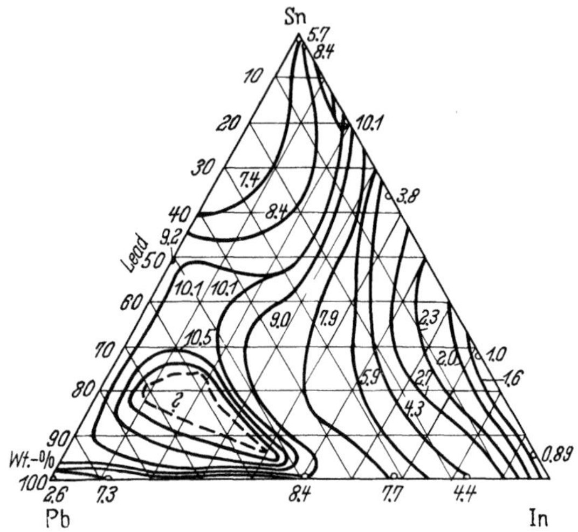

Fig. 166. Hardness of Lead-Indium-Tin alloys. (According to CAMPBELL).

lattice. A deeper investigation of the diagram was made by including hardness measurements (Fig. 166). The hardness maximum of over 10 BHN in the indium-tin system is due to the γ-phase formed in the solid state. The maximum traverses the ternary system until near to the lead corner of the lead-indium system. The alloys are of interest for low-melting solders; and in this connection attention may be drawn to the binary lead-indium system.

GRYMKO [447] investigated the effect of indium additions on the strength and corrosion resistance of lead-tin solders. He found a solder with high strength, good wetting and good corrosion resistance at the composition 37.5% Pb, 37.5% Sn and 25% In.

34. Pb–Mg–Sb. Lead-Magnesium-Antimony

ABEL [3] showed in an incomplete examination of the system that only the boundaring system phases Pb, Sb, Mg, Mg_2Pb and Mg_3Sb_2 occurred in the alloys. The Pb–Mg_3Sb_2 section is regarded as quasi-binary, and the eutectic Pb–Mg_3Sb_2 has 3.17% Mg, 10.6% Sb at 248 °C. As well as the quasi-binary section, the partial triangle Pb–Sb–Mg_3Sb_2

has been investigated. It contains three binary eutectic troughs arising from the boundary systems, and a ternary eutectic E_T, 2.7% Mg, 8.9% Sb at 248 °C (Fig. 167). No solubility determinations in the lead corner or age-hardening tests were carried out. They would perhaps be

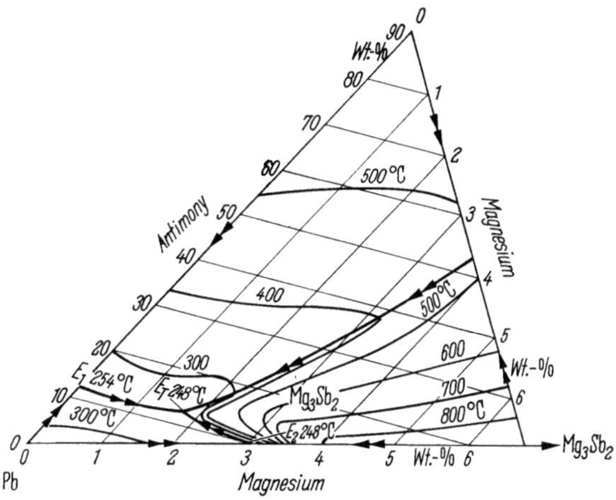

Fig. 167. Lead-Antimony-Mg_3Sb_2 [3]. Redrawn in Wt.-%.

of interest, since in alloys between the boundary lead-antimony system and the section Pb–Mg_3Sb_2, only antimony and Mg_3Sb_2 can occur as precipitates in the solid state — not, however, the readily decomposable Mg_2Pb.

35. Pb–Mg–Sn. Lead-Magnesium-Tin

The ternary system only contains the phases of the boundary systems, together with the intermetallic compounds Mg_2Pb and Mg_2Sn. Fig. 168 shows a spatial representation of the ternary system in atomic per cent, while Fig. 169 gives weight per cent. The Mg_2Sn–Pb section is quasi-binary and has a eutectic E_6 at 1.0 Wt.-% Mg, 2.5 Wt.-% Sn and 300 °C. The Mg_2Pb–Mg_2Sn section is also quasi-binary. The two phases have considerable mutual solubility at high temperatures, so it is feasible that the point P (Fig. 168) represents a peritectic decomposition of the melt with Mg_2Sn-solid solution to Mg_2Pb-solid solution. The troughs from E_6 and P, meet at the point E_{T1}, which, with 3.0% Mg, 97.0% Pb at 248 °C, practically corresponds to the binary eutectic point E_1 of Pb–Mg_2Pb. The trough PE_{T1} is, in its lower temperature region, regarded as eutectic as the solubility of Mg_2Pb in Mg_2Sn may practically disappear here. This slightly conflicts with the views of v. VEGESACK [1218].

A second ternary eutectic, Mg_2Sn–Pb–Sn, solidifies at E_{T2} with the composition 1.5% Mg, 30% Pb, 68.5% Sn at 166 °C. The ternary system is such

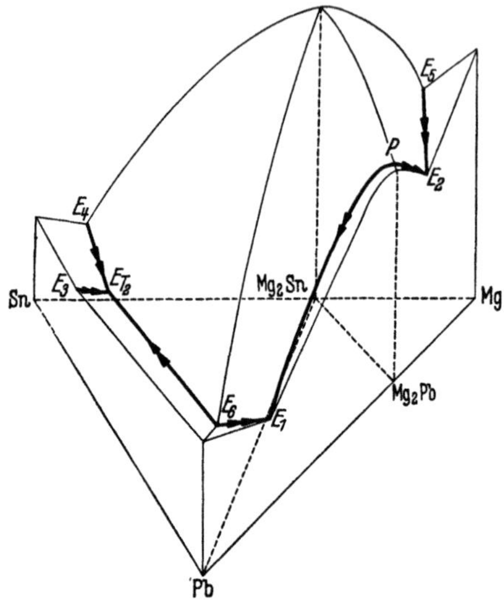

Fig. 168. Lead-Magnesium-Tin. Solid diagram in At.-%. (According to v. VEGESACK [1218]).

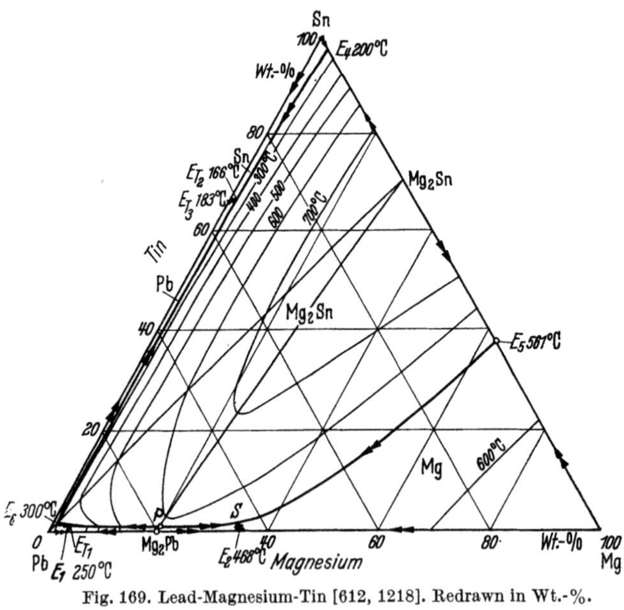

Fig. 169. Lead-Magnesium-Tin [612, 1218]. Redrawn in Wt.-%.

that the Mg_2Pb and Mg_2Sn precipitates are to be expected in the lead solid solution on cooling but both of these compounds easilydecompose.

36. Pb–Na–Sn. Lead-Sodium-Tin

Like the lead-sodium-mercury system, a eutectic trough runs from one boundary system parallel to the other (Fig. 170). The ability of the

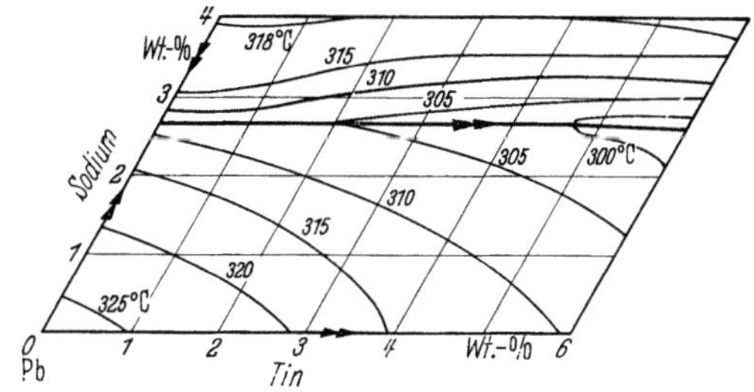

Fig. 170. Lead-Sodium-Tin. (According to GOEBEL).

ternary alloys to age-harden is less than that of the binary lead-sodium alloys, but bending and compressions tests show an increase in ductility, particularly with high tin contents.

37. Pb–S–Sn. Lead-Sulphur-Tin

This ternary system within the Pb–PbS–SnS–Sn quadrilateral was investigated by VOGEL and ZASTERA [1230], see Fig. 171. A redetermination of the PbS–SnS boundary system indicated the presence of two crystalline phases; PbS, which takes up a certain amount of SnS in solid

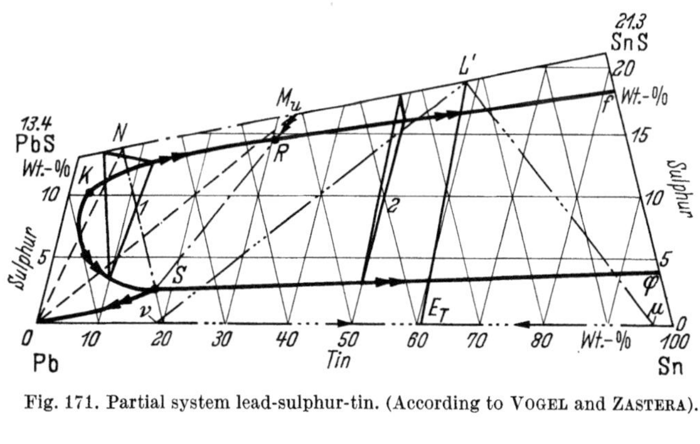

Fig. 171. Partial system lead-sulphur-tin. (According to VOGEL and ZASTERA).

solution, and the solid solution SnS, whose region of homogeneity extends practically to the composition $PbSnS_2$. The determination of this solid solution region agrees with earlier work (unknown to VOGEL and ZASTERA) by the present author [553], who investigated the crystalline structures of SnS and $PbSnS_2$ (which occur naturally in the mineral teallite) and established the same structure for the two substances. The liquidus and solidus curves on the right hand side of the boundary system (Fig. 172) practically coincide, and the horizontal in the left-hand part of the diagram represents a peritectic three-phase equilibrium. The composition of the melt, which reacts with PbS, is practically identical with the composition M_u of the SnS solid solution formed by the reaction. The miscibility gap in the liquid state from the Sn–SnS side of the ternary

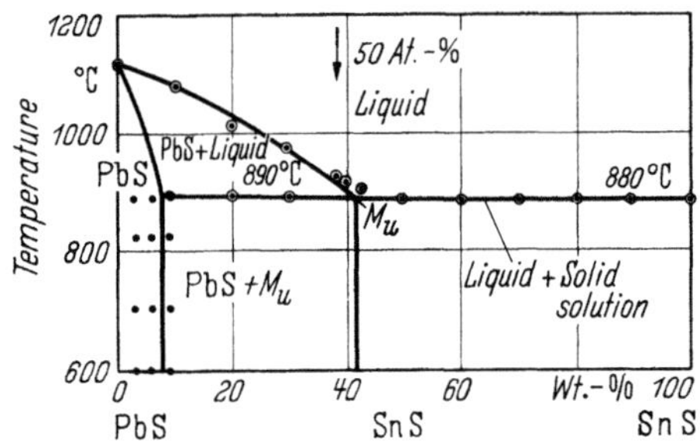

Fig. 172. Binary PbS–SnS section. (According to VOGEL and ZASTERA).

system, extends up to the critical point K with the composition 87% Pb, 3% Sn, 10% S at 1060 °C. A transformation curve $M_u - R$ runs from the PbS–SnS side to the boundary curve of the miscibility gap, and on the other side of the miscibility gap from the point S to the lead corner. The two branches of this transformation curve represent the peritectic three-phase equilibrium, Melt + PbS solid solution ⇋ SnS solid solution. The stratification curves KR and KS (entering at points R and S) signify monotectic three-phase equilibria, e.g., "Sulphide-rich melt ⇋ PbS + Lead-rich melt". Where the peritectic and the monotectic three-phase areas intersect, a four-phase plane M_u–R–S–N is formed, which is practically a triangle. The reaction:
Melt R + Lead sulphide solid solution $N \rightleftharpoons$ Melt S + Tin sulphide solid solution M_u occurs here at 860 °C.

The temperatures of the points R and S indicate that the section $f - R$ and $\varphi - S$ of the stratification curve are very flat. The main rise in temperature of the stratification curve occurs between R and S and the critical point K. It is worth noting that in this ternary system the solidification surface of the SnS solid solution nearly extends to the lead-tin boundary system. Towards the end of the solidification of sulphur-containing Pb–Sn alloys, alloys practically free from sulphur solidify. In the ideal diagram the peritectic trough running from the point S to the lead corner enters a eutectic trough which extends from the lead-lead sulphide side, parallel to the lead-tin side, to the tin-tin sulphide eutectic. Theoretically two four-phase equilibria result, first, a peritectic four-phase plane in the lead corner "Melt + PbS $\rightleftharpoons$ SnS solid solution + Pb". Two of the points of this four-phase plane practically coincide with the lead corner point. The ternary eutectic E_T is quite close to the binary Pb–Sn eutectic, in fact practically coincident with it. However, the proportion of SnS solid solution of composition L' in the ternary eutectic is extremely small.

38. Pb–Sb–Sn. Lead-Antimony-Tin

a) Structure of Alloys. This ternary system is the basis of a series of important lead alloys, notably type metals (p. 357) and white bearing metals (p. 381). Soft solders also belong here, for, in addition to lead and tin as the main constituents, they always have a certain amount of

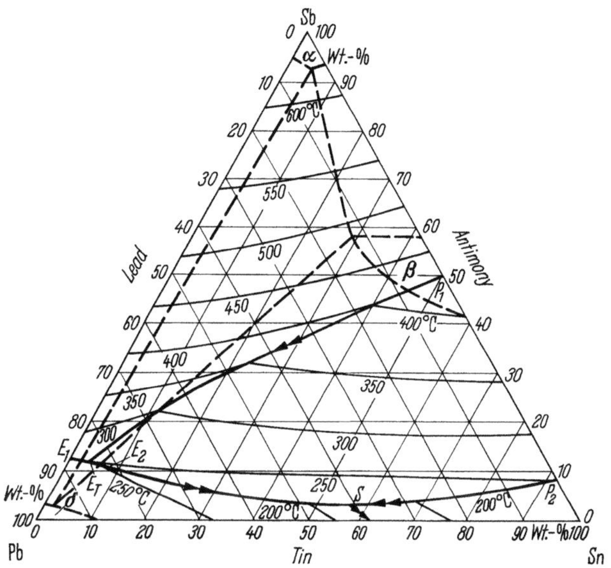

Fig. 173. Lead-Antimony-Tin. (According to WEAVER, and to IWASÉ and AOKI).

antimony (p. 470). The alloys with lower concentrations of tin and antimony have significance in cable sheathing (p. 414).

The diagram devised by v. LOEBE [756] is still approximately valid in the region of the primary crystallisation fields, but not in the concept of three- and four-phase equilibria in the neighbourhood of the lead corner. The ternary system now regarded as authentic is based on an investigation by IWASÉ and AOKI [608] and later by WEAVER [1246]. As no ternary compound exists, the diagram only contains the crystallisation fields of the boundary system phases, i.e. of lead, antimony, tin and of their solid solutions in one another (including the two phases β and β' of the antimony-tin-system). As β and β' correspond approximately to the formula SbSn, with an excess of antimony or tin respectively, and presumably have very similar crystal structures, they can be regarded at a first approximation to be one single phase. According to Fig. 173 the liquidus surface of the diagram is crossed by a eutectic trough which runs from point E_1 fo the lead-antimony system to the lead-tin eutectic, E_3.

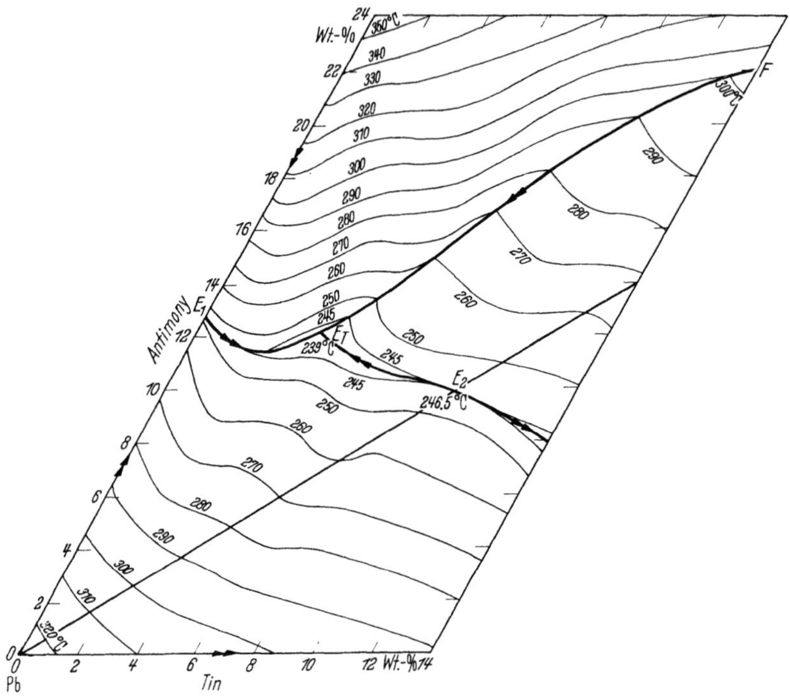

Fig. 174. Lead-Antimony-Tin. (Lead corner according to WEAVER).

It is necessary to consider the significance of the individual sections of this trough. A peritectic trough corresponding to the equilibrium, Melt $+ \beta \leftrightarrows \gamma$, starts from point P_2 of the antimony-tin system and meets the eutectic trough mentioned above at S with 57.5% Sn, 2.5% Sb, 40% Pb and 189 °C. S is the corner point of a peritectic four-phase plane with the reaction, Melt S $+ \beta \leftrightarrows \gamma$ (tin) $+ \delta$. The trough $P_1 - E_T$ largely represents, according to Fig. 173, the reaction, Melt $+ \alpha = \beta$, but in the neighbourhood of E_T assumes a eutectic character, corresponding to the reaction Melt $\leftrightarrows \alpha + \beta$, which can be recognised by a change in the direction of the tangent. The lead corner is divided into two partial systems by a ridge-like elevation of the liquidus surface, shown in Fig. 174 by a straight line. The trough $E_T E_2$ rises up at E_2 and then falls steadily to a point S and E_3 resp. (Fig. 173). E_2 heaving the composition

10% Sb, 10% Sn at the temperature 246.5 °C, signifies a quasi-binary eutectic of lead with SbSn. The ternary eutectic E_T of lead with antimony and β, found by IWASÉ and AOKI [608], is at the composition 11.5% Sb and 3.5% Sn and the temperature 240 °C. Fig. 173 contains, in addition to the liquidus surfaces, a horizontal section through the eutectic four-phase plane and the related phase fields. The solid solution area in the lead corner, as shown in Fig. 175, was evaluated on the basis of resistance measurements. KOGAN [692] recently confirmed the existence of the quasi-binary section Pb–SbSn.

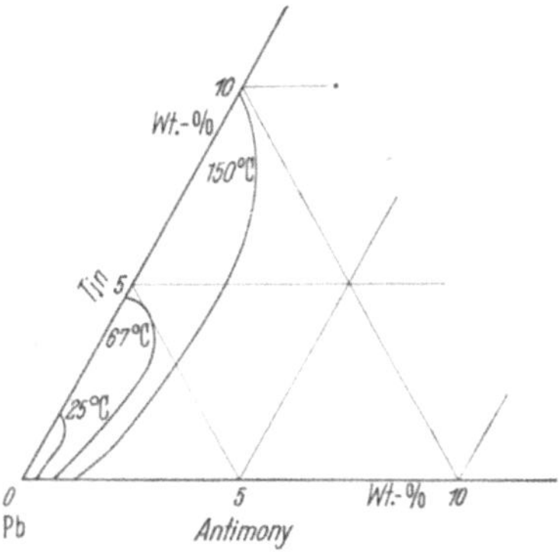

Fig. 175. Solubility of antimony and tin in solid lead. (According to MORGEN and coworkers [869]). An improvement in the drawing corresponding to the theory of ternary systems was not made.

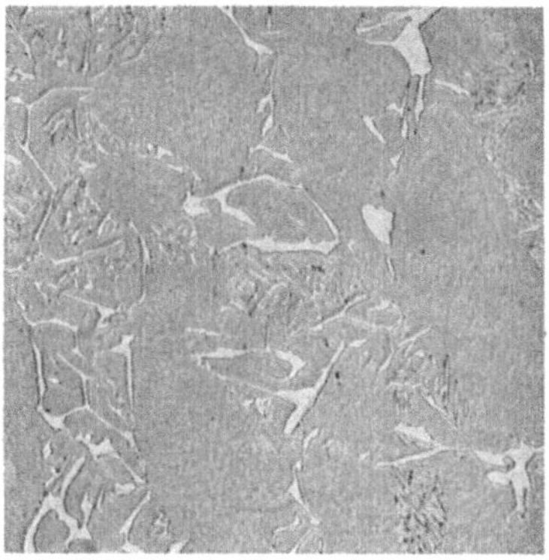

Fig. 176. 8% Sb, 2% Sn. Ground mass: lead solid solution. White: antimony. Fine dark lamellae: SbSn. Binary eutectic of lead with antimony coarsely developed, ternary eutectic very finely developed. ×500

In order to throw some light on this ternary system, some alloys were cast in a mould, which was heated to 180 °C, and examined metallographically. The sections were etched with the ammonium persulphate-tartaric acid solution mentioned on p. 28. If the section tarnishes on etching, the thin layer of lead sulphate can be removed by quick immersion in ammoniacal tartaric acid solution. Etching allows Sb and SbSn to be distinguished. If the two constituents exhibit a coarse structure, Sb remains white on etching; SbSn becomes yellowish, and under some circumstances develops etch pits. In a fine eutectic structure, the lamellae of SbSn acquire a very marked dark edge on etching whereas the lamellae of Sb remain unchanged. Even without etching SbSn can sometimes be distinguished from the pure white Sb by its light grey shade. Phase determination is also possible in polarized light. SbSn is pseudo-cubic and so between crossed Nicol's gives a much feebler change than the kind of layered lattices formed by antimony.

Fig. 176 shows an alloy from the primary crystallisation field of lead, and the succession of primary, binary and ternary eutectic solidifications are clearly recognisable. The pure ternary eutectic alloy is shown in Fig. 177, and Fig. 178 shows an alloy with primary antimony. SbSn of the binary Sb–SbSn eutectic is present in large amounts. Alloys with primary SbSn, the cubic form of which is characteristic, are re-

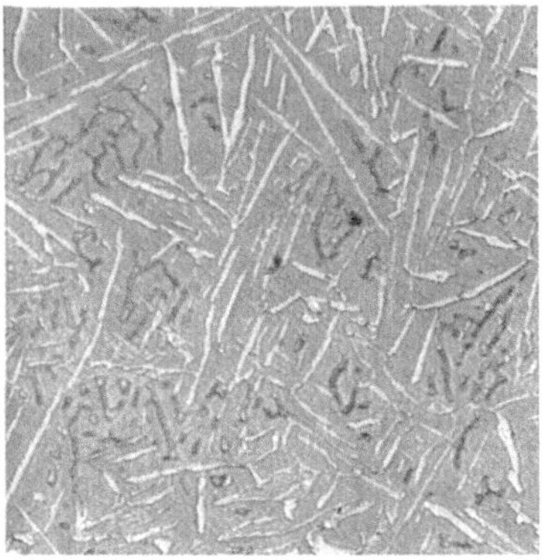

Fig. 177. 12% Sb, 4% Sn. Ternary lead-antimony eutectic (light lamellae) – SbSn (dark lamellae). ×500

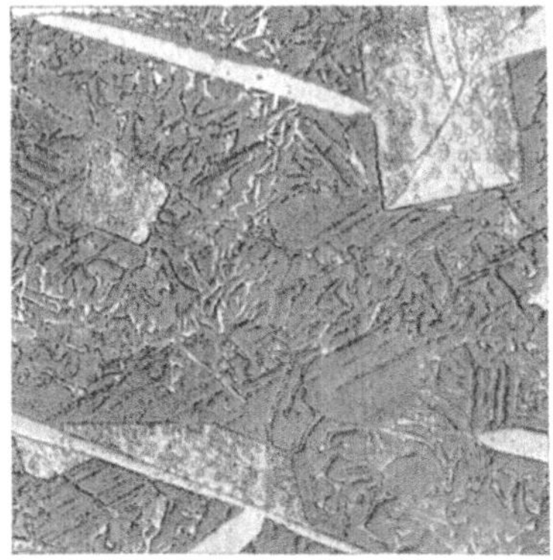

Fig. 178. 28.5% Sb, 8.8% Sn. White: antimony. Grey with etch-pits: SbSn. Ground mass: ternary eutectic. ×500.

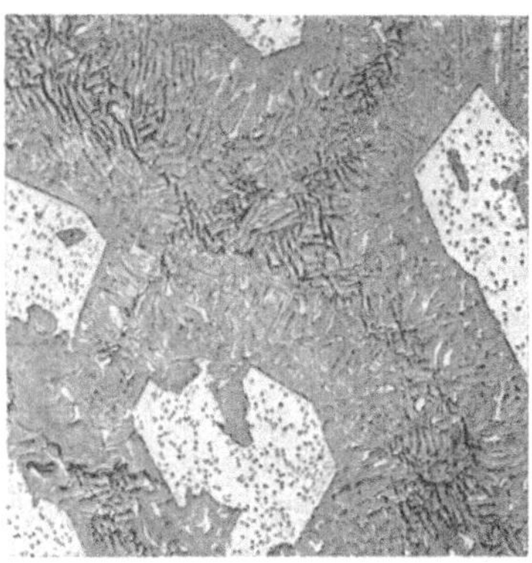

Fig. 179. 20% Sb, 9.5 Sn. Large primary crystals of SbSn. Ground mass: ternary eutectic, as a net of lead + SbSn, formed round the lead + antimony region. ×500

Fig. 180. 12% Sb, 12% Sn. Primary crystals SbSn, in part in skeleton form. Ground mass: quasibinary eutectic lead-SbSn. ×150

produced in Figs. 179 and 180. The ground mass in the first case is ternary eutectic, in the second the quasi-binary eutectic Pb–SbSn.

Löhberg and Schulz [758], in a study of type metals, observed the surprising influence sulphur had on the crystalline development of the primarily-solidifying antimony, independent of its tin content. If the alloys contain more than 0.05% sulphur, antimony always occurs in pseudo-cubic crystals, but apart from this, a refinement of the structure is observed. In sulphur-free alloys the antimony crystals exhibit coarser development—the extent depending on the tin content. Antimony, free from tin, solidifies in dendrites in the rhombohedral class, while antimony

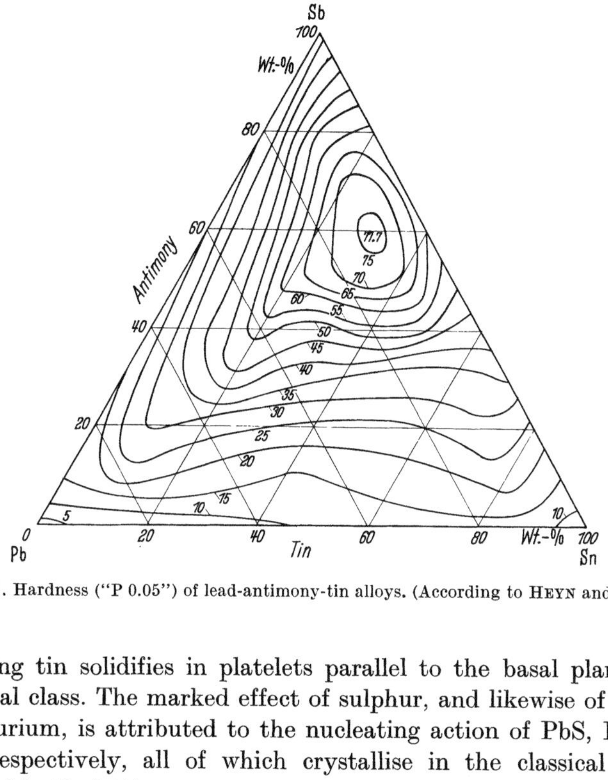

Fig. 181. Hardness ("P 0.05") of lead-antimony-tin alloys. (According to HEYN and BAUER).

containing tin solidifies in platelets parallel to the basal plane in the hexagonal class. The marked effect of sulphur, and likewise of selenium and tellurium, is attributed to the nucleating action of PbS, PbSe and PbTe, respectively, all of which crystallise in the classical B1-type sodium chloride lattice.

b) Hardness and Strength. The mechanical properties of the ternary alloys, in so far as they are of importance for bearing metal applications, were examined in some fundamental research by HEYN and BAUER [520] which is still worth noting. Hardnesses of the alloys were determined on specimens which had been cast in a water-cooled chill mould.

Fig. 181 indicates, as a hardness measure, the load in kilogrammes necessary to produce an impression with a depth of 0.05 mm. Since this value is not identical to the Brinell hardness, the diagram can only be used for comparisons. Fig. 182, from the work by WEAVER [1246], gives Brinell hardnesses for the lead corner of the phase diagram. It is noteworthy that the hardness maximum is in the neighbourhood of the quasi-binary eutectic E_2 (Fig. 174). As alloys containing primary antimony or primary SbSn are of significance as bearing metals and type metals, respectively, it is of interest which of the two structural constituents is the harder. A determination of scratch hardnesses gave the

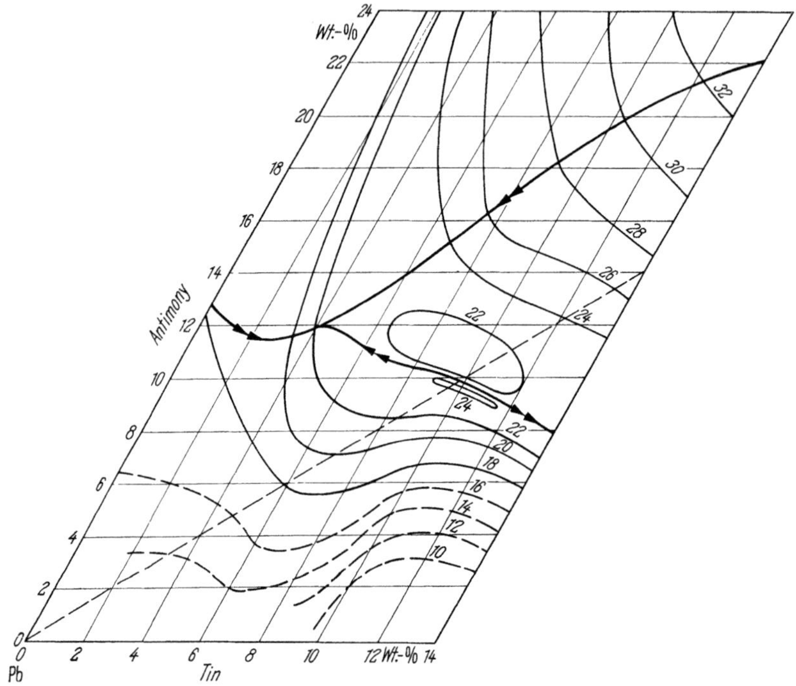

Fig. 182. Brinell hardness in the lead corner of the ternary system lead-antimony-tin. (According to WEAVER).

values, 200 for antimony and 280 kg/mm² for SbSn. The more accurate microhardness measurements showed only insignificant differences between the two phases (Fig. 321).

The change in hardnesses of the alloys with increased temperature is shown by comparing Fig. 183 with Fig. 181. The heat treatment of the specimens in Fig. 183 consisted of twenty-four repetitions of: slow heating to 150 °C, holding at this temperature for between 2 and 6 hours, and subsequent slow cooling. The lead-rich alloys, apart from a narrow strip near the Pb–Sn system, were softened by this heat treatment, which can be attributed to the formation of precipitates from solid solution and granulation of the eutectic material.

The tensile strength of the lead-rich alloys and possible changes through age-hardening were studied by MORGEN and co-workers [869].

A series of alloys, having 0.5% Sb and Sn contents of 0.5 to 14.7%, exhibited no age-hardening on quenching from 150 °C; alloys with the higher tin content in fact softened. A second series of alloys with 0.5% Sn and antimony contents of up to 2.5% underwent age-hardening, but

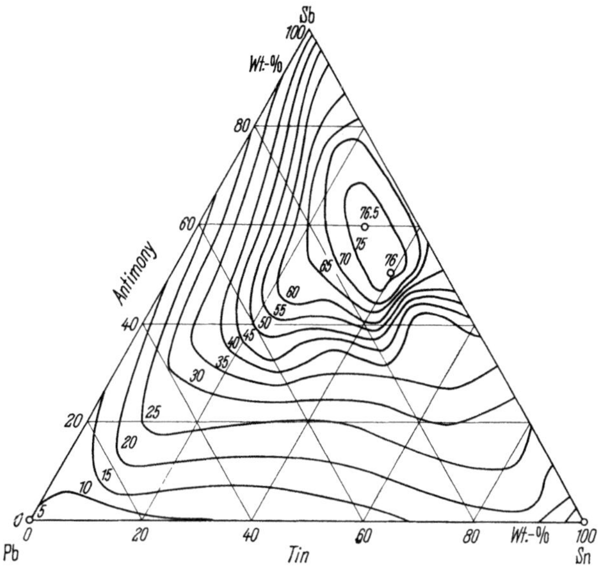

Fig. 183. Hardness ("P 0.05") of lead-antimony-tin alloys after annealing at 150 °C. (According to HEYN and BAUER).

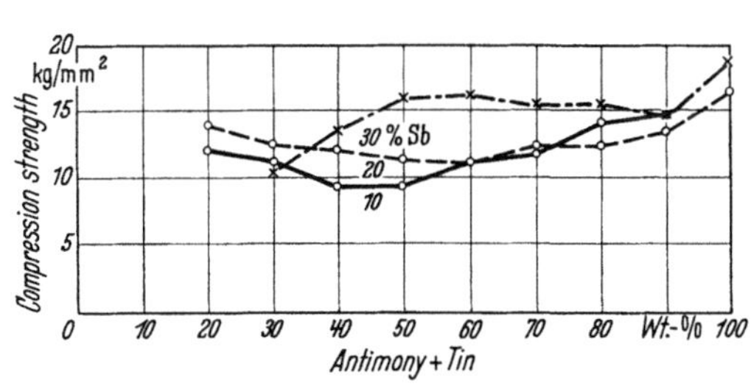

Fig. 184. Lead-Antimony-Tin. Compressive strength = stress to initiation of crack formation. (According to HEYN and BAUER).

of a lesser degree than in the tin-free alloys. This result is, however, not convincing as the quenching temperature was only 207 °C. In a further series of experiments on an alloy with 3% Sb and 3% Sn the effect of the quenching temperature on age-hardening was investigated. It was found that marked age-hardening was only to be expected when quenching temperatures were above 200 °C. According to PELZEL (see p. 87) the age-hardening of lead-antimony alloys is increased by tin additions. The tensile strength of an alloy with 1% Sb, after homogenising at 220 °C and subsequent quenching, rose in 26 days from 2.4 kg/mm² to 4.4 kg/mm². However, the corresponding values for an alloy with 1% Sb and 1% Sn were 2.5 and 6.6 kg/mm².

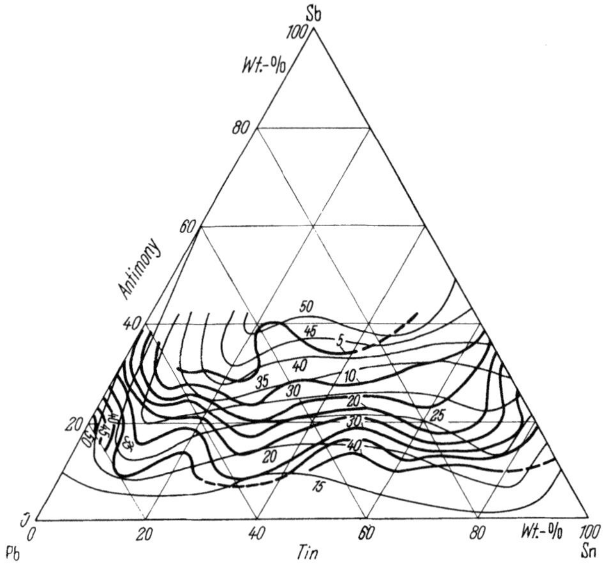

Fig. 185. Lead-Antimony-Tin. Thin curves: hardness ("P 0.05"). Thick curves: percentage reduction to fracture in the compression test. (According to HEYN and BAUER).

c) Static and Dynamic Compression Tests [520]. The test cylinder, of 2 cm diameter and height 1.77 cm, was loaded in steps of 80 kg/cm² for 30 seconds. After each load was released, the reduction in thickness was measured. The occurrence of cracks and the development of the fractures was recorded as a measure of the compressive strength of the alloys (KÖRBER [690]). The diagram in Fig. 184 corresponds to three sections parallel to the Pb–Sn system. The test pieces from the boundary system showed no cracking, even after a 50% reduction in height after which the experiment was concluded.

The percentage reduction in height of the test pieces up to the inception of cracking is a measure of the workability of the alloys. In Fig. 185 this is compared with the hardness. By and large the two values are divergent but some intersections of the curves do occur.

The deformation limit and the compressive stress corresponding to a specific height reduction (2%) were also derived from the compression tests. As the shape of these curves is very similar to the hardness curves, they have not been reproduced.

Impact tests at temperatures of —20 °C, 20 °C and 93 °C used the same form of test specimen. The specific impact energy per blow amounted to 37.4 cm kg/cm³. The reduction in thickness after individual blows was measured and the inception of crack and fracture observed. The specific impact energy at the inception of cracking is shown in

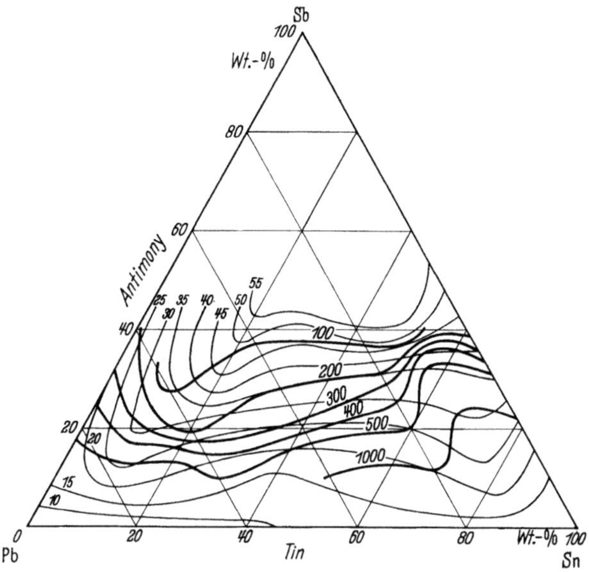

Fig. 186. Lead-Antimony-Tin. Thin curves: hardness ("P 0.05"). Thick curves: specific work of impact at beginning of crack formation in cmkg/cm³. (According to HEYN and BAUER).

Fig. 186 together with the hardness. The diagram shows many intersections of the two curves, so it is possible to select alloys with minimum brittleness at desired hardnesses. In this respect the tin-rich alloys are particularly favourable.

39. Pb–Sb–Zn. Lead-Antimony-Zinc

TAMMANN and DAHL [1165] fixed the broad outlines of this diagram by thermal analysis and structure investigations (Fig. 187). The extent of the miscibility gap was given by WRIGHT's work [1289], which included the boundary curves at 650 °C. Only two intermediate phases, Sb_2Zn_3 and SbZn were attributed to the antimony-zinc system in place of the three known now. While Sb_2Zn_3 has a melting point maximum, Sb_3Zn_4 and SbZn form peritectically. The other two boundary systems contain one eutectic each—the boundary lead-zinc system has in addition a monotectic. The formation of SbZn in the binary and ternary alloys only takes place after considerable undercooling, with a sudden emission of heat. Consequently it is not possible to follow the peritectic trough, Melt + $Sb_3Zn_4 \leftrightarrows$ SbZn, emerging from the antimony-zinc system into the interior. The course of the other peritectic trough, emerging from the antimony-zinc system, was not determined. The Pb–Sb_2Zn_3 section was

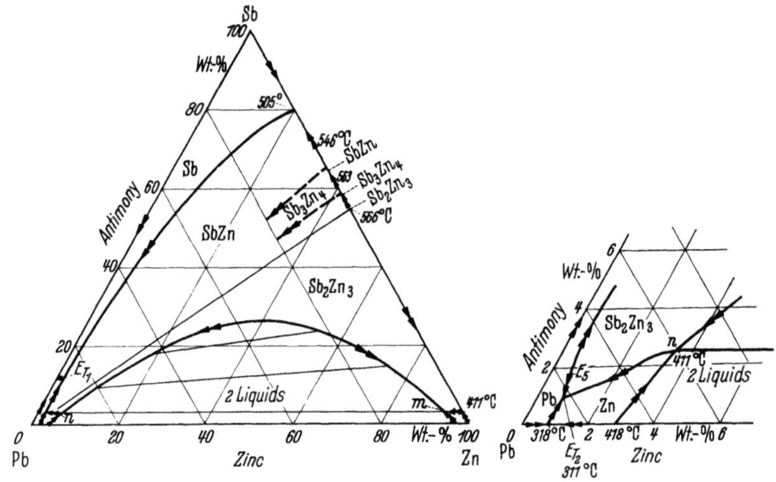

Fig. 187. Lead-Antimony-Zinc. (According to TAMMANN and DAHL).

found to be quasi-binary; it contained a eutectic E_5 of Pb with Sb_2Zn_3 at 98% Pb, 2% Sb_2Zn_3 and 312 °C. The quasi-binary section divides the ternary system into two partial systems. The partial Pb–Sb–Sb_2Zn_3 system has a ternary eutectic E_{T1} of lead, antimony and SbZn but the greater part of the second partial system is filled by the miscibility gap. The bounding curve of the miscibility gap towards the liquidus surface meets the eutectic trough emerging from the zinc side of the antimony-zinc system at point *m* at *a* temperature of 411 °C. This practically coincides with the binary eutectic Zn–Sb_2Zn_3. A second invariant point n at the same temperature is near the lead corner. The four-phase plane with corners n, Zn, Sb_2Zn_3 corresponds to the reaction, Melt m $\leftrightarrows$ Melt n + Sb_2Zn_3 + Zn. A eutectic trough runs from *n* in the direction of the lead corner. It unites with the eutectic troughs emerging from the lead-zinc system and the quasi-binary section, to a ternary eutectic E_{T2}, which practically coincides with the quasi-binary eutectic E_5. The diagram given by PONOMAREW [970a] does not differ essentially from the one described here. The ternary system needs reinvestigation since no attention has heen paid to the Sb_3Zn_4 phase or the transformation of the Sb–Zn phases in the solid state which can, under some circumstances, have effects up to the lead corner.

In a wider investigation, for developing lead-based bearing metals, ZUNKER [1308] looked into the technical properties of lead-antimony-zinc alloys. With higher antimony contents (in part above 30%) and lead contents around 50%, he was able to attain Brinell hardness values of above 30 kg/mm², but the alloys were very brittle and their compressive strength, at about 12 kg/mm², was too low for use as a bearing metal. Further, a partial replacement of the zinc by iron resulted in no definite improvement.

40. Pb–Sn–Zn. Lead-Tin-Zinc

As early as 1891, WRIGHT [1288] gave an approximate boundary to the miscibility gap in the molten state on the basis of stratification at 650 °C and 800 °C, but a more precise investigation of the ternary system was carried out by LEVI-MALVANO and CECCARELLI [745] by

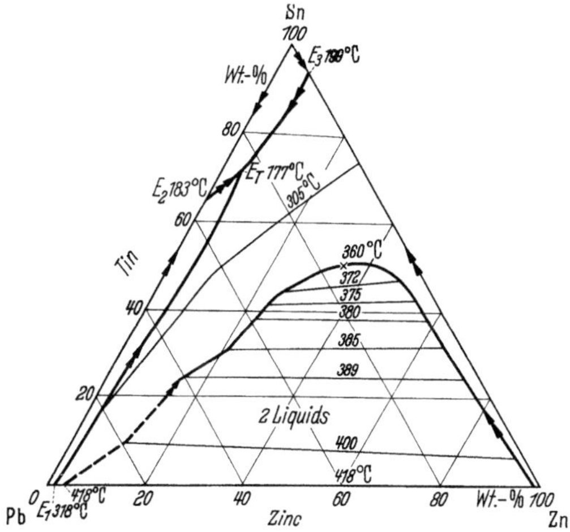

Fig. 188. Lead-Zinc-Tin. (According to LEVI-MALVANO and CECCARELLI).

thermal analysis (Fig. 188). Since from this work, the miscibility gap in the lead corner enters the lead-zinc boundary system at a zinc content which is too high, it is shown dashed. The miscibility gap only falls in the solidification field of zinc, so the structure. of the ternary system is simple. The binary eutectic troughs, emerging from the three boundary systems, unite at the ternary eutectic, E_T, with the composition 24% Pb, 71% Sn, 5% Zn at 177 °C. A section through the miscibility gap at 520 °C was given by MONDAIN-MONVAL and GABRIEL [863].

41. Comparative Investigations of the Mechanical Properties of 37 Ternary Alloys

In addition to the binary alloys mentioned on page 31, 37 ternary alloys of lead were made at the Armour Research Foundation [24a], from the purest starting materials (Pb 99.9999%). The dilute ternary lead-based alloys had the following additions,

Ag with Ba, Cd, Cu, Te; As with Cu, Sb, Sn, Te;
Ba with Ca, Cu, Sn, Te; Ca with Sn, Cd with Sb;
Sb with Sn; Sn with Te.

As described for the binary alloys, tensile test specimens were prepared from rolled sheet and heated for one hour at 100 °C. The tensile strengths of the dilute ternary alloys were calculated from those of the corresponding binary alloys, by assuming that the action of the alloying elements in the ternary alloys was additive. In fact, the calculated tensile strengths agreed to some extent with those determined experimentally. There were particularly marked differences between calculated and experimental results in the age-hardening alloys with 0.03 Wt.-% Ba + 0.4% Ca and 0.01% Ca + 0.1% Sn, where the tensile strengths were far higher than calculated. Most of the ternary alloys with larger amounts of alloying elements were extruded into wire (obviously at room temperature). The following combinations of alloying elements being selected:

Ag with Ba, Ca, Cd, Te; As with Ba, Ca, Cd, Sb, Sn, Te;
Ba with Bi, Ca, Cd, Sn, Te; Ca with Cd, Sb, Sn, Te;
Cd with Sb, Li; Cu with Ni, Sn; Li with Sb, Sn, Te, Zn;
Mg with Sn; Na with Sn; Ni with Sb, Te; Sn with Sb, Te.

After extrusion and subsequent heat treatment at 100 °C, the tensile strengths of the wires were determined at room temperature. The heat treatment at 100 °C should provide information on the stability of the alloys on storage and slight heating. Some of the alloys largely retained their strength after heat treatment, whereas others, for example:

1.72% Sn + 0.35% Ba; 2.90% Sn + 2.90% Sb;
1.88% Sn + 0.91% Na; 1.77% Cd + 0.45% Ba;
2.0% Cd + 2.0% Sb,

underwent a considerable loss in strength. Presumably this is due to coarse breakdown of supersaturated solid solutions. A third group of alloys showed an increased strength through age-hardening, for example (in addition to most of the alloys of the alkali metals and the alkaline earth metals), the alloy with 0.5% Ag + 0.5% Te.

A further series of experiments was specially devoted to a study of age-hardening. The alloys were solution treated at 230 °C, quenched in water and heat treated to various temperatures. As expected, the highest strengths after age-hardening were attained in the alloys containing alkali metals and alkaline earth metals. The alloy with 1.93% Sn + 0.61% Si after age-hardening gave a strength of 10.8 kg/mm^2 but at 125 °C it had fallen to 6.5 kg/mm^2. On the other hand, the alloys with Ba and Ca retained their strengths up to heat treatment temperatures of 150 °C, as did also an alloy with 0.2% Cd + 0.5% Ag (3.5 kg/mm^2). Higher heat treatment temperatures up to the commencement of melting increase the strength of the age-hardening alloys. The highest strength values are obtained when the alloys are quenchcast from the molten state. This is in agreement with earlier results. It would be worthwhile to continue the age-hardening experiments on dilute ternary alloys.

42. Concluding Remarks

In the foregoing section the ternary systems, which were principally considered, were those whose structure could be represented on a diagram. Minor observations, made on other ternary systems, however, are scattered throughout the literature. For example, reference may be made to JOLLIVET's data [630] on the lead-arsenic-zinc system, and to the lead-antimony-sodium and lead-arsenic-sodium systems, the study of which was stimulated by the metallurgical researches of KROLL [710] and BETTERTON and LEBEDEFF [77] (p. 8). SILINA and PONOMAREW [1121 a] investigated the bismuth corner of the ternary diagram lead-magnesium-bismuth. The section Pb–Mg_3Bi_2 is semi-binary and contains a eutectic $\alpha + Mg_3Bi_2$ at 294 °C. The temperature of the ternary eutectic $\alpha + Mg_3Bi_2 + Mg_2Pb$ is 248 °C. For the ternary system WEINSTEIN and ELLIOT gave the isotherms of the liquidus surface as ranging from 650 to 1000 °C [1254 a]. The results of RAYNOR and GRAHAM [997] on the lead-indium-thallium ternary system may also be mentioned, or the communication by STEIDLITZ [1144] on the formation of a nickel-germanium phase when nickel and germanium are soldered. Finally, a thermodynamic study of the lead-oxygen-sulphur system by KELLOG and BASU [653] can be reported. Only a small number of the ternary systems covered, represent technically important alloys, but most of the systems are of interest because they contain combinations of elements which can occur as deliberate or unintentional additions to the technical alloys. In the absence of more accurate knowledge of multi-alloy systems the relevant ternary systems can give an indication. Further, the ternary systems are of significance in understanding lead smelting, the best example being the PARKES zinc desilverisation process. A further example is provided by DAVEY [235a], who used the ternary lead-antimony-copper, lead-arsenic-copper, lead-copper-sulphur and lead-copper-tin systems as a substitute for the sexenary system lead-antimony-arsenic-copper-sulphur-tin, and related the ternary system to the copper-drossing of lead.

FEDEROW and co-workers [307a] prepared a series of alloys of lead, bismuth and calcium in a closed vessel and brought the alloy to equilibrium at 400 °C, i.e. the crystals which formed were floating on a melt. The structure was frozen by quenching the vessel and the composition of the upper and lower layers were analytically determined. The tie lines of the related compound of the upper and lower layers showed graphically the composition of the crystals on the bismuth-calcium side of the system. The composition of the melts (lower alloy layer) and the crystals which were formed ($CaPb_3$, Ca_3Bi_2, $CaBi_3$) could be distinguished on the isothermic section at 400 °C. For example, if calcium, to an extent of 0.21%, is added to a melt of lead containing 0.2% Bi, then, at 400 °C, a residual melt E_1 having only 0.065% Bi, results. The work does not show any exact ternary system but illustrates the removal of bismuth from the lead by means of calcium.

FEDEROV and co-workers [307b] set up a ternary system lead-magnesium-bismuth in broad terms. It breaks up into three ternary partial systems, Pb–Bi–$BiMg_3$, Pb–$PbMg_2$–Bi_2Mg_3, and $PbMg_2$–Mg–Bi_2Mg_3. A ternary phase was found in the section Pb–Mg_2–Bi_2Mg_3. The section Pb–Bi_2Mg_3 contains a quasi binary eutectic at 98% Pb at about 300 °C. The lead corner of the diagram was not closely examined. In view of this, therefore, further details of the work are not given here. No section on quaternary systems is provided, but the lead–cadmium-bismuth-tin (PARRAVANO and SIROVICH [936]) system investigated is mentioned in the section on low-melting alloys (p. 388). Indications of the presence of a quaternary eutectic are given by JÄNECKE [615] for the systems lead-cadmium-zinc-tin (137.0 °C), lead-bismuth-zinc-tin (93.5 °C) and lead-cadmium-bismuth-zinc (90.9 °C). Since the zinc content of the quaternary eutectic is only small, the solidification temperature is only a few degrees below that of the ternary eutectic without zinc. The quaternary eutectic lead-cadmium-zinc-tin was confirmed by LINDNER [751], obviously without knowledge of the work of JÄNECKE [615].

In concluding this section dealing with alloys, it should also be mentioned that lead is dissolved in small quantities by silicate slags. It is suspected that neutral lead atoms fit into vacancies in the silicate network (MEYER [849a]). The importance of the quinary system Bi–Cd–Pb–Sn–Zn for recovering wastes of low melting alloys by fractionation has been shown by SPENGLER [1141a].

IV. Specific Properties of Lead and its Alloys

1. Straining and Recrystallisation

a) Deformation of Lead: General. The deformation of cubic face-centred metals, to which group lead belongs, takes place by slip on the octahedral planes {111}. There are six directions of slip ⟨110⟩ in each of four possible octahedral slip planes. The shear stress necessary for the inception of slip, SCHMID's [1070] critical shear stress τ_0, is in

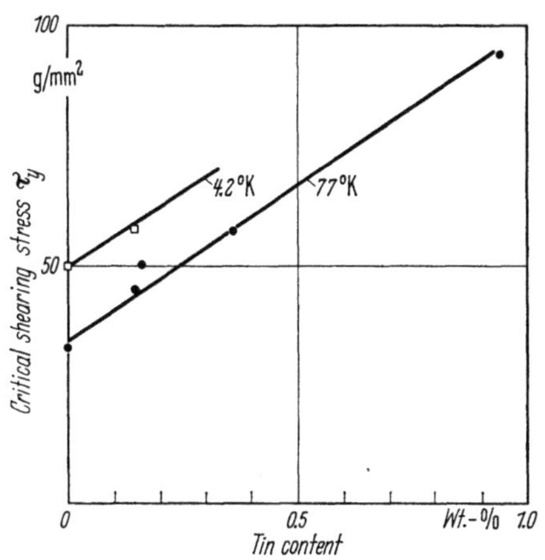

Fig. 189. Influence of a dissolved tin content on the critical shear stress of lead single crystals at 4.2 °K and 77 °K. (According to FLEISCHER [327]).

the region of 20 to 500 g/mm² (SEEGER [1100]), in metals with cubic or hexagonal close packing. In the creep tests on lead single crystals described below, the present author and LEY [566] found critical shear stresses between 56 and 71 g/mm² at 20 °C; a communication by DEMING [244] completely agrees with this. At −190 °C, NEURATH and KOEHLER [895] determined a value of 96 ± 8 g/mm² for τ_0. From very divergent results of measurement on 99.99% lead, FELTHAM and MEAKIN [313] determined the temperature dependence of the critical shear stress. According to the latest measurements by FLEISCHER [327] on very pure lead with only 2 ppm by weight of impurities, the critical shear stress amounts at 77 °K to only 28 g/mm², at 4.2 °K to 51 g/mm². Dissolved tin raises it as shown in Fig. 189.

According to the present state of knowledge, the slip process consists of a migration of dislocations (dislocation lines) on the slip planes. While the dislocation line sweeps over a certain surface, the atoms above and below are displaced with respect to one another by an elementary slip step (Burgers vector). The Burgers vector is, in the simplest case, one lattice vector of the atoms. In that section of the dislocation line where it is perpendicular to the Burgers vector, the dislocation appears as an edge; where the dislocation line runs parallel with the Burgers vector, there is a screw dislocation. The other parts of the dislocation line are combinations of edge and screw dislocations. Pure edge and screw dislocations are shown in Figs. 190 and 191.

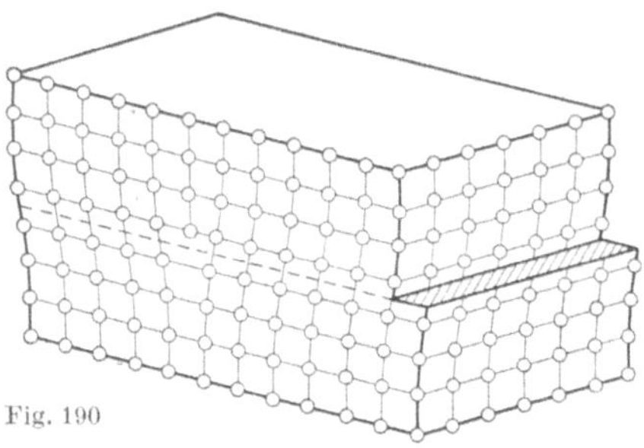

Fig. 190

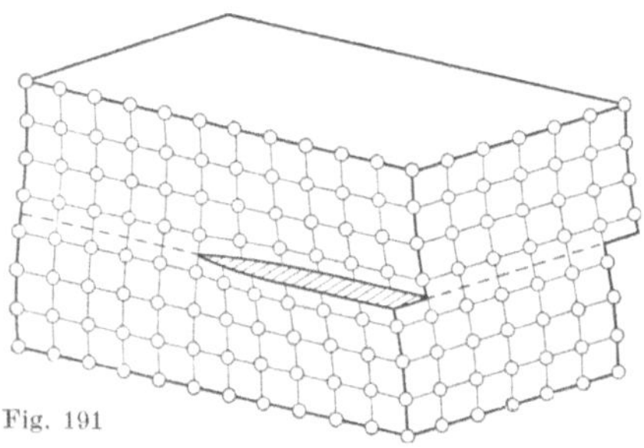

Fig. 191

Fig. 190/191. Edge dislocation (190) and Screw dislocation (191).

The edge dislocation can be regarded as though the crystals were cut to a certain depth and an additional half lattice plane pushed into the forcibly widened gap. The line of lattice points at the end of the half plane is the dislocation line. Apart from the great disturbance of the lattice structure in the immediate neighbourhood of the dislocation line, the upper region in Fig. 190 is under internal compressive stresses, the lower part under internal tensile stresses. The plane formed by the dislocation line and the slip vector is the slip or translation plane. According to whether the above-mentioned half plane is above or below the slip plane, reference is made to a positive ($\perp$) or to a negative ($\top$) dislocation. From their initial formation the metal crystals contain a spatial system of branching dislocation lines, which is the basic structure. The dislocation density is 10^4 to 10^8 dislocations/cm^2.

While a dislocation loop, under the action of applied shear stress, spreads to the crystal boundaries and is terminated there, new dislocations (which also spread and end) are formed inside the original dislocation line from the Frank-Read sources. The Frank-Read sources are dislocation lines which issue from the slip plane at two fixed points. This mechanism therefore makes it possible that in one single slip plane larger slip paths are traversed, which can be made visible on the crystal surface. For more detailed study of dislocation theory and its experimental foundations, attention is particularly drawn to the surveys by

COTTRELL [217], READ [998], SEEGER [1099, 1100], FRIEDEL [341] and VAN BUEREN [141].

In the cubic face-centred metals, the curves of stress-strain and of shear stress-glide respectively, show the three sections indicated in Fig. 192. In the first section, the region of easy glide, almost only one

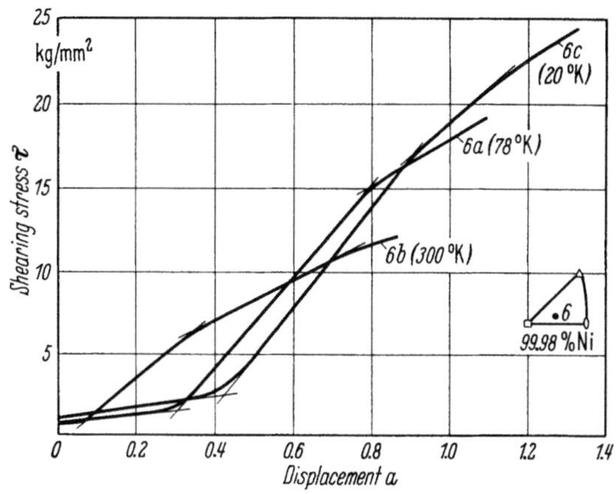

Fig. 192. The temperature dependence of the strain-hardening curve of nickel crystals of uniform orientation. (According to HAASEN [1100]).

slip system is active. The magnitude of region I is therefore very dependent on the orientation of the crystal. The dislocation loops move without much hindrance through the network of dislocations. This explains the small increase of shear stress. Electron microscope observations of the surface are not available for lead; in the other face-centred metals they show uniform fine slip.

Only suggestions of easy glide were found by FELTHAM and MEAKIN [313] in lead crystals. FLEISCHER [327] attributed this to the insufficient purity of the crystals used. He found a marked region of easy glide (Fig. 193) in the stress-strain curves of lead crystals with only 2 ppm of impurities. This region is enlarged by dissolved impurities and the strain-hardening in the region of easy glide reduced. Insoluble impurities such as copper or aluminium increase the strain-hardening and, as shown in Fig. 193, even in very small amounts will promote the disappearance of easy glide. Light and electron-microscopic traces of secondary slip systems were observed by various authors in region II of the strain-hardening curve of face-centred metals. Dislocations on intersecting slip planes of the cubic face-centred lattice react with one another and become immovable (sessile) LOMER-COTTRELL dislocations. The LOMER-COTTRELL dislocations of different orientations therefore act as obstacles to the movement of dislocations on the octahedral slip planes and are the main cause of the marked strain-hardening in part II of the strain-hardening curve, independent of soluble impurities. The rise of the strain-hardening curve in this region amounts to about 1/300 of the shear modulus, independent of the temperature. In part III the increase of strain-hardening is less than in part II. On the strain-hardening, ac-

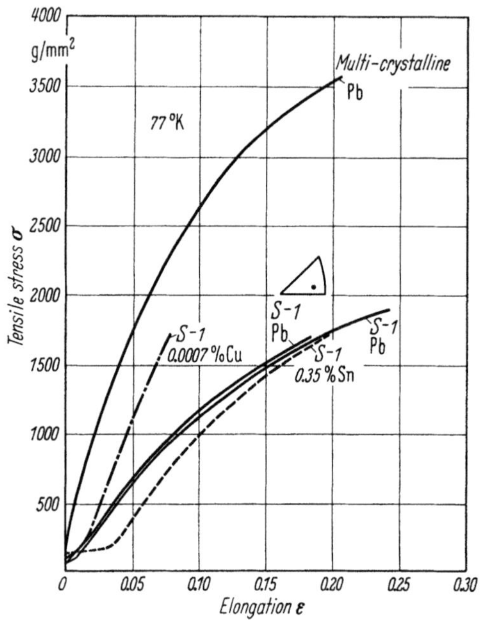

Fig. 193. Strain hardening curves of crystals of very pure lead and of lead with additions of tin and copper at 77 °K. (According to FLEISCHER).

cording to the above-described mechanism, there is now superposed a dynamic recovery due to cross slip or climbing (p. 184). By the former, screw dislocations are by a double change of the slip plane, displaced into a new slip plane, which is parallel to the original one; in this way they can avoid obstacles (circles in Fig. 194). Cross slip leads to an accumulation of fine slip lines into the so-called slip bands. The remaining edge dislocations can arrange themselves in cell walls by climbing, which results in fragmentation of the grains. The transverse slip is thermally activated and therefore occurs, with rising temperature, at lower degrees of deformation. The fact that the transverse slip of lead is very marked at room temperature, is connected with the magnitude of the temperature of deformation relative to the melting temperature, and, further, with the relationship of the shear modulus to the stacking fault energy. Stacking fault energy means the energy of one of the sequence of close-packed octahedral planes deviating from the cubic close-packing, e.g., at the twinning joint of recrystallisation twins and between two partial dislocations of an octahedral slip plane. According to the most recent investigations by BOLLING and co-workers [102c, 103] the stacking fault energy of lead is not so high as was originally assumed on the basis of the polyvalency of lead. The value of $\gamma = 25$ erg/cm^2 is of the order of magnitude of the stacking fault energy of silver. The energy of twin boundaries for lead amounts to 10, for silver less than 15, but, on the other hand, for aluminium to approximately 120 erg/cm^2.

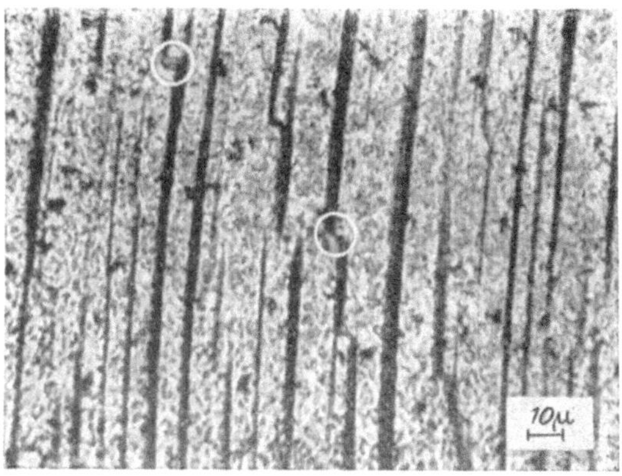

Fig. 194. A single crystal of lead after a tensile test at room temperature. Slip bands and cross slips. Stress = 86 kg/cm^2; elongation = 18%. (According to FELTHAM and MEAKIN).

That the stacking fault energy of lead is lower than that of aluminium explains the fact that in aluminium hardly any recrystallisation twins are found, but these are quite abundant in lead. The views hitherto expressed relate to the behaviour of single crystals. In the plastic flow of a polycrystalline material, the individual grains must conform to the change of shape of the macroscopic body. Simultaneously the coherence of grains in the grain boundaries must be preserved. This is only possible if several slip systems were active in the crystallites from the beginning of flow. There is therefore no region of easy glide in polycrystals. The conditions laid down can, further, only be fulfilled if additional deformations occur in the neighbourhood of the grain boundaries. Therefore part of the yield stress of a polycrystal, σ, is derived from the individual crystals σ_c, and another part σ_b from the grain boundaries: $\sigma = \sigma_c + \sigma_b$. The part from the grain boundaries is the greater, the smaller the grain diameter. Therefore the yield stress of a fine-grained material is higher than that of a coarse-grained. However the yield point is not of significance in the application of lead as a constructional material; the creep strength is of greater importance. This depends on the grain size in a manner different to the yield point given in tensile tests.

In addition to the surface occurrence of slip bands, the deformation of lead single crystals, on the basis of crystal translation, is evident by the crystals assuming an elliptical cross section. Slip lines in deformed polycrystalline lead were revealed on the surface by gentle bending of cast layers on glass plates (VOGEL [1228]). They also appear in etched sections of highly deformed specimens (stable to recrystallisation) and bent in the usual manner (bending slip) (Fig. 195). KHOTKEVICH and coworkers [660] measured the potential energy accumulated in the deformation of lead at about −196 °C. At 65.8% deformation the accumulated energy amounted to 0.53 cal/g. The ratio of accumulated energy to work of deformation was surprisingly almost 1:1 at low degrees of deformation, i.e., the total work applied was accumulated as potential energy of defects of the crystal structure. These results do not agree with general experience; compare, for instance, CLAREBROUGH and coworkers [198]. After adiabatic compression at 30 °C, WILLIAMS [1274 b] found, within a time of 0.01 sec, accumulated energy of more than 10 cal/gram atom; that is, up to 0.05 cal/g. After one minute only one-third of it still remained. These results, too, should be applied only with reservations.

Fig. 195. 2% Sb. Rolled to 50% after being homogenised and quenched. Sample taken immediately. Stretched structure with slip lines. ×160

b) Textures. In extruded wires and rolled foils of age-hardened lead alloys, which were stable to recrystallisation, it was possible to identify textures which largely corresponded to those known in metals with face-centred structures (HOFMANN [554, 555]). Thus in extruded antimonial lead wires a fibre texture was found in which generally the space diagonal was in the direction of the wire axis, less frequently the edge of the elementary cube. In recrystallisation of the wires, a recrystallisation texture developed which can be described in the same way. In antimonial lead foils a rolling structure was found, as illustrated by the pole figures of the cubic, octahedral and dodecahedral planes in Fig. 196. It is very similar to the corresponding texture of aluminium foils (WASSERMANN [1240]) and corresponds as a first approximation to the alignment of a cube surface in the rolling plane and of a surface diagonal of the elementary cube in the direction of rolling (Fig. 197).

Another, barely age-hardening, type of antimonial lead gave a completely different rolling texture, as is perhaps shown by comparison of

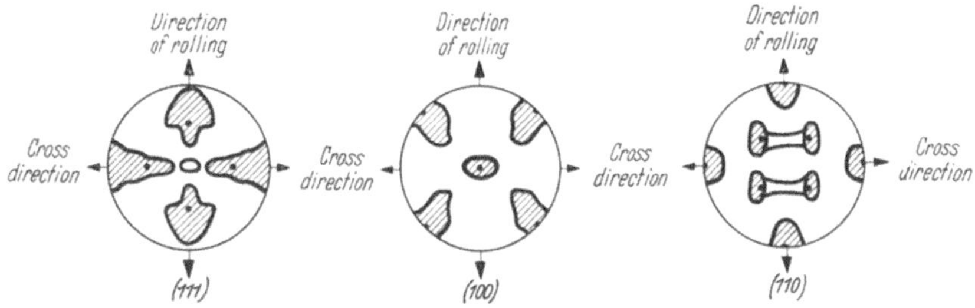

Fig. 196. Pole figures of rolled antimonial lead foil.

the two X-ray pictures in Figs. 198a and b. The second rolling texture, which after a short time changed to a similar recrystallisation texture, might perhaps correspond to the so-called cube position, which is obtained from Fig. 197 by rotation of the cube round the perpendicular to the foil at 45°. The rolling textures of other lead alloys appear to be combinations of the two orientation rules described. The lead-tellurium (0.05 Wt.-%) and lead-calcium (0.05 Wt.-%) alloys, after rolling reductions of 99.7%, showed the {011} ⟨211⟩ texture usual in cubic face-centred metals (Hirst [531]). A recently published investigation by Toman and Simerska [1203a] on textures of lead and their derivation from the elements of plastic deformation was not available to the present author before publication.

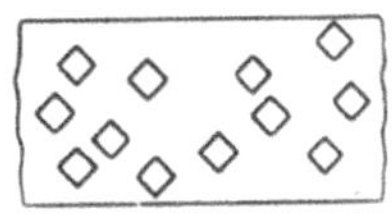

Fig. 197. Ideal representation of the crystals in a newly rolled antimonial lead foil.

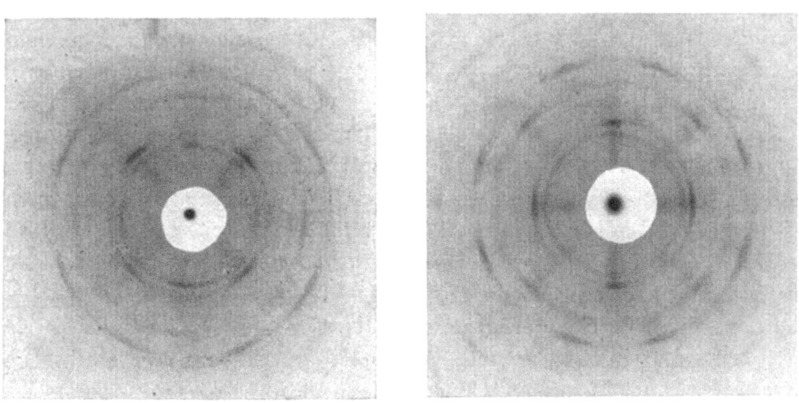

a b

Fig. 198a and b. Debye-Scherrer photographs of antimonial lead foil by transmitted X-rays (Molybdenum source).
a) Arsenical, very hard antimonial lead; b) arsenic free, slightly hardened antimonial lead.

c) The Recovery and Recrystallisation of Lead: General. The reappearance on annealing of the physical and mechanical properties of a metal altered by cold deformation takes place in several stages. In general a distinction is drawn between recovery and recrystallisation, according to whether the change in properties takes place with new grain development or not. LUNDQUIST [783 c] measured hardness versus temperature. The recrystallisation properties give an estimate of the purity of the lead. BOESONO [101] reported on the lowest stage of recovery hitherto observed on lead. He measured the increase of electrical resistance due to the rolling of lead at 90 °K, and its decrease with rising temperature of anneal. Ordinary commercial lead showed recovery stages at —130 to —100 °C and at —40 to —10 °C. In pure lead the stages were not clearly separated. The energy of activation for the first stage of recovery amounts to 0.48 eV (11,100 cal/gram atom) for both kinds of lead. It is attributed to the disappearance of vacancies. The next stage with an activation energy of 23,040 cal/gram atom has already been associated with recrystallisation.

Two processes are principally involved in the recovery of mechanical properties. If the amount of deformation is small and the deformation, therefore, very uniform, positive and negative dislocations are present together in about equal number. Pairs consisting of a positive and a negative dislocation mutually attract one another and thereby disappear (BURGERS [153]).

The process is made possible by thermally activated climbing, i.e., the inserted semi-plane illustrated in Fig. 190 is elongated or shortened by addition or loss of atoms. A further elementary process of crystal recovery is polygonisation. If, in a deformed volume, dislocations of one sign are heaped up by impinging on an obstruction, they leave their original slip plane by thermally activated climbing, and arrange themselves upon one another in grain sub-boundaries, as this arrangement is favourable and stable from the energy point of view (SEEGER [1100]; Figs. 199a and b). Only in single crystals does recovery result in complete loss of strain-hardening; in polycrystalline materials complete loss of strength occurs only on recrystallisation. Of course the ability to recover is pronounced in certain lead alloys even in the polycrystalline condition.

Recrystallisation is nowadays defined in general according to BECK [67a] as the formation and movement of large angle grain boundaries (LÜCKE [776a]). This definition embraces:

1. Formation and growth of nuclei in primary recrystallisation. Growth of nuclei is ended when the deformation structure is completely absorbed.

2. Grain growth after completed primary recrystallisation. Normal grain growth takes place particularly on increase of temperature.

Sometimes grain growth occurs, not as a rather uniform increase of size of crystallites, but as a remarkably intense crystal growth occurring in individual places. The primary recrystallised structure first remains fine-grained, until it is absorbed by the few growing grains. This process is described as discontinuous grain growth or secondary recrystallisation. It is not typical of lead and lead alloys.

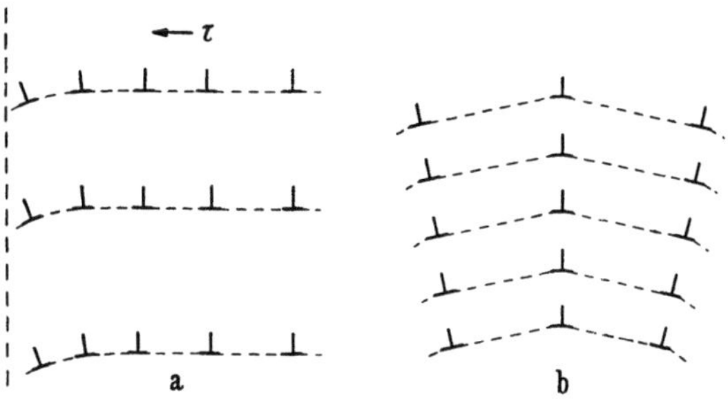

Fig. 199a and b. Atomic interpretation of polygonization.
a) The piling up of edge displacements in separate slip planes as a result of plastic deformation.
b) "Polygonization" of the slip planes by climbing of the dislocations shown in a). (According to SEEGER).

3. Grain boundary movement in slightly deformed or undeformed materials.

The problems of nucleus formation are very difficult and only partly elucidated. A very good review is given by BURGERS [153a]. According to BECKER [153a] the critical nucleus of an undeformed lattice region is of such a magnitude that equilibrium exists between the required interfacial energy and the gain in volume energy. An estimate of the activation energy necessary for such a process, on the basis of known data, gave the improbably high value of 10^5 kcal/gram atom. This explains the widely held view that the orientations of the lattice regions, acting as nuclei, are already present in the deformed metal, and indeed, in the places of highest deformation (BURGERS [153]). However there is also the opposite opinion that the sections of lattice acting as nuclei are the regions of minimum deformation. The basic concepts for the quantitative comprehension of recrystallisation are the number "N" of nuclei formed in the unit of time and in the unit of volume of unrecrystallised metal, and the linear rate of nuclear growth G in cm/sec. Measurements and calculations, with a view to obtaining ideas about these magnitudes, were made, particularly on aluminium, by KORNFELD [698] as well as by JOHNSON and MEHL [626]. LÜCKE [775] summarised these studies and in particular amplified the calculations made by JOHNSON and MEHL, in which a relationship was sought between the basic magnitudes mentioned and the development of structure. Here LÜCKE, in agreement with the experiments, assumed constancy of time in the rate of nuclear growth, G; on the other hand, the rate of nucleus formation N depends on the time in the manner discussed below. The calculation by Lücke [775] shows that there is a mathematical relationship between measurable structural magnitudes (e.g., the fraction of the structure already recrystallised) and the value $G \cdot N$. A separation of the basic magnitudes G and N is therefore not directly possible. The determination of $G \cdot N$ from structural magnitudes is however so inexact that it possesses no great practical value. Direct experimental determinations of the basic magnitudes were made on aluminium by Kornfeld [698], and recently in particular by Anderson and Mehl [19], and thoroughly discussed by Lücke [775]. According to this, the rate of formation of nuclei is dependent on time; it sets in after the expiration of a certain time (the period of incubation), reaches a maximum, then fades away to the value 0. The period of incubation sinks exponentially with rising temperature according to the equation $\frac{1}{\tau} = A_\tau \cdot \exp \frac{-U_\tau}{RT}$ and falls further with increasing degree of deformation. In the experiments with aluminium it was in the range of 2 minutes to 3.6 hours. Formation of nuclei and growth of nuclei are temperature-activated processes and their velocities therefore obey respectively the equations

$$N = A_N \cdot \exp \frac{-Q_N}{RT} \quad \text{and} \quad G = A_G \cdot \exp \frac{-Q_G}{RT}$$

As is to be expected, both activation energies fall with increasing degree of deformation. According to Lücke [775], the growth of nuclei does not take place as a sort of diffusion, but on the contrary, whole regions of lattice can be built into the new lattice with very great speed by one single activation process.

Kennedy [655] carried out creep tests on pure lead with interruptions. In the rest periods, part of the material recovered, another part recrystallised. He developed a theory of recrystallisation, according to which nuclei are formed by a recovery process, and derived from his measurements and calculations an activation energy of grain growth of 31 kcal/gram atom. He gave 19.8 kcal/gram atom as the activation energy for crystal recovery.

d) Movement of Large Angle Grain Boundaries in Recrystallisation. The velocity of grain boundary migration G can, according to Beck [67a], be equated to the product $B \cdot K$. B represents the grain boundary mobility and K the driving force. In primary recrystallisation, K is given by the line stress of the dislocations, on subsequent grain growth by the grain boundary stress. The grain boundary energy falls with increasing grain size. It can therefore be understood that with increasing grain size at constant temperature the velocity of grain boundary migration also falls.

BECK [69] proposed for the relationship of grain growth with time a law of the form $D = C \cdot t^n$ in which D signifies the grain diameter, C is a constant dependent on temperature, t the time and n an exponent of magnitude 0.5 approximately.

BOLLING and WINEGARD [104] carried out recrystallisation experiments on zone-refined lead. The specimens were deformed with a drop hammer. After annealing at various temperatures, the grain diameter was determined optically. The authors derived from their measurements a value of 0.4 for n. Further they evolved the expression $K = K_0 \exp -Q/RT$ in which K_0 signifies a constant of the dimensions of a diffusion constant, and, as a result of their experiments, obtained an activation energy of grain growth of 6.7 ± 0.7 kcal/gram atom. This value agrees well with that of 6 kcal/gram atom determined by AUST and RUTTER [30a] by another method. AUST and RUTTER found the velocity of grain boundary motion to be independent of time.

The beginning and the course of recrystallisation are very strongly influenced by impurities. According to measurements on copper, contents of a few multiples of 0.01% can change the rate of recrystallisation by many orders of magnitude. MASING, LÜCKE and NÖLTING [802a] have plausibly suggested that the greatest delay in recrystallisation is to be expected from elements whose atomic radius differs markedly from that of the basis metal. These addition elements become enriched in the grain boundaries (equilibrium segregations). In certain cases the equilibrium segregations can be detected by means of radioactive isotopes (THOMAS and CHALMERS [1187]). The grain boundaries become less mobile, owing to the accumulation of foreign atoms. LÜCKE and DETERT [777] developed this idea into a quantitative theory. According to this the rate of movement v of a grain boundary is proportional to C^{-1} if C signifies the concentration of the dissolved alloying metal.

AUST and RUTTER [30] tested this theory in the case of lead crystals with additions of tin. Single crystals from the melt were not uniform and contained, in the cell, structure differences of orientation of individual "streaks" of several angular degrees. By a local deformation, recrystallisation was induced at one end. From here, a grain boundary moved into the undeformed "streak" structure and absorbed it. Measurements of the rate of movement of the grain boundary in relation to the concentration of the dissolved tin showed that v is proportional to $C^{-5/2}$. By increasing the tin content from 0.0004 to 0.006 Wt.-%, v decreased not by 1/10 but by 1/1,000. The experiments further showed that special orientation relationships, between the growing crystal and the one to be absorbed, resulted in a rate of grain boundary movement multiplied many times. The activation energy of grain boundary movement amounted in this case to only 6 kcal/gram atom, independent of the tin content (between 0.0005 and 0.02 Wt.-%) [30a]. In arbitrarily orientation relationships, the activation energy increased from 15 (0.0001% Sn) to 43 (at 0.001% Sn) kcal/gram atom. According to the authors, the low value of 6 kcal/gram atom for the activation energy of grain boundary movement of the purest lead indicates that the grain boundary movement is due to the activation of individual atoms (single process theory) and not to the movement of groups of atoms. The special orientation relationships observed could be expressed by a rotation of the two crystals about a common ⟨111⟩ axis by 36 to 42°, by a rotation of the two crystals about a common ⟨111⟩ axis by 23°, and by a rotation about a common ⟨100⟩ axis by 26 to 28°. In the purest lead, as well as in lead with additions of silver and gold [1039a], the mutual orientation of the grains had only a small effect on the rate of movement of the grain boundaries; on the other hand, the special orientations mentioned led, at a tin content between 5 and 40 ppm Wt., to a rate of movement multiplied one hundredfold [31]. In experiments with copper, Kronberg and Wilson [713] found that similar orientation relationships were active. The large angle grain boundaries mentioned, with special orientation relationships between the neighbouring grains (coinciding boundaries), are distinguished by a low value of the grain boundary energy in comparison with non-special large angle grain boundaries (random boundaries). It may be assumed that dissolved atoms in random grain boundaries are more highly enriched (equilibrium segregation) than in special grain boundaries. Thus the greater mobility of the special grain boundaries can be fundamentally understood. The enrichment of dissolved atoms in the grain boundaries and the difference in enrichment in random and in special grain boundaries were demonstrated by Westbrook and Aust [1262a] by microhardness measurements. The elements dissolved in lead were In, Sn, Ag, Au.

The enrichment of dissolved atoms in oriented grain boundaries is not yet incorporated in the theory of Lücke and Detert [777]. The theory ought to be tested by further experiments on lead alloys of the highest purity. A fundamental understanding of recovery textures should also emerge from the new work. On page 183 it is shown that the recovery texture of antimonial lead with 1% Sb changes completely if the lead-antimony alloy contains an addition of a few multiples of 0.01% As. Presumably here a mutual action similarly takes place between the arsenic and certain grain boundaries, as a result of which their rate of movement is influenced. Bolling [102a] found that there was no longer any grain boundary migration in zone-refined lead below 168 °C, but on the other hand observed it, if this lead was subsequently remelted under hydrogen. From this he concludes that grain boundary migration is inhibited by oxygen (cf. GIFKINS [378]).

e) Observations on Strain Hardening and Softening. The increase of strength of lead by compression was investigated by MOLNAR [862] in the temperature range —75 to 50 °C. The degree of compression amounted to 25 and 50%. Strain-hardening, detected by hardness measurements, still occurred at the highest temperatures used, but disappeared after

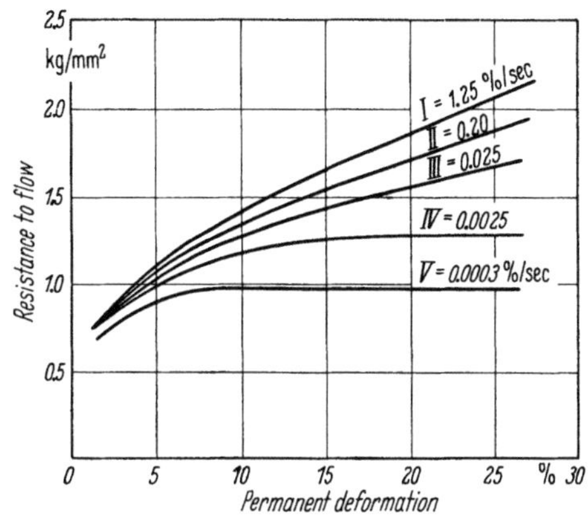

Fig. 200. The flow of soft lead plotted against the variation of the rate of change. (According to HOFF and DAHL [550]).

5 and 10 minutes respectively, in the weakly and strongly compressed specimens. At room temperature, softening required 25 and 64 hours respectively, at —20 °C it was still incomplete after 96 hours, at —40 °C and below, it was completely absent. The dependence of softening on time has the result that the resistance to flow of lead in the compression test increases with increasing rate of deformation (Fig. 200).

Similar experiments were carried out by NORBURY [904] on 99.99% lead at room temperature. Softening occurred far more rapidly than in the tests described above, for instance, in a few minutes at 25% compression. This may have been particularly the result of the greater purity of the lead under investigation (cf. below).

MIMA and co-workers [857] examined the strain hardening and the softening of lead-tin alloys of varied tin content in the bending test. In alloys containing the eutectic, they found a loss of hardness with increasing degree of deformation, whereas lead and its solid solutions first showed a certain hardening. Strain-hardening and softening were also demonstrated on twisted wires by recording torsion curves with intermediate periods of no load (LUDWIK [772]) and by electrical resistance measurements (TAMMANN and DREYER [1168]).

The strain hardening of lead is also seen in the tensile test, through the rise of the load-elongation curve. The softening which simultaneously occurs at room temperature results in the rate of test considerably affecting the tensile strength and the elongation (p. 198). The occurrence of recrystallisation can also be discerned, in part, in the tensile test diagrams of Fig. 210. Interrupted tensile tests by CHASTON [192] were particularly clear; here after attaining an elongation of, say, 15% a rest period of, say, 12 minutes was successively applied. In this way elongations at fracture above 300% were obtained. Rest periods of less than 4 minutes duration had little effect; by prolonging the rest period above 12 minutes the elongation was not increased any further.

f) The Recrystallisation Threshold. Additions which Inhibit Recrystallisation. In this section test results are described (disregarding exceptions) in which the lead used was not of the degree of purity now available. The tests therefore (in part) require repetition. LOOFS-RASSOW [765] determined the appearance of recrystallisation from measurements of the softening of lead (cf. MAYER [813a]). As the effect of crystal recovery was neglected here, the curve shown in Fig. 201 can only be regarded as a rough approximation. Moreover, such a curve is strictly valid only for a specified grade of lead and a specified grain size. On increasing the degree of purity, the beginning of recrystallisation is displaced to lower temperatures and less amounts of work.

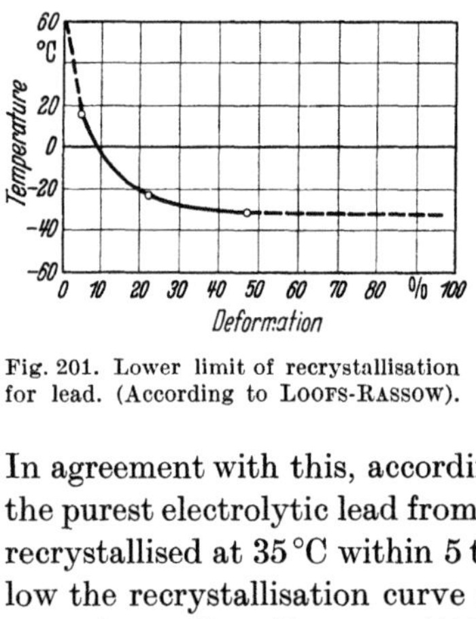

Fig. 201. Lower limit of recrystallisation for lead. (According to LOOFS-RASSOW).

In agreement with this, according to observations under the microscope, the purest electrolytic lead from Melbourne University, strained by 2.5%, recrystallised at 35 °C within 5 to 10 minutes, but also at 12 °C (i.e. far below the recrystallisation curve shown in Fig. 201) within 45 to 120 minutes after rolling (RUSSELL [1039]). Recrystallisation was complete after five minutes at a rolling reduction of 5%. A content of 0.05% antimony delayed the end of recrystallisation to 21 days. If this alloy was homogenised, then the delay, as compared with pure lead, was essentially less. A decrease in grain size favours the occurrence of recrystallisation, as is known also in other metals (JONES [632]).

JENCKEL and HAMMES [618] tested the effect of a large number of alloying elements, mainly in the concentration of 0.1%, on the softening temperature of rolled alloys (61% reduction). They assumed that here the softening and recrystallisation temperatures were approximately the same. An increase of softening temperature, as compared with pure lead, was found for additions which with lead formed intermetallic compounds of brittle character, that is, platinum, palladium, selenium, gold, barium, lithium, potassium, sodium, magnesium, calcium and tellurium. The action of the additions was in general the more marked, the higher the melting point of the intermediate phase. Additions of zinc, mercury, thallium, tin, antimony and bismuth, which with lead form no intermediate phase of brittle character, did not raise the recrystallisation temperature. The effect of the first-named alloying elements was not increased further by contents above the solubility limit.

Even though this regularity is generally true, some observations may be mentioned which do not fit into the pattern. Thus silver, which was not included in the investigation, delays the recrystallisation of lead in a particularly marked way (Fig. 202). According to BOLLING and WINEGARD [105], the activation energy of grain growth of zone-melted lead is increased from the value 6.7 to 9.7 kcal/gram atom by a silver content of 0.005%: according to RUTTER and AUST [1039a] the activation energy of grain boundary migration of the purest lead (6 kcal/gram atom) increases to the value of 33 kcal/gram atom at a gold or silver content of about 1 atom ppm. This discrepancy in the quantitative results is discussed by BOLLING [102b]. In this connection attention should be given to the investigations of HEIDENREICH [503] on the lead-silver diagram. The grain-growth of lead is even more delayed by addition of gold than by silver (Fig. 203). Further, the recrystallisation is greatly inhibited by larger concentrations of antimony, so that in these alloys the course of recrystallisation with time could be demonstrated particularly well (HOFMANN, SCHRADER and HANEMANN [577]). The photomicrograph of a specimen, rolled to 50% reduction, showed, two hours afterwards, only a worked structure with clearly visible slip lines. After one day recrystallisation began, particularly at the grain boundaries and slip lines. After twelve days it had covered almost the whole structure (Figs. 204, 205, 206). It is noteworthy that the two last-named metals, which delay the recrystallisation of lead strongly, have higher melting points than the less active additions.

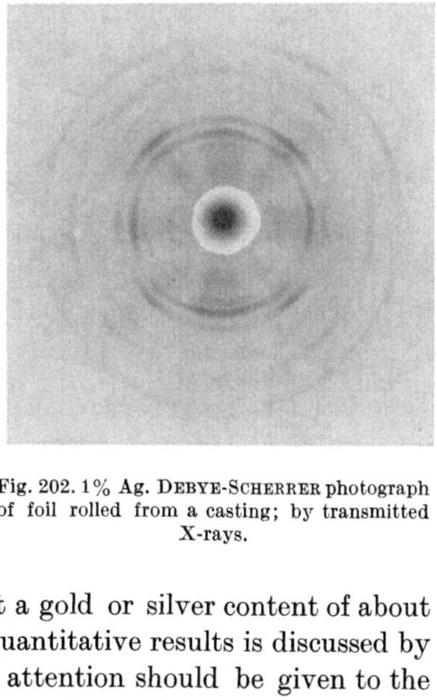

Fig. 202. 1% Ag. DEBYE-SCHERRER photograph of foil rolled from a casting; by transmitted X-rays.

The recrystallisation of the purest binary alloys mentioned on page 31 were studied at the Armour Research Foundation [24a]. The specimens were compressed 50% at the temperature of −75 °C. Subsequently the hardnesses were tested after annealing at various temperatures. The steep fall of hardness is ascribed to recrystallisation. This work confirmed most of the existing results. 99.9999% lead recrystallises at —59 °C which is similar to lead with bismuth. The recrystallisation of lead is delayed in the known way by alloying with silver and tellurium, and in a smaller degree also by alloying with tin. Lead alloyed with

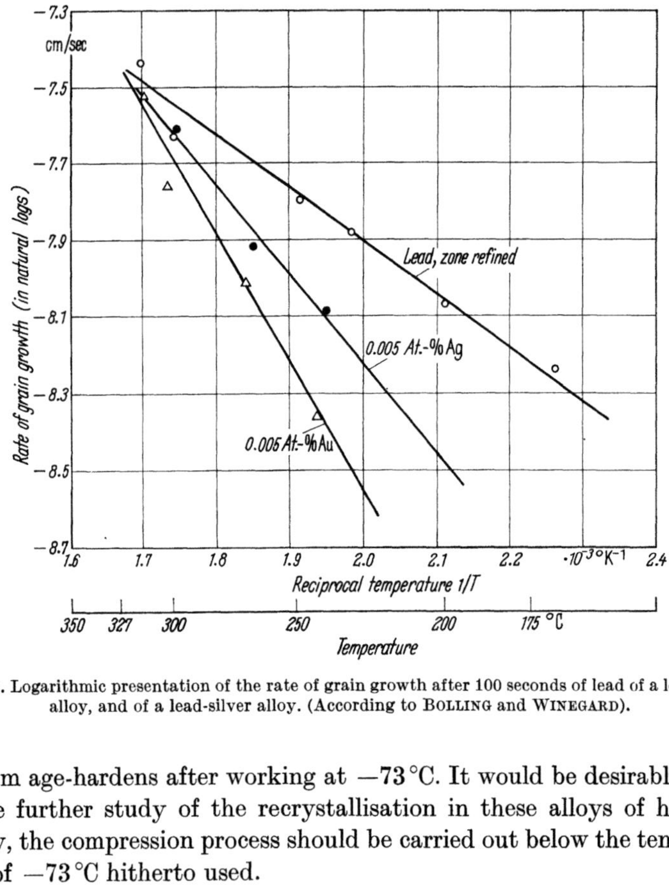

Fig. 203. Logarithmic presentation of the rate of grain growth after 100 seconds of lead of a lead-gold alloy, and of a lead-silver alloy. (According to BOLLING and WINEGARD).

calcium age-hardens after working at —73 °C. It would be desirable that in the further study of the recrystallisation in these alloys of highest purity, the compression process should be carried out below the temperature of —73 °C hitherto used.

Further information was obtained by including age-hardening in the investigation. A particularly pronounced delay in recrystallisation occurs if the alloys are deformed in the age-hardened condition (HOFMANN, SCHRADER and HANEMANN [577], HOFMANN [555], HOFMANN and HANEMANN [561]). In this case a reduction of hardness by crystal recovery takes place during rolling (Fig. 207). It was thus possible to roll foils of age-hardened lead-antimony and lead-calcium to 99.9% reduction, which were stable to recrystallisation for days and months respectively. As they were transparent to molybdenum-radiation, they were useful for the X-ray determination of rolling texture. The behaviour of lead-tellurium, lead-lithium and lead-sodium was similar, though not so marked. If the supersaturation of the alloys and therewith the age-hardening was obviated by slow cooling, then on rolling the specimens, recrystallisation set in immediately. The behaviour of age-hardened alloys is of practical significance, as owing to recrystallisation after slight deformations, such as occur, for instance, in the bending of pipes, a breakdown of the age-hardened condition would occur, combined with softening and formation of coarse precipitates in the structure. Fig. 86 showed this in a non-age-hardened, highly deformed, alloy.

In extruded rods of pure and of alloyed lead a reduction of strength and elongation was observed after storage for several months, which was described as "aging". It is mainly attributed to recrystallisation (BURKHARDT [155]). This assumption is very probable as wires (particularly if extruded at lower temperature) are very fine-grained, and in some circumstances, recrystallise or become coarse-grained after slight subsequent deformation (JONES [632]). The view put forward is supported by the fact that 24 hours annealing at 100 °C gave the same strength values as 340 days of storage at room temperature, and that on further storage only slight changes of the mechanical values occur. In addition to recrystallisation, slight intercrystalline corrosion

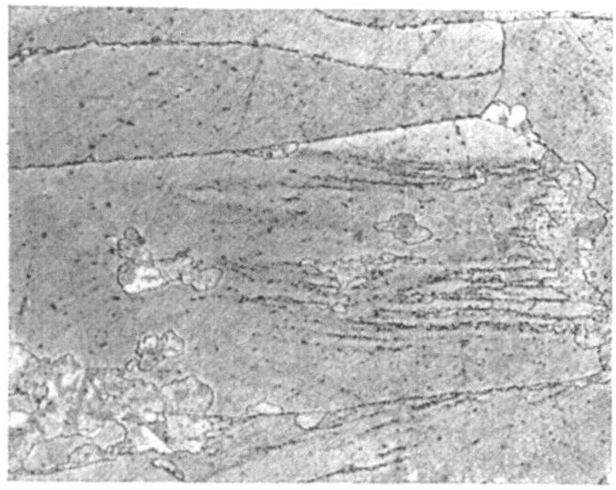

Fig. 204

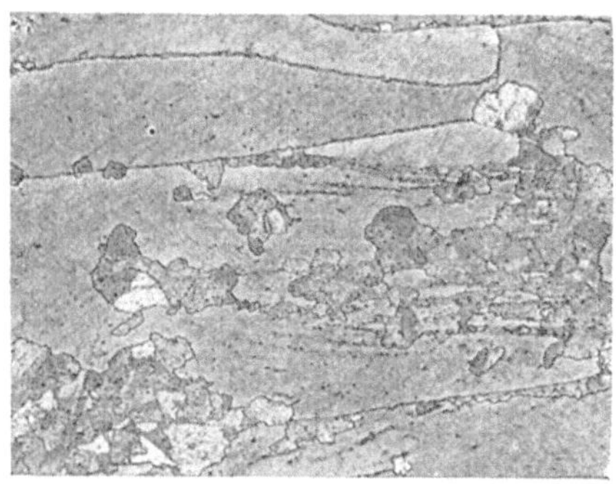

Fig. 205

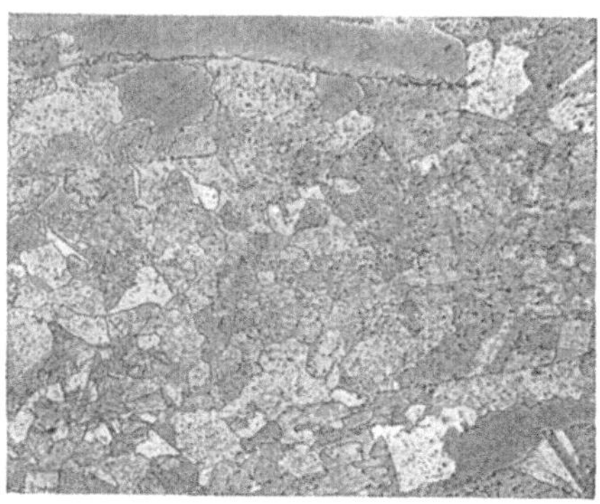

Fig. 206

Fig. 204–206. Microsection as Fig. 195. Same region after 1, 2 and 12 days. ×160

was also taken into consideration, as the aging — obvious, in particular, for pure lead — is slowed down by coating with asphalt. Fundamentally, in supersaturated solid solutions, a breakdown on storage must be taken into consideration as a further cause of the fall in tensile strength, as described on p. 267. The cause of the so-called aging must accordingly be checked individually from alloy to alloy.

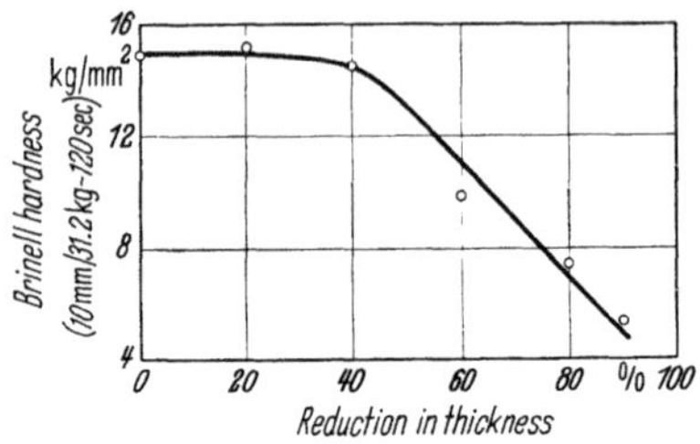

Fig. 207. Alteration in hardness by rolling age-hardened lead-calcium alloys.

g) The Course of Grain Growth with Time. Recrystallisation Diagrams. Grain Coarsening. The duration of recrystallisation at room temperature after various degrees of working, was determined by BECK [68] in an original manner, in that he regarded the time of returning sensitivity towards critical deformation as the end of recrystallisation. The test was carried out by slight local working of the specimen at the time in question, and immediately annealing. If coarse grain occurred, it showed that recrystallisation had ended (see above). In this way curves were prepared for the purest lead (99.999%) and on five other grades of lead. Throughout, increasing amount of work gave increasing rate of recrystallisation (BECK [68]). TAMMANN and CRONE [1164] examined the course of recrystallisation with time at a rolling reduction of 82% by measuring the isothermal grain growth at three different temperatures.

To establish the recrystallisation diagrams of lead, the specimens were in general worked at room temperature and then brought to the desired heat-treatment temperature, as soon as possible. In so doing, room temperature recrystallisation could not always be entirely avoided. This, however, does not affect the grain size after heat-treatment, as was shown by rolling at —70 °C (LOOFS-RASSOW [765]). Neither did longer storage between rolling and heat treatment, except at very small amounts of work, affect the shape of the recrystallisation diagram. Complete recrystallisation diagrams were prepared for the "purest" lead, as well as technical 99.9% lead with copper, iron, antimony, arsenic, zinc and silver as additions (GARRE and MÜLLER [356]), and for electrolytic lead (NA), Parkes lead and Pattinson lead (LOOFS-RASSOW [765]). The recrystallisation diagrams of the three last-named types of lead are reproduced in Figs. 208a—c. In their preparation the investigator proceeded from the cast grain instead of a uniform recrystallisation grain. It may be recorded as a noteworthy result, that Pattinson lead (owing to its copper content of about 0.06%) still shows no grain growth at 200 °C (TAMMANN and DREYER [1166]). However grain growth sets in at higher temperatures, so that at the heat-treatment temperature of 310 °C the differences between the three types of lead disappear. The grain-refining action of a copper addition of 0.04 to 0.05% in cable sheathing (24mm diameter and 2 mm wall thickness) of refined lead and bismuth-containing lead, extruded on the works scale, was further convincingly demonstrated by EMMERICH and BECKMANN [278].

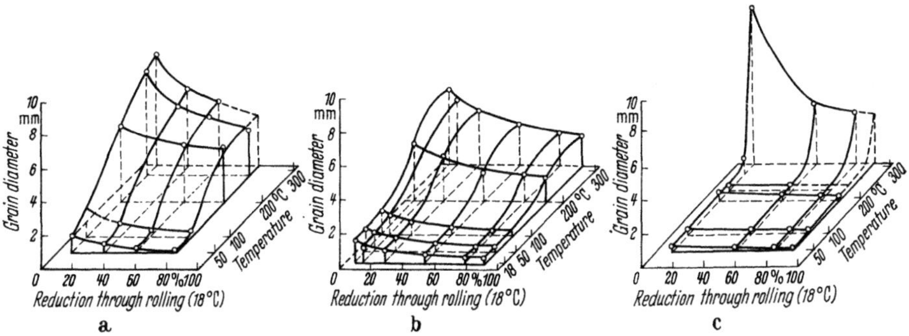

Fig. 208. Recrystallisation diagrams of different types of lead.
a) Electrolytic; b) PARKES; c) PATTINSON. (According to LOOFS-RASSOW).

Complete investigations of the kind given above are not available for other lead alloys. However there are many observations scattered through the literature which concern the effect of impurities on the size of the recrystallised grain. They are of particular importance with regard to grain-coarsening.

In a table which is not reproduced, data were collected from three different laboratories on the grain size of lead and lead alloys after rolling (50 and 60% reduction) as well as after annealing at 160 to 200 °C and in part on cast grain (LOOFS-RASSOW [765], KRÖNER [709], JENCKEL and HAMMES [618], JENCKEL and THIERER [619]). Disregarding a few exceptions (potassium), the cast grain throughout was more or less coarse as long as the content of alloying constituents was only moderate (JENCKEL and HAMMES [618]). The grain sizes of alloys including unalloyed lead, rolled, and recrystallised at room temperature, also showed few peculiarities. Only in lead-calcium and lead-lithium was the grain diameter essentially below 0.1 to 0.2 mm, and only in lead-gold was this exceeded. On the other hand, on heat-treatment at 160 to 200 °C larger differences of grain diameter occurred, which must be explained by a specific action of the different alloying elements. A grain diameter, substantially above 1 mm, was obviously present in the purest types of lead, whereas in many technical types of lead it was around 1 mm. The grain size is only slightly affected by alloying elements which go into solid solution in lead, such as bismuth, a few multiples of 0.1% of antimony, presumably also small contents of tin, cadmium and thallium.

The view has also been advanced that by such dissolved elements the grain is actually coarsened (WERNER [1260]). In order to test this assumption, EMMERICH and BECKMANN [278] carried out groups of tests under practical conditions of cable extrusion. In this case the bismuth content of the press charges was raised to the value 0.18% in nine steps.

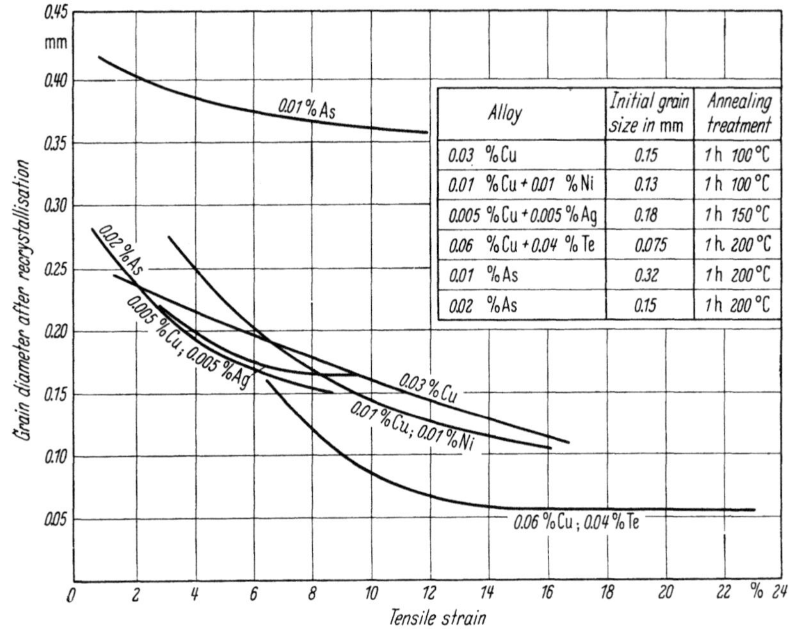

Fig. 209. The influence of the degree of deformation on the grain size of different lead alloys after recrystallisation. (According to BUTLER).

Up to bismuth contents of about 0.02%, the grain size of the cable sheath was practically the same as in the refined lead sheath. When the bismuth content was increased further, a certain refinement of grain was unmistakable. The same was determined on lead-antimony alloys. These observations also correspond to the theory put forward by LÜCKE and DETERT [777] (p. 187).

Impurities which do not go into solution, or only in part, refine the grain, so that even after heat treatment the diameter is considerably below 1 mm. According to ZENER [1302] grain growth should cease, in the case of the presence of inclusions, at an average grain size $D_m = d/F$ where d signifies the average diameter of an inclusion, F the fraction of the volume in the alloy. Among grain-refining impurities particular mention may be made of a few hundredths of 1% of copper, nickel, tellurium and calcium. Antimony and tin also refine the grain if they are added in larger amounts. On the other hand, copper, nickel and tellurium are without action if their concentration is within the very low solubility limit (0.01%) at the temperature of heat treatment. BUTLER [163] gives extensive observation data on the effects of small amounts of impurities in lead on recrystallisation and grain growth (Fig. 209). Details on this subject will be found in the section "Lead Pipe" (p. 434).

In commercial lead, coarse grain occurs even at room temperature; owing to uncontrollable influences, it is often impossible to indicate the cause with certainty. However in most cases, critical straining may be the basic cause, as is shown, for instance, by the behaviour of sligthly bent lead sheet. Rolled, very fine grained lead sheet is particularly regarded as susceptible towards grain-coarsening of this kind. As this danger should not exist at medium grain size, it is recommended that efforts should be made to attain the latter in technical working by extrusion and rolling (JONES [632]).

2. Mechanical Properties (in the Short-Time Test)

a) Measured Values from the Tensile Test. Data on the elastic limit or the yield point of lead, that is the stress which in the short-time test brings about a permanent elongation of a fixed small amount, have only limited value, as they are very dependent on the duration of test. As a result of the tendency of lead to creep, permanent elongations will take place (in certain cases of considerable magnitude) at stresses below the elastic limit determined by the short-time test, if only the stress is applied for long enough. The following few data can be evaluated accordingly.

FAUST and TAMMANN [306] determined the compressive yield point as the stress at which the crystallites in a section emerged from the polished surface; it amounted to $25\ \mathrm{kg/cm^2} \pm 8\%$. CHALMERS [188] determined the limit of proportionality in material free from stress, by means of the interference method, at $9\ \mathrm{kg/cm^2}$. This value is of the order of magnitude of the creep strength of pure lead. In deformed metal deviations from HOOKE's law occur. A natural yield point of the kind shown by carbon steel is not observed in the stress/strain diagrams of lead and slightly alloyed lead.

As expected, the tensile strength falls with decreasing rate of strain, as the softening which opposes the strain-hardening, depends on time. Fig. 210 shows the relationship according to measurements by ERDMANN-JESNITZER and HANEMANN [286]. With a longer duration of test, the tensile strength changes smoothly into the creep strength whereby the value of σ_B, the tensile strength, sinks to the value (many times smaller) of the creep strength or creep limit. It would therefore be desirable to specifiy a definite rate of strain for tensile tests of lead and lead alloys. In Germany a rate of strain $\Delta l/(l \cdot \text{min}) = 0.25$/minutes is applied in tensile tests of zinc and its alloys. Fortunately the effect of the duration of the test first makes itself clearly noticeable at very low rates of strain

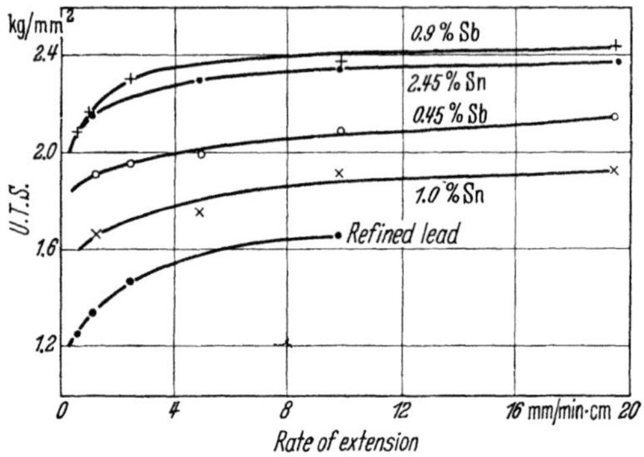

Fig. 210. The influence of the rate of extension on the tensile strength of lead and lead alloys. (According to ERDMANN-JESNITZER and HANEMANN).

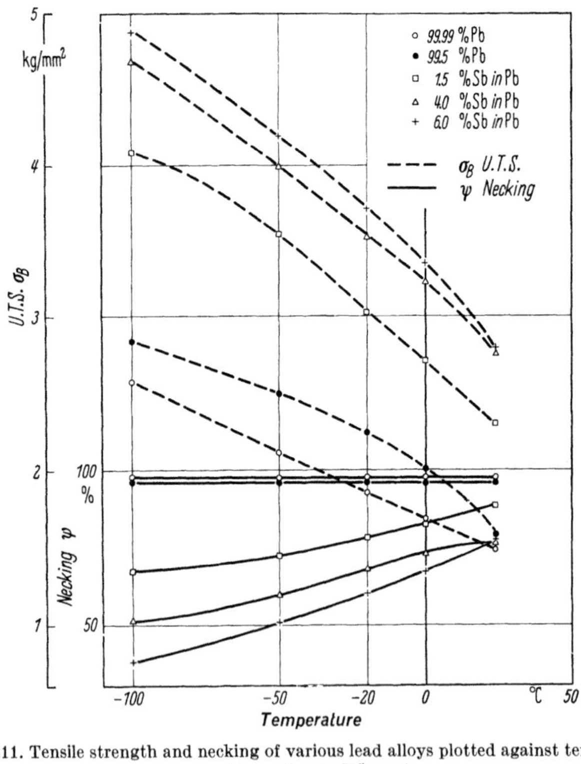

Fig. 211. Tensile strength and necking of various lead alloys plotted against temperature. (According to KÖLSCH).

in practice which are not used in tensile tests, so that data on tensile strengths at unknown rates of strain, while certainly inexact, are not however valueless (SIEGLERSCHMIDT and FIEK [1119]).

The considerable dependence of tensile strength on the duration of test is paralleled by a similar relationship of temperature. Fig. 211 shows this for lead and some lead alloys, according to measurements by KÖLSCH [688]. According to determinations by POMP and co-workers [970], the tensile strength at −183 °C is about twice as great as at 20 °C. BURKHARDT [155] gives the following data for the range between room temperature and 265 °C.

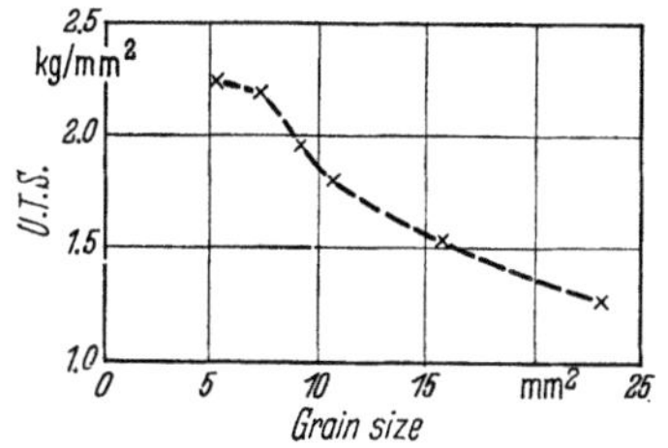

Fig. 212. Grain size and tensile strength of lead. (According to GARRE and MÜLLER).

Temperature °C	20	82	150	195	265
Tensile strength (σ_B)kg/mm²	1.35	0.80	0.50	0.40	0.20
Elongation (δ) %	31	24	33	20	20

According to GARRE and MÜLLER [357], the tensile strength increases (as in other metals) with decreasing grain size (Fig. 212). The tests related to grain sizes between 23 and 5 mm², and therefore need to be extended to smaller grain sizes.

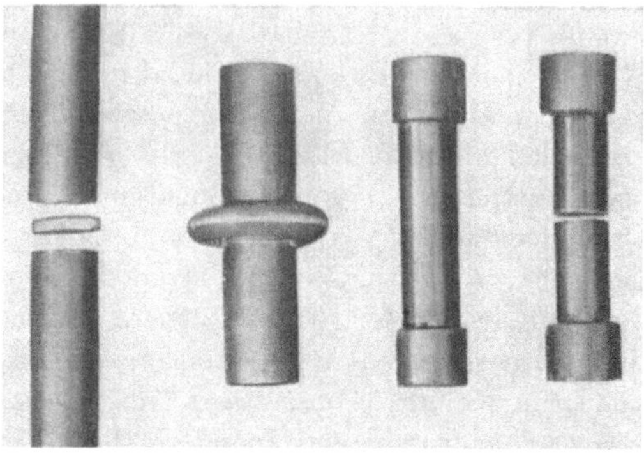

Fig. 213. Stages in a cold pressure weld. (According to BURAT and HOFMANN).

Values for the cohesive strength = separation strength of lead, can be obtained if the plastic elongation and contraction in the tensile strength are hindered. This was possible by the use of cold pressure welding (p. 57). Thin discs of lead could be welded on both sides to cylindrical rods of aluminium or steel (Fig. 213). After welding, the thickness of the lead layer was a few multiples of 0.1 mm. On applying a tensile stress, an inhomogeneous triaxial state of tensile stress is present in the lead. Fracture occurs in the lead. The value of the cohesive strength, about 4 kg/mm², thus determined can only be regarded as a guide.

An evaluation of the data in the literature on elongation at fracture is unsatisfactory because it is generally not clear whether the test piece was flat or round and whether, in the latter case, δ_5 or δ_{10} was measured. For commercial lead, 35 values from four different authors (Harris [494], Burkhardt [155], Garre and Müller [357]) were taken from the literature. They varied between 21 and 73%, the corresponding tensile strengths between 1.12 and 2.22 kg/mm². The wide scatter is explained by the factors described above, on which are superposed the actions of different grain sizes and impurities. Apart from this, the strength values also alter in some circumstances in storage (p. 193). In commercial lead the reduction of area amounts as a rule to 100%, i.e., the fracture is a point or an edge (Sachs and Fiek [1043]). This phenomenon, still to be observed at a temperature of —100 °C, is a particularly good expression of the high workability of lead. Only in the creep test in coarse-grained commercial lead, slight reduction of area and intercrystalline fracture are observed in some circumstances.

According to measurements by Mayer-Wegelin [814], the elongation at fracture of lead is reduced by only a few per cent on increase of the rate of strain in the region of 0.7 to 6.1 mm/min · cm. Further, in commercial lead, elongation at fracture and reduction of area show no marked dependece on temperature in the range —100 to 150 °C, but, on the other hand, in lead alloys they show a tendency to fall if below 0 °C. For the purpose of investigating the rule, quite generally valid for alloys, that by increasing the concentration of the alloying element the strength is increased, but elongation at fracture and a reduction of area are reduced, the tensile strength (σ_B) and elongation at fracture (δ) were determined for commercial lead and a large number of arbitrarily chosen alloys. The specimens were in the form of extruded strip of cross-section 2 · 10 mm²; they were in part heat-treated. There resulted the region of correlated values of tensile strength (σ_B) and (δ) elongation (referred to $l_0 = 50$ mm) shown in Fig. 214. Those alloys and pretreatments will be preferred whose pairs of values correspond to the upper boundary of the region. As examples of application, lead-tellurium with a good elongation level, commercial lead and antimonial lead (1.1% Sb + 0.04% As) with different pretreatment are particularly significant in the diagram. The antimonial lead with high elongation at fracture was extruded at 240 °C and immediately quenched in water; that with lower elongation at fracture was homogenised for sixteen hours at 250 °C (which resulted in grain coarsening) and quenched. The differences in the commercial lead specimens are brought about by similar pretreatments.

b) Types of Stress-Strain Diagrams. Haigh and Jones [471] put forward the view that the value of the tensile strength and of the elongation at fracture do not suffice for the characterisation of different

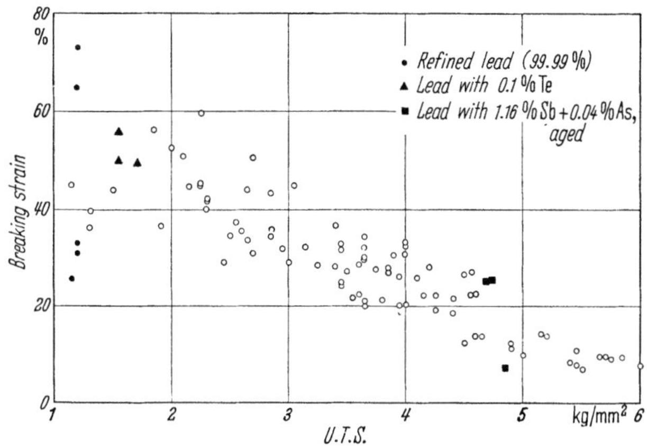

Fig. 214. Relationship between strength and elongation of lead and several lead alloys. (Unpublished measurements by v. HANFFSTENGEL).

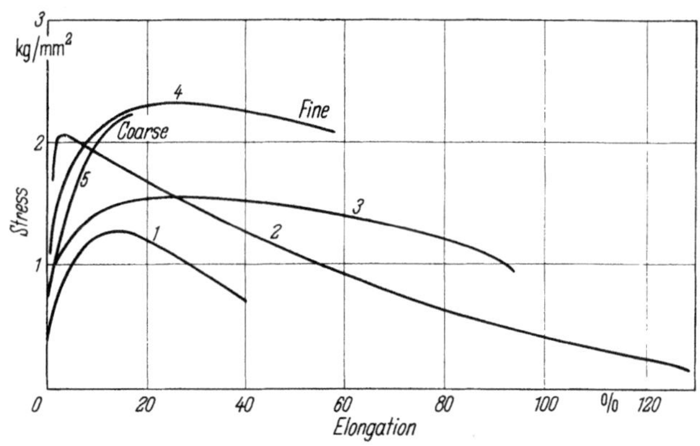

Fig. 215. Types of stress-strain diagrams for some lead alloys. 1. Refined lead. For various designations on the curves, see the text. (According to HAIGH and JONES).

lead alloys, and that really complete stress-strain diagrams must be provided. Moreover they referred to the known fact that the elongation at fracture is composed of the uniform elongation and of a contribution to the elongation arising from necking. Both parts can be recognised to a certain extent in the stress-strain diagram. Stress-strain diagrams from the investigation on commercial lead and on three unnamed lead alloys are reproduced in Fig. 215. In comparison with alloy 2, alloy 3 shows a marked uniform elongation, alloy 4 in the same way shows necking elongation and uniform elongation, but there is evidence of intercrystalline fracture which has occurred in coarse grained specimen 5. Alloys with high uniform elongation should be favourable when there is risk of frost, as can occur, for example, in water pipes. The dip in the load immediately behind the maximum in curves 1 to 3 is attributed to spontaneous recrystallisation. That the recrystallisation of lead is not hindered by additions of bismuth or thallium, is in agreement with other experience. In the lead-calcium alloy, age-hardening obviously occurs during the tensile test.

High transverse tensile stresses occur in the middle of a necked rod (Fig. 216); hence a triaxial condition of stress is present. In some lead

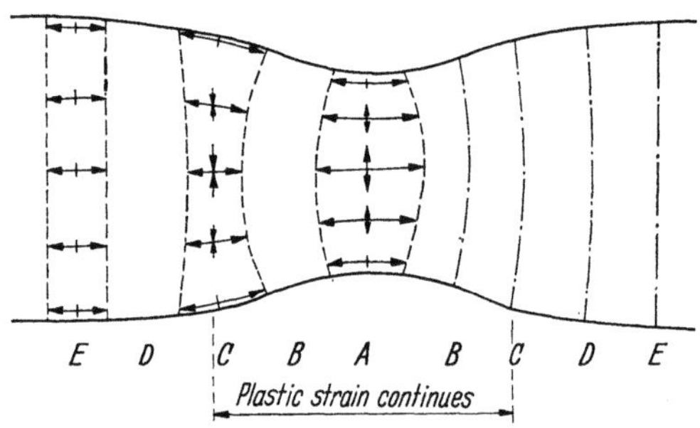

Fig. 216. Stress-distribution in a necked tensile test piece. (According to HAIGH and JONES).

alloys it results in cracking from within, whereby the broken halves of the rod only touch externally when put together. Lead in strip form often shows another kind of fracture, which results from shearing (HAIGH and JONES [471], KÖRBER and KRISCH [690]).

c) Hardness. The hardness of lead on the MOHS scale amounts to about 1.5 (KOHLRAUSCH [695], LUDWIG [771]).

α) Brinell Hardness. Since in the case of lead, in some circumstances, a considerable grain size must be expected, a ball diameter of 10 mm is recommended for the determination of the Brinell hardness. In the usual way the load should be so chosen that the diameter of the impression of the ball is between 0.2 and 0.7 times the diameter of the ball. In this way, loads are derived for testing the hardness of lead, according to the hardness value to be expected, the loads being 15.6; 31.2; 62.5; 125 and 250 kg. The test loads quoted apply, according to the above condition, to the following regions:

Test Load:	15.6 kg	31.2 kg	62.5 kg
Hardness HB:	0.38 to 4.9 kg/mm²	0.75 to 9.75 kg/mm²	1.5 to 19.5 kg/mm²
Test Load:	125 kg	250 kg	
Hardness HB:	3 to 39 kg/mm²	5.6 to 78.8 kg/mm²	

As a cessation of flow cannot be realised, even on tests of long duration the hardness numbers of lead are only relative values, valid for the duration of testing. The use of a uniform duration of test would therefore be desirable. The testing time of two minutes, formerly usual in the Bleiforschungsstelle, could be shortened, bearing in mind practical requirements. Frequently a testing time of thirty seconds is used. In the literature these conditions and notations are not uniform. As far as possible any quoted Brinell hardness in this book is supplemented with: Ball diameter (mm)/load (kg) or load per diameter square (kg/mm^2) — time (sec).

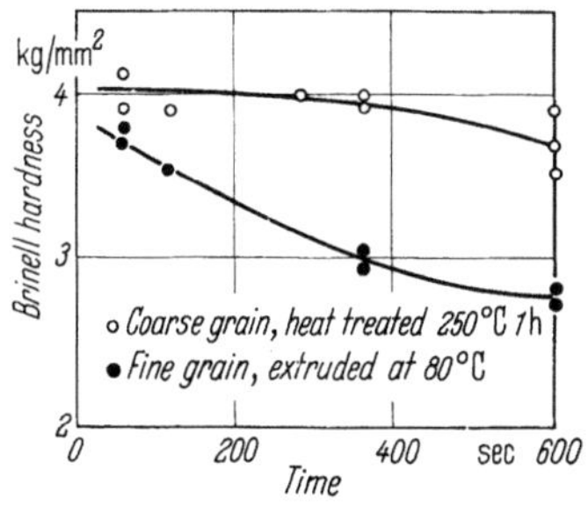

Fig. 217. Hardness creep curves of fine— and coarse — grained lead. (According to ERDMANN-JESNITZER and HANEMANN [285]).

On the other hand, it is perhaps not possible to set up a generally valid monogram for the recalculation of hardness values for a uniform testing time, as the dependence of hardness on the duration of testing alters, corresponding to the laws of creep, with the nature of the alloy, the grain size and so forth (KENNEFORD [657]). In particular there is a greater time-dependence in fine-grained lead than in coarse-grained (BALLAY [39] and ERDMANN-JESNITZER [285]), so that for the usual testing times the latter can, curiously enough, appear harder than the former (Fig. 217). The Brinell hardness of high purity lead 99.9999 (Cominco) was found to be H $10/100/30 = 2.74 \pm 0.04$ kg/mm^2 at $20° \pm 0.5$ °C with a grain size of 0.2 mm^2, and the Brinell value for pure lead (99.99) was determined as being H $10/100/30 = 3.57 \pm 0.05$ kg/mm^2 at $20° \pm 0.5$ °C with a grain size of 0.5 mm^2 (KRYSKO [714d]).

At room temperature the hardness of commercial lead can be quoted at 2.5 to 3 Brinell units, and corrected by about 0.5% of the value at 20 °C for each degree rise of temperature (HOFMANN, SCHRADER and HANEMANN [577]). WORNER [1285] says that the Brinell hardness of very pure polycrystalline lead falls by about 0.027 units for a temperature increase of 1 °C in the range 0 to 60 °C. The temperature-dependence of the Brinell hardness was further measured from −253 to 0 °C (HOLM and MEISSNER [583]), and from −40 to 125 °C, although with a high test load (ITO [607]). At −183 °C the Brinell hardness, like the tensile strength, has 2.1 times the value at 20 °C (POMP, KRISCH and HAUPT [970]). MULHEARN and TABOR [881a] investigated the dependence of Brinell hardness (or yield pressure) on the duration of test (between 10^{-4} and 10^{+3} sec) and temperature (between −183 and 287 °C). If the duration of load is less than one second, the hardness increases very much, at a duration of test of less than 10^{-3} sec an upper value of the hardness, p_{max}, of about 8 kg/mm² is attained, independent of the temperature of test. At the temperature of liquid air, the value of the hardness ist nearly independent of the duration of load and about equal to p_{max}.

β) Relationship between Hardness and Tensile Strength. In unpublished investigations, the Brinell hardness was also determined (under the conditions of test set out above) on the tensile test pieces of lead alloys

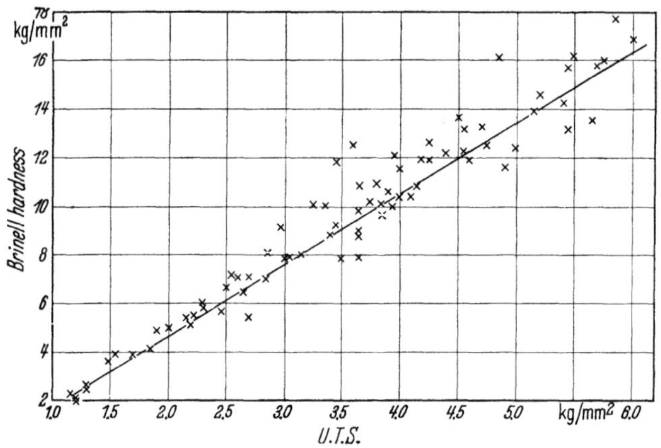

Fig. 218. Relationship between hardness and tensile strength of lead and lead alloys. (According to v. HANFFSTENGEL).

chosen arbitrarily (cf. Fig. 214) which had already been examined for the relationship between tensile strength and elongation at fracture. The diagram of the correlated values of HB and tensile strength (σ_B) (Fig. 218) shows an approximately linear relationship of the two quantities.

γ) Diamond Cone Hardness. The diamond cone hardness of KAHLBAUM lead was measured by LUDWIK [773, 774] over a large range of temperature and for two different durations of test. A selection of values is given in Table 15.

Table 15. *Diamond Cone Hardness in kg/mm² of Lead at Different Temperatures*

°C	15	45	62	102	115	150	153
Duration of test 300 sec	3.95	2.68			1.39		1.10
Duration of test 15 sec	4.98		3.30	2.16		1.44	
°C	213	214	272	282	323	324	
Duration of test 300 sec		0.67	0.43			0.27	
Duration of test 15 sec	1.01			0.54	0.40		

The diamond cone hardness is referred to the surface of the circular impression. If, corresponding to the Brinell hardness, it is desired to refer it to the convex cone surface, then the figures given must be divided by $\sqrt{2}$. The values so calculated would approach nearer to the Brinell hardness.

δ) Dynamic Hardness. The dynamic hardness of lead, which had been annealed at 270 °C, was determined by SAUERWALD and KNEHANS [1051] over a large range of temperature in the course of a comparative investigation of different metals. They found (similarly as in other metals) a linear decrease of the dynamic hardness with rising temperature. The value of the absolute dynamic hardness, which can be regarded as the specific work of displacement (HENGEMÜHLE [508]), sinks from 7.44 at 25 °C to 4.24 mkg/cm³ at 267 °C. At higher temperatures no measurements could be carried out, owing to the formation of cracks as a result of grain coarsening.

d) Compression. Impact and Notch Tests. As regards the compressive yield point of lead, attention is drawn to what was said under the tensile tests. No value can be given for the compressive strength (σ_{dB}) of lead,

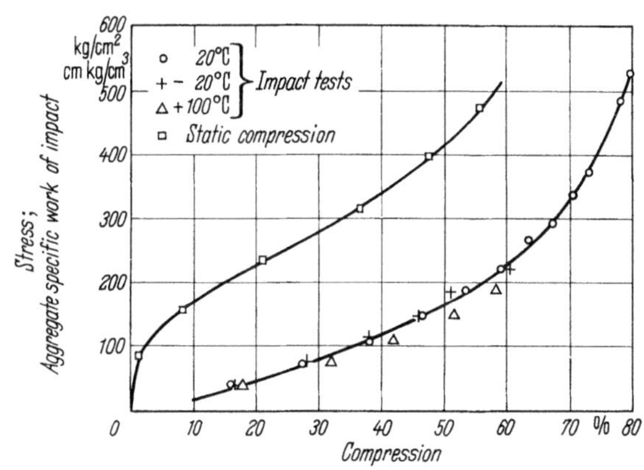

Fig. 219. Diagrams of compression and impact tests on commercial lead. (According to HEYN and BAUER [520]).

as no fracture occurs under static pressure. The course of flow in the static test is given by the upper curve in Fig. 219. where the stress is referred to the initial cross section. The dimensions of the test pieces in the static and in the dynamic test were dia. 20 mm, height 17.7 mm. The specific work of impact per blow in the dynamic test amounted to 0.374 mkg/cm³. The values of the impact test in the temperature range investigated (−20 to 95 °C) showed that the resistance to impact was only slightly dependent on temperature. PELZEL [949] carried out static compression tests on cast test cylinders of dia. 16 mm and height 16 mm at room temperature. The rate of compression was 50 mm/min. For the comparison of different alloys he used the resistance to impact, in kg/mm² referred to the initial cross-section at a 50% reduction of ineight of the specimens. The result (Fig. 220) is particularly noteworthy hor the high strengthening effects of small contents of silver and tellurium.

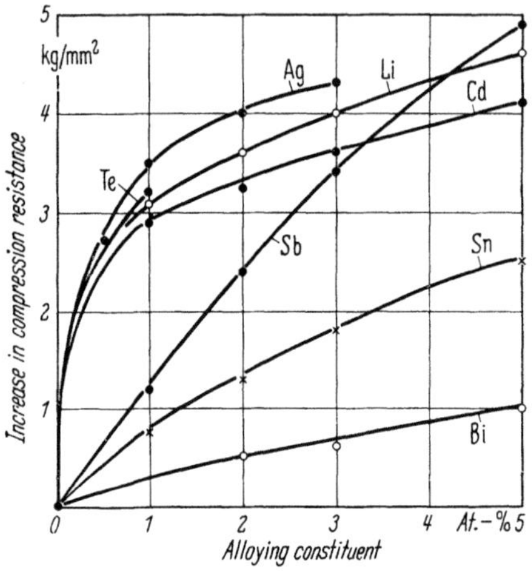

Fig. 220. Increase of compression resistance of lead alloys over that of unalloyed lead. (According to PELZEL).

This is also apparent in the effect of these elements on the recrystallisation of lead (p. 190). BAILEY and SINGER [33c] developed a constant strain rate plane strain plastometer which can apply very high compression rates between 0.4/sec and 311/sec and thus simulates practical working processes. Among other things they provide experimental results on lead between 22 and 300 °C. The relationship between yield stress σ and strain rate $\dot{\varepsilon}$ obeys the equation $\sigma = \sigma_0 \cdot \dot{\varepsilon}^n$, that is, on a log-log scale the values of σ and $\dot{\varepsilon}$ lie on straight lines (cf. Fig, 328). σ_0 and n are constants, which depend on the degree of deformation, $\varepsilon = \ln h_0/h$ (h_0 = height before compression, h = height after compression) and the absolute temperature T. According to the experimental conditions n lies between 0.04 ($\dot{\varepsilon} = 0.1$ at 22 °C) and 0.26 ($\varepsilon = 0.5$ at 300 °C). Notched bar impact tests on coarse- and fine-grained lead specimens of dimensions 15·15·180 mm³ and notched to a greatest depth of 10 mm, resulted in brittle fracture (GARRE [357]). The impact resistance on coarse grained specimens was lower than with fine grained material.

The impact resistance of commercial lead was determined at 20, −183 and −253 °C as >2.3 and >3.7 and >4.4 mkg/cm² [970]. During this work ductile fracture occurred at all temperatures of test. Therefore lead is still malleable at the temperature of liquid hydrogen.

3. Creep Strength

a) Basic Phenomena. Creep is regarded as the time-dependent flow of a metallic or non-metallic material under load. It becomes of greater significance as the temperature approaches the melting point of a metal. The creep behaviour is therefore of particular importance for lead and lead alloys, The knowledge of creep strength is not merely a basis for design in lead and lead alloys. As lead behaves, at room temperature, in

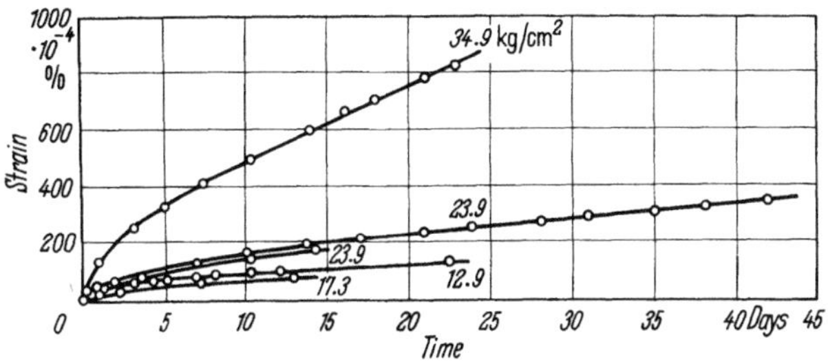

Fig. 221. Creep curves of antimonial lead with 1.0% Sb and 0.04% As [563].

a manner similar to high-melting materials in a much higher temperature range, strength studies on alloys for use at high temperature can be simulated by model tests on lead at room temperature. For example, Sachs [1041] carried out model tests on lead shells in order to apply the results to Inconel. Model tests on lead have also proved suitable for the development of forging technique for high-melting metals.

Creep tests are usually carried out under constant load by weights or springs. The increase of stress which occurs owing to the reduction of cross-section with time, can be neglected for small amounts of elongation. Higher creep rates require the application of constant stress and therewith the use of a special regulating mechanism, if more than a qualitative evaluation of the results is desired. The principles of such regulating mechanisms are described by Sully [1155]. Relaxation tests are of less significance for lead (p. 220).

Creep curves of lead usually exhibit the course shown in Fig. 221. The elongation occurring at the instant of loading, the so-called instantaneous creep (which is part elastic and part plastic) is followed by the time-dependent flow. Thus the transient creep, which is characterised by reduction of the creep rate with time, is distinguished from the steady-state creep (quasi-viscous creep or secondary creep) which represents the condition of lowest rate of strain. Sometimes, creep with constant rate of strain is completely absent, that is, after the transient creep no further elongation occurs. This is particularly the case for steel at room temperature (MEYER [849]; it was also found, however, in lead single crystals at low temperatures (NEURATH and KOEHLER [894]). On the other hand, an absence of transient creep was observed in age-hardened lead-antimony-arsenic (0.001%) alloys (HOPKIN and THWAITES [590]). Here the instantaneous creep is immediately followed by secondary creep (see p. 234). If a period of increasing creep rate follows on the secondary creep, it is described as tertiary creep. The cause of this is not (or not only), as has often been said, the increase of stress due to the reduction of cross-section at constant load. As the increase of creep rate is also observed on application of constant tensile stress and even in the compression test (SULLY, CALE and WILLOUGHBY [1156]), it must be assumed that it is to be ascribed to changes in the materials due to creep (see below).

The ideas developed by BECKER [72] and OROWAN [925] form the basis for the understanding of creep. While the plastic deformation of metals, when the yield point is exceeded, occurs suddenly, and the flow assumes high values, creep occurs slowly. In the first case the deformation takes place solely through the applied stress, without the essential contributions of thermal energy; on the other hand, in the case of creep, the difference between the applied stress and the yield point beyond it, the so-called activation stress σ_a is introduced by fluctuations of thermal energy.

b) Time Laws of Transient Creep. The transient creep is of technical significance insofar as only, after it has finished, can a constant rate of creep be seen which permits extrapolation to longer periods of test. According to COTTRELL [217], the time law of transient creep can often be expressed by the power law $a = da/dt = At^{-n}$ in which a represents the gliding off, A and n are constants depending on the stress and the temperature; n lies between 0 and 1. The extreme case $n = 1$ gives the logarithmic creep law $a = \alpha \ln t$, which is valid for a number of metals, for instance, approximately for steel at room temperature and for lead single crystals at -190 °C (NEURATH and KOEHLER [894]). At higher creep rates and amounts of strain a value of n less than 1 is observed, frequently $n = 2/3$. The gliding off then appears as $a = \beta t^{1/3}$: Andrade's creep law. It is found for lead, copper, cadmium, tin and lead-tin alloys, if the arrangement for the maintenance of constant stress, mentioned at the outset, is applied.

The evaluation of many results provided in the literature gives the impression that in creep tests with different loads on a material, the inflection of the creep curves in the rectilinear part takes place after approximately equal times of test; it is assumed that the temperature is constant. This rule can usefully be applied in planning creep tests. In the creep curves in Fig. 221, the transient creep has ended after about seven days. On the basis of years of experience with lead, at least eighty days has been proposed as the duration of practical creep tests (MOORE, BETTY and DOLLINS [866]); steady state creep then takes up the greater part of the creep curves.

c) Processes in Steady State Creep. In addition to the elementary processes of plastic deformation, there are further mechanisms of deformation present in creep phenomena. Important service in their elucidation was rendered not only by light microscopy but particularly by electron microscopy and *X*-ray diffraction.

The prevention of crystal translation by the neighbouring grains leads to rotation of the slip planes within the crystals and therewith to a fragmentation of the grains. McLEAN [825] has drawn particular

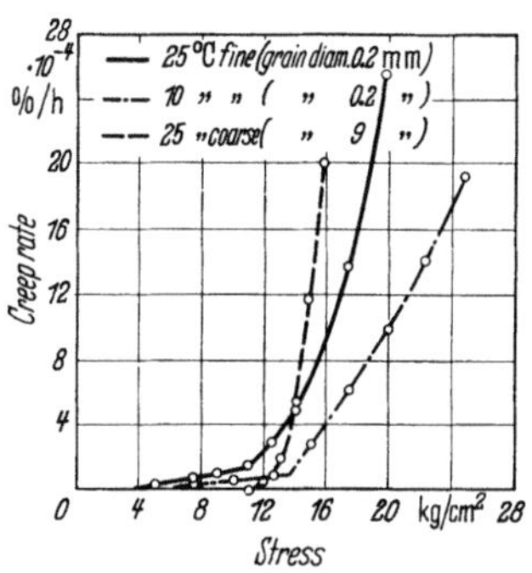

Fig. 222.
Creep rate of commercial lead in relation to load. (According to v. HANFFSTENGEL and HANEMANN).

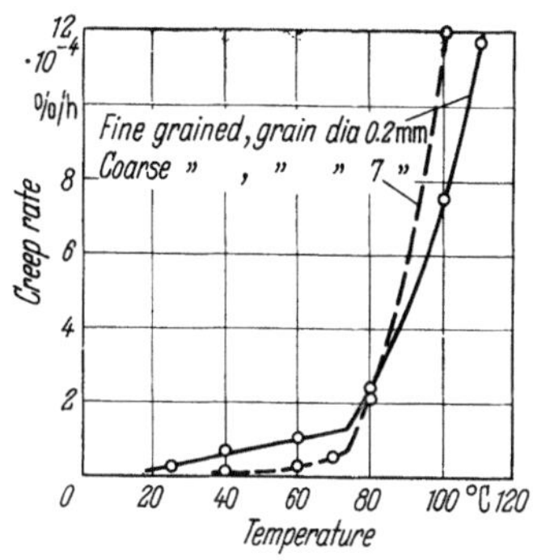

Fig. 223.
Creep rate of commercial lead in relation to temperature. Stress 5 kg/cm². (According to v. HANFFSTENGEL and HANEMANN).

attention to the formation of subgrains in creep owing to the occurrence of polygonisation (Fig. 199). They form preferentially at the location of break planes. He sees the process of polygonisation as an essential part of secondary creep. The activation energy derived from the temperature-dependence of the creep rate, and its close relationship to the activation-energy of self-diffusion (see below), speak for the accuracy of this assumption.

It has been known for decades that the grain boundaries exert a great influence on the rate of creep of a material. Here mention may be made of experiments on pure lead (99.99%) which were carried out thirty years ago at the former Bleiforschungsstelle by v. HANFFSTENGEL [482]. The creep curves were systematically followed into the region of constant strain rates. In the experimental series, the temperature was altered at constant load and the load at constant temperature. Each group of experiments was carried out for coarse- and for fine-grained lead. All the curves in Figs. 222 and 223, in which the creep rate is plotted against the stress or against the temperature, show a break after passing which the rates of strain rise very steeply. In addition the curves for coarse- and for fine-grained lead intersect so that at low stresses and temperatures fine-grained lead, at high stresses and temperatures coarse-grained lead, creeps more quickly. The break in the diagrams of stress/strain rate is also indicated in many curves given by MOORE and co-workers [865] though here it is not marked as such.

At that time the results were interpreted in the sense that the creep process to the right and left of the break arose from different causes. The creep to the left of the break was said to be due to a mutual displacement of the grains along the grain boundaries. The greater creep of fine-grained lead, as compared with coarse-grained, which is observed to the left of the break, would be understood in this light. The creep above the stress or temperature, indicated by the break, should be due to crystal plasticity as well as grain boundary slip.

The views developed at that time are now no longer entirely defensible. In particular, the assumed sudden occurrence of crystal translation in polycrystalline specimens on an increase of temperature or of load above the value at the "break", is a simplification which is not justified. In spite of this the interpretations of that time are qualitatively true in broad outline. The view that creep is composed of crystal slip and a mutual displacement of grain on both sides of the grain boundaries (grain boundary sliding), has been further supported in the interim by a number of observations. For example, a screen was placed on aluminium specimens consisting of only two crystals, and the displacement of the crystals measured along the grain boundaries lying diagonally to the direction of tension (RHINES [1006]; p. 212). Similar measurements of the grain boundary sliding in lead in comparison with the general creep were carried out by STRUTT and co-workers [1152a]. They used a bicrystal with a small angle grain boundary 45° to the direction of tension. The energy of activation of the grain boundary sliding was between 12.8 and 15.5 kcal/gram atom, the energy of activation of the general creep between 17.8 and 22 kcal/gram atom. Extensive measurements of grain boundary sliding in creep were made at the National Physical Laboratory in Teddington (McLEAN [826, 827]). At the beginning of the experiment the specimens had polished surfaces. In the course of creep, owing to grain boundary sliding, differences of height of the surfaces of neighbouring grains occurred, which could be measured with an interference microscope. The amounts of grain boundary displacement calculated from this were plotted against the total elongation (in %). For all the metals hitherto investigated (Al, Sn, Zn, Cd, Fe) straight lines were obtained of different slopes, that is, the relationship $\frac{\text{grain boundary sliding}(\mu)}{\text{elongation}(\%)}$, appeared constant for one creep test. From the respective average value of the grain boundary sliding (in μ) a corresponding elongation (in %) was calculated. The quotient, $\frac{\text{total elongation}}{\text{elongation due to grain boundary sliding}}$, which up to a constant factor is the reciprocal value of the above-mentioned coefficient, is higher for coarse-grained material than for fine-grained. Moreover it increases with the applied stress. So far the result corresponds qualitatively with the concepts put forward by v. HANFFSTENGEL [481, 483]. The comparatively small fraction of the total elongation formed by the elongation due to grain boundary sliding is unexpected. The measurements showed in one case that the fine-grained specimen, in the region of steady state creep, crept twice as fast as the coarse-grained, while the additional effect of the grain boundaries could only (according to the interference microscope measurements) give an increase in the ratio 5:4. The greater creep rate of the fine-grained specimen must therefore have arisen not only from a larger amount of grain boundary sliding, but also from a higher rate of slip inside the crystal. The unexpectedly small fraction, according to McLEAN [826], of the total creep strain formed by the grain boundary sliding, could partly be explained on the basis of the observations by RACHINGER [989] that the surface of a specimen shows a much smaller displacement of grain boundaries than the inside. According to measurements by GIFKINS, RACHINGER's observation is true only at large amounts of strain ($\gg 10\%$). GIFKINS [381 b] says that at small elongations G (that is the ratio of the strain due to sliding E_{gb} to the overall extension E_t) is the same on the surface and inside the specimen. In this case G reached the high value of 0.88. If equiaxed grains (length L = breadth B) stretch in the direction of load during creep, so that L is greater than B, then the strain due to crystal slip $E_s = (L/B)^{2/3} - 1$. HENSLER and GIFKINS [512a] *suggest* an improved method of making such measurements.

PARKER [931] gives another possible explanation for the effect of grain boundaries on the creep rate. As was shown by McLEAN's [823] experiments, the flow of grain boundaries explains to a small extent only the increase of creep rate with decreasing grain size. The grain boundaries can however (in an indirect way) strongly influence the rate of creep inside the grains by favouring the movement of vacancies. According to SEEGER [1100] the diffusion of vacancies is the rate-determining factor for the climbing of dislocations and consequently for creep (see below). If, in fine-grained lead, many large-angle grain boundaries are present, the movement of vacancies will take place for the greater part on the grain boundaries, and only to a small extent within the grains. In this way there is a greater rate of diffusion of vacancies in fine-grained than in coarse-grained lead, where after recrystallisation-anneal only a few large-angle grain boundaries, rich in energy, have remained.

Above, the concept has been presented that in general the creep rate decreases with increasing grain size. With very coarse grain, an increase of creep rate will again result, owing to the small delay in crystal translation by the grain boundaries. There then results a minimum creep rate at a certain grain size, as is shown, for example, in the curves reproduced in Fig. 240.

In further investigations, McLean [824] added measurements of grain boundary displacements parallel to the surface, to those of the differences of level perpendicular to the surface. A part of this displace-

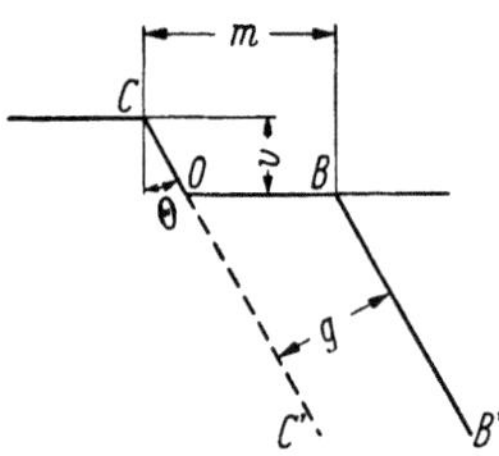

Fig. 224

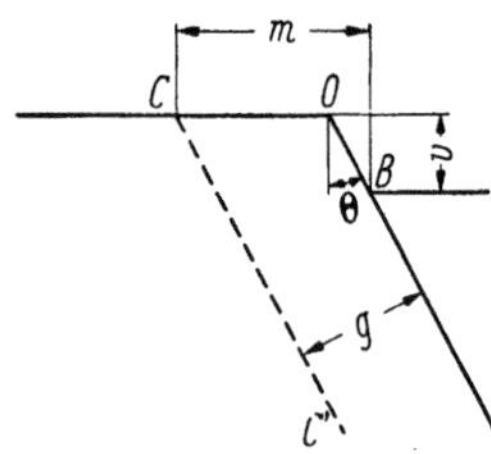

Fig. 225

Fig. 224 and 225. Grain-boundary displacement (m) by grain boundary gliding ($v \cdot \text{tg}\,\theta$) + grain boundary migration ($g/\cos\theta$). BB' Position of grain boundary in initial condition. CC' Position of grain boundary in final condition. Fig. 224. Sequence of migration and gliding. Fig. 225. Sequence of gliding and migration.

ment of grain boundaries arises from the fact that, in boundaries inclined to the surface, the grain boundary slip contains a component parallel to the surface (Figs. 224, 225). The other part consists in a grain boundary migration. The measurements showed that, with increasing temperature, the grain boundary migration becomes more important than the grain boundary sliding. According to measurements by Gifkins [378] there is also, in a creep test on a lead specimen, a linear relation between grain boundary migration and extension in creep.

As shown on page 11, at high temperatures there is a certain concentration of vacancies (holes) in thermal equilibrium with the crystal lattice. The diffusion of vacancies in a certain direction is equivalent to the transport of matter in the opposite direction. Kauzmann [648] and Nabarro [887] have shown that such a process will take place under load and will make a contribution to steady state creep. Creep by volume diffusion of vacancies should indeed play a part in the creep of lead primarily in the neighbourhood of the melting point (Seeger [1101]), but quantitative investigations on this point are not yet available (cf. Balluffi and Seigle [40]).

d) Dependence of Steady State Creep on Temperature and Stress. The occurrence of a constant creep rate in the secondary part of the creep curve has long been interpreted as an equilibrium between strain hardening and softening. This view implies that at constant load the yield point is certainly exceeded in small regions for a short time, but that the applied stress is below the yield point when the latter is raised by strain hardening. Consequently, the energy of activation, required for occurrence of a unit amount of slip, increases during creep. These higher values of the energy of activation rarely occur: that is, the creep rate decreases. When the strain hardening is in equilibrium with the softening and has therefore taken up a constant value, the energy of activation and with it the rate of creep remain constant.

Various mechanisms have been developed in order to be able to calculate the dependence of creep rate on temperature and stress. Most of these equations form a kind of Arrhenius equation with an activation energy H. The relationship derived by MOTT [874] (later replaced by improved equations) may be given as an example: creep rate $= \dot{a} = da/dt = A/\theta \exp \frac{-(H - q\sigma)}{kT}$. Here A and q are constants. Data for their magnitude are given by FELTHAM [312]. The negative sign before $q\sigma$ signifies that an increase of stress decreases by $q\sigma$ the activation energy necessary for the occurrence of a unit (elementary) process in creep. θ is the slope of the stress/strain curve at the applied stress σ, and therefore a measure of the strain hardening in the short-time tensile test. The energy of activation H is obtained by measurements of the creep rate at various temperatures T under the constant stress σ. The natural logarithm of $\dot{a}$ plotted against $1/T$ gives (if the above relationship is valid) a straight line, from the slope of which the value of H follows,

Such determinations of the energy of activation from their own measurements and from data in the literature were carried out in particular by FELTHAM [312], DORN [255], WISEMAN, SHERBY and DORN [1281]. They found that creep curves, that is, curves of strain against time, which had been taken at constant stress but at different temperatures, could be converted into one another by a parallel displacement in the direction of the axis of ordinates, if a double logarithmic scale was used. This state of affairs can be expressed thus, that the creep strain is regarded as a function of the stress and of the "temperature compensated time" θ, namely $\theta = te^{-H/RT}$; here t is the time expressed in a usual unit, R the gas constant, T the temperature in °K. The activation energy of creep H can be determined from two creep tests at different temperatures T_1, T_2 and constant stress. If the times which, at these temperatures, give the same creep strain are t_1 and t_2 then:

$$t_1 \exp - H/RT_1 = t_2 \exp - H/RT_2 ,$$

from which the single unknown H can be calculated.

Dorn [255] found for most metals a surprising coincidence of energy of activation for creep and for self-diffusion (Fig. 226). From this it was concluded that the process which controls creep rate at high temperatures (which is, already, room temperature for lead) must be a diffusion process. Mott [873] in 1951 had already regarded the climb of a dislocation as such, as in this way a transport of matter

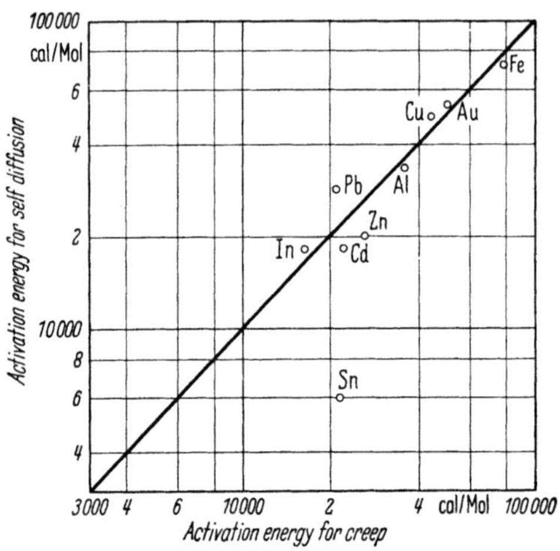

Fig. 226. Comparison of activation energy for creep and for self diffusion in various metals. (According to Dorn).

takes place. The full treatment of these ideas by Weertman [1250] gave a creep law of the form: Rate of strain $\dot{\varepsilon} = C(\sigma^{\alpha/kT}) \cdot e^{-E/kT}$ where C and α are constants, σ = stress, k = Boltzmann's constant and T = absolute temperature. The discrepancies between the energy of activation of creep and of self-diffusion were exhaustively discussed by Seeger [1100]. Butcher and Ruoff [160a] carried out a further determination of the activation energy of creep of lead (99.9999%). In the tests the temperature was cycled. In analogous tests the hydrostatic pressure was also varied and the activation volume was determined similar to the activation energy. The value of the activation energy amounted to 1.18 ± 0.03 eV, and the activation volume was found to be $(24.2 \pm 0.6) \cdot 10^{-24}\text{cm}^3 = 0.80$.

Some values of the creep rate given in the literature were evaluated by the method described. Thus, for instance, the natural logarithms of the values of $\dot{\varepsilon}$ in pure lead, according to Fig. 227 (v. Hanffstengel [483]) showed a nearly rectilinear dependence of $1/T$ as required by the equation given by Weertman. The graphs were, however, clearly curved mainly at low creep rates. This shows that in the equation given we are dealing with a very useful approximation.

From the formula of the creep rate there also follows a linear relationship of the logarithm of the rate of creep $\dot{\varepsilon}$ with the logarithm of the stress σ, if all other magnitudes in the equation, including the temperature, are made constant. In fact, on double logarithmic plotting ($\log \dot{\varepsilon}$ against $\log \sigma$) all values of creep rate available in the literature fall on to an almost rectilinear graph, an indication of the very useful

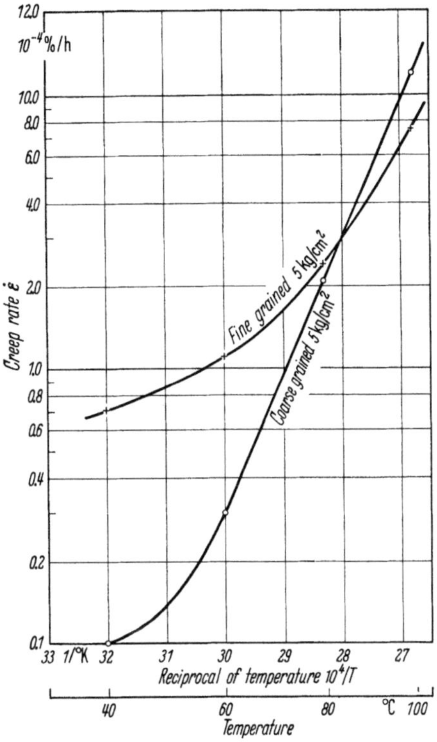

Fig. 227. Creep rate ε of commercial lead (on logarithmic scale) in relation to 1/T. (Redrawn from v. HANFFSTENGEL and HANEMANN).

nature of the assumed equation (Fig. 246). It is self-evident that all the above views are valid only for materials which do not undergo any phase changes (such as precipitation processes) in creep, that is, in the first place, for pure metals.

In a later investigation, WEERTMAN [1251] deals exhaustively with the dependence of creep rate on stress. From the assumption that the climb of dislocations is the rate-determining factor, he derives a formula which gives the proportionality of the creep rate as $\sigma^{4.5}$, that is, when the stress is doubled the creep rate is multiplied by 22. This relationship only applies for the range of low stresses, which are of primary importance for lead. In alloys, in addition to climb, other mechanisms (microcreep mechanisms) participate in the progress of creep, for example the dragging of clouds of dissolved atoms by migrating dislocations according to the COTTRELL-JASWON [218] mechanism.

If the creep rate is determined by such microcreep mechanisms instead of by climbing, then, according to WEERTMAN, the power law applies in the form that the creep rate is proportional to σ^3, that is, on doubling the stress the creep rate grows eight-fold. In a series of experiments on lead single crystals and polycrystalline lead alloys of the solid solution type, he showed that actually the exponent n in the power law σ^n, beginning with the value 4.5, approaches the value 3 on passing from lead to solid solutions with higher concentrations of alloying elements. It is however remarkable that lead-bismuth solid solutions behave like pure lead, an indication of the great mobility of bismuth atoms, which also follows from other observations (p. 42). The diagrams of the dependence of creep rate of lead and some lead alloys on stress (Figs. 246, 247, 250, 253), set up by the present author twenty-five years ago, are quantitatively in good agreement with the deductions of WEERTMAN, as well as the results of measurements on lead strip by v. MALOTKI [569a]. Variations in the magnitude of the exponent are discussed by GIFKINS [381a].

e) The Tertiary Stage of Creep and the Related Embrittlement Phenomena. As far as the creep tests take place under constant load, and the stress at high strains grows perceptibly as a result of reduction of area (necking), the increase of creep rate of the tertiary stage of test can be attributed to this in a trivial way. In addition, there is a true tertiary creep which is characterised by an increase of rate of strain at constant stress. This is attributed to the formation of microcracks at the grain boundaries, among other things. They lie chiefly perpendicular to the direction of the applied stress (HANSON and WHEELER [489]). If intercrystalline cracks occur, they will exert an internal notch action. In addition to the increase of creep rate due to reduction of the supporting cross-section, the micro-cracks thus also explain the reduction of toughness, that is, the occurrence of fractures at comparatively small strains. While such premature fractures are particularly, observed in high-strength high-temperature alloys, for instance in chromium-nickel steels, they also occur in certain lead alloys (cf. p. 412). CRUSSARD and FRIEDEL [226], following on previous work by GREENWOOD [424], have developed the theory that the cracks arise by diffusion of unoccupied places in the crystal lattice (vacancies) to the grain boundaries and subsequent combination there of the holes (condensation). According to FRANK [337], impurities in the metal, which reduce the surface tension of the holes, may exert an influence on the course of the event.

Eborall [267], from ideas taken from Zener [1301], gave a clear model for the origin of cracks by local accumulation of stress (Fig. 228). However, he also takes a thermally activated fracture process into account. Moreover in lead the possibility of a certain intercrystalline oxidation in the course of a creep test cannot be entirely ruled out.

A considerable increase of creep rate during creep can also be brought about by recrystallisation (Gifkins [379]). In such curves, periods of increased rate of strain, due to recrystallisation, alternate with those of lower rate of strain (Andrade [20]). A particularly clear example of this is given by Brock [135]. Such observations occur preferentially at stresses which are not to be expected with lead in practice.

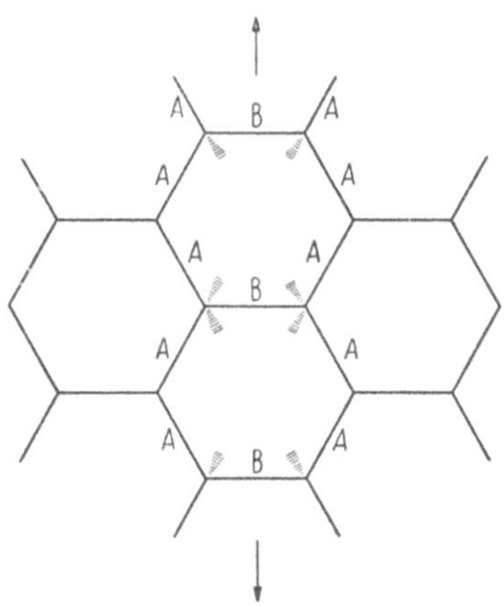

Fig. 228. Model for the origin of intercrystalline cracks in creep. By gliding of the grain boundaries A, stress accumulations occur at grain corners, with subsequent fracture at the grain boundaries B, or plastic deformation in the hatched regions. (According to Eborall).

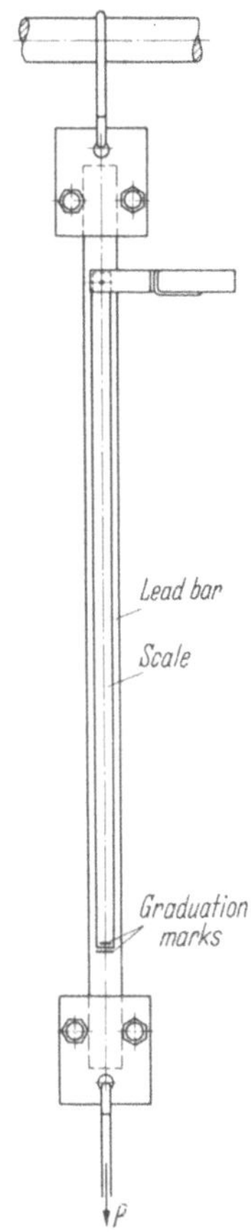

Fig. 229. Simple measuring arrangement for creep tests on lead rods. →

f) Performance and Evaluation of Creep Tests. In general the measurements are carried out on lead rods under tensile load. The stress is, in the usual way, referred to the initial cross-section. If a sufficiently long gauge length, such as 200 mm, is used, then the elongation can be determined with a cathetometer with sufficient accuracy for practical purposes. If only a measuring microscope with an ocular micrometer is available, then, nevertheless, a gauge length of any desired size may be used by clamping a stiff measuring bar on the rod and measuring the small distance between the mark on the end of the bar and a mark on the rod. (Fig. 229).

The distance between two measuring marks can also be determined by unloading the rod from time to time and reloading it after measurement. If the duration of unloading is taken as vanishingly small, then in the ideal case, on reimposition of the load, the previous length of specimen and the former creep rate set in. In the degree in which the duration of unloading is increased, the instantaneous creep and the transient creep return. In extensive creep tests on lead with intermediate periods of recovery, KENNEDY [655] determined that the creep curves after reloading could be interpreted as if a fraction n of the material recovers completely, while the part $1 - n$ behaves as though no unloading had taken place. On application of rest periods for the measurements, greater strains are therefore to be expected than if the creep tests had proceeded without interruption.

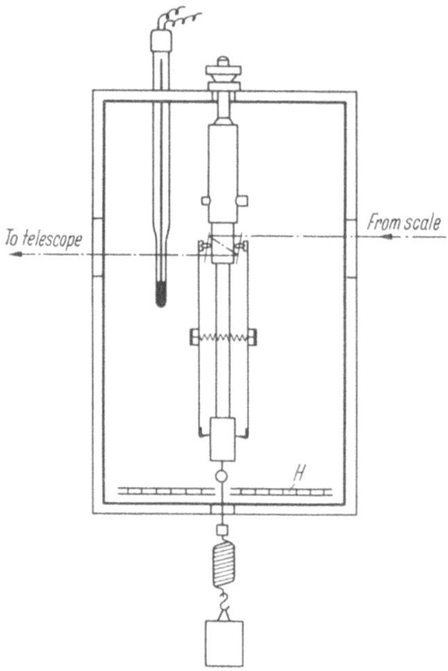

Fig. 230. Measuring arrangement with Martens Mirror Extensometer for creep tests. H = Heating.

A sufficiently high accuracy of measurement of about $10^{-4}\%$ can also be attained for shorter durations of test by using a Martens mirror extensometer in the arrangement in Fig. 230 [483, 463]. The specimen is in a thermally insulated box which is kept at constant temperature. If a room with temperature regulation is not available, then a cellar with additional heating is often a sufficient substitute for tests of industrial accuracy.

It would be desirable, as a matter of principle, for all creep curves to be carried through to fracture of the test piece. As the necessary duration of test would (particularly for the low stresses of interest in practice) amount to many years, it is usually regarded as satisfactory to follow the creep curve into the region of minimum creep rate. The rates of strain on a logarithmic scale determined from various creep tests are plotted against stress on the linear or logarithmic scale, and in both cases approximately straight lines are obtained (see p. 237). From the diagram the stress relating to a permissible rate of strain can be derived. This stress is, in accordance with the German standard specification DIN 50119 "Standversuch", described as the limiting creep rate. As a permissible creep rate, for instance for use as water pipe, the value of 0.1 to 1 · 10^{-4}%/hour has been proposed for lead (v. HANFFSTENGEL and HANEMANN [482]). It may be pointed out that a creep rate of 10^{-4}%/hour closely corresponds to a strain of 1% per year. If only one bar or one testing apparatus is available, the test is begun at the lowest load, and the latter increased from time to time when a minimum creep rate has set in.

Lead alloy										
Nominal alloy	Chemical composition (Wt.-%)									
	Ag	Cu	Bi	Sb	Sn	As	Zn	Ni	Fe	Ca
Calcium-Lead	0.0022	0.051	0.003	Sb+Sn+As=0.325			0.001	Trace	0.007	0.030
Calcium-Lead	0.0016	0.052	0.001	Sb+Sn+As=Trace			0.001	0.0029	0.0009	0.037
Arsenic-Lead	0.003	0.045	0.023	0.031	0.033	0.059	0.0006	0.0004	0.001	
Arsenic-Lead	0.001	0.038	0.033	0.021	0.143	0.111	0.0003		0.0003	
Arsenic-Lead		0.06	0.012	0.029	0.113	0.16	0.001	0.001		
„Corroding Lead"	0.0003	0.0001	0.005							
Antimony-Lead				1.0						
Desilverized Lead	0.0007		0.073	0.008	Trace		0.0002		0.0006	
Chemical Lead	0.0012	0.0585	0.0009	Sb+Sn+As=0.0009			Trace	0.0048	Trace	
Copper-Lead	0.0015	0.0424	0.0290	0.003	Trace		0.0001	0.0032	0.0013	
Tin-Lead					2.5					

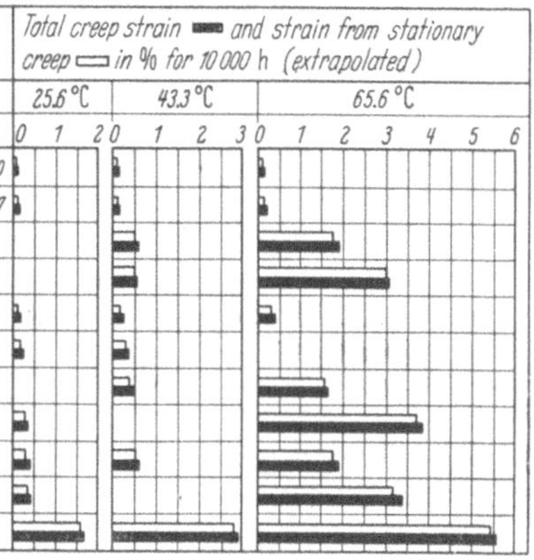

Fig. 231. Comparison of creep strains in some lead alloys (cable-sheathings) at various temperatures after 10,000 hours under 14 kg/cm² tensile stress; extrapolated from tests of 2,000 hours duration. (According to DOLLINS [866]).

Often the creep rate is not plotted in the diagram but instead the time in which the strain reaches a fixed amount, say, 1%. In the determination of the limiting creep rate the instantaneous creep and the initial strain are neglected. They can be taken into account when the linear part of the creep curve is extended backwards to the axis and the value thus obtained, having regard to the time in which a strain of 1% occurs: the procedure of MACVETTY [791], POMP [969]). According to the German standard specification DIN 50119 "Standversuch" and DIN 50118 "Zeitstandversuch", the limiting creep stress is defined as the load, referred to the initial cross-section, which brings about a fixed amount (e.g., 0.1%) of creep after a fixed time (e.g. 10,000 hours). In this figure, instantaneous creep and initial strain are included. As shown by numerous measurements, with or without regard to instantaneous creep and to initial strain (MOORE, BETTY and DOLLINS [866]), the difference is of no great significance for the assessment of an alloy (Fig. 231).

For some alloys, particularly lead-tellurium and lead-silver, creep curves have been given which, in contrast to those described, are concave upwards (GREENWOOD and WORNER [432]). This characteristic shows itself noticeably at very high creep rates (which are not allowed in constructional units). Therefore it is not particularly necessary to consider this type of creep curve at the usual rate of strain. Another type of curve, designated as B in the same investigation (GREENWOOD and WORNER [432]), which exhibits an increase of creep rate in the first stages of the creep test and then normal course of strain, was only found in commercial lead when it was quenched from about 125 °C and the test begun within two days. If the test was delayed longer, creep curves of the normal type were obtained. This observation leads to the conjecture that the slight deformation connected with quenching is to be held responsible for the initial increase in the creep rate. If the specimen was stored for some time before applying the load, normal creep curves were obtained, presumably owing to crystal recovery which had occurred. Thus there is apparently no question of a special type of creep curve, but of a side-phenomenon determined by the nature of the test. Various peculiarities in published creep curves can be attributed to similar unintentional occurrences, such as deformation and recrystallisation, or formation of precipitates from supersaturated solid solutions. Perhaps this is the reason for the peculiar creep curves for commercial lead in an American investigation, where the creep rate decreased considerably about one year after the beginning of the test (PHILLIPS [959]).

A further observation was made on extruded specimens for creep tests (GIFKINS and COE [382]). In X-ray back-reflection photographs, which were used as the criterion for complete recrystallisation, it was shown that in a surface layer of about 0.025 mm, an essentially finer structure (perhaps a portion of deformed structure) was present than inside the specimen. Even a long period of heat treatment brought hardly any change of structure of the surface zone. If the die and the lead ingot were lubricated, then this effect did not occur in a small number of specimens. Neither did specimens, produced by rolling, show this effect. From existing experiments it may be concluded that the observed skin effect had no influence on creep behaviour.

Basically, views on the creep strength of a material may also be obtained if the test piece is not subjected to creep under constant stress, but given a single permanent deformation, and the gradual fall of stress at constant temperature (relaxation) is observed. This principle became known, for instance, by torsion tests carried out by WELLINGER [1256] on steel, which were discussed by a committee of the International Institute of Welding (IIW). On the whole, relaxation experiments on lead have been carried out much less frequently than creep tests, with the determination of strain.

g) Effect of Prestrain on the Creep Rate. In the course of an extensive study of the creep behaviour of lead and lead alloys, HOPKIN and THWAITES

[589] investigated the creep rates of a very pure lead-copper (0.1%) and lead-tin (1%) alloy after prestraining by various amounts. They found that, owing to prestraining, the type of creep curve changed. The instantaneous

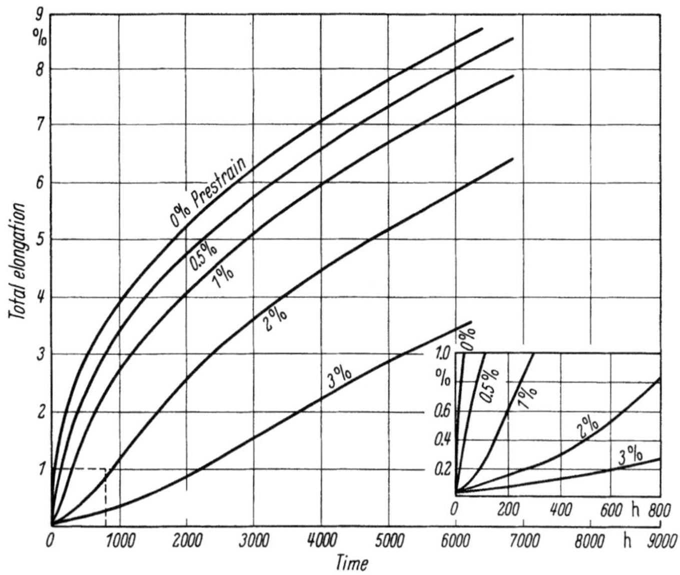

Fig. 232. Effect of prestraining on the creep rate of lead with 0.1% Cu at 35 kg/cm². (According to HOPKIN and THWAITES).

creep was not followed by a period of high creep rate which gradually died away (transient creep), but on the contrary, the creep rate, after the instantaneous creep, was at first very low and then gradually increased (Fig. 232).

Fig. 233. Effect of straining on the strain behaviour of coarse-grained lead under a load of 14 kg/cm². Grain diameter 9 mm. (According to v. HANFF-STENGEL and HANEMANN).

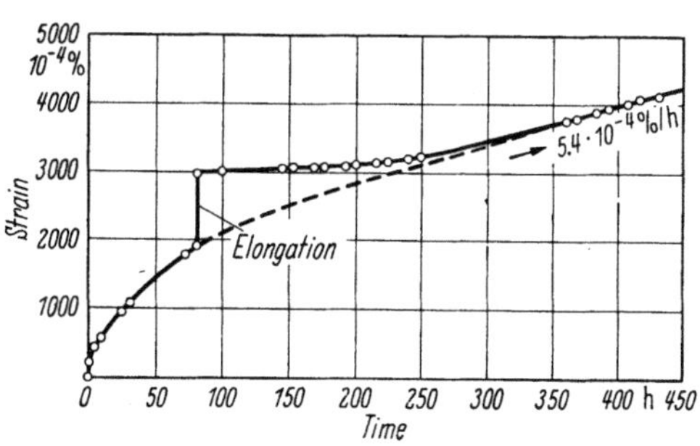

This increased creep rate then fell in the course of time to the value for the non-prestrained specimen (cf. Fig. 233.) At the end of the tests all the rods in fact had the same creep rate, but the total strain was the lower, the higher the amount of prestraining. Recrystallisation does not seem to have occurred in these tests; the initial acceleration of strain was attributed to polygonisation (a phenomenon which accompanies crystal recovery), as it was combined with a sharpening of the originally diffuse X-ray reflections. A curve similar to the one described, with points of inflexion, was also found by v. HANFFSTENGEL and HANEMANN [483] with commercial lead after very slight prestraining. Here, too, no recrystallisation took place, but only crystal recovery. If the degree of prestraining is such that recrystallisation occurs at the beginning of the following creep test, considerable increase in creep resistance can be expected during recrystallisation (Fig. 234).

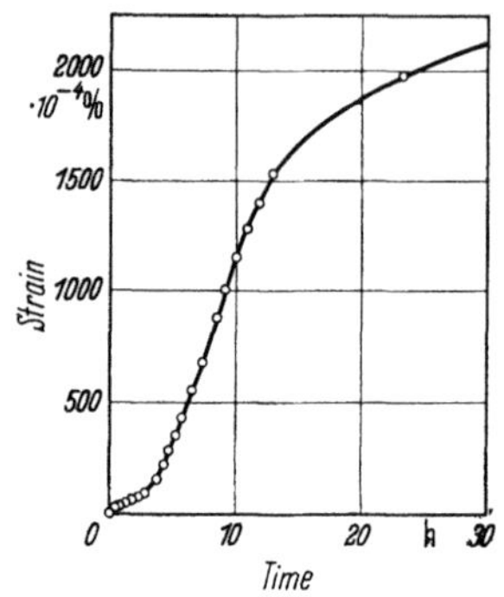

Fig. 234. Creep of lead under a load of 16 kg/cm². Prestrain 2.3%. (According to v. HANFFSTENGEL and HANEMANN).

h) Creep under Multi-Axial Tension and Pressure and under Bending Stress. Most creep tests are carried out under uniaxial tensile stress. In the use of lead as a constructional material, however, multiaxial stressing conditions occur. The treatment of creep under multiaxial stress is based on simple hypotheses (FINNIE and HELLER [321], ODQVIST and HULT [910a]).

1. For the uniaxial stress condition a relationship $\dot{\varepsilon} = f(\sigma)$ is known in the static part of the creep curve; e.g. $\dot{\varepsilon} = B \cdot \sigma^n$.

2. The creep takes place under constant volume, i.e. $\dot{\varepsilon}_1 + \dot{\varepsilon}_2 + \dot{\varepsilon}_3 = 0$.

3. The rate of deformation is not affected by a superimposed hydrostatic pressure.

4. The material is isotropic.

According to the theory of HUBER-v. MISES-HENCKY a reference stress σ^* can be defined from the three principal stresses, and a corresponding value $\dot{\varepsilon}^*$ can be defined from the creep rates.

$$\sigma^* = \frac{1}{\sqrt{2}} [(\sigma_1 - \sigma_2)^2 + (\sigma_2 - \sigma_3)^2 + (\sigma_3 - \sigma_1)^2]^{1/2}$$

$$\dot{\varepsilon}^* = \frac{1}{\sqrt{2}} [(\dot{\varepsilon}_1 - \dot{\varepsilon}_2)^2 + (\dot{\varepsilon}_2 - \dot{\varepsilon}_3)^2 + (\dot{\varepsilon}_3 - \dot{\varepsilon}_1)^2]^{1/2}$$

If it is assumed that σ^* and $\dot{\varepsilon}^*$ are related in the same way as σ and $\dot{\varepsilon}$ in the steady state part of the uniaxial creep test, that is $\dot{\varepsilon}^* = B \cdot \sigma^{*n}$ for example, then the stress distribution in a body under multiaxial creep stress can be calculated, in that the above hypotheses 1 to 4 lead to calculation of the stress by conventional means. If the distribution of stress is known, then the pattern of the creep rate can be calculated. The equations for creep rates reported initially by SODERBERG [1134] and MARIN [795] were:

$$\dot{\varepsilon}_1 = B \cdot \sigma^{*n-1} \left[\sigma_1 - \frac{1}{2}(\sigma_2 + \sigma_3)\right]$$

$$\dot{\varepsilon}_2 = B \cdot \sigma^{*n-1} \left[\sigma_2 - \frac{1}{2}(\sigma_3 + \sigma_1)\right]$$

$$\dot{\varepsilon}_3 = B \cdot \sigma^{*n-1} \left[\sigma_3 - \frac{1}{2}(\sigma_1 + \sigma_2)\right]$$

The quantities B and n are determined from the steady state part of the uniaxial creep test.

With regard to the practical significance of the multiaxial stress in the constructional use of lead for pipes and washers (pressure and friction), for example, it is desirable that the practicability of the determinations outlined above should be examined further on a broader basis.

In the case of the creep of lead pipes under stress due to uniaxial tension, internal pressure and various combinations of internal pressure and tension, both calculations and tests were carried out, (RIECHE [1011a]). In this way the tests of HOFMANN and v. MALOTKI [569a] were continued. These authors conducted comparative creep tests on pipes of various lead alloys under stress due to internal pressure and tension and they correlated a specific stress in the uniaxial tension test with the internal pressure which led to a creep rate of the same order. For the stress calculation of the pipe under internal pressure they used various methods which assumed either an elastic or a completely plastic behaviour of the material. On the basis of these considerations it seemed most appropriate, in designing lead pipes, to start with the equations for the completely plastic condition. Hence the maximum allowable internal pressure is

$$p_{i,zul} = \frac{2}{\sqrt{3}} \cdot \sigma_F \cdot \ln\left(\frac{r_a}{r_i}\right)$$

if σ_F is substituted by a creep rate limit from the uniaxial creep test. The stress distribution in the pipe under internal pressure is reported by ODQVIST and HULT [910a] for the case of steady state creep. Corresponding equations for the simultaneous stressing by internal pressure and tension were derived by RIECHE [1011a], and the stress distribution was calculated. From this it appeared that, with increasing ratio of the internal pressure to the axial load, the slope of the reference stress σ^*

over the cross-section of the pipe is steeper. In all cases the maximum value for the reference stress is on the inner wall of the pipe.

At the same loading the reference stresses, determined in this way, are always smaller on the inner wall of the pipe and always larger on the outer wall than the stresses calculated for the elastic condition; that is to say, the distribution of the reference stresses over the cross-section of the pipe is more uniform than in the elastic condition.

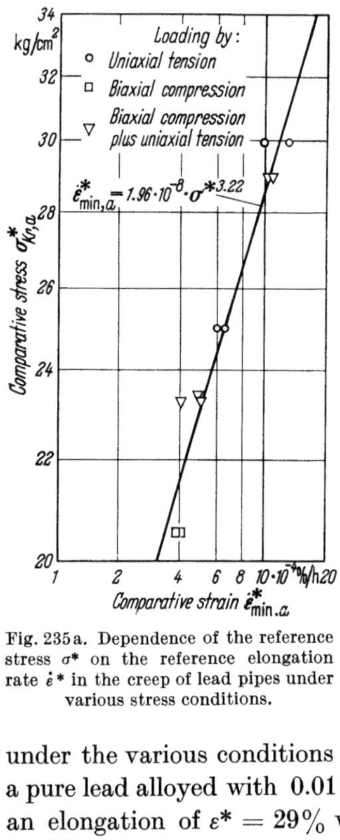

Fig. 235a. Dependence of the reference stress σ^* on the reference elongation rate $\dot{\varepsilon}^*$ in the creep of lead pipes under various stress conditions.

If the reference stress σ^* at a point on the pipe wall, calculated for the various stress conditions, is plotted on a 2-log scale against the reference elongation rate $\dot{\varepsilon}^*$ determined at this point in the test, then, in accordance with the equation $\dot{\varepsilon}^* = B \cdot \sigma^{*n}$, the points must lie on a straight line of slope n. Fig. 235a shows that this condition is fulfilled for a pure lead alloyed with 0.05% by weight of copper, within the scatter of the test values. The test results show that the dependence of the creep rate on the stress determined in the uniaxial test is valid also, in general terms, for the triaxial stress condition if the reference stress is used in accordance with the hypothesis of HUBER-V. MISES-HENCKY.

As the tests were not carried out to the fracture of the samples, only little can be said on the elongation of the lead under the various conditions of stress. It should be noted, however, that a pure lead alloyed with 0.01% by weight of silver could be loaded to an elongation of $\varepsilon^* = 29\%$ without signs of fracture.

When lead is used as a pressure-equalizing material, the frictional forces, as well as the pressure forces, play an outstanding part. In this case lead should transfer the pressure forces, e.g. the weight of a structure, uniformly onto the foundation. Thus it is expected that the lead will flow at the points at which the surface pressure would be inconveniently high for the foundation, and accommodate itself in the regions of low pressure. In this way a uniform surface pressure is imposed upon the entire foundation. On the other hand, however, the lead should not be pressed out completely from the equalizing joint after a long time, i.e. the creep of the lead should come to a halt after the uniform surface pressure has been set up.

However it is not possible to make a statement here and now, in advance, on the load-carrying capacity of lead under stresses of this type because the effect of the frictional forces between the inserted lead and the contact surfaces is largely unknown and is of varying extent, according to the size and thickness of the pressure pad. As the frictional force acts on the lead in a direction opposite to that of flow, a tri-axial compressive stress will be set up at an adequate distance from the narrow part of the slit, so that the creep rate is reduced appreciably or eliminated completely. Due to the extremely large dimensions of this type of pressure pad it is advisable to obtain, by means of tests on models, the reference data for use in practice.

For this purpose an arrangement was constructed in which lead sheets of various thicknesses and having a size of 40 · 60 sq.mm were

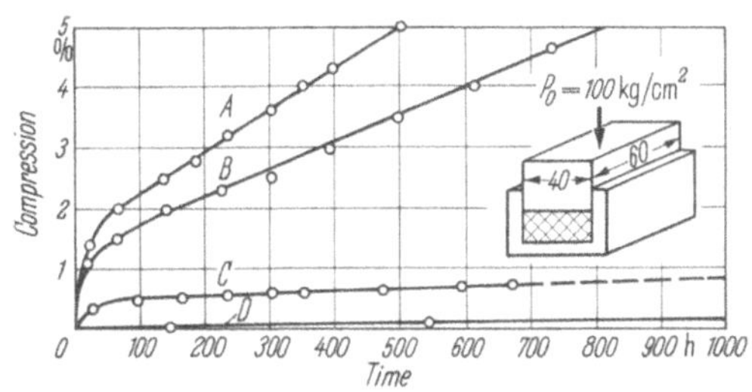

Fig. 235b. Time compression lines of rolled refined lead 99.90 (DIN 1719) of various thicknesses at compression stress of 100 kg/cm² in the apparatus illustrated above. Extrusion is possible on two sides. (The value of the compression rate $\dot{\varepsilon}$ has been extrapolated).

Curve	Sheet thickness (mm)	$\dot{\varepsilon}$ 10^{-4}%/h
A	30	70
B	20	45
C	10	4
D	5	1

subjected to a pressure stressing (Hofmann and v. Malotki [569b]). The result of the creep tests at a constant pressure stressing of 100 kg per sq.cm is shown in Fig. 235b. In this way the lead could spread out without hindrance from the front and rear openings in the apparatus shown in Fig. 235b, whilst the sides of the unit were closed. The sheet thicknesses and the compression rates taken at 10,000 hours are reported in the caption for the individual creep curves. Despite the variations in creep strength, due to differences in grain size it can be seen clearly from the curve that the creep rate drops very markedly with the sheet thickness.

According to experience with pressure welding (p. 57) it is known that it will not be possible to press the lead completely out of the apparatus even at very high surface pressures. With decreasing thickness the creep rate will fall finally to zero. If the creep rate from Fig. 235b is plotted in relation to the sheet thickness, then the curve shown in Fig. 235c is produced. The scatter of the values is to be attributed to the effect of the grain size already mentioned. An exact comparison of the creep properties of different sheet thicknesses would be possible only with identical grain sizes.

If it is desired to vary the compression rate with constant dimensions, pressure stresses and friction ratios, then the spreading of the lead must be restrained by applying a counter pressure in the openings in the apparatus or by reducing the cross-sections of the openings. Such tests were carried out by HOFMANN and v. MALOTKI [569b]. With the tests described here, with pressure and frictional forces, a calculation of the stress and elongation rate in accordance with the laws of creep mechanics should encounter very great mathematical difficulties. Moreover, the boundary conditions will be very difficult to determine.

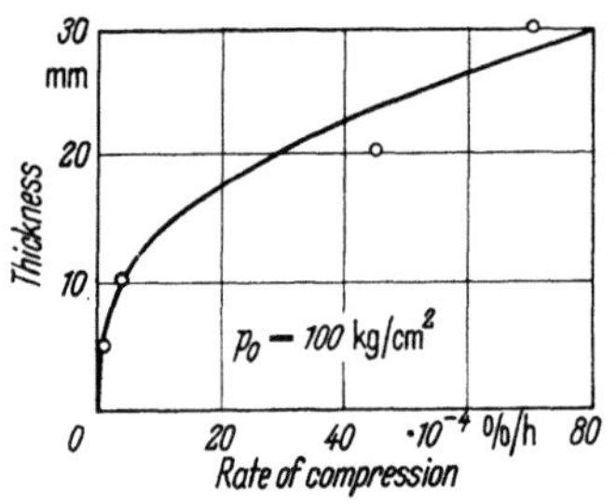

Fig. 235c. Influence of the lead sheet thicknesses on the compression rate at a constant pressure (values extrapolated). For dimensions of the specimen see Fig. 235b.

In connection with these tests rough determinations of the friction between steel and lead were also made. A lead block was placed between two steel plates and the force required to withdraw the block was measured. A friction value μ between the oxidised steel and lead of 0.8 to 0.9 was determined in this way. In this the effect of creep also became noticeable after a short period of test. After approximately 3 hours the friction value dropped to $\mu = 0.4$ to 0.5, i.e. a very slow sliding of the lead block took place with a smaller drawing force. It was observed that the surface of the steel plate was coated very quickly with lead which entered into the surface irregularities and filled them in. Thus the lead block slid entirely on the lead coating adhering to the steel, and in fact, the friction between lead and lead was being measured.

An appreciable reduction in the friction value to $\mu = 0.2$ could be achieved by coating the steel surface with graphite. Moreover a drop in the friction value to $\mu = 0.1$ occurred after a short time although particles of lead no longer adhered to the surface. If it is desired to reduce the flow resistance of the lead under pressure stressing, then a lubricant must be introduced between the lead coating and the contact surface.

A strip of lead was clamped edge uppermost and was subjected to bending moments which, in the case of elastic stress, would correspond to stresses in the range of 57.45 or 73.8 kg/cm² (TAPSELL and JOHNSON [1173]). The elongation or compression was measured continuously at various distances from the neutral axis during the test period of 7 days or 30 hours. Both tests showed that the elongation or compression of the various strands behaved at any time in accordance with the distance from the neutral axis, i.e. that the creep curves of the individual strands have the same dependence on time, and, by compression in the direction of the ordinate axis, can be related to the distances between the strands and the neutral axis. Similarly, as for elastic deformation, planes, during creep under bending strains, remained as planes. In order to determine the longitudinal strands, and to follow up their time change, creep curves of the same material were obtained by tensile stresses at 21.1 to 49.2 kg/cm², and creep rates were determined in relation to stress and to the duration of test. By equalizing the stresses, to which the same creep rates correspond in tensile and bending tests, the distribution of stress was obtained in the bending specimens. The distribution of stress that resulted was apparent 12 minutes after the test had begun and did not change any further. This means the same time relationship exists in creep rate for the longitudinal strands in the bending test and the tensile specimens. An earlier publication (MacCullough [788]) had led to similar conclusions. Further tests must first show whether the results can be generalised.

i) Tests on Single Crystals. The grain boundary slip should not apply as a component part of the creep stress on single crystals. Creep tests on single crystals therefore attract special interest and consequently should be given more consideration in future investigations into the effect of alloying elements on creep resistance. The work described in the first edition of this book on single crystals (Baker, Betty and Moore [36], Moore, Betty and Dollins [865]) was not quite satisfactory, particularly as the stresses applied, between 24.9 and 39.5 kg/cm², were rather high. Ley [566] and the present author carried out creep tests on 10 to 11 mm thick zone-refined crystals of a purity of 99.99%. The rods were very carefully installed in order to avoid bending stress, and the strain was measured at 25°C by a Martens mirror apparatus (Fig. 230). The creep curves taken on six different carefully centred crystals showed a very speedy reduction of creep to complete stand-still, on loads which were above the creep limit of polycrystalline lead. After exceeding the creep limit, creep curves of normal appearance were obtained so that the initially rapid flow changed into one with constant strain rate. The resulting critical shear stresses in the octahedral slip system amounted, on an average, to 6.3 kg/cm². Two further crystals, that were particularly carefully handled, were soldered into brass sockets. The crystals behaved in a purely elastic way up to the creep limit (Fig. 236). Their critical shear stress was almost the same as the value given above. The values of the elastic modulus were, for the one crystal, slightly above, for the other slightly below, the value for quasi-isotropic lead.

More recent investigations were carried out at the Carnegie Institute of Technology at Pittsburgh on lead single crystals of 99.999% purity. In the tests at $-190\,^\circ C$ the lead crystals were found "hard". By this it is understood that every increase of load is combined with a corresponding increase of length, which ceases after 10 to 20 minutes. Steady state creep practically did not occur, so that the yield point could be determined in the usual way. The critical shear stress in the octahedral slip system amounted to 9.6 ± 0.8 kg/cm². At 25 °C, as expected, lower values of the critical shear stress were found, namely 7 kg/cm² in the non-heat-treated specimens and 3.4 kg/cm² in those which had been heated to 90 to 120 °C. From this and other observations it was concluded that at room temperature lead can retain a certain strain-hardening for

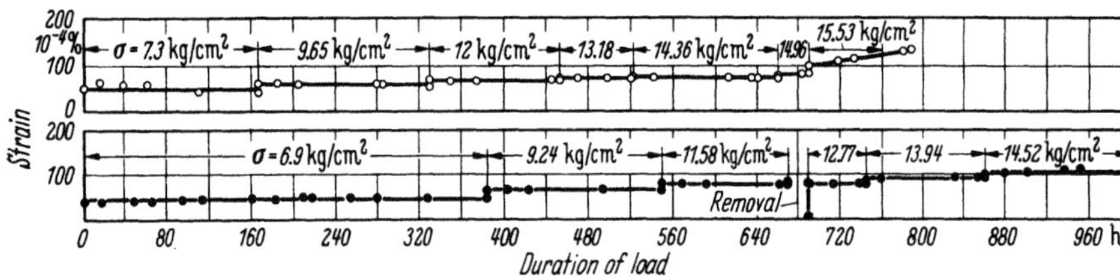

Fig. 236. Behaviour of single crystals of lead in a creep test. (According to HOFMANN and LEY [566]).

weeks. The creep limit still amounted to 1.8 ± 0.4 kg/cm² at 110 °C. The values agree throughout, in order of magnitude, with the results of the work described above; unfortunately no long-time tests were carried out. A continuation of these tests is desirable for these and other reasons. Creep tests on single crystals were also undertaken in Australia by HIRST [532, 533, 534]. These were, by and large, at higher creep rates than in the two investigations previously discussed, so that the results are different. The change of orientation of the crystals in creep was observed and thereby proof obtained that slip on the octahedral planes {111} takes place in the ⟨110⟩ direction. According to HIRST's observations [532, 533], a single crystal specimen creeps faster than a polycrystalline one of the same dimensions under the same tensile stress. This does not exclude the fact that at very low tensile stresses the reverse is true (p. 209). v. MALOTKI [569a] set himself the task of studying the creep of single crystals of the same orientation in unalloyed lead and in lead alloys. The initial 99.99% lead has been so purified to exclude traces of noble metals. The alloy single crystals with copper had a heterogeneous structure; those with indium, bismuth or tin at a concentration of 1 At.-% were single phase. Other single crystals contained 0.035 Wt.-% Te or 0.01 Wt.-% Ag. The latter alloys in the main, formed solid solutions. The silver-containing single crystals exhibited marked honeycomb segregation. In all the alloy single crystals the creep strength was increased as compared with that of unalloyed lead, least in those containing copper, most in those containing silver (Fig. 237). The lead-tellurium single crystals were distinguished by an exceptional capacity for strain hardening.

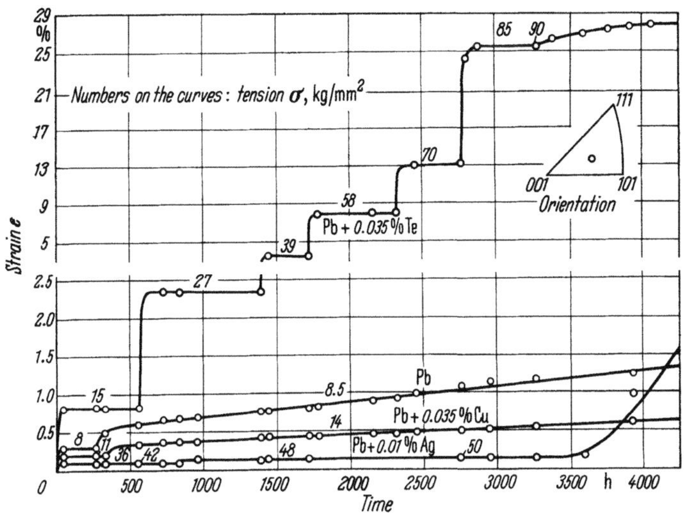

Fig. 237. Creep of single crystals of alloyed lead.

The alloys with indium, tin, bismuth (atomic radii 1.57; 1.58; 1.82 Å; atomic radius of lead: 1.75 Å) behave almost alike in the creep test. According to GEGUZIN [364a] another result was to be expected. He says that in a solid solution the creep rate depends only on the degree of saturation of the solid solution. However, in v. Malotki's experiments, the degree of saturation was far higher in the lead-tin alloy than in lead-indium.

A theoretical formulation for the creep of single crystals is given by FOLBERTH and KOCHENDÖRFER [331].

j) Is There a True Creep Limit? The investigation of creep processes has led in recent decades to finer and finer methods of measurement. The question arose whether there is really a true lower limit creep (SPÄTH [1139], v. HANFFSTENGEL and HANEMANN [483]). DEHLINGER [241] considered this problem thermodynamically. If a very small load is applied to a polycrystalline specimen, it momentarily undergoes a slight elastic strain. If, now, in very long periods of time, this is followed by a measurable elongation due to creep, then the latter would be combined with an increase of the elastic energy of the specimen owing to the necessary deformation of the grains. As the duration of this imaginary experiment can last as long as desired, and therefore the load as small as desired, then the assumed increase of energy would (below a certain load) exceed

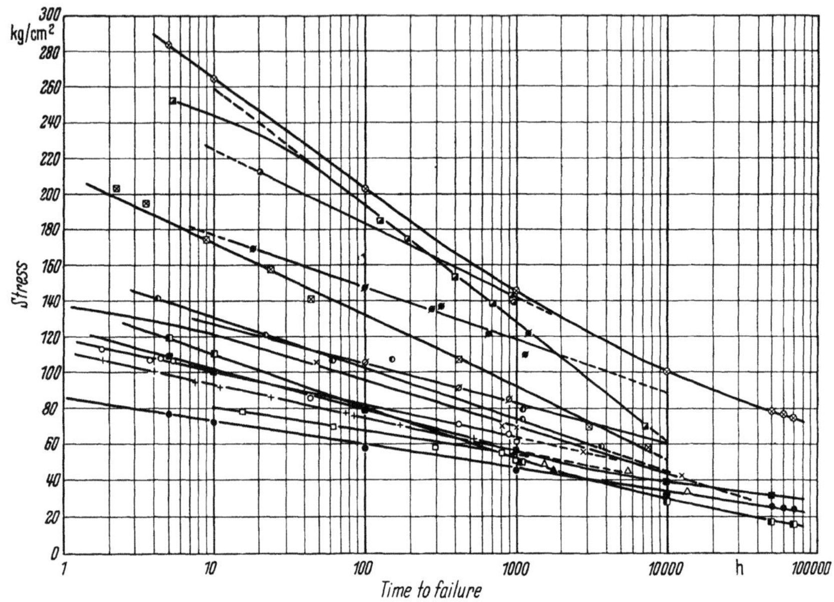

□ Lead (99.89%) with 0.09% Bi
+ Commercial Lead 25.8 °C
○ Commercial Lead 21.1 °C
◐ Cable Sheathing 2% Sn
⊠ Cable Sheathing 0.75% Sb
Cable Sheathing 0.03% Ca
◪ 0.04% Ca Material 1
Material 2
(According to Moore and co-workers [866]).

⌀ Cable Sheathing 0.16% As + 0.113% Sn. (According to Dollins [254]).

● Refined Lead (stamped)
◧ Refined Lead + 0.06% Te (stamped)
■ Chemical Lead 0.059% Cu
⊗ Chemical Lead 0.05% Ca
(According to Gohn and co-workers [402]).

× Lead Pipe 0.018% Ag (According to Gohn and Ellis [403]).

△ Commercial Lead, Coarse Grained
▲ Commercial Lead, Fine Grained
(According to McKeown [817]).

Fig. 238. Creep-to-rupture of lead and lead alloys.

the product load · distance of creep = work of creep. This means that now the increase of elastic energy is supplied by absorption of heat from the surroundings. As this, according to the second law of thermodynamics, is impossible, then below the given stress even this slow creep must cease, i.e., there must be a true creep limit.

k) Creep-to-Rupture. The stress which leads to fracture after a given time, e.g., 10,000 hours, is in the terms laid down in German standard specification DIN 50119, designated as creep-to-rupture. According to MOORE and co-workers [866], the creep-to-rupture of lead alloys is obtained when the stress referred to the initial cross-section is plotted against the time to occurrence of fracture on the logarithmic scale. By this means of representation, straight lines are obtained according to Fig. 238. In addition to the values by MOORE, recent ones by other workers [402, 254, 403] are considered in Fig. 238. As the tests extended to 300 days at the most, the extrapolation of the straight lines to longer times is uncertain.

In connection with the question of creep-to-rupture, attention must again be drawn to the phenomenon considered under "tertiary creep", namely that the elongation-at-fracture is lowered by increasing the duration of the experiment in the creep test. Under certain circumstances, the creep fracture, contrary to the tensile fracture, therefore shows no necking. This was observed in the alloys of lead with antimony and lead with tin (MOORE, BETTY and DOLLINS [865]), and further, in the lead-tellurium alloys (GREENWOOD and WORNER [432] (c.f. p. 411). In general, commercial lead shows necking in the creep-to-rupture test. Only in specimens which were quenched from about 125 °C and immediately loaded, did brittle fracture occur, combined with cracks at the grain boundaries (GREENWOOD and WORNER [434]). The picture of a specimen showed very coarse grain. Intercrystalline fracture as a result of coarse grain was also observed by other investigators, particularly at stresses of about 30 kg/cm².

l) Dynamic Creep Strength. The effect, on the rate of creep of commercial lead of the superposition of a dynamic stress on the static stress was, in particular, investigated by BERNHARDT [76]. A machine for tension-compression stressing allowed variations of the static preload between 0 and 50 kg, the dynamic portion of the load between 0 and $\pm$ 20 kg, and the frequency between 1.5 and 25 Hz. By means of a mirror apparatus, the elongation could be exactly measured from 10^{-4} to 10^{-5}%. As shown in Fig. 239, at an equal high stress (σ_0) a growing rate of creep was found, as the frequency increased. The break in the curves, assumed by BERNHARDT [76] in agreement with v. HANFFSTENGEL [483], is of no significance for the present problem, and will therefore not be treated further. It may be assumed that the increase of creep rate with the frequency, as shown in Fig. 239, can be mainly attributed to an increase of the temperature of the specimen as a consequence of damping (p. 13). Measurements of the temperature of the specimens were not carried out.

According to the results in Fig. 239, the rate of creep in the static test is lower than in the dynamic test with the same upper stress. Investigations by LEY [747] with similar apparatus, and by KENNEDY

[656], led to similar results (cf. VASIL'EV[1216a]) MEYER [849] established in careful tests on steel that creep under a purely static stress of 90 kg/mm² took place no more slowly than under a dynamic stress of 89 $\pm$ 1 kg/mm² at a frequency of 9 Hertz. Further experiments must be awaited

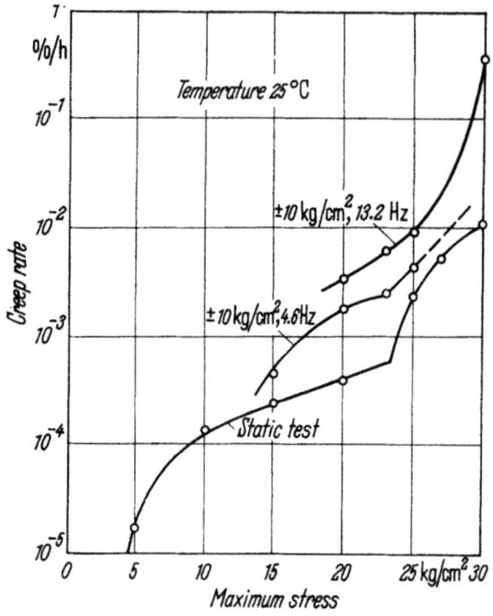

Fig. 239. Creep rate of commercial lead under static and dynamic stress. (According to BERNHARDT and HANEMANN).

in order to decide whether there is a fundamental difference in the dynamic creep test between steel and lead. It may however be expected with certainty that the creep rate will be increased if a stress vibration is superposed above and below a static average stress that is, if the dynamic creep test is not compared with a static creep test at the upper stress σ_0, but with a creep test at the stress σ_m.

Experiments to determine the effect of a dynamic stress on the rate of creep were also carried out by MOORE and co-workers [866]. The stress was $\sigma_0 = \pm 2.11$ kg/cm², the frequency 16.7 Hz. Only the time to fracture and the elongation at fracture were determined for commercial lead and for lead with 2% Sn, lead with 1% Sb, and lead with 0.03% Ca. The life was considerably reduced as compared with the values shown in Fig. 238 and indeed for commercial lead, according to the magnitude of the average stress, to 1/30 th–1/100 th; for lead-tin to 1/6 th–1/7 th; on the other hand, for lead-antimony and lead-calcium to only about half. Only in lead-calcium was the elongation at fracture notably reduced. The experiments were also extended to single crystals. New observations by GREENWOOD and COLE [428] and by GREENWOOD [423] on this problem are also available. The latter work throws no particular light on the fundamental questions, as practical points of view predominated. The rate of creep of specimens under tensile stress was increased by superposed vibrations. A survey on creep under dynamic stress for different materials is given by VITOVEC [1227].

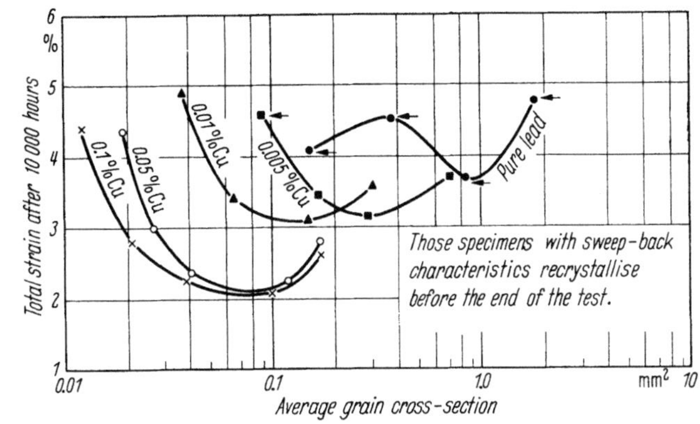

Fig. 240. Effect of grain size on creep rate of lead-copper alloys at a stress of 21 kg/cm². (According to HOPKIN and THWAITES [589]).

m) Alloy Structure and Creep Strength. Some general points of view on this theme are set out by ALLEN [16]. Fundamental investigations on the relationship between alloy structure and creep strength were carried out by HOPKIN and THWAITES [589] at the British Non-Ferrous Metals Research Association. The purest lead was converted into single phase alloys by additions of tin, cadmium, indium, antimony and thallium; and into 2-phase alloys by additions of copper. The alloys were worked in an extrusion press. By extruding at various temperatures and speeds it was possible to study the effect of grain size and grain fragmentation on the creep behaviour. By and large, there was a decrease in the creep rate with increasing grain size as in Fig. 240. The curves of the lead-copper alloys illustrated however possess a minimum of creep rate at a certain grain size, above which the creep rate again increases. Further observations on this point will be found below. If the curves for the alloys in Fig. 240 are compared with those of pure lead, the reduction of the creep rate with increasing copper content can be recognised. The creep rate is also reduced (apart from inclusions) by dissolved atoms, as the example of tin additions in Fig. 241 illustrates. On this point attention is drawn to experiments on alloy single crystals. In a further series of experiments, cadmium, indium, tin, thallium and antimony were added to lead at a content of 0.09 At.-%. At this concentration all the binary alloys mentioned are still single phase at room temperature. By extrusion and subsequent tensile straining and annealing two specimens of each alloy with different grain sizes were obtained. In creep tests under a stress of 21 kg/cm², the lead-cadmium and the lead-indium alloys had the

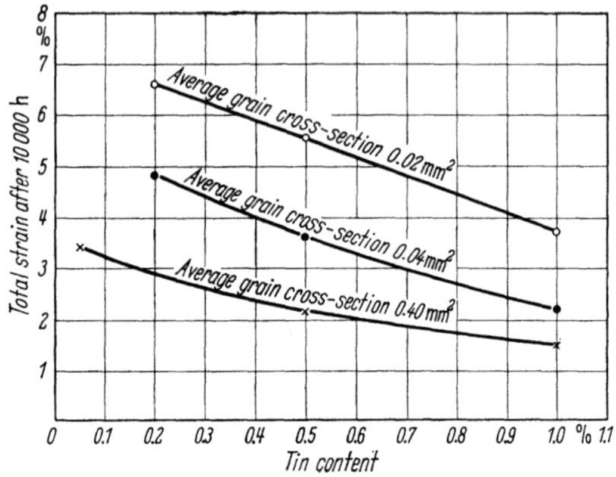

Fig. 241. Effect of tin content on creep strength of specimens of similar grain size under a stress of 21 kg/cm². (According to HOPKIN and THWAITES [589]).

lowest creep rate. It is noteworthy that among the alloying metals considered, cadmium and indium have the greatest difference in atomic radius with lead. According to COTTRELL's theory [217] it may be assumed

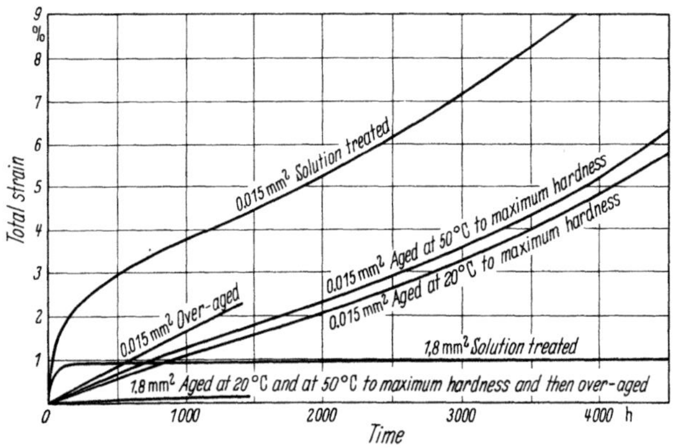

Fig. 242. Effect of age-hardening on creep behaviour of lead with 0.9% Sb and 0.001% As. (According to HOPKIN and THWAITES [589]).

that such elements preferably remain in the neighbourhood of dislocations and reduce the mobility of the latter (equilibrium segregation). This would provide a certain explanation for the favourable creep behaviour of these alloys. At the same time a relationship between the reduction of the creep rate and the decreased tendency to recrystallisation was found. The authors adduce several observations which point towards a recrystallisation-inhibiting action of both elements. There are however alloys with much greater inhibition of recrystallisation, e.g., lead with silver.

Further, alloys with 0.9% Sb and 0.001% As were age-hardened by storage at 20° and 50 °C (Fig. 242). The precipitates thus formed were at the limit of microscopic resolution. In part, these alloys were also stored beyond the maximum of hardness increase. The effect of age-hardening consisted, particularly, in a reduction of the amount of transient creep. The considerable increase of the creep strength of age-hardened lead-antimony-arsenic alloys and other age-hardened alloys, such as lead-calcium, has already been thoroughly treated in earlier investigations at the Bleiforschungsstelle in Berlin (p. 238).

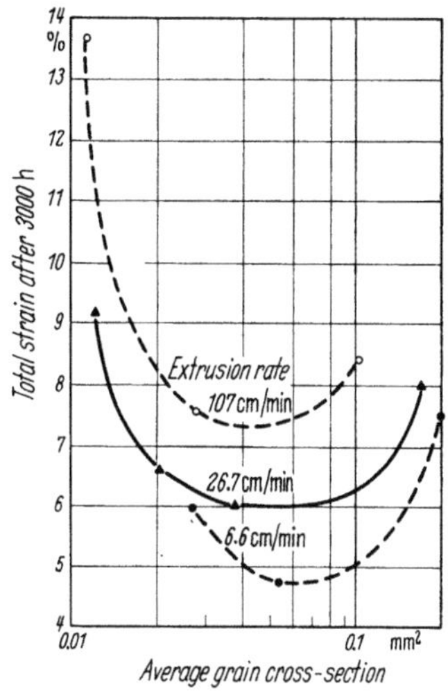

Fig. 243. Effect of the rate of extrusion on the creep strength of lead with 0.1% Cu at 35.1 kg/cm². The curve of lowest rate of extrusion indicates greatest grain fragmentation. (According to HOPKIN and THWAITES [589]).

The grain fragmentation of the specimens in dependence on the temperature and the rate of extrusion was investigated by means of X-ray photographs with filtered copper radiation. According to the arrangement by BARRET [46], specimen and film were synchronously swivelled through an angle of 10°. The grain fragmentation can thus be recognised by an increase in size of the interference spots. It was found that the amount of grain fragmentation increases with falling temperature and with reduced rate of extrusion. Apart from this, the grain fragmentation in lead-copper alloys increased with the copper content. Grain fragmentation lowers the creep rate, as shown in Fig. 243.

By working and annealing (100 °C) a series of lead-copper alloys were obtained with a recrystallised grain without grain fragmentation. Metallographic observation of the specimens showed that grains of very irregular size had been formed with numerous twins. The representation of the creep rate fo these specimens in relation to the grain size and the copper content in Fig. 244 offered an unequivocal conclusion. As expected, the creep rate fell with increasing grain size and with increasing copper content. The creep rate of these specimens was in general below those in Fig. 240. On the basis of these results, a recrystallised structure of very variable grain size with twins would be regarded as exceptionally favourable in respect of creep properties.

The comparison of the creep behaviour of wrought and cast alloys of the same composition deserves special interest. It is known that cast

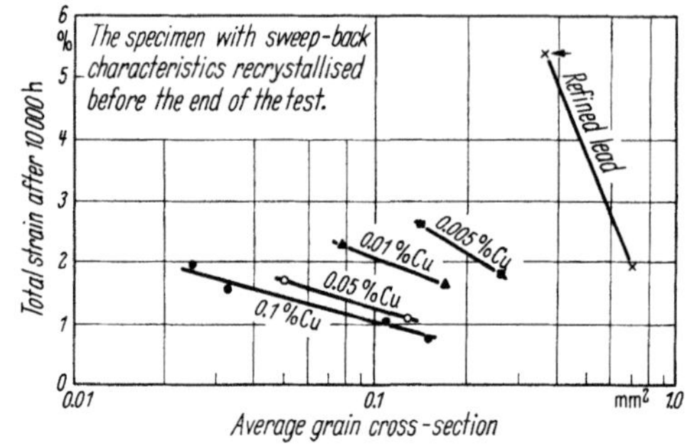

Fig. 244. Effect of grain size on the creep strength of recrystallised lead-copper alloys at a stress of 21 kg/cm². (According to HOPKIN and THWAITES [589]).

alloys exhibit a higher creep strength than wrought alloys. This is in general attributed to the coarser structure of the cast state. Other factors certainly play their part here, for instance, the number of crystal structure defects in the cast and in the recrystallised material. The difference

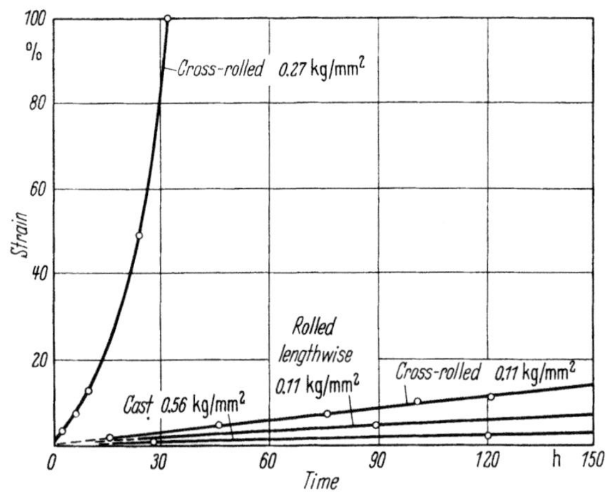

Fig. 245. Lead with 64.5% Sn. Creep curves of cast and rolled sheets under tensile stress. (According to HOFMANN and ENGEL [560]).

between wrought and cast alloys was especially clear in the present author's own investigations with the eutectic lead-tin alloy, see Fig. 245. The creep rates of the alloy in various conditions of work differ here by orders of magnitude.

n) Results for Individual Lead Grades and Alloys. As measurements of creep strength in different laboratories often diverge strongly from one another, results of various origin for a number of lead grades and lead alloys were compiled to provide a general view. In the following curves,

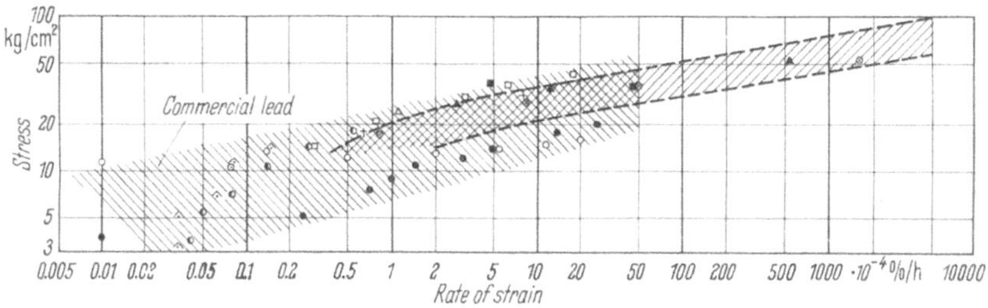

(According to PHILLIPS [959]).
○ coarse-grained } Pure lead (99.99%). (According to v. HANFFSTENGEL
● fine-grained } and HANEMANN [483]).
+ determined 1935 [478]
△ U-lead (99.99%). (According to GREENWOOD and WORNER [434]).
□ coarse-grained } (According to MCKEOWN [817]).
■ fine-grained }
◮ pure lead } (According to MOORE among others [866])
△ chemical lead } at 30 °C. (According to GOHN and others [402]).

Fig. 246. Creep rates of commercial lead from measurements in various laboratories.

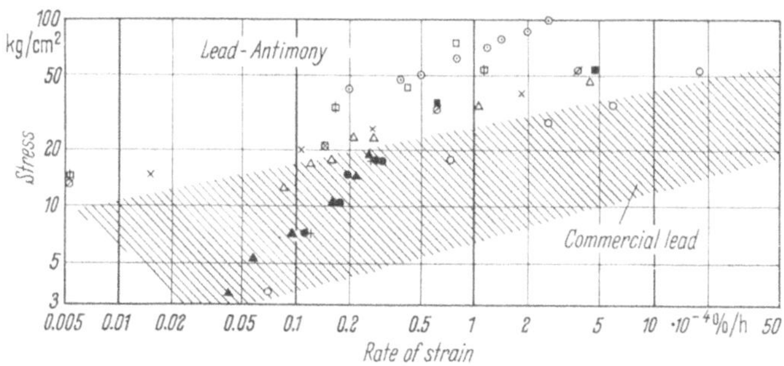

+ 0.75% Sb }
● 0.75% Sb } (According to MOORE, BETTY and DOLLINS [866]).
▲ 1.0% Sb }
■ Antimonial lead A, determined 1935.
⊙ 1% Sb + 0.05% As, determined 1937 [477].
× 0.8% Sb + 0.02% As }
⌀ 1.0% Sb + 0.02% As } (According to Stockmeyer).
⦶ 1.2% Sb + 0.02% As }
△ Arsenical lead from Bleiind. A. G.
⊠ Antimonial lead A, tested after 4 1/2 years storage [563].
□ 1% Sb, artificially aged. (According to PHILLIPS [959]).
○ 1% Sb at 30 °C (According to GOHN among others [402]).

Fig. 247. Creep rates of antimonal lead from measurements in various laboratories.

the stress is plotted on the double logarithmic scale against the rate of strain in steady state creep. In this type of plotting, the results for commercial lead (see Fig. 246) fall on to a band showing slight curvature (p. 215).

The evaluation of lead-antimony alloys with about 1% antimony as used, for instance, for pipes, was long in dispute. At different times a superiority in creep was found, compared with commercial lead at higher stresses (which in practical operation are no longer permissible) but on the other hand, not at lower stresses (MOORE and ALLEMANN [864], v. GÖLER [392]). In several investigations (HANEMANN and HOFMANN [478], HANEMANN, v. HANFFSTENGEL and HOFMANN [477]) it was shown that the behaviour described applies to soft or only slightly age-hardened antimonial lead, but on the other hand, in age-hardened lead-antimony, particularly if arsenic is present, there is superiority over commercial lead even at low stresses. These results were confirmed in investigations by the British School (HOPKIN and THWAITES [589]). Fig. 247 shows clearly the two groups of alloys and (particularly in the left-hand side of the diagram) their creep

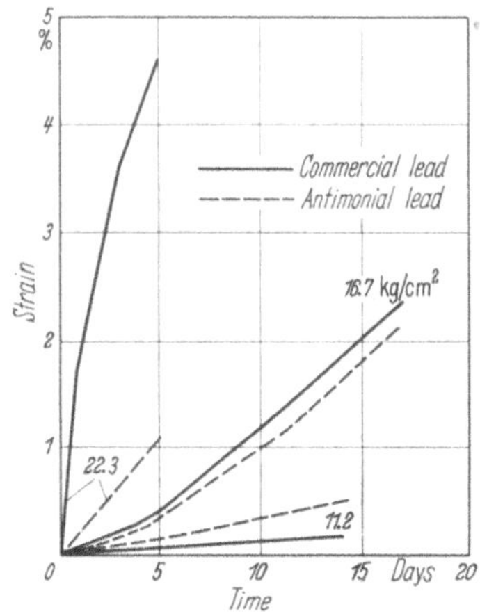

Fig. 248. Creep curves of commercial lead and antimonial lead with 1.16% Sb and 0.04% As at 85 °C [556].

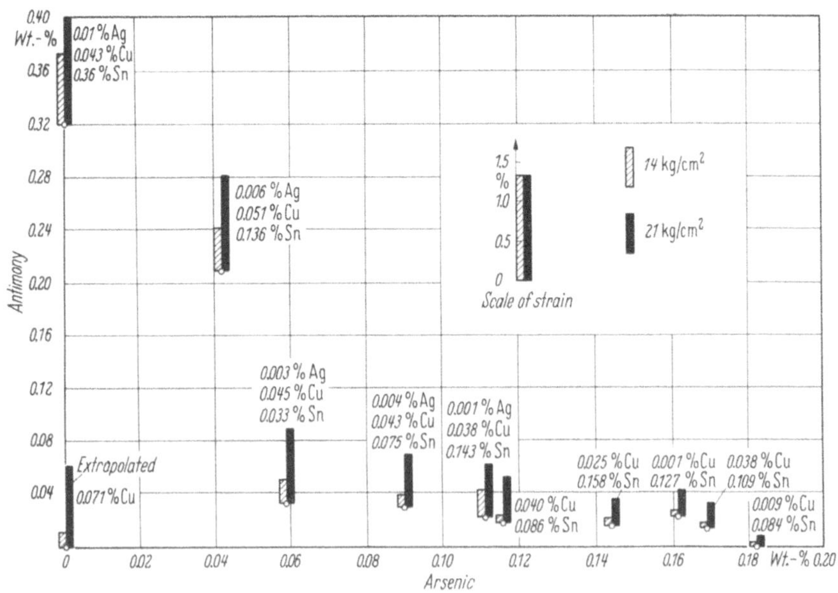

Fig. 249. Creep strain of lead alloys after 5,000 hours at 43.3 °C under various stresses, plotted in relation to the antimony and arsenic contents. (According to DOLLINS [254]).

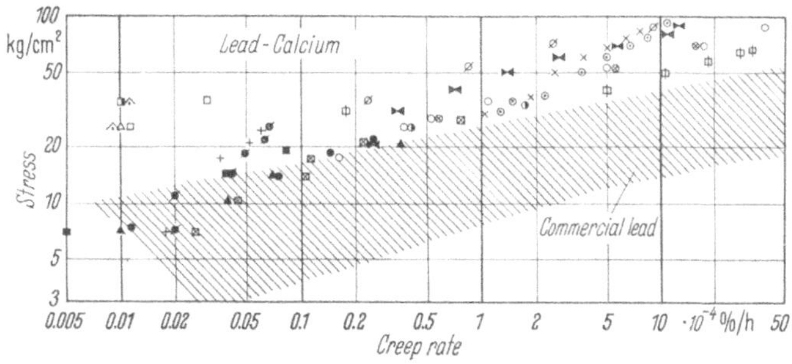

- ● 0.03% Ca, ▲ 0.03% Ca, ■ 0.04% Ca, ◑ special (According to MOORE and others [866]).
- 0.01% Ca, ◧ 0.05% Ca, △ 0.1% Ca (According to GREENWOOD and WORNER [434]).
- × 0.10% Ca quenched, ⊙ 0.10% Ca air cooled, ▶◀ 0.12% Ca quenched, [|] 0.13% Ca air cooled — Cast. Bleiforschungsstelle [502].
- ⌀ 0.027% Ca, ○ 0.043% Ca, ⊗ 0.05% Ca at 30 °C. (According to GOHN among others [402]).
- + 0.03% Ca, ⊠ 0.037% Ca (According to DOLLINS [254]).
- □ 0.055% Ca, ⩓ 0.105% Ca (According to GREENWOOD and COLE [429]).

Fig. 250. Creep rates of lead-calcium alloys. From measurements in various laboratories.

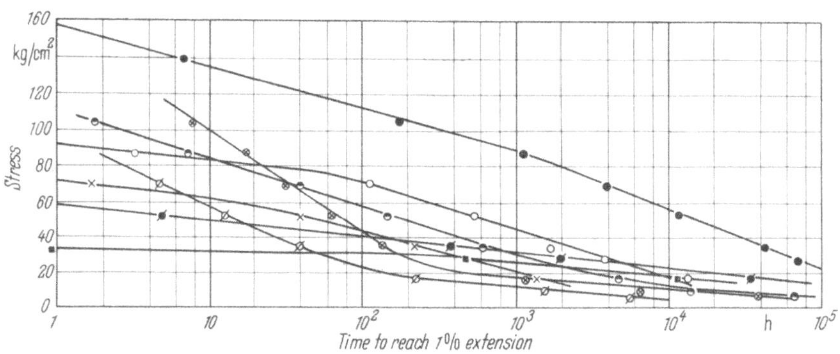

- ● Pure lead + 0.027% Ca, ○ Chemical lead + 1% Sb, × Pure lead + 0.06% Te, Chemical Lead, ■ Pure Lead — Extruded in laboratory.
- ⊗ Chemical Lead + 0.9% Sb, ◓ Secondary Lead + 3% Sn, ⌀ Secondary Lead + 0.7% Sb — Cable Sleeves.

Fig. 251. Time yield limits of lead and lead alloys. (According to GOHN and co-workers).

rates differing almost by an order of magnitude. The region of values for commercial lead is shown hatched. The superiority of antimonial lead over commercial lead is lost at higher temperatures (Fig. 248).

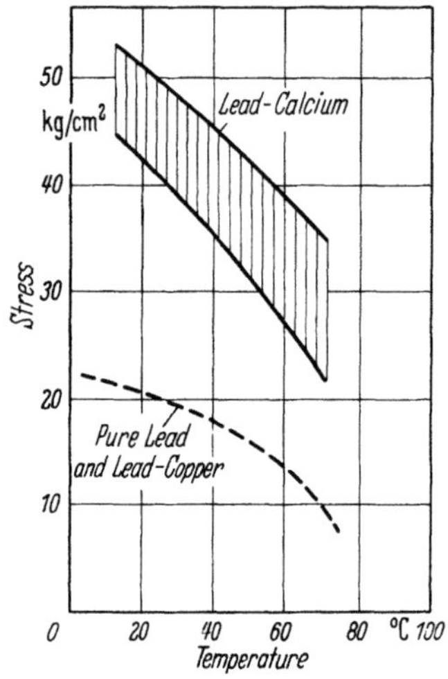

Fig. 252. Temperature-dependence of stress which causes 1% strain per year. Comparison of pure lead, lead-calcium (0.08 to 0.05%) and lead-copper. Extrapolated values. (According to GREENWOOD and COLE [429]).

The lead-arsenic alloys, and the polymetal alloys containing arsenic, also have a certain significance in respect of their creep strength, particularly from the results by DOLLINS [254]. This behaviour can certainly be related to the marked difference between the atomic radii of lead and arsenic. Beyond this there must be present an effect of the inclusions of arsenic in lead, as the alloys considered in Fig. 249 are, in part, far beyond the solubility limit at room temperature.

In Fig. 250 a considerable superiority of lead-calcium alloys over commercial lead for all stresses can be noted. The range of values on the right contains cast alloys of higher calcium content, that on the left, extruded alloys of lower calcium content. The remarkably small creep rates of lead-calcium alloys were also confirmed in recent publications by GOHN and co-workers [402]. Their results are shown in Fig. 251 together with those for some cable-sheathing alloys. The superiority of lead-calcium (Fig. 252) still persists in full extent even at 65°C (GREENWOOD and COLE [429]).

The values of the creep rates of lead-copper alloys available when the first edition of this book was written, did not yet suffice to provide a clear picture of the effect of copper. Since then further values have become available which are considered in Fig. 253. According to this, no essential improvement of the creep strength of lead is present owing to copper additions. Perhaps the result is to be interpreted in the sense that the copper addition does indeed increase the creep strength of lead, the effect being however again neutralised by the finer grain of the specimens containing copper. Creep tests on commercial lead and on lead-copper which have been extruded at two different temperatures, support this interpretation (Fig. 254). According to this, lead-copper extruded at 120°C is markedly superior to commercial lead. On the other hand, the effect of the added copper is precisely reversed after extrusion at 180°C. In contrast with the lead-copper, commercial lead becomes very coarse-grained at this extrusion temperature. The explanation offered is confirmed by experiments on single crystals of lead-copper. According to Greenwood [423], a copper content only increases the creep strength of

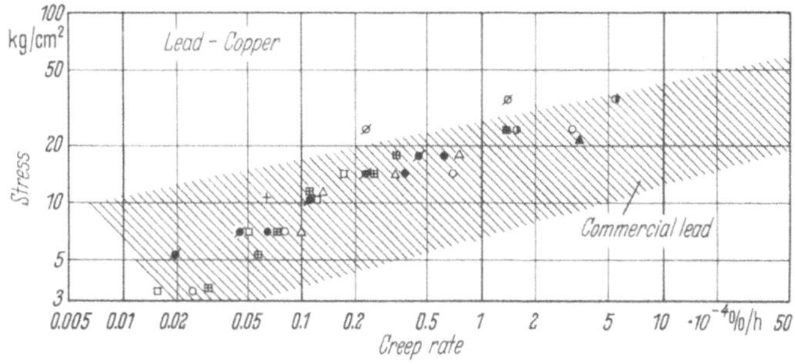

○ 0.04% Cu, □ 0.055% Cu (According to Moore among others [866]).
◑ 0.05% Cu, ⌀ 0.1% Cu (According to Greenwood and Worner [434]).
■ 0.014% Cu (According to Greenwood and Cole [427]).
⊞ 0.037% Cu, • 0.042% Cu, ● 0.052% Cu, △ 0.059% Cu (According to Dollins [253]).
▲ 0.1% Cu (According to Hopkin and Thwaites [589]).

Fig. 253. Creep rates of copper-alloyed lead.

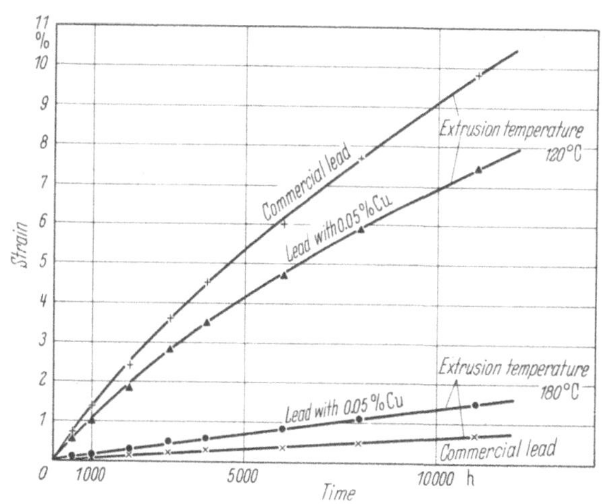

Fig. 254. Effect of extrusion temperature on creep strain of commercial lead and of copper-alloyed lead, at a tensile stress of 15 kg/cm². (According to v. Malotki).

lead if the copper is present in fine sub-division; copper in coarser distribution has no effect on the creep strength. The high creep strength found by McKeown and Hopkin [821] for an alloy of very pure lead with

0.005% silver and 0.005% copper, as also the high creep strength of other lead-silver-copper alloys in Fig. 255 is (according to the single crystal experiments) mainly attributed to the action of silver.

The evaluation of lead-tellurium alloys with regard to their creep strength is not quite unanimous. While GREENWOOD and WORNER [432], found a small creep strength in these alloys, GÜRTLER and SCHMID [454] put forward an opposite opinion. On an extruded alloy with 0.015% Te,

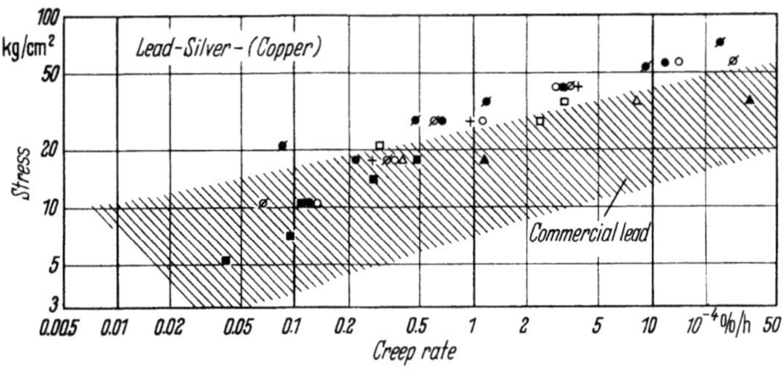

+ 0.003% Ag + 0.004% Cu
∅ 0.010% Ag + 0.004% Cu
○ 0.017% Ag + 0.061% Cu
● 0.018% Ag + 0.004% Cu
} (According to GOHN and ELLIS (at 30 °C) [403]).

□ 0.005% Ag + 0.005% Cu
1% Ag
} (According to McKEOWN and HOPKIN [822]).

■ 0.0024% Ag + 0.06% Cu (According to DOLLINS [253]).

△ 0.005% Ag + 0.064% Cu
▲ 0.0054% Ag + 0.061% Cu
} (According to GOHN among others (at 30 °C) [402]).

0.01% Ag
0.03% Ag
0.05% Ag
} (According to GREENWOOD and WORNER [434]). 3 points (not shown) lie inside the upper edge of the hatched area.

Fig. 255. Creep rates of silver-alloyed lead and of copper-alloyed lead.

of grain size 0.20 mm², under a load of 21 kg/cm², McKEOWN and HOPKIN [821] found a strain of only 0.4% in 17,000 hours. According to the experiments on single crystals, very large initial strains are to be expected in lead-tellurium alloys in the creep test, but the creep soon comes to a standstill. These experiments should be continued and extended to polycrystalline material.

For practical purposes, the temperature-dependence of the creep strength of lead and its alloys is of great significance. Measurements on five grades of commercial lead and four alloys are represented by solid diagrams in Figs. 256a—i. The relatively small effect of temperature in both lead-calcium specimens is noteworthy. It is presumably related to the high stability with temperature of the age-hardening of these alloys. Values for the creep strength of antimonial lead of higher antimony content up to 5.4% are shown in Fig. 257. Disregarding the alloy with 2.95% antimony (which is not considered in the diagram and which falls out of line) the great increase of creep strength at high antimony contents is especially remarkable. The data of SMITH [1128] on lead-antimony alloys at room temperature and at 100 °C agree with this. However it seems as if

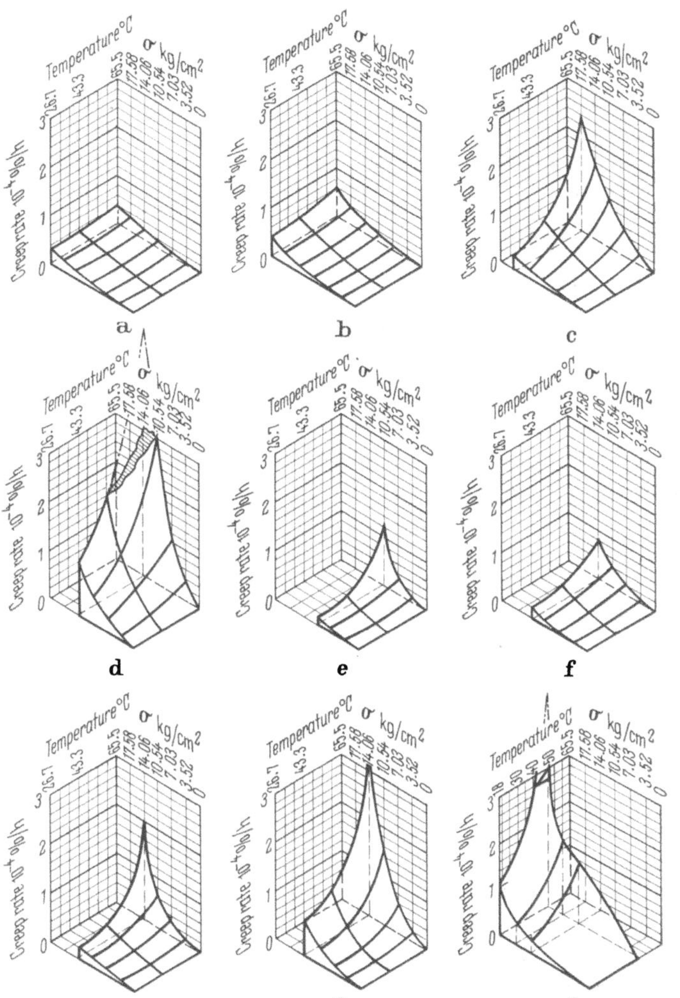

Fig. 256a – i. Temperature-dependence of creep rate of lead and lead alloys. (a) 0.03% Ca + 0.04% Cu; (b) 0.03% Ca + 0.05% Cu; (c) 1.0% Sb; (d) 2.0% Sn; (e) ASTM Grade III with 0.07% Bi; (f) Grade II with 0.06% Cu + 0.04% Bi; (g) Grade III with 0.09% Bi: (h) Grade II with 0.04% Cu + 0.03% Bi; (i) U-Lead (Port Pirie) with a total of 0.009% of impurities. (a) to (h): (According to MOORE and co-workers). (i): (According to GREENWOOD and COLE).

this superiority weakens with increasing temperature (cf. Fig. 248; GREENWOOD [425]). Individual observations on the creep of lead alloys are also scattered through other parts of this book, particularly in the section on technical working. Mention remains to be made of creep tests by GIFKINS [381] on specimens of lead-thallium alloys, which in preceding creep tests were stressed to fracture. The new creep curves

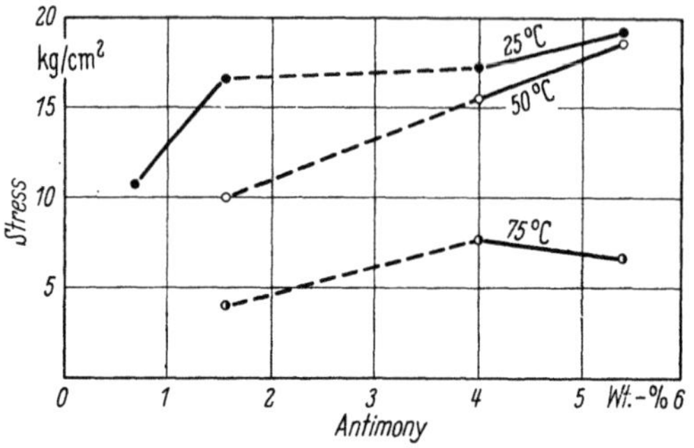

Fig. 257. Load bearing behaviour of lead-antimony alloys at a strain rate of $0.1 \cdot 10^{-4}$%/hour at various temperatures. (According to HILLEN and HOFMANN [528]).

were similar to the original ones. The ratio of the old minimum creep rate to the new one varied in a range $^1/_2$ to $2^1/_2$. According to SISTIAGA [1126 a] the creep rate in lead-mercury alloys with less than 2% Hg is increased in comparison with pure lead.

4. Endurance (Fatigue Strength)

In cable-sheathing made from commercial lead ("soft lead"), cracks were frequently observed, the origin of which was not understood for a long time (ARCHBUTT [24]). It was shown, particularly by the observations of HAEHNEL [464], that the cracks in question had been caused by fatigue: BECKINSALE and WATERHOUSE [73] arrived at the same result. Over 48 fractured cables were investigated. They had been installed near railways or bridges, and only a few were underground cables. Overhead cables, and cables which had been carried by ship, or prolonged railway transport, were well represented.

Almost without exception the cable sheaths were commercial lead. Fracture was mainly intercrystalline. Therefore fatigue fracture of lead is frequently designated as intercrystalline brittleness or intercrystalline corrosion. The fact that cracks mostly start from the inside of the cable shows that, primarily, the problem is not one of corrosion. Nevertheless, the investigations described below allow the conclusion that air is involved in the process of fatigue fracture. The observations mentioned gave impetus to extensive investigations on the fatigue strength of commercial lead and its improvement by alloying additions.

According to the older view, the beginning of fatigue fracture is attributed to the fact that the crystallites of the metallic material (owing to their directional elastic properties) are stressed to different amounts on the application of an external load, so that, in individual grains, plastic flow and consequent strain hardening set in far below the conventional elastic limit. Owing to the strain hardening the micro-stress increases further. When it exceeds the cohesive strength of the material, the first crack occurs, which, owing to notch action, is gradually propagated. Recent theories emphasise local slip processes on the surface

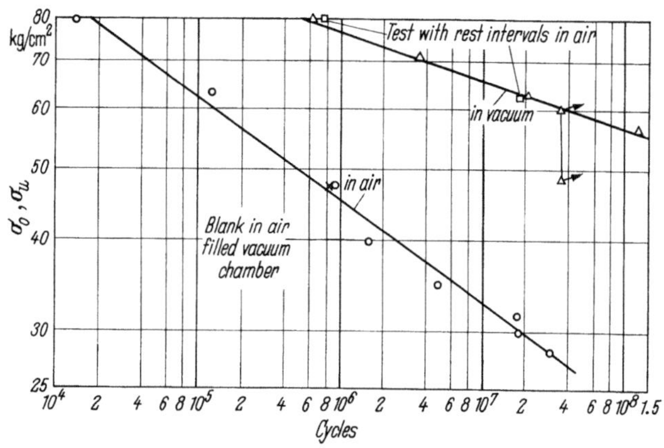

Fig. 258. Wöhler curves of lead from experiments in air and in vacuo. (According to GOUGH and SOPWITH [411]).

of the specimen due to alternating loads. If a steplike structure due to slip is formed on the surface, atoms formerly situated within the crystal are exposed and transformed to an active state. A return of these atoms to the inside of the crystal by backward deformation will then be impossible if, say, in the interim, oxygen has been absorbed (SHANLEY, SCHAUB and LISSNER [1109]). Damage of the surface by the mechanism described is increased by further cycles of stress, until a crack forms. Some support for this view is the fact that with a number of metals (among them lead) the influence of the surrounding medium on the fatigue strength has been established.

An increase of the fatigue strength of lead (at 10^7 cycles of stress) was found in oil and even in acetic acid, this phenomenon being attributed to the absence of air (HAIGH and JONES [470]). Experiments with a Haigh direct-stress-machine with high stress-frequency in air and in vacuo indicated (Fig. 258) quite a considerable atmospheric effect (GOUGH and SOPWITH [411]). The difference of fatigue strength in air and in vacuo increased with the duration of the test. MCKEOWN [820] gives data, shown in Table 16, on the fatigue strength in relation to the environment.

Table 16. *Effect of Surrounding Media and Protective Coatings on the Fatigue Resistance of Lead and Lead Alloys*

Materials	Surrounding medium or protective coating	Semi-range of stress, ± kg/mm²	Endurance cycles, ÷10⁶
Lead . . .	Air	0.055	1.3
	Normal acetic acid	0.055	8.5 U
	Rape oil	0.055	7.9
	Vaseline	0.063	9.8 U
Lead + 1.5 per cent tin + 0.25 per cent cadmium	Air	0.103	1.6
	Petroleum bitumen	0.126	9.3 U
Lead + 0.5 per cent antimony + 0.25 per cent cadmium	Air	0.126	1.3
	Rape oil	0.142	9.6
	Vaseline	0.142	6.4

U = unbroken.

Further data on this matter are given by McKeown and Hopkin [822]. On the other hand, according to Chaston [191], the fatigue strength of lead is greatly decreased by etching. In this case, there should be notch action, as scratches also reduce the fatigue strength (Haigh and Jones [470]), as do abrupt changes of stress at fixing clamps. Snowden [1132] compared the appearance of the fracture and the structure in fatigue specimens stressed in air and vacuum resp. The rods broken in air showed preferentially intercrystalline cracks; on the other hand, in the vacuum-broken specimens, Lüders bands resulted in the 45° directions, i.e. in the directions of the greatest shear stresses. Snowden [1131b] concludes from the results of his investigations that during the alternating stressing, oxygen migrates into the grain boundaries, and favours the formation and broadening of cracks. The fatigue strength of lead in air is only raised to the value in vacuo if the air pressure is reduced from $1.5 \cdot 10^{-1}$ to $5 \cdot 10^{-3}$ mm Hg. In the first stages of the fatigue test, Snowden [1131a], observed grain boundary migration and recrystallisation in a narrow zone on both sides of the grain boundaries. In the course of the fatigue test, the grain boundaries took up stable positions at ± 45° to the direction of stress. This observation was confirmed, in the present author's institute, on specimens of pure lead containing nickel or copper (Fig. 259); [577a].

Haasen explained the 45° direction of grain boundaries (which corresponds to an angle of 90° between grain boundary and the slip plane active at the time) by polygonisation. The small-angle grain boundaries thus formed gradually transform into large-angle grain boundaries.

It cannot yet be decided whether the most recent views, given below, are fruitful for understanding the fatigue of lead. In a discussion of the latest investigations on metal fatigue, HARTMANN [495a] first describes strain hardening observed in the fatigue tests. Strain hardening can also be shown in fatigue tests on lead, since here the amplitude of strain gradually decreases notwithstanding a constant amplitude of stress. In various cubic face-centred metals (Al, Cu, Ni) a distinction is drawn between strain hardening I and strain hardening II. In strain hardening I virtually the same slip processes occur as in the static stress test, but on the other hand in stage II, the so-called "Fatigue striations" or "persistent slip bands" are formed (*F*-bands). In *F*-bands, extrusions and intrusions frequently occur, as observed by FORSYTHE [336a] and explained by MOTT [875] in a dislocation model. A dislocation line opening on to the surface of a specimen, with a component of the BURGERS vector perpendicular to the surface, is forced by the alternating stress into a rotating motion. If therefore the end point of the dislocation line describes a closed curve on the surface, then at each circuit the volume circumscribed by the dislocation line completes a step of slip in the direction towards the surface. In this way a cavity is produced in the interior. The material above it is pressed out of the surface as an extrusion. According to WOOD [1284a], the formation of cracks takes place thus: within the *F*-bands pores are formed, which grow by aggregation and finally unite with surface channels (intrusions) to a crack. On the other hand, HAASEN [461b] puts forward the view that crack formation is not a volume effect, but a pure surface phenomenon which is not fundamentally connected with chemical reactions. The spreading of cracks in polycrystals with low stacking fault energy is said to take place frequently along the grain boundaries [496b]. This would agree with the observations on lead, which has a low stacking fault energy. It may be imagined that the spreading of cracks along the grain boundaries, observed in lead, may be attributed not only to the low stacking fault energy but also to the penetration of oxygen.

Fig. 259. Grain boundary migration of fatigued specimens of copper-lead alloy.

alternating stress	± 48 kg/cm²
alternating strain	± 0.076%
number of cycles to failure	604,000
temperature	53 °C
H_2O_2 + CH_3COOH	× 60

At this point attention may be drawn to some comprehensive surveys of metal fatigue, for example, the book by THOMPSON and WADSWORTH [1193].

Lead exhibits some peculiarities with regard to its fatigue strength. First there is the intercrystalline character of the fatigue fracture (already mentioned), which is not usual in other metals. Further, GÜRTLER and SCHMID [454] drew attention to the fact that in lead the fatigue strength is greater than the creep strength. In the ordinary construction metals with a higher melting point, the reverse is true at

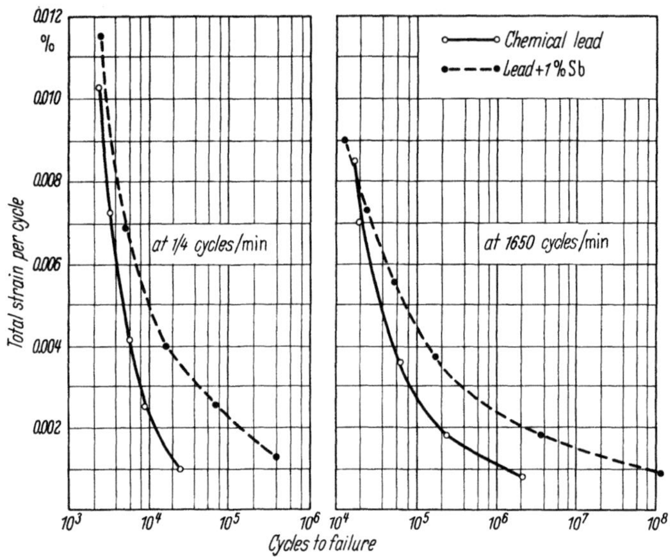

Fig. 260. Effect of frequency on the number of strain cycles to failure. (According to GOHN and ELLIS).

room temperature: Fatigue strength is less than creep strength (yield point). Also Wöhler-curves of face-centred metals do not, as a rule, show an inflection to the horizontal, but according to the mode of presentation, either a general curvature or even a completely straight line (Fig. 258). Even at the highest numbers of cycles of stress yet used, (100 millions) a horizontal asymptote to the Wöhler curve of lead was not reached (SCHUMACHER and BOUTON [1090], HOFMANN and MÜLLER [572]). In determining the fatigue strength of lead, the number of cycles of stress must therefore be set high, and furthermore, stated in the results. Moreover, in lead, fatigue strength is markably dependent on the frequency of vibrations (MOORE, BETTY and DOLLINS [866]). GOHN and ELLIS [404] referred to extensive tests with one cycle per day as the lowest frequency. In Fig. 260 the number of cycles at frequencies of $^1/_4$ and 1,650 per 1 min. for two types of lead are compared. At an alternating strain pure lead of $\pm$ 0.2%, withstands 10,000 and 200,000 cycles respectively (lives of respectively 700 and 2 hours); in lead with 1% Sb, the numbers of cycles are respectively 130,000 and 2 million (lives of 8,670 and 20 hours). The life expressed in cycles to fracture therefore increases with increasing frequency. If the duration of a vibration exceeds 4 to 6 minutes, then, according to SNYDER'S observations [1133], the fatigue strength does not depend on the frequency. It seems that with the harder alloys the dependence of fatigue strength on frequency is less marked.

On the contrary the life, expressed in units of time, decreases in all known cases (at constant amplitude) with increasing frequency of vibration. McKEOWN gives the following further example [820].

Table 17. *Effect of Cyclic Speed on Endurance of Lead Alloys Under Conditions of Rotating Flexure*

Material	Strain, per cent	Cycles to failure at		Time to failure, hours, at	
		3,000 cycles per min.	1.35 cycles per min.	3,000 cycles per min.	1.35 cycles per min.
Pure lead	0.10	$0.09 \cdot 10^6$	4,700	0.5	58
(1) Pure lead + 0.2 per cent antimony + 0.4 per cent tin	0.10	$0.2 \cdot 10^6$	16,600	1.1	205
(2) Pure lead + 0.85 per cent antimony	0.10	$1.0 \cdot 10^6$	100,000	5.5	1,230

According to Table 17, the ranking order of alloys with regard to their fatigue strength is practically independent of the frequency. For the region of low frequency and high amplitude of vibration attention is drawn to a discussion by ECKEL [268]; (Fig. 261).

The heat evolution in alternating stress was measured on various metals including lead (HAIGH [469]). The heat is the equivalent of the content of the hysteresis loop given by the stress-strain curve in alternating stress. The primary stage present in some metals, characterised by high damping as a result of the commencement of plastic deformation, is absent in lead. For the rest, it shows normal behaviour, i.e., a slow increase of damping in the secondary stage. In the tertiary stage of the test, in which fatigue fracture occurs and proceeds, the evolution of heat was, as usual, greatly increased.

The effect of grain size on the fatigue strength was tested by HOPKIN and THWAITES [589] on an alloy with 0.85% Sb, which did not recrystallise in the tests. Wöhler-curves were taken on rotating bending fatigue specimens at 3,000 stress cycles per minute. The stress, resisted up to 20 million cycles, was designated the endurance limit. A superiority (though only moderate) of the fine-grained material can clearly be seen (Table 18).

Table 18. *Effect of Grain-Size on the Fatigue-Resistance of the Lead — 0.85% Antimony Alloy*

Extrusion Temp., °C	Average Grain Area, mm.²	Endurance Limit, ± kg/mm²
160	0.0039	98.5
200	0.012	88.0
250	0.043	82.6
300	0.19	72.1

The curves obtained with lead-tin alloys (Fig. 262) show simultaneously the effect of grain size and of alloy concentration. Very fine-grained specimens of commercial lead and of lead with less than 1% Sn, recrystallised with coarse grains during fatigue testing at a stress

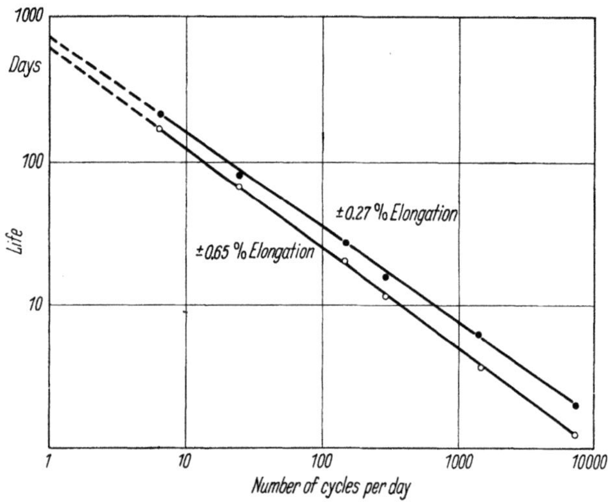

Fig. 261. Relation between the life of the specimen and the frequency in the rotating bending test. (From experiments on acid lead by ECKEL).

closely above the fatigue strength. This grain-growth presumably led to a reduction of the fatigue strength and masked the dependence on the grain size shown above. Lead with coarser initial grain, and alloys more stable to recrystallisation, ought to maintain their initial grain size in the fatigue tests.

Tests with one single stress cycle per day are of significance for evaluating the effect of daily temperature fluctuations on the durability of lead-sheathed telephone cables, lead pipes and so forth. The nightly cooling causes a longitudinal contraction of the lead, which is followed in the daytime by an expansion due to warming. In heavy current cables the longitudinal movements of the lead sheath result from the fluctuations in the current load. This alteration indeed repeats itself more rapidly than the sequence of day and night, but is however still much slower than the rhythm of the ordinary fatigue testing machines. GOHN and ELLIS [404] proposed, for imitating the longitudinal movement of

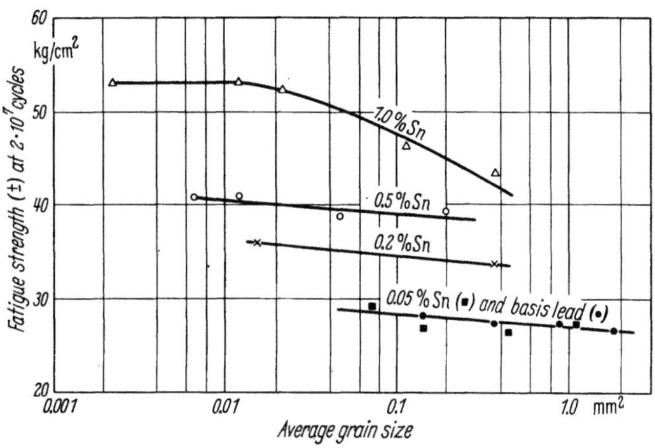

Fig. 262. Effect of grain size on fatigue strength of lead-tin alloys. (According to HOPKIN and THWAITES).

cable sheaths due to temperature fluctuations, and having regard to the possible duration of test, a frequency of 15 per hour. Testing machines of usual high frequencies are, on the other hand, of importance for investigating the effect of short-period vibrations in lead.

When it is desired to arrive rapidly at a comparison of the fatigue strength of different alloys, the use of simple test pieces, e.g., in the form of extruded flat rods, is recommended. If, on the other hand, predictions of the durability of lead cable sheaths in practical service are required, the tests should be made on corresponding test pieces and the service conditions imitated as closely as possible.

Generally, in the Wöhler-curves of lead, the stress is plotted as ordinate, the cycles as abscissa. However according to PFENDER and SCHULZE [957], the alternating bending stress is of less significance in the behaviour of cable sheathing materials than the alternating bending strain, since as a rule, cable sheaths in strain would not absorb any energy worth mentioning. In view of its fundamental character, this work may be described in somewhat more detail. It was carried out on flat test pieces in a plane bending machine. The "soft lead" containing 0.025% Sb, 0.046% Sn, 0.001% As, 0.002% Zn, 0.006% Cu, 0.001% Ag, 0.02% Si, trace Cd, was compared with a series of lead alloys. As an example, Fig. 263 shows the Wöhler-curves of "soft lead" and of lead with 0.10% Sb, 0.10% Sn, 0.08% As, plotting respectively the alternating strains (left) and the alternating stresses (right).

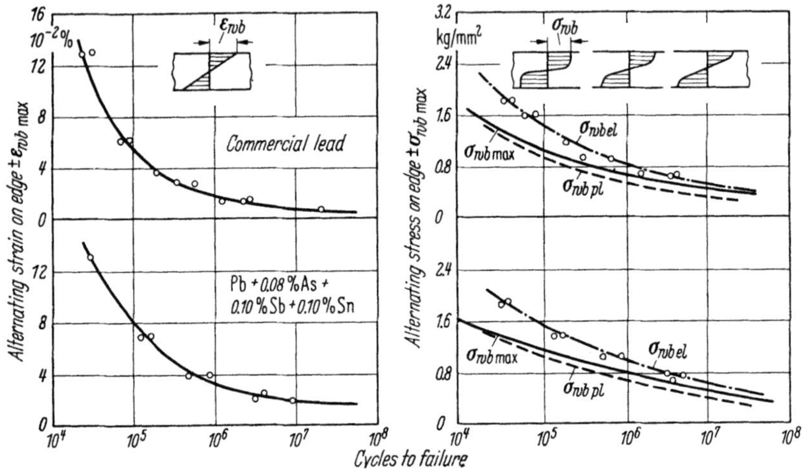

Fig. 263. Wöhler curves of two types of lead, plotting respectively (left) the alternating strains, and (right) the alternating stresses calculated from the measured bending moment by two methods. (According to PFENDER and SCHULZE).

The edge stress was separately calculated for the fully elastic and the fully plastic behaviour of the bending-test rod. The values of the actual edge stress should lie between these two cases, so that at high number of cycles the elastic, at low number of cycles the plastic strain predominates (continuous line). Fig. 264 shows a series of Wöhler-curves of all the alloys tested, both forms of representation being given side by side. It can be seen that at high alternating strains the effect of alloying largely disappears, and that on the other hand at small alternating strains the favourable effect of the alloying elements is more and more evident. If the alternating stress is used as a criterion, then at high number of cycles a somewhat similar ranking order of alloys is obtained as at low number of cycles of alternating strains. However here the particularly high resistance to alternating stresses of the alloy with 5.5% Sb is noteworthy, since this is not evident when plotting against alternating strain. This example exhibits, with especial clarity, the importance of adapting the testing of cable-sheathing lead to practical requirements. Recent investigations also emphasize the use of the strain rather than the stress when plotting Wöhler-curves, as being more adapted to the practical problems (GOHN and ELLIS [404], ECKEL [268]). The representation of stress in relation to cycles is useful if it is desired to observe the behaviour of one and the same specimen. The decrease of stress indicates the beginning of damage to the material (Fig. 265). The edge stress is

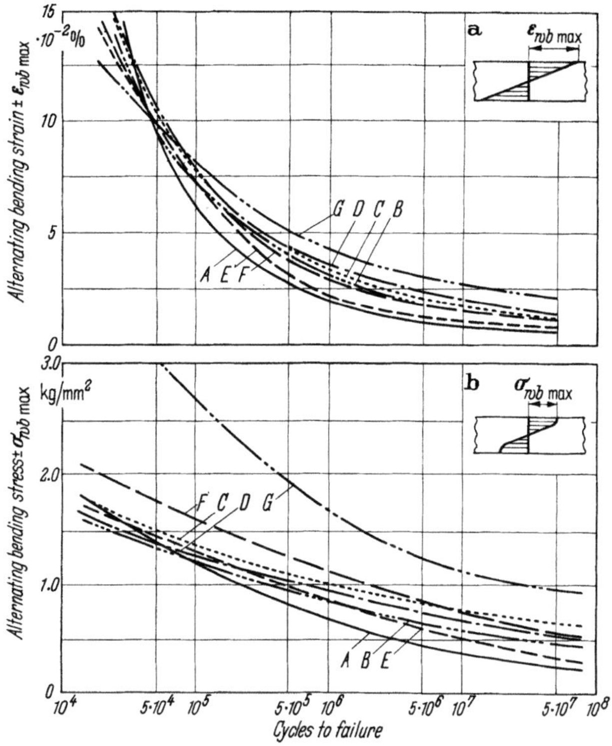

Fig. 264a and b. Wöhler curves of various leads, plotting respectively (upper) the alternating strain, and (lower) the alternating stress, calculated from the measured bending moment. (According to PFENDER and SCHULZE).
A commercial lead; *B* 0.08% As; *C* 0.8% Zn; *D* 1.5% Zn; *E* 0.64% Sb; *F* 0.47% Sb + 0.18% Cd; *G* 5.5% Sb.

calculated from the measured bending moment and the moment of intertia. Thus fully elastic behaviour is assumed, which is certainly not true in the case of high alternating stress. The measurement of strain is made by a measurement of curvature or by means of strain gauges.

The temperature-dependence of the fatigue strength of lead is significant for its use at higher temperatures. By plotting the stress against the number of cycles, a decrease of fatigue strength with rising temperature will be expected in all cases. As an example, values by MCKEOWN [820] (Table 19) are quoted.

Data on the fatigue strength of homogeneous linings of pure lead alone or with addition of 0.02% Ni, or 0.06% Cu, or 0.05% Te are shown elsewhere (see p. 459).

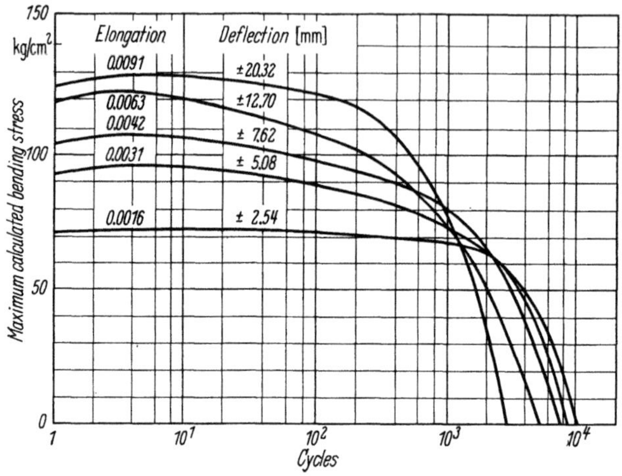

Fig. 265. Decrease of edge stress to fracture at constant bending strain in the course of the bending fatigue test on the purest lead. (According to GOHN and ELLIS [404]).

Table 19. *Fatigue Resistance of Extruded Lead and Lead Alloys*
Direct-stress tests

Material	Endurance limit at $10 \cdot 10^6$ cycles, ± kg/cm²	
	At room tem-perature	At 100 deg. C.
Pure lead	28.1	12.65
Lead + 0.06% Te	77.4	52.0
Lead + 1.5% Sn + 0.25% Cd	90.0	44.3
Lead + 0.5% Sb + 0.25% Cd	116.8	63.2
Lead + 0.06% Te	77.4	52.0
Lead + 1.5% Sn + 0.25% Cd	90.0	44.3
Lead + 0.5% Sb + 0.25% Cd	116.8	63.2

Attention has already been drawn above to some fundamental investigations concerning the effect of structure — e.g. grain size. Further observations on the relationship between alloy structure and fatigue strength may be added.

"Soft lead" in comparison with lead alloys, has the lowest values in respect both alternating strength and alternating strains. It has already been noted above, that in taking the alternating strain behaviour as a criterion, the differences between unalloyed and alloyed lead become progressively smaller with increasing amount of strain. According to PFENDER and SCHULZE [957], the curves of various alloys as shown by Fig. 264 intersect at an alternating strain of 0.1% and a number of cycles of $5 \cdot 10^4$, but according to GOHN and ELLIS [404] (Fig. 266) the

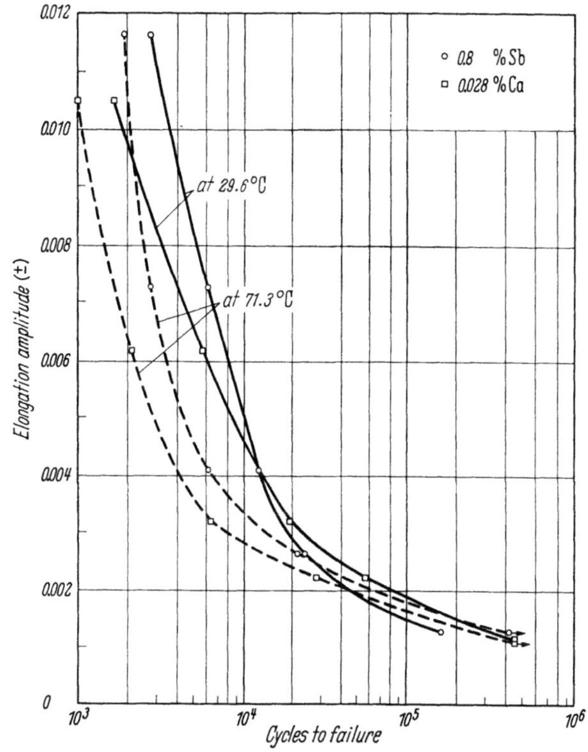

Fig. 266. Effect of temperature on the Wöhler curves. Alternating bending strain of lead with 0.8% Sb and with 0.028% Ca. Tested at 1/4 cycle per minute. (According to GOHN and ELLIS [404]).

"levelling" of the types of lead with respect to their alternating strain behaviour sets in at a strain of about 1% and a number of cycles of $2 \cdot 10^4$. (The two diagrams ought not to be strictly compared, as curves for pure lead are lacking.) At alternating strains below 0.4%, such as are significant for telephone cables, the life of cable sheaths (according to the last-named authors) can be considerably increased by alloying. In Fig. 267 the alloys with 1% Sb, 0.15% As + 0.10% Sn + 0.10% Bi, 0.65% Sb + 0.25% Zn can be regarded as about equal. According to PFENDER and SCHULZE, the curves for the very impure lead (which would be better described as a polymetal alloy) and for lead with 0.64%

Sb, almost coincide. Higher values of the alternating strain endurance (almost equal among themselves) are given by the alloys designated in Fig. 264 as *B*, *C*, *D* and *F*. According to GOHN and ELLIS, additions of silver of the order of 0.001—0.01% produce a small improvement in the

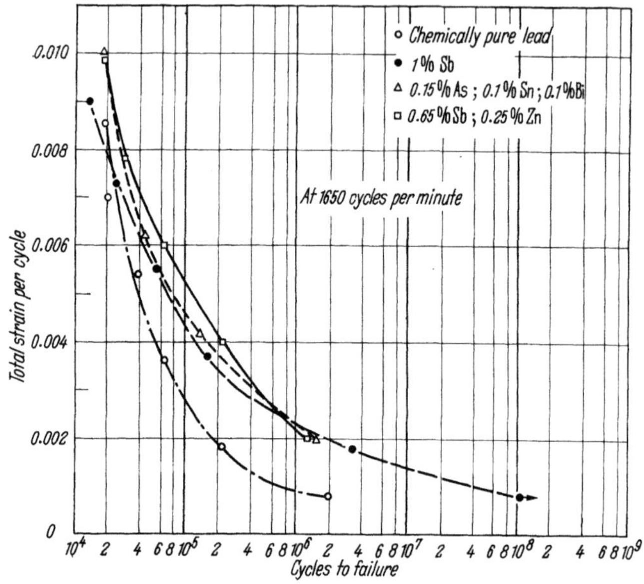

Fig. 267. Wöhler curves of alternating strain (±) for some lead cable sheathing alloys. (According to GOHN and ELLIS [404]).

alternating strain behaviour, as does an extra addition of 0.06% Cu. According to the results in the literature, the alternating strain behaviour of lead is improved more or less, by all kinds of alloying elements, whether they go into solid solution or form precipitates. In age-hardenable alloys, the presence or absence of age-hardening had no special effects on the alternating strain behaviour. Doubtless, specific effects of the alloying elements are present; they are however veiled in part by other influences, so that no definite ranking order for their alternating strain behaviour can be given.

This is also why most of the investigations and comparisons have been made on the basis of alternating stress (i.e. stress amplitude as the ordinate of the Wöhler-curve). Some results of such investigations will be briefly given. According to HOPKIN and THWAITES [589], antimony in lead increases the fatigue strength as long as the alloys are single phase, but if the antimony content is raised from e.g., 0.5 to 0.85% in two-phase alloys, the fatigue strength increases only insignificantly (from 70.3 to 79 kg/cm²). Additions of tin show a similar effect. In alloys with 0.9% Sb and 0.001% As, a further increase of fatigue strength occurs if the homogenised single phase alloy is allowed to age-harden before the fatigue test. A clear increase in fatigue strength in the direct

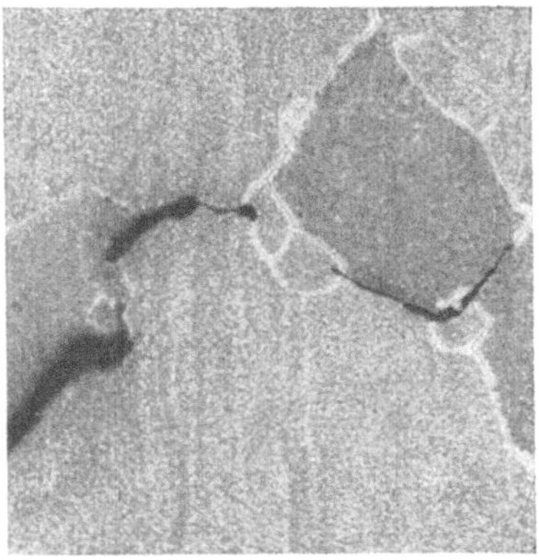

Fig. 268. Intercrystalline cracks in the water pipe, shown in Fig. 366, damaged by fatigue fracture. ×50

Fig. 269. 1% Sb, 0.05% As. A rod of the age-hardened alloy broken on a plane bending machine [485]. Intercrystalline fatigue fracture. ×200

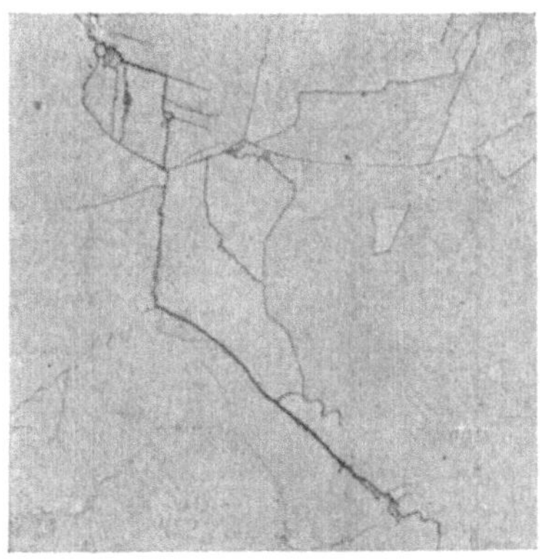

Fig. 270. 0.07% Ca. A rod of the age-hardened alloy broken on the plane bending machine [485]. Cracks also within the grain. ×50

tension-compression test was given in pure lead [572] by small additions of copper, which were mainly present as a second phase.

In conclusion, values of the fatigue strength of lead (Table 20) and of lead alloys (Table 21), and some photomicrographs of fatigue fractures are given. Alloys of lead with cadmium, antimony or tin, alone or combined, exhibit considerably increased strength, as do also alloys with

Table 20. *Fatigue Strength of Lead*

Grade	Treatment	Machine	Author	Number of Cycles	Frequency 1/min	Fatigue Strength kg/cm²
99.99	Extruded	Haigh Push Pull	Beckinsale and Waterhouse [73]	10^7	2,000	±27.5
99.99	Extruded, 100 h, 250 °C	,,	,,	10^7	2,000	27.5
99.99	Cold Rolled	,,	,,	10^7	2,000	26,6
99.99	Cold Rolled 1 h, 250 °C	,,	,,	10^7	2,000	30,5
Broken Hill (99.99%)	Extruded	Haigh Push Pull in air	Gough and Sopwith [411]	$3 \cdot 10^7$	2,200	25.9
(99.99%)	,,	Haigh Push Pull in vacuum	,,	$3 \cdot 10^7$	2,200	58.0
Commercial Lead	,,	Rotating	Townsend and Greenall [1205]	$5 \cdot 10^7$	700	15.1
,, ,,	,,	Illinois Plane bending	Moore, Betty and Dollins [866]	10^7	700	49.2
with 0.09% Bi Commercial Lead	,,	Illinois Plane bending	,,	10^7	2,500	45,8
,, ,,	—	Haigh Push Pull	Waterhouse [1243]	10^7	2,000	21.1—29.5
	,,	,,	Jones [633]	$3{,}6 \cdot 10^7$	3,000	28.4
99.99	,,	DVL Plane Bending	v. Hanffstengel and Hanemann [485]	10^7 extrapol.	740	32.0
99.99	,,	Rotating	Schumacher and Bouton [1090]	$5 \cdot 10^7$	800	13.0
99.99	,,	,,	,,	$1 \cdot 10^7$	800	22.6
Commercial Lead	,,	,,	Burkhardt [155]	$2 \cdot 10^7$	—	20.0

Table 21. *Fatigue Strength of Lead Alloys*

Addition	Wt.-%	Treatment	Machine	Author	Number of Cycles	Frequency 1/min	Fatigue Strength kg/cm²	UTS kg/cm²
Ca	0.04	Extruded	—	Moore, Betty and Dollins [*866*]	10^7	2,500	109	—
	0.04	,,	—	,,	10^7	700	105.5	—
	0.04	—	Haigh Push Pull	Waterhouse [1243]	10^7	2,000	112—116	—
	0.03	—	,,	,,	10^7	2,000	102	—
	0.028 to 0.039	Extruded	Rotating	Townsend and Greenall [1205]	$5 \cdot 10^7$	700	46.4—59.1	191—243
	0.07	Extruded and age hardened	DVL Plane bending	v. Hanffstengel and Hanemann [485]	10^7 extrapolated	740	100.0	—
	0.04	Extruded and stored 2 weeks	Rotating	Schumacher and Bouton [1090]	$5 \cdot 10^7$	800	60.5	244
					$5 \cdot 10^7$	800	73.1	292
	0.04	Extruded and stored 6 months	,,	,,	$5 \cdot 10^7$	800	80.3	307
	0.06	,,	,,	,,	10^7	1,700	105.5	316
	0.04	Extruded	Illinois Plane bending	Dean and Ryjord [238]	extrapolated			
	0.04	,,	,,	,,	,,	1,700	84.5	268
Cd	0.3	Rolled	—	Beckinsale and Waterhouse [73]	—	—	71.8	—
	0.3	1 h 250°C	—		—	—	64.0	—
	0.5	Rolled	—	,,	—	—	99.0	—
	0,5	1 h 250°C	—	,,	—	—	90.0	—

Table 21 *(continued)*

Addition	Wt.-%	Treatment	Machine	Author	Number of Cycles	Frequency 1/min	Fatigue Strength kg/cm²	UTS kg/cm²
Cu	0.06	Extruded	Haigh Push Pull	Jones [633]	6.3 · 10.7 extrapolated	3,000	44.1	—
Sb	0.25	Extruded in laboratory	"	J. McKeown (BNFMRA) [818]	10^7	2,200	58.9	—
	0.50	"	"	"	10^7	2,200	77.1	—
	0.75	"	"	"	10^7	2,200	92.5	—
	0.75	Extruded		Moore, Betty and Dollins	10^7	2,500	91.5	—
	0.75	"	—					
	0.75	—	—	"	10^7	700	84.4	—
	1.00	Extruded	Haigh Push Pull	Waterhouse	10^7	2,000	83.0	—
			Rotating	Townsend and Greenall	$5 \cdot 10^7$	700	21.1—31.7	194—281
Sb	1.0	Rolled	—	Beckinsale and	—	—	99.0	—
	1.0	1 h 250 °C	—	Waterhouse	—	—	94.6	—
	1.0	Extruded and stored 6 months	Rotating	Schumacher and Bouton	$5 \cdot 10^7$	800	37.0	211
	1.0	Extruded	Illinois Plane bending	Dean and Ryjord	10^7	1,700	70.3	225
	1.0	Extruded in laboratory	Haigh Push Pull	J. McKeown (BNFMRA)	10^7	2,200	97.5	—
Sn	1.0	"	"	"	10^7	2,200	51.8	—
	2.0	"	"	"	10^7	2,200	69.1	—
	2.00	Extruded	"	Moore, Betty and Dollins	10^7	2,500	66.9	—

Table 21 (*continued*)

Addition	Wt.-%	Treatment	Machine	Author	Number of Cycles	Frequency 1/min	Fatigue Strength kg/cm²	UTS kg/cm²
Sn	2.00	Extruded	Haigh Push Pull	Moore, Betty and Dollins	10^7	700	49.2	—
	2.00	—	„	Waterhouse	10^7	2,000	61.9	—
	3.00	Rolled	„	Beckinsale and	—	—	82.3	—
	3.00	1 h 250 °C	„	Waterhouse	—	—	73.1	—
	3.00	Extruded in laboratory	„	J. McKeown (BNFMRA)	10^7	2,200	75.7	—
Te	0.05	Extruded	„	Singleton and Jones [1126]	$2 \cdot 10^7$	—	78	—
	0.05	„	Rotating	Burkhardt [155]	$2 \cdot 10^7$	—	60	—
{ Cd	0.25	Extruded	—	Beckinsale and	—	—	87.0	—
{ Sn	+1.5	1 h 250 °C	—	Waterhouse	—	—	77.9	—
{ Cd	0.25	Extruded	—	„	—	—	113	—
{ Sb	+0.5	1 h 250 °C	—	„	—	—	108	—
{ Sn	1.2	Extruded	DVL Plane bending	v. Hanffstengel and Hanemann	10^7 extrapolated	740	70.0	—
{ Cd	+0.2							
{ Sb	+0.1							
{ Sb	1.0	„	„	„	„	740	125.0	—
{ As	+0.05							

tellurium, lithium, calcium, copper. Even though many alloys of higher fatigue strength also exhibit higher hardness (v. HANFFSTENGEL and HANEMANN [485]), there is no simple relationship between hardness and fatigue strength. This is shown for instance, by the lead-tellurium alloys, which exhibit a slight increase of hardness and a much improved fatigue strength. Nineteen binary and ternary lead alloys, suitable for use in cable-sheathing, were tested for their resistance to vibration. The best results were shown by the compositions lead + 0.5% Sb + 0.01% Te, lead + 0.5% Sb + 0.15% As, lead + 0.15% As + 0.1% Sn + 0.1% Bi. The alloy with tellurium is particularly recommended. It has finer grain than the other alloys; possibly the tellurium addition is enriched at the grain boundaries (BOL'SHANINA [108a]).

Figs. 268 and 269 clearly show the intercrystalline character of the fatigue fracture of commercial lead and of arsenical lead-antimony. In lead-calcium alloys, on the other hand, the cracks also pass through the interior of the grains (Fig. 270).

Alternating bend tests, with a bending angle of 90° round a mandrel of 6.4 mm dia. were carried out on lead and lead alloys (MOORE, BETTY and DOLLINS [866]). While in commercial lead bending values of 60 to 100 were attained, lead-cable-sheathing with antimony, tin and calcium only survived 20 to 35 bends. A reduction of the bending value may be due to alloying effects, but also to the presence of coarse grain, oxide inclusions (through defective welding) and similar causes.

5. Hardening and Heat Treatment of Lead Various Conditions of a Lead Alloy

Basically a metal's resistance to flow can be increased in several ways:

1. By reduction of grain size (p. 181). This effect cannot be practically utilised in lead, as fine-grained lead recrystallises very easily, and creeps more than a lead of medium grain size.

2. By strain hardening as a result of cold working. This possibility is excluded with lead and lead alloys, as lead (in contrast with metals of higher melting point) is already in the region of crystal recovery and recrystallisation at room temperature.

3. By formation of solid solution. The increase of the yield point of a metal by the introduction of dissolved atoms into the crystal lattice is now attributed to the mutual action of dislocations and foreign atoms. HAASEN [461a] gives a survey of the various forms of this mutual action. The effect of the different atomic radii of the dissolved atoms and of the

atoms of the solvent metal is well known. This effect was recently tested at the Armour Research Foundation [24a] on fifteen binary alloys of lead of the highest purity. In log — log graphs an approximately linear relationship was found between the differences of the atomic radii and the solid solution strengthening rate ds/dc (s = strength, c = concentration in atomic %). The rate of solution strengthening increased in the order Tl, Bi, Sn, Cd, Sb, Li, As, Ca, Zn, Cu, Ba. There was also a relationship between solid solution strengthening and solid state solubility as the amount of solubility in the solid state decreased with increasing difference of the atomic radii. The investigation should be extended to other alloying elements such as mercury or indium. The relationships mentioned do not apply strictly, i.e., solid solution strengthening does not depend solely on the difference between the atomic radii. A mutual action between the dissolved atoms and the atoms of the basic metals also takes place when the atomic radii are of equal size, if the dissolved atoms alter the shear modulus of the metal or if the dissolved atoms have a valency different from the atoms of the basic lattice, or if a chemical interchange takes place between a stacking fault in a split dislocation and a foreign atom (the Suzuki effect). If the amount of the alloying constituent exceeds its solubility limit in lead at room temperature, then generally changes take place on storage. If the supersaturated solid solution breaks down coarsely, softening results, for example in lead-tin alloys (see p. 98). The flow stress of the softened alloy will in general increase a little above that of commercial lead.

4. If the breakdown due to storage takes place in a very fine form, then, on the contrary, increase of hardness can occur, which is designated as age-hardening. It depends on the mutual action of the dislocations and the stress yield of the precipitated particles. The optimum age-hardening occurs at a definite degree of dispersion of the precipitates (critical dispersion), i.e. a definite average distance between the precipitated particles. Age-hardenability presupposes a region of solubility which narrows with falling temperature. This assumption is realised in many lead phase diagrams, for example with lead-antimony and lead-calcium alloys [562, 502] (Figs. 271 and 28). At the beginning of the storage period the solid solution must be supersaturated and this is attained by rapidly cooling of an alloy saturated at a higher temperature. In lead alloys, quenching is, in the main, only necessary if the work pieces have large dimensions. In the case of duralumin age-hardening is not, as described, connected with the formation of precipitates, but with a single phase breakdown of the solid solution. Age-hardening here leads to the formation of ordered distribution of copper atoms, of the nature of Guinier-Preston zones (Dehlinger [242]). This kind of age-hardening is usually called cold age-hardening, in contrast with the hot age-

hardening described above for lead alloys (precipitation-hardening). The single phase breakdown must however also be taken into account as a partial process even in the age-hardening of lead-antimony alloys; indications of this are given in investigations by BORCHERS and co-workers [109, 109a, 109b, 146] (p. 93).

5. By heterogeneous structure. In the age-hardening of an originally homogeneous solid solution there is the possibility of attaining a heterogeneous structure. This can be brought about by using a sufficiently

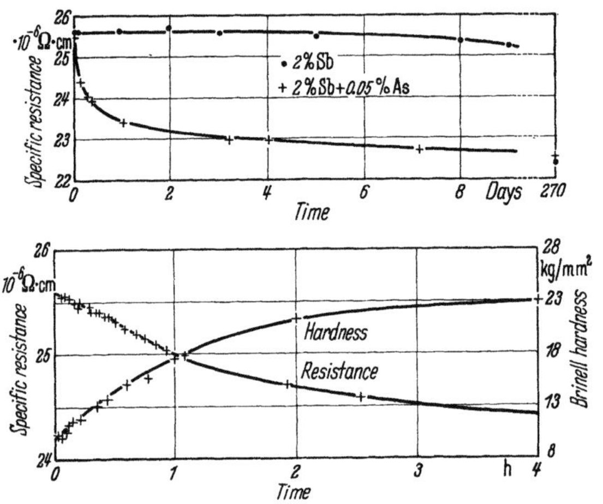

Fig. 271. (Top) Change of electrical resistance on storage of quenched alloys with 2% Sb, and 2% Sb + 0.05% As. (Below) Change of hardness and electrical resistance of the latter alloy immediately after quenching.

large amount of alloying element. Metals such as copper or nickel which are almost insoluble in solid lead are particularly efficacious in this system. Such alloys can be produced by melting or by powder metallurgy methods (p. 450). Examples of the first case are given by experimental alloys of lead with larger contents of copper, nickel and cobalt, (WILLJAMS and co-workers [1273a, 1274] (Fig. 272)). In further experiments of this kind an alloy with 1.5% Cd and 0.2% Ni was developed. The increase of strength was attained here by a combination of solid solution formation (Principle 3) and dispersion hardening (Principle 5); [1274b]. Hardening by this method, which, for instance is widely used for storage battery alloys, has the advantage of great stability, while hardening according to Principles 3 and 4 is sometimes diminished or lost in course of time by softening. The increase of flow stress by inclusion of metals and metalloids is particularly marked if these form a stiffening framework in the lead solid solution, perhaps in the form of a finely striated eutectic (Figs. 66, 177).

6. A hardening by formation of superstructures, as in the copper-gold system, has not yet been found in lead alloys, but there are hints in silver-lead alloys (see p. 32).

7. A hardening by transformations, similar to the hardening of steel, is unknown in lead alloys.

For further details, reference may be made to the sections on lead-antimony, lead-calcium and lead-tin. Some basic observations on the age-

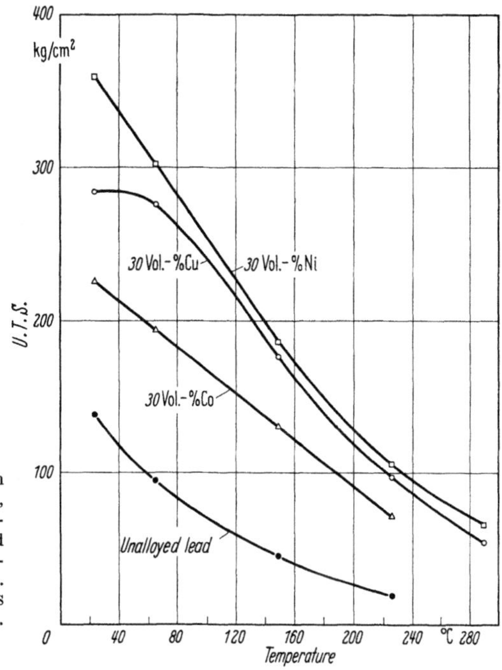

Fig. 272. Tensile strength of various 'lead alloys', i. e. cast lead with infiltrated finely-divided metal powders, in relation to the temperature. (According to WILLIAMS and co-workers [1274]).

hardening and softening of lead alloys will now be discussed. Age-hardening frequently occurs in casting alloys if they cool rapidly after solidification. The most favourable compositions are those slightly below the saturation concentration at the temperature of the end of solidification. At higher concentrations too many inclusions are formed in lead solid solutions by crystallisation; as a result of their nucleating action they favour the breakdown of the solid solution in cooling. For this reason the increase of hardness on storage is greater, for instance, in a lead-calcium alloy with 0.1% Ca than in one with 0.2% Ca (see p. 48). The crystal segregation present in the casting is in general in no way detrimental to age-hardening. Thus the degree of age-hardening in the alloy mentioned with 0.1% Ca is not increased by homogenising.

Extruded alloys also age-harden on storage, assuming that they did not cool too slowly in the container. The temperature should, if possible, be so high that in extrusion the alloy is above the temperature of the onset of breakdown. If extrusion is carried out below the solid solubility line, a coarse breakdown occurs on working and the age-hardenability of the alloy is reduced.

Rolled alloys hardly age-harden. Rolling of alloys generally takes place at such a low temperature, that a coarser breakdown occurs in alloys which are in themselves age-hardenable, corresponding to the solubility limit at the rolling temperature (p. 90). If it is desired to impart age-hardenability again to such an alloy, then the precipitates must be brought into solution by heat-treatment at higher temperatures (Fig. 83). Age-hardening of lead alloys can be accelerated by heat-treatment to moderately high temperatures. It is then however necessary to make allowance for a lower final hardness (see p. 89).

The formation of precipitates involved in the age-hardening of lead alloys was investigated in earlier researches at the Bleiforschungsstelle. The precipitates are often below the limit of visibility under the microscope, and can then only be detected by the lowering of the electrical resistance (Fig. 271), or by the change in the lattice parameters of the lead solid solution. In lead-antimony alloys which have been stored for a long time, the precipitates become visible under the optical microscope if strong objectives are used, as shown in Fig. 273. In such age-hardened specimens, the antimony precipitates always appear in the form of small rods and are oriented. The antimony lamellae lie parallel to the octahedral planes of the lead[1] and further lattice planes. Oriented precipitates were also established in age-hardened lead-magnesium and lead-calcium alloys (Fig. 52). Such observations are also of practical value, as in some circumstances they permit of conclusions on the pre-treatment of a material.

While therefore, the age-hardening of lead alloys at room temperature involves the formation of the finest microscopic or sub-microscopic precipitates from the supersaturated solid solution (critical dispersion), coarse breakdown can however still occur in storage without being conditioned by recrystallisation due to working. Such a coarse breakdown is often observed in age-hardened lead-antimony alloys which have been stored for a long time, such as cable-sheaths or water pipes. The regions attacked form fringes at the grain boundaries. In a microscope section they appear white, and contain round antimony particles (Fig. 273). A slight formation of fringes only involves a small decrease

[1] This is occasionally recognised by the parallelism between the antimony lamellae and the twinning planes of the lead.

of hardness, as is shown by Fig. 274 from the investigation quoted [562]. It is not certain whether the process gradually comes to a complete standstill. In an antimonial lead pipe with 1.15% Sb and 0.002% As, 17 years old, the same structure as in Fig. 273 was observed under high magnification. The fringes mentioned had remained narrow. The Brinell hardness at 9.3 kg/mm^2 may have fallen somewhat below the initial value. The high creep strength of the age-hardened lead-antimony-

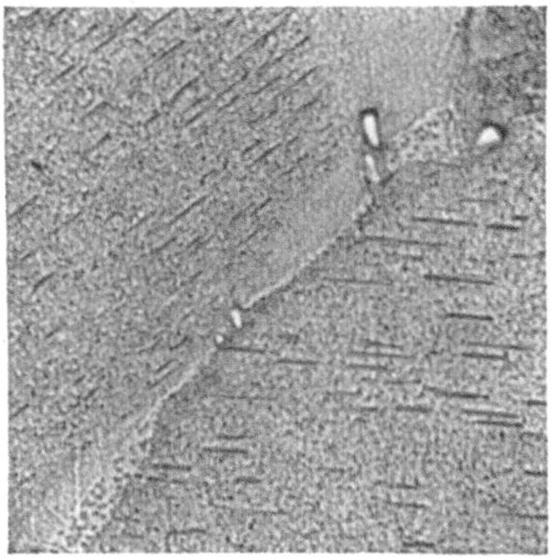

Fig. 273. Lead pipe with 0.95% Sb after storage for 3 years. Oriented precipitates of antimony. Coarse breakdown of solid solution at the grain boundaries. ×1,200

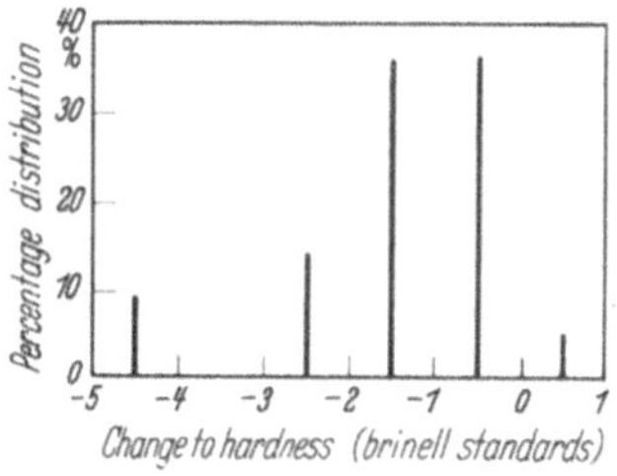

Fig. 274. Statistical distribution of change of hardness of age-hardened antimonial lead in storage for years [562].

arsenic alloy still persisted. Lead-calcium alloys do not exhibit the breakdown at the grain boundaries. Their age-hardening should therefore be more stable, which is perhaps connected with the high melting point of calcium.

In lead-tin and lead-cadmium alloys, assuming suitable pretreatment, temporary age-hardening occurs in storage. It is soon overcome by a complete softening (Fig. 38). Observation under the microscope shows that the softening is connected with the formation of very coarse, partly oriented precipitates (Fig. 35). It is natural to relate this behaviour with the low melting points of tin and cadmium.

In alloys with practically stable age-hardening, such as lead-antimony and lead-calcium, a softening of the age-hardened material also sets in on heavier working and on heat treatment at moderately high temperatures (Figs. 32, 84). The softening is probably connected with a coarsening of the precipitates.

V. Lead as a Corrosion-Resistant Material[1]

In co-operation with Dr.-Ing. J. Maatsch, Essen.

The generalities on the corrosion behaviour of lead collected in this section, and the individual observations quoted, form a selection and can provide a guide only for the appropriate applications of the material. If need be, further information on the occurrence of corrosion must be obtained by laboratory tests. In each case the final decision on the correctness of the action taken can only be the behaviour under service conditions. The reason for this is the large number of factors which are concerned with the progress of corrosion. It is therefore not surprising that, in part, observations recorded in the literature differ markedly from one another and are even contradictory.

1. Fundamental Properties

Lead is very stable in air of any composition, as it becomes covered with a protective film. Vernon [1219] found that at room temperature the oxide film grows very quickly at first, and that the curve showing the increase in weight flattened off after seven days. Up to this time the increase amounted to 10 to 15 mg/m^2, but hardly rose to more than double this value in 180 days. Under the action of moisture, the oxide film on lead absorbs carbon dioxide and changes into basic carbonate (Falk [303]). In perfectly dry air the rate of oxidation is less by orders of magnitude. Single crystals of lead grown in vacuo retain their silver-bright appearance for months (Seith and Keil [1106]). According to the present author's own experiments, pure lead through which hydrogen was passed while it was melted in a closed vessel, and which solidified polycrystalline under hydrogen, behaved in a similar way. Tests over a period of ten years confirmed the good durability of lead in industrial, marine and country air (Finkeldey [320]). The growth law of the oxide film has been determined several times in the range of temperature 100 to 800°C (Weber and Baldwin [1247], Hofmann and Mahlich [568], Stahl [1143]).

The most important factor affecting the corrosion of lead is the formation of protective films of difficultly-soluble compounds. Table 22 gives the solubilities of various lead compounds (Hodgman [538]).

As an easily soluble salt, lead nitrite may be mentioned. Further, the salts of various organic acids are also somewhat soluble. Lead lactate, formate, tartrate and citrate, for example, dissolve to the extent of

[1] Data on the corrosion-behaviour of specific alloys, or on corrosion phenomena arising in certain fields of application, will be found in the appropriate sections.

more than 0.1 g/litre of water. The compounds at the bottom of table 22 will act as protective films if they form on the lead surface in the respective media. These surface films may not form if the electrolyte has great mechanical activity, e. g., high rate of flow.

Among the electronegative elements in the electropotential series, lead is next to hydrogen, as its potential against the normal hydrogen electrode in a solution of normal ion concentration amounts to only − 0.126 V (LINGANE [752], LANDOLT-BÖRNSTEIN [729]). Accordingly, lead should displace hydrogen from acids at normal concentration. Owing to the high hydrogen overpotential, this only takes place to an insignificant extent with pure lead. The hydrogen overpotential depends upon many factors and for lead it can amount to 1 Volt or more (MILAZZO [853]). Further investigations on the hydrogen overpotential of polycrystalline lead and of lead single crystals were carried out by PIONTELLI and POLI [964].

The conditions are different if the acids under consideration are oxidising acids, such as nitric acid, by which the nascent hydrogen is consumed; or if oxidation of the hydrogen takes place by oxygen dissolved in the liquid and replenished from air. The latter case is of fundamental importance in the corrosion of lead. Two areas, in a metal otherwise homogeneous which are distinguished only by different aeration, will form a short-circuited galvanic element, in which the less aerated parts are usually the anodes, and the parts better supplied with oxygen, the cathodes (EVANS [297]). Metal goes into solution at the anode. The cathode is depolarised by oxygen. Simultaneously, the alkalinity of the solution increases here owing to the consumption of hydrogen ions or through direct formation of hydroxyl ions from water, oxygen and electrons. The rate of solution of the metal at the anode depends on the provision of the cathode with oxygen, unless, as is often the case, the reaction product forms a protective film, which inhibits further anodic attack.

The oxygen-absorption type of corrosion described above, occurs, for instance, in water pipes. As the potential differences between parts of the metallic material (which, according to the electrochemical theory of corrosion are decisive for the initiation and the progress of corrosion) occur here mainly through differential aeration, it is to be expected that the purity of the metal will be of minor influence on its durability. By and large, this assumption is confirmed by experience. On the other hand, the composition of the water is decisive, as on this factor depends whether or not a surface film of sufficient protective action is formed.

Often, in corrosion with consumption of oxygen, no clear separation of anodic and cathodic areas is present. In the course of time, areas of the surface can behave alternately as anodes or as cathodes. The conditions

Table 22. *Solubility of Lead Salts in Water*
Abbreviations: S = soluble, Ak = alkalines

Name	Formula	Temperature °C		Solubility in grams in 100 ccs: Water, Cold	Water, Warm	Alcohol, Acids, etc.
Lead Perchlorate	$Pb(ClO_4)_2 \cdot 3H_2O$	25	—	499.7	—	s in alcohol
Lead Chlorate	$Pb(ClO_3)_2 \cdot H_2O$	18	80	151.3	171	s in alcohol
Lead Nitrate	$Pb(NO_3)_2$	20	100	56.5	127.3	s in NH_3, alkali
Lead Acetate	$Pb(C_2H_3O_2)_2$	20	50	30.7	221	s in glycerin
Lead Chloride	$PbCl_2$	20	100	0.99	3.34	—
Lead Bromate	$Pb(BrO_3)_2$	20	—	1.34	—	—
Lead Bromide	$PbBr_2$	20	100	0.844	4.71	s in acids, KBr
Lead Chlorite	$Pb(ClO_2)_2$	20	100	$9.5 \cdot 10^{-2}$	0.42	s in KOH
Lead Fluoride	PbF_2	20	—	$6.4 \cdot 10^{-2}$	—	s in HNO_3
Lead Iodide	PbI_2	20	100	$6.3 \cdot 10^{-2}$	0.41	s in alkali, KI
Lead Dioxide (hydrate)	$PbO_2 \cdot H_2O$	25	—	$1.48 \cdot 10^{-2}$	—	—
Lead Oxide	PbO	20	—	$1.7 \cdot 10^{-3}$	—	s in HNO_3, alkali, lead-acetate, NH_4Cl, $CaCl_2$, $SrCl_2$
Lead Sulphate	$PbSO_4$	25	40	$4.25 \cdot 10^{-3}$	$5.6 \cdot 10^{-3}$	s in (NH_4) salts
Lead Iodate	$Pb(IO_3)_2$	20	—	$1.83 \cdot 10^{-3}$	—	s in dilute HNO_3
Lead Carbonate	$PbCO_3$	20	—	$1.1—1.75 \cdot 10^{-4}$	decomposed	s in acids, alkali
Lead Sulphide	PbS (precipitated)	18		$8.6 \cdot 10^{-5}$		s in acids
Lead Dioxide	PbO_2			$2.3 \cdot 10^{-5}$		s in dilute HCl hot
Lead Phosphate	$Pb_3(PO_4)_2$	20		$1.4 \cdot 10^{-5}$		s in HNO_3, alkali
Lead Chromate	$PbCrO_4$	25		$5.8 \cdot 10^{-6}$		s in acids, alkali

are different if, for instance, inclusions of lead compounds or other metals are present, when anodic and cathodic areas are locally fixed. Coarse and non-uniform distribution of inclusions can give rise to so-called pitting. The potential differences are then often much larger than in an aeration cell, and, furthermore, these inclusions can have a lower hydrogen overpotential than lead. Corrosion with evolution of hydrogen (which, with lead, is, in general, only to be expected in the stronger acids) is thus facilitated. Corrosion with consumption of oxygen, and corrosion with hydrogen evolution, cannot be strictly separated. Sometimes both kinds of corrosion exist side by side, that is to say, corrosion with hydrogen evolution is intensified by the influence of dissolved oxygen [1094a].

Apart from the question of formation of protective films, the effect of electrolytes on corrosion consists, on the one hand, in that with decreasing pH value (increasing degree of acidity), corrosion with hydrogen evolution is facilitated, and so the hydrogen electrode becomes nobler, while on the other hand, an increase of concentration of the lead ions in the electrolyte makes the lead potential nobler. These relationships play a part in the behaviour of lead towards alkalis (see p. 294). Als lead is amphoteric, in the presence of oxygen, attack with formation of unstable plumbites occurs in caustic alkalis (Laugen) (SCHIKORR [1061]). Here lead is a constituent of the anions; consequently the concentration of lead cations becomes extremely low and the lead electrode less noble.

The corrosion of lead in strong hydrochloric acid proceeds essentially according to the hydrogen evolution pattern, and is therefore accelerated by nobler inclusions in lead. A contributory factor here ist that lead dissolved in strong hydrochloric acid forms a constituent of complex anions, as a result of which a low concentration of lead cations and so a low value of the lead potential is permanently maintained. The formation of complex ions is especially marked in the strongest hydrobromic acid, owing to its high concentration, about 65% by weight. For this reason, a hydrogen evolution, fifteen times as high as that in concentrated hydrochloric acid, was found here (SCHIKORR [1060]). In highly concentrated sulphuric acid, complex formation likewise occurred. Further data, particularly on formation of protective films, will follow in the individual sections. A comprehensive interpretation of these questions was attempted by means of the thermodynamic equilibria of reactions in the lead-water system (DELAHAY, POURBAIX and VAN RYSSELBERGHE [243]). The results could be represented in a potential pH diagram. Further diagrams deal with the effect of carbon dioxide, bicarbonate and carbonate ions, and sulphate ions respectively.

A special type of attack is represented by intercrystalline corrosion. It is promoted by the presence of easily-decomposable precipitates on the grain boundaries, and was observed above all in alloys of lead-magnesium (p. 65) and of lead-calcium-lithium (v. HANFFSTENGEL and HANEMANN [484]) (Fig. 275). It already occurs in commercial lead in certain solutions (p. 10) and is often a contributory factor in the fatigue fracture of lead (p. 245).

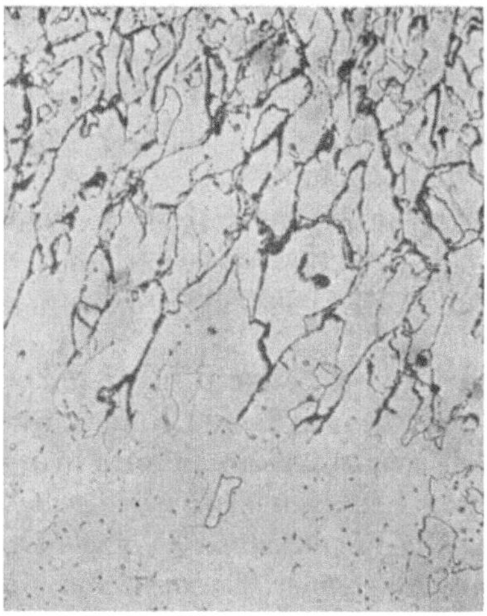

Fig. 275. 0.10% Ca, 0.07% Li Cast, 5 months exposure to air. Peripheral layer with intercrystalline corrosion. ×100

The rate of corrosion in cases of uniform attack should, according to draft German specification DIN 50901, be measured either as loss of weight in g/m² per day or as loss of thickness in mm/year. In the case of pure lead and of high-lead alloys respectively, a loss of thickness of 1 mm/year corresponds to a weight loss of about 31 g/m² per day. The limit of acceptability is likely in many cases to be a decrease in wall-thickness of about 1 mm/year. An evaluation of corrosion-resistance in this manner however, is sensible only in the case of uniform attack and not of pitting, intercrystalline corrosion or the like.

In this connection, the comparatively large electrochemical equivalent of lead is noteworthy. With lead, a given quantity of electricity corresponds to a weight of 3.7 times, and a volume about 2.6 times, that of iron (ROBINSON and FEATHERLY [1015]).

2. Resistance to Sulphuric Acid

a) Fundamentals. Apparatus, containers, pipe lines and fittings in which sulphuric acid is to be kept or used are preferably made of lead. Usually, working temperatures below 60°C, and only in special cases up to 130° and above, are in consideration. The resistance of lead to sulphuric acid depends on the low solubility of lead sulphate in water and the considerable further reduction of the solubility due to the presence of sulphate ions (c.f. Table 23).

Lead sulphate formed by the attack of the sulphuric acid forms a protective film on lead; as a result, further attack at moderately high temperatures practically ceases (Fig. 276) (HECKLER and HANEMANN [501];

Table 23. *Solubility of Lead Sulphate in Sulphuric Acid of Various Concentrations and Temperatures*

(CROCKFORD and BRAWLEY [223])

% H_2SO_4		0	0.005	0.01	0.10	1.0	10.0	30.0	60.0	70.0	75.0	80.0
mg $PbSO_4$	0%	33.0	8.0	7.0	4.6	1.8	1.2	0.4	0.4	1.2	2.8	6.5
in 1 litre	25%	44.5	10.0	8.0	5.2	2.2	1.6	1.2	1.2	1.8	3.0	11.5
of solution	50%	57.7	24.0	21.0	13.0	11.3	9.6	4.6	2.8	3.0	6.6	42.0

TÖDT [1203]). The formation of the protective film is therefore of decisive importance for the question of corrosion resistance. The sulphated lead surface exhibits semi-conductor properties, which may be attributed to the probable inclusion of oxygen atoms in the surface layer (CASEY and CAMPNEY [183]).

In acids of very high concentration, an increase of the solubility of lead sulphate again occurs, owing to the formation of complex anions. This explains the lower resistance of lead in sulphuric acid of high concentration. The coarsely crystalline condition of the protective film and the increasing solubility of oxygen in strong acids (p. 280) have also been held responsible for the lower corrosion resistance. In these cases the attack proceeds almost proportionally to time (SCHIKORR [1060]), which is an indication that the protective action of the film is no longer complete. For sulphuric acid of above 80% by weight (especially if hot) high-silicon cast iron is used instead of lead. Fig. 277 shows the effect of temperature and concentration on the corrosion of lead in sulphuric acid (ROLL [1028]). As the extent of corrosion depends on numerous factors, not least on the method of test, the numerical values given here and later are only of limited significance.

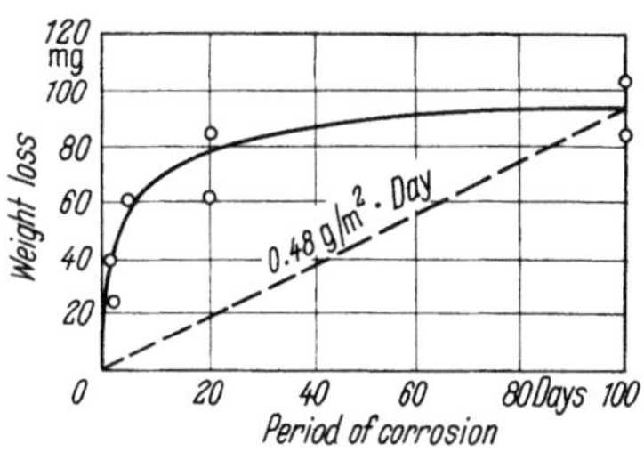

Fig. 276. Course of corrosion vs. time of lead in 80% sulphuric acid at room temperature. (According to HECKLER and HANEMANN [501]).

b) Effect of Small Additions to Lead. On the basis of an extensive literature, good resistance to sulphuric acid can be credited to very pure technical lead. This statement of course needs qualification. For a long time, special resistance has been attributed to lead containing copper, nickel, silver or tellurium (see p. 56). As regards the latter, favourable experience (particularly at elevated temperatures) has been reported (Chem. Met. Eng. [196], HIERS and STEERS [525]). Particularly if, in addition to corrosive attack, mechanical effects occur due to vibrations or to rapid changes of temperature, slightly alloyed lead is to be preferred to pure lead (LAIDLER [725]).

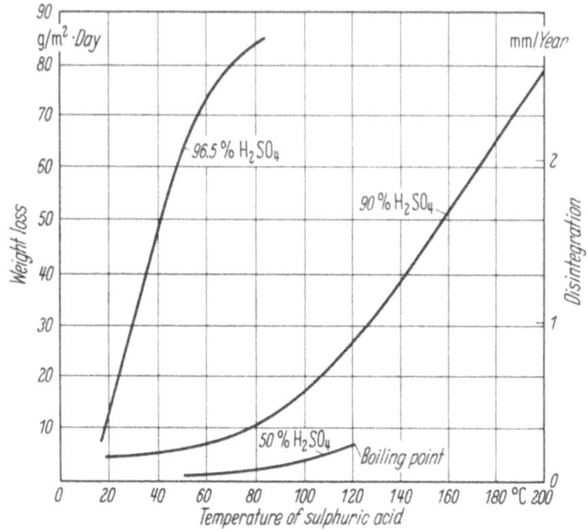

Fig. 277. Effect of temperature and concentration on corrosion of lead in sulphuric acid. (According to ROLL [1028]).

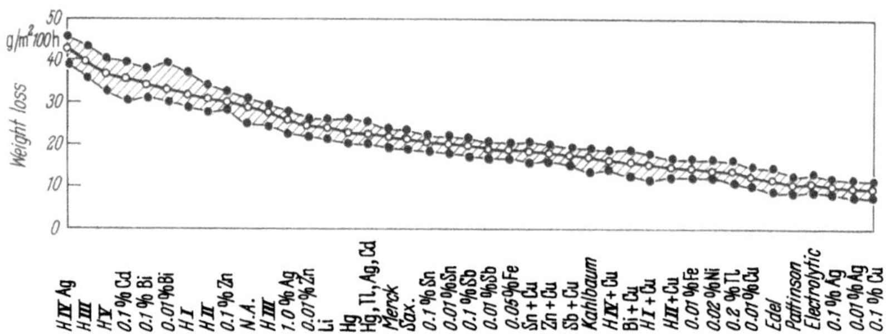

Fig. 278. Effect of additions in lead on corrosion in 70% sulphuric acid at 90 °C. Duration of test — 10 days. (According to BURKHARDT.)

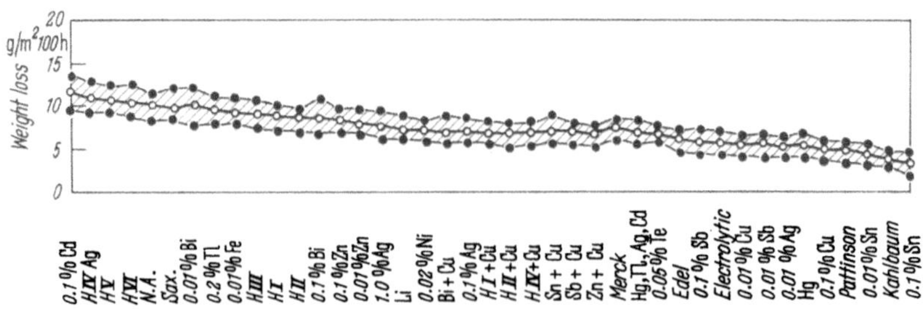

Fig. 279. As Fig. 278. 20 °C. Duration of test — 10 or 20 days. (According to BURKHARDT).

In the extensive researches at the Metallgesellschaft (BURKHARDT [154]), the duration of tests at 170 °C was 4 h; at 130 °C, 2 days; at 90 °C, 10 days; at room temperature 10 or 20 days. The sulphuric acid was 70%. The compositions of the lead specimens on the basis of spectroscopic analysis, are shown in Table 24. The alloys were prepared from electrolytic lead of the composition indicated.

It may be mentioned, as particularly noteworthy, that the differences between the various leads are basically greater at elevated temperatures than at room temperature. At 170°C, the copper- and nickel-containing leads behaved best. The bad behaviour of lead-bismuth here is confirmed by decades of experience. At 170 °C, a content of zinc, cadmium or silver was also markedly disadvantageous. The effect of silver, however, in agreement with other investigations, seemed to be reversed at small contents and somewhat lower temperatures. At 90°C there is, as before, the detrimental effect of bismuth, cadmium and zinc. It is also still present at room temperature, but here only of small significance (Figs. 278 and 279). At elevated temperatures, e.g., 90°C, the effect of small tin and antimony contents is neither markedly favourable nor unfavourable, but at room temperature it is slightly advantageous. Tellurium must be reckoned among the additions with favourable action. The injurious effect of bismuth and zinc can be

Table 24. *The Compositions of Leads Used by* BURKHARDT *[154] in Corrosion Tests*

Impurity	"Kahlbaum" i.e. very pure	Electrolytic	Harris	Pattinson	High Grade Lead (p. 69)	Commercial Lead I	Commercial Lead IV	Saxonia	Commercial Lead III
Cu	0.001	0.004	0.004	0.068	0.001	0.002	0.004	0.006	0.006
Ag	—	—	0.0003	0.002	0.0005	0.0003	0.0005	0.002	0.0005
Bi	0.0005	0.003	0.015	0.032	0.002	0.010	0.028	0.032	0.020
Ni	—	—	—	—	0.035	—	—	—	—
Sb	—	0.001	0.001	0.001	0.003	0.0005	0.001	0.001	0.001
Cd	—	0.0005	0.0005	0.0005	0.0005	0.001	0.0005	0.001	0.0005
Sn	—	0.0005	0.0005	0.020	—	0.0005	0.0005	0.0005	0.0005
Tl	—	—	—	0.001	—	—	—	—	—
Fe	0.001	0.001	0.002	0.001	0.001	0.001	0.001	0.001	0.001

largely offset by the addition of small copper contents, as well as by a few multiples of 0.01% tellurium (HOHLSTEIN [579a]).

Before proceeding to other work, consideration will be given to the nature of the alloy structures formed with lead by the various additions. Small contents of antimony, tin, thallium, mercury, cadmium and bismuth are entirely in solid solution, while iron, tellurium, copper, nickel, silver and zinc, if not present in too small quantities, form inclusions of a second structural constituent. The latter consequently give rise to the formation of local elements (cells), whereby lead forms the anode in the case of copper, silver and other additions discussed below. In these cases the resistance to sulphuric acid is particularly good, if the behaviour of silver at very high temperatures is disregarded. Owing to the anodic dissolution, an insoluble corrosion product (lead sulphate) is formed on the lead, which hinders or stops further attack (self-passivation). The formation of a continuous protective film is said to take place within 0.1 sec. (MÜLLER and MACHU [880]). By stirring the electrolyte, by increasing the oxygen concentration or by addition of oxidising agents to the sulphuric acid, the process is accelerated. The presence of the above mentioned nobler metals on the lead surface has the same effect (MÜLLER and MACHU [881]).

HECKLER and HANEMANN [501] carried out further experiments on the subject of protective film formation. The lead which they used as a starting material had a purity of 99.997 to 99.998%. The alloying additions of silver, arsenic, bismuth, calcium, cadmium, copper, nickel, antimony, selenium, tin, tellurium and zinc were in the region of 0.005 to 0.16%. The alloys were homogenised, rolled at 100°C and stored. Their condition therefore approximately corresponded to equilibrium at room temperature. The corrosion experiments were carried out in static 80% sulphuric acid at room temperature, in concentrated acid at 120°C for fourteen days and at 180°C for two days. At room temperature, in accordance with the tests described above, only slight differences were found between the individual alloys. On the other hand, at 120°C the specimens behaved very differently. By the appearance of the specimens it was possible to distinguish between two types of attack:

Corrosion Type A: The alloys were superficially blackened, but otherwise hardly attacked (self-passivation). An under-layer of lead sulphate had not formed. The black film consisted of small leaflet-shaped transparent crystals of lead sulphate (anglesite). The blackish colour is presumably due to the presence of traces of lead sulphide which had been formed by reduction of lead sulphate. This type of corrosion also occurred in lead with silver, copper, nickel, tellurium, bismuth (0.02%) + copper (0.05%), and bismuth (0.02%) + nickel (0.05%).

Corrosion Type B: The alloys were more or less intensely decomposed into white, fine-grained lead sulphate. The acid had in part broken down with separation of elementary sulphur. In addition to pure lead, the alloys with arsenic, bismuth, antimony, selenium, tin and zinc behaved basically in this way. Individually, however, differences were exhibited by these alloys. Lead-bismuth was completely decomposed; and so, to a large extent, was lead-arsenic, lead-zinc as well as high-percentage lead-calcium. Lead-tin, lead-cadmium and pure lead were partly decomposed, while the loss of weight of lead-antimony was the smallest.

At 180°C, the two types of corrosion *A* and *B* again appeared in similar distribution in the individual alloys as at 120°C.

In addition to HECKLER [501], a contribution to the elucidation of corrosion types *A* and *B* has recently been offered by HOHLSTEIN and PELZEL [579], by corrosion tests on commercial lead in boiling sulphuric acid. The acid concentrations were 10, 30, 50, 60, 70 and 80% H_2SO_4. While the lead specimens underwent only small and very uniform weight-losses in the two lowest acid concentrations, considerable variations of experimental results occurred in the concentrated acids; in this case the specimens corroded according to either *A* or *B*. Corrosion type *A* could be arbitrarily brought about by an artificial oxidation of the lead specimens before the boiling experiments. PELZEL assumes that the surface oxide acted in the corrosion experiments as oxide-cathode and, together with the lead anode, provided the current density necessary for the passivation. The protective action of the oxide is lost to the extent in which it is converted to sulphate. Lead behaved passively, if its potential towards the normal hydrogen electrode was between —0.20 and +2.0 V. This corresponded to losses of 40 to 200 g/m² per day. On the other hand, lead became active at a potential of —0.36 V. The weight-losses amounted to 5000 to 15000 g/m² per day. The change from the active to the passive state could also be attained (apart from the means mentioned) by impressing an external voltage or by additions of oxidising acids (nitric acid, chromic acid).

Current-voltage curves, such as are the basis of PELZEL's results, were obtained by WEHR for pure lead (alone or alloyed with nickel or copper) and for antimonial lead with 4% Sb at 53°C in 85% sulphuric acid and at 80°C in 93% and in 80% acid (Figs. 280, 281). The region of passivity lay between —0.25 and +2 volts, referred to the normal hydrogen electrode. Under the action of an external voltage, passivity suddenly occurs if the current density exceeds the value of $0.5 \cdot 10^{-4}$ Amp/cm². The current density then falls to a very low value, which corresponds to the rate of corrosion in the region of passivity. In the absence of an external current, the current density necessary for the inception of passivity is only attained if either an oxide-cathode, or (as described by HECKLER) inclusions of metals of low hydrogen overvoltage (Cu, Ni, Te, Ag) are present or if an oxidising agent is added to the acid. In corrosion type *B*, the current density necessary for passivation is obviously not attained. According to HECKLER [501], the film

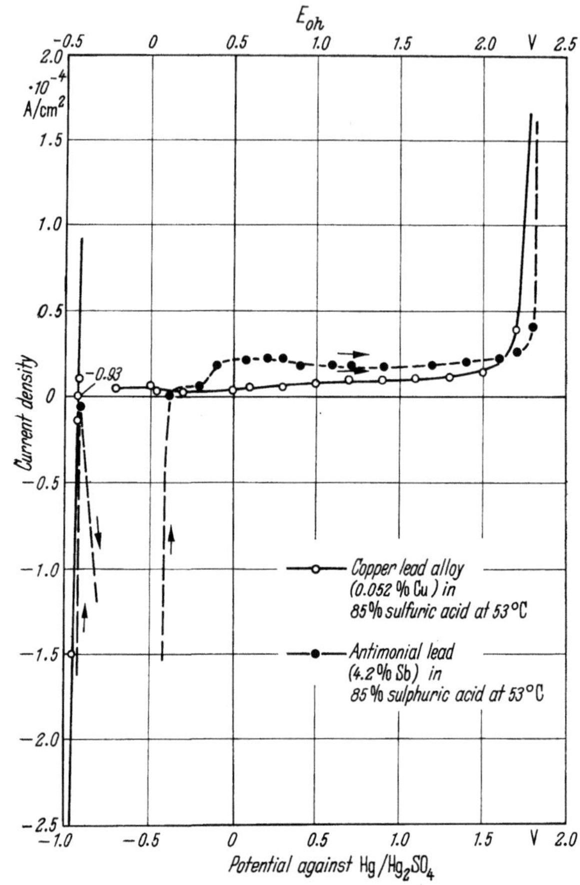

Fig. 280. Polarization curves of copper lead alloy and of antimonial lead.

of white corrosion-product which is formed adheres with sufficient firmness, up to temperatures of about 100°C, to protect the underlying metal from further attack. At higher temperatures, on the other hand, the influence of this protective film is no longer sufficient. Here only the protective film formed according to type *A* corrosion offers sufficient resistance. Some further experiments by HECKLER which PELZEL [579] confirmed, fit well into the present view of the type *A* corrosion. A lead-bismuth alloy, whose normal temperature of decomposition was

100 to 120°C, was stable up to 260°C, if metallic bismuth powder was mechanically forced into the surface, or if the specimen was connected by a conducting wire with an auxiliary bismuth cathode. A current flowed in the wire in a direction indicating that the lead was the anode, passing into solution. If at this high temperature, electrical connection was broken then sudden decomposition occurred. Auxiliary cathodes of graphite, copper, nickel or platinum gave equal service. Lead-bismuth alloys, connected to the auxiliary cathode, were stable for days at 180°C. The appearance of the specimens corresponded to corrosion type *A*. In a similar way, lead-arsenic and lead-cadmium alloys could be made resistant to sulphuric acid at elevated temperatures. Even a small addition of copper sulphate (SCHIKORR and SCHALLER [1064]) to boiling 55% sulphuric acid (which produced about 0.06 g of copper per square metre of lead surface) changed the corrosion from the dangerous type *B* to the practically harmless type *A*. The particularly poor corrosion resistance of lead containing bismuth was explained by a low local current activity due to the remarkably great homogeneity of this alloy. This was concluded from the fact that the alloys showed the lowest rate of solution in nitric acid.

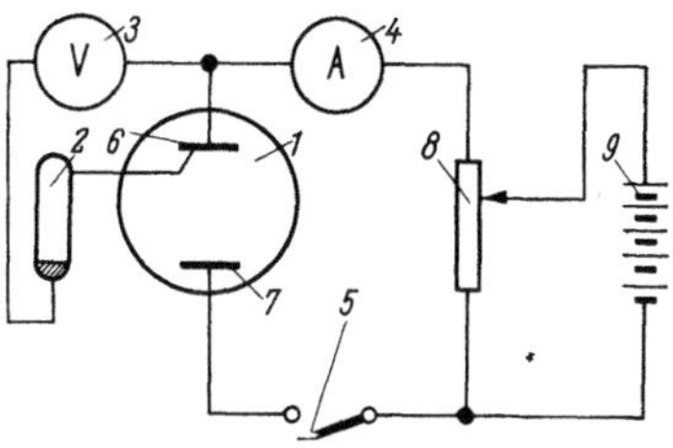

Fig. 281. Wiring diagram.
1 Analyzer; *2* Standard cell; *3* Valve voltmeter; *4* Ammeter; *5* Switch; *6* Electrode (e.g .Pb); *7* Counter-electrode; *8* Resistor; *9* Direct-current supply.

Further publications which deal with the behaviour of lead alloys as anode material in the electrolysis of sulphate solutions, report similar corrosion-protective effects by other metals which were conductively connected to the lead anodes (KIR'YAKOV and STENDER [668]).

It should be repeated that, fundamentally, oxygen dissolved in sulphuric acid may be concerned in both types of corrosion (CASEY and CAMPNEY [183]). It is possible, for instance, that in type *A* corrosion, local currents between anodic lead and cathodic nickel arise only from the depolarising action of the dissolved oxygen, without the release of hydrogen at the nickel. Iron inclusions may have a similar effect. In type *B* corrosion of homogeneous lead, the oxygen can, in the case of variable aeration, contribute to the formation of cathodic areas, as a result of which the rate of corrosion can in some circumstances be increased. On the other hand, in type *A* corrosion the formation of the protective film is often accelerated by dissolved oxygen. That aeration plays a part in sulphuric acid corrosion is also shown by older corrosion tests in nitrosylsulphuric acid with and without access of air. In the first case a more intense attack could regularly be noted (LUNGE and

SCHMID [785]). The general nature of the dependence of weight-loss of lead on the concentration of the acid used (Fig. 282) has also been brought into relationship with the action of the dissolved oxygen, as the solubility curve of oxygen in sulphuric acid follows a course similar to the weight-loss curve. The relationship has not however been firmly established, as the solubility of lead sulphate in sulphuric acid of different

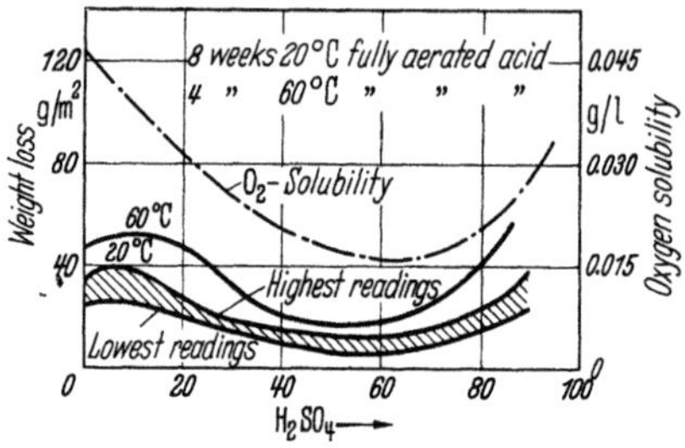

Fig. 282. Solubility of oxygen and lead in sulphuric acid of various concentrations [196].

concentrations exhibits the same course (Table 25). In contrast with Fig. 282, HOHLSTEIN [579a]) found that at the boiling temperature of sulphuric acid, a 50% concentration of acid is more aggressive than one of 30% or 70%. Perhaps variable aeration plays a part in the inception of pitting corrosion in Figs. 283 and 284. The arrangement of the pits indicates an origin in scratches, which are less well aerated and therefore anodic at the bottom.

A far more favourable behaviour of pure lead came to light in the investigation of BURKHARDT [154] than in that of HECKLER. It may be that the variable degree of purity of the purest leads still plays an important part. The lead used by HECKLER contained only a few tenthousandths of 1% of Cu, whereas the electrolytic lead used by BURKHARDT had a copper content of 0.004%. WEISSBACH [1254b], on the basis of his experiments, regarded pure lead as a material unsuitable for sulphuric acid above 20% for temperature above 100°C.

WICKERT [1266]), gives further results of corrosion tests. The lead specimens were heated for 48 hours at 90°C in a glass tube containing 80% sulphuric acid. Among other things an increase of resistance was observed as compared with pure lead, by alloying with zinc, aluminium and silicon, while an alloy with 0.1% nickel behaved unfavourably. Lead alloys which, in addition to nickel, also contained tin, aluminium, silicon, zinc or copper, showed a smaller attack. The observations reported here are, in part, at variance to the results quoted above. As a criticism to the work ot WICKERT [1266] it may be pointed out that very short-time tests are of limited value, and that no information is given about the degree of purity of the lead employed. Further extensive investigations on the corrosion resistance of lead and lead alloys in sulphuric acid were carried out by HIRAMA and WATANABE [529].

In the works discussed above, the nature of the corrosion was considered in relation to the alloying constituents; the investigation which is next quoted is mainly concerned with the effect of the concentration of the sulphuric acid (SCHIKORR and SCHALLER [1064]). Observations

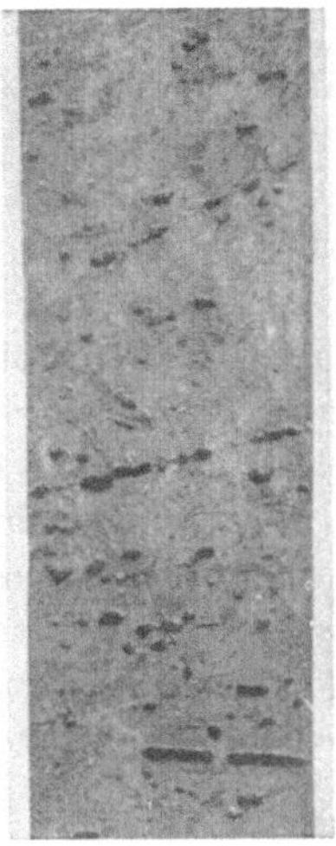

Fig. 283. Lead sheet with pitting corrosion. Bottom of a container in which fat was split by sulphuric acid. 0.5 ×

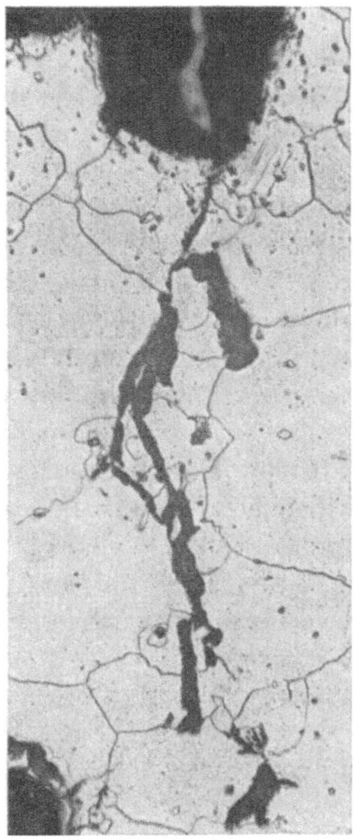

Fig. 284. As previous figure. Section through a pit. Corrosion proceeding on grain boundaries and inside grains. × 150

were made on the attack of hot sulphuric acid of medium concentration on three leads with different degree of purity (99.99 and 99.9%). Apart from the type *A* and type *B* corrosion described above, a further, type *C*, was noted. It corresponds to the sudden decomposition of lead in hot sulphuric acid in the flash test, described below. The duration of these tests, in general, was only two hours.

THOMPSON [1190] suspected that the iron content of lead was the black sheep in its corrosion by sulphuric acid. Even the adhesion of fine iron particles (derived from the casting mould or from working) is said to prevent the formation of a well-protective lead sulphate film (RABALD [987]). Contrary to this, however, it is conceivable that inclusions of iron, similar to those of nickel, improve the resistance to corrosion by sulphuric acid. The measurements mentioned above, on circuits of lead/sulphuric acid/iron showed that iron can be passive and so become the cathode (WERNER [1261]). Similar behaviour was found for circuits of lead/oleic acid/iron with access of air (HEYER [519]).

If the resistance of lead alloys to corrosion by sulphuric acid is to be determined in the laboratory, it is more realistic to carry out long-term tests at the temperatures in question. Moreover, the possibility of corrosion at the liquid-gas phase boundary should be investigated. Often the flash test is used for rapid evaluation of a grade of lead (BARRS [47]). The flash point is determined as the temperature at which a bright strip of lead in sulphuric acid spontaneously changes to lead-sulphate, while the decomposition of the acid to sulphur dioxide and sulphur occurs simultaneously. The flash test must be carried out carefully and under precisely controlled conditions, if reproducible results are required. For the rest, it is of merely limited utility, as it merely provides information on the behaviour of lead at the highest temperatures, which in practice are rarely encountered, and not in the important medium range of temperature. For testing the applicability of lead alloys in strong sulphuric acid systems, the flash test was found unsuitable (WICKERT [1266]). Here particular attention must be drawn to the wide scatter of the results of corrosion tests, which requires caution in application to practice.

Among other factors, the flow of liquid plays a part in the formation of protective films on chemical equipment. While with 20% sulphuric acid at 25°C a rate of flow up to 0.2 m/sec was without effect, at rates up to 1.6 m/sec the corrosion increased about proportionately to the rate and at higher rates more than proportionately (ROLL [1028]). Lining the equipment with acid-resistant bricks can largely prevent erosion, and further serve as a protection against temperature and provide mechanical support to the lead (UHLIG [1211]). It is also advisable to expose a lead-lined stirring vessel for a few days to static sulphuric acid before first use (RABALD [987]).

c) Corrosion-Resistance of Antimonial Lead. The resistance of lead-antimony alloys to sulphuric acid is of interest since parts of pumps, valves, stirrers and the like are preferably made of antimonial lead on account of its mechanical strength. Extensive tests in dilute (5%, 20%) and in concentrated (96%) sulphuric acid were carried out (WERNER [1261]). The antimony contents were stepped from 0.1 to 25%, the temperatures of test were 20, 50 and 100 °C. It is difficult to bring the results on to a common scale, especially as the measurements showed considerable scatter.

In older investigations, the behaviour of commercial lead (99.983%) was compared with that of lead-antimony (0.2% Sb) in sulphuric acid and in nitrosyl-sulphuric acid. At room temperature the alloy showed somewhat smaller losses in weight than did commercial lead but at higher temperatures by and large the reverse applied. Further, commercial lead (0.04% Bi as the most important impurity) was compared with two antimonial leads, respectively having 1.8% Sb, 0.10 As, 0.05% Cu, and 18.1 to 18.3% Sb, 1 to 3.1% As, 0.1 to 0.3% Cu. At room temperature, in 96% sulphuric acid alone or with 1% N_2O_3 no great difference was found between commercial lead and antimonial lead and similarly at 100 °C, in acids with 83 and 79% H_2SO_4. On the other hand, in sulphuric and in nitrosylsulphuric acids of densities 1.84 (96%), at 100 °C, antimonial lead behaved significantly worse than commercial lead althought the latter contained bismuth.

The comparison of the rate of dissolution of pure lead, antimonial lead with 8.2% Sb and a lead-antimony-tin alloy with 4.78% Sb, 1.12% Sn, 0.08% As, 0.01% Cu, 0.01% Bi, 0.002% Ag, 0.005% Fe, in accumulator acid (density 1.24) at room temperature, indicated that pure lead showed the smallest attack. The attack was about doubled in the case of the 8.2% antimonial lead. The highest losses were shown by the tin-containing antimonial lead in which the gain in weight still persisted after 27 days of test, whereas in pure lead and in tin-free antimonial lead corrosion practically ceased after 3 and 16 days respectively (Katz [647]). The effects of elevated temperatures, and longer periods of test at room temperature, are not given in this investigation. According to newer tests by Weissbach [1254b], pure antimonial lead with 3, 5 to 6% antimony behaves very well in the region of 20 to 60% H_2SO_4 at temperatures between 80 °C and boiling, in some respects more favourably than pure lead, and at boiling temperature definitely better than copper lead containing 0.04 to 0.08% Cu. Using antimonial lead with 4% antimony, in 80% sulphuric acid at 80 °C, Wehr (1252a] observed type *A* corrosion and only a slight decrease of resistance compared with copper lead. Thus Weissbach and Wehr confirmed the recommendations of a Canadian investigation (Turner [1208a]).

At concentrations of sulphuric acid which are not too high, antimonial lead thus appears to be a material equivalent to copper lead. On the other hand, increased attack must be expected in concentrated acids at elevated temperatures. Here Weissbach recommends a copper lead with a small antimony or tin (0.1%) content. The application of lead-antimony-tellurium should also offer advantages [196]. For sulphuric acid, hydrochloric acid and their mixture with sulphates and chlorides, antimonial lead (0.8 to 1.1% Sb) with an addition of copper is recommended [735a].

The superior resistance of a lead alloy with only 0.05% Ag in sulphuric acid and in ammonium sulphate solutions, as compared with antimonial lead with up to 7% antimony, and with leads higher in silver, is shown in a Japanese investigation (FUKE [352]). EMICKE [275] proposes replacing lead with up to 3% Sb by an alloy with 0.1% As. The latter was found to be superior to the lead antimony alloy in sulphuric acid.

d) Stress-corrosion and corrosion-Fatigue in Sulphuric Acid. Lead, while in contact with sulphuric acid, is often exposed to static or dynamic stresses. In such cases a type of corrosion can arise, which is not known in other circumstances. Materials which corrode under tensile stress can even crack without the formation of visible corrosion products. This particularly dangerous form of damage is known as stress-corrosion.

The first indication of the behaviour of lead towards such a combined attack is given by GOUGH [411a]. In a comprehensive work on corrosion-fatigue, he describes the formation of cracks in a lead pipe through which 70% sulphuric acid was passed at 140 °C and which was set vibrating by a motor running in its immediate vicinity. Gough was able to show that, in the absence of vibration, cracks no longer occurred. Mack devoted a systematic investigation to the corrosion-fatigues of some hard lead alloys in sulphuric acid [790b]. He describes alternating bending tests on pure lead, on lead-tellurium and on lead-antimony alloys in 38% sulphuric acid. MACK interprets the mechanism of corrosion-fatigue in the sense that as a result of continuous vibrating stress the lead sulphate film formed by the acid cracks in various places on the surface of the specimen, thus continuously exposing the underlying bright metallic lead to an increased corrosive attack.

In his dissertation, WEHR [1252a] treats the corrosion-fatigue of some kinds of lead used in the chemical industry, (pure lead, copper lead, pure lead with 0.025% Ni and antimonial lead with 4% Sb). The application of an alternating tensile-compressive stress provided a constant distribution of stress over the cross-section of the specimen. The frequency of cycling of the testing machine was 50 Hz, the concentration of sulphuric acid 30 and 85% and the temperature of test 22 and 53 °C. The curve diagram (Fig. 284a) gives a survey of the majority of the experimental results. In this "Wöhler Diagram", the introduction of lines of constant corrosion loss is an innovation. Their slope makes it clear that the corrosive attack increases with rising alternating stress. The "damage line" (beginning of cracking) shown by dashes was derived from potential-time curves which were taken simultaneously (Fig. 284b). The beginning of passivation can be recognised in the potential-time curves by a clear potential-jump after about 16 hours. However, owing to the vibrating stress the potential of the passivated lead constantly falls in the subsequent period, until after 22 hours the first cracks appear in the protective film. From this point the course of the potential curve is unsteady, but from its up-and-down nature it can be concluded that the cracks formed are temporarily healed by the renewed formation of a protective film. In the instant of fracture, after a fatigue-testing time of about 29 hours, the potential sinks to the original value of non-passivated lead.

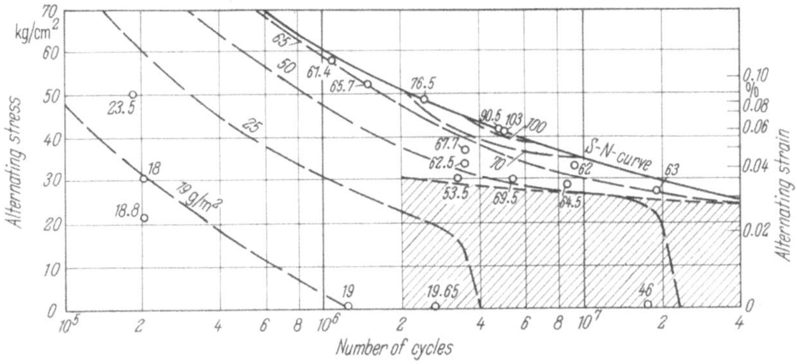

Fig. 284a. Loss in weight by corrosion under alternating stress. Copper lead alloy (0.052% Cu)/85% sulphuric acid at 53 °C.

– – – – Damage line of the sulphate layer. Frequency of cycles 2,970/min.
/////////// Lead passivated

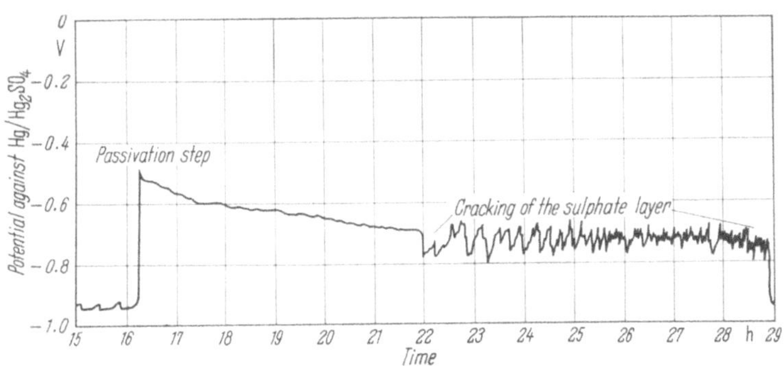

Fig. 284b. Potential time diagrams of copper lead alloy in 85% sulphuric acid at 53 °C. Magnitude of stressing $\sigma = \pm 30.5$ kg/cm²; $\varepsilon = \pm 0.037\%$.

The experiments showed that the preparation and knowledge of Wöhler curves are not by themselves sufficient in many cases for the evaluation of corrosion-fatigue. Thus, for example, the Wöhler curve of copper lead rods tested in acid was superior than the corresponding tests in air. The fact that, in spite of this, a considerable corrosion loss occurred, could only be established by weighing and by plotting the lines of equal corrosion-loss in the Wöhler diagram.

In the dissertation mentioned, the effect of creep on the corrosion of lead was also discussed. In one series of experiments, specimen rods of the leads enumerated above were statically stressed in tension and at the same time immersed in 93% sulphuric acid (80% for the antimonial lead) at 80 °C. After 30 days of test the specimens were removed, washed with ammonium acetate and reweighed. The curves for the corrosion-

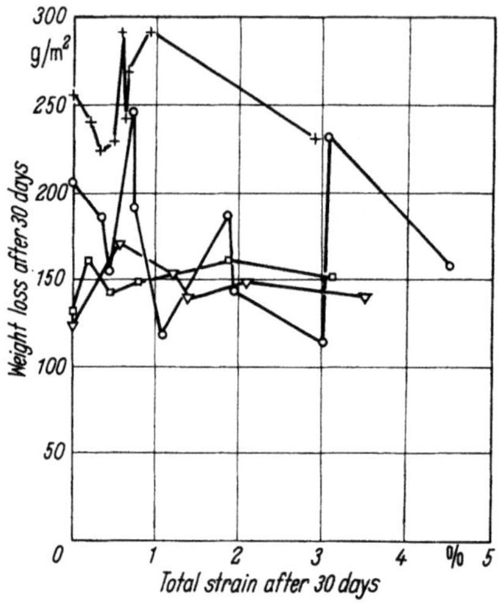

Fig. 284c. Losses in weight versus creep strain of lead in sulphuric acid after 30 days. Temperature of the acid: 80 °C.

○ Pure lead; ▽ Copper lead alloy (0.052% Cu); □ Nickel lead alloy (0.025% Ni) — in 93% sulphuric acid
+ Antimonial lead (4.2% Sb) — in 80% H_2SO_4

loss show that there is no unequivocal effect of creep on corrosion (Fig. 284c). The large fluctuations in corrosion-loss in pure lead are duc to the fact that in this unalloyed lead large variations in potential occurred during part of the corrosion process in sulphuric acid, which

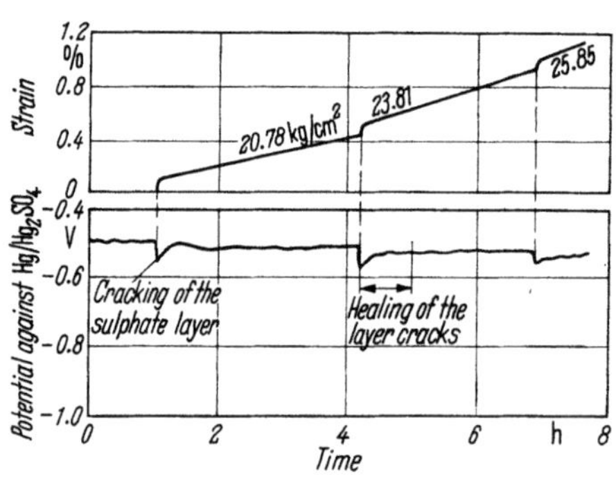

Fig. 284d. Creep curve and potential time curve of pure lead in case of discontinuous loading. Corroding agent: 93% H_2SO_4; temperature: 80 °C.

delayed a passivation of the lead. This behaviour did not occur in the alloyed lead. When the corrosion potential in these cases was plotted against time, it was shown that the protective film of lead sulphate, once formed, remained undamaged even at very high rates of creep, such as rarely occurred in practical use. In creep tests on pure lead with stepwise increase of load it was found that only in the instant of loading, a transient fall of potential occurred (Fig. 284d) but that the cracked protective film was at once restored.

e) Corrosion in Nitrosylsulphuric Acid. In the lead chamber process, sulphur dioxide is oxidised by air and water vapour (with the participation of N_2O_3) to sulphuric acid[1]. N_2O_3 dissolves in strong sulphuric acid to form nitrosulphuric acid HSO_4NO. Conversely nitroylsulphuric acid, under the action of water, splits off N_2O_3. As the latter goes into solution in weaker acids with separation of NO and formation of nitric acid, the acid of lead chambers forms a mixture whose quantitative composition depends principally on the water content. In an old, exhaustive investigation, nitrosylsulphuric acids with densities of 1.30 to 1.77, and a nitrogen content of 0.1% were prepared by adding nitrosylsulphuric acid or nitric acid or a mixture of both to sulphuric acid. A definite ratio N_2O_3:N_2O_5 is set up in the endproduct, N_2O_5 predominating in dilute acids, N_2O_3 in concentrated solutions and this is largely independent of the nature of the acids which were added. This means that nitrosylsulphuric acid is stable only in concentrated acids, and decomposes in dilute acids progressively with dilution, to form nitric acid (LUNGE and SCHMID [785]).

The corrosion tests [785] were carried out at temperatures up to 70 °C. A minimum of attack was found at densities between 1.44 and 1.58; these are the concentrations at which only a little nitrosylsulphuric acid is still present in equilibrium. The attack was more intense with decreasing and with increasing density. In the first case the relative increase of free nitric acid made itself noticeable, in the second, the action of the concentrated sulphuric acid added to that of the nitrosylsulphuric acid. The concentrations of chamber acids which occur in practice approximately correspond to the indicated minimum of corrosive attack.

In other test series [785], investigations were made, at temperatures between 20 and 200 °C, of the corrosion of antimonial lead and commercial lead in pure sulphuric acid of densities between 1.84 and 1.725 and in sulphuric acid of equal concentration with addition of 1% N_2O_3. While corrosion in the strongest acid was in general increased by a content of N_2O_3, in the less concentrated acids the reverse occasionally occurred. However, according to the series of tests first described above, an increase of attack due to N_2O_3 is to be expected at still higher dilution.

Corrosion tests by JONES [634] in sulphuric acid of density 1.72, with or without addition of 2.5% N_2O_3 at 52 °C indicated a lower degree of adhesion of the lead sulphate formed in the nitrosylsulphuric acid.

[1] The mechanism of this process can be disregarded here.

Further, sheets of lead containing copper showed higher weight-losses than commercial lead sheet. In pure acid the corrosive attack on the two kinds of sheet was exactly reversed (p. 273). In nitrosylsulphuric acid of the composition indicated it was found that a copper content of only 0.01% was permissible. Also, in the tests by SCHÜNEMANN [1084] at the Hüttenwerk Oker in the Harz mountains, leads containing copper behaved unfavourably, if the period of observation was sufficiently long. Further tests of slabs of various types of lead, carried out for 6 months at the hottest place in the Glover tower in a sulphuric acid works, at 130 °C, indicated no advantage of a copper content and other additions. Here the injurious effect of bismuth was hardly noticed. On the basis of these results, and above all, of the behaviour in nitrosylsulphuric acid in the Gay Lussac tower, lead containing copper is only recommended in places where high fatigue strength is required. Otherwise pure Parkes lead is preferred (SCHÜNEMANN [1084]).

Various investigators have treated the corrosion of lead in sulphuric acid manufacture particularly with reference to the most favourable working conditions of the process (WICKERT [1266], BURBANK [151]). The suitability of lead and iron as material have been compared (JUSCHMANOW and co-workers [639]). Some investigations were concerned with the origin of local pitting in lead chambers. It was attributed to the action of variable quantities of free nitric acid in the chamber acid, whereby concentration cells occur; $^{+}Pb/H_2SO_4$ or nitrosylsulphuric acid/$HNO_3 + H_2SO_4$ or HNO_3 + nitrosylsulphuric acid/^{-}Pb (PERSCHKE [953]).

Such differences in the composition of the acid might occur, for instance, owing to mechanical imperfections in the lead lining such as folds, seams and the like. The nitric acid dissolves lead anodically and is constantly reformed by precipitation of lead sulphate. If the lead sulphate formed is of loose texture and falls to the bottom, the process can continue unhindered. Corrosion tests in artificial cells of the above kind, as well as some orienting potential measurements, provide support for the views put forward. The observation made by JONES [634], that pitting occurs at a higher nitric acid content of the sulphuric acid, may also be adduced here. In further studies, investigation was made of problems of lead corrosion which arose in relation to intensifying the lead chamber process (WICKERT and ZENKER [1267], WICKERT [1266]). Here, in addition to the action of sulphuric acid and of nitrosylsulphuric acid, an effect of chloride ions, which could reach the sulphuric acid system as chlorides in the water, was assumed. In order to test the behaviour of various lead alloys, the specimens were heated in a mixture of sulphuric and hydrochloric acids at 90 °C for 48 hours. The numerous experimental results could only be regarded as of relative value, because in general the formation of the protective film had not ceased in these short periods of test. Apart from this, an increased susceptibility of lead at the boundary between the sulphuric acid and the nitric oxide gas phase, in the presence of chloride ions, was established.

f) Further Phenomena, Summary. In addition to the metallic impurities in the lead, special significance is also ascribed to the oxygen content of the metal when considering resistance to sulphuric acid. SCHIKORR [1064] observed a favourable effect of oxides which had been rolled into the lead surface in the form of dross. On the other hand, WERNER [1261] challenges the effectiveness of any oxygen which is detectable under the microscope and in particular any oxygen content of above 0.006%. He attributes the unfavourable effect of inclusions of lead monoxide in lead to the preferred dissolution of the oxide, whereby the lead sulphate developed does not form a sound protective film. If lead monoxide is present in continuous filaments, perhaps in welds or in homogeneous lead linings, the sulphuric acid is given the opportunity to penetrate deeper into the lead. The effect of surface oxide must be interpreted differently.

BARTELD [48] confirmed these observations in their essentials. He found that only oxygen contents above 0.002% were injurious in corrosion by hot concentrated sulphuric acid. In nitrosylsulphuric acid only a slight increase of corrosion could be determined even in specimens with 0.04% oxygen; that is, an oxygen content below 0.002%, such as is present in the usual types of lead (see under binary alloys) should be of no significance in this connection.

Course-grained structure of lead is frequently regarded as unfavourable for resistance to sulphuric acid (BRENTHEL [130]). The occurrence of grooves at the grain boundaries—particularly at high temperatures—has been described, and these, in some circumstances are so deep that the lead crystals fall apart (WERNER [1261]). It is further reported that lead of very variable structure is more quickly destroyed in sulphuric acid evaporators than lead of uniform grain size (COTTON [216]). The development of eutectic structures is also regarded as unfavourable (WERNER [1259, 1261]), as a more readily corrodible eutectic between the grain boundaries can lead to inter-crystalline corrosion. While lead with uniformly distributed silver inclusions behaves favourable, this was not the case in a cast alloy, owing to the presence of eutectic. An increase in the number of pits was observed in the upper parts of suspended sheets. The lead here is under mechanical stress (McKELLAR [816], LOVELESS and co-workers [768]). In contrast with this, WEHR [1252a] found that creep had no effect on the resistance of lead to sulphuric acid.

The resistance of lead to sulphuric acid is greatly reduced by a content of hydrochloric acid (CALCOTT and co-workers [167]). Also alternate filling of lead containers with hydrochloric acid and with sulphuric acid is disadvantageous, as the lead sulphate formed in sulphuric acid is dissolved by strong hydrochloric acid. According to LEVIN [745a], additions of various organic substances to sulphuric acid do not impair the corrosion-resistance of lead.

Summarising, it can be said on the question of the resistance of lead to sulphuric acid, that the points still open are concerned less with room temperature than with higher temperatures and with acids with further additions. The impurities in the acids can be due to the course of the reaction itself, or they can be derived from impurities in the raw materials, or reach the sulphuric acid from the air. Thus in one case the waste gases from a steel works were recognised as the cause of corrosion (COTTON [216]). Impurities in lead play a far smaller part at room temperature than at higher temperature. The low corrosion resistance of pure, very dilute lead-bismuth alloy even at moderately high temperatures is beyond doubt (see also p. 279). For the rest, whether the advantage lies with pure lead or with lead with small contents of copper, silver, nickel or tellurium or several additions together cannot, in general, be decided. This question must be settled from case to case. In the production of rayon by the viscose process only 99.99% lead is suitable as lining material for the containers. Copper inclusions in the lead would react with the hydrogen sulphide in the solution, forming black cupric sulphide and giving rise to an undesired coloration of the rayon (WESSON [1261a]). The alloys mentioned provide, in part, improved corrosion resistance at very high temperatures, and they further offer increased fatigue resistance and a slightly higher hardness. On the other hand, very good results were also obtained with pure lead. A comparison of nine types of lead, which proved good in lead chambers for 5 to 20 years, lists, in addition to pure lead specimens, only two with copper contents of 0.033 and 0.04% respectively (JONES [634], compare SCHOPPER [1078]).

3. Resistance to Inorganic and Organic Acids and other Chemicals

a) Halogen Acids. At low temperatures, lead has a certain resistance to dilute hydrochloric acid, while concentrated acid, particularly at high temperatures, attacks strongly (WICKERT [1265], CALCOTT and co-workers [167]). At room temperature, attack up to a concentration of 30% HCl remains within tolerable limits, at 100 °C up to HCl-content of 20%. Impurities in the acid increase the attack, sometimes to a considerable extent (ROLL [1025]). The attack is roughly proportional to time, an indication of the incompleteness of the surface film (cf. Table 22). In cold concentrated hydrochloric acid the loss of weight amounts to 10 g/m² per day. Single lead crystals behave similarly in one mole/litre-hydrochloric acid (BOLOGNESI [108]). Corrosion is considerably accelerated by stirring and by access of oxygen (WHITMAN and RUSSELL [1264]). Lead is therefore only used to a limited extent as a material for hydrochloric acid. Pure lead is preferred over all lead alloys with the exception of lead-antimony alloys of high antimony content (RABALD [986]). On the other hand, lead is sufficiently resistant towards dry and gaseous hydrogen chloride (ROLL [1025]). According to an American patent, lead in diluted hydrochloric acid is said to be suitable as sacrificial anode for the cathodic protection of steel (ROBINSON and FEATHERLY [1015]).

In concentrated hydrobromic acid, lead is attacked still more strongly than in hydrochloric acid. Commercial lead is accepted as a material for hydrofluoric acid up to about 65% at room temperature. Use at higher temperature (up to about 85 °C) is said to have limited possibilities but only in hydrofluoric acid which has been well deaerated. Weight losses of about 2.5 g/m² per day were observed in 40 to 60% hydrofluoric acid at room temperature. The attack increased to about 10 g/m² per day in mixtures of 60 to 65% hydrofluoric acid with 12 to 25% fluosilicic acid, 0.3 to 1.25% sulphuric acid and 0.01 to 0.03% iron. Lead is also established in the production of anhydrous hydrofluoric acid (ROLL [1026], LINGNAU [754], FRIEND and TEEPLE [348]). In corrosion tests lead was found to be insufficiently resistant towards 5 and 48% hydrofluoric acid at about 66 °C (SCHUSSLER and co-workers [1091]). The corrosion rate of lead in hydrofluosilicic acid and other media is reported elsewhere (KATZ [647]).

b) Nitric Acid and Mixed Acids. Corrosion in nitric acid decreases with increasing concentration to a minimum at 65 to 70% HNO_3. At still higher concentrations sometimes an increased attack was observed. In no case, however, was it higher than that of 96% sulphuric acid (LUNGE and SCHMID [785]), SCHIKORR [1060]).

Potential measurements in acids of various concentrations confirmed this result (GUITTON [455]). It is concluded that lead could therefore be used for storage of nitric acid of about 60% upwards in the cold (JONES [635]). On the other hand, lead vessels are not suitable for dilute acids or for higher temperatures.

The attack of a mixture of equal volumes of the strongest sulphuric acid (98.85%) and the strongest nitric acid of density 1.50 or fuming nitric acid of density 1.52 on lead is astonishingly small (LUNGE and SCHMID [785]). On the other hand, lead cannot be used if the water-content of the mixed acids is above 25% [195].

c) Phosphoric Acid. Lead was recommended as material for saturators for the production of ammonium phosphate from ammonia and concentrated phosphoric acid, although the tests concerned did not show complete resistance (Clarkson and Hetherington [199]). Corrosion tests in pure 75 and 40% phosphoric acid prepared in the dry way, showed a considerable attack on lead. On the other hand, it was completely resistant in 40% acid prepared in the wet way, from calcium phosphate and sulphuric acid (Portevin and Sanfourche [974]). This is due to the fact that the technical acid produced in the wet way had a small content of sulphuric acid, corresponding to 0.44% $CaSO_4$. The protective film formed in phosphoric acid containing sulphuric acid, however, offered no protection against pure phosphoric acid, as it obviously dissolved. Lead is regarded above all as a material for phosphoric acid produced in the wet way in the synthetic fertiliser industry (Weber [1249]). If this acid contains much fluorine, lead is considered unsuitable (Rohrmann [1022]). In phosphoric acid plants, lead-tellurium is said to have advantages over commercial lead (Chem. Met. Eng. [196]).

Lead with 6% Sb is proposed for evaporators. However, this material is sensitive to erosion by agitated gypsum sludge (Porter and Lowrison [971]). With certain restrictions, the use of lead is possible in the production of phosphoric acid by the electric arc process (Hartford [495]).

d) Sulphurous Acid. Lead is resistant to sulphur dioxide and aqueous sulphurous acid (Chem. Met. Eng. [195]). It is used as a material for pipelines, wash towers and pans handling this acid, as well as in sulphite-pulp equipment (Ullmann [1212, 197]). The resistance of lead towards sulphur dioxide is not reduced by a content of fluorine, such, for instance, as can exist in gases from zinc roasting plants (Roll [1027]). Moreover lead was only slightly attacked in liquid sulphur dioxide with 1% of water and 0.21% of oxygen at 20 °C (Bollinger [107]).

e) Organic Acids. Lead is attacked by most weak organic acids in the presence of air or organic oxidants, which fact can be deduced from the considerable solubility of the corresponding salts (p. 268). Lead should therefore not be used in contact with foods. Some organic acids, as constituents of tap water, prevent the precipitation of lead as carbonate and thus the formation of a protective film, as is indicated in the discussion on water pipes (p. 304). The attack on lead by a series of organic acids exhibits an intercrystalline character (Coles and co-workers [208]).

Corrosion tests in 0.1 mole/litre oxalic, tartaric, citric and benzoic acids showed, in that order, increasing attack of 0.17; 3.4; 9.6; and 7.7 g/m^2 per day. As this order is also that of the solubility of the corresponding lead salts, a clear relationship thus exists between the latter and the rate of corrosion (SCHIKORR [1060]). As regards the corrosion of lead in mole/litre-butyric acid, see PUTILOVA [983a]. While lead is strongly attacked by dilute acetic acid, corrosion in glacial acetic acid is slight. Therefore lead is considered for the storage of glacial acetic acid (CALCOTT and co-workers [167], BECKINSALE and WATERHOUSE [73]; Chem. Met. Eng. [195]). On the other hand, tests by WHITMAN [1264] at 20 °C in dilute acetic and in glacial acetic acid through which oxygen or hydrogen was bubbled, showed a very strong attack in the glacial acid. Here unfavourable accompanying circumstances certainly contributed. In view of its resistance towards higher fatty acids, lead is also used in the fatty acid industry (WIEDERHOLT [1269]). Interesting cases of corrosion of porous sheet lead claddings have been traced to galvanic lead/fatty acid/iron cells (p. 282); (HEYER [519]).

The so-called phenol-corrosion has been observed on lead claddings wrapped with jute soaked in bitumen. The corrosive attack follows the screw line of the jute strings. The catalytic action of the phenol content of the tar product was formerly regarded as the effective cause, and therefore this kind of corrosion received its name (DA FANO [305]). In investigations of the next decade, the cause of corrosion was seen as the chemical, and particularly the bacteriological, decomposition of the jute and of the impregnating hydrocarbons, phenol also being thus formed (HESS [515]; COLES [207]; SENEZ and PICHINOTY [1108]; BONDE and LUNN [108b]). There is however, no connection between the phenol-content and the corrosion. According to VÖGTLI [1227a] even this view of phenol-corrosion is incorrect. He regards the intercrystalline character of this type of corrosion as its determining characteristic, and also finds a "phenol-corrosion" in bright lead cable sheaths which were laid directly in the ground. From his experiments in the ground and in the laboratory he draws the conclusion that the nitrate content of the soil is a necessary, even if not sufficient, condition for intercrystalline corrosion of cable sheathing. This view of the origin of phenol-corrosion cannot however be reconciled with the observations of BONDE and LUNN [108b], which indicate a mutual action of organic substances, soil bacteria and oxygen at somewhat raised temperatures in ducts lying at insufficient depth. The Danish and British specialists suggest to avoid the description "Phenol-corrosion" and replace it by "Compounding-corrosion". A further investigation would be desirable.

In this connection, mention may also be made of the attack on lead by some hard woods, particularly oak, which must be kept in mind in building operations (BRADY [124]; SOUTHERDEN [1138]), or in the storage of printing type in wooden boxes (SCHULZE [1087]). The supporting action of moisture is decisive. Soft woods, such as pine wood, are harmless. However, all rotting wood attacks lead (SANDMEIER [1046]).

f) Alkalis and Ammonia. As lead oxides have an amphoteric character, and form soluble plumbites or plumbates with alkalis, no considerable resistance of lead toward alkalis is to be expected (p. 271). However the attack only becomes considerable if the alkali concentration is above 10 to 15% and the temperature is raised [195]. Alternating treatment with sulphuric acid and with alkaline solutions, such as is used in petroleum refining, can take place in one and the same lead-lined container (UHLIG [1211]). On the other hand, lead corrodes strongly in quicklime and should therefore not be brought into contact with fresh mortar, cement or concrete, Infiltrated water from fresh concrete, if it reaches a lead surface in the presence of oxygen, behaves similarly (UHLIG [1211]). Soil-corrosion of lead is to be expected if the pH-value of the soil is above 10 [1175]. The special action of quicklime on lead, as compared with caustic soda solution, was cleared up by experiments by SCHIKORR [1060].

He concluded from the reduction in the rate of attack in 0.044 mole/litre NaOH that a protective film was formed. The protective film of lead carbonate forms from the dissolved lead at the surface of the specimen by reaction with sodium carbonate, which, for its part, is formed by the absorption in the caustic solution of carbon dioxide from the air. In calcium hydroxide solution of the same concentration, the attack was more intense. In particular, the curve of weight-loss against time showed no flattening, i.e., no protective film is formed here. This is because in the absorption of CO_2 by the lime solution practically insoluble calcium carbonate is formed, and this prevents the supply of carbonate ions to the lead surface. Thus similar conditions are present as were described in type II attack of lead in distilled water (p. 301). In fact, almost corresponding curves of loss of weight could be obtained in distilled water and in lime solution. Corrosion tests by KATZ [647] also showed the production of a protective film in 0.01 mole/litre NaOH and in 0.01 mole/litre $Ba(OH)_2$ after 3 or 4 days. In 0.04 mole/litre $Ca(OH)_2$, 0.1 mole/litre and 0.5 mole/litre $Ba(OH)_2$ a continous corrosion of about 10 g/m^2 per day was observed. The attack in mole/litre NaOH solution was only about half as intense, but also showed no reduction in the course of sixteen days.

The action on lead of quicklime or of building materials containing it, was thoroughly discussed in some older investigations (DITZ [251], BRAME [125]). The characteristic of this type of attack is the formation of red and of yellow PbO (BAUER and WETZEL [65]; BRCIC and SIFTAR [128]). Which of the two modifications of PbO is formed, obviously depends on the humidity, because in a lead pipe which had corroded in mortar, rings of alternate yellow and red layers, corresponding to the seasons, had been formed (Fig. 285). The corrosion decreases with the setting of the lime.

The attack in fresh ferroconcrete (DODERO [252], GOSDEN [410]) is intensified by an electrochemical interaction between iron and lead. This occurs without metallic conducting connection between lead and iron. Additions to the concrete (FRIEDLI [343]), mainly alkali or alkaline earth chlorides for accelerating the setting process and preventing frost damage, increase the attack.

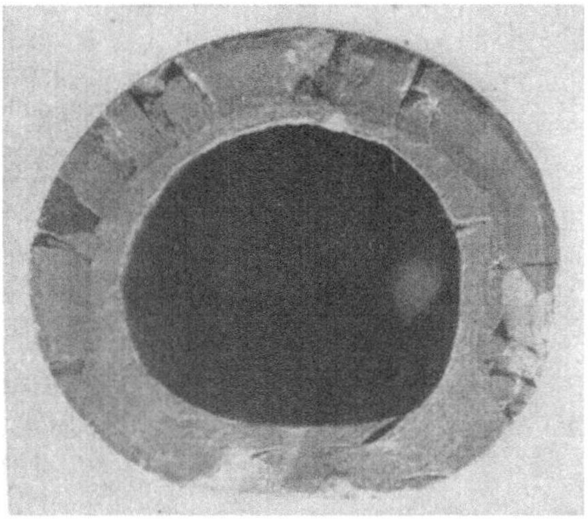

Fig. 285. Lead pipe corroded by lime mortar. Very fine annual rings of yellow and red PbO. (According to KOHLMEYER [693]).

In order to avoid such destruction, it has been proposed, on the one hand, to use gypsum mortar or gypsum-sand mortar (BAUER and WETZEL [65]) or cement high in Al_2O_3 [280] in contact with lead. On the other hand, direct contact of lead with lime-mortar can be avoided by interposing asbestos paper (MULLARKEY [882]), by coating the lead pipes with asphalt or with tar paint or embedding in loam (OBST [909]). According to a Belgian patent specification, lead pipes in cement are also protected by grease with addition of K_2CO_3 and of $KHCO_3$ [185]. In order to protect cable sheathing in concrete pipes against infiltrated water containing $Ca(OH)_2$, carbon dioxide is passed through the pipes (HIERS [524]). Further methods are proposed by PERRY [952].

Dilute ammonia solutions, as well as pure gaseous ammonia, at temperatures up to 300 °C, hardly attack lead [195]. A Japanese investigation of the resistance, towards the attack of alkaline solutions, of lead with addition of Zn, Te, Sb, Sn, Ca, Hg, or Cd, indicated that alloys with 5 to 15% Sn, or 0.5 to 1% Zn, or also 1% Cd, were relatively resistant (HAYAKAWA and co-workers [499]).

g) Salt Solutions. The resistance of lead in salt solutions depends on the development of protective film and therewith on the solubility of the corresponding lead salt. The resistance of lead to sodium carbonate or bicarbonate in fire extinguishers (HEWLETT [518]) and to phosphates can thus be understood (p. 269). There is especial resistance to ammonium carbonate (BOSCH [116]). The rate of corrosion of lead was investigated in solutions of various concentrations of sodium chloride, sodium sulphate, ammonium sulphate, ammonium acetate, sodium carbonate, sodium fluoride and other substances. A considerable decrease of the rate of corrosion after a period in the sulphate, carbonate and fluoride solutions was exhibited, whereas in the sodium chloride and ammonium acetate solutions the attack remained about constant (KATZ [647]). Data on the anodic polarisation of lead in hydrochloric acid and in halide solutions, and on the related surface phenomena, are given by BRIGGS and WYNNE-JONES [134].

Sulphate solutions can generally be regarded as harmless (Chem.Met. Eng. [195]), e.g., aluminium sulphate, acid or neutral solutions of ammonium sulphate (BOSCH [116], Chem.Met.Eng. [195]), copper sulphate for electroplating baths. With potassium sulphate in certain concentrations, an attack has been observed in laboratory experiments (THAU [1178]), which indicates an incomplete condition of the lead sulphate protective film. In a comparison of the resistance of lead in potassium sulphate with that in distilled water, the CO_2 content of the solution plays a decisive part. In two series of tests (SCHIKORR [1060]) the concentration of potassium sulphate was increased from zero to saturation. The carbon dioxide content of the distilled water used was in one case zero, in the other 51 mg/l, and the lead was attacked by the distilled water according to corrosion type II or I (p. 303). An arbitrary potassium sulphate content reduced the attack in water free from carbon dioxide, whereas the attack in water containing carbon dioxide had been less pronounced from the beginning, decreased further only on saturation with potassium sulphate. The nature of the tests did not permit a strict comparison between potassium sulphate solution of equal sulphate concentration but of different carbon dioxide content; however the solutions containing carbon dioxide seemed to be less corrosive. Corresponding tests were carried out with potassium chloride and gave similar results. The attack of distilled water free from carbon dioxide on lead was reduced by all chloride concentrations, that of distilled water containing carbon dioxide increased by all chloride concentrations. Contradictions in the literature (FRIEND and TIDMUS [347]) are now explained by the fact that in one case carbon dioxide-rich, in the other carbon dioxide-poor water was used, i.e., the lead corroded in the distilled water used according to corrosion type I or II (p. 303). Particularly intense corrosion was found in magnesium chloride (Gummi-Ztg. [457]) as well as in chloride solutions which find application as cooling brines (MIETHKE and WITT [851]). However, lead was only slightly attacked in a 10% $FeCl_3 \cdot 6H_2O$ solution at room temperature (WICKERT [1265]). In 5, 10 and 20% solutions of potassium fluoride, potassium hydrogen fluoride and ammonium hydrogen fluoride at 20 °C, lead shows good resistance. On the other hand, a strong attack is to be noted at 80 °C (Köhler [687]). Addition of about 0.1% sodium bichromate to distilled water or to dilute sodium chloride solution decreases the corrosion of lead to a certain extent; a protective action as in iron is however not attained (Roetheli and Cox [1020]). On the other hand, a content of 10^{-4}% sodium silicate in distilled water is said to reduce corrosion to 1/10th. as compared with pure distilled water (Albano [12]).

h) Gases. Lead is very resistant towards many gases as well as air. Data on its behaviour in industrial atmospheres are given by Schikorr [1062]. Mechanical stresses by static loads, vibrations and erosion can reduce its resistance. The favourable behaviour of lead towards chlorine can be used, for instance, in chloride of lime chambers.

The attack of dry bromine vapour on lead is very slight, that of moist bromine vapour comparatively low; on the other hand, the corrosion is very intense in saturated bromine vapour (bromine water) in the region of the water-line (Haines [472]; Bloch [90]).

Fluorine forms a protective lead fluoride film particularly in the presence of water vapour. For this reason, lead is suitable as a material for fluorine up to about 100 °C. Lead fluoride is soluble in nitric acid and in sulphuric acid, but, on the other hand, is stable towards acetic acid, hydrogen fluoride and ammonia (Roll [1027]; Lingnau [753]). The favourable behaviour of lead towards sulphur dioxide, also in the presence of fluorine, has been mentioned earlier (p. 292 sulphurous acid). It also behaves similarly in dry and in moist hydrogen sulphide (Uhlig [1211]).

Illuminating gas can attack lead strongly (Sandmeier [1046]). The carbon monoxide mechanically trapped in the corrosion product can be chemically detected years later (Vögtli [1227 b, c]). Lead-coated sheet iron showed itself to be insufficiently resistant towards the combustion products of illuminating gas (Witt [1282]). Further, lead is sensitive towards gases which contain volatile organic acids (Schikorr [1061]).

i) General. There is not complete agreement regarding the resistance of lead to liquid fuels as they often contain agressive sulphur, and since the formation of lead sulphide can cause operational difficulties, the use of lead-lined tanks is generally not advised (Grempe [439] Marder and Farnow [794]; Öl und Kohle [911]). On the other hand, hot-lead-coated fuel cans, and lead-lined tanks for small Diesel motors, have been successfully employed. Mercaptans and disulphides have shown themselves to be particularly aggressive towards lead. However, lead can be considerably attacked by vapourizing fuels which do not contain aggressive sulphur. Water and acid present in fuel mixtures can increase the attack (SCHLÄPFER [1065]). The rules governing the corrosion of lead by numerous organic acids dissolved in hydrocarbons, e.g., benzene or xylene, were investigated. In the absence of oxidising media, the corrosion was often insignificant (TURNBULL and FREY [1207]); (PRUTTON and DAY [982]). In corrosion tests in lubricating oils at 150 °C, it was the peroxide content and not the acidity of the oil which proved decisive (WILSON and GARNER [1276]). One case of corrosion benzene could be traced back to the presence of nitric acid. Lead was also strongly attacked in a solution about 0.22 mole/litre of iodine in benzene or iso-octane (GINDIN and PAVLOVA [384]). Vessels for the acid washing of light oils in the production of coke-plant by-products are often clad with lead. While its corrosion properties here are quite favourable, it is unsatisfactory in mechanical respects (TICE [1196]). Lead is also recommended for containers for insulating oil (ORNSTEIN and co-workers [923]). Lead-coated iron is said to be stable in spirit and in absolute alcohol (DIETRICH [249]). Guiding tests at the Bleiforschungsstelle also showed stability to 96% alcohol, but elsewhere experience was worse (EISENSTECKEN and ROTERS [273]). Alcoholates are formed by the attack of methyl or ethyl alcohol on lead. As regards further references on the resistance of lead, attention is drawn to the "Korrosionstabellen metallischer Werkstoffe" (RITTER [1012]) and UHLIG [1211].

4. Lead Anodes

The positive plates of lead storage batteries will be disregarded at this point, as they are treated elsewhere (p. 341). In the first place, mention may be made of the use of lead as anode material in the production of electrolytic zinc. A hypoeutectic (arsenical) lead-silver alloy, the so-called TAINTON lead has proved the most favourable (TAINTON and co-workers [1162], EGER [270]; p. 33). In addition, among 28 alloys tested in the laboratory, lead-thallium showed good resistance, although it gave no reduced anode potential. Lead-calcium-silver alloys displayed an exceptionally favourable behaviour (HANLEY and co-workers [486]).

The corrosion of anodes in zinc electrolysis was thoroughly investigated in further work. Consideration was given to cast and rolled commercial lead (99.965%), to lead with 0.96% and with 2% Ag, and further to an alloy with 0.22% Cd + 0.05% Ag (CAMBI and PIONTELLI [168]). The corrosion was investigated microscopically; the lead peroxide sludge, the lead content of the cathodically-deposited zinc and the anode potential after the current had been briefly switched off, were determined. Lead-silver alloys showed the most favourable behaviour i.e. little contamination of the cathodes, little corrosion of the anodes and low anode potential, and therefore reduced oxygen-overpotential. The PbO_2 layer on the anodes here had particularly good cohesion. Cast electrolytic lead showed intercrystalline attack at current densities above 1,000 A/m^2 in contrast with commercial lead which had not been recast. A small amount of manganese or cobalt salts in the bath had a favourable effect in the electrolysis. Lead contents appear in the cathodes at current densities below 300 A/m^2 owing to lead sulphate in solution. At higher current densities suspended PbO_2 gets into the cathodes mechanically, or by electrophoresis. In a zinc sulphate/sulphuric acid electrolyte, at a $ZnSO_4$ content of 30 to 120 g/l and a H_2SO_4 content of 50 to 300 g/l, the solubility of lead depends on the sulphuric acid concentration and the temperature, but very little on the zinc sulphate content (WATANABE and FUKUSHIMA [1242]). The solubility falls with increasing content of sulphuric acid, but, on the other hand, rises by 0.11 to 0.16 mg/l for a rise of temperature of 1 degree.

REY and co-workers [1005] in their tests considered only pure lead (99.994%) and lead-silver. The total corrosion process of the anodes was broken down into partial processes; for instance, the dissolved and the suspended lead were determined, with and without current passing. Fresh and used anodes were tested as well as anodes made of platinum, coated with lead oxide. Lead-silver alloys in fact showed great superiority over pure lead. In particular, even in the new state, they showed hardly any attack, the formation of suspended PbO_2 being completely stopped. Small additions of cobalt in the electrolyte reduced the formation of this sludge even from pure lead. Cobalt is said, in this way, to effect an easier release of oxygen, so that the latter does not reach the lead. From the dependence of the time of passivation on the current density, SMIALOWSKI [1127] draws similar conclusions as regards the favourable effect of 1% Ag as an alloying element and of cobalt ions in the electrolyte. According to this work, chloride ions in the electrolytic delay passivation.

KIR'YAKOV [667] carried out extensive tests on the stability of numerous lead alloys in the electrolysis of sulphate solutions. The specimens were first anodically treated in 2 mole/litre sulphuric acid at 25 °C, with about 400 A/m^2 for about 3 to 4 weeks. Then current densities up to 10,000 A/m^2 temperatures up to 50 °C and concentrations of 2 to 6 mole/litre sulphuric acid were applied. Pure lead served as cathode material. Anodes of lead with 1% Ag behaved quite favourably. By alloying with further elements such as thallium or tin, which form a solid solution with lead and improve the distribution of silver, the durability of the anodes was further increased. Particularly good properties were shown by an alloy with 1% Ag, 0.3% Sn and 0.02% Co. Calcium and selenium only give temporary protection. Gold and mercury had an injurious effect. Tin, calcium, barium and strontium facilitated the formation of protective films; cobalt and arsenic increased the effect of these films, as did chloride ions. The grain size of the anode material had no effect worth mentioning on the corrosive attack. Increase of current density, of temperature and of acid concentration only lowered the durability of lead silver alloys slightly, but it increased the amount of anode slime and decreased the thickness of the anode film.

TSUKI [1205a] reports on good experiences with a lead 5.5% silver, a lead 30% tin or a lead 60% tin alloy. An addition of cobalt to the alloy, likewise additions of cobalt sulphate to the electrolytic system (10% H_2SO_4, 15% Na_2SO_4) improved the durability.

The relationship between the durability and the potential of numerous anode alloys was also investigated (KIR'YAKOV and STENDER [668]). The potentials of the alloys which were more durable than pure lead were more negative than the potential of the latter. For less durable alloys the opposite holds good. The presence of chloride and cobalt ions in the electrolyte lowered the potential of all anodes. Manganese ions had practically no effect. Similarly as in the investigations on the effect of alloying elements on the durability of lead in sulphuric acid (p. 279), short-circuited local cells of pure lead and the alloying element concerned were prepared and loaded anodically. Measurements of potential and of current density were carried out using sulphamic acid and not sulphuric acid as the electrolyte, so that no film of lead oxide could be formed. Silver and thallium protect lead in spite of their high potential, but cobalt only in part. Calcium and other electronegative metals only acted as long as they could go into solution in the electrolyte. Gold and mercury promoted the attack on lead even under these experimental conditions. The superposition of 4% alternating current on a direct current of 25 to 150 mA/cm² in 20% sulphuric acid did not provide any improvement in the durability of the lead anodes (COSTA [215a]). KIR'YAKOV [666] found that corrosion does not depend on the thickness and porosity of the protective film.

Lead and antimonial lead anodes are used in the electrodeposition of copper from sulphuric acid solutions for the purpose of the wet production of copper; lead anodes with 1% silver are preferably used for electrolytic production of cadmium. The use of lead raises difficulties if the liquors contain chlorine or nitric acid (FINK and ELDRIDGE [318]; TAFEL [1159]). Rolled anodes of lead or antimonial lead are also used in the chromium plating industry. The electrolyte consists of chromic acid with various additions. The lead anodes usually become covered with a layer of lead peroxide. Previous anodic treatment proved beneficial except in the case of Pb–Sb (CARTER [181]). Commercial lead is said to behave most favourably with continuous operation, lead-antimony with interrupted operation (BAKER and MERKUS [34]). Lead-tin alloys, tellurium-containing alloys [603]), MORRAL and BRAY [871], etc. are also used. Data on the shape and the installation of anodes are given by FRIEDBURG [340]. Lead has been proposed (HOTHERSALL and GARDAM [597]) as anodes for the internal nickel plating of boreholes in a nickel sulphate/boric acid electrolyte. In the electrolysis of manganese sulphate, the deposition of manganese dioxide on the lead anode can be prevented by mechanical treatment of the anode surface, by previous electrolysis of cobalt sulphate with the same anode, by addition of cobalt salts to the solution and by using an anode with 1% Ag (NISHIHARA [899)].

For the cathodic protection of steel using an applied voltage MORGAN (868) proposed anodes of an alloy of 93% Pb, 6% Sb, 1% Ag instead of platinum, ferrosilicon or graphite. The rate of solution of this anode is very small. In tests on destroyers, anodes of lead with 2% Ag behaved very favourably (BARNARD [45a]). In further tests over a period of a year PEPLOW [951a] found that the deposition of PbO_2 on the lead anodes was reduced by adding tellurium or bismuth to the lead – 1% Ag alloy. An alloy with 0.5% Bi + 0.5% Sb was superior to the lead – 1% Ag alloy. Further, the combination of lead and platinum as the so-called Bi electrode has been discussed (cf. ROYUELA and SERRA [1036b]). If a slight alternating current is superposed on the impressed current, the life of anodes is in general reduced, except in the case of lead anodes.

5. Behaviour Lead towards Water

a) Distilled Water without Access of Air. According to the theory of electromotive forces, lead should release hydrogen from the purest water. Here, however, the water becomes enriched in lead ions and hydroxyl ions; as a result the "metal electrode" becomes increasingly nobler, the "hydrogen electrode" increasingly less noble, until the hydrogen evolution and the dissolution of lead come to a stop at equal potential of both electrodes. If the over-potential of the hydrogen at lead is ignored, then, according to a calculation by WERNER [1261], this equilibrium is attained at a concentration of about $10^{-9.5}$ gram atoms of lead ions per litre of water. In this process the pH-value of the water should increase from 7 to about 7.0015. At a slight acidity of the water, such as is frequently brought about by dissolved CO_2, the solubility of lead increases rapidly. Thus the above-mentioned equilibrium is only reached at pH = 5 at a concentration of about $10^{-5,5}$ gram atoms of lead ions per litre, corresponding to about 0.6 mg of lead per litre of water. The experimental results, which can perhaps be taken from an old survey of literature (GMELIN-KRAUT [388]), by and large confirm the predicted slight attack of lead by the purest air-free water. MARKOVIC [797] investigated the mode of origin of lead hydrides (such as Pb_2H, PbH and PbH_2) in distilled water by means of polarisation curves. The measurements were carried out in a nitrogen atmosphere. According to this, the lead hydrides are to be regarded as compounds chemically or physically adsorbed on the lead electrode.

b) Distilled Water with Access of Air. Lead is strongly attacked by distilled water containing air. Potential-time measurements by MARKOVIC [797] showed that the potential of a lead electrode in distilled water containing air becomes nobler with increasing depth of immersion, i.e. the decomposition becomes the smaller the greater the distance between lead and the surface of the water. This fact can be understood as the effect of the diffusion of oxygen on the corrosion. Together with the oxygen, a decisive role is played by the carbon dioxide content of the air and of the water (BAUER and WETZEL [65], BAUER and SCHIKORR [61]). The attack on the lead is the most intense, and proceeds unchecked, if air free from (or poor in) carbon dioxide is in contact with distilled water free from carbon dioxide (attack type II). A white turbidity forms in the liquid immediately after the suspension of the specimens. A white skin of basic lead carbonate of composition $PbCO_3 \cdot Pb(OH)_2$ forms on the surface of the liquid. OŠIS [928] observed a basic lead carbonate of composition $2\,PbCO_3 \cdot Pb(OH)_2$ as the corrosion product in aerated distilled water at a pH value of 8.6. In the complete absence of carbon dioxide the solution contains about 100 mg/l of lead (cf. table 22, ZINC [1305]). It reacts alkaline, corresponding to a pH-value 9.5 to 9.7. As a result of this alkaline reaction, the surface skin absorbs the small amount of carbon dioxide in the air and prevents its penetration into the solution. The skin, with increasing thickness, sinks to the bottom from time to time and is repeatedly replaced. A protective film cannot form on the metal. Its surface appears etched and it is covered with small crystals of lead hydroxide and lead oxide. In the majority of cases, the mechanically stressed parts of the lead are first attacked. If the water is richer in carbon dioxide, the attack is weaker, as a protective film forms on the lead (attack type I). The curve of loss of weight in relation to time correspondingly shows a flattening (Fig. 286). The grey-tarnished lead exhibits white efflorescences, presumably of basic lead carbonate. The deposits settled preferably on places which are mechanically stressed.

Attack type II can be converted to attack type I by increasing the water surface. The lead hydroxide which goes into solution is not then sufficient to produce the surface skin on the water and to maintain it continuously. A similar effect would be expected from a movement of the water (SCHIKORR [1060]).

Attack type I can also occur in water poor in carbon dioxide if the lead already possesses a protective film before it is put into the liquid (BAUER and SCHIKORR [61]). This is formed by the action of moisture in the presence of the normal carbon dioxide content of air. The protective film is disrupted by mechanical stress and cannot then be reformed in water free from carbon dioxide, so that in these places corrosion according to type II takes place vigorously.

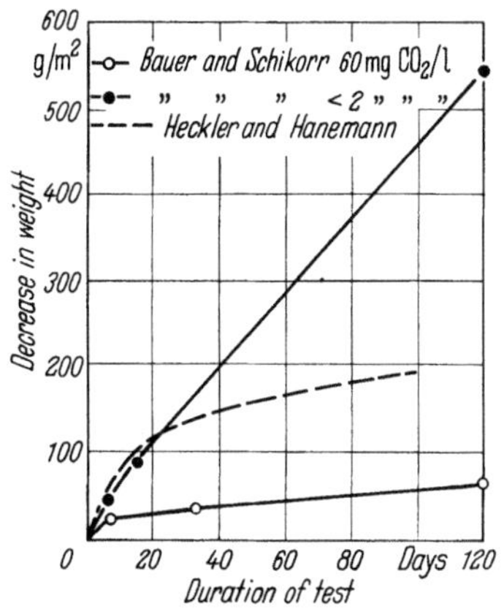

Fig. 286. Lead attacked by distilled water. (According to various investigations).

The presence of a protective film due to various pretreatments could be tested by potential measurements in distilled water poor in carbon dioxide. The calculated potential of lead in saturated $Pb(OH)_2$ solution against the hydrogen electrode was reported as -0.23 V. Measurements on lead specimens without a protective film or with a protective film disrupted by bending, gave values which were permanently less noble. The reason could be a lower solubility of $Pb(OH)_2$ than that assumed. The specimens with a protective film, on the other hand, permanently had a potential nobler by about 0.09 V. Other observers also confirmed the above inhibiting action of carbon dioxide on the attack of lead by distilled water (ZINC [1305]; LIDDIARD and co-workers [748]). It was also found that at high carbon dioxide content the attack again increases, as thereafter lead carbonate and basic lead carbonate go into solution as bicarbonate (GMELIN-KRAUT [388]). Heckler's corrosion tests [501] also essentially confirmed these results. In ordinary distilled water, the purest lead exhibited behaviour between corrosion type I and II (cf. Fig. 286). The attack was reduced by a higher carbon dioxide content of the water. Tests of the various alloys enumerated in the section on "sulphuric acid corrosion" indicated no special differences. Only the lead-bismuth alloys displayed a somewhat heavier. the lead-calcium a significantly slighter attack. Additional experiments however showed that this favourable behaviour of lead-calcium no longer prevails at very low CO_2 contents.

The effect of additions to distilled water on the solubility of lead was dealt with in the section on "salt solutions".

c) Tap Water, Sea Water. There is an extensive literature on the hygienic aspects of the attack of lead pipes by the tap water. Therefore only the results of some selected investigations will be described and references given to further literature. Lead pipes for cold water are, according to Klut [679], attacked principally:

1. By soft waters containing air. The softer the water and the richer in oxygen, the stronger the attack, which is further promoted by organic acids which have water-soluble salts. These occur especially in water from moorlands (Heap [500]). The root fibres of bilberries and of heather contain quinic acid, which prevents the precipitation of lead carbonate and the consequent formation of a protective film on the lead pipes (Evans [297]).

2. By all waters containing aggressive carbonic acid, i.e. carbonic acid which dissolves lime stone. Aggressive carbonic acid is not identical with free carbonic acid, as the latter (up to a certain amount) does not attack lime stone. In order to explain this concept, attention is drawn to the fact that the dissolution of calcium carbonate in carbonic acid leads to the formation of soluble bicarbonate:

$$CaCO_3 + H_2CO_3 = Ca(HCO_3)_2$$

The process is reversible and in the presence of excess calcium carbonate does not proceed to the complete utilisation of the carbonic acid. A certain amount of H_2CO_3 or CO_2 is present as free, so-called proper carbonic acid. Only the amount which exceeds the quantity of proper carbonic acid is aggressive carbonic acid, and therefore has the ability to dissolve more calcium carbonate[1] (Carius and Schulz [180]). The injurious action of aggressive carbonic acid on lead pipes depends on the disso-

---

[1] The full equation is

$$\begin{array}{ccccc} & & CO_2 + H_2O & & \\ & & \upharpoonleft\downharpoonright & & \\ CaCO_3\ (\text{in solution}) & + & H_2CO_3 & \rightleftharpoons & Ca(HCO_3)_2 \\ \updownarrow & & & & \\ CaCO_3\ (\text{in soluble}) & & & & \end{array}$$

In any ordinary balanced reaction, the position of equilibrium is determined by the concentrations of the ingredients concerned, which are variable. However, in the above main (i. e. horizontal) equation the concentration of the sparsely soluble $CaCO_3$ (in solution) is not a variable as long as insoluble $CaCO_3$ is present; it is a constant at any given temperature. Thus the position of the above system is determined only by the concentration of $Ca(HCO_3)_2$ in the main equation and the CO_2 in the upper equation. This CO_2 is designated "proper carbonic acid" by Prof. Hofmann.

If now excess CO_2 be added, the whole system will swing over to the right; i.e. insoluble $CaCO_3$ (the protective film on the inner walls of water pipes) will dissolve and soluble $Ca(HCO_3)_2$ will be produced. It is this excess CO_2 which is described as "aggressive carbonic acid" (Wesson [1261a]).

lution of calcium carbonate, since the protective film, if formed, consists mainly of this compound. In addition, excess carbonic acid is also able to redissolve lead carbonate or basic carbonate which has deposited on the lead (p. 303). The subject is not complete if consideration is given only to the carbonate equilibrium of a water, if the latter also contains organic constituents (MILES [854]). In the case of lead, the matter is further complicated by the fact that not only the resistance of the metal involved, but the question of the minimum lead content of the water is also of concern. It should also be noted that points 1 and 2 above in part overlap.

3. Quite widely by waters which do not gradually produce an adherent, dense, wall-deposit of calcium carbonate. In general this deposit forms at a carbonate hardness of the water above 7 degrees of hardness. However, if the water has an alkaline character it can still be produced at lower degrees of hardness (p. 306). A carbonate content corresponding to about 7.5 German degrees of hardness is shown, for instance, by Vienna High Spring water (MEYER [848]). As it contains some aggressive carbonic acid, it represents a limiting case for the use of lead pipe. The lead taken up from the pipes does not exceed the permissible amount, as the investigation showed (p. 306).

The formation of a calcium carbonate/lead carbonate protective film in water with sufficient carbonate hardness can be imagined, according to the electrochemical theory of corrosion (EVANS [297]) as follows. The lead surface owing, for instance, to varying rate of flow of water, is not uniformly supplied with air. As a result, aeration cells are formed, in which the areas richer in oxygen are cathodes. At the anodic places lead goes into solution. An equivalent quantity of hydrogen ions is discharged at the cathode, and oxidised to water by the dissolved oxygen. As the alkalinity of the solution consequently rises at the cathode, the calcium bicarbonate/carbonic acid equilibrium described above is disturbed at these points and calcium carbonate precipitated. The cathodic reaction is thereby pushed to other places and so gradually the whole internal wall of the pipe becomes covered with calcium carbonate. The precipitation of lead carbonate from dissolved lead bicarbonate can be imagined in a similar manner. The formation of calcium carbonate by the electrochemical process described above was demonstrated by quantitative experiments for an aerated platinum electrode (SCHIKORR [1059]). There is a great probability that the processes in a lead pipe take place similarly.

4. All waters which do not react alkaline towards the indicators litmus and rosolic acid. This condition is implied essentially, in the points considered already. For example, water with a great excess of carbonic acid, according to point 2 above, rarely reacts alkaline. Thus

four waters which dissolved lead strongly and were injurious to health (NACHTIGALL [889]) had pH-values of between 4.7 and 6.8. The above formulation by KLUT [679] which is synonymous with a pH-value of 8, may, however, be somewhat too strict. NACHTIGALL [889] requires a pH-value of 8 only for water with 0 to 3 degrees of carbonate hardness. The minimum pH-value for precipitation of calcium carbonate decreases with increasing carbonate hardness, and at above 7 degrees of hardness amounts only to 7.4 to 7.5. It may be mentioned that an alkaline reaction (i.e. pH greater than 7) is quoted for the above mentioned Vienna water (MEYER [848], p. 305).

According to HAASE [460], the solubility of lead (as far as no sound protective film of calcium carbonate is formed) is favoured by contents of chloride, sulphate, H_2S, above all nitrates, nitrites and ammonium salts in the water. As the data obviously derive from older investigations which were done with distilled water, HÖLL [545] carried out a re-examination, using tap waters of widely differing compositions. Neither chlorides, nitrites, nitrates nor ammonium salts in quantities, such as occur in drinking water in practice, increased the solubility of lead. At very high gypsum contents of the water (above 250 mg SO_4/l) a slightly increased attack occurred (cf. BAUER and WETZEL [65]). The earthing of wireless aerials on lead pipes had no effect on the solubility. The tests were carried out in a lead pipe which was filled (obviously successively) for a day with waters of different composition. This method is not wholly free from objection, as the lead pipe perhaps already had a protective film.

There is not complete agreement on the part played by oxygen in the attack of water on lead pipes. As oxygen is necessary in any case for the dissolution of lead, it can promote the attack in aggressive waters; for this reason, pipes are often not permitted to stand empty on account of the entry of air (NACHTIGALL [889]). On the other hand, as shown, for instance, by the electrochemical conceptions mentioned above, oxygen is required for the formation of a protective film. Attention has already been drawn to the varying opinions regarding carbon dioxide.

Remedies against the lead-solvent action of drinking water can be applied to the water or to the pipe material. In the first place the aggressive character of the water can be obviated by deacidification. This is done by treatment with calcium carbonate and quicklime (in which the latter should not come into contact with the lead) and, in hard waters also by deaeration (HAASE [460]). In soft water with organic constituents, as from moorlands, treatment with aluminium sulphate combined with alkalis should be advantageous (MILES [854]). Formerly, lead pipes were frequently treated with sulphur to increase their resistance. There is, however, the risk that the sparingly soluble lead sulphide will be converted by oxygen to sulphate, which has a significant solubility in water (p. 273). The increased resistance of lead by alloying additions is insignificant. At best, the lead-calcium alloys (according to HECKLER [501]) have greater resistance in certain cases. According to this investigation the corrosion resistance (in Berlin tap water) is hardly reduced by small amounts of other metals in lead. Similar results were obtained for antimonial lead pipes with 1% Sb (BAUER and SCHIKORR [60]; NAUMANN [891]), with 0.5% Sb + 0.25% Cd and with 0.8% Sb + 0.2% Sn + 0.03% Na (BRADY [123]), and with lead-sodium and lead-calcium pipes[1], as well as with lead-arsenic-cadmium alloys (SORRENTINO and INTONTI [1137]).

A sure method of preventing lead from being taken up by the water, is the use of lead pipes internally coated with tin (p. 430). However, the coating should not have any defects, as otherwise the lead will suffer pitting attack (HAASE [460]). So far, the permissible limit of lead content has been stated as 0.3 mg/l, approximately "after standing for nine hours in a lead pipe of the usual diameter through which flow has already taken place for a long time". (Gas und Wasserfach [360]). Nowadays the view is often expressed that the permanent content of lead permitted in water should be a maximum of 0.1 mg (MÜLLER [878]).

According to SCHMELING and RÖSCHENBLECK [1068], marked pitting is occasionally observed in lead pipes for waste water. This is connected with the formation of hydrogen sulphide from refuse and the consequent coating of the lead surface with lead sulphide by solution action. Lead sulphide has good electrical conductivity and acts as a cathode. On the other hand, the uncoated parts of the surface, or those coated with carbonate, form the anode and can be corroded to perforation. With a loss of 0.4 g/m^2 per day, lead is comparatively stable in sea water, as has been shown by direct tests over many years (FRIEND [346]) and also by laboratory tests (BAUER and WETZEL [65]). Protective films of lead chloride, oxychloride, sulphate or chlorocarbonate are formed, which prevent further attack (HAEHNEL [467]). Lead is therefore used for submarine cable and in ship building. Nevertheless, the resistance of lead to sea water for secular periods of time cannot be counted upon, as HAEHNEL [464] showed for various historical examples; on the other hand, a lead cross-beam (about 315 kg) of the anchor of a ROMAN galley in the Mediterranean was not corroded.

HAEHNEL describes the corrosion of an armoured water pipe in the Bay of Rio de Janeiro, which had to be removed after 20 years owing to local perforation. Here however, it appeared that a part had been played by remarkably large fluctuations in the tin content of the alloyed pipe,

[1] Information kindly provided by Bleiwerk Goslar K. G.

as numerous small corrosive attacks had occurred in succession on the inside, which had been in contact with fresh water; this is a rarity in water pipes. According to this investigation, extensive protection towards corrosion by sea water is provided by covering the lead with sand or mud. Otherwise the risk of corrosion in the upper layers of the sea is particularly great in lead pipes which are not covered with mud, as the development of the protective coating is impaired by air bubbles and the passage of waves.

While pure lead is very sensitive towards erosion in moving sea water, good resistance is attributed to an alloy with 40% tin as well as to pure tin. For this reason, sufficiently thick coatings of lead-tin alloys are recommended for copper. to protect it against erosion in sea water (Uhlig [1211]).

6. The Behaviour of Lead in the Ground

a) Soil Corrosion. The question of soil corrosion is of importance principally for buried cable (p. 404) and for water pipes. Somewhat other conditions apply to cable drawn through conduits, which are treated in section b). The discussion of soil corrosion combined with stray currents follows in section c).

A series of fundamental investigations on soil corrosion are available. At the National Bureau of Standards, following earlier investigations in progress since 1922 (Logan and co-workers [761 to 764]), in fifteen analysed soils of various kinds, pipe specimens of three types of lead were buried in 1937, followed, in 1941, by further pipe specimens of another type (Denison and Romanoff [246]). Laboratory investigations on corrosion cells with a series of these soils as electrolytes showed that predominantly the control of the corrosion of the lead was anodic. The conditions only changed in a reducing environment, which contained many organic products of decomposition. If however, much sulphate and chloride were simultaneously present in such soils, then the corrosion was again controlled anodically (Denison [245]). Among the specimens buried since 1937 are chemical lead (see p. 6 with 0.056% Cu, 0.002% Bi, 0.0011% Sb, lead-tellurium with 0.08% Cu, 0.0011% Sb, 0.043% Te and an antimonial lead with 0.036% Cu, 0.016% Bi, 5.31% Sb. The fourth type provided no fresh evidence and therefore will not be considered here. The losses of weight of the specimens and the depth of the corroded places were investigated at periods of a few years.

It appeared fundamentally that the differing compositions of the lead have practically no effect on its corrosion, and that a satisfactory resistance of lead in direct contact with soil cannot be expected everywhere. The risk of corrosion is however significantly less than in iron.

Many lead specimens showed pitting-type corrosion. However only in two cases did the pits reach more than half-way through the pipe wall. In the first case the specimen was buried in cinders, in the second in a feebly aerated soil. In both cases there was only a moderate sulphate content. In alkaline soils, which were fairly well aerated, a certain in-

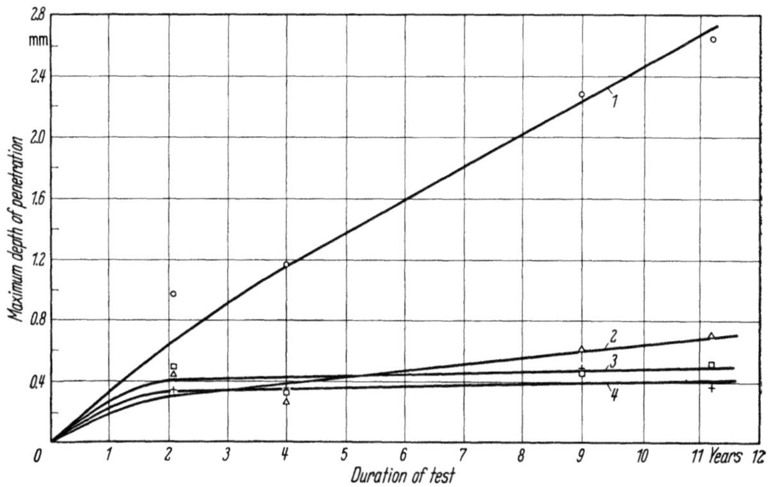

Fig. 287. Development of depth of pits in lead in various soils. (According to DENISON and ROMANOFF [246]):
1. Lake Charles Clay, with verry little aeration; 2. Rifle Peat, slightly aerated; 3. Docas Clay, slightly aerated; 4. Cecil Clay Loam, well aerated.

hibition of corrosion by bicarbonate, chloride and sulphate ions could be recognised. These soils were superior to others with equal aeration but with a smaller content of soluble salts. In general it appeared that with increasing aeration the corrosion attack, and particularly the tendency to pitting, diminished (Fig. 287).

According to the earlier investigations of the National Bureau of Standards (LOGAN [761]), the growth of the pits takes place without hindrance in sulphate-free soil, therefore roughly proportionately with time, whereas in sulphate-containing soil it gradually comes to a stop owing to the development of a protective film. Observations on a large number of soils can be interpreted in the sense that the corrosion decreases with increasing salt content as expressed by the anion content $Cl^{1-} + (HCO_3)^{1-} + (SO_4)^{2-}$ (1). To consider the corrosion only in relation to the total of the salt contents of a soil, would, however, be too great a simplification.

(1) The effect of a chloride content is sometimes disputed. For instance in museum objects derived from soils containing salt, strong intercrystalline breakdown was observed (MATIGNON [807]).

In the older investigations of the National Bureau of Standards (LOGAN and co-workers [761 to 764]) a somewhat stronger attack on antimonial lead was observed, by and large, than on commercial lead. The differences between the two kinds of lead were however far smaller than the differences in various soils.

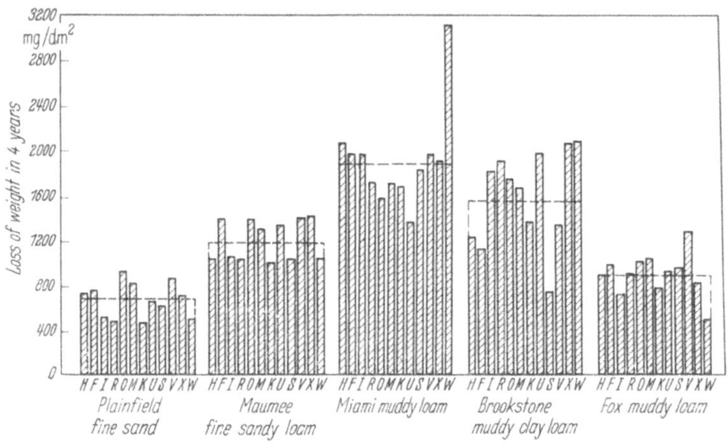

Fig. 288. Corrosion of lead alloys in various types of soils. Hatched areas show average values for a soil. (According to BURNS). *H* Lead (99.94%); *F* "Chemical Lead" (0.06% Cu); *T* Lead (99.85% Pb, remainder Bi); *I* H + 0.8% Sb; *R* T + 0.8% Sb, *O* F + 1.0% Sb; *M* F + 1% Sb, 24 hours at 107 °C; *K* H + 1% Sb; *U* H + 1% Sb + 0.06% Cu; *S* T + 1% Sb; *V* T + 1% Sb + 0.06% Cu; *X* F + 2.5% Sb, quenched behind the press; *W* F + 3% Sn.

A somewhat stronger attack on lead with 1% Sb, as compared with commercial lead, was also shown in the tests by ANDEREGG and ACHATZ [18]. An alloy with 3% Sn here behaved better than commercial lead.

Tests by the Bell Telephone Co. [156] only extended to five soils for four years, but numerous alloys were considered. The weight losses given in Fig. 288 show that the soil has a far greater effect on the corrosion than has the nature of the alloy. A similar conclusion followed from the observations of pits. Only the lead-tin alloy was completely penetrated (cf. however above).

Of the soils investigated by BURNS [156] at Bell Telephone, the sandy ones showed the most favourable behaviour, the clayey ones the most unfavourable. As corrosive constituents of the soil the following are reported: nitrate and chloride ions (these in higher concentrations), alkalis, organic acids; and as protective constituents: silicates, sulphates, carbonates, colloids and certain organic compounds. In both of the latter cases cathodic polarisation may occur. According to VÖGTLI [1227d], moorland soil behaves far less aggressively towards lead than does loamy soil. These data do not, however, make it possible to foretell with certainty the effect of a given soil, as several conditions always work together. According to a report from the British Non-Ferrous Metals Research Association, copper and lead show the best behaviour, in comparison with aluminium, in 15-years' exposure in a clay marl containing gypsum, at a depth of 0.61 m. Lead suffered only an insignificant uniform loss. The maximum depth of pitting was 1 mm [134a].

Similar data on the action of various soils were available in the investigations of the Deutsches Reichspost-Zentralamt, where further observations are given (Haehnel [466 to 468]). A low moisture-content of the soil is particularly favourable, as moisture is indeed one of the fundamental prerequisites of (electrochemical) corrosion. That lead is much less at risk in sandy soil than in clay is thus based in the first place on the good porosity for water of the former as compared with the latter. Marl and limestone soils favour the attack. The destruction of lead by lime mortar and concrete is discussed elsewhere (p. 294). A high carbon dioxide content of the moist air in the soil, humus content of the soil, and contact with coke and cinders (Haehnel [466], Anderegg and Achatz [18]), as well as with graphite (Titoo [1201a]), are disadvantageous.

Markovič [796, 799] carried out electrochemical measurements on synthetic soils. For example, quartz or montmorillonite (clay mineral) was dried at 105 °C and brought to the desired moisture-content by the addition of distilled water or sodium chloride solution. It was found that, in a soil containing clay, a lead electrode formed a double layer, similar to that in an electrolyte. In general lead is passive in moist soils. Corrosion can be activated by additions of suitable electrolytes. The water content of the soil affects the corrosion, particularly in connection with the transport of oxygen (Marković and Dugi [798]). In a further investigation Marković [799a] points particularly to the decrease of corrosion of lead with increasing pH-value of the soil moisture. However the measurements show considerable scatter. Soils in which frequent change of water can occur seem less aggressive than those with little movement of water, as here hydroxyl ions become enriched near to points of corrosion, owing to the solution of metal and the consequent discharge of hydrogen (Haase [461]). The development of pits (which in spite of the usual light-coloured corrosion products contained a black mass, predominantly lead sulphide) was explained as due to sulphate-reducing bacteria (Reinitz [1004]). Sulphates and sulphites are reduced to hydrogen sulphide or metal sulphide by the metabolic processes of these micro-organisms. These processes occur particularly in moist environments poor in air, such as, for instance, mud or waste waters. The metabolic processes of other micro-organisms can also be concerned, directly or indirectly, in the destruction of lead in soil (Uhlig [1211], v. Wolzogen-Kühr [1283]).

The corrosion of lead in the ground can often be interpreted by the action of aeration cells. Air has limited access to the points of contact of lead with the grains of the soil. These points are anodic and are attacked. This explains the irregular, sometimes pitting-hollowed surface of such lead specimens (Fig. 289). Once pits are formed, the possibility of a further increase in depth is explained by the fact that the bottom of the pit is worst provided with oxygen, particularly if a protective film forms on its walls. The attack was more severe in coarse-grained than in fine-grained soil (Burns and Salley [157]). The observation that more deeply embedded parts of experimental strips were more strongly attacked than the parts higher up, is of the same kind (Logan and co-workers [763]). In addition to cells due to differential aeration, concentration cells must also be taken into account. The latter occur because the cables or pipes pass through soils of varying composition and differing moistness and aeration. Concentration cells may cause corrosion in a greater degree than has hitherto been assumed. The recognition of such cases is rendered difficult by the fact that circumstances can change considerably according to the season and the weather. Thus the severe corrosion of lead in a limestone-clay mixture was attributed to concentration cells (Brandt [126]), as well as the corrosion of a bare telephone cable which lay in the neighbourhood of mineral springs (Kaja [641]). The cell here was of the kind lead/soil containing salt/ordinary soil/lead. Laboratory investigations (Robson and Taylor [1016]) showed that, particularly between lead in clay and lead in cinders containing coke, a strong concentration cell develops if the lead specimens have metallic connection, and the contact between the cinder and the clay is electrically conductive. While lead in the neighbourhood of the clay was strongly attacked, the half of the specimen in the cinders remained practically undamaged over a period of a year. A lead pipe, which passes from clay to a layer of cinders is therefore severely damaged in the clay close to the interface. Lead specimens which lay only in clay or only in cinders showed no pronounced attack. As a protective measure it has been suggested that cables in cinder areas should be buried in sand, gravel or chalk. If there is risk of flooding by water, the only remedy is a water-tight, insulating coating. Anodic behaviour and consequent destruction of a lead cable in cinders has however also been reported (Thompson [1188]). A steel gas pipe also

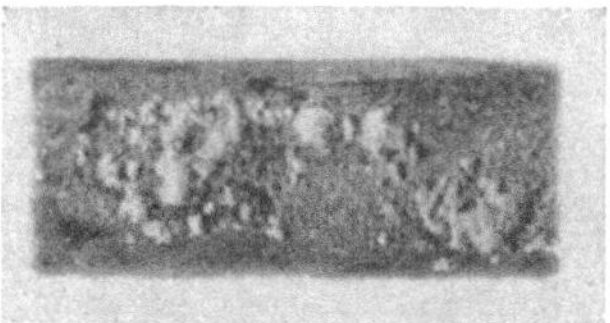

Fig. 289. Lead pipe corroded in soil. Pitting attack. White corrosion product of basic lead carbonate. × 0.8

passed through the cinder layer. When the cinders were replaced by sand, and a film formed from the corrosion products and cinders on the gas pipe were removed, the cable showed a cathodic potential. Waste waters of all kinds, scattered salt and synthetic fertilisers can promote the formation of concentration cells.

The behaviour of wrapped cable (Parkway cable) in the ground was also investigated by the National Bureau of Standards (Logan and Ewing [762)]. Here the lead sheaths showed good behaviour throughout. After many years they were, at most, only lightly attacked.

b) The Behaviour of Ducted Cables. The corrosion of ducted cable was thoroughly investigated on the basis of the experience of the Deutsches Reichspost-Zentralamt with the Rheinland cable (Haehnel [463]). The cable was laid in concrete ducts, which had the disadvantage of being permeable to water. Five years after the completion of the cable, serious damage showed itself here and there. Investigation indicated that the corrosion was exclusively accompanied by the exit of current at the places in question, and was therefore of an anodic nature. This conclusion corresponds exactly with modern views. If stray currents (which are discussed in the next section) are disregarded, then primarily here also electrolytic cells of the type "lead/water of one composition/water of another composition/lead" are involved. Such cells occur preferentially at the boundaries of geological formations, in which the entry of lead into limestone zones is particularly hazardous (Haehnel [463]). Lead is thus the anode in calcareous water and here goes into solution. The distribution of anodic and cathodic areas shows great variation with the season and with the position of the ground water level. If the cable duct is at the upper margin of a fluctuating ground water level, and is often alternately moist and dry, then the lead sheath is more endangered than when the cable lies permanently in ground water and access of oxygen is thus impeded (Haehnel [467]).

Apart from the processes mentioned, the following can also contribute to corrosion, irregular distribution of alloying elements in lead, contact of the lead sheath with the iron coil-cases and stray currents which use the sheath as conductor.

On the basis of experience, cement fittings are only regarded as permissible in towns where access of water through the road surface and the pavement has largely been blocked. The cement must be sufficiently set, otherwise it contains free lime (Haehnel [467]).

The conditions of ducted cables in towns are distinguished from those of buried lead pipes by the fact that in the former moisture has in general less access, and air greater access, than in the case of lead pipes. Soil air contains far more carbon dioxide, and less oxygen, than the atmosphere. Moreover the environment of ducted cables is, in general, more uniform than that of a buried naked lead pipe, where, on the basis of a more or less intimate contact with the surroundings, the presence of aeration cells must be assumed. The greater durability of ducted cables in comparison with bare underground cables (which are hardly ever laid) has been connected with this uniform environment of ducted cables, which only permits aeration cells under special conditions (BURNS [156]). As far as ducted cables come into contact with earth, it is only the fine mud which settles from water which need be considered. However, the ingress of small amounts of moisture to ducted cables seems more important.

The work of Bell Telephone Co. already mentioned (BURNS [156]) includes further aspects of the corrosion of ducted cables. The duct materials mentioned are glazed clay or pinewood impregnated with creosote from coal tar. Owing to its acetic content, creosote from wood tar is not recommended; acetic acid can indeed originate in unfavourable circumstances, for instance, on heating, from wood itself. Cement ducts are not used by Bell Telephone. They are said to be more suitable for power cables than for telegraph cables, as there heating is to be expected. Little significance is attributed to alloy inclusions in lead, or soldered joints, or oxide accumulations as possible electrodes of galvanic cells resulting in corrosion.

c) Stray Current Corrosion and Electrical Methods of Protection. While in normal soil corrosion, the EMF predominantly originates from the reciprocal action of lead with its varied environment, stray current corrosion (MICHALKE [850]) has its origin in external sources of potential such as high-tension cables with defects of insulation, or uninsulated conductor rails of direct-current tramways. Some of the current flowing back from the conductor rails to the power station leaves this conductor and uses instead lead sheaths of buried cables lying adjacent to the tramways (Fig. 290); (EVANS [297]). Thus cathodic places, where the current enters, and anodic places of current exit can be found on the lead sheaths. In general, corrosion only occurs at anodic regions and the degree of corrosion depends on the current density at these places. Lead is considered to be endangered if the current density is more than 25 mA/m^2 (HAEHNEL [467]). Under otherwise similar circumstances the current density is the greater the better the ground conducts, (which is determined largely by the moisture content) and the worse the conductive connection between the rail section joints. Scattering of salt in winter is disadvantageous. Corrosion need not be feared at a distance of more than 100 metres from the current source. This limit is better taken too high than too low (HAEHNEL [467]).

The appearance of the corroded areas depends on the size of the surfaces of contact of the anodic regions with the electrolyte. According to conditions, either a smooth pitted surface or a more general rough etching formed. The corroded areas are covered with a corrosion-product of a mostly whitish colour.

The detection of stray current corrosion is carried out, on the one hand, by electrical measurements (MICHALKE [850]; German Standard Specification DIN 50910 [905]), on the other by chemical investigation of the corrosion product (HAEHNEL [464-5-7], GLANDER [386]). Almost

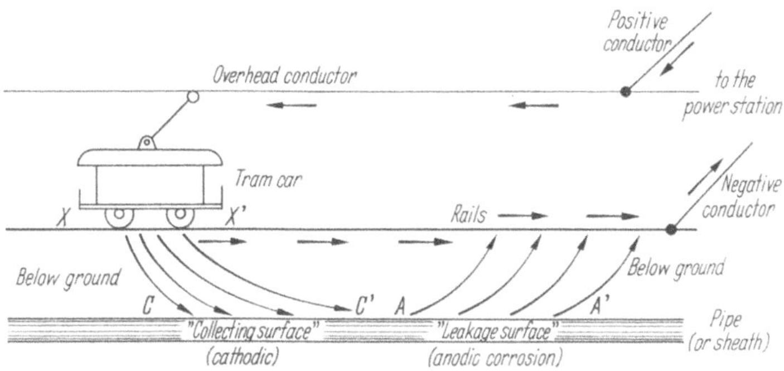

Fig. 290. Corrosion by stray currents. (According to EVANS [297]).

without exception the latter contains lead chloride in considerable amounts even when the content of soluble chloride in the soil is as low as 0.01%. Lead chloride can be detected with silver nitrate after dissolving the corrosion product in dilute nitric acid. In addition, according to the composition of the soil, lead nitrate, sulphate and dioxide can occur as corrosion products. Lead carbonate, which forms the chief product of the direct corrosion of lead in soil, on the contrary, is not a principal feature in stray current corrosion. Under American conditions, stray current corrosion was approximately equal to direct corrosion (CHASTON [190]). whereas, according to older experiences of the Deutsche Reichspost, cases of direct corrosion were more numerous. When stray currents are to be feared, wrapped cables are used instead of asphalted cables. Replacement of asphalt by a synthetic tar preparation is said greatly to increase the insulation and therewith the protection against stray current corrosion (GLANDER [386]). A series of counter-measures is described in the literature (HAEHNEL [467], EVANS [297], REINER [1001], BOREL [110], GOSDEN [410]). Attention may be drawn here to the experience of the British Post Office (RADLEY [990]). The best corrosion-protection against attacks of any kind should be sought in a coating which shows mechanical strength and good electrical insulation. The protective measures should however also be applied to the sources of stray current themselves. By suitable design of conductors, the danger of the passage of current to the ground can be reduced. In this connection attention may also be drawn to electrical drainage. A well-conducting metallic connection is made between the source of the disturbing current and the objects to be protected after paying regard to a whole series of precautionary measures (SCHMID [1072], FORETAY [333]).

Even in the absence of stray currents the cathodic protection of lead also plays an increasing part (UHLIG [1211]). Lead is regarded as completely protected in most cases if the potential against the standard hydrogen electrode amounts to —0.8 volts (HORNUNG [594]). According to another report, lead must be made about 0.1 volts more negative than its equilibrium potential in the type of soil in question (COMPTON [211]). The requisite potential and the necessary protective current are produced either by means of carbon anodes and a source of direct current (DOYLE [257]) or by means of sacrificial anodes. A magnesium alloy with 6% Al and 3% Zn has been proposed as a material for sacrificial anodes for the protection of lead cable sheaths (ROBINSON and FEATHERLY [1015]). An excessive cathodic load can render the environment alkaline and thus bring about corrosive attack (MUYLDER and POURBAIX [885]). A cathodic attack was observed in the presence of ions of sodium, potassium, lithium, magnesium and aluminium in a sulphuric acid electrolyte (ANGERSTEIN [21]). It is assumed that, for example, a lead-sodium alloy forms which easily decomposes in water. Ammonium ions caused no attack.

As an example of the application of cathodic protection, mention may be made of the measures taken by the British Post Office for protection of telephone cables against corrosion (GERRARD and WALTERS [368]), and also of the report of an American committee [215]; (cf. PLYM [967]).

In cable damage, particularly in high-tension networks, larger or smaller circular sharp-edged craters or holes with traces of melting in the armour and the lead sheath can occasionally be seen in the neighbourhood of the actual penetration, (BUSS and MÜLLER [160]). Such phenomena occur particularly where the soil is moist and well-conducting; they must not be confused with the corrosion damage hitherto described. These holes and craters are connected with the arcing of current which arises from sudden potential impulses between the lead sheath and the armour. Lightening flashes can act as sources of potential; for instance, those which strike telephone masts and find the shortest way to earth through the cable sheath. The holes and other cable damage which occur are caused by the prior damage to the cable sheath, and are not themselves the origin of the damage.

One more kind of electrolytic corrosion, which occurs on the inner side of cable sheaths, may be mentioned. It is caused by currents which pass from the cable sheath to the cable wire. Here lead peroxide occurs as corrosion product (HAEHNEL [467], GLANDER [386]).

While alternating currents of different frequency and strength do not noticeably affect lead up to 40 °C, at temperatures above this, thin film of finely-crystalline lead is said to form on the metal surface (BECK [70]). SANDMEIER [1046] also discusses the possibility of corrosion by alternating currents in relation to a rectifying action by the soil or by layers of corrosion products.

7. Damage by Animals

Damage to lead objects by animals, which is possible owing to the softness of the metal, is not merely a peculiar natural phenomenon but owing to the severity of the damage, a serious problem in some places, especially for the cable industry. For instance, in view of erosion of ducted cables by rats, special measures have been taken to prevent

Fig. 291. Water pipe, damaged by gnawing rats. 1:1

access of the rats to the ducts. An example of damage by rats is shown by Fig. 291. Similar damage is caused by other rodents, such as mice and hamsters.

Very varying kinds of injury are caused by insects. Observations here go back many decades. The final evidence for the correct interpretation of the phenomena was produced by direct observation of the creatures' activity (BAUER and VOLLENBRUCK [64]). Fundamental investigations were carried out particularly by the Deutsches Entomologisches Institut of the Kaiser-Wilhelm-Gesellschaft (HORN [591 to 593], BÖRSIG [99], GREFF and LÖHBERG [437]).

The greatest damage of this kind to underground cables has been caused, in warmer countries, by termites. Direct damage by insects is not involved, but rather a chemical corrosion brought about by the formic acid of the creatures (ALLEMAND [15]). Attack on water pipes in cellars by beetles may also be mentioned here (BAUER and VOLLENBRUCK [64]).

Another aspect concerns "secondary" damage, such as that arising from injury to neighbouring wooden structures by insects. Examples are lead chambers supported by wood, overhead cables suspended on wooden posts, and lead cable wound on wooden drums. Here larvae or pupae of insects develop in the wooden objects and eat their way to the surface; if they chance to meet a lead object then they have no option but to bore through it (VALLAND and SALMON [1216]). The damage takes the form either of channels which continue from the wood directly on to the lead, or larger irregular serrated cavities. In Germany, wood wasps for example are responsible in the former case, the house beetle and the dust beetle in the latter. It is a relatively simple matter to prevent damage of this kind by, say, impregnation of the wood.

Damage by insects to overhead cables, particularly in warmer countries, has often been described. The incidence here is purely random. As the pests in question are accustomed to gnaw in wood, they attempt to do the same on lead. Roundish holes occur, which are produced by fully-developed beetles according to their body diameter; or holes which are gnawed by beetle larvae after they are hatched from eggs laid on the surface of the overhead cable; or holes which are gnawed for the purpose of laying eggs inside the cable. Recently the cause of such damage, observed in Germany, was also attributed to beaver beetle larvae (LAPKAMP and MAGNUS [731]).

The detection of insect damage in lead should, in general, offer no difficulties, if the phenomenon is at all known, particularly if an inspection of the cable, together with its surroundings, is possible. Generally grooves will be perceived on the inner wall of the hole which correspond with the impressions of the chewing organs of the insects. The rim of the hole is not deformed, the walls are often bright and unoxidised.

There is no universal remedy for insect damage. In addition to the indications already given the following are relevant: armouring of the cable, oxidation of the surface of the cable sheath, suitable design of the suspension appliance for overhead cable, painting with fatty and sticky substances (to which, if necessary, a contact insecticide may be added), and further measures described in the investigations mentioned (cf. GIBLIN and KING [377]). On the other hand, alloying of the lead (perhaps with poisonous arsenic) has not proved successful as the lead is not taken up by the insects into their alimentary canals.

8. Indications of Further Causes of Corrosion

Many indications of the behaviour of lead in contact with other metals has already been given above. To supplement this some further experiences will be mentioned. According to a Dutch investigation (VAN DUIJN [261]), no severe attack on either metal occurs in a conducting combination of lead and monel metal in sulphuric acid. Coupling with nickel causes an increase of corrosive attack on lead in alkaline and acid solutions (pH >10 and <5), with the exception of sulphuric acid. This also holds for lead alloys, such as lead-tin solders. Solders joined to monel underwent an attack of 33.5 g/m^2 day in hot water, whereas in the absence of monel the attack amounted to only 0.22g/m^2 day. An antimonial lead alloy showed increased attack in a boiling mixture of sulphuric and nitric acids, when coupled with passive chromium-nickel steel. Under atmospheric conditions, electroplated lead coatings on steel often caused pitting in sheet steel (COMPTON and co-workers [212]). Magnesium or zinc were likewise attacked when in conducting combination with lead. On the other hand, when in contact with copper or nickel under atmospheric conditions, lead corroded, In a magnesium chloride solution, iron was at first nobler than lead (MÜLLER and LÖW [879]), but the potential relationships were soon reversed. Lead-coated sheet iron was found to be unsuitable for the manufacture of wet gas meters.

If lead is to be thoroughly protected against contact corrosion, and if such damage is to be recognised and eliminated, then as the first criteria the electrochemical series of the elements should be consulted together with the observations made elsewhere. Caution is of course advisable, as shown, for example, by the investigations on corrosion cells of stainless steel and lead in sulphuric acid (PRATT and COLLINSWORTH [977]). Normally, chromium-nickel steel, owing to its passive condition, here behaves as a noble metal towards lead. The stainless steel, when acting as a cathode, promotes the formation of the protective sulphate film on the lead, as described above, and neither the lead nor the alloy steel is attacked. The conditions can however change completely, if the oxygen content of the sulphuric acid is decreased; for example air can be expelled from the acid by an increase of temperature. The oxygen content is now too low to maintain the passivating oxide film on the stainless steel. This was observed, for example, in hot 50% sulphuric acid, in which lead behaved as a noble metal, owing to the formation of the sulphate film, whereas most of the chromium-nickel steels lost their passivity owing to lack of oxidising media. Under these conditions lead became the cathode, while the stainless steel was strongly corroded anodically. These observations show that the mutual electrochemical behaviour of different metals is very dependent on the conditions of corrosion.

Mention may also be made of the contact corrosion of an antimonial lead sheet with a welded seam, in a chromium electrolyte. The antimonial lead sheet had a composition of 5% Sb and 0.004% Bi, the welded seam an average content of about 4% Sb and 0.011% Bi, but 2.75% Sb and 0.026% Bi at the surface. A further difference between the original material and the welded seam was also revealed in the crystal structure. The sheet exhibited a worked structure, the welded seam, on the other hand, a cast formation. On both sides of the seam, the sheet showed corrosion over a large area and was penetrated by pits. The seam, on the other hand, showed no damage. In Fig. 292 there is shown a place where severe contact corrosion has occurred between the original metals and weld metal. While the weld metal exhibits only a slight attack, a thick layer has been destroyed in the original metal. In order to prevent corrosion in this case, it is necessary that an effort should be made to match the characteristics of the weld metal to the chemical composition and structure of the basis metal.

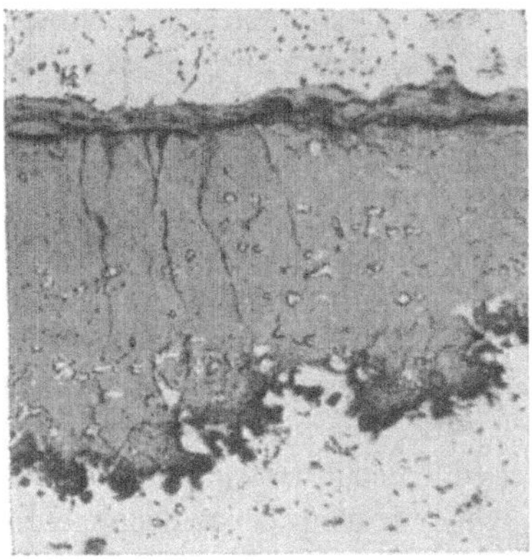

Fig. 292. Contact corrosion in a chromium electrolyte, between an antimonial lead sheet (below) and a welded seam in it. More intense attack in the original sheet. Unetched. ×300

In all cases in which lead is directly in conducting contact with other metals or alloys, the possibility of contact corrosion must be considered. A simple remedy consists in interposing an electrically-insulating layer.

C. The Technical Processing of Lead

I. Foundry Alloys

1. Melting and Casting

a) General Properties of Lead Melts. Data on the physical properties of pure lead melts are given in section A III. The physical properties of lead alloy melts were largely treated in connection with the corresponding alloys (sections B II and III). Comparative investigations of the properties of various molten alloys will be found in a number of places in this book (p. 362). In addition, attention may be drawn here to an investigation by Patterson and co-workers [938] on the viscosity of molten lead alloys. The authors found, among other things, that the viscosity was the more increased by an alloying element, the lower the solubility limit of the latter in solid lead.

Furthermore, relationships were found between these results and the elements' position in the Periodic Table.

b) Melting. The melting of lead is nowadays carried out mainly in oil- and gas-fired furnaces. Coal-heating is only rarely used. Oil- and gas-heated furnaces are equipped with automatic temperature regulation. The furnaces are operated at full power in melting down the charge; for temperature maintenance a small supply of gas or oil suffices.

Electrical heating of lead melts has also been applied widely. Electrically heated lead melting furnaces of 75 kW, and a holding capacity of about 3,600 kg of lead are described, particularly for cable presses (Höltje [546]). Of particular significance is the subdivision of the heating coil, over height and length, in three separate groups with individual automatic regulation, so that, on the one hand, less heat is provided to the upper marginal zone when the bath surface sinks, and, on the other hand, more to the loading side in charging. When the desired bath temperature of 450 °C is exceeded, an additional pyrometer in the molten lead cuts out the entire heating current. By these measures local overheating and oxidation occurring there, which is undesirable particularly in cable extrusion, are avoided. Lead baths such as are used in the patenting, annealing and hardening of steel, are also often heated electrically (Höltje [547]).

Where the alloying elements themselves melt at low temperatures, the melting of lead alloys presents no difficulties. If the alloying elements have high melting points (such as copper, nickel, tellurium, calcium), it is more expedient to begin with a hardener alloy of known composition. This is further to be recommended with the alkali and alkaline earth metals in view of their ease of oxidation and the consequent difficulty of achieving the desired composition by direct alloying with lead. Even by the use of hardener alloys, it is difficult to obtain a composition corresponding to the amounts originally weighed. Exact information on this point is available for lead-calcium (v. GÖLER [393]). According to this, the loss of calcium is made up of loss by burning off, by segregation of Pb_3Ca (consequent on melting at too low a temperature) and drossing of the complete alloy (lead and calcium) together.

c) Attack of Lead Melts on the Crucible Materials. The melting of lead takes place almost entirely in iron pots. These may be made of sheet steel, grey cast iron or cast steel. They are emptied by pumps, by siphons, and only occasionally by discharge outlets at the bottom.

If the furnaces are to be used for alloying, then stirrers or hand-stirring are provided. RIEMANN [1011 b] describes two kinds of mechanical stirrers which can be installed in the melting pot, and mentions the possibility of bath-mixing by introducing chemicals. Wear of the melting pots is to be expected. In melting unalloyed lead, this is to be attributed in the first place to the reaction with the iron of the PbO formed, or of the oxygen in solution in the lead (HÖLTJE [546], KRAUTMACHER and PÜNGEL [702]). In extensive tests TIMMERHOFF [1199] established that molten antimonial lead attacked much more severely than did commercial lead. Presumably in the case of antimonial lead there is an effect due to the antimony, which forms an intermetallic compound with iron. The greatest resistance to antimonial lead was shown by a low strength cast iron, Ge 12.91. TIMMERHOFF further found in his tests that specimens of Armco iron, steel with 0.13% C, Sicromal 6, Sicromal 12, Ge 12.91 and Stg 45.81 S, were much more severely attacked when rotated in melts of pure lead, lead + 8% Sb and lead + 8% Sn than when at rest. In melts at 750 °C, a certain amount of carburisation was observed in Armco iron and in steel specimens, i.e., only the ferrite dissolved, the carbon diffusing into the inside of the specimen.

In addition to the wear of the melting pots brought about by the attack of the lead, considerable distortion can occur owing to the creep of steel at high temperatures. This is particularly observed in sheet steel pots. As is known, the creep resistance of cast steel is higher than that of rolled or wrought steel. KRAUTMACHER and PÜNGEL [702] dealt with the life of pans for lead for patenting and annealing steel wire.

Pans of open construction with walls of sheet steel of 25 to 30 mm thickness, lasted for several years at the usual patenting temperatures of 500 to 550 °C. On the other hand, in the actual annealing baths, which were operated at 700 to 740 °C, a weakening of the wall took place, sometimes after a comparatively short time, which necessitated changing the pans. As expected, the laboratory tests showed that corrosion is significantly more increased by raising the bath temperature than by prolonging the duration of the operation. Killed steel lasted longer than rimmed steel. Here a direct effect of the low silicon content of 0.09% in the rimmed steel is much less likely than the absence of segregations. It is suspected that attack on rimmed steel proceeds more strongly after exposure of the inner zones rich in sulphur. In works corrosion tests at 720 °C, steels *A* and *B*, resistant to scaling, were compared with unalloyed steel. The ferritic steel *A* contained 0.09% C, 1.80% Si, 0.85% Al, 23.1% Cr; and the austenitic steel *B* 0.10% C, 19.5% Cr, 8.86% Ni, 1.11% Mn. Surprisingly, the attack on steels *A* and *B* was only 1/20 to 1/10 of that on unalloyed steel. However, for various reasons (price, thermal conductivity), the steels resistant to scaling could only be considered in the form of a roll-cladding on carbon steel. Furthermore, some caution is necessary in the use of steel resistant to scaling; TIMMERHOFF [1199] in fact noticed intercrystalline corrosion and a carbide network at the grain boundaries in Sicromal 12 with 3% Si and 24% Cr after exposure to lead at 750 °C.

In the region of still higher temperatures, namely between 850 and 900 °C, lead patenting is operated according to the Double Lead process. The pans, "lead retorts", are cast from a steel G40CrNi 108 (resistant to scaling) with 0.3 to 0.5% C, 26 to 28% Cr, 3.5 to 4.5% Ni, closed up to the openings for entry and exit of the wire; the wall-thickness amounts to 40 mm, the life is estimated as 3 to 4 years (HAUG [497]). Sometimes a particularly intense attack is to be expected in kettles for zinc desilverisation of lead. It arises from a migration of zinc along the grain boundaries, which may give rise to cracking without substantial wear of the kettle wall.

In order to elucidate this type of damage, several kinds of steel sheet (including also two high-alloy steels) were exposed to attack by zinc-containing lead at temperatures between 350 and 550 °C (HERRNKIND [513a]). For a degree of wear x, there was found for all specimens and at all temperatures the parabolic time law: $x^2 = at$. The values x^2/t of the various types of steel did not essentially diverge from one another. Even the high-alloy sheets underwent no smaller losses. The constant a depends little on the material, but much on the temperature (Fig. 292a). Among the unalloyed materials, the rimmed steels exhibited the smallest reductions of thickness. No effect of the grain size on the wear could be found. Moreover tests in zinc-containing lead did not show an increased attack in the temperature range around 500 °C.

Furthermore, creep-to-fracture tests were carried out in lead, in zinc and in zinc-containing lead at temperatures of 425, 450, 475 and 500 °C. Fig. 292b shows (on the single logarithmic scale) the results of creep

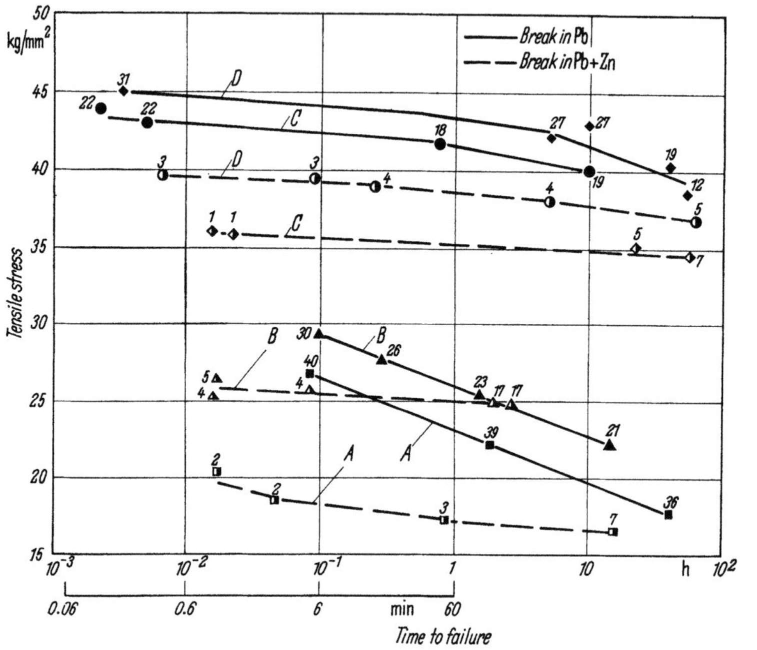

No	Type of Steel Abbreviation in DIN 17006	Chemical Composition in Wt.-% C	Si	Mn	Cr	Mo	Al	Cu	Grain Size in μm²
A	HI	0.12	0.20	0.70			0.015	0.18	1,010
B	HI	0.12	0.22	0.44			0.006	0.11	301
C	15 Mo 3	0.12	0.26	0.69		0.38	< 0.001	0.17	
D	13 CrMo 44	0.15	0.23	0.61	0.88	0.52	0.005	0.18	

Fig. 292a. Endurance of alloyed and unalloyed vessel steel sheets (DIN 17155) in normal heating conditions at 475 °C in melts of lead and of lead with zinc contents at about 2.5 Wt.-% zinc. The numerical values of the measuring points represent the elongation at fracture as percentages. (According to HERRNKIND).

tests on Armco iron and on a kettle steel of quality HI (German Standard Specification DIN 17155) from which most of the melting kettles are made. While molten lead did not effect the mechanical properties of the steel, molten zinc considerably reduced the time to fracture and particularly the elongation at fracture. An astonishing reduction of life and almost brittle fracture occurred in contact with melts of zinc-containing lead, if a minimum zinc content in lead (depending on the material) and a minimum tensile stress in the material were exceeded. For example, in Armco iron at 475 °C, a tensile stress of 11.8 kg/mm² already led to fracture about one minute after loading, in lead containing 1.3 Wt.-% of

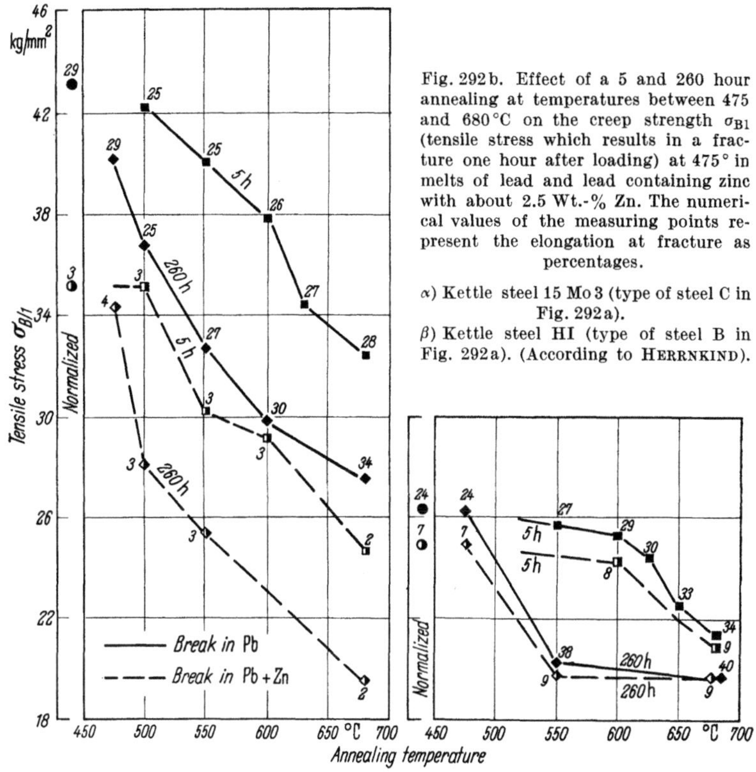

Fig. 292b. Effect of a 5 and 260 hour annealing at temperatures between 475 and 680 °C on the creep strength σ_{B1} (tensile stress which results in a fracture one hour after loading) at 475° in melts of lead and lead containing zinc with about 2.5 Wt.-% Zn. The numerical values of the measuring points represent the elongation at fracture as percentages.

α) Kettle steel 15 Mo 3 (type of steel C in Fig. 292a).

β) Kettle steel HI (type of steel B in Fig. 292a). (According to HERRNKIND).

zinc. The elongation here amounted to hardly 2 to 3%. Under otherwise similar conditions, a rod lasted 12 hours in pure zinc, 24 hours in pure lead. Tests on various deliveries of kettle sheet of quality HI showed that at 475 °C, tensile stresses between 14 and 22 kg/mm² could bring about fracture in zinc-containing lead 0.1 hours after loading. Zinc contents of 0.5 to 0.6 Wt.-% in lead are not yet hazardous for this type of steel. This is the reason why in general no cracking is observed in kettles for vacuum dezincing in spite of the higher operating temperature of almost 600 °C. The maximum zinc content at the beginning of dezincing amounts to only about 0.57 Wt.-%. The considerable reduction in life and the complete embrittlement of the material which are observed, presumably depend on the phenomenon of "liquid metal embrittlement" brought about by zinc-containing lead melts. Pure lead melts do not react with steel. Zinc melts indeed react with the steel, but at the point of contact a solid alloy layer is formed which strongly inhibits further reaction. In the case of lead melts containing zinc, after the zinc (in contact with the walls) has reacted with the steel, the denuded metal can take up fresh zinc from the bulk of the melt, and the mixture will react once more. In normal heating conditions, of all the sheets of standard "HI" used for sides of vessels, those which have the best liquid metal embrittlement are those which have the larger grain size (Fig. 292a). Steels killed with aluminium (fine grain steels) have poor resistance to lead-containing zinc. The low alloyed vessel steels, 15 Mo 3 and 13 CrMo 44, as unalloyed coarse grain steels show much better liquid metal embrittlement resistance. Liquid metal embrittlement resistance of all types of steel is reduced by heating at temperatures over 500 °C (Fig. 292b). Desilvering vessels therefore must not be maintained for an extended period above this temperature limit, neither during preparation nor in operation. The tensile stresses occurring on the innerside of vessels are caused mainly by heat stresses during the heating-up periods.

Attention is directed elsewhere to corrosion by molten lead alloys used for heat transfer (see p. 41, compare Epstein [283], Gurinsky [458]). Tantalum is resistant in lead melts up to 1,000 °C (Erben [284a]). Tungsten is not appreciably attacked up to 1,740 °C (Wurms [1293a]). Iridium and Ruthenium are not attacked at 200 °C above M. P. of the lead (Rhys [1007a]).

d) Drossing (Oxidation) of Lead Melts. In air, molten lead becomes covered with an oxide layer. *X*-ray investigation of the surface film with a high-temperature camera at 400 °C by O'Neill and Farnham [921] showed the presence of the rhombic modification of PbO (yellow lead monoxide). Bircumshaw and Preston [86] showed by *X*-ray diffraction that the part of the film against the metal was pseudo-tetragonal (red) PbO, whereas the layer facing the air was rhombic (yellow) PbO. The (001)-plane of the crystals is parallel to the metal surface. The surface oxide thus forms a kind of single crystal. In agreement with this, Gruhl [446] found that after cooling, the oxide formed in 8 hours at 400 °C was coloured dark red on the metal side and light yellow on the upper side. Thus below 488.5 °C, (the transition temperature red PbO $\rightleftharpoons$ yellow PbO), rhombic PbO can be first formed as a metastable phase, which in the course of hours transforms to pseudo-tetragonal PbO. On the other hand, red PbO was still found far above the transition temperature, i.e., at 750 °C (Stahl [496, 1143]). In long-time experiments between 475 and 525 °C, Pb_3O_4 (minium, red lead) occurred as a further oxide. Weber and Baldwin [1247] likewise established, by *X*-rays, the oxides already mentioned. However, they noticed a sooty black colour of the surface film after the disappearance of the interference colours. On this account the observations of KATZ [646] on the colours of lead oxide might be noted in future investigations of drossing (cf. further [1110a]).

Various investigations were concerned with the growth of oxide layers on lead melts and the effect of impurities and alloying elements in the lead. Not all the observations combine to give a uniform picture, as methods of experiment were very different, and the degree of purity of the initial lead was to some extent, by present views, insufficient. Individually, the results depend on whether the melt was at rest or agitated, whether the processes were observed in the first stage of oxidation, or whether tests of hours or days were carried out. Further it is important in the long-time tests whether the melts were allowed to cool and solidify for weighing, and then being heated up again. If it is desired to avoid this and to weigh at constant temperature, then the crucible, with the melt, which is in a vertical furnace, must be suspended on one beam of an analytical balance. In addition to the weighing method, there is the possibility of following the thickness of oxide layers by means of interference colours on parts of the melt which have become bright after scraping off the dross.

At temperatures which were not too high (see below) investigations carried out at rest gave the parabolic time-law of increase of thickness. It held not only for the first minutes of the formation of the tarnish colours (STAHL [496, 1143]) but was in part followed for days (GRUHL [446]). As is known, the parabolic law, $g = k\sqrt{t}$, where g is the increase of weight and k the so-called scaling (Zunder) constant, is an expression of the fact that the growth of the oxide film takes place on the oxide-air interface and that metal ions from the melt migrate through the oxide layer and at its surface form metal oxide with the oxygen of the air (HAUFFE [496d], WAGNER [1234a]). The process of diffusion, as the slowest partial process of drossing, determines the rate of the total event. More recent measurements of the rate of oxidation of 99.999% lead in the temperature range 453 to 643 °C also showed the parabolic time-law (ARCHBOLD and GRACE [23]). From measurements at various temperatures, the activation energy of the oxidation process is given as about 13 kcal/mole. According to measurements by THOMPSON [1187a], the oxidation of lead may, on the contrary, be controlled by the diffusion of oxygen in lead oxide.

The oxidation of solid lead takes place according to the same law (WEBER and BALDWIN [1247]). Certainly the parabolic law should be valid if, according to the rule of PILLING and BEDWORTH [963], the volume of the oxide is greater than the volume of the metal from which it is formed.

It is then possible to expect a crack-free surface film. A linear dependence of oxide layer thickness on time, which was established in some cases, was, on the other hand, explained by a cracked or loosened condition of the dross. The cracks permit immediate access of air to the surface of the melt, so that oxidation can take place below the surface film. However, in addition to the PILLING and BEDWORTH volume quotient [963], still other influences determine the loose condition of the surface film, such as local overheating of the melt surface by spontaneous oxidation of impurities with high affinity for oxygen (Li, Na, Mg) or deposits of oxide of accompanying elements (CaO, CdO) on the grain boundaries of the PbO crystals (GRUHL [446]). Deviations from the parabolic time-law in the direction of the linear time-law were observed not only in the alloys with Li, Na, Mg, Ca and Cd, but also in pure lead at temperatures about 600 °C (GRUHL [446], MAHLICH [792]). Changes of the time-law can happen in the course of long-time tests through the occurrence of secondary reactions, for example, the oxidation of PbO to Pb_3O_4 in the temperature region around 500 °C. The rate of diffusion of lead ions in Pb_3O_4 is lower than in PbO, so that the formation of Pb_3O_4 expresses itself in a reduction of the scaling constant.

The effect of individual additions will now be considered on the basis of various investigations. By double electrolysis of assay lead from the Harz district, which is itself very pure, STAHL [496, 1143] prepared about 500 g of lead in which foreign metals could no longer be detected spectroscopically. This lead, at temperatures up to 400 °C, showed no oxydation colours. This confirmed an observation of the former Kaiser-Wilhelm-Institut für Physikalische Chemie, according to which the purest lead, a little above its melting point, shows the lowest rate of oxidation. This result also applies to the crystalline state, as bars of 50 g weight, produced by STAHL [496, 1143], still showed no change in their mirror-bright surfaces after storage for two years in a desiccator. According to the results of short-time tarnishing tests by MAHLICH [792], which were discussed by Stahl, bismuth, nickel and cadmium in amounts of 0.01% have no effect on the rate of oxidation, whereas 0.01% of zinc, silver, copper, arsenic, antimony and tellurium considerably increase the rate of oxidation, not only at 500 °C (Fig. 293) but also at 400 °C. However, against these results by MAHLICH [792], there is the objection that the lead used, 99.99% lead from LAUTENTHAL, was not of the extreme degree of purity necessary for such investigations. Therefore the results of the two-hour experiments carried out at 520 °C, must also be regarded with some reservation. The considerable inhibition of drossing, found with additions of aluminium and of tin, agreed with the results obtained by GRUHL [446]. By small additions of these metals the drossing of technical lead alloys can also be greatly reduced, e.g., of lead-calcium (by 0.005% Al), lead with 6% Sb (by 0.02 to 0.05% Sn), and lead with 0.05% Te (by 0.02% Sn). As was to be expected, enrichment of the air with water vapour or with pure oxygen affected the oxidation only slightly.

Pelzel [951] showed that the more rapid occurrence of interference colours on molten lead, due to small contents of additional elements, can

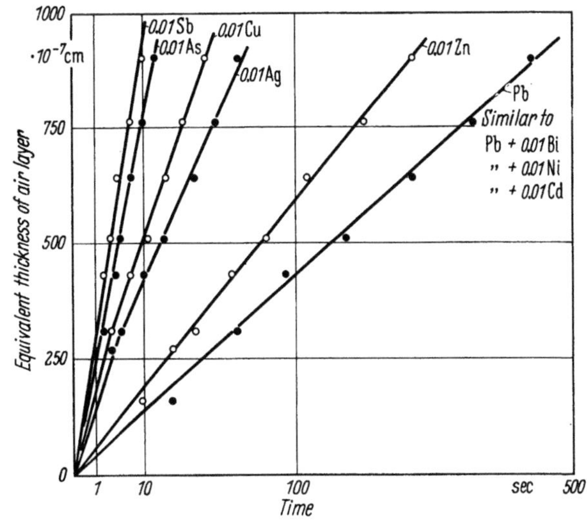

Fig. 293. Effect of various additions in lead on the rate of growth of oxide layers at 500 °C. Thickness of oxide layer expressed in terms of equivalent thickness of air layer. Mahlich's results plotted parabolically. (According to Stahl).

provide estimates of the diffusion coefficients. In this way he studied the diffusion of antimony, arsenic, cadmium, tin and zinc in molten lead.

Gruhl [446] conducted an experiment for one hundred hours, which was interrupted by repeated cooling and solidification for the purpose of weighing. He used 99.993% electrolytic lead, containing as chief impurities 0.006% Bi, and 0.0004% Ag. He found that the majority of additions increased the drossing of the lead which he used (Fig. 294). He referred the action of nobler metals, such as copper, in promoting oxidation, and their enrichment in the drosses, to the higher rate of diffusion of copper ions as compared with lead ions.

Mahlich's observation (792) on the so-called delay is not yet fully elucidated. By holding the lead melt at 500 to 700 °C with frequent, thorough, stirring, the rate of oxidation after removal of the dross is much retarded. The delay can be obviated, and the original higher rate of oxidation recovered, by reducing-melting and expecially by gassing with hydrogen.

A well-known investigation by BURKHARDT [155] was devoted to the drossing of agitated lead melts. The lead used for the production of the alloys was Braubach Commercial Lead of unspecified composition, which forms little dross. Air at the rate of 4 litre/min was blown through

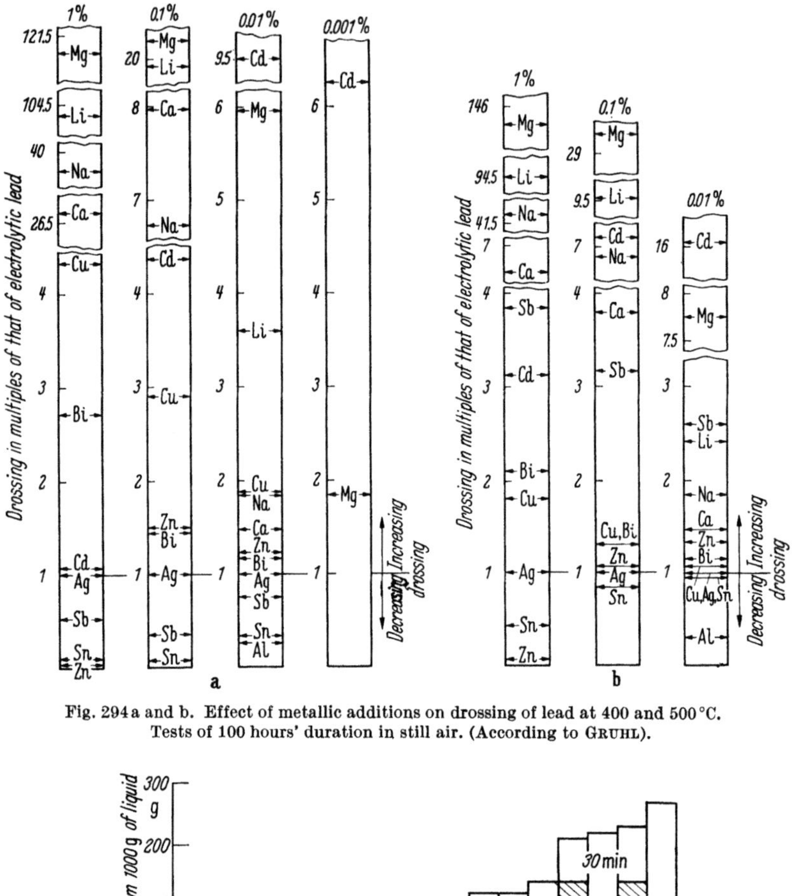

Fig. 294a and b. Effect of metallic additions on drossing of lead at 400 and 500 °C. Tests of 100 hours' duration in still air. (According to GRUHL).

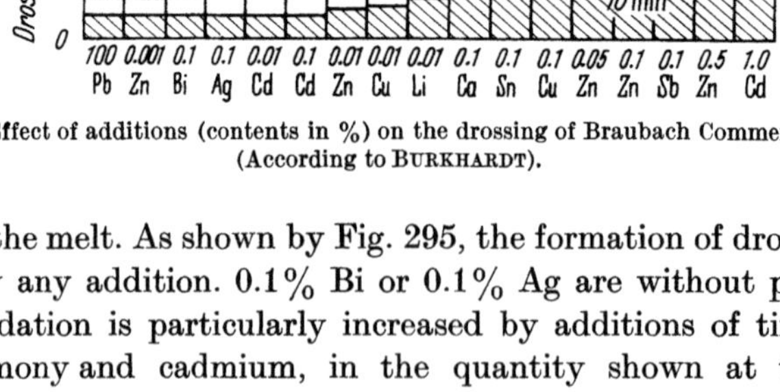

Fig. 295. Effect of additions (contents in %) on the drossing of Braubach Commercial Lead. (According to BURKHARDT).

1,000 g of the melt. As shown by Fig. 295, the formation of dross was not lowered by any addition. 0.1% Bi or 0.1% Ag are without perceptible effect. Oxidation is particularly increased by additions of tin, copper, zinc, antimony and cadmium, in the quantity shown at the right-hand side of the diagram. With larger amounts of antimony, the drossing again declined. In further experiments, additions of various kinds and quantities were made to 99.999% lead. A noticeable decrease of dross formation was effected by a few multiples of 0.001% Ag. Of the other additions to the high-purity lead, only 0.1% Bi caused an insignificant decrease of drossing. DREWS [258a] found, on the other hand, with agitated melts that silver as well as copper (0 to 0.1%) raised the oxidation rate to some extent, whilst with quiescent melts the oxidation is controlled by copper but not by silver (0 to 0.05%). (See also [366c]). It seems noteworthy that in the oxidation of bismuth-containing lead, the bismuth itself is hardly oxidised. DAVEY [235] found that at temperatures of 900 to 1,000 °C and bismuth contents of 0.06 and 0.14%, the ratio

$$\frac{\%\ \text{Bi in unoxidised lead}}{\%\ \text{Bi in reduced lead oxide}}$$

was 19.2:1. The method of passing, for determination of drossing was so improved by STAHL [496a], that reproducible results can be obtained. During weighings, the crucible with the melt and the tube for injecting air remains in the furnace. Over a large range, proportionality was established between the weight-increase of the melt and the volume of air passed through, between the rate of air flow and the frequency of bubbles, as well as the constancy of bubble size. By means of model experiments, the varying bubble frequency, observed in some type metals, was brought into relation with the surface tensions at the surfaces of the bubbles.

In a series of alloys with antimony contents up to 15%, a sharp maximum of the rate of oxidation was revealed by this method at 0.04% Sb, in agreement with earlier investigations (Fig. 296). Only GRUHL

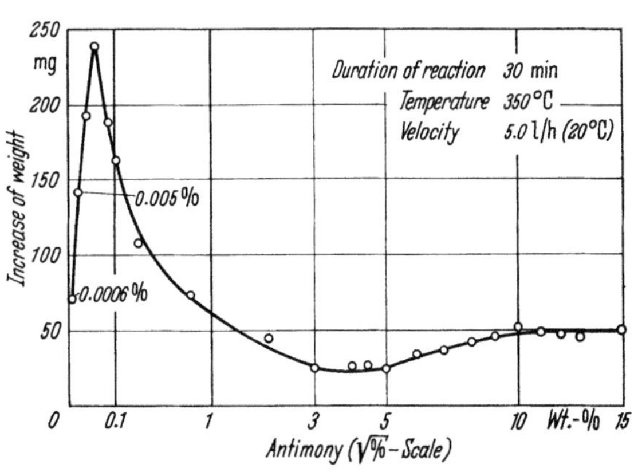

Fig. 296. Weight-increases of lead-antimony melts on passing-through of air. (According to HARTMANN, HOFMANN and STAHL).

[446], using his method of experiment described above, was unable to establish a maximum. According to an evaluation of the literature by STAHL [496, 1143], and his own experiments, the position of the maximum (disregarding the temperature) depends on the intensity of oxidation, as shown by the following data:

No movement of bath: maximum at approx. 0.01% Sb (STAHL [1143] at 350 °C).

Stirring of melts: maximum at 0.013% Sb (WILLIAMS [1274a] at 420 and 500 °C).

Passing air in: maximum at 0.04 to 0.05% Sb (BURKHARDT [155], STAHL [496a] at 350 °C).

In contrast with the investigations mentioned, which were concerned with the total amount of oxides formed ($PbO + Sb_2O_3$), RÖNTGEN and co-workers [1019] investigated only the decrease of the antimony content of a melt by selective oxidation at refining temperature. PELZEL [946], in a study of the drossing of lead-antimony melts, also provides data on the selective oxidation of antimony. RÖNTGEN [1019] found that the speed with which the antimony is oxidised out of the melt, decreases with falling concentration of antimony, and sinks to very low values when the antimony content falls below 0.1%. A maximum in the rate of oxidation of antimony at a given antimony content was only found if the lead contained, in addition to antimony, additions of more electro-negative metals such as arsenic, tin or zinc. In oxidising treatments the accompanying base metals burn first, with only slow decrease of the antimony content. The oxidation of antimony only sets in more intensely after removal of the associated base metal. A parallel to this phenomenon is provided, for instance, by the oxidation of phosphorus in the Thomas process. The results obtained by ERMISCH [289] can be interpreted in the same way. GERLACH and HERRMANN find a maximum of oxidation between 0.002 and 0.05% Sb; here chiefly lead is oxidised. A second maximum of oxidation at 0.3% Sb represents the maximum removal of antimony ([366b] and BARTHEL [49b]).

STAHL [496, 1143] also considered the oxygen take-up of lead-antimony melts at 600° and 750 °C. For this purpose he developed a special test apparatus. He brought the specimens to the temperature of reaction under nitrogen, completely free from oxygen, and determined the beginning of the reaction by rapid replacement of the nitrogen by air. The end of the reaction could be enforced just as exactly by replacing the air by an inert gas and the simultaneous freezing of the reaction.

No marked maximum of the rate of oxidation in relation to the antimony content was found at 600° and 750 °C in these experiments (Fig. 297). Both at 600 °C and also at 750 °C, antimony contents up to 0.01% cause, at first, a steep increase of the scaling (Zunder) constants. Then at 600 °C the rate of oxidation remained about constant up to an antimony content of 0.1 %, at 750 °C up to 0.03 %. With further rise of the antimony content a considerable increase of the rate of oxidation set in at these temperatures. The parabolic law now no longer holds. As the reason for this, STAHL suggests that above the last-mentioned content and temperature, fusion of the layer of scale takes place. The melting of the oxide layer at certain antimony contents could be confirmed by surface tension

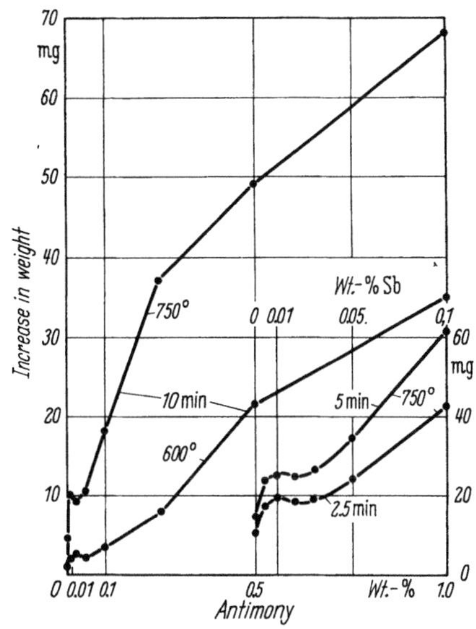

Fig. 297. Weight-increases of purest lead-antimony alloys at 600 and 750 °C. After oxidation for 10, 15 and 2.5 minutes. (According to HARTMANN, HOFMANN and STAHL).

measurements. The observed effect of antimony on the rate of oxidation of lead was interpreted on the basis of WAGNER's theory of oxidation ([1234a], HAUFFE [496d]), and supported by additional experiments. According to the parabolic time-law, the rate of reaction is given by the diffusion of the reaction constituent in the oxide layer as the slowest partial process. Again the rate of diffusion depends on the amount of disorder in the oxide lattice. For PbO as the primary oxidation product, a disorder with cation vacancies is regarded as very probable. By entry of Sb^{3+} or Sb^{5+} ions into the PbO lattice, the concentration of cation vacancies, and consequently the rate of diffusion of cations into the lattice, increases. This explains the acceleration of the oxidation of lead by the smallest antimony content. The fall in the rate of oxidation beyond the maximum in the temperature range 350 to 500 °C (Fig. 296) is interpreted by the appearance of lead antimonite in the oxide layer. Above 500 °C these high antimony contents of the oxide cause a lowering of the melting point until fusion of the oxide layers. The transport of material in molten oxide layers however takes place more quickly than in solid, so that the rate of oxidation increases further with the antimony content and no marked maximum is formed. In practice, the tendency of a grade of lead to drossing is judged by the crucible test. A weighed amount of lead (e.g., 600 g) is melted at 400 °C in a porcelain crucible and the surface made bright. The time to occurrence of the various interference colours, and to general clouding of the surface, is found. In grades of lead with little tendency to drossing, the surface should remain bright for at least 30 to 40 sec. This crucible test has not yet been standardised. FRITZE [349a] describes experiments on the statistical analysis of the bright period, that is the time to inception of tarnishing, and experiments on the increase of this period.

In connection with the drossing of lead, mention may be made of a proposal by KRYSKO [714], to deoxidise lead melts before casting, by addition of aluminium.

e) Casting. Mention may first be made of the casting properties of commercial lead. The linear shrinkage was determined at 0.75% for sand casting and 0.94% for chill casting (BAUER [57]). Another determination gave values of 0.90 to 0.99% (HONDA and KIKUCHI [585]). It was thus demonstrated that the linear shrinkage approximately corresponds with the thermal contraction in the cooling of the casting from the temperature of solidification to room temperature. The volume, or linear change (ENDO [279]) of 1.15%, which occurred on solidification is not included in measurement of the shrinkage. Data on the shrinkage of lead-antimony alloys are given on page 79.

The castability of a metal, that is its ability to fill a mould completely, can be determined either with moulds used in practice (p. 338) or in a spiral chill mould, in which measurement is taken of the length of the spiral formed.

PATTERSON and KÜMMERLE [939] criticised the equating of castability with the length of the cast spiral. They describe this property as the fluidity. By ability to fill the mould they understand, on the contrary, the ability of the melt to reproduce fine contours of the mould cavity. While the fluidity is principally influenced by the heat content of the melt, the ability to fill the mould is connected with the surface tension, with the solidification behaviour and the rate of cooling. In future investigations on the castibility of lead alloys this distinction should be kept in mind.

A comparative investigation of low-melting metals including aluminium, by means of a spiral chill mould of initial temperature 18 °C, showed an approximately linear dependence of length of the cast spiral on the degree of superheat, i.e., the difference between casting temperature and melting temperature (PORTEVIN and BASTIEN [973]). The upward slope of the curve was less for lead and tin than for the higher-melting metals, zinc and aluminium. The curve for lead flattened towards the ordinate, so that the length of the spiral only altered by 10% between 400 °C (the lowest casting temperature investigated) and 520 °C, but on the other hand, changed more noticeably above this. An analytical treatment of the curves showed a simultaneous dependence of the fluidity, both on various properties of the cast metal (such as specific heat, latent heat of melting, density and viscosity) and on the mould material.

An understanding of the general relationship between fluidity and course of solidification is of great importance for the knowledge of lead alloys. Thus in binary alloy series, with eutectics, there was found in general a change of the fluidity, as shown in Fig. 298a, for lead-antimony. Maxima of fluidity occurred with the pure metals and the eutectic alloy. A minimum of fluidity is situated in the neighbourhood of the solubility limit in the solid state, i.e., where the solidification range is greatest (shown in the diagram by the length of the lines in the hatched part) (cf. Mashovets and Daragan-Sushchov [800]). Corresponding results were obtained for the lead-bismuth-tin alloys. The ternary system exhibits three binary eutectics and one ternary eutectic (Fig. 137). The absolute maximum of the fluidity corresponds to the ternary eutectic; other maxima are provided by the pure metals and the eutectics of the boundary systems. The latter maxima are each connected through a ridge with the peak of ternary eutectic composition. The ridges have minima at the places where the temperature range of binary eutectic solidification is greatest. Absolute minima of castability are present near the lines which join the vertices representing the pure metals to the ternary eutectic, at the place where the primary solidification range is greatest, i.e., in the corner points of the eutectic 4-phase plane. A similar pattern of the fluidity was found in the technically important lead-antimony-tin ternary system and is illustrated in Fig. 298b. Kogan [692a] also expresses views on the castability of the lead-antimony-tin alloys. He disputes the view that tin-contents increase the fluidity of these alloys.

Casting is carried out either as ingots or as shaped castings. In casting ingots in smelters, chill moulds are frequently arranged on a turntable or a continuous belt [836]. In the production of ingots for lead extrusion presses (as distinct from other metals) the melt is poured directly into the container and the ingot then extruded in a hot condition (see p. 389). The cross-section of the solidified contents of the container of a cable press (Fig. 299) shows columnar crystals, arranged and distributed in the manner which is usual in cast metals. The long axis of the crystals is parallel to a cube-edge of the elementary cell, i.e., the direction of greatest crystal growth-rate (NIX and SCHMID [902], cf. WALTON and CHALMERS [1236]). Here we are concerned with the cast texture usual in cubic metals. On the other hand, ROSENBERG and TILLER [1033] succeeded in producing the ⟨111⟩ texture by solidification of zone-refined lead with less than 10^{-4} Wt.-% Ag. If the solidification of a lead melt is

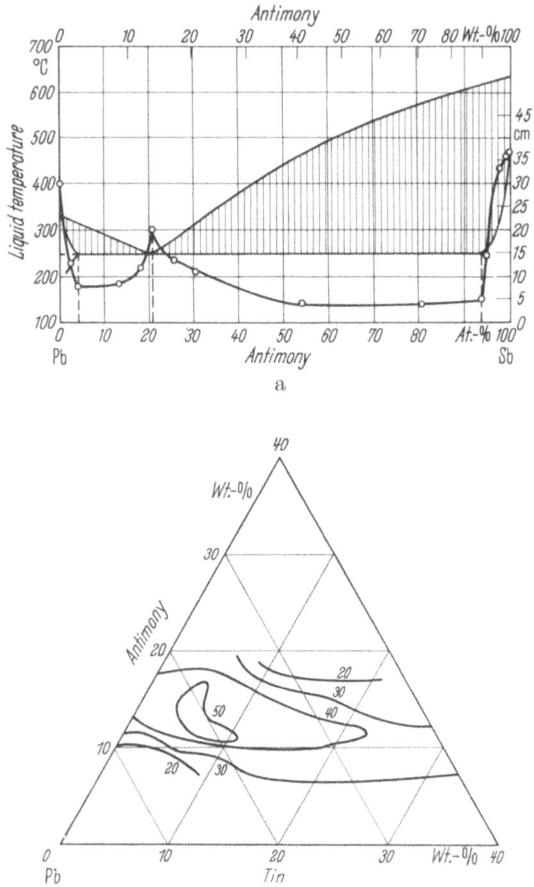

Fig. 298a and b. Fluidity of (a) lead-antimony melts, and (b) lead-antimony-tin melts. Arbitrary scale in (b). (According to (a) PORTEVIN and BASTIEN, and (b) CARTLAND [182]).

carried out in such a way that a single crystal forms from the middle of the surface of the liquid, then its octahedral plane lies parallel to the latter (ATWATER and CHALMERS [29]). BOLLING and co-workers [102d] similarly found that on quenching a melt of zone-refined lead a {111} surface texture of the crystals is formed.

In ingots of alloys the composition is not uniform, owing to segregation. For example, in lead-antimony the regions which have solidified last in the interior of the ingot are richer in antimony (STOCKMEYER and HANEMANN [1147]). An indication of the occurrence of weak inverse segregation is given by JACOBY [609].

Fig. 299. Section through the solidified container-charge of a cable press. (According to PRÜMM [981]).

Comparative investigations of grain size of cast lead alloys showed in most cases coarse cast grains (Fig. 299). Therefore in general greater grain refinement in the cast state is only to be expected at higher concentrations of the alloying elements (JENCKEL and HAMMES [618], JENCKEL and THIERER [619]). Low casting temperature favours the formation of fine grain (SCHEIL [1055]).

Shape casting of lead is carried out both in chill and in sand moulds. Lead bullets, plugs, and keyboard lead are, for example, cast in chill moulds. A further example of this is lead packing glands which are cast on steel moulds. (They prevent loss of oil in surface grinding machines if they are substituted for leather packing sleeves, which wear out rapidly) (Bassow [54]). Lead plugs are made of commercial lead or antimonial lead (Kühnel and Pusch [718]). The casting of memorials, souvenirs, ornaments and fancy goods in antimonial lead is widespread. Examples are caskets and bowls from Japan, in which amazingly good reproductions of all details of the mould surface have been achieved. Hollow-cast figures are produced, where, after solidification of the surface layers, the internal metal, still molten, is poured out. Flat castings are produced in superheated moulds and controlled solidification takes place from below upwards by application of wet cloths. The castings are generally given electroplated coatings of copper or brass and then gold, (Snelling and Thews [1131]), or chemically coloured (Krause [700, 701]). A continuous casting machine for lead strip of cross-section 50 · 5 mm is described in [839a]. The machine does not operate with a chill mould usual in continuous casting, but with a cooling wheel, which has a profile cut into its outer circumference and is cooled by water in the interior. The main casting wheel, together with small rollers, shapes the cross section of the strip. As the cooling zone is short, the temperature of the melt, the peripheral velocity and the cooling of the casting wheel must be carefully balanced with one another. In this plant, for instance, pure lead required a casting temperature of 360 °C, and a 60/40 lead-tin alloy, 300 °C. In this case a rate of 1.2 m/min was attained.

A similar kind of continuous casting is described by Kettler [657b]. A slowly rotating internally water-cooled steel casting roll dips into the lead melt. The layer of lead which solidifies on the steel surface runs as strip over various rollers and through pairs of rolls and is reeled on to a drum. The thickness of the strip depends on the depth of immersion of the casting roll. The most favourable rate of production of strip lies between 9 and 30.5 m/min; on the average a strip 460 mm wide, at a casting rate of 113 kg/min, uses about 180 litres of water per minute at 20 °C.

Krysko [714b] describes a chill-mould for casting blocks for radiation-protection in 4% antimonial lead. The length of the block is 10 cm, the thickness 5 cm, the height is optional. The mould consists of steel sheet 4 mm thick. It is heated to a temperature near the liquidus curve; the melt is at a temperature no higher than 20 to 30° above the liquidus. By water-cooling at the base, solidification from below upwards is achieved and the formation of pipes is prevented.

The casting of lead-silver alloy anode plates for zinc electrolysis may be mentioned as a further example of chill casting. The casting of large plates takes place by top-pouring. Commercial lead starting sheets for lead electrolysis are cast without a complete mould, the melt being allowed to run down an inclined plane. Cable sockets can be produced by dipping a steel core into molten lead when the lead solidifies as a coat on the core. Lead-tin alloy sheet of about eutectic composition, of thickness a few mm, is cast on felt cloth. After milling, the sheets are bent into the shape of organ pipes and soldered.

The manufacture of shot, as an example of casting without a mould, may also be mentioned in this connection. So-called soft shot consists of lead containing up to 0.5% As; hard shot contains in addition up to about 2% Sb. The action of the arsenic does not depend, as is usually assumed, on an increase of the surface tension, but on the oxide-dissolving property of the arsenious acid formed on the surface (TAMMANN and DREYER [1167], WHITE [1263]). Casting takes place from the top of a tower or the upper part of a mining shaft. The alloy is in a pasty form and put into a gas-heated iron pot, the base of which is perforated with a series of holes, corresponding to the desired diameter of shot. On stirring the pasty metal, spheres of molten alloy flow through the sieve, or coherent streams of melt are formed, which however immediately break into individual drops. The spheres, after solidifying in free fall and cooling, are caught in water at the lower end of the shaft. After drying the shot, the twins (that is, grains joined together) are removed by using the ability of the spherical grains to roll down an inclined plane. The shot is now sorted for diameter in cylindrical sieves and finally polished with graphite in a drum. The method is fast and economic. It is possible to produce in one plant several tons daily of smaller-diameter shot, so-called fine bird-shot.

The lead-coating of fittings of cast iron, malleable cast iron, cast steel, gunmetal, bronze, special brass or steel is often carried out by casting in or round with antimonial lead. A low-stress joint is made possible by heating the part to be coated to 100 to 180 °C. The details of the process are fully described (BECKER [71]).

Further applications of chill-casting, namely the production of bearings, storage battery grids, pressure die-cast parts and printing type, are described in special sections.

Parts for acid-resistant evaporating vessels, valves, pumps, stirrers etc., for the chemical industry are produced by sand-casting. Casting is carried out in moist ("green") or dry sand [302]. Antimonial lead with 2 to 10% antimony is usual for this purpose. The casting of an evaporator for titanium sulphate, of total weight 16,000 kg, whose individual parts weighed up to 4,000 kg, has been exhaustively described (MACGRAIL [789]). The mould and cores were prepared from dry sand. The mould was just strong enough to bear the pressure of the heavy metal, and beyond that, efforts were directed to obtaining the highest possible permeability of the mould material to allow for escape of the gases. At a casting temperature of 340 °C, the shrinkage amounted to 1.04%.

In connection with the melting and casting of lead, a reference to certain health hazards is also relevant. Within the compass of this book, the problem of lead poisoning can only be outlined by a few remarks. According to modern knowledge, danger from lead is present only from the inhalation of lead fumes or lead dust. On the other hand, contact with metallic lead and with lead compounds is largely harmless, as for example in the work of typesetters, plumbers or fitters. TUFTS [1206] reports on the determination of lead particles in air.

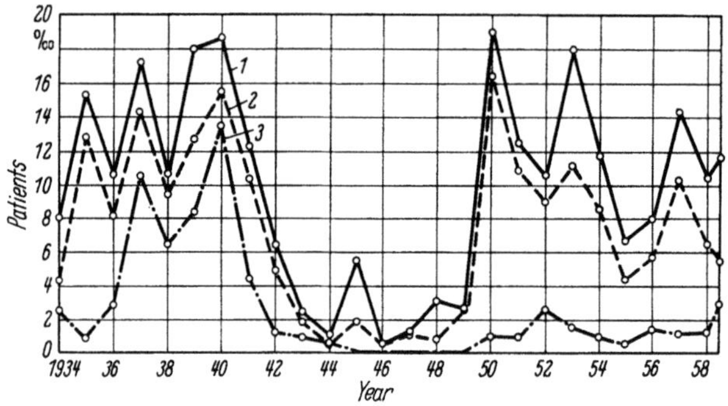

Fig. 300. Lead illnesses in German metal smelting works per 1,000 men employed. *1.* Notified. *2.* Recognised but not yet compensated. *3.* Compensated for the first time. Period 1934 to 1959. (According to BÖRGER [98]).

The essential protective measures against lead poinsoning are already contained in the Lead Smelter Order (Bleihütten-Verordnung) of the German Government, 16. 6. 1905. The preventive measures relate principally to the question of the cleanliness and ventilation of the working areas, but also to the behaviour of the employees and their medical supervision. Masks must be worn in especially dangerous areas. The personal cleanliness and healthy mode of life of the lead worker appear particularly important, but his constitution also plays a part.

A survey of the frequency of lead illnesses in the years 1934 to 1959 (Fig. 300) shows a particularly low point of lead illnesses in the last years of the war and afterwards. This fact has been connected with the nutrition of those years, which was particularly poor in fat. Objective confirmation of the protective action of milk has not yet been possible.

As regards the details of lead poisoning, the protective measures and the position of medical research, attention may be drawn to the publications of the Gesellschaft Deutscher Metallhütten- und Bergleute, whose committee on lead smelting held two symposia on lead poisoning after the war[1].

[1] Gesellschaft Deutscher Metallhütten- und Bergleute Publications H. 3 and H. 7, "Papers and Discussions at the Symposia on Lead Poisoning", Clausthal-Zellerfeld 1952 and 1960.

2. Lead Storage Batteries

a) General Development. The principle of the lead storage battery was first discovered in 1850 by WILHELM SIEMENS, and, independently of him, by SINSTEDEN. However, the credit of having, in 1859, introduced the battery as a means of storing energy belongs to PLANTÉ. The plates were formed by two principles; in one case the electrochemically active mass (see below) is formed from a commercial lead plate itself PLANTÉ's principle), in the other it is introduced mechanically as paste in the antimonial lead grid or as powder in so-called reinforced plates (FAURE's principle).

The plates produced in accordance with PLANTÉ's principle are called large area plates. The total area of such a plate is about eight times the projected area. Fig. 301 shows a section of such a plate. The large area plates are employed only as positive plates. At the top they have two lugs with which they rest on the edge of the container or on supporting members. The one lug serves for suspension, the other also for conducting the current. The latter lug is extended to enable it to be welded to the bars which carry the current. The formation of the porous active mass from the solid lead takes place by anodic oxidation of the plates in about 12% sulphuric acid, containing an addition of an acid which dissolves lead. Nowadays perchloric acid is generally used as the additive. It is necessary to remove the lead-solvent acid completely in order to avoid corrosion later. This is effected by cathodically polarising the plates in the same electrolyte until the lead peroxide is completely transformed to spongy lead. After removal, the plates are thoroughly washed and dried. The renewed transformation of the active mass into lead peroxide takes place in pure sulphuric acid. The formation could be described as deliberate anodic corrosion of lead. The lead peroxide thus formed is a good electrical conductor. Its specific electrical resistance is about 130 micro-ohm/cm, which is only 6 or 7 times that of commercial lead (THOMAS [1185], THIRSK and WYNNE-JONES [1183]).

Plates prepared in accordance with Faure's principle are called pasted plates. According to the proposed use, the grids have fine to coarse patterns. Fine-meshed grids are necessary for very high discharge currents. Coarse or wide-mesh pattern is suitable when the currents are small and importance is attached to low self-discharge. Plates with large-meshed grids are called frame plates (Rahmenplatten). If the grid is covered on both sides with finely-perforated lead sheets, the plates are called box plates. They serve only as negatives in the battery, in combination with positive large area plates. Reinforced plates have insulating hollow bodies permeable to electrolyte, usually in the form of small tubes which contain the paste and a current conductor (lead core) of antimonial lead (Fig. 302). They are produced only as positives. The pasted plates other than the box plates are used both as positives and negatives (GARTEN [359]).

The grids of the pasted plates are made of 6 to 12% antimonial lead. The apertures are then filled with a spreadable paste consisting of lead oxides (PbO, Pb_3O_4), grey oxide and an addition of sulphuric acid.

Fig. 301. Large area plate AFA.

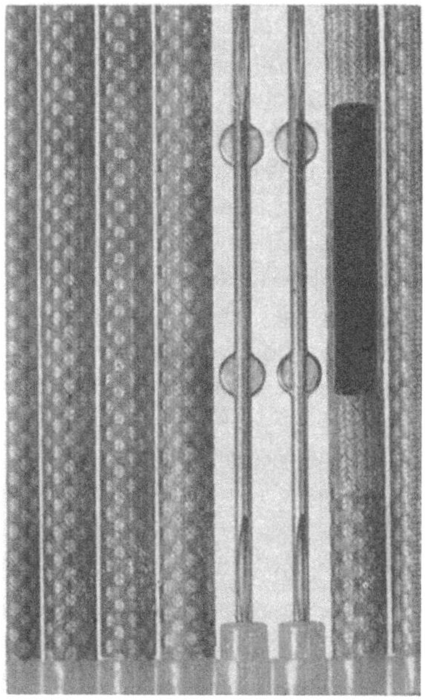

Fg. 302. Structure of tube plate with slit hard rubber tubes and antimonial lead supporting rods. (Akkumulatorenwerk Hoppecke.)

The active paste of the negatives must contain expanding additives, in order to prevent premature sintering of the fine spongy lead particles to coarse, less active ones (shrinkage of the negatives). The grey oxide is mostly produced at the storage battery factories by pulverising cast lead shot in drum mills [259] or by blowing air or steam into molten lead. It consists of metallic lead and monoxide (PbO). Lead contents smaller than 5% are usual for positive plates, and contents of 20 to 25% for negatives. Apart from this, essential characteristics of grey oxide are grain size, density and bulk density, and water-and-acid absorption (REUTEL [1004a]). The porous structure of the active mass is ensured by introducing powdered lead compounds. The conversion of the paste to lead peroxide and spongy lead respectively (formation) can then take place in pure 10 to 20% sulphuric acid by anodic or cathodic polarisation respectively. Formation takes 20 to 50 hours, according to the thickness of the plate. After charging, the paste of the positive has approximately the composition 90% PbO_2, 7% PbO, 3% $PbSO_4$; that of the negatives

95% Pb, 3% PbO, 2% $PbSO_4$. After formation the plates are soaked and dried. As chemical reactions occur during the drying of the plates, particularly an oxidation of the spongy lead of the negatives, a first charge is necessary to bring the assembled batteries into serviceable condition. The oxidation of the negatives can be partly prevented by drying in vacuo or in superheated steam. This is done particularly in starter batteries, as the latter, after being filled with sulphuric acid already have a certain capacity without special charging, and are in working condition ("precharged") for a certain limited time. Those plates which were not assembled can be stored dry as long as desired, and can, without risk, be packed, despatched and, if necessary, quickly assembled. For combining several plates of the same polarity into sets, the pole lugs are fixed to a connector, i.e., welded to it or cast to a set (Hoehne [540]). Contact between positive and negative plates in storage batteries is prevented by suitable interposed separators. These consist of wood, microporous vulcanite, plastic or glass fibre.

The working capacity of lead storage batteries, calculated on the weight, are approximately:

Cells with positive large area plates and negative box plates	8 to 10 Wh/kg
Cells with pasted plates	28 Wh/kg
Cells with positive reinforced plates	28 Wh/kg

The capacity of lead storage batteries depends on the magnitude of the discharge current, the density and temperature of the sulphuric acid. The discharge is limited by the final discharge voltage which, according to specification, or to specific instructions from the manufacturer, is about 10 to 15% below the initial discharge voltage.

Details on the design, metal contents, capacities etc. of stationary lead storage batteries are given in the new German standard specifications DIN 407 30 to 407 37.

According to the kind of use, lead storage batteries are classified as follows (cf. VDE Specification [1217]):

1. Traction batteries, for driving vehicles or in power plants,
2. Lighting batteries, e.g. for train lighting or emergency lighting,
3. Batteries for communication purposes, e.g. for telephone systems,
4. Batteries for auxiliary services, for providing current for electrical plant, e.g., for control and recording systems, braking systems,
5. Starter batteries, for starting and ignition of automobiles as well as for lighting.

In stationary batteries for electrical power stations, domestic lighting plant, emergency lighting and telephone installations and much else, positive large area-, and negative box plates were originally used, but more recently reinforced and grid plates have been employed to save weight and space. Batteries for accumulator-driven rail vehicles and for train lighting use large area plates as well as the combination of reinforced plates and negative grids. Grid batteries and, recently, the combination of positive reinforced plates and negative grids, have been introduced in large numbers as vehicle storage batteries, particularly for electric trucks and mine locomotives. Starter batteries should withstand, in particular, high-current outputs. They are therefore equipped with very thin grid plates which have a large surface as well as a large contact surface between grid and paste. For small portable storage batteries use is made of large area, reinforced and grid plates.

b) Production and Properties of Storage Battery Plates. Large area plates and the grids of pasted plates are produced almost exclusively by casting. Grid plates contain two systems of ribs, enclosed in a rectangular frame, crossing one another to form either straight lines parallel to the edges of the plate or diagonal lines. These systems comprise two planes, parallel to each other, of alternatingly spaced ribs. By this arrangement, strong internal coherence of the paste is obtained.

The large area plates and the grids of pasted plates are cast by top pouring, many types even now being made by hand. For large numbers, for instance grids for starter batteries, machines are used which effect casting and cutting mechanically. At each operation molten metal, delivered by a pump, runs into the mould from a heated casting trough above and parallel to the ingate. Pressure die casting is often used for casting the lead cores of the small tubes of reinforced plates.

Before each cast, the moulds for large surface plates are powdered, for instance with talcum or lycopodium. The moulds for the grids of pasted plates are dusted with, say, a cork powder/water glass suspension (Drotschmann [259]). In casting by hand an additional powdering is occasionally necessary. In the mould, air holes or ventricles in the individual panels, which are connected by very fine channels with the troughs for the molten lead, are of particular significance.

According to type and alloy, the mould temperature is between 180° and 220°C; it is either attained during the course of casting (in which case the first plates or grids poured in the cold mould are scrapped) or is provided by heating. According to the heat content of the lead and the mould, as well as the rate of operation, the mould temperature is maintained constant, by cooling with air or water, or by heating. The ingate must be so high that the molten lead in the mould is under a certain pressure.

For the production of box plates, a perforated sheet of lead is laid in the mould and a large frame of antimonial lead is cast round it as a stiffener. Of two such plates, one is perforated and the other is fitted with a set of spines. The whole of the paste is put into one plate, the other placed over it, the spines are pushed through the perforations, and the complete unit is riveted together by hammering.

A report in *Industrial Heating* [604] gives details of the casting of grids in Pb-9% Sb in a Californian battery factory. The ingots of weight 23 kg are melted in a steel kettle of 1,135 kg capacity and 9.5 mm wall-thickness. The kettle is heated with gas burners; regulators maintain a bath temperature between 450 and 510 °C. The round casting table with eight fine-grained grey cast iron moulds, each for two plates, make 5 to 6 revolutions per minute. In the event of a stoppage of the machine, burners under each mould serve to maintain a temperature of at least 200 °C. The solidification of the plates, their cooling and ejection take place during three-quarters of a revolution of the casting table, equal to a time of about 9 seconds. The ingate ends, after cutting, go straight back into the kettle. For casting larger plates, a kettle of about 500 kg capacity is available. The melt is pumped from there to a single mould which provides two identical plates. Small parts, such as connectors of lead with 3% Sb, are also produced in a casting machine. The corresponding 500 kg kettle, of height and diameter each 40.6 cm and wall thickness 9.5 mm, is provided with gas burners at 3 different heights. The temperature is automatically regulated.

Refined lead is used as a material for large area plates. Some works make a choice from types of lead known to be very pure, or subject the lead from the smelter to an extra refining process (Evers [298]).

The usual material for grid plates and for other parts in which a certain strength is required (e.g., for connectors) is antimonial lead with between 3 and 12% antimony as already mentioned above. The hardness of antimonial lead increases greatly with increasing antimony content up to the eutectic composition (Fig. 75). For this reason, alloys with 5 to 6% Sb are usual for the casting of thicker cross-sections, such as connectors of framed plates; on the other hand, alloys with 8 to 9% Sb are employed for grids for starter batteries. The alloys of higher antimony content also have the advantage of better casting fluidity (see p. 335) with, however, the disadvantage of a higher self-discharge of the batteries made from them. The structure of an alloy with 6% Sb taken from a starter battery shows (according to Fig. 303) dendritic primary crystals and lead-antimony eutectic.

Pipes in the casting are recognised externally by the sunken surface and the mat appearance. The structure here usually shows a marked microporosity (Fig. 304). The cause is the local delay in solidification, which can arise, for instance, because a cross-section is too large or the melt was poured too hot, or by heat accumulation. Cavities reduce the life of plates particularly if they are cut when the edge of the grid is stamped out. In order to prevent porosity, a steady progress of solidification in the mould (among other things) is encouraged. The mould can be provided, in certain places, with cooling ribs, or cooled by airblast, thus displacing the pipes into the feeder head which is cut off (WILLIHNGANZ [1275]). In manual casting there is a risk of oxide in-

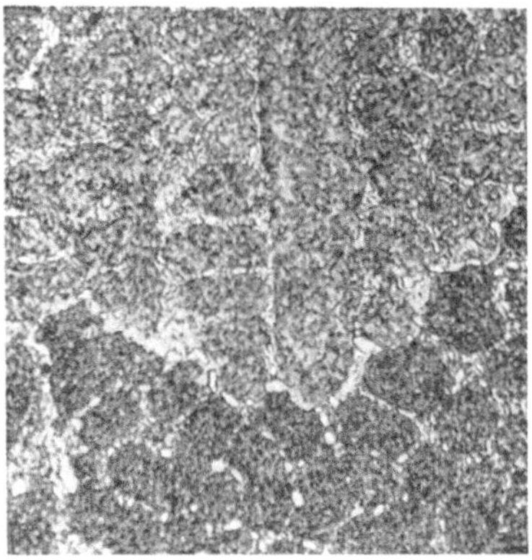

Fig. 303. 6% Sb. Cast. Dendrites of lead with fine antimony segregate. Lead-antimony eutectic in the residual fields. ×1,200

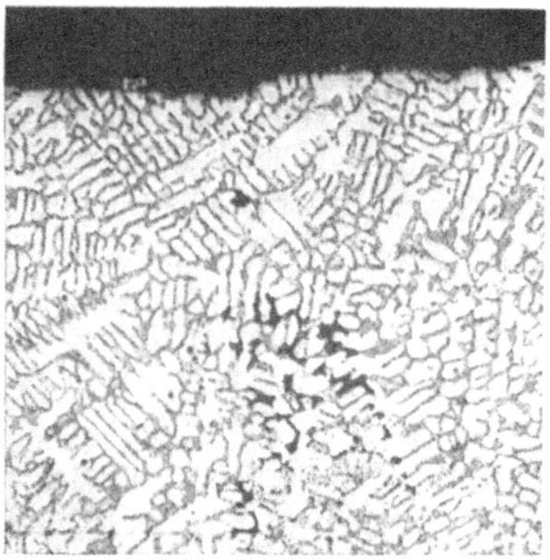

Fig. 304. Section through the lug of a battery plate at a depressed area. Micro-pipes below the surface. × 125

clusions owing to the oxidation of the surface of the melt, but in machine casting the lead near the bottom of the melting kettle is pumped through a pipe into the casting trough and the introduction of oxide largely avoided (ZAHN [1299, 1300]).

SIMON and JONES [1122a, 1123] found a great enrichment in antimony at the surface in cast alloys. They traced this inverse segregation to a delay of the crystallisation of antimony. The residual melt between the lead dendrites therefore becomes hypereutectic. Owing to the contraction of the cooling casting, it flows into the gap between the mould and the casting and there solidifies.

The fluidity of antimonial lead in relation to the antimony content has already been treated, above (p. 335). According to practical experience, alloys with about 3% Sb in particular tend to develop edge cracks. This may be connected with the bad casting behaviour of the alloy. Edge cracks can also arise from too rapid removal of the grid from the mould. In regard to the hardness of the alloys, attention is drawn to a preceding section (p. 83). From the account given there it can be seen that in battery grids which have solidified rapidly, a certain increase of hardness on storage, due to age-hardening, is to be expected. The increase is however trifling, if it is compared with that of heat-treated and quenched alloy with 2 to 3% Sb (p. 84). Additions of arsenic (p. 87) can increase the age-hardening. The use of additions to improve the quality of the grid will be treated in the next section.

Attempts have also been made to produce thinner grids by stamping. An alloy for this purpose with 5 to 6% Sb was, after heating to a casting temperature of 320 to 380 °C, brought on to a rotating cast iron cylinder. The ribs and then the frame of the grid were stamped from the cast strip in two operations [259].

Lead-antimony alloys for storage batteries are produced in part by stirring pure antimony regulus into molten lead, in part by diluting antimonial lead with commercial lead. Apart from this, a large proportion of antimonial lead is recovered from storage battery residues. Liquation furnaces with sloping bottoms are customary for this purpose in small and medium-sized works. At hearth temperatures of 500 to 600 °C, about 50% of the metal content of the antimonial lead grids can be melted out, and again cast into grids after a brief refinement. The metal yield can be increased by using higher temperatures and a reducing furnace atmosphere. In recent times, rotary furnaces are used which can give yields of over 90% of the metal content. Depending on the material provided, a lead more or less enriched with antimony is obtained. Usually, the low-percentage antimonial lead which is recovered is raised to a desired antimony content by means of antimony regulus. In the treatment of storage battery residues by the Harris process, the antimony is converted to sodium antimonate, from which a very pure antimony regulus is obtained by reduction and subsequent refining (Evers [298], Liebscher [749]).

c) The Durability of Storage Battery Plates. Special Phenomena. The user's interest is that his storage batteries should have the longest possible life, i.e. that they should retain a sufficient capacity as long as possible. The composition of the paste after charging is indicated above. Considerable changes of volume of the active paste occur in discharging; the paste of the negative grids expands by 164%, corresponding to the change from lead to lead sulphate, that of the positives by 82%, following the change from lead peroxide to lead sulphate. During this process, the concentration of acid in the pores of the paste (the so-called internal acid) decreases; moreover, as the paste particles become covered with a layer of lead sulphate, their conversion and utilisation remain incomplete. In the calculation of the capacity of starter batteries, an efficiency of the active paste between 25% (Drotschmann [259]) and 50% is regarded as normal. The conversion can be driven further in weak discharges over long periods, as the internal and external acids can equalise their concentrations during the resting times.

The volume changes in charging and discharging cause considerable mechanical stress in the grid and, particularly, in the paste itself. Experience shows that the volume changes in charging and discharging are more marked in the positives than in the negatives, so that normally the positives fail by sedimentation of the paste (and grid corrosion). The life of pasted plates is less than that of large area plates and reinforced plates. In large area plates a life of 12 years is expected. Starter batteries are expected to last 2 to 4 years; the author has just removed from his car a battery which has been in service for six years.

In the operation of storage batteries a certain amount of anodic corrosion of the positives is to be expected. It sometimes occurs when the batteries are being charged. In consequence, the metallic lead of the grid is superficially converted to peroxide, that is, to active paste (see p. 298). The increase of capacity thus attained can initially compensate the loss of capacity due to sedimentation of the active paste, thus to a certain extent having a favourable effect, particularly in large area plates of commercial lead; apart from this, however, anodic corrosion is undesirable because of the risk of fracture of the grid ribs. The corrosion of the positives in certain types of cells is compensated by making the positive grid of thicker dimensions than the negative (Drotschmann [259]).

In discussing investigations of anodic corrosion, a recent study by Lander [727] may first be mentioned. He kept lead specimens in sulphuric acid of various concentrations, at potential constant with time, and determined the losses of weight with time. Potentials near to the equilibrium potential of + 1.684 V[1] of the reversible reaction $PbSO_4 + 2H_2O \rightleftharpoons PbO_2 + H_2SO_4 + 2H^{1+} + 2e$, which takes place at the positive plate of the storage battery, were especially important. The potential − 0.355 V of the reaction at the negative plate, $Pb + H_2SO_4 \rightleftharpoons PbSO_4 + 2H^{1+} + 2e$, may be quoted for the purpose of completeness, though it has nothing to do with the anodic corrosion. The lead peroxide formed at the equilibrium potential $PbO_2/PbSO_4$ becomes unstable at lower potentials. The reaction in the solid state $Pb + PbO_2 \leftrightharpoons 2\,PbO$ can only take place below a potential of 1.58 V. At this point the corrosion in relation to potential shows a sharp maximum, which coincides with the occurrence of PbO_2 as corrosion-product in addition to $PbSO_4$ and PbO Corrosion increases with rising temperature and with decreasing acid density. Therefore, Lander [727] suggests that the life of floating cells can perhaps be extended by increase of the acid density. Burbank [149] allowed the thinnest layer of corrosion product to be formed on a single crystal of lead, and by means of electron diffraction established an orientation relationship between the crystal lattices of lead and lead

[1] Related to the normal hydrogen electrode.

peroxide: $(100)_{PbO_2}//(110)_{Pb}$ and $(001)_{PbO_2}//(001)_{Pb}$. In discharging the positives, various intermediate products occur between the PbO_2 and the end product $PbSO_4$, such as lead monoxide, lead hydroxide and basic lead sulphate, which became evident in the electron diffraction and also in the time-potential curves. Studies by FEITKNECHT and co-workers [310, 311] contain electron micrographs of the transitions of lead sulphate to lead peroxide at the positives.

Recently BODE and VOSS [97] established a rhombic modification of PbO_2 ($\beta - PbO_2$) — found by SASLAWSKI and co-workers [1047] under quite different conditions — in the positive electrode of the charged storage battery. According to ASTAKHOV [26b] is is formed during the anodic charging of lead, first as a thin layer, through which oxygen diffuses. The latter forms, with the underlying lead, a layer of $\alpha - PbO_2$. Beneath this, a layer of PbO occurs in the further course of anodic corrosion.

The positive plates of used storage batteries were examined by SIMON [1122a]. The plates, which still possessed their nominal capacity, consisted of hard branched crystals of (tetragonal) α–PbO_2. Amorphous[1] β–PbO_2 was embedded between these crystals. The crystalline structure of α–PbO_2 determines the strength of the plates, while the active β–PbO_2 principally controls the capacity of the battery. In batteries with a capacity below 50% of the nominal, the plates consisted only of β–PbO_2. BURBANK [150a] found that crystalline structure of the lead peroxide is advantageous as compared with an amorphous[1] structure.

BÜCKLE and HANEMANN [139] investigated the microscopic development of the anodic corrosion. They found that in antimonial lead it advances preferentially along the eutectic. During this process the lead is first converted to a corrosion product, while the antimony is left unchanged. Later, the antimony inclusions also change to a corrosion product, which, as a pseudomorph, still reveals the eutectic structure. The corrosion of a positive large area plate (GOUGH and SOPWITH [411]), cf. Fig. 305, favoured the grain boundaries in its advance. The growth of plates in use is explained primarily by this intercrystalline attack (PARR and co-workers [932]). BURBANK and SIMON [150, 151] confirmed these details. They also found in commercial lead and in lead with 0,5% Sb a preferential attack on the grain boundaries, while in alloys of higher antimony content there was a selective corrosion of the eutectic. Owing to the large proportion of eutectic, the latter type of attack is more uniform, and is therefore regarded as less dangerous. In the initial charging (formation) of the pasted plates, the antimony, enriched on the surface in the solidification process, is preferentially

[1] Designation in [1122a] and [150a].

dissolved anodically and precipitated on the counter-electrode. The disappearance of the antimony from the surface and from the related interdendritic cavities should have a favourable effect on the quality of the plates, as space is thus provided for the corrosion products formed, and the tendency of the plates to grow is lessened.

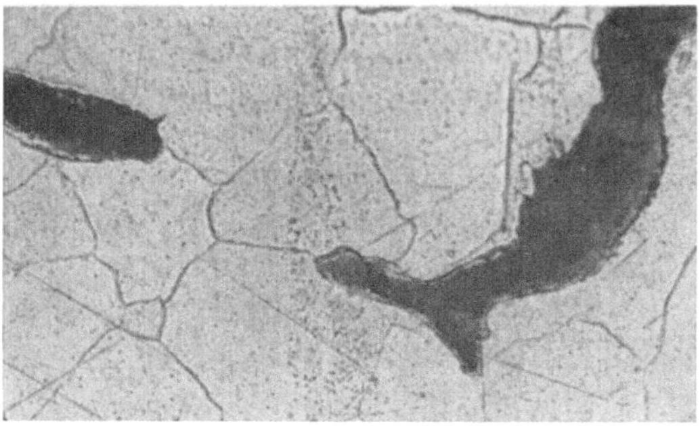

Fig. 305. Corrosion of a large area plate. Intercrystalline attack is indicated. (According to BÜCKLE and HANEMANN). ×500

LANDER [726] investigated the anodic corrosion and the growth of test strips, both at a constant potential of 1.05 V against the $Hg/HgSO_4$ electrode (corresponding to 1.67 V against the normal hydrogen electrode), as well as on cyclic charging and discharging. Here the effect of the nature of the alloy on corrosion and growth was prominent. If

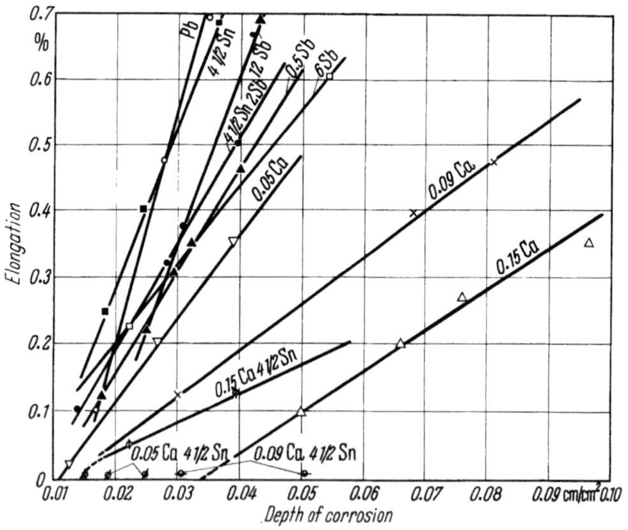

Fig. 306. Relationship between the longitudinal growth and the depth of corrosion of battery grid ribs (cross-section 1.27 mm²) of various alloys. Tests of 20 days duration at about 50 °C. (According to LANDER [726]).
The dimension giving the depth of corrosion is taken from LANDER.

variations in LANDER'S results are disregarded, the tendency to growth seems to decrease with increasing tensile strength of the alloy. This relationship is plausible, as the cause of growth is to be seen in the force which an oxide film of suitable thickness exerts on the grid; nevertheless, consideration of creep strength instead of tensile strength would have been more appropriate. In addition, LANDER found a certain proportionality between the depth of corrosion and the amount of growth (Fig. 306). The growth of the plates is disadvantageous, insofar as it spoils the contact between the paste and the grid and between the individual particles of the paste. This is prejudicial in the charging of the plates, and gassing leads more easily to a deterioration of the active paste. In a plate with 8% Sb the beginning of growth was seen by a fall in capacity; at a growth of 1%, a marked defect was apparent. Alloys with about 4.5% Sn and strengthening additions of calcium are particularly recommended in respect of corrosion stability. According to tests (casting, mechanical and corrosion) conducted by PARR, MUSCOTT and CROCKER [932] on alloys with antimony, tin, calcium, barium, tin + calcium, and tin + barium, an alloy with 3 to 3.5% Sn and 0.07 to 0.1% Ba exhibits particularly good prospects for battery grids. Apart from this, the work of LANDER [726] contains data on the depths of corrosion to be expected in different types of cells. They are in the region of $2 \cdot 10^{-3}$ to $16 \cdot 10^{-3}$ cm annually.

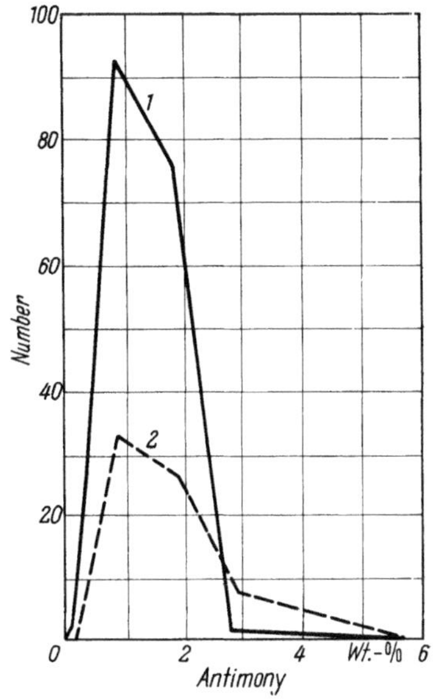

Fig. 307. Curve *1*: Number of visible cracks after stamping in the frames of each of 17 grids in relationship to the antimony content. *2*. Number of frame-fractures after 170 discharges on each of 5 plates. (According to HOEHNE and v. SCHWEINITZ).

GROSHEIM-KRISKO [442, 443] investigated the effect of the antimony content in antimonial lead grids. He found a region of particularly low durability at antimony contents between 2 and 4%. Antimony contents above 10% were also found unfavourable. Plates with antimony content between 6 and 8% behaved best in service. HOEHNE and v. SCHWEINITZ [544] investigated the region of alloy composition between 0 and 6% Sb and found, according to Fig. 307, particularly bad behaviour of the grid around 1 to 2%. DROTSCHMANN [259] gives data on the distortion and growth of lattice grids with 5, 8 and 12% Sb. The grid with 5% Sb comes out far the worst. According to the data of Dassojan [233], the corrosion-resistance of antimonial lead grids decrease uniformly with the antimony content (cf. p. 283). According to Grosheim-Krisko [442, 443], bismuth additions of 0.047% in antimonial lead grids with 6% Sb did not have any injurious effect, but 0.1 and 0.5% Bi reduced the durability of the positive plates. Mashovets and Lyandres [801] established an increase of corrosion by addition of bismuth and of zinc in lead-antimony anodes.

Grosheim-Krisko [442, 443] also investigated the effect of bismuth additions in large area plates, and found a marked increase of corrosion and of growth at a bismuth content of 0.047%. Hoehne [541] established similar facts regarding the effect of bismuth on distortion and growth, corrosion and sedimentation of large area plates. Under his experimental conditions, he found a detrimental effect with 0.036% Bi. At low loads the detrimental effects were not so marked as at higher demands. Moreover, the plates containing bismuth at times showed (after very deep discharges on subsequent charging with high current) an increased charging voltage during the charging operation.

Hofmann and Petri [573] investigated the behaviour of large area plates of refined lead with small additions of silver, tin, arsenic, copper, zinc, mercury, nickel, cobalt and antimony in respect of capacity, sedimentation, tendency to buckling and growth, and corrosion, Dassojan [233] tested large area plates with 0.05 to 0.1% tellurium. As compared with commercial lead plates, the mechanical properties are improved, but the capacity somewhat reduced. Hoehne[1] likewise found a smaller growth in large area plates of commercial lead with 0.05% Te. As regards capacity, no difference could be established in comparison with commercial lead plates; however cells with positive plates of commercial lead containing tellurium showed a decrease of the charging voltage on the negatives of 0.14 V.

Self-discharge normally occurs by formation of short-circuit elements in the positive and negative plates. The short-circuit element in the positive is formed by the lead framework and the PbO_2 of the paste, which possesses metallic conductivity. It therefore produces, so to speak, a small-scale accumulator. Current arises from the same reaction as in the normal discharge: $PbO_2 + Pb + 2H_2SO_4 = 2PbSO_4 + 2H_2O$. This is one cause of the corrosion affecting the lead framework of the positives. The process is the more active, the greater the contact-surface between the paste and the grid. The contact-surface is smallest in framed plates, greatest in large area plates.

The short-circuit element in the negatives is formed on the one hand by the lead of the paste, and on the other by the inclusions of antimony

[1] Private communication.

and other metals which are in the paste electropositive to lead. In this process lead is converted to lead sulphate with evolution of hydrogen, the latter escaping at the electropositive metal or at inclusions with low hydrogen overvoltage (p. 278). The process is accelerated if the hydrogen is oxidised by access of air (THOMAS [1184]). The inclusions in the lead paste which have been mentioned may be considered, with regard to the evolution of hydrogen, as gassing metals. Insofar as they derive from impurities in the acid, e.g., copper sulphate, they are avoidable. On the other hand, the formation of antimony inclusions in the negative paste of pasted plates cannot be prevented, as antimony is anodically dissolved from the positives and in part precipitates on the negatives (CRENNELL and MILLIGAN [222], HARING and THOMAS [491], VINAL [1223]). The path of antimony in the storage battery was ascertained by HERRMANN and PRÖPSTL [513] by addition of a radioactive isotope. The antimony found in the paste of the plates derives in greater part from the grid of the positive plates. The main quantity of the antimony is found in the positive paste, a smaller part also gets into the negative paste, and into the separators and the acid. Probably primary antimony hydride forms on the negative grid. The self-discharge was investigated by weighing the plates (VINAL and RITCHIE [1225]).

The following proportions of elements, in ppm in the acid, double the rate of self-discharge (VINAL and SCHRAMM [1226], GILETTE [383]): 0.3 of platinum, 10 of antimony, 15 of arsenic, 30 of tungsten, 150 of copper, 200 of silver, 1,000 of tin, 1,000 of bismuth. A content of iron in the acid is particularly harmful in respect of self-discharge and sedimentation. The effect is due to the fact that the iron ions are alternately oxidised at the positives and reduced at the negatives. Further reports on self-discharge in accumulators are given by RÜETSCHI and ANGSTADT [1037] and by GABRIELSON [354].

AGRUSS and co-workers [11] added bismuth, silver, copper, nickel, antimony and tellurium to electrolytic lead, made lead oxides from it, and pasted grid plates of starter batteries with the latter. The final charging voltage of the cells was considerably reduced by additions of tellurium and nickel of about 0.001%. At higher contents, copper, silver and antimony also exerted a noticeable influence. Self-discharge measured by the fall of acid density during a period of 28 days at 32°C, was greatest in the case of additions of tellurium and nickel, and also antimony. Here silver and copper produced no effect of any kind. It was also found that the effect of a combination of several additions always exceeded that of the individual ones. In all the investigations, bismuth was without effect.

HOEHNE [543] made a full examination of the condition of pasted lead storage batteries during and after long periods of rest. In practice, the most important process is the fall in metal content and in the strength of the positive grid which occur in batteries at rest. On starting after a rest period of up to 400 days, other unfavourable changes were found in the batteries. It is therefore advisable to charge and discharge during the rest period. If a rest period of more than six months is in prospect, then valuable batteries should be taken to pieces and the parts stored.

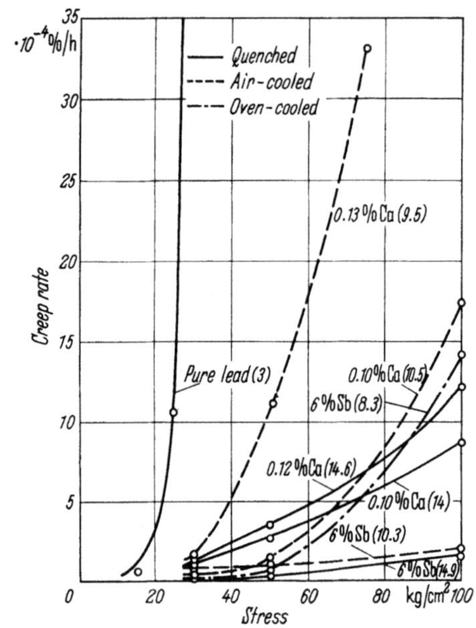

Fig. 308. Creep strength of cast battery alloys. Hardness shown in brackets. (According to Heckler, Hofmann and Hanemann [502]).

In the years before the last war, tests were carried out on the reduction of self-discharge by the use of alloys with electronegative constituents. The alloy with calcium seemed particularly promising. Indeed investigations by the Bell Telephone Laboratories [74] showed that the self-discharge amounts to only one-fifth of that of antimonial lead grids and that the alloy is of sufficiently high hardness [491]. Practical experience was not at first wholly satisfactory, as buckling of the plates frequently occurred on long use (cf. [540]). In fact, investigations of the creep strength of such alloys in the cast state showed lower values than in cast lead-6% antimony (Fig. 308) [562]. Moreover, the hardness did not reach the values usual here. Further additions of lead-calcium have been recommended [17, 14], latterly, e.g., 1.5% Sn, 0.05% Al in lead with 0.08% Ca [1235].

Other results with grids of lead-calcium and with lead-calcium ternary alloys are available in Hoehne's papers [542]. According to him, positive grids of lead-calcium are unsatisfactory in charging and discharging operations. They exhibited increased corrosion, the active paste gave more sediment and after deep discharges abnormally high charging voltages occurred. These disadvantages could be mitigated by further constituents. Apart from this, abnormally high voltages after deep discharges were observed by HOEHNE and v. SCHWEINITZ [544] in positive plates of commercial lead and of antimonial lead with less than 3% Sb. This phenomenon was traced back to the formation of a badly-conducting surface film on the grid. THOMAS and co-workers [1186] reported results with lead-calcium alloy in storage batteries for floating operations. The life of such plates is limited by the growth of the positives. If their expansion reaches 5%, the active paste is lost and the grid framework often breaks to pieces. Lead-antimony alloy cells reach this point in 9 to 10 years. If the above value for the expansion is considered as the permissible maximum with which good output is still maintained, then for cells with lead-calcium (0.065 to 0.090% Ca) grids a life of 20 to 30 years seems not unattainable. Drs. SCHUHMACHER and BOUTON of Bell Telephone Laboratories have told the author that batteries of this kind have been successfully introduced in the internal operation of the Bell Telephone System.

A series of developments concern the improvement of the lead-antimony grid by further additions to the basic alloy.

MASHOVETS and LYANDRES [801] mention that addition of silver up to 1% reduce the attack on lead-antimony. A similar effect of an addition of 0.1% Ag was established by FINK and DORNBLATT [317]. STOERTZ [1148] obtained a patent for improvement of the alloy with 6% Sb by addition of about 0.5% As + 0.04 to 0.6% Ag (compare KAWABA [649]). According to VINAL [1223], tin improves the casting properties of the alloy, presumably by retarding oxidation (p. 328), and is added in amount 0.15 to 0.50%. Traces of copper are regarded as favourable [1223].

As regards low self-discharge and low hydrogen evolution, an alloy with, for instance, 5% Sn and 0,075% Se is said to be favourable [174].

DASSOJAN [233], in a comprehensive investigation, confirms most of the facts stated above. He also directs attention to an alloy with 2% Sb and 1% Cd for grids. He offers his own experimental results on the reduction of the corrosion of antimonial lead grids by silver contents up to 1%. Such an addition can only be tolerated by industry when the life of storage batteries is limited by an insufficient corrosion-resistance of the positive grids. The corrosion-resistance of storage battery grids is also said to be increased by an addition of cobalt sulphate to the acid; however the addition has an unfavourable effect on the wooden separators (KRIVOLAPOVA and KABANOV [708]). For the grids of storage batteries an alloy with 4% Sb and additions of tin, silver and arsenic

(ZIEGFELD [1304a]) is used instead of the lead-antimony alloy with 7 to 12% Sb which was customary. Similarly, ZAHN [1299a] recommends a lead alloy with 1.5 to 4.5% Sb, 0.10 to 0.50% As, 0.1 to 0.50% Sn and 0.005 to 0.05% Cu. By this composition, self-discharge is reduced and life improved, without partial sacrifice of hardness, strength and castability.

Finally, BAGSHAW and HUGHES [33a] found that lead strengthened with lead-oxide dispersion was suitable as grid material.

There are specifications for the purity of storage battery acid and for water used for topping (VDE [1217]). Small contents of acetic acid become anodically oxidised to CO_2 in charging, so that only larger quantities have a permanently harmful effect [259].

HOEHNE [539] makes a survey of additions to sulphuric acid and their effect in corrosion and sedimentation of the positives, shrinkage of the negatives, self-discharge, capacity, voltages, evaporation and spraying.

Additions of cobalt sulphate and silver sulphate considerably reduce the corrosion of the positive plates in starter batteries in the overcharging test. LANDER [728] showed that this effect did not occur in practice and attributed it to the fact that only 6% of the grid-corrosion arises from the overcharge, while, on the other hand, 89% of the corrosion processes occur during the rest period, and the latter are not affected by these additions.

Finally, attention may be drawn to the measures necessary in storage battery manufacture for protection against lead poisoning. A series of practical precautions are recommended in a report by DAUBENSPECK and McCLURE [234].

3. Pressure Die Casting

The pressure die casting of lead alloys is easy, as the melts hardly attack iron kettles and show only slight shrinkage in solidifying and cooling. Pressure die casting was therefore first carried out with tin alloys and lead alloys. The most important application, the casting of printing type, is treated in a separate section. For this reason, it is merely mentioned here that the alloys for general applications have the same alloying elements. According to the German Standard Specification DIN 1741. "Lead Die-Casting Alloys", still valid in Germany, they contain lead with 4 to 41% Sn, 2 to 14% Sb, 1.5 to 3.5% Cu. Two of the alloys mentioned in the specification are practically binary alloys with about 3 and 13% Sb respectively; one alloy (with 10% Sb and 5% Sn) is almost the lead-antimony-tin ternary eutectic. Alloys with high tin contents and additions of copper are also listed in the specification. ASTM Designation B 102-52 contain a binary alloy with 15% Sb and a ternary with 15% Sb amd 5% Sn. The contents of aluminium and zinc are limited to 0.01% of arsenic to 0.15% and of copper to 0.50% (p. 363).

Lead pressure die casting can be used, for instance, for parts of measuring apparatus or meters, where low strength but high accuracy is required. The German standards also contain strength values and foundry data as well as specifications for the production of test-pieces. As regards pressure die-casting technique in general, attention is drawn to Frommer [350], Richter [1010], AWF [32] and [301]. Halliday [474] offers an example of the casting of a conical stopper.

4. Lead Alloys in the Printing Industry

a) Working Methods in General. Until a century ago, letters were exclusively cast by hand and assembled to form sentences. The casting of individual printing types for hand setting is still carried out by so-called automatic casting in type foundries or in foundries of printing works. It is done on typesetting machines, that is, single-letter casting machines which produce a series of castings from an applied matrix. The production of the casting moulds, their dimensional testing and the mode of operation of a single-letter-setting machine is exactly described in [837]. The matrix, the most important part of the mould, which produces the letter, is made, for instance, by applying an electroplate of nickel followed by copper to a positive made of low-melting alloy. The positive itself is obtained by casting into a brass negative. The dimensional accuracy of the moulds is given as 5 μm; the temperature of the melt is 340 to 400 °C. A usual machine (Bruce type) treats about 30 kg of metal daily, the rate of casting amounting to one shot per 5 to 10 seconds, depending on the size of the type.

A large proportion of type metal is converted in the printing establishments themselves to printing type on composing machines. A distinction is drawn between machines for setting and casting lines (linotype, intertype and typograph), and those for individual letters (monotype) (Bretag [132]). In the first case the brass matrices are in a magazine. They are assembled to a line by pressing a key. From this a casting is immediately produced, the molten metal being injected into the mould under pressure. In the monotype process, the key and the casting machine are separated. By means of the key, holes are first punched in a paper strip, the position of the holes corresponding to a certain series of letters. The strip of paper is inserted into the casting machine. The holes permit the passage of compressed air which actuates the movement of the matrix frame in two directions at right angles to one another. The matrices appear individually before the casting orifice. The cast letters are assembled in the machine to lines and the latter to columns.

If it is desired to produce many copies in a very short time (as in printing newspapers, periodicals, forms etc.) then stereotyping is appropriate. The original setting, together with the blocks contained in it, is moulded in a specially prepared paper pulp under a hydraulic or mechanical press either at room temperature or with moderate heat and a matrix made of the whole area. The plates of the press are heated electrically or steam-electrically whereby, it is claimed, very small differences of temperature can be attained (MAN)[1].

The matrix forms the one side of a mould. Printing plates corresponding to the original, so-called flat stereos, are cast in it. For rotary printing semi-circular stereos are made instead of flat stereos. For this purpose the matrix is curved and put in a sector of a cylindrical mould. In semi-automatic or in automatic casting machines, the metal solidifies under its own weight and the additional pressure is produced by the machine. Up to four plates a minute can be cast. Separate mouldings are saved because the units are integral with the machine[1].

In this connection electrotyping must also be considered. Half-tones are used for printing pictures and for high demands on the quality of print, such as, for illustrated papers, advertising, securities etc., instead of stereos, or of original type. The original, that is, a type-setting, wood-engraving, half-tone block or line block is moulded in wax, celluloid or thermo-plastic foil, or, by the method of Albert and Fischer (Gygax [459]), in lead. Prior to moulding the lead surface is sensitised with graphite and non-conducting matrices are made conductive by the silver-spray method, and copper (and sometimes subsequently nickel) applied by electrodeposition. The finished half-tone block is produced by peeling the copper deposits from the matrix, tinning it on the reverse side and strengthening it with backing metal.

b) Composition, Structure and Properties of Alloys. Type metals are almost exclusively lead-antimony-tin alloys. For ease of reference, common compositions of the most important kind of alloys (DIN 16512, "Lead Alloys for the Printing Industry", October 1954) are recorded in the co-ordinate system of the ternary diagram (Fig. 309). By comparison of this presentation with the diagrams of the liquidus temperature (Fig. 174) and of hardness (Fig. 182), the upper melting points of the individual alloys and their hardness values can be read off. The solidus temperatures, that is, the temperatures of the end of solidification, coincide for nearly all the alloys with the ternary eutectic temperature

[1] Information kindly provided by Maschinenfabrik Augsburg-Nürnberg.

of 239 °C. Only in the backing metals, the end of solidification was established in the neighbourhood of the quasibinary eutectic, at 246,5 °C. The fields of application of the different alloys can be derived principally from the melting and casting properties, the microstructure and the hardness values. The determination of hardness is most suitable as a

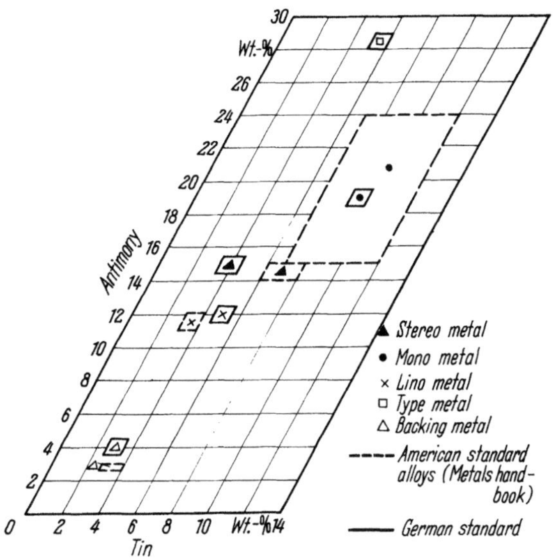

Fig. 309. Composition of the most important type metal alloys in the coordinate system of the lead-antimony-tin ternary diagram.

method of testing the demands of pressure and friction on the type, and it has the advantage of being easily ascertained. As shown in Figs. 182 and 183, in the region of small antimony contents, the hardness is, in the first place, raised by an increase of the antimony concentration, on the other hand, at higher antimony contents (from about 10 to 15% Sb) mainly by an increase in the tin concentration.

Owing to the fineness of grain, it is difficult to study the microstructure of cast type. For this purpose it is better to use alloys which have not been cooled too quickly as was done in the construction of the ternary lead-antimony-tin system (p. 163). By means of the photomicrographs given there, the structure of type metals can also be easily understood. The group of backing metals lies in the primary solidification region of lead. Their structure is therefore similar to that illustrated in Fig. 176, although the proportion of the constituents is different. The low hardness of the alloys causes no difficulty, as the printing surface consists of copper or nickel.

The linotype alloys are in the vicinity of the ternary eutectic, which is illustrated in Fig. 177. They are therefore distinguished by a particularly low melting point and good fluidity as the result of a small solidification range (p. 335 and Fig. 298b). Furthermore, segregations can never occur in the eutectic alloy. Similar compositions with a tin content about 3% lower are used for blank material with which spaces between letters, words and lines in the printing mould are filled.

Of the alloys for the monotype process and for stereos, those with tin contents of 5 or 6% lie in the primary solidification field of antimony. Their structure therefore corresponds approximately to that illustrated in Fig. 310, if the details of the course of solidification are disregarded. The alloys with higher tin content lie, on the other hand, in the primary solidification field of SbSn (Fig. 179). The stereotype alloys, and especially the monotype alloys, are more highly alloyed and therefore harder than the linotype alloys. In spite of the greater solidification range of the monotype alloys, there is no risk of harmful segregation, as cast individual letters solidify more quickly than lines. The higher hardness of the monotype alloys is necessary for the reason that the mechanical stress (particularly owing to the moulding of the matrices) is greater in individual type than in whole lines. In stereos the stress is particularly great in the printing process with regard to the large editions of newspaper, etc. Moreover, as far as flat intermediate stereos are concerned, they must withstand the lead moulding in the production of half-tone blocks. Use is sometimes also made, in stereotype alloys, of the effect of a small copper or nickel addition for increasing strength and reducing segregation, as in bearing metals (Bargilliat [43]). The same applies to backing and to founder's type metal and music printing metal, yet to be described. The German Standard specification DIN 16512 of October 1954, "Lead Alloys for the Printing Industry", allows in these four groups of alloys, maximum contents of copper plus nickel up to 0.3% whereas the copper plus nickel content in the other alloys described is limited to 0.05%. The alloys of greatest hardness are specified for casting by hand-setting, as after setting the type is not remelted but stored and used again and again. The usual German founder's type metal,

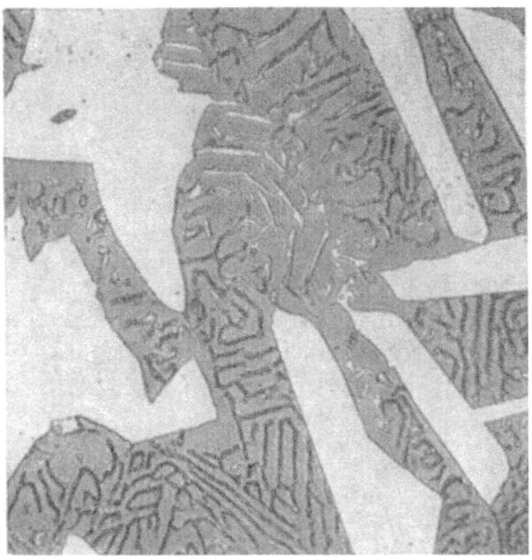

Fig. 310. 28.5% Sb, 5.5% Sn. Primary antimony, but predominantly ternary eutectic, whose antimony component is mainly crystallised on the primary antimony. ×500

with about 28% Sb and 5.5% Sn, is in the primary solidification field of antimony. Its structure corresponds with that shown in Fig. 310.

For printing music, engraving plates of a few mm thickness are first rolled. The notes are hand-engraved, and the text punched in with a die. Defects are corrected by flattening the appropriate places with a hammer. Having regard to the necessary good rolling properties, an alloy of high tin content is used, for example the German standardised alloy with 15.5 to 16% Sn and 4 to 5% Sb. The surface of the engraved note plate is transferred to a zinc etching plate [1200].

In hardness measurements on type metals, a small degree of age-hardening must also be expected, which in the main takes place in the first two days after casting. Thus in this period the hardness of a stereotype metal with 13.1% Sb and 4.4% Sn increased from 29.1 to 30.1 Brinell units (SCHWARZ and WINKLER [1094]). The age-hardening is proportionately greater in alloys with lower antimony and tin contents, such as, in backing metal. For an alloy of composition 3.1% Sb and 2.7% Sn a heat treatment, consisting of 15 minutes at 235 °C and water-quenching, has been proposed. While the hardness on the back of a plate which had merely been cast increased in two days from 15.9 to 18.2 Brinell units, a value of about 25.9 Brinell units was attained in the same period after heat treatment. The view has also been put forward that it is the microhardness of hard inclusions, and not the Brinell hardness of type metal, which is decisive for its wear resistance (CARTLAND [182]). As the microhardnesses of antimony and of SbSn have proved to be about the same (p. 385), this view can hardly be correct.

With regard to its stressing in the moulding of matrices, compression tests with fine measurement were made on type metal (NICOLAUS [897]). Here pressures up to 100 to 200 kg/cm^2 are said to occur, whereas the pressure in the high speed printing press is estimated at only 20 kg/cm^2. Pressure/compression diagrams showed that no elastic limit is present in type metal, so that permanent strains readily occur at low loads. As the behaviour when hot is still more unfavourable, a hardness as high as possible is required for types which are copied hot. If this is not possible, then only the production of so-called hand-struck matrices by means of brush beating can be considered.

c) Casting Properties and the Occurrence of Defects. The "castability" of type metal in relation to its composition was determined by CARTLAND [182] by means of a spiral chill-mould. According to Fig. 298b, a saddle of good fluidity (see p. 336) passes up from the binary lead-antimony eutectic to higher tin contents, there gradually flattening out. The absolute maximum coincides approximately though not exactly, with the ternary eutectic. Thus the highest castability is present in linotype alloys (p. 360).

The exact reproduction of fine detail of the mould has frequently been attributed to an alleged expansion of type metal melts on solidification. Measurements by MATSUYAMA [810] of the change of volume on solidification show that this is not the case. The contraction was determined as 2.04 to 2.06% for linotype, monotype and stereotype metal, and as 2.61% for backing metal (4% each of Sb and Sn). The values for the first-named alloys are nevertheless essentially lower than the value

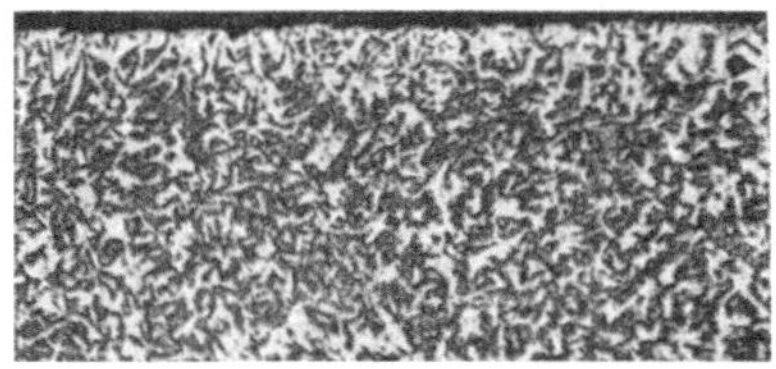

Fig. 312

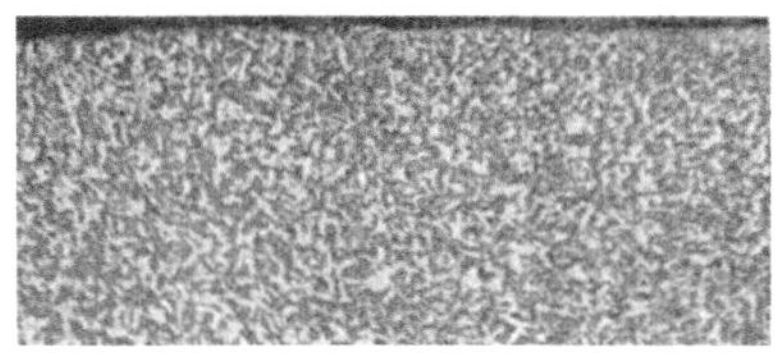

Fig. 311

Fig. 313

Fig. 311. 27.5% Sb, 6.11% Sn. Section parallel to the spotty surface of a letter. Bright and dark places. ×100

Fig. 312. Same specimen as Fig. 311. Section perpendicular to the surface of a bright place. ×300

Fig. 313. Same specimen as Fig. 311. Section perpendicular to the surface of a dark place. ×300

of 3.44% for pure lead. Therefore the cast types necessarily show a certain slight porosity as in casting the skin solidifies first and the inside, which solidifies subsequently, has no opportunity to compensate the loss of volume in solidification by sucking in melt from a reservoir which still remains molten. The small pipes are without significance, as they are generally in the interior and therefore do not interfere with the printing process. At the surface of the hypereutectic alloy, particularly the high antimony founder's type, the specifically lighter primary crystals (in this case the crystals of antimony) are heaped up, and form a firmly coherent hard crust. A similar phenomenon is produced in the centrifugal casting of white metal — that is bearing metals of similar composition — thus achieving an especially hard surface. An occasional defect in type is a spotted appearance, the normally bright surface being broken by places of darker colour (Fig. 311). Figs. 312 and 313 contain cross-sections through a bright and a dark place in the type surface. The antimony enriched portions (mentioned above) do not occur in the dark places. Obviously the cooling in the mould was not uniform, perhaps as a result of impurities in the mould surface.

The casting temperature of type metals is given by the liquidus temperatures which can be derived from Fig. 174. In the region of a composition particularly in question, they are somewhat raised by an increase in the antimony content, and on the contrary lowered or a little affected by an increase in the proportion of tin. The casting

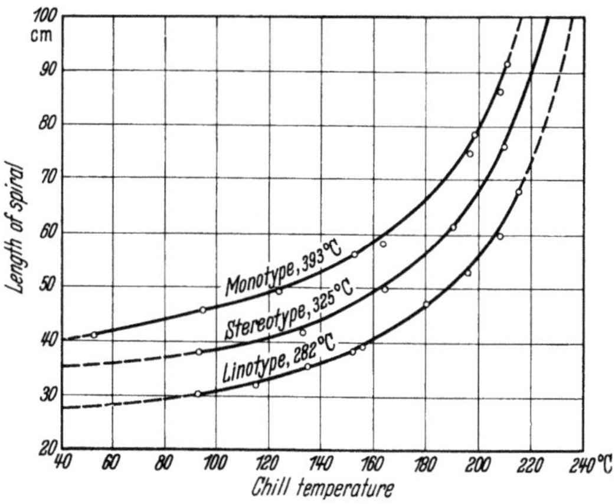

Fig. 314. Relationship between mould temperature and castability in different type metals. (According to Epstein).

temperature should be about 50 °C above the upper melting temperature (Epstein [284]), and carefully checked or automatically regulated. For example, in the linotype setting machine, it is 285 ± 4 °C. High casting temperatures certainly improve the fluidity, as measured in a spiral chill-mould, but have the disadvantage of greater wear of the casting machines, and cause dross-formation, absorption of gas and spongy castings. As a result of its tin and antimony content, type metal slightly attacks the iron of casting machines. The wear of cast iron is less than that of steel (Timmerhoff [1199]). The relationship between mould temperature and "castability" (p. 334) is shown in Fig. 314.

Type metal must endure repeated remelting, as the cycle of casting, printing and remelting often recurs daily. Contact of the surface of the melt with air should therefore be restricted as far as possible. Remelting often takes place with application of fluxes which dissolve oxides. The impoverishment in tin and antimony which nevertheless unavoidably occurs in time, is compensated by occasional additions of rich lead-antimony and lead-tin alloys. Aluminium and zinc rank as particularly injurious impurities in type metal, as they give rise to considerable drossing and therefore bad castings. Consequently the maximum content of the alloys in both these impurities is limited to 0.01% in German standard specification DIN 16512. Stahl [496, 1143] investigated the effect of temperature and alloying additions on the drossing of type metals of the usual commercial compositions (linotype, monotype and founder's type). He used two different methods of experiment:

1. Weighing of the oxygen absorption of tranquil melts,

2. Weighing of the oxygen absorption on passage of air into the melts.

Long-time tests of 50 hours on tranquil melts showed that up to 400 °C no oxidation of the alloys occurs. This was traced back to the immediate formation of a compact oxide film probably consisting of SnO_2. As the temperatures usual in practice are below 400 °C, the oxide film ought to protect the molten type metals against oxidation. Not until 450 °C could oxidation of the alloys be detected gravimetrically. The increase in weight, that is, the amount of oxide formed, was proportional to the square root of the time, and indeed the same proportionality constant was given by all the alloys. It follows from the parabolic time law that the oxidation is controlled by a diffusion process (p. 327). Most of the experiments were carried out by method (2) above. By investigations on monotype metals 6/15 and linotype 5/12 at 350 °C, the accuracy of measurement of the method was tested and it was proved that the maximum thickness of the oxide skin formed by passing in air was 10 µm. At 300 °C the alloys mentioned already oxidised as strongly as at 350 °C. However at 400 °C a fundamentally more intense oxidation sets in.

Alloying additions of 0.03% Ag; 0.1% Bi; 0.02 and 0.05% Al were found to be without effect on the oxidation of monotype metal 6/15 at 350 °C. Contents of 0.08 and 0.4% Al brought about a trifling increase of the absorption of oxygen. Addition of 0.1% Zn brought about a doubling of the oxygen absorption. From the volume of dross formed on passing in air, it could be concluded that 0.05% Zn already greatly increased the tensile strength of the oxide film on monotype metal 6/15. The detrimental effect of small zinc contents was thus confirmed, whereas under the experimental conditions chosen, aluminium impurities were found to be scarcely disadvantageous.

Arsenic in small amounts increases the hardness and improves the fluidity of the alloys. Consequently the German standard specification permits a maximum content of 0.3% As. Whether arsenic promotes the corrosion of casting machines (Epstein [284]), is not demonstrated. Bismuth is also said to improve the castability of type metal [147]. After earlier investigations by White [1263], Stahl [496, 1143] also considered the effect of additions on the surface tension of type metal. The frequency of bubbles on passage of air through the melt served as a unit. According to his (partly published) results, the surface tension of molten monotype metal is increased by a zinc addition, but, on the other hand, reduced by phosphorus. Although freedom from sulphur was formerly required in all type metals, this requirement cannot be maintained according to present experience. According to the investigations of LÖHBERG and SCHULZ [758] (lead-antimony-tin), as confirmed by STAHL in an unpublished study, small sulphur contents of the order of 0.01% promote the fine-grained solidification of printing metals with primarily solidifying antimony. For this reason the question of the sulphur content of stereo, mono and founder's type metal has been left open in the German standard specification (p. 166).

The literature contains a few further hints on the casting of type metal. In typesetting machines the melt enters the mould through the feed tubes slightly above the liquidus temperature (MUNDEY, BISSETT and CARTLAND [883]). If cooling occurs, owing to air Currents, crystals are said to deposit in the small cross sections, and result in piping, owing to reduction in the supply of melts. Piping can also result from too slow escape of air from the mould.

5. Bearing Metals

a) Mode of Operation of Plain Bearings. The following account closely follows that given by PEKRUN [943].

The bearing metal is generally understood to be the softer material of any pair of metals which comprise a bearing. In a journal bearing it is, therefore, the material forming the shell or bushing. However the duties and the performances of bearing materials cannot be considered by themselves, but only in conjunction with the other bearing surface and with the lubricant between them. The characteristics of the bearing metal must be regarded as depending, to some extent, on the properties of the shaft material. In the latter case, however, other considerations predominate, such as the transmission of torsional moments, the resistance to bending moments and shearing forces within the limits of allowable deformation and without the risk of a sudden failure or fatigue fracture. A plain bearing which is to run permanently must be designed and lubricated so that the bearing surfaces are completely separated in service by a film of lubricant. Sliding with partial, or quite completely, solid contact generally leads to high wear, usually to high friction and to a considerable temperatuie rise. Disregarding hydrostatic bearings, in which the pressures of the lubricant, required for carrying the load, are produced outside the bearing by a pump (BEUERLEIN [79]), plain bearings must themselves produce the pressure of the lubricating film in the bearing space by hydrodynamic action. The region between the bearing surfaces, the bearing space, must have a shape suitable for this purpose. Frequently, use is made of a wedge shape, narrowing in the direction of motion (Fig. 315). The attainable hydrodynamic pressures, and consequently the bearing capacity, are the higher, the smaller the narrowest gap of the bearing space. With regard to the shape in the axial direction, the most favourable proportions are obtained if the width of the gap is the same over the whole breadth of the bearing. Deviations from this ideal shape occur, not only if the sliding surfaces are oblique to the axis or are curved, but also if the method of production gives rise to wavy surfaces or roughness of the bush and journal.

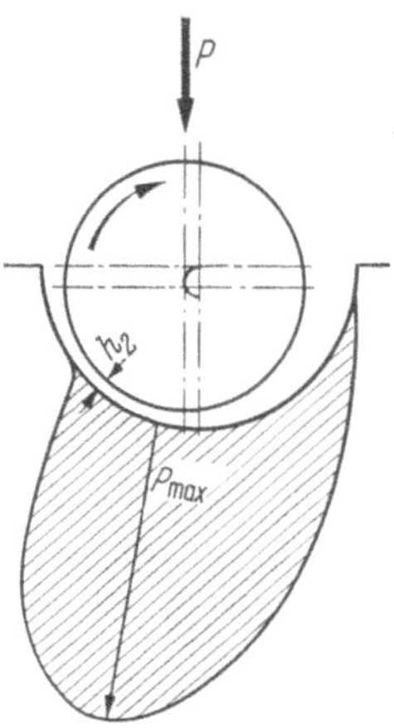

Fig. 315. Characteristic pressure distribution in the lubrication-gap of a cylindrical bearing.

In journal bearings, the wedge-form of the gap is usually attained by the eccentric position of the journal in the bush. For this bearing, BUSKE [158], among others, has shown experimentally the remarkably great effect of configuration on the bearing capacity. In a series of experiments with a rigid bearing body, the bearing capacity of the bushing decreased in the following order: leaded bronze (750 kg/cm²), aluminium alloy, tin bronze, white metal, grey cast iron, steel (50 kg/cm²). With medium stiffness of the bearing body, all the bearing materials showed an increase of loading capacity—the hard bearing materials, however, much more than the soft—thus in a certain example the sequence was: steel (1,200 kg/cm²), cast iron, leaded bronze, aluminium alloy, tin bronze, white metal. With a very flexible bearing body, in respect of loading capacity, the same sequence of bearing metals, as with the rigid bodies, occurred again. Here the loading capacity of the leaded bronze was 500 kg/cm². By measuring the distribution of pressure in the bearing space, and confirming traces of galling, it was established that, with the rigid bearing body, the lubricating gap was narrower at the end of the bearing than in the middle, due to bending of the shaft. With very flexible bearing bodies (small wall thickness of boss), the ends of the bosses deflected too far under the pressure of the lubricant, so that the bushing was more bent (in the axial direction) than the shaft and the lubricating gap was wider at the ends than in the middle. With medium rigidity of the bearing body there was about constant width of gap over the whole breadth of the bearing. The hard bearing materials with high modulus of elasticity, especially grey cast iron and steel, only gave the high loading capacity in an occasional, very narrow, region of thickness of the boss wall. Consequently the effect of the shape of the bearing is particularly important in such a case. The soft bearing materials provide better compensation for local overloads by elastic and plastic yield. It is here too, that highest possible adjustment in the microgeometric region is possible, that is the compensation of waviness, ridges and roughness. Indeed, the loading capacity of soft bearing metals also increases with more favourable bearing configuration, but owing to limited strength, does not reach the high values of the hard materials. For white metal and leaded bronze, the loading capacity can be increased by using thin linings (see below); this, however, reduces the conformability.

As well as the shape of the bearing space, it is also important, for the development and maintainance of hydrodynamic pressures, that the bearing surfaces should be impervious to the lubricant under pressure. A reduction of loading capacity owing to porous bearing surfaces is to be expected, for example, in porous sintered materials (WIEMER [1271]), in flame-sprayed bearing layers, if the pores intercommunicate (VOGELPOHL [1233]), and in extruded plastics, which become permeable at high pressures (WENGER [1257]).

Raised bearing temperature not only reduces the strength of the bearing material but also lowers the viscosity and the chemical stability of the oil. The working temperature should, therefore, not exceed certain limits. In general, bearing materials must dissipate at least a part of the frictional heat by thermal conduction. Here, apart from the thermal conductivity, the thickness of the lining may be of importance.

Mixed friction is present when the hydrodynamic pressure has not kept the bearing surfaces completely separated, and, as a consequence, contact of the solid bodies takes place at different points. It occurs on starting and on stopping the shaft, on overloading, in the running-in process, and in deficiency of lubrication. It has frequently been pointed out (see particularly VOGELPOHL [1231]) that at the commencement of mixed friction (in the hydrodynamic region) there is not perhaps a "breakdown" of the film of lubricant, but that the hydrodynamic pressures continue to operate and that only a part (and at first, only a very small part) of the load is carried by solid body contact. As the coefficient of solid body friction is much higher than that of fluid friction, the solid body contact soon makes itself perceptible by the fact that the friction and consequently the heating become higher than should be the case according to the laws of hydrodynamics.

For journal bearings, VOGELPOHL [1231] gives a survey of the frictional behaviour and formulae for the coefficients of friction both in the hydrodynamic region and especially in the region of mixed friction. According to this, the course of the Stribeck curve (STRIBECK [1152]) —which represents the coefficient of friction in relation to the rate of sliding—for the region of mixed friction and for the immediately following hydrodynamic region simultaneously can be represented as a first approximation by a mathematical formula, which is based on physical considerations:

$$f = f_0 \left[1 - \frac{1}{S_0}\left(\frac{s}{h_{\text{eff}}} - 1\right)\right] + 3.0 \sqrt{\frac{\eta \cdot \omega}{\bar{p}}} \text{ for } S_0 > 1.$$

The symbols here have the following meanings:

f Coefficient of friction of the bearing
f_0 Coefficient of friction at rest
η Viscosity of the lubricant in the bearing space
ω Angular velocity
$\bar{p}$ Average surface pressure
ψ Relative play of bearing (= the difference of diameter between the bushing and the journal, based on the journal diameter)
s Radial play of bearing (= half the difference of diameter between the bushing and the journal)
h_{eff} Smallest effective thickness of lubricant film (the average value corresponding to the actual bearing space)
$S_0 = \frac{\bar{p} \cdot \psi^2}{\eta \cdot \omega}$ SOMMERFELD number (a dimensionless number).

The condition $S_0 > 1$, for the region of mixed friction and for the hydrodynamic region immediately following it, is always fulfilled in practice. Thus the bearing material has far-reaching influence on the values of f_0 and h_{eff}. Having regard to low coefficients of friction, both should be as small as possible. The surface roughness also plays a part here. For h_{eff}, the configuration of the bearing is important as well as the conformability of the bearing material, as was shown in the above-mentioned experiments by BUSKE [158].

In bearings which have already been run in, the main function of the bearing material is to ensure a safe transition through the region of mixed friction in starting and stopping. The bearing surfaces should, therefore, be free from scratches—which reduce the hydrodynamic loading capacity—and suffer no deformation. The temperature should not rise beyond a certain point. If the region of mixed friction is quickly traversed, the frictional heat plays only a minor part. The coefficient of friction is then not actually the deciding factor; in many cases, however, high coefficient of friction and surface damage go hand in hand.

In the running-in process, adjustments occur of the sliding surfaces to one another, both in regard to the macrogeometric shape (adjustment of the bearing material to an inclination or bending of the shaft, elimination of waviness), as well as in regard to the microgeometric shape (see above). Running-in can take place by plastic deformation and by mutual attrition of the bearing surfaces. In general, running-in occurs in the region of mixed friction with, therefore, partial solid body contact. An adjustment of the bearing surfaces to one another is, however, also possible without solid body contact, by the hydrodynamic pressure alone. In the direction of motion—that is, in journal bearings in a circumferential direction—adjustment of the bearing surfaces must not go too far. This could result in the partial or complete transformation of the wedge-shape of the bearing space into a parallel gap, in which a sufficient increase of pressure by hydrodynamic action would not be possible. This can occur in soft bearing materials under a load which is too high.

b) Properties of Bearing Metals and Manufacture of Bearings. Good surveys of the properties important for bearing metals are found, for example, in Metals Handbook [835a] and in Weber [1248]. They are described below.

Hardness, Plasticity Limits (yield point, compressive yield point). The loading capacity of a bearing depends on the compressive yield point of the bearing material, the relative magnitude of which can be estimated as a first approximation from the Brinell hardness. Here, however, not only is room-temperature hardness significant, but still greater importance should be attached to the hardness at elevated temperatures. The point of inception of melting is, therefore, significant not only for the manufacture, but, equally, for the evaluation of the bearing capacity of the bearing at high loads and speeds, that is, at high heat development. In testing the hardness and resistance to flow of alloys high in lead and tin, attention must be given to the time factor, that is to say, the duration of test must be kept constant, if results obtained with different materials are to be compared with one another (cf. page 203). The flow of a bearing material is rendered more difficult, the thinner the layer in which it is used. Bearing metals of low strength are, therefore, only used as a lining in a supporting shell. This point is again mentioned in the discussion of dynamic strength (below). Cuthbertson [227] gives data on the elastic properties of lead and of tin-base bearing metals.

The hardness of the bearing material and of its individual structural constituents is of further importance with respect to contingent ridge-formation and wear of the shaft material. For this reason unhardened shafts can only be used in combination with soft bearing metals, less with lead-copper alloys or tin bronzes.

Deformability. The elastic deformation of a bearing metal under a given surface load will be the more marked, the lower its elastic modulus. Likewise, its plastic deformation will be all the greater, the lower its yield point, that is, the more the latter is exceeded by the load. The deformability is, therefore, roughly inversely proportional to the hardness. Adequate deformability favours the adjustment of the bushing and the shaft to one another, as described above, and, therefore, the running-in, and reduces the sensitivity of the bearing to edge pressures. Simultaneously good deformability of the bearing material favours the embedment of hard wear and dust particles, so that wear of the shaft is reduced. This function of the bearing metal presupposes its use in sections of sufficient thickness. Embeddability is so much more important, if little reliance can be put on clean filtration of the oil and on exclusion of external dust and sand.

Dynamic Strength (fatigue resistance). Fatigue fractures, such as can occur as a result of the type of stress in a reciprocating engine, begin at the bearing surface and penetrate the bearing material until near to the bond between the lining and the steel backing, then proceed parallel to this suface, so that a thin layer of bearing metal continues to adhere to the steel. The lining becomes more and more brittle and finally shatters.

The low melting-point of lead gives rise to a reduction of the fatigue strength with rising temperature. A similar behaviour is also to be expected of the white metals based on lead and tin, while the effect of temperature on the fatigue strength of the alloys of high melting point, such as the leaded bronzes and tin bronzes, is less. The life of white metal linings under dynamic stress is increased by a reduction of wall-thickness. LOVE and co-workers [767] provide estimated values of the fatigue strength of bearing metal linings. DUCKWORTH and WALTER in [334a] determined a fatigue strength of 1.53 kg/mm^2 for a steel-backed, tin-base white metal bearing of 0.2 mm thickness. On reduction of the layer thickness of the white metal to 0.04 mm, the fatigue strength increased to 2.82 kg/mm^2.

In most cases, white metal is put on to a backing of steel, bronze or cast iron. To attain adhesion, the backings are cleaned and tinned. Cleaning can take place chemically, by pickling, or mechanically, for instance, by shot blasting. THWAITES (in [334b]) has shown that pickling leads to a stronger bond between bearing metal and the backing. The bearing metal is applied by casting in the space between the backing and a core, by pressure die-casting or by centrifugal casting. The latter method, in general, gives rise to considerable segregation in lead alloys, as the light phases such as SbSn rise to the surface.

The strength of the bond between a thick cast lining and the backing is tested either qualitatively by opening a half-shell or quantitatively by the method indicated by CHALMERS (in [334a]; p. 197).

In repetition work a continuous casting method is frequently used [1018a]. Thus a flat steel band runs through successive different baths for cleaning, electrolytic etching and tinning, before the white metal is cast on. Water-cooling provides for rapid solidification of the melt. In the continuous production method the excess bearing metal is machined off. For the finished bimetal strip a thickness tolerance of 0.03 to 0.04 mm is attained. Segments of strip are formed to half-shells under a press. These tolerances still permit, even in continuous production, the maintenance of a layer thickness of 0.1 mm of white metal. Frequently, the thin layers of a white metal are not applied directly to steel, but use is made of an intermediate layer of leaded bronze, which prevents damage to the shaft in the event of wear or of fusion of the white metal (three-layer bearings). In the so-called electrodeposited bearings, the white metal (lead with 8 to 12% Sn or with 8 to 12% Sn + 2 to 4% Cu) is electrodeposited in a thickness of 0.02 to 0.06 mm on leaded bronze. In order to prevent the diffusion of tin into the leaded bronze below it, a layer of nickel, the so-called nickel dam, is applied below the white metal layer. A three-layer bearing made in this fashion has, for instance, the following succession of layers: steel backing, leaded bronze layer 0.5 mm thick, nickel dam 1 to 1.5 μm thick, running layer of white metal 0.02 to 0.025 mm thick. (For lead-indium bearings, see p. 60).

Efforts to improve further the dynamic strength of white metal by alloying brought little success. For increased values of the fatigue strength, a change must, therefore, be made from white metals to copper alloys. For determination of the fatigue strength, bearing testing machines have been developed which imitate service stresses, or alternatively the service trial itself is used as a test method.

In addition to the fatigue strength, the impact strength of bearing materials, that is, the ability to endure a limited number of blows without fracture, is of importance. The performance of this test can take place in different ways, such as a dynamic compression test or on a Krupp repeated impact machine.

Thermal Expansion. The thermal expansion of the bearing material and of the shaft material is responsible for the alteration of the bearing play on change of temperature. Usually a low melting-point and a low elastic modulus are related to a high value of the coefficient of thermal expansion. The difference in coefficient of thermal expansion between the bearing material and steel also determines the magnitude of the heat stresses which set in between the bushing and the backing on change of temperature.

Corrosion-Resistance. If acid constituents are present in the oil, or are formed in service, corrosive attack on the bearing metal is to be expected. WILSON [1277] gives data on the mechanism of corrosion. In this respect, tin-base bearing metals are considered to be superior to lead-base metals. In the event of corrosion in the copper-lead alloys, a preferential dissolution of the lead portion is to be expected, so that in the surface layers, which are important for the running properties, an impoverishment in lead occurs.

Running Properties. A number of properties can be grouped under this heading, such as anti-seizure characteristics, low value of wear, absence of production of ridges on the shaft, good running-in properties, good emergency running properties—that is, no excessive heating-up with deficient or absent lubrication. These questions are treated in the next section.

c) Running Behaviour of the Bearing Material. Pekrun [943] gives a full treatment of the specific effect of the bearing material on the running behaviour. First of all, the different theories on this subject, as reported in the literature, will be considered.

The *hydrodynamic theory,* which was referred to earlier, makes no mention of bearing materials. Nevertheless these have, as shown above, an essential influence on the hydrodynamic behaviour owing to their contribution to the macro- and micro-geometric shape of the bearing space. In this respect, there may be some effect of the bearing metal on the formation of layers of increased viscosity which can play a part at very small gap widths, and, as a consequence, particularly in the region of mixed friction (Deryaguin [247]). A mathematical treatment of bearings is only possible on the basis of the hydrodynamic theory. An exhaustive presentation of bearing calculations, appropriate to practice, is given by Vogelpohl [1232]; cf. p. 367.

The *supporting crystal theory* is based on the structure of high-tin white metals, which were introduced by Babbitt as cladding of backings. The structure of the alloys includes harder crystals, chiefly of SbSn, in a somewhat softer eutectic matrix. The concept has been developed, on the lines of the work of Schmid and Weber [1071], that under the high pressures in the lubrication-gap, the soft matrix yields more than the hard included crystals, so that these appear in relief on the running surface. The result is that in the condition of mixed friction the areas of solid contact and of boundary lubrication respectively, are given by the supporting crystals, as a consequence of which wear is reduced. The formation of cavities further improves the lubrication, as they form additional oil pockets of microscopic dimensions.

Even if most bearing alloys have a heterogeneous structure and consist of soft and hard phases, there are however exceptions to this rule, such as the silver bearings. For this reason, the supporting crystal theory has now largely been abandoned (Holligan [581]). A direct confirmation of the supporting crystal theory was recently put forward in the above investigation by Pekrun [943]. The very sensitive interference micrographs of the running surface of a Weissmetall 80 bearing showed no indication of cavities (Fig. 316a) around possible "supporting crystals" of SbSn. In a metallographic section of WM 80, by contrast with this, the crystals appeared clearly after etching (Fig. 316b).

The welding theory was put forward and developed by Bowden and Tabor [119]. In their book the authors first show that the actual surface of contact between solid bodies is in general practically independent of

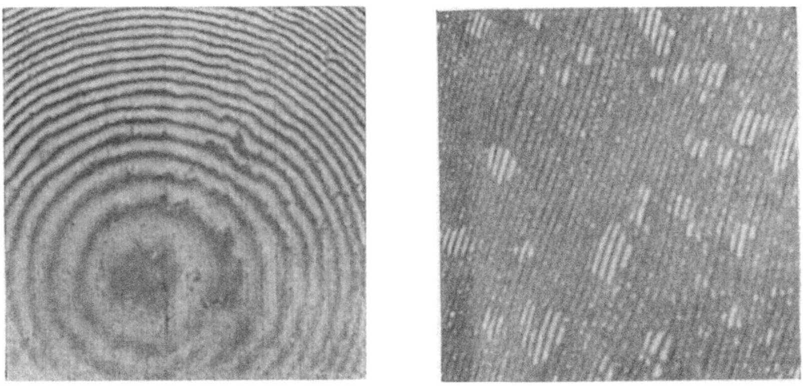

a b

Fig. 316. (a) Interference micrograph of the running surface of a rotating bearing on WM 80. Extension by the width of one band corresponds to a roughening-depth of 0.27 μm. ×150
(b) Microsection WM 80. Interference micrograph. SbSn crystals light. Etched. (According to Pekrun). ×150

the apparent surface of contact. Contact always takes place only at some projecting parts of the surface (asperitics). In general the consequent stresses far exceed the elastic limit. The average surface pressure at the points of contact, insofar as no work hardening takes place under pressure, is porportional to the yield stress of the softer material and independent of the load. Therefore the actual surface of contact is then proportional to the load and inversely proportional to the yield stress of the softer material. Bowden and Tabor [119] arrived at the following formula, which should apply to most practical cases and to all types of roughnesses:

$$A = \frac{W}{p}$$

with A = actual surface of contact
W = load
p = average yield-compressive stress of the roughnesses.

It has been concluded, from observations on metal transfer in the sliding of metals on one another, that local welding of the metal surfaces occurs at the asperities. Then work must be applied in sliding in order to shear off the welds and also to "plough off" the metal. The frictional energy of sliding without a lubricant is accordingly divided into a "shearing fraction" and a "ploughing fraction". The ploughing fraction depends on the ratio of the hardness of the metals sliding on one another and on the shape of the asperities. In general, it is small for hard metals. The shearing fraction S is equal to the product of the actual surface of contact A and the shearing stress required for shearing s; thus $S = A \cdot s$. It has been deduced from experiments that s is proportional to the shearing stress which is required to shear a cylinder out of the softer material. If the ploughing fraction is negligible, then we have:

$$F = S = A \cdot s \quad \text{with} \quad A = \frac{W}{p}, \qquad \text{therefore} \quad F = W \cdot \frac{s}{p}$$

(where F is the energy of friction; W = the load; s = shearing strength of the welds; p = yield-compression stress = compressive yield point of the softer metal). As the shearing usually proceeds inside the softer metal, the shear strength of the latter can (as already mentioned) be approximately substituted for s. Then the coefficient of friction becomes:

$$f = \frac{F}{W} = \frac{s}{p} = \frac{\text{Shear strength of softer metal}}{\text{Compressive yield point of softer metal}}$$

In soft metals s and p are both small, in hard metals both large. The coefficient of friction is therefore of the same order of magnitude for most metals; it lies between 0.6 and 1.2. A small value of s with a large value of p, i.e., a small coefficient of friction, can be achieved by putting a thin layer of soft metal on a hard basis metal. The opposite sliding member should be at least as hard as the basis metal. Then the high yield-compression stress p of the hard basis metal primarily determines the magnitude of the actual surfaces of contact, and the more so, the thinner the surface layer. In leaded bronze, over which a body had been slid, traces of lead were found smeared over the surface. Bowden and Tabor [119] assume that these give rise to a reduction of friction according to the above mechanism. If a lubricant is used, the behaviour of a bearing metal is essentially determined by the extent of absorption of the lubricant on the surface layer of metal. As metallic contact between the sliding surfaces easily occurs even if a lubricant is used, the basic frictional behaviour of the alloys is of the first order of significance for the running properties. From experiments on lead-base and tin-base white metals, with and without hard inclusions, it is concluded that the hard crystals contributed only little to the average hardness of the alloy and to the sliding properties. On the other hand—according to an investigation by Boas and Honeycombe [95]—it is held to be possible that hard particles in a soft ground mass make the alloy "more rigid" and that cavities and rifts are formed in the ground mass which can serve as minimal spaces for oil storage.

According to BOAS and HONEYCOMBE [95], the following phenomenon should be kept in mind: tin, cadmium and zinc crystallise in non-cubic systems. Their thermal expansion is therefore different in the direction of the individual crystal axes. Usually the orientation of two crystals meeting at one grain-boundary is not the same. Consequently, on heating, stresses occur which generally suffice to bring about plastic deformation and slip in the crystals. The extent of the deformations increases with the number of temperature changes. This phenomenon can be moderated by the hard crystal inclusions in tin-base white metals, as the hard constituents have a coefficient of thermal expansion which lies between the two main coefficients of thermal expansion of the anisotropic ground mass. These considerations are not valid for lead-base alloys, as the cubic lead crystals have the same thermal expansion in all directions.

FENG [314] contrasts the *welding theory* of BOWDEN and TABOR [119] with his own *toothing theory*. Atomic movements along certain crystallographic planes are connected with the plastic deformations of asperities which occur at contacts. The surfaces of two such contacting asperities must thereby adjust themselves to one another. Thus movement in various slip-planes occurs and there ensues a roughening and simultaneously toothing of the surfaces at the points of contact. Slip in the touching surfaces is thus prevented. On movement, tearing occurs in the place where the strength of the cross-section of the asperities, that is, the product of the cross-sectional surface and the shear strength, is smallest. In general the shear strength is increased by work hardening near the surface. Shearing-off therefore takes place at a distance from the surface of contact. The energy of shearing off is mainly converted to heat and can—but must not—lead to welding of the surfaces of contact. If no welding occurs, then any particle sheared off can remain suspended by adhesion or become a loose wear particle. While according to the welding theory, cold welds are the cause of friction, here, on the contrary, metal transfer is regarded as the result of the heat of friction.

In many papers, particularly in the English literature, the concept of *boundary lubrication* was introduced. However it is not always defined in the same way. CLAYTON [204] explains boundary lubrication as the region of friction in which two solid surfaces slide on one another and there is no longer present a liquid lubricating film to carry part of the load. Thus, not only metallic surfaces are to be regarded as solid surfaces, but all absorbed layers (oxides, gases, moisture etc.) should be considered in this light. For the explanation of friction, CLAYTON refers to the theory of adhesion. According to this, the energy of friction should be a function of the molecular adhesion between the solid surfaces. If a lubricant is present it becomes absorbed on the surfaces or combined with them, and thus reduces, or completely prevents, direct solid body contact. Adhesion between the lubricant films should be less than that between the clean surfaces; shearing movements in the surfaces in contact would therefore occur more easily. A layer of a molecular thickness, which will be directly present on each solid surface, should effect lubrication under these boundary conditions. Barwell [50] introduces the concept of thin film lubrication. By this term there is understood an intermediate state of affairs, which is not a hydrodynamic type but, on the other hand, is not quite boundary lubrication.

According to Finch [315] the function of the boundary layer is that it replaces the coefficient of friction of the oxide film by a still lower value. According to him, lubricants consist in general of mixtures of long-chain hydrocarbons (paraffins) or the corresponding alcohols, esters, ketones and acids. At temperatures below the melting point of the hydrocarbons, the chain molecules are grouped in bundles like twigs in faggots. These bundles form kinds of crystals. If such films are heated, as occurs under practical condition of sliding, the faggots break up at a temperature near the melting point leaving an actual boundary layer of molecules. The molecules directly attached to the surface stand perpendicular or slightly inclined to it. Above these, there are films which are less precisely aligned. The maximum thickness of the boundary layer is of the order of less than 200 Å.

Shooter [1112] emphasises the great influence of chemical attack on boundary friction. According to Lunn's theory [786], the function of a bearing depends on the formation of adherent layers which prevent metallic contact between the sliding surfaces. These layers are said to result from chemical reaction between the sliding metals, the lubricant and the surrounding atmosphere. Pekrun [943] indicates that the coefficients of friction for boundary lubrication quoted in the English literature were largely measured with apparatus with elastic, lightly-built arrangements of the sliding piece without any damping. Thus, if the coefficient of friction at rest was essentially greater than that in movement, so-called "stick-slip" easily occurs. Considerations were then based on the coefficient of friction at rest observed at the "stick". The friction during "slip", that is the friction in sliding, was usually not measured. Moreover, in most cases the work only extended over short periods and short sliding paths. Forrester [334] carried out investigations with various bearing metals in which hemispherical sliding bodies of the bearing metal pressed against a rotating steel disc. He came to the very noteworthy conclusion that the differences in the sliding properties of the individual bearing metals are due to the fact that the latter, under given conditions, allow different degrees of fluid friction to develop. Lead-base and tin-base white metals favour the development of fluid friction much more than do leaded bronzes. Further details of the extensive literature on boundary lubrication and the related problems can be found in the report by PEKRUN [943], mentioned above.

WENGER [1258] carried out running tests, at constant load and constant rate of rotation, at the Institut für Reibungsforschung (Institute

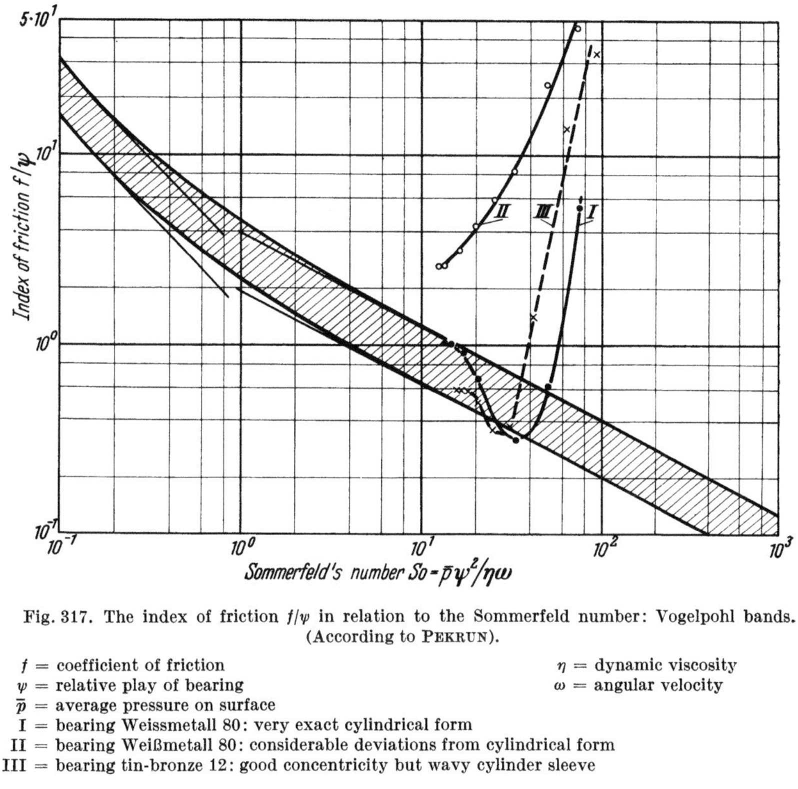

Fig. 317. The index of friction f/ψ in relation to the Sommerfeld number: Vogelpohl bands. (According to PEKRUN).

f = coefficient of friction
ψ = relative play of bearing
$\bar{p}$ = average pressure on surface
I = bearing Weissmetall 80: very exact cylindrical form
II = bearing Weißmetall 80: considerable deviations from cylindrical form
III = bearing tin-bronze 12: good concentricity but wavy cylinder sleeve
η = dynamic viscosity
ω = angular velocity

for Research on Friction) of the Max-Planck-Gesellschaft in Göttingen. The state of running-in was estimated by measuring the friction and comparing it with the value to be expected according to hydrodynamic theory $\left(f = 3\sqrt{\eta \cdot \omega / p}\right)$ A rough calculation of the narrowest width of gap indicated that under the above conditions, and with a bearing of exact cylinder form, this amounted to about 1 μm. The deviations from the ideal cylindrical form were up to 10 μm, that is up to ten times the calculated narrowest gap, and in some cases still higher. Correspondingly very high scatter of results was found. The indexes of friction f/ψ determined by constructing the Stribeck curve, are plotted in Fig. 317 against the SOMMERFELD number S_0. The VOGELPOHL band represents the coefficient of friction at full lubrication. In machine operation in practice a coefficient of friction $f \cong 0.003$ is accepted (VOGELPOHL [1232a]). The comparison of the results given by the three bearings shows the preponderant effect of the macrogeometric form on the running-in behaviour. The properties of the bearing material, in particular its hardness, played no appreciable role in these tests. All three curves, show, with increasing running-in, entry into the VOGELPOHL band, which represents the state of fluid friction. The points situated above are therefore an expression of the presence of mixed friction. Curves I and III are closer together than I and II, which relate to the same alloy. In order to exclude the effects of the macrogeometric form on the running behaviour, PEKRUN [943] developed an apparatus which can be regarded as the simplest type of machine for testing bearings. A rotating ball continually moistened with oil is forced with constant force against the bearing metal and thus presses out an indentation of suitable diameter. Measurement of the force of friction and of the electrical current passing enables the state of running-in to be followed. At the end of the test the ball is carried by the developed hydrodynamic pressure alone. The width of the lubricating gap is estimated at 0.1 μ and below. The low coefficients of friction which were found indicate that the load was then being carried mainly hydrodynamically when the insulating action of the film of lubricant was still incomplete. It could be concluded from interference pictures that smooth surfaces favour the development of hydrodynamic lubricating films. The results of PEKRUN's calculations [943] show that the radius of curvature of an asperity exerts a great influence on the hydrodynamic processes when it is pressed against a surface. The time of pressing to a given distance at constant force, and the force for pressing with constant speed, grow as the square of the radius of curvature of the asperity. At very narrow gap widths, remarkably high pressures can occur. At a constant pressing force, the pressure of liquid is inversely proportional to the gap width, at constant rate of pressing inversely proportional to the square of the gap width. As the asperity and the opposite surface are however not ideally rigid, they are deformed at first elastically and then plastically, and thus adjust to one another.

If, over a certain region, asperities and opposite surfaces assume the same curvature, so that a parallel gap is formed, and if the film of lubricant is not torn perhaps by the deformation, then further pressing is impeded to an extraordinary degree. Calculation for a parallel gap indicates that the time required for pressing together to a given distance at constant force increases with the fourth power of the radius of curvature of the parallel gap. The better the adjustment of the asperities to the opposite surface, and therefore the closer the approximation to the ideal parallel gap, the greater becomes the load-bearing capacity of the film of lubricant. The production on deformation of sharp edges and points is unfavourable for the persistence of the film of lubricant. Hydrodynamic forces occur not only in the macro-region in "thick" films of lubricant, but also in con tinuous transition down to the thinnest layers, where molecules are still present between the sliding surfaces. These molecules have ready ease of displacement of molecules of liquid.

The first prerequisite for the cessation of wear and the increase of the running surface is that the sliding surfaces should be completely separated by a hydrodynamic film of lubricant. The second prerequisite is that only elastic but not plastic deformations should be brought about by the pressures and shearing stresses exerted by the film of lubricant on the sliding surfaces. Low modulus of elasticity (at the temperature of operation!) largely favours elastic deformations and therewith speedy cessation of wear. Good plastic deformability permits an adjustment without formation of wear particles. Examples of this are lead, tin and cadmium, and at higher temperatures, aluminium. In the leaded bronzes, a thin layer of lead presumably takes over the adjustment to the asperities on the opposite surface (p. 375).

It is certain that the wettability of the running surface by the lubricant only affects the process of lubrication for very narrow gaps. The forces due to capilliary action (interfacial tension), which draw the lubricant into the gap, have in this case the same order of magnitude as the hydrodynamic pressure forces of the surrounding lubricant acting on the cross-section of the gap. With full lubrication and sufficient supply of lubricant, wettability no longer has any significance, as the film of lubricant is no longer broken. Wettability is not a property of the bearing metal itself, but is characteristic of the combination bearing material/lubricant. According to PEKRUN [943], the effect of the attachment of polar molecules to the sliding surfaces can be seen in the fact that it gives rise to an increase in the radius of curvature of the asperities, which raises the hydrodynamic bearing capacity. Apart from this, the attachment of lubricant molecules can perhaps bring about an improvement in the wettability. In the light of what has been said above, good wettability and attachment of polar molecules to the surfaces improve the process of lubrication only by facilitating the development of hydrodynamic pressure. At this point it may be mentioned that practical experience ascribes to lead-base bearing metals a better wettability for oil than to tin-base bearing metals.

As was to be expected, the experiments indicated a considerable dependence of the frictional behaviour on the fine surface structure of the running surface arising from the sliding process. It was very different

Table 25. *Composition of White Metals According to DIN 1703*

Description	Sn %	Sb %	Cu %	As %	Others %
LgPbSb 12		10.5 to 13.0	0.3 to 1.5	up to 1.5	up to 0.3 Ni
LgPbSn 5	4.5 to 5.5	14.5 „ 16.5	0.5 „ 1.5		
LgPbSn 10	9.5 „ 10.5	14.5 „ 16.5	0.5 „ 1.5		
LgPbSn 6 Cd	5 „ 7	14 „ 16	0.8 „ 1.2	0.3 to 1.0	0.6 to 1.0 Cd 0.2 „ 0.6 Ni
LgPbSn 9 Cd	8 „ 10	13 „ 15	0.8 „ 1.2	0.3 to 1.0	0.3 „ 0.7 Cd 0.2 „ 0.6 Ni
LgSn 80	79 „ 81	11 „ 13	5 „ 7		
LgSn 80 F	79 „ 81	10 „ 12	8 „ 10		
Nr. 11		14.0 „ 16.0	up to 0,50	up to 0.25	
Nr. 12		9.3 „ 10.7	„ „ 0.50	„ „ 0.25	
Nr. 15	0.75 „ 1.25	14.5 „ 17.5	„ „ 0.60	0.80 to 1.40	
Nr. 10	1.75 „ 2.25	14.0 „ 16.0	„ „ 0.50	up to 0.20	
Nr. 8	4.5 „ 5.5	14.0 „ 16.0	„ „ 0.50	„ „ 0.20	
Nr. 19	4.0 „ 6.0	8,0 „ 10,0	„ „ 0.50	„ „ 0.20	
Nr. 16	9.0 „ 11.0	11.5 „ 13.5	0.40 to 0.60	„ „ 0.20	
Nr. 7	9.3 „ 10.7	14.0 „ 16.0	up to 0.50	„ „ 0.60	
Nr. 6	19.0 „ 21.0	14.0 „ 16.0	1.25 to 1.75	„ „ 0.15	
Nr. 2	88.0 „ 90.0	7.0 „ 8.0	3.0 to 4.0	„ „ 0.10	

The permitted iron content is in the German Standard in general 0.1% and in the American Standard 0.08%; the permitted content of Al and Zn is 0.05% each, in the American Standard 0.005% each.

in the individual alloys. However, no relationship with the micrograph could be established.

d) Composition, Structure and Properties of White Metals and Leaded Bronzes. The type of alloy known as white metal was introduced into technology by BABBITT in 1838. The composition then chosen of 89 to 90% Sn, 8 to 9% Sb and 1 to 2% Cu (BUNGARDT [145], BLANDERER [89]), can still be regarded as modern (Table 25); in Germany it has been altered to that of Weißmetall 80 F. The lead-free alloy WM 80 F has a higher hardness at elevated temperatures than Weißmetall 80, which contains lead.

After the tin-rich white metals, the lead-rich white metals came forward prominently at the time of the first world war; in the English literature they are often described also as babbitts. They are essentially alloys of the ternary lead-antimony-tin system (p. 163 or of the binary

and According to ASTM Designation B 23—49 (No. 11 to No. 2)

Pb %	Melting Range	Casting Range or Casting Temperature	Brinell Hardness[1] kg/mm²			Remarks
			20° C	50° C	100 °C	
	254 to 380°C	380 to 550°C	18	14	8	Good attachment to steel and chilled castings for thicknesses up to 1.5 mm
77.5 to 79.5			22	13	6	
72.5 to 74.5	235 „ 370°C	420 „ 450°C	23	16	9	Attachment to bronze, steel and chilled casting with an intermediate layer, by a special process also to grey iron
	245 „ 420°C	480 „ 520°C	26	21	15	
	240 „ 400°C	450 „ 520°C	28	23	15	
1 „ 3	230 „ 400°C	440 „ 460°C	27	20	10	Best attachment and castability
up to 0.5						
84.0 to 86.0	244 „ 262°C	332°C	15.0		7.0	
89.0 to 91.0	245 „ 259°C	329°C	14.5		6.5	
Remainder	248 „ 281°C	350°C	21.0		13.0	
82.0 to 84.0	242 „ 264°C	332°C	17.5		9.0	
79.0 to 81.0	237 „ 272°C	341°C	20.0		9.5	
Remainder	239 „ 257°C	327°C	17.7		8.0	
Remainder	244 „ 257°C	327°C	27.5		13.6	
74.0 to 76.0	240 „ 268°C	338°C	22.5		10.5	
62.5 to 64,5	181 „ 277°C	346°C	21.0		10.5	
0.35	241 „ 354°C	424°C	24.5		12.0	

[1] Definition of Brinell Hardness in DIN 1703, $P = 2.5\,D^2$, Time 180 secs. In the American Standard the recommendation is $P = 500$ kg, ball diameter $= 10$ mm i.e. $P = 5\,D^2$ for a time of 30 secs.

lead-antimony system also (p. 75) with additions of copper and of other metals, and thus have some similarity, structurally speaking, to some printing metals. The lead-alkali alloys, e.g., Bahnmetall, or the lead-tin-alkali alloys, have a limited significance. Guertler [450] and Zunker [1308] tested the strength properties of lead alloys containing several metals, with regard to bearing problems. The basic result was that, owing to the simultaneous requirements of hardness and toughness, the range of alloys was limited.

As examples of the composition of lead-rich white metals, data are provided from German and American standard specifications. The values of the compressive yield point such as are given, for example, in the ASTM standard, are not included in Table 25, as the variation of Brinell hardness with temperature gives a criterion for the loading capacity of the bearing with rising temperatures of operation.

The generally higher position of the upper melting point of the alloys in the German standard specification as compared with those of the American can be related to the content of copper and nickel. Larger contents of these high-melting metals considerably extend the melting range upwards, as can be observed for instance by consideration of the ternary-lead-antimony-copper system (p. 149). On cooling such an

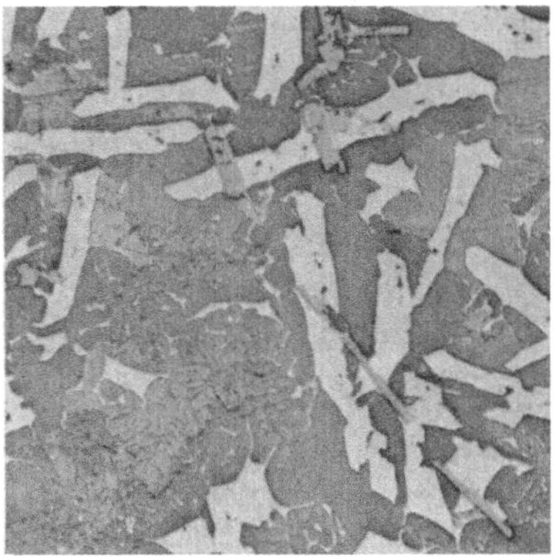

Fig. 318. Alloy "PbSn10" with 73.5% Pb, 10% Sn, 15.5% Sb, 1% Cu. Grey rods (of violet colour): Cu_2Sb. White: "primary" antimony. Ground mass: ternary eutectic lead + antimony + SbSn. ×150

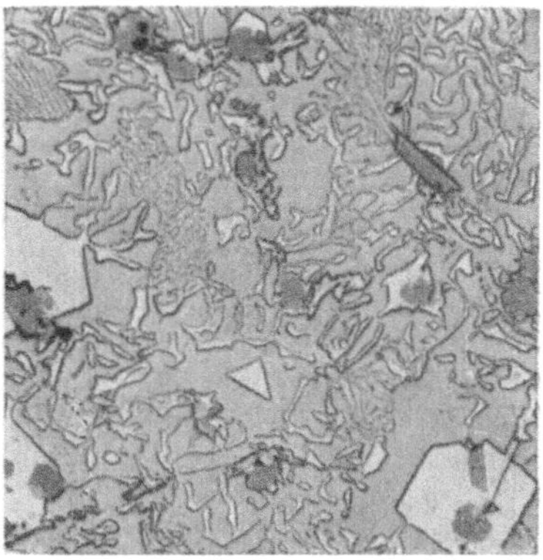

Fig. 319. Thermit bearing metal from Th. Goldschmidt A. G., Essen. Several coloured (reproduced as dark) phases containing copper and nickel, in part in primary antimony. Antimony and SbSn of the ternary eutectic can be distinguished even without etching. ×150

alloy from the region of complete liquid miscibility, the amount of copper- and nickel-containing phases first solidifying is so small that it is difficult to recognise the beginning of solidification by thermal analysis. The same holds in reverse for the heating curve. The indication of the upper melting point should therefore be regarded with some reserve.

Most of the alloys, if only the contents of lead, antimony and tin are considered, lie in the primary crystallisation field of antimony or of SbSn of the ternary system in question, and accordingly contain—apart from the high-melting phases of copper already mentioned—primary crystals of antimony or SbSn in a binary or ternary eutectic matrix. Examples of this are shown in Figs. 318 and 319 as well as in Figs. 177 to 180 discussed in connection with the ternary system. In Fig. 318, the platelike shape of the crystals regarded by Müller [877] as Cu_2Sb, may be observed. Forrester [334a] considers them to be the Cu_6Sn_5 phase. According to the general view, the framework of this phase, first formed by the solidification of the melt, impedes the segregation of the ensuing antimony or SbSn crystals from the specifically heavy melt (HEYN and BAUER [520], REICHENECKER [999]). Copper contents above 1.5% also increase the hardness of the alloys (v. GÖLER and SCHEUER [396]). According to the investigations quoted in [396], nickel contents should likewise reduce the tendency to segregation, and moreover have a grain-refining action. Arsenic does not make itself evident as such in the structure, as it is in solid solution in lead, in antimony and in antimony-containing phases such as SbSn. According to PHILLIPS, SMITH, BECK and co-workers [960] it should improve the mechanical properties (particularly the hot-hardness) and refine the structure. Consequently, alloys containing arsenic should be particularly suitable for the continuous production of steel/white metal strip in the casting process. The fine-grained alloy layer tolerates the high stress involved in bending the strip to form bushings. On the other hand, should white metals containing arsenic, having regard to combination with copper alloys, give rise to difficulties in three-layer bearings. In the interest of good impact strength of the alloys it is recommended that the arsenic content should not be raised above 0.8% [1300c]. Cadmium, in the small amounts occurring in white metals, is dissolved in the lead (p. 50) or precipitated in the solid state. Other metals added in lead-antimony-tin alloys brought no special advantages (ACKERMANN [6], v. GÖLER and SCHEUER [396]). According to the references quoted in [396], the effect of aluminium is said to be due to the fact that it promotes segregation and spoils the bond and the running properties. The chief disadvantage of zinc, according to ACKERMANN [6], is that it causes drossing, lowers the castability and coarsens the structure. With regard to the fundamental mechanical properties of the alloys, attention is drawn to the section on lead-antimony-tin. Some further investigations may be discussed. Determinations of the hardness of the alloys in relation to tin content and to temperature showed that an improvement only results up to a tin content of about 10% (v. GÖLER and SCHEUER [396]). This result of hardness determination at room temperature agrees with that obtained by HEYN and BAUER [520]; (p. 67). This can be seen by comparing the section at 15% Sb in the diagram given in Fig. 181. The hot-hardness, especially at high temperatures, lies below that of the high-tin white metals (Fig. 320). According to v. GÖLER and PFISTER [395], the lead-base white metals are also inferior to tin-base in respect of fatigue strength, at least as far as solid test-pieces are concerned. On the other hand, the relationships are different for thin bearing linings (p. 370). Further, repeated impact testing of various white metals at elevated temperatures indicated a superiority of lead-base over tin-base alloys (HERSCHMANN and BASIL [514]).

Like the hardness, the tensile strength shows a pronounced increase only up to a tin content of 10% (v. GÖLER and SCHEUER [396]). This, considering the relationship between hardness and tensile strength, (p. 204) is plausible. The fatigue strength of the alloy with 10% tin is also comparatively high. With an additional copper content of 1 to 1,5% it amounts to 2 kg/mm². The investigation described therefore provides a proof for the fact that white metals of medium tin content, between about 10 and 80% Sn, are rarely used in technology. The elongation at

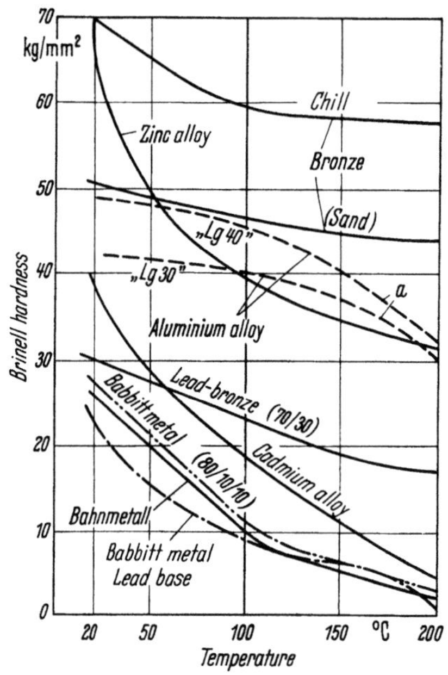

Fig. 320. Hot-hardness of various bearing metals. (According to KÜHNEL [717]).

fracture falls strongly up to tin contents of about 10%, then remains about the same. A considerable fall of ductility in the region of 0 to 10% Sn was also found in the bend test and in the compression test (cf. Fig. 185). The effect of the copper content on the ductility was determined in the impact-compression test. With increasing copper content, there was a reduction in the number of blows and a reduction in height at fracture (HEYN and BAUER [520], v. GÖLER and SCHEUER [394]). Finally attention may be drawn to the determination of the microhardness of the individual structural constituents (Fig. 321). The thermal conductivity of bearing metals based on lead-antimony-tin is less than that of high-tin white metals. Since the lining thicknesses are, in general, below 10 mm, this fact is of minoı importance (BUNGARDT [144]).

Casting temperature and mould temperature have great influence on the mechanical properties of bearing metals, as was closely investigated on an alloy with 14.6% Sb, 5.4% Sn, 0.04% Cu, 0.06% As

(ARROWSMITH [25]). The casting temperature was varied between 350 and 450 °C, the temperature of the chill mould between 20 and 200 °C. With increasing mould temperature there was found a decrease of tensile strength and of yield point, while in the region indicated the casting temperature had little influence. The elongation at fracture decreased slightly with increasing temperature of the chill mould, while

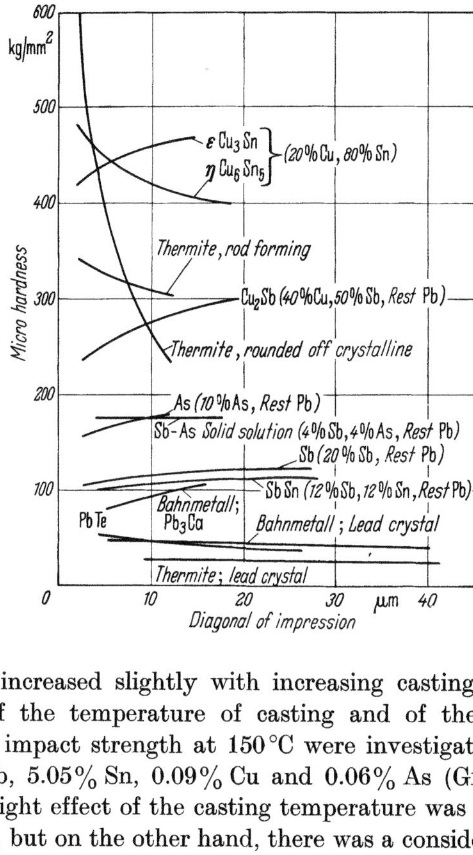

Fig. 321. Microhardness of structural constituents of lead-base bearing metals. (According to RAPP and HANEMANN [992]).

on the other hand, it increased slightly with increasing casting temperature. The effect of the temperature of casting and of the chill mould on the repeated impact strength at 150 °C were investigated on an alloy with 14.9% Sb, 5.05% Sn, 0.09% Cu and 0.06% As (GREENWOOD [421]). Again a slight effect of the casting temperature was found between 350 and 450 °C, but on the other hand, there was a considerable reduction of the impact strength if the chill mould temperature rose from 70 to 120 °C. By and large, the casting temperature will be chosen as low as possible, partly on economic grounds, and partly to avoid the risk of segregation and of producing coarse-grained solidification.

Of the lead-alkali bearing metals or precipitation-hardened bearing metals, no doubt the best-known representative is Bahnmetall with 0.69% Ca, 0.62% Na, 0.04% Li and 0.02% Al. In the United States similar alloys with 0.5 to 0.75% Ca, smaller contents of further elements and an addition of about 1.5% Sn (e.g., Satco metal), have been developed. The structure of Bahnmetall (Fig. 322) exhibits essentially primary crystals of $CaPb_3$ in lead solid solution.

The melting range of Bahnmetall is stated as 312 to 425 °C, the casting temperature as 500 to 550 °C (v. GÖLER and WEBER [398]). The temperature of the chill mould must also be higher than for white metals. The tendency of the melts to oxidise is reduced by an addition of aluminium (cf. p. 328). As impurities of arsenic, antimony and bismuth form high-melting compounds (with calcium, for example) which segregate out (p. 48), they are harmful when present in hardening constituents. The precipitation-hardening of Bahnmetall after casting is connected with the formation of highly-supersaturated solid solutions on solidification (p. 45). The rate of cooling affects the rate and degree of precipitation-hardening. Fig. 320 shows the temperature-dependence of the hardness of some types of bearing metals. According to this, Bahnmetall still has a considerable hardness at 150 °C. Long heating to above 60 °C leads however to a certain reduction of hardness, measured at room temperature. This phenomenon is fundamental and can be observed in all precipitation-hardening lead alloys. The softening of Bahnmetall takes place most markedly at 100 °C. On holding material at this temperature for three days, its hardness at 20 °C fell from 36 to 21 BHN; longer annealing had no further action. The softening is not noticeably disadvantageous in the running properties. Over and above this, the reverse process again sets in on heating to higher temperatures (above 120 °C), and on subsequent storage the alloys again undergo more or less precipitation-hardening (v. GÖLER and WEBER [398]).

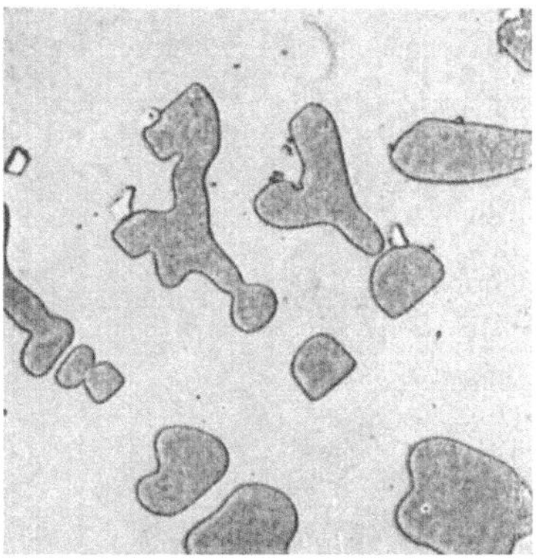

Fig. 322. Bahnmetall. The white constituent, mostly crystallised on to $CaPb_3$, is presumably aluminium. × 500

SCHMID and co-workers [1069] further improved Bahnmetall during the last war. In an altered alloy, the sodium content in particular was lowered to 0.2% and 0.4% barium added. According to SCHMID [1069] the alloy formerly used—which had otherwise very good properties—exhibited two defects, namely partial softening on running hot, and loss of alloying metals in melting and in casting. The decrease of hardness is prevented by reducing the sodium and the lithium contents down to the solubility-limit at room temperature, which ensures that the safety-factor of the alloy against loss of alloying metals is increased at the same time. The barium additions compensate the loss of hardness which occurs and the otherwise favourable properties of Bahnmetall are retained in the new alloy (p. 63).

Today at a time when raw materials are no longer scarce, Bahnmetall is used only in small amounts in railway operations, especially as bearings made of it were not wholly satisfactory for increasing the speed of goods trains; and the precautions to be observed in remelting the alloys seemed a disadvantage. Nevertheless this interesting group of alloys could not remain unmentioned.

Leaded Bronzes and Tin Leaded Bronzes. In the narrow sense, these alloys are not a subject for this book, as their major constituent is copper. German specification DIN 1716 "Cast Leaded Bronze and Cast Tin Leaded Bronze" includes five cast alloys, namely a binary leaded bronze essentially with 25% Pb, as well as four tin leaded bronzes with about 10 to 5% Sn and increasing contents of lead of 5, 10, 15, 20%, remainder copper. In England and the United States lead contents up to 40% are used. All the alloys consist mainly of a matrix of copper or a copper-tin solid solution, and inclusions of lead. They are preferably attached to the steel supporting shell by the centrifugal casting process or combined in thin layers with steel strip, which is subsequently bent into a bearing. Sintering of the alloy on to steel is also common. Thus lead contents up to 40% can be used (FORRESTER [334b, 335], DUCKWORTH [260]). The steel strips are copper-plated in order to provide a good bond of the applied alloy to the basis metal, the alloy powder loosely poured on, and sintering then carried out in a protective atmosphere. The strips thus prepared are rolled to bring about compaction of the applied layer, and in some cases again sintered or heated. They are then cut into suitable lengths from which the bearing shells or bushings are formed by bending. The leaded bronze bearings also frequently receive electrodeposited coatings of soft metals, for instance, of lead-indium or lead-tin (MOHLER [860], ROGGENDORF [1021], JUNGE [638], AZZOLINO [33]) (p. 60). The characteristics of the leaded bronzes and tin leaded bronzes are again described in the general sections of this book and compared with those of the white metals (p. 362). As described by EISENKOLB [272a], sintered iron with lead inclusions forms a certain parallel with the leaded bronzes.

6. Fusible Alloys

If it is desired to reduce the melting point of lead-tin solders below the eutectic of about 180 °C, cadmium is the first metal to suggest itself as a further alloying element. The lead-cadmium-tin alloys have good soldering properties, as the melting point of the ternary eutectic (p. 140) is at only 145 °C. By addition of zinc, a quaternary eutectic can be obtained of melting point 138 °C and composition 28.6% Pb, 16.7 % Cd, 52.45% Sn, and 2.25% Zn (p. 176). Solders containing cadmium of course tend to somewhat increased drossing, which must be watched particularly in dip-soldering baths (KEIL [651]). According to WASSERMANN and GRUHL [1241], drossing of the alloys is least along the binary eutectic channel emerging from the lead-tin eutectic, including the ternary eutectic. The composition of solder should be chosen from this point of view. A further effective depression of the melting point is brought about by additions of bismuth to the lead-cadmium-tin alloys. In this way the optimum alloy is obtained, the quaternary lead-cadmium-bismuth-tin eutectic of melting point about 70 °C, the composition of which is usually given (PARRAVANO and SIROVICH [936]) as 50% Bi, 12.5% Cd, 25% Pb and 12.5% Sn (Wood's metal). According to FRENCH [338] the quaternary eutectic is said to have the composition 50% Bi, 10% Cd, 27% Pb, 13% Sn (LIPOWITZ metal) and the melting point of 71.7 °C. The above investigation of the quaternary system (PARRAVANO and SIROVICH [936]) is fundamental for a survey of the solidification range of any combination of the metals in question. It contains detailed tables, by means of which the composition suitable for specific purposes can easily be determined. The melting point of the quaternary eutectic alloy can be further effectively lowered only by additions of mercury (WAEHLERT [1234]) or of indium (FRENCH [338], JÄNECKE [615]). However mercury, on account of its high vapour pressure and its toxicity, is excluded for most applications. The phases of the quaternary eutectic are the lead-cadmium-tin solid solution and the β-phase of the lead-bismuth system. The quaternary eutectic alloy is brittle when cast, becoming plastic on storage for two or three hours. A similar phenomenon has already been described for the lead-bismuth-tin alloys (p. 134). Non-eutectic alloys of higher tin content do not lose part of their brittleness even after storage for 24 hours. The eutectic or near-eutectic alloys are always used where a melting point below the boiling of water is desired, e.g., as fillings (which can be easily removed by subsequent melting out) for bending thin-walled tubes, as solders for heat-sensitive parts, as second solders after first soldering with lead-cadmium-tin, for safety melting devices, as casting patterns and for fire alarm appliances. Further examples of application are given by KEIL [650] and SEEDS [1097]. Fig. 323 gives a clear representation of the binary and ternary systems on which the quaternary system is based.

In addition to the alloys described, those of the ternary-lead-bismuth-tin system are also of great technical importance (p. 133). Some of these are specified by name, such as Lichtenberg- or Newton-metal with 50% Bi, 30% Pb, 20% Sn which approximately represents the ternary eutectic melting at 90°C, and similarly Rose's metal with 50% Bi, 25% Pb, 25% Sn. A standardisation of nomenclature would be desirable. Alloys of the quaternary lead-antimony-bismuth-tin system may also be mentioned here, for instance, "Matrix alloy" (COURNOT [220]).

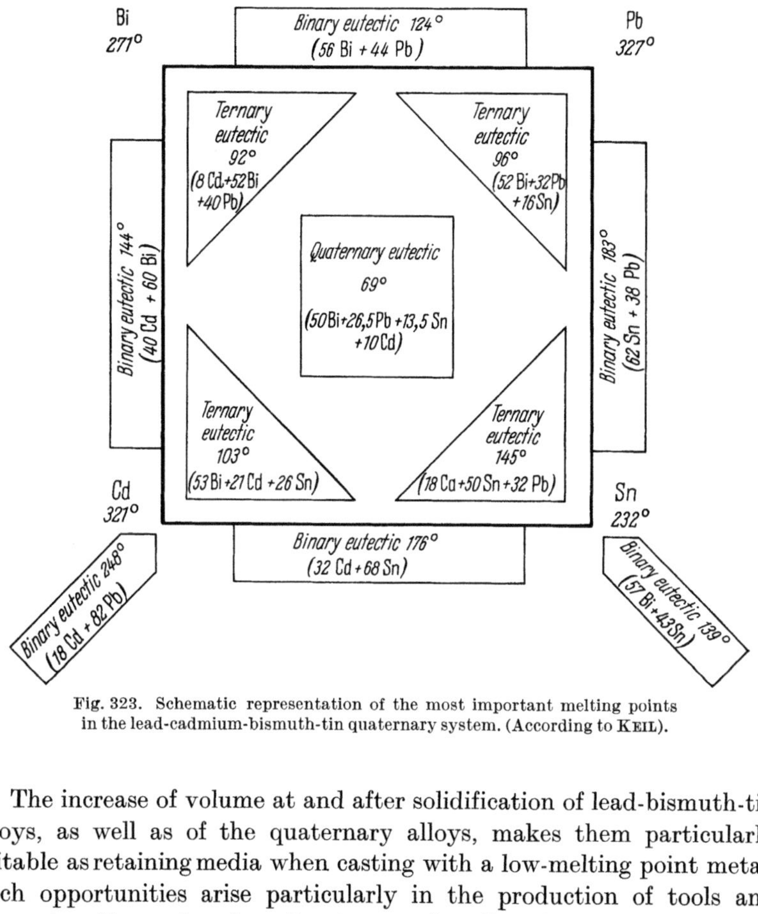

Fig. 323. Schematic representation of the most important melting points in the lead-cadmium-bismuth-tin quaternary system. (According to KEIL).

The increase of volume at and after solidification of lead-bismuth-tin alloys, as well as of the quaternary alloys, makes them particularly suitable as retaining media when casting with a low-melting point metal. Such opportunities arise particularly in the production of tools and apparatus. Examples of application are described by SEEDS [1097] and by KEIL [650]. Here some guidance may be given on the effect of the bismuth content on the changes of volume on solidification and storage. Bismuth content 35 to 45%: slight contraction on solidification, more or less compensated by expansion on storage. Bismuth content above 48%: expansion on solidification and to a slight extent during a few hours of storage. Lead-free alloys with more than 50% Bi: expansion on solidification and slight contraction on cooling to room temperature. In using solders which contain bismuth it should be noticed that the metal surfaces wet less well than with the usual soft solders. The surfaces to be soldered should therefore first be tinned. Moreover, according to KEIL [650], the use of an acid flux is generally to be recommended.

SPENGLER [1141] provides an extensive list of low-melting metals and alloys.

Wider fields of application for the alloys of the ternary lead-cadmium-bismuth and cadmium-bismuth-tin systems are not known.

II. Plastic Working (Technology)

1. General Discussion of the Extrusion of Lead

In antiquity, lead pipes were produced by casting. The fabrication of lead by extrusion was reserved for the machine age. The first patent for a lead press was granted in 1797 to BRAMAH in London. The first hydraulic press was built by BURR in England in 1820. Effective im-

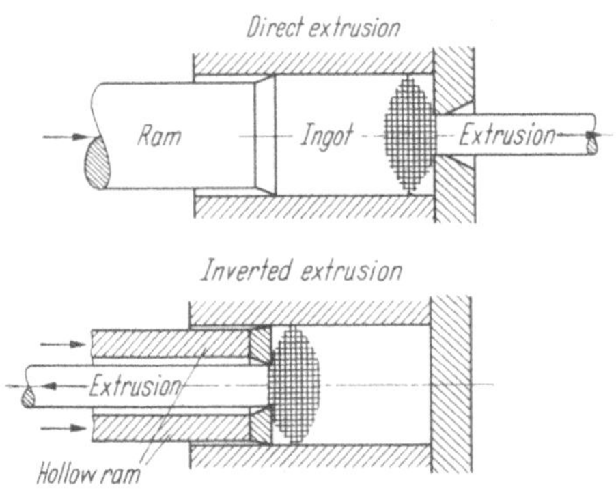

Fig. 324. Schematic illustration of direct and inverted extrusion. (According to SIEBEL and FANGMEIER [1116]).

provements, such as a container which could be heated and a hydraulic accumulator for more powerful extrusion techniques, were provided in HAMMON's patent in France in 1869. The first lead cable-sheathing presses with vertical rams were built around 1880 from ideas by WESSLAU in Germany, BOREL in FRANCE and EATON in the United States. HUBER invented the horizontal lead cable-sheathing press (PEARSON [940[1], 282], SCHNEEBERGER [1075] and KLEIN [669]).

The extrusion of metals, as shown in Fig. 324, takes place either forwards or backwards. The terms direct and indirect presses, or forward

[1] [940] is a book on "The Extrusion of Metals" by C. E. PEARSON, 1944. In 1960 there appeared the second edition (revised) by C. E. PEARSON and R. N. PARKINS.

and inverted presses, are also used[1]. In inverted presses, the billet of metal is at rest relative to the wall of the container, whereas in forward presses it is moved relative to the container. Here, therefore, the frictional forces between billet and container must be overcome. PEARSON and SMYTHE [941] made a comparison between the two methods in experiments with lead, cadmium, bismuth and tin. The internal diameter of the heatable container was 31 mm. An important result of measurement was, that for equal pressures, greater rates of flow occurred in the inverted process than in the direct. In the latter, the rate of flow rose progressively only towards the end of the experiment to the value which had been constant during the whole process in the former. In billets of the diameter indicated and of length 63.5 mm the forward movement of the billet, relative to the container and necessary in the extrusion, expressed itself as a 25 to 30% pressure-loss. The pressure-loss was lower with a shorter billet and with use of a lubricant. The pronounced effect of a lubricant, emphasized by PEARSON, [940] seems to relate more to the experiments with bismuth. In experiments on lead and on lead-tellurium alloys, JOHNSON [625] achieved only a slight reduction of the pressure by the use of a lubricant. On the other hand, SACHS and DRAPER [1042] found, in the case of extrusion of lead pipe, a notable effect of lubrication, particularly on the maximum pressure, which appears at the beginning of the extrusion process.

SIEBEL and FANGMEIER [1116], in the extrusion of lead in a container of internal diameter 35.7 mm, kept the movement of the ram constant at 6 mm/min and observed the course of pressure with time. The representation of the results (Figs. 325 and 326) contains curves for various ratios of the cross-section of billet and wire. The rise in the right hand part of the upper curve in Fig. 325 is due to lead being squeezed out between the die and the wall of the container. The two lower curves have an approximately horizontal course. The curves in Fig. 326 show a somewhat greater reduction of pressure during extrusion, but essentially still a horizontal course. The pressures are scarcely higher than in the inverted press. That the wall-friction in direct extrusion was so little in evidence here arises from the fact that the wall of the container was cleanly ground. When this was not the case, the ram-forces exceeded by far those occurring in the inverted process, and only fell at the end of the extrusion process to those for a ground container-wall. The curves of pressure versus stroke obtained by PUGH [982a] for direct and inverted extrusion with good lubrication are identical among themselves and in complete agreement with those of SIEBEL in Fig. 326.

---

[1] PEARSON and PARKINS in the 1960 edition of the book "The Extrusion of Metal", speak only of direct and inverted.

EISBEIN and SACHS [272], as well as SIEBEL and FANGMEIER [1116], investigated the effect of different apertures on the extrusion pressure: the results were substantially in agreement. According to Siebel's equations [1113] for the calculation of the work of extrusion, the ex-

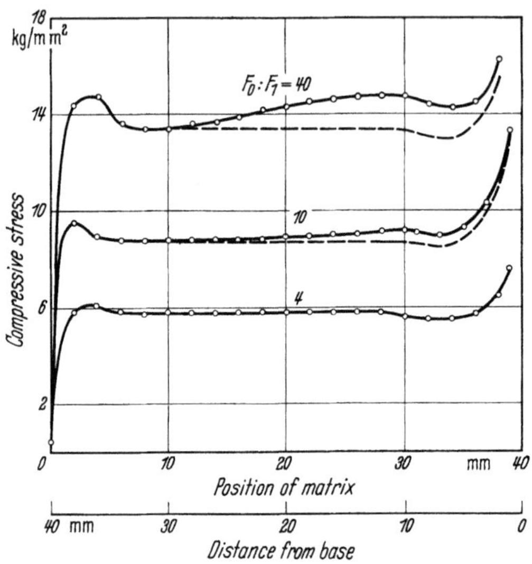

Fig. 325. Pressure variation during inverted extrusion. (According to SIEBEL and FANGMEIER).

trusion pressure p necessary for extruding a round billet of diameter D to a round rod of diameter d (neglecting external forces of friction, and in the absence of hardening) is:

$$p = k \cdot \ln \frac{D^2}{d^2} \; [kg/mm^2] .$$

The yield stress k was determined from compression tests (EISBEIN and SACHS [272]), and inserted in the above equation as 2.8 kg/mm². In order to test the formula, measurements of the work needed for various ratios D^2/d^2 were then carried out. On plotting against D/d on the semi-logarithmic scale, the values fall on a straight line. Owing to the effects of friction, its slope was steeper than is to be expected for $k = 2.8$ kg/mm², and corresponded to a value $k = 2.8 \cdot 1.6$ kg/mm². The measurements supported the concept of the effect of friction on the container-wall in direct extrusion. The law of similitude derived by KICK [661] from the above formula was broadly confirmed.

Conversely SIEBEL [1116], by means of the formula given, calculated the yield stress k from the extrusion pressure and, in good agreement with the preceding investigation, obtained values between 3.6 and 4.7 kg/mm², which are not far from the Brinell hardness values for lead.

In multi-hole dies, which for the extrusion of thin wires, may have up to 12 individual orifices [835a], the necessary extrusion pressure can be derived as a first approximation (BISHOP [88]) by regarding the individual round orifices as replaced by a single hole of equal cross-section and

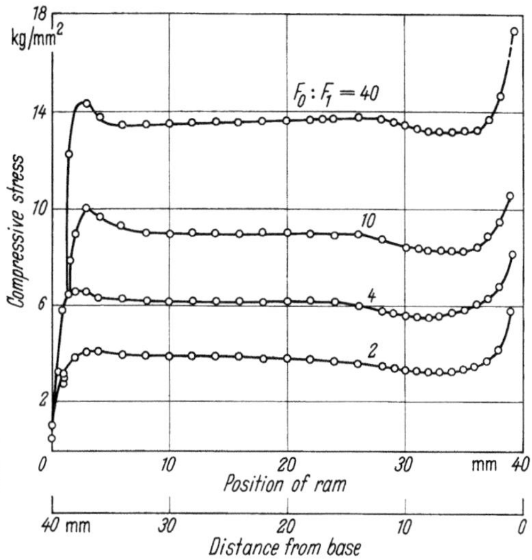

Fig. 326. Pressure variation in direct extrusion. (According to SIEBEL and FANGMEIER [1116]).

putting this diameter into the above equation. In this case, however, a certain addition must be made in view of the greater frictional losses in several small holes as compared with single larger ones. For example, the addition amounts to 22% for a 4-hole die (PEARSON [940]).

The question of the dependence of the extrusion pressure on the temperature is of great practical significance. PEARSON [940], on the basis of experiments on a series of low-melting metals, gives the curves reproduced in Fig. 327. According to PEARSON [940] they almost correspond to the equation—extrusion pressure $P = Ae^{-\alpha t}$—derived by SHISHOKIN [1111] from experiments on aluminium alloys. Here t signifies temperature and α a material constant which for lead has the value 0.0035. BOUTON and PHIPPS [118] found the same rule in lead and lead alloys for the temperature-dependence of an extrusion pressure associated with a given degree of compression. A linear relationship between extrusion pressure and temperature, such as BUTLER [162] assumes for lead, should only be valid for a narrow range of temperature.

As is known, the work used in the deformation of metal is mainly converted to heat. According to an evaluation by MASING [802] of the literature on various metals, an average fraction of 10% of the work of deformation remains as internal energy in the material. Disregarding heat losses, a temperature increase of 114 °C (91 °C) can be calculated for the extrusion of a lead billet of cross-section F_0 to cross-section $F_0/100$ ($F_0/40$) at room temperature. WATKINS [1245, 1245a] measured the temperature distribution actually occurring in lead and aluminium billets during extrusion, and represented it in the form of diagrams.

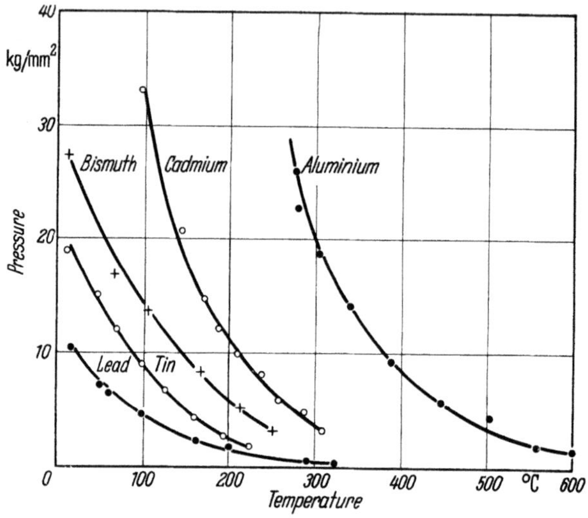

Fig. 327. Effect of temperature on the pressure in extrusion.
Ram velocities: Aluminium v = 5.08 mm/min, other metals v = 2.54 mm/min.
Percentage changes of cross-section: Aluminium 90%, other metals 96%. (According to PEARSON).

PEARSON [940] also gives numerical values for the dependence of the extrusion pressure on the velocity of the ram. On the double logarithmic scale, with the temperature as parameter, straight lines were obtained as shown in Fig. 328. This also agrees with the results of compression tests (BAILEY [33c], p. 206). SINGER and co-workers [1125a, 1125b] measured for lead the exit-temperature of extrusion without lubrication. The rise of temperature during the extrusion process can be divided into three regions. Region 1 comprises the rapid rise of temperature during the first 20% of the extrusion length; in region 2 the exit temperatures increase only slightly with the extrusion length. In the last 10% of the extrusion length, a greater rise in temperature occurs (region 3).

In Fig. 328a the rises of temperature of the slug are shown after leaving the die for an extrusion length of 63.5 mm. The lead billet had an original diameter of 50.8 mm and a length of 88.9 mm. It is found that the rises of temperature against the rates of extrusion plotted logarithmically give straight lines with the cross-section-ratio as parameter.

The linear relationship between the extrusion pressure and the rate of flow of the extrusion, which was found by FRISCH and THOMSEN [349] for low rates of extrusion, cannot be used in practice.

The extrusion pressure further depends on the shape and surface quality of the die. The die entry is often given a slight lip; however lips

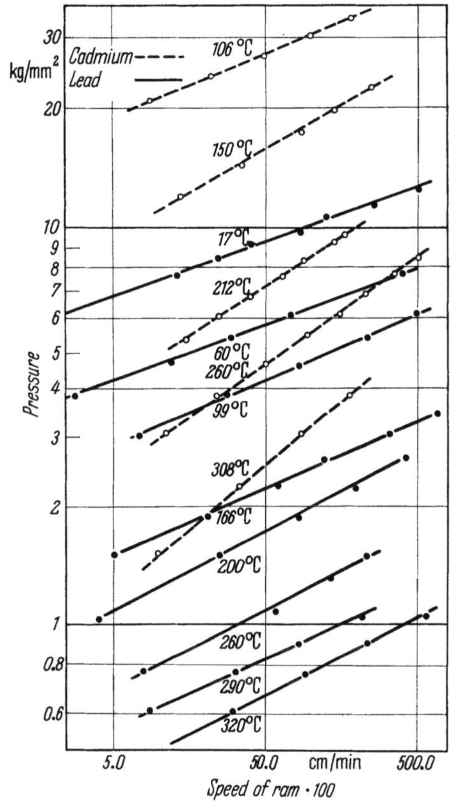

Fig. 328. Variation of pressure with the ram speed at various temperatures. (According to PEARSON [764a]).

of too large radius are disadvantageous since—according to the experiments of EISBEIN and SACHS [272]—they increase the extrusion pressure. The evaluation of the effect of a conical die has given different results. While LÖHBERG [757] found for zinc alloys an increase of extrusion pressure arising from the conical inlet, EISBEIN and SACHS [272] found, on the contrary, a decrease in the extrusion of lead. For lead, the length of the cylindrical part of the die affects the extrusion pressure only slightly if the hole is made cleanly. The effect of alloying elements on the extrusion pressure will be treated in the next section, on the extrusion of cable-sheathing. Further practical questions, such as the appearance of oxide-enriched zones in the extruded product, will also be discussed there. SACHS and DRAPER [1042] carried out exhaustive investigations on the special case of the direct extrusion of lead pipe. They put solid, unpierced lead billets in the container of the press. In

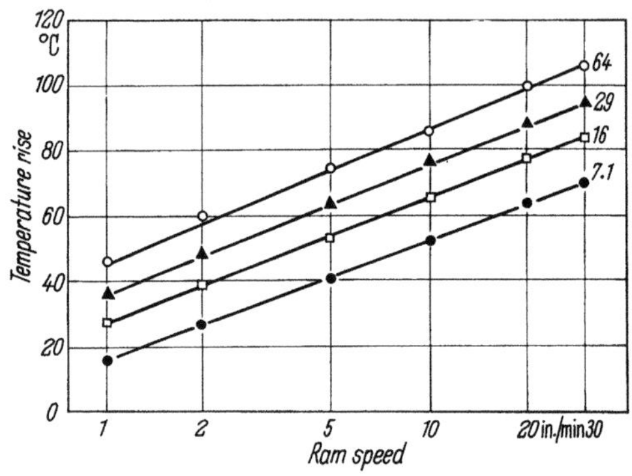

Fig. 328a. The relationship between temperature rise and ram speed for the extrusion of lead at various extrusion ratios shown on the curves. [1125a].

pipe extrusion, the forces depend essentially on the same conditions —that is, on the geometrical proportions—as in the extrusion of solid rods. In the extrusion of lead, as distinct from other nonferrous metals, only a slight tensile force is exerted on the mandrel. Uniform wall-thickness of the pipe depends to a great degree on the shape of the mandrel and its coaxial position in the container. Loss of material due to variable sizes of the extrusion residue can be reduced by effecting the piercing of the solid lead billet and extrusion to pipe in two separate operations.

The flow process in extrusion can be made visible in different ways, for instance, by using, as model materials, wax billets having layers of different colours, or alloy billets having layers of somewhat varied composition, which contrast with one another on etching. For the most part, use is made of billets cut longitudinally, the two halves having a rectangular network (grid lines, Fig. 329) inscribed on the cut surfaces, and being (after the two halves have been put together) then extruded like an undivided billet (e.g. EISBEIN and SACHS [272], THOMSEN [1192a], [1296]). Welding together of the two parts is prevented by a lubricant such as graphite, or white lead in oil. The distortion of the network consists, in the simplest case, in an extension in the axial direction and a contraction perpendicular to it, that is, it is similar to the deformation of the billet as a whole. This applies particularly to that part of the billet in the neighbourhood of the axis. In the outlying regions, on the other hand, the sides which were originally parallel become curved and displaced towards one another. On the basis of such experiments, PEARSON [940] distinguishes three kinds of flow in the extrusion process. Type *A* is characterised by the fact that friction between the outside of the billet and the container is excluded or at least extensively reduced by using inverted extrusion or, in the case of direct extrusion, by a lubricant. Type *B* consists of those extrusion processes in which friction works itself out fully on the skin of the billet. In that case the skin remains, in part, adherent to the container and is sheared off the bulk of the billet. The meshes of the network on the periphery of the emerging extrusion are therefore much more distorted than in Type *A*. Types *A* and *B* occur principally in low-melting metals. Type C is, on the contrary, found in extrusion processes which take place at high temperatures, for instance in the extrusion of copper. Here the outside of the billet cools considerably; first of all therefore, the warmer internal zones flow within a thick peripheral layer. Only afterwards does the latter take part in the flow process.

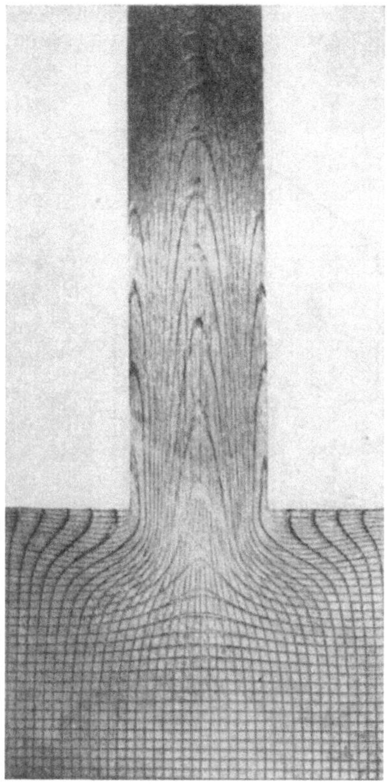

Fig. 329.
The flow pattern in inverted extrusion, revealed in the middle plane of the lead billet by grid lines. (According to YANG and THOMSEN [1296]).

In these experiments, designed to render the deformation process visible, the velocities of extrusion are lower than those usual in practice. Owing to the absence of permanent hardening, lead is a suitable model material for experiments at room temperature. It should however be borne in mind that in high-speed deformation processes, the duration of deformation is not sufficient for a complete regression of hardening.

According to HADDOW [461c], an almost ideal plastic isotropic material is obtained by 40% working of a lead alloy containing 0.065% Te. If cylindrical specimens pretreated in this way are compressed, thus obtaining the diagram of "true stress against the normal elongation $\varphi = \ln h_0/h$", then in the region of φ between 0.1 and 1.8 (h_0/h between 1.1 and 6), and a stress of 4.5 kg/mm², an almost horizontal part of the curve is obtained. The worked alloy is said still to exhibit this almost ideal plastic behaviour after storage for a year. If on the contrary use is made not of the cold-worked alloy but of that annealed at 220 °C, the course of the curve obtained is similar to that illustrated in Fig. 219 (static compression test).

The theoretical treatment of extrusion has achieved great advances in recent years, and is at present making much progress. However it is first of all limited to the case of plane strain, that is, the assumption is made that the average principal rate of elongation is zero: $\dot{\varepsilon}_2 = 0$, $\dot{\varepsilon}_3 = -\dot{\varepsilon}_1$ (on account of constancy of volume). This assumption is permissible if the billet extends very far (in theory, infinitely far) at right angles to the line of pressure and so has the same overall cross-section in this direction. In this case, the same process of conversion takes place in all cross-sections (e.g. plane A in Fig. 330), and is regarded as a stationary flow process. The concept of an infinitesimal parallelepipedal conversion, that is, the conversion of squares into equal sided parallelograms, leads to the view that the conversion is considered as essentially brought about by shearing stresses; the lines of maximum shear stress are called "slip lines"; the slip lines form an orthogonal network. Therefore at every place in the plastic region an infinitely small square can be inscribed in such a way that in the process of yielding its sides undergo no changes of length but only twists (Fig. 331a, b). On the contrary, another square, at an angle of 45° to the former one, shortens along one side and lengthens in the direction perpendicular to this. Owing to the condition of constancy of volume, the amounts of shortening and of lengthening are equal. Owing to the condition of continuity, the first kind of square forms the network of two systems of curves which intersect below 90°, the so-called slip line field. This should not be confused with the slip lines of crystals, as in plasticity theory, isotropic (that is non-crystalline) bodies are assumed.

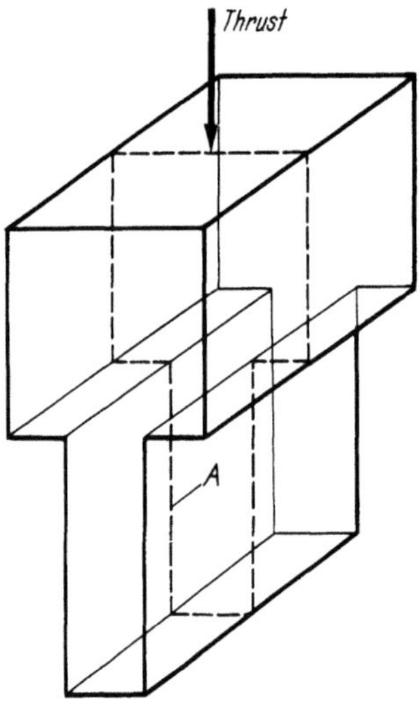

Fig. 330.
Model of plane strain in extrusion.

According to the theory of plastic yielding, the inception of yielding is connected with a condition for yielding which the state of stress must fulfil. Various such conditions for yielding have been put forward. However, only two of them—which differ only slightly from one another in their results (maximum 15% for the amount of the comparative stress in the uniaxial condition of stress)—have much significance today. According to the shear stress theory (TRESCA, ST. VENANT) the inception of yield is assumed when the greatest shear stress $\tau = \frac{\sigma_1 - \sigma_3}{2}$ exceeds a critical value. In practice this value is derived from the yield point of the uniaxial tensile test ($\sigma_1 = \sigma_F$, $\sigma_3 = 0$). According to the hypo-

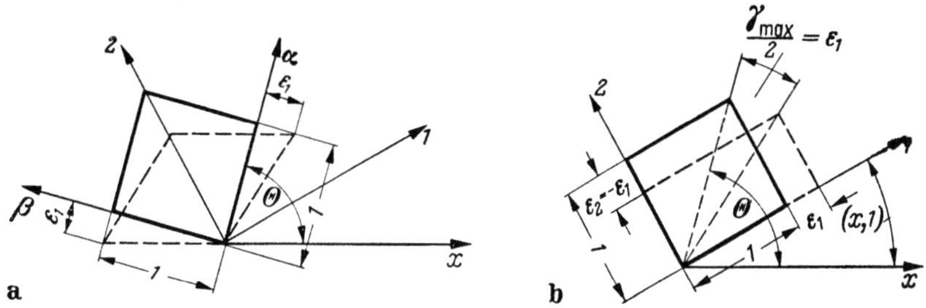

Fig. 331a and b. Strains of volume elements in plane strain. (a) Sides of elements parallel to slipline field. (b) Sides of elements parallel to directions of main stress. (According to HOFFMANN and SACHS [551]).

thesis of the energy of change of form put forward by v. MISES [858] and HENKY [511], the change of shape is regarded as divided into a volume change due to the average hydrostatic pressure $\frac{\sigma_1 + \sigma_2 + \sigma_3}{3}$ and a change of form due to shearing. Flow should occur when the energy necessary for change of form exceeds a critical value. From this, the yield stress is derived as

$$\sigma_0 = \frac{1}{\sqrt{2}} \sqrt{(\sigma_1 - \sigma_2)^2 + (\sigma_2 - \sigma_3)^2 + (\sigma_3 - \sigma_1)^2}.$$

HENKY [510] and PRANDTL [975] have brought the concept of the slip line field into relation with the conditions of flow and have treated a series of practical cases by means of mathematical-graphical methods. Use was made here of various geometrical properties of the slip lines. They proceed—for instance, to bounding surfaces, on which operate only forces at right angles and no frictional forces—at an angle of 45°. No flow occurs in regions without slip lines: these regions behave as though they were rigid (Fig. 332). In the recent development of the theory attempts are made to bring the speed of the process of deformation also into consideration, but this changes the fundamental structure of the theory. Some problems could be solved in the framework of the theory of plastic flow by means of the concept of slip lines (HILL [527]). Parallel with the theoretical advances, the experimental methods have also been further developed. For example, the velocity field during deformation was determined by means of the networks described above, on surfaces of separation of laminated blocks, the network being photographically checked in various successive stages of the process. Comparison of calculation and experiment gave satisfactory agreement. An example is given in Fig. 333, from a comprehensive survey by BISHOP [88]. The assumption of a plane state of deformation scarcely fits the

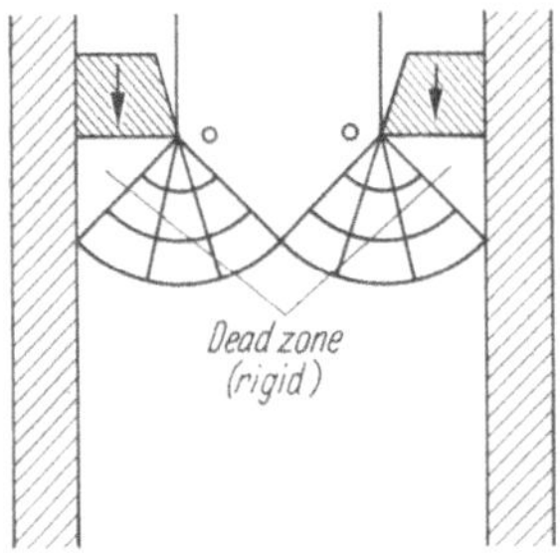

Fig. 332. Representation of slip line region for the case of inverted extrusion through a square die orifice with 50% reduction of cross-section. (According to HILL [527]).

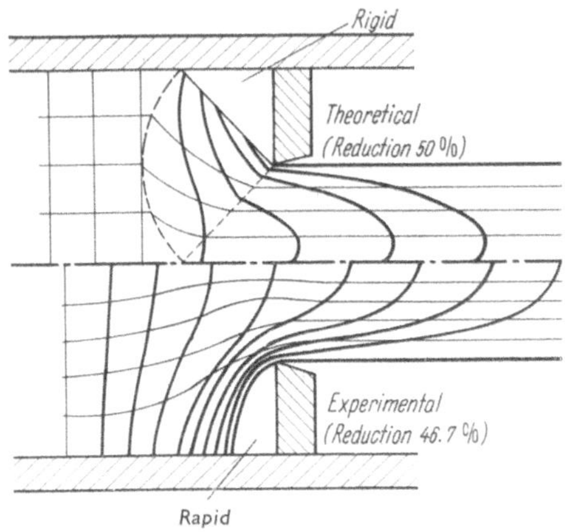

Fig. 333. Deformation of a grid network. Comparison with the theoretical solution (50% reduction of cross-section) for extrusion with plane deformation. (According to BISHOP).

practical cases. Fortunately it has been determined that the results thus obtained can be transferred as a first approximation to the axially-symmetric case which applies in the extrusion of round cross-sections. In addition to the determination of the state of deformation in the above cases, the relationships in the extrusion of tubes were treated in a further investigation by THOMSEN [1192]. The lead is inserted as a solid unpierced billet and extruded by the inverted process. The velocity fields and the stress fields, as well as the wall pressures of the billet against the container, are obtained.

2. Cable Sheathing[1]

a) General Design of Cables. A fundamental distinction is drawn between cable for power installations and for communications. The commonest type of communications cable is the paper-insulated type with a lead sheath. Here the insulation of the conductor is carried out—having regard to the significance of capacity and of dielectric

[1] For an exhaustive description of cable technology: KLEIN [669], Telegraphenbauordnung [1176] and EHLERS and LAU [271].

losses for the transmission properties—with the smallest possible amount of solids. Accordingly the conductors are encircled with paper and there is an air-space between the conductor and the paper (paper-air space insulation). The core conductors are stranded to the cable core on special machines. The core receives a wrapping of paper, and after careful drying, has a lead sheath extruded over the whole.

Nowadays—in addition to lead—aluminium, copper and steel (the latter two exclusively in the form of corrugated tube from welded strip) are used as metallic cable sheathing materials. Aluminium and steel are less corrosion-resistant than lead and therefore require additional protection. In addition to the metallic cable sheathing materials, plastics are also coming into prominence for sheathing, but they are not absolutely impervious to water.

Above all, lead-sheathed cables in power installations differ from communication cables in the limited number of conductors, in the size of the cross-sections of the conductors as well as the greater insulation against one another and the outside. Heating occurs during operation owing to the transmission of current through the conductor. Heat is also produced owing to the dielectric losses of the insulation, to eddy current losses in the lead sheath, and eddy current and hysteresis losses in the steel armouring. The loading tables are determined on an approximate temperature rise of 25 to 45 °C, according to the nominal voltage, the higher rise of temperature being associated with low voltage cables. Power cables are produced as single conductor or multi-conductor cables. Paper-lead cables are still very common. Here the insulation consists of paper which, after application to the cable, is carefully dried in vacuo and impregnated with a mineral oil-resin composition. In multi-core cables, either the stranded core conductors have a mutual insulation cover, the so-called belt-insulation which has a lower thickness than the insulation between the conductors; or, on the other hand, the insulation cover of the individual core conductor is (in high-voltage cables) surrounded with perforated metal-packed paper, the belt-insulation being omitted, and a conducting combination of the lead sheath with the metal coverings produced through a textile strip with webbed-in metal filaments (Höchstädter cable, or briefly H-cable); or, on the other hand, each individual conductor has a separate lead sheath (three-sheath cable).

The Höchstädter type of metal covering is also often retained here in order to exclude from the electrical field hollow spaces between the conductor insulation and the lead sheath.

In addition to the paper-lead cables, mention must be made of lead-sheathed power cables, with rubber or plastic insulation, for operating voltages up to 1 kV. The conductors are surrounded by a vulcanised rubber or plastic cover. In multi-core cables the individual conductors are stranded to one another and this bundle surrounded by a common cover, on which the lead sheath is laid. Paper-lead cables are not suitable as solid-type cables for the highest voltages. Fluctuations of temperature produce, in time, hollow spaces in the dielectric, in which ionisation occurs, because the dielectric has a high coefficient of thermal expansion in relation to the lead sheath. This reduces the quality of the insulation. In order to remedy this, oil-filled cable is predominantly used for voltages above 60 kV. The first such installation for 130 kV was provided in 1927 in New York and Chicago (Emanueli [274]). There next followed in 1927/28 in Nürnberg, Germany, an installation for 100 kV (Coninx [213], Held and Gasser [505]). As far as single-conductor cables are concerned, a hollow copper or aluminium conductor is used (Fig. 334); the paper insulation is impregnated with a thin oil. On loading the cable, heating and therewith expansion of the oil occurs. It passes from the paper insulation into the hollow conductor (between the individual strands comprising the latter) and flows off towards the ends into closed compensating vessels. On reduction of the load and cooling of the cable, the reverse process occurs. As the oil is always kept under a certain excess internal pressure, air or moisture cannot immediately penetrate the insulation even on mechanical damage. In oil-filled cables the insulation can be thinner than in solid-type cables. While the latter reach the limit of their stability at about 50 °C, an oil-filled cable can be permanently operated up to 80 °C. In this way, for the same voltage and conductor cross-section, an increase of about 50% in the transmission can be obtained as compared with solid type cables, and for the same thickness of insulation—owing to the greater ohmic resistance—an increase of 200%. A fundamentally similar means of avoiding ionisation is provided by the so-called "diaphragm" cables. The cable sheath here works like a diaphragm, that is, on heating it permits the cable-core to expand (again on cooling reverting to its original position under the action of an external pressure). The oldest type of diaphragm cable is the

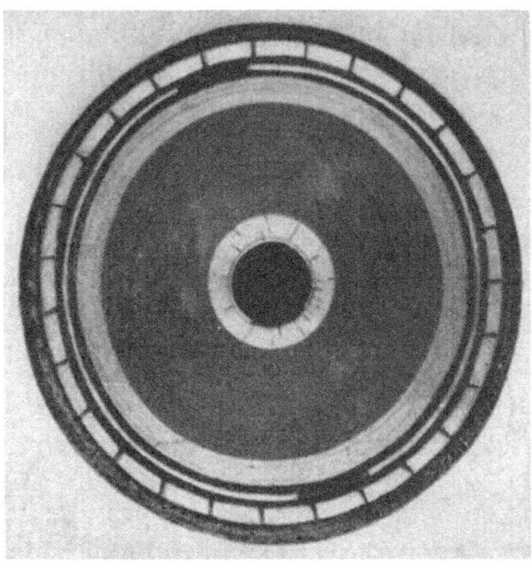

Fig. 334. Single core oil isolated cable for 125 kV. Composition; Aluminium core, oil impregnated paper insulation, lead sheathing, pressure-protective spiral, flat wire announing, asphalt coating (Siemens).

pressure cable. The lead-sheathed solid type cables are in a steel or lead tube, reinforced with copper or bronze strip. In this tube the cables are kept under an increased external gas pressure. In the flat cables, which are made as oil-filled cable without external excess pressure, the conductors, as shown in Fig. 335, lie close together. The flat lead sheath is reinforced with springy strips of bronze, the flat sides of the cable behaving like a springy membrane or diaphragm. Further, the so-called internal pressure cables have been developed in the regions of medium high voltage. The oxygen-free gas is inside the cable sheath which must therefore be given protection against pressure. The type known as the internal pressure cable has found most application. The path through the cable is kept free for the pressure gas, by keeping the spacers in the cables, the jointing sockets and the terminals free from compound.

Fig. 335. High-voltage flat cable 3 × 95 mm² for 66 kV service voltage according to JOHS. MØLLERHØJ (Nordiske Kabel- og Traadfabriker, Copenhagen). The numbers denote: *1* to *3* cores wrapped with metallised paper; *4* flat lead-sheathing; *5* padding (asphalt-coated fabric or paper); *6* two thin overlapped bronze strips; *7* wave-type springy bronze strip reinforcement; *8* bronze wire binding. (According to EHLERS [271]).

The design of the high-frequency cable departs from that of the communications cable (briefly sketched) and of the power cable. As a rule the two conductors are arranged coaxially; thus the one conductor forms the axis of the tubular second conductor. The transmission of high frequencies requires the use of insulating materials with particularly low dielectric losses. If certain ceramic materials are disregarded, paramount consideration should be given to those plastics the molecules of which contain no polar groups. Use is made, for instance, of polystyrene or polythene.

Further, as cable-type conductors, the so-called lead cable sheath conductors sometimes used in domestic installations, are distinguished from the lead-sheathed power cables. They are designed for use in a corrosive environment.

Bare lead-sheathed cables are also used as overhead cables. For this purpose they are suspended at short intervals by steel supports. In towns, there is widespread use of bare greased lead-sheathed cables drawn into hollow asphalted cement bricks or into glazed clay tubes: the so-called "ducted" cables. Buried cables are mostly laid under stone covers in order to guard against mechanical damage, from pickaxes, for example. They are preferably laid on sand. In the simplest case, the so-called bituminised cables, the lead sheath is given a protective covering of alternate layers of viscous compound and pre-impregnated paper, and over this a layer of pre-impregnated jute which is covered with hard compound. This protective coating should, above all, exclude the chemical effects of the environment. If an additional mechanical reinforcement is required, the cable is armoured. Armour is used on buried cables, river and submarine cables as well as mining shaft cables. Here the lead sheath receives—over the protective covering (already mentioned) of alternating layers of viscous compound and pre-impregnated paper—further layers of tar-coated steel strip, or flat or round or profiled wire, and over this an outer protective covering of asphalted jute. The type of armouring is determined by the magnitude of the tensile stress to be expected. Cables with galvanized steel wire armouring are used among others, as self-supporting cables (Thiel [1182]) or as ducted cables if increased tensile stresses occur as a result of greater length of cable to be drawn in, or of a distortion of the tube channel. If necessary, single conductor, alternating current cables can be armoured with non-magnetic material, for instance copper or aluminium alloys (ETZ [292] and Otten [929]) on account of magnetisation losses which could otherwise take place.

As indicated in what has been said, no very high demands can be made on the strength properties of the lead sheath. The individual significance of the mechanical properties of lead cable sheaths will be discussed elsewhere (p. 410).

The most important task of the lead sheath is to protect the cable against the penetration of moisture. Owing to the numerous environmental influences, special significance is attached to the corrosion-resistance of the lead. Attention is drawn to the appropriate section of this book, particularly to the chapter on soil corrosion (p. 308).

b) Production of Cable Sheathing. The production of cable sheathing and the choice of lead alloys (section *c*) have been comprehensively surveyed by Ehlers and Lau [271] and by Hiscock [534a], and attention is drawn to the reports in Lead 62 [735b]. Also of interest are older reports on the occurrence of failures by v. Göler [394], Prümm [981], Sherman [1110]. Now, as before, there is wide use of the vertical lead cable sheathing press, one design of which is illustrated in Fig. 336. First, its mode of operation will be described in broad outline. The water-cooled container shown in the middle of the illustration is charged with molten lead at a temperature of 375 to 400 °C. The melting plant necessary for this purpose will be discussed later. In a medium cable sheathing press a lead charge of 500 kg is usual. Solidification is complete in 6 to 9 minutes (ZICKRICK [1303]). When solidification is complete, the ram (which has been pressed onto the lead previously to prevent con-

Fig. 336. Vertical lead cable sheathing press. (Krupp).

traction cavities) is driven in by means of a hydraulic cylinder in the upper part of the press. The pressure water is delivered during operation of the press by a reciprocating pump; the incorporation of a pressure water accumulator is not usual. Below the container is the die-block. The cable core enters this from behind through the hollow mandrel holder. As the cable core passes through the space between the mandrel and the die, lead is formed round it as a continuous tube, the length of cable behind being pulled along by the lead. For very thin lead sheathing some further aid must be given. The finished cable runs over the rollers visible in the foreground, is cooled with water by means of a sprinkler and immediately reeled on to a drum.

The flow processes of the lead may now be described with reference to a die-box of older design, schematically illustrated in Fig. 337. It can be seen that on leaving the container by a bridge which rests on the mandrel holder, the metal is separated into two independent streams. These, with sharp deflections to left and right, enter into the lead chamber surrounding the mandrel, and weld below and above the mandrel

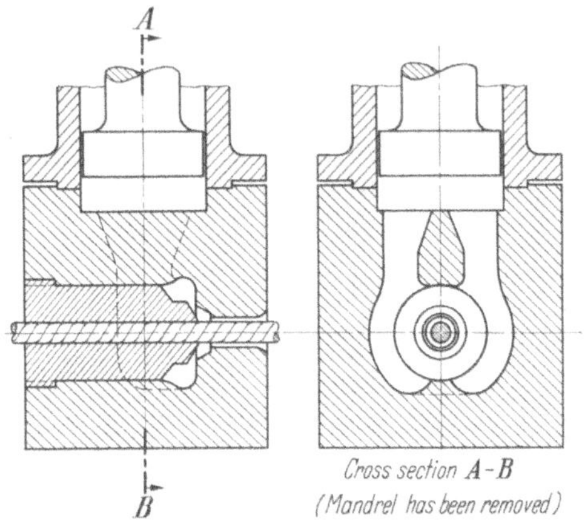

Fig. 337. Die box for cable sheathing with two welded seams. (According to v. Göler and Schmid [397]).

to a thick-walled tube. This is further extruded through the space between the mandrel and the die, and forms round the cable core and attains its final dimension. Before the introduction of the cable core into the die-box, a section of sheathing is extruded as an empty tube for the purpose of adjusting exact wall thickness. Only a moderate pressure is exerted on the cable core while sheathing.

The vertical lead cable sheathing press in Fig. 336 works with a die-box which produces only one welded seam. Division of the lead into two streams by a bridge is not provided for here. Rather the lead is pressed from above on to the mandrel, flows round it on both sides, to weld on the lower side and then be further extruded between the mandrel and the die (Fig. 338a and b). As the lead must travel different paths in order to form the lower and the upper side of the cable sheath, there is the possibility of a non-uniform distribution of pressure in the lead tube entering the die. Uniform pressure is ensured (among other things) by the bridge over the mandrel holder. In addition to the vertical press described, horizontal lead cable sheathing presses of the Huber type are still in use. They work with two horizontal containers, which are symmetrically arranged to the left and right of the cable. The lead streams enter on both sides, horizontally, into the die-box and weld here, as described above, with an upper and a lower seam. Vertical two-ram presses work on a similar principle (Prümm [981]).

The ram pressure on the lead varies with the temperature, the dimensions and the alloy. 3,000 kg/cm² is given as an average value. Ordinary industrial lead cable sheathing presses, with charges of between 120 kg

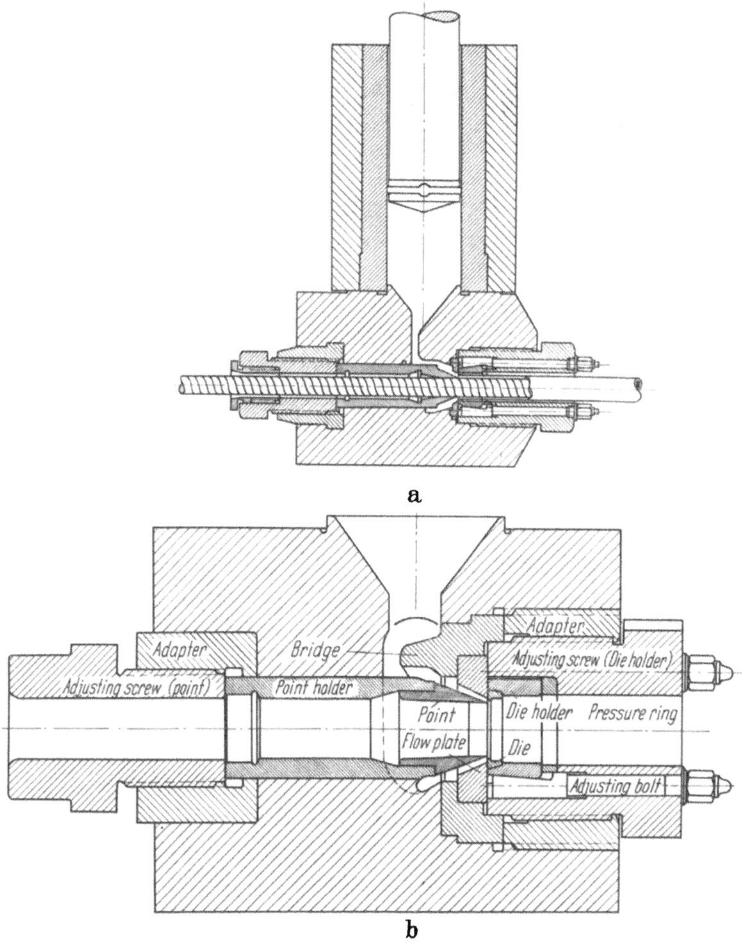

Fig. 338a and b. Die box of a lead cable sheathing press. (Krupp). (a) Schematic representation; (b) Section diagram.

and 1,000 kg of lead, are designed for thrusts of 800 to 3,800 tons[1]. The rate of extrusion amounts to 15 to 60 m/minutes (v. Göler [394]). The extrusion temperatures for lead cable sheathing are in general between 200 and 220 °C in the die-box, which must consequently be heated electrically or with gas. The temperature of the container and the die-box

[1] Information kindly provided by Hydraulik GmbH, Duisburg.

are checked by thermoelements. The extrusion of cable sheathing is discontinuous. After extrusion of a charge, the container is recharged with molten lead. On solidification, extrusion is resumed. In this way cables of any desired length can be sheathed.

In addition to hydraulically-operated cable sheathing presses, mechanical screw-operated presses have been increasingly introduced. They have the advantage of continuous operation, in which the diffi-

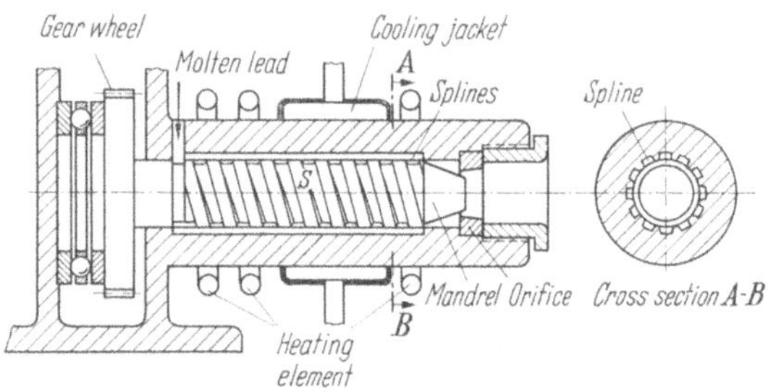

Fig. 339. Diagram of a continuous cable sheathing press.

culties of cable-sheath extrusion with intermittent charging (described below) are obviated [838]. The basic principle of the process is illustrated in Fig. 339, which roughly corresponds to a design by Henley [512] Molten lead is introduced at the rear of the press. It passes forward through the spaces between the central screw and the longitudinal grooves of the cylindrical inner wall of the housing, where it cools and solidifies. The solid lead is driven forward by the screw and extruded between the mandrel and the die as a tube; it adheres to the cable core, which passes through the hollow screw. Rotation of the entire lead content is prevented by the anchoring action of the longitudinal grooves. The Pirelli presses [965], developed in Milan and Southampton, are particularly well known. The Pirelli-General (Southampton) design differs in numerous respects from the diagram in Fig. 339. The cable is delivered through the stationary inner component. Around this, a hollow cylindrical screw, with an internal and an external thread, rotates in a stationary steel cylinder. There are longitudinal grooves on the inner part and on the steel cylinder. The molten lead enters between the inner part and the inner side of the screw, as well as between the outer side of the screw and the cylinder. On rotation of the screw, a conveying action occurs as soon as the lead in the middle and at the front solidifies. These thermal conditions are attained either by previously heating the machine by means of an induction coil round the cylinder or by inserting a cooling device. The solidified lead emerges between the die mounted at the front part of the cylinder and the mandrel set in the front end of the inner part. It forms, as a sheath, round the cable core delivered through the hollow inner part. BUNCE [143a] discusses the construction and some operation parameters of the Pirelli machine. While, basically, cable sheathing without a welded joint can be produced in the Pirelli press, in the more recent design the absence of a seam has been relinquished for the purpose of better centring of the mandrel. The screw has a front bearing which gives rise to radial distributions of the lead strands and rewelding. The good centrability offers the possibility of producing very thin lead sheathing.

While the continuous extrusion process described was hitherto introduced chiefly for the treatment of unalloyed and low-alloyed lead, it has also—according to RADTKE [991]—recently been applied to lead-antimony and lead-arsenic alloys.

Pirelli of Milan returned to the single cylinder screw press which improved their output per power unit. DAVINI [236a] proposes deoxidation by a lead-sodium alloy to obviate oxide inclusions in the lead. This measure reduces the number of interruptions for cleaning the screw. The Hansson-Robertson design placed the driving screw at right angles to the direction of the cable movement and thus returned to the ram press design, replacing the discontinuous ram action by a continuous screw drive (Fig. 339a). As an advantage of his arrangement, HANSSON [489a] mentions, among other things, the robust design of the screw. He reports on experiences in the extrusion of cable sheaths in unalloyed and alloyed lead. For purification of the lead melt before extrusion, he bubbles vaporised stearine through it. This should prevent disturbances in the operation of the press due to residues. Owing to the fact that the extrusion of some lead alloys creates difficulties in continuous screw presses, attempts were made to design continuous ram presses. The first step in this direction was mady by the firm of GLOVER who placed two vertical ram presses side by side and used them alternatively to extrude lead in a common mixing chamber.

The two channels for filling the mixing chamber extend in wedge shape in the direction of the lead stream, so that on descent of one piston only a little lead passes back into the other container, which is recharged. The Hydraulic Company introduced the non-return valve to stop the backflow in their design of a continuous ram press of 3,800 tons, which was originally designed to extrude aluminium cable sheathing [108c] but, when converted, could be used to extrude lead and lead alloys. During the downstroke (Fig. 339b) the antechamber filled with material and the surplus of the material was extruded through the die-box round the cable. During the upstroke, conducted by the bottom ram, the material from the antechamber was extruded, and at the same time the chamber was recharged for the next cycle. Another development in continuous ram press design was made by Krupp-Rheinhausen, whose press works with smaller capacities and a faster cycling operation, which decreases the size of the press without losing productivity. This press, KBK 1,500, is designed to operate with lead (1,500 tons). The principal features of this press are the main cylinder, two side cylinders, and the non-return valve, all of which are placed

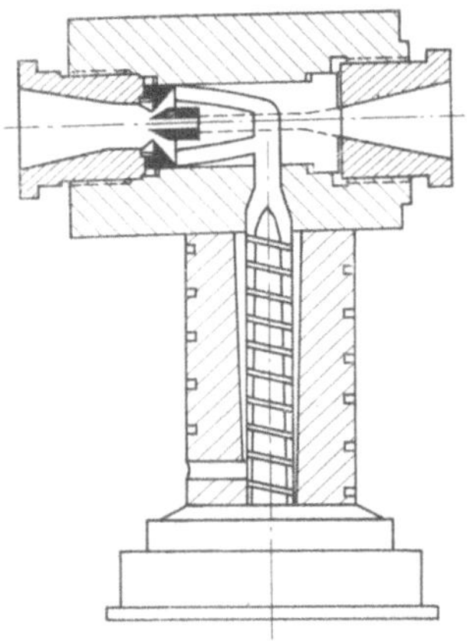

Fig. 339a. The Hansson-Robertson screw driven extrusion machine.

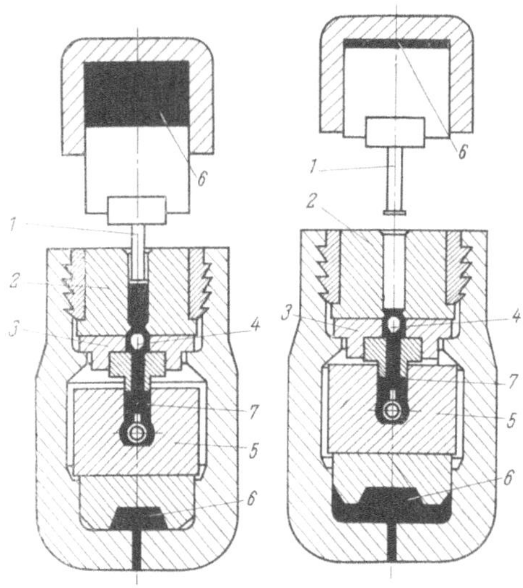

Fig. 339b. The hydraulic continuous ram extrusion press. (Downstroke position and upstroke position).

on a conventional die-box. During the downstroke of the main ram, extrusion takes place and at the same time two side cylinders are filled with material. After this the side rams take over the continuity of extrusion while the main cylinder is recharged for the next operation. This recharging takes place under vacuum protection. A Glover tray (see p. 424) with five times the capacity of the press is placed on top. The small capacity of the main cylinder decreases the tendency for any segregation, particularly as faster cooling can be achieved as a result.

c) Types of Lead, and Lead Alloys for Cable Sheathing. A cable must be easy to coil and uncoil; the required low rigidity of the sheath in the elastic and in the plastic range is synonymous with low values of the elastic modulus and of the yield point, as given for lead. In general, the creep limit given in the long-time test is not of great significance for the preservation of the cable sheath in practice; where higher permanent stresses of the cable covering are found, as in oil-filled or gas pressure cable, the internal pressure is taken up by the armour. Since, as indicated, deformations of the sheath can always be expected in manipulating and laying cable, a further region of plastic deformability is essential for good service. Criteria for that are the elongation and the reduction-in-area at fracture in the tensile test, and the value in the alternate bending test. LOESCHMANN [760] gives data on this subject for a number of lead cable-sheathing materials. Further, for cables which are subject to vibration — whether it be in transport, or additionally, in operation — an improved endurance or better alternating deformability as compared with commercial lead is to be desired. Moreover, a certain stability to recrystallisation, that is, absence of coarse grain as a consequence of deformation following frequent rises of temperature, should be present. The investigation by LOESCHMANN [760], already mentioned, provides ample data on the changes in mechanical properties and of grain size of lead cable sheathing materials, owing to deformation and heating. After cold working (10%) and heating to 150 °C, the alloys with 0.04% Cu, 0.04% Te, 0.6% Sb and 1% Sb, that is, alloys which still have heterogeneous structures at this temperature, remained fine-grained. The sheathing should show as little deterioration with time as possible and, in particular, it should not embrittle as a result of age-hardening phenomena. Furthermore, lead for cable sheathing should suffer as little drossing as possible when molten, it should give sound welded seams in the press at pressures as low as possible, and have good solderability. Finally, any alloying elements should be favourably priced.

A series of lead alloys are considered in papers by LATIN [733] as suitable for cable sheathing, embracing the above requirements. Among other things, he extruded pipes of different composition with external diameters of 50 to 75 mm and wall thicknesses of 2.5 to 3.3 mm in works presses or in the laboratory, and subjected them to the internal pressure test. The short-time tests lasted 30 to 100 hours, the long-time tests up to over one year. The short-time tests generally showed the knife-edge type of fracture (flowing wedge), except in alloy E (0.2% Sb + 0.4% Sn), which, in spite of this, showed a high elongation at fracture. Pipes which were left straight after extrusion gave higher values of elongation than those which had been bent on a drum. No significant importance is attached to this result, with cables in use. A summary of the results of the short-time internal pressure tests is provided by the following Table 26, according to LATIN [733].

When the results of the long-time tests were plotted, they showed in general "a diminishing extension characteristic" such as is undesirable in cable sheathing, that is, the expansion of the pipe until fracture decreased with decreasing stress (Fig. 340). In this respect, the only exceptions were commercial lead and alloys which formed dilute solid solutions (lead with 0.1% Sn, with 0.1% Sb). All the types of lead and

Table 26. *Internal Pressure, Short Time Tests on Pipes. According to* LATIN

Material	S = Straight Pipes C = Coiled Pipes	Greatest Deformation				Least Deformation			
		Hoop Stress kg/cm²	Time to Fracture in Hours	Average Strain Outside the Fractured Zone in %	Breaking Strain in %	Hoop Stress kg/cm²	Time to Fracture in Hours	Average Strain Outside the Fractured Zone in %	Breaking Strain in %
Pure Lead	S	58.7	26	12.85	17.4	58.0	26.5	10.1	13.3
	C	60.7	33	8.0	10.9	55.0	79.25	5.05	7.4
Lead with 0.1% Sn	S	61.2	43.5	10.4	17.8	67.2	26	7.7	12.45
	C	70.3	111	6.4	8.65	69.2	33	3.45	7.4
Lead with 0.06% Cu	S	64.0	66.5	10.95	14.45	64.7	112.5	9.5	14.75
	C	64.7	148	6.5	12.45	69.6	50	4.45	9.45
Lead with 0.1% Sb	S	61.9	80	10.85	12.85	61.9	83	9.45	11.9
Alloy E (0.4% Sn and 0.2% Sb)	C	102.7	40	19.2	23.4	99.0	59	14.6	18.95

lead alloys which were investigated, showed, with increasing duration of test, a transition of the knife-edge type of failure to the intercrystalline type of fracture, more or less without yield; the harder alloys fractured in this manner even in the short duration test. The diminishing extension characteristic of alloy *E* (0.2% Sb + 0.4% Sn) and of other harder alloys was connected with the processes occurring in age-hardening, but also with the creep process as such. Grain size also plays an important part. If, for example, alloy *E* which normally exhibited a grain size of 0.3 mm, were extruded to strip at low temperatures, the grain size was only just 0.05 mm, and the alloy showed now an increasing extension characteristic and a yield fracture. Creep tests on pipes under internal pressure with simultaneous bending vibration, showed not only the expected increase of creep rate, but also a considerable decrease of expansion at fracture. The "disc test" described in this investigation, in which alloys for pressure-cable sheathing can be tested, should be noted. The disc, of 57.2 mm dia, is subjected, on one side, to a gas pressure through an aperture of

25.4 mm dia, so that it slowly bulges, and bursts at the end of the creep test. BAILEY [33b] gives a good review of the development and properties of lead alloys for cable sheathing.

Until after the last war, commercial lead was the cable sheathing material most used in Germany. A prerequisite for the use of commercial lead is that it should be laid in an environment free from vibration. Its

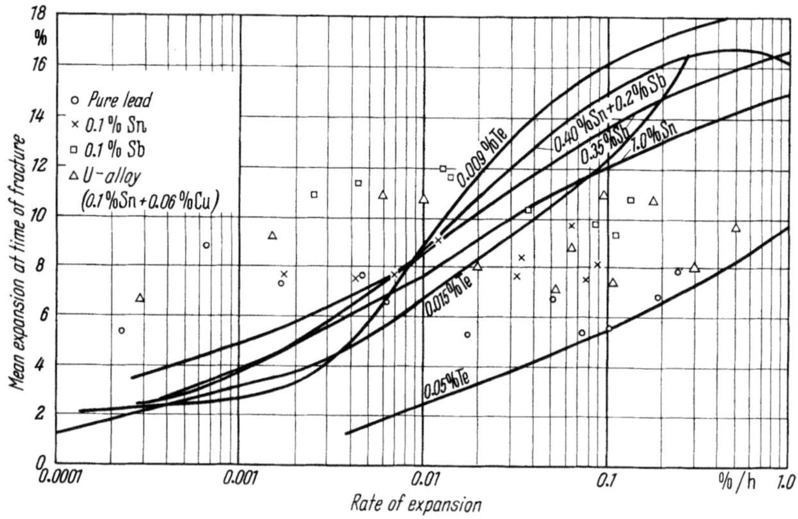

Fig. 340. Expansion (at fracture) of lead pipe under internal pressure. (According to LATIN). ○ Commercial lead; × 0.1% Sn; □ 0.1= Sb; △ U-alloy (0.1% Sn + 0.06% Cu).

use has declined strongly in favour of alloyed cable sheathing. The reason for this development is, first, the fact that with practically every cable, vibration is to be expected during transport, and second, the increasing purity of smelter lead, which simultaneously results in an increasing susceptibility to endurance fracture. EMMERICH and BECKMANN [278] described such damage due to transport in a lorry over several hundred kilometres. It concerned a bare telephone cable, which, owing to its paper-air insulation and absence of armour, had a greater tendency to vibration in transport than perhaps a solid filled power cable. The cable sheathing lead used was of high purity but with a bismuth content of 0.05%; the bismuth impurity had not brought about the occurrence of cracks. In supplementary experiments no disadvantageous effect of bismuth up to 0.1% could be demonstrated. An unalloyed lead with maximum contents of antimony up to 0.15% (for certain purposes a minimum content of 0.05% Sb is specified), tin up to 0.1%, copper up to 0.06% is included in the British Specification BS 801 of 1953. The minimum lead content is deliberately fixed at only 99.8%. Here we are therefore virtually dealing with a low-alloyed lead (alloy *F*).

The main advantage of alloys is their increased fatigue strength. Alloys with 3% Sn were common in America before the first World War; from considerations of price, an alloy with 0.75% Sb has been used there for overhead cables since 1913, for power cables a softer alloy with 1 and 2% Sn. This was known in England as alloy *A* (2% Sn). In Germany also, after the First World War, the alloy with 3% Sn predominated; its tin content was later reduced. The alloy with 0.5 to 0.95% Sb (alloy *B*) is widely used, and laid down both in the German Standard Specification DIN 17640 of August 1959 and in the British Standard Specification BS 801. It age-hardens on storage so that with the increased fatigue strength there is simultaneously a greater rigidity of the lead-sheath. The use of this age-hardening alloy is not advised for power cables, where daily cycles of heating and cooling are to be expected [1209]. Fig. 341 shows the longitudinal section through a cable sheath of lead with antimony. The line structure due to the coring of the cast billet is characteristic, the lines with antimony inclusions originating from the eutectic residual melt. If the extrusion temperature is above the eutectic melting temperature of 251 °C, the alloy becomes molten here and there. Regions with eutectic cast structure then occur in the cable sheath, as can be clearly recognised in Fig. 342. The extrusion temperature here was demonstrably too high. Owing to intercrystalline cracks in the sheath (Fig. 343) the cable had to be removed. It may be suspected that the cracks emanated from the antimony-rich parts with the eutectic structure. Loeschmann [760] also concludes from alternating moistening tests that cable sheathing lead, alloyed with antimony, has increased safety towards atmospheric corrosion, as compared with unalloyed lead.

The directional precipitates, formed after long storage, were described on page 266 in connection with age-hardening. There, the explanation is also given of the appearance of coarse segregation at the grain boundaries. This edge-development is still more strongly marked in alloys with antimony plus tin. In supersaturated lead-tin alloys, coarse precipitates occur in the entire structure following storage (Fig. 93). Whether the technological properties of lead cable sheathing alter in the course of time, owing to these processes, has not been investigated. Of the ternary alloys developed in England, the alloy with 0.25% Cd + 1.5% Sn has disappeared from British Standard Specification BS 801 of 1953. A similar alloy with 0.4% Sn + 0.15% Cd (alloy *C*) and an alloy with 0.2% Sn + 0.075% Cd (alloy $^1/_2$ *C*) are however still in general use, particularly in cases where no very high fatigue strength is required. BS 801 now lists, in addition to 99.8% commercial lead and alloy *B*, only alloy *E* with 0.4% Sn + 0.2% Sb and alloy *D* with 0.5% Sb + 0.25% Cd; the latter age-hardens. A general disadvantage of alloys, particularly alloys of lead and antimony, is the higher extrusion pressure as compared with commercial lead and low-alloyed lead cable sheathing. According to

Fig. 341. Longitudinal section of a lead-antimony cable sheath. ×500

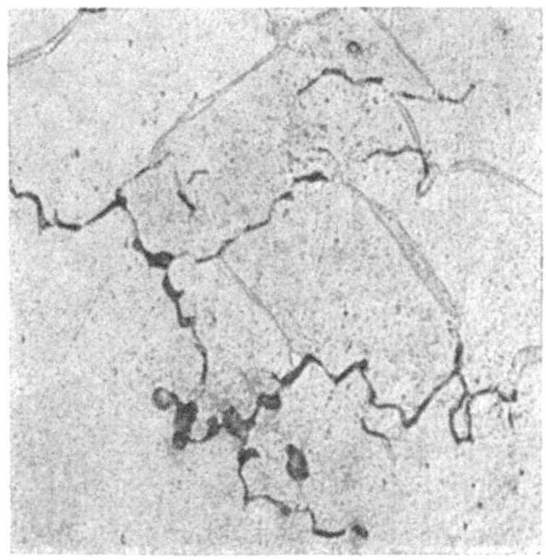

Fig. 343. The lead-antimony cable sheath of Fig. 342. Intercrystalline cracks as a result of transport by ship. ×500

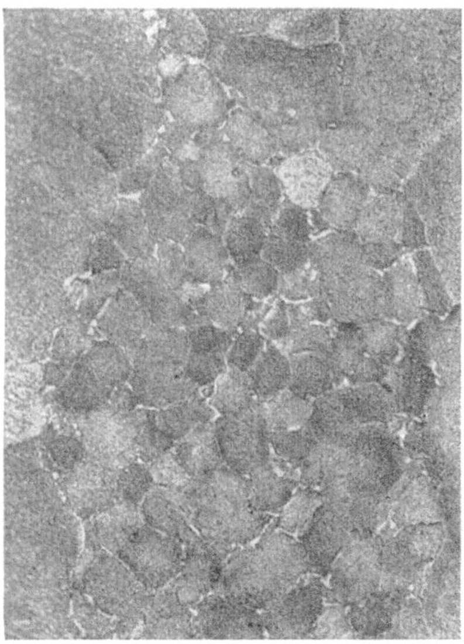

Fig. 342. Lead-antimony cable sheath. Eutectic on grain-boundaries as a result of remelting. ×500

LOESCHMANN [759], elements (such as antimony and tin) which form solid solutions, give rise to increase of extrusion pressure; on the other hand, insoluble or slightly soluble alloying additions such as copper, tellurium, aluminium and calcium increase the required pressure only insignificantly. Increase of fatigue strength, with hardly any augmentation of hardness and rigidity, is emphasised as a special characteristic of lead-tellurium as compared with lead-antimony (SINGLETON and JONES [1126]). Lead-tellurium alloys were specially recommended for submarine cable (BUSS and MEYER [159]). The usual tellurium content is about 0.04%. Further data on submarine cable are provided by GIBLIN [376a]. He gives preference to alloy *E* over alloy $^1/_2\,C$ as the former exhibits higher fatigue strength, although the creep ductility is somewhat lower. TELLIER [1176a] mentions alloy $^1/_2\,C$ in connection with the Cross-Channel line (± 100 kV). A certain increase of endurance without increase of extrusion pressure, with simultaneous grain-refinement, is also brought about by a copper addition up to about 0.06% — that is, the content of the eutetic lead-copper alloy (WEHR [570]). This alloy has been introduced in Germany instead of commercial lead (DIN 17640, "Lead and Lead Alloys for Cable Sheathing"), particularly having regard to damage in transport. Some cable works provide for a tin addition of a few multiples of 0.01%, or a small antimony content, in addition to copper. The tin-content is said to reduce the drossing tendency of the melt. The antimony content is chosen at 0.1 to 0.15% so that in any case the drossing maximum at about 0.01 to 0.05% Sb (p. 331) is exceeded (BOESCHE [100]). The maximum antimony content of 0.15% in unalloyed lead in the British Standard is to be interpreted similarly. The 0.1 to 0.15% antimony addition, which remains completely dissolved in the lead, also contributes somewhat to increasing the endurance of the cable sheathing. The advantage of the copper addition — apart from an increase in the endurance — is expressed in considerable grain refinement. This eliminates the situation in which large single grains can extend across the entire wall-thickness of the cable sheathing. In view of the favourable effect of this small addition of copper, provision has been made in Germany for a copper content of 0.04% in most cable sheathing alloys in addition to the main alloying element, e.g., 0.5 to 1% Sb. In this way a desirable reduction of the extrusion pressure is effected in this group of alloys (GLANDER [387]). The British Standard Specification also provides for the possibility of a copper addition in all alloys, since the maximum copper content is uniformly fixed at 0.06%. Apart from the alloys mentioned, EMICKE [275] has recommended for cable sheathing a lead-arsenic alloy with about 0.1% As. The lead-arsenic alloys found particular favour in the United States, where their good creep strength was recognised (p. 240). In general use is made not of binary lead-arsenic alloys, but of more complicated alloys, for example the compositions 0.15 to 0.2% As + copper and bismuth (*F*-3 alloy), 0.2% Sb + 0.015% As or 0.4% Sb + 0.03% As (HISCOCK [534a]). Alloys with small contents of antimony (0.2 to 0.4%) + arsenic (0.02 to 0.05%) + tellurium (0.02 to

0.05%) are also of interest. Tellurium produces the desired structural stability instead of copper. Copper is avoided in this alloy, as it combines with arsenic (BROCK and MCKEOWN [135a]). Such an alloy, described in a patent specification, containing 99.65% lead, in addition to the arsenic content, has a little tin and bismuth in solid solution. It is said to withstand an internal pressure 50% higher than the usual lead-copper alloy, and not to age-harden, as is shown by the constancy of the tensile strength over a period of many years. Zinc contents (e.g., 0.8%) were also tried as an addition to cable sheathing lead in Germany during the war. The conclusions of technical investigations by PFENDER and SCHULZE [957] throughout indicated promising results. Lead-calcium alloys, and those with calcium + copper, also have very good mechanical properties, for example, high creep strength as well as high endurance which makes them seem suitable for cable sheathing. The evident fact that they have not come into use may be related to difficulties in melting and casting.

Table 27 provides data on some mechanical properties of cable sheathing materials, extracted from a compilation by BASSETT [51], and from v. GÖLER and GREFF [394].

Table 27. *Composition and Properties of cable sheathing Alloys*

	Tensile Strength in kg/mm^2	Elongation in %	Bending fatigue strength in kg/mm^2
Commercial Lead (99.85)[3]	1.34	50	0.28[1]
Lead with 0.06% Cu	1.62	40	0.42[1]
„ „ 1% Sn	1.7	40	0.50[2]
„ „ 2% Sn	2.11	45	0.56[1]
„ „ 3% Sn	2.2	35	0.55[2]
„ „ 0.6% Sb	2.0	30	0.70[2]
„ „ 0.06% Cu + 0.85% Sb	2.96	10	0.85[1]
„ „ 0.03% Ca	1.8	40	0.60[2]
„ „ 0.035% Ca + 0.6% Cu	2.82	30	0.99[1]
„ „ 0.05% Te	2.0	45	0.70[2]
„ „ 0.07% Te	1.97	45	0.63[1]
„ „ 0.25% Cd + 0.5% Sb	2.96	15	0.85[1]
„ „ 0.25% Cd + 1.5% Sn	1.97	30	0.63[1]

[1] 10^7 Alternations of load [51].
[2] $2 \cdot 10^7$ Alternations of load. 3,000 rotations/min [394].
[3] ASTM Tab. 3.

Further technical data are given in the work of Loeschmann [760] already mentioned, by Jungkönig [638a], by Bolshanina [108a] and particularly in Lead 62 [735b]. A survey of the endurance and the compressive fatigue strength of lead cable sheathing alloys is given in

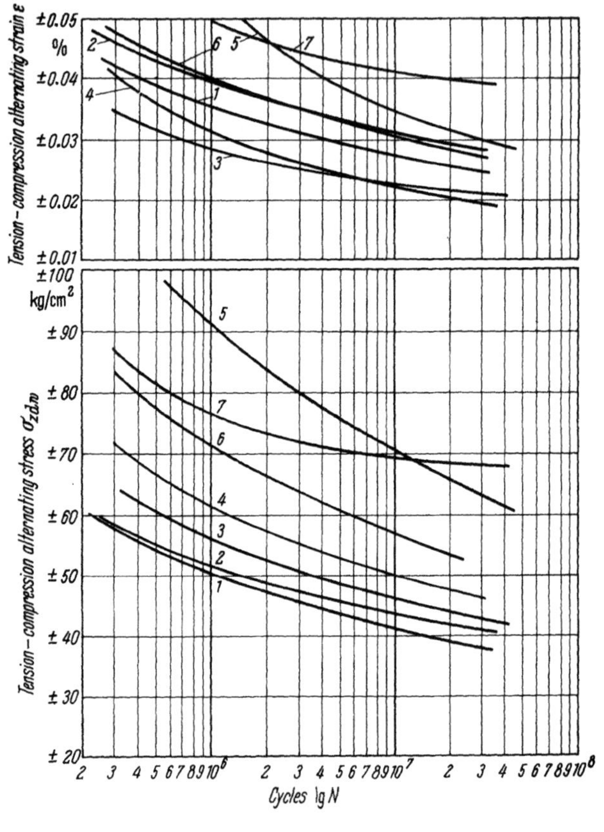

Fig. 344. Endurance and fatigue strength of lead cable sheathing alloys in the tensile-compression test Load cycles 100/sec [570].

1 Commercial lead 99.94%
2 0.03 to 0.05% Cu
3 0.03 to 0.05% Cu, 0.13% Sb
4 0.002% Cu, 0.13% Sb
5 0.03 to 0.05% Cu, 0.56% Sb
6 0.03 to 0.05% Cu, 0.04% Te
7 0.03 to 0.05% Cu, 2.5% Sn

Fig. 344 from measurements by Wehr [570] on a tensile-compression testing machine with magnetic drive [572]. The alloys were produced, as pipes of alternating larger and smaller wall-thickness, on an experimental press. By cutting up the pipes, a number of specimens was obtained with thickened ends to prevent fractures due to clamping.

In addition to the cable-sheathing materials in the table, other alloys have been used, for example in Sweden lead with 0.2% Ag (HANSSON [489a]).

As already mentioned, allowance must in general be made in lead alloys for a deterioration in extrudability with increase of alternating impact strength. On this point, static compression tests were carried out at 200 °C on various types of lead and lead alloys (ZICKRICK [1303]).

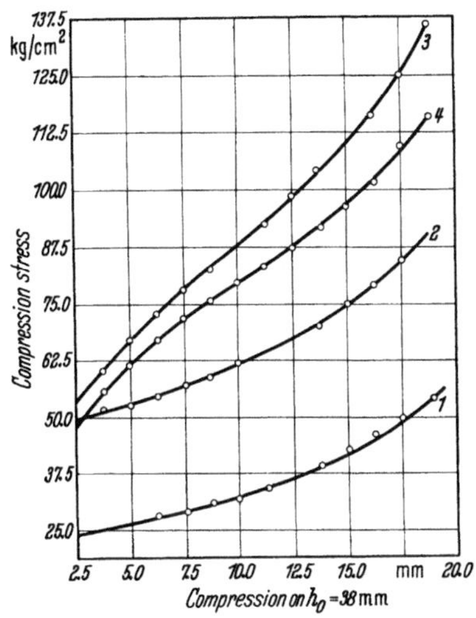

Fig. 345. Compression-diagrams of cable sheathing alloys at 200 °C. Dimensions of test-pieces 19 × 19 × 38 mm. Rate of compression 0.85 mm/min.
1 Lead A (99.99%); *2* Lead A with 0.06% Cu; *3* Lead A with 1% Sb; *4* Lead B (0.06% Cu) with 1% Sb. (According to ZICKRICK [1303]).

Fig. 345 shows some of the compression curves obtained. The most important results are as follows: the "crushing pressure" of the purest lead is almost doubled by 0.06% Cu. Further increase of the copper content beyond this eutectic concentration has only slight effect. The crushing pressure of commercial lead is also greatly increased by 1% Sb (likewise by 2.25% Sn and 1.5% Sn + 0.25% Cd). The effects of antimony or tin and of copper are not additive. The crushing pressure of an alloy with 1% Sb is in fact slightly reduced by addition of 0.06% Cu. This also agrees with the experience of the cable companies and with the later work of GLANDER [387] which has been mentioned. Bismuth contents in cable-sheathing lead only raise the crushing pressure slightly. BOUTON and PHIPPS [118] carried out similar tests. They found, among other things, that at a deformation temperature of 220 °C the compression curves of lead with 1% Sb + 0.06% Cu, with 0.015% As + 0.1% Sn, and with 0.06% Te, lay close together. At a deformation temperature of 265 °C, the lead-tellurium alloy mentioned has relatively the highest resistance. It is obvious that at this temperature the entire tellurium content is in solution.

d) Special Phenomena and Defects. Imperfect welding in the joint of cable sheath signifies a mechanical weakening and therewith the possibility of longitudinal cracks. Good welding presupposes an extrusion temperature which is not too low. In order to clear up this question, strips were extruded in a small press at various temperatures (ERDMANN-JESNITZER and HANEMANN [286]). The lead strand was divided on the way from the container to the die by a crosspiece and welding occurred behind this crosspiece. The quality of the welded joint was tested by tensile tests on flat rods. They were taken diagonally to the strip, so that they were completely crossed in the middle by the welded seam over the whole cross-section. After extrusion at 150 °C, satisfactory welding was ascertained in all the cable-sheathing alloys investigated, without exception. Having regard to the output of the press, only commercial lead could be extruded at room temperature. The welding was defective. This is purely a question of temperature, as with the experimental arrangement indicated, oxides could not occur in the welded joint. DUNSHEATH [263] carried out similar experiments.

In practice, conditions are more severe because oxides can also occur in the welded joint. This is connected with the mode of operation of ram presses. The oxides, as far as they are not included within the lead, have a tendency to be on the surface of the billet after solidification. After the ram has been driven down, they therefore form a part of the lead residue, which is not extruded. When the container is recharged with molten lead, this welds up with the lead residue, without completely melting it. Thus the oxides have no opportunity of rising to the surface of the bath and are therefore enriched at the boundary of the two charges. On extrusion of the new charge, the boundary oxide zone now moves into the cable sheathing, making itself more or less noticeable over long stretches. Cross-sections through the cable show, at various distances, a structure such as is reproduced in Fig. 346 (BASSETT and SNYDER [53]). "Tongues" can be observed on both sides near the upper — and at somewhat greater distance from the lower — welded joints. The lead of the new charge is separated from the first charge by an upper and a lower tongue. The detailed structure of a tongue is shown in Figs. 347 and 348. The oxide inclusions and cavities, as well as the refined grain in the boundary zone, are noteworthy. The latter is a result of many impurities, which prevent crystal growth on recrystallisation. The cavities can, in some circumstances, extend through the entire thickness of the sheath, leading to permeability. As far as they get into the welded joint, the oxide inclusions give further weak spots, since they constitute a reduction of cross-section. In cables with defective welding, longitudinal cracks often occur near to the weld under internal pressure in the short-time tests; on the other hand, on stressing for several days with a quarter of

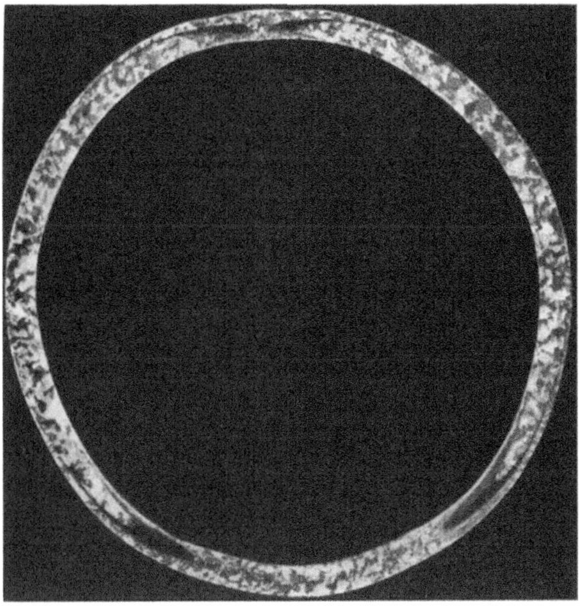

Fig. 346. Cable sheath with portions of different container charges. (According to BASSETT and SNYDER [53]).

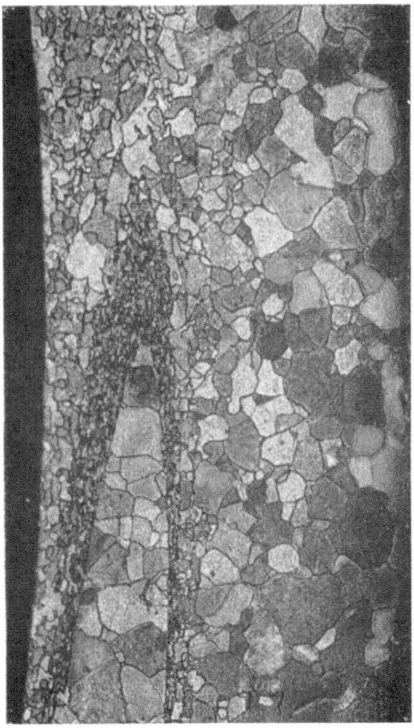

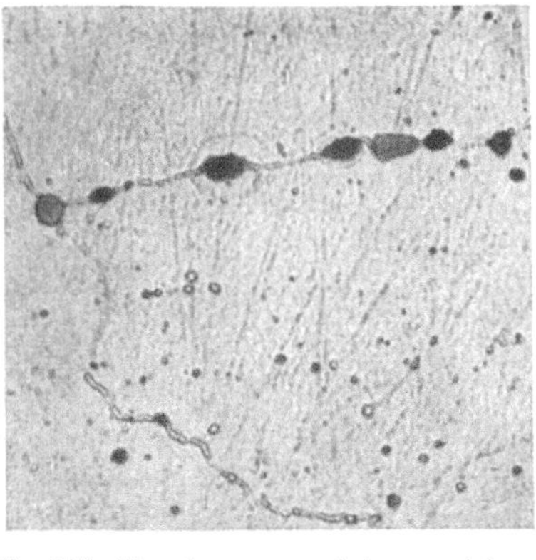

Fig. 348. Boundary zone of two container charges from Fig. 347 with oxide inclusions. ×1,200

Fig. 347. Commercial lead cable sheath. Cross-section with portions of two container charges. ×10

the normal load to fracture (similar to practical operation) they occurred even in the weld (SACHS [1040]). This indicates that not only the strength in the short-time tests but also the creep of the joint is of importance. It is possible that creep under internal pressure is promoted in the fine-grained joint zone.

Several investigations have been concerned with the formation of the tongues described above (DUNSHEATH and TUNSTALL [264], v. GÖLER and SCHMID [397]). In particular, the latter reference will be considered as it is closely concerned with the flow processes in the die box. Model experiments with wax were carried out in the die box, illustrated in Fig. 337, which produces two welded joints. The container was first charged with white wax which was extruded. Then alternate discs of black and grey plasticine were inserted, and the contents of the container were extruded. Cross-sections through the extruded tube at various distances, as well as sections through the extruded residue, led to conclusions on the flow process. The principal finding was that the different areas of a horizontal layer in the container flow with very different speeds corresponding to the different flow-paths. Consequently, zones from several charges are distributed on a cross-section through the complete tube. Figs. 349 and 350 show two such sections, which reveal the residue of the first charge and a further four or five layers of the second charge. A new layer at first appears in the right and left upper halves under the inscribed angles, then spreading from there upwards and downwards. The upper and certainly the lower welded joint still contain residues of the first charge. A cross-section through the extrusion residue, which contains the wax strands coming from the container, both from the left and the right of the bridge (Fig. 351), and in which (below) the strands were diverted forwards, allows the flow process to be clearly seen. It is apparent that the regions near the wall are very far behind in the rate of flow. On both sides, the grey plasticine last extruded has only advanced above and been diverted in the direction of the cable. It therefore appears first in the cable at the places described above. In the lower part of the extrusion chamber an almost dead space has formed, containing the residues of the first charge. A considerable length of the lower weld is fed from here with white plasticine.

The results are very important for the understanding of the production and structure of lead cable-sheaths, although a complete transfer of the results to other presses and to lead, as the material, is not justifiable, in view of the different wall-friction and viscosity. The fact that the upper weld is usually narrow, the lower broad and composed of several zones, is explained by the preferred formation of the dead space on the lower side (Fig. 351). More favourable flow relationships could be attained in the model experiments by a somewhat different arrangement of the die box. As parts from different charges must be expected in the cross-section of a cable, then on change of alloy the die box must be completely emptied (SHERMAN [1110]). If from any reasons a charge shows particularly heavy drossing, there is the risk that the welds will be fed with this dross for a long time.

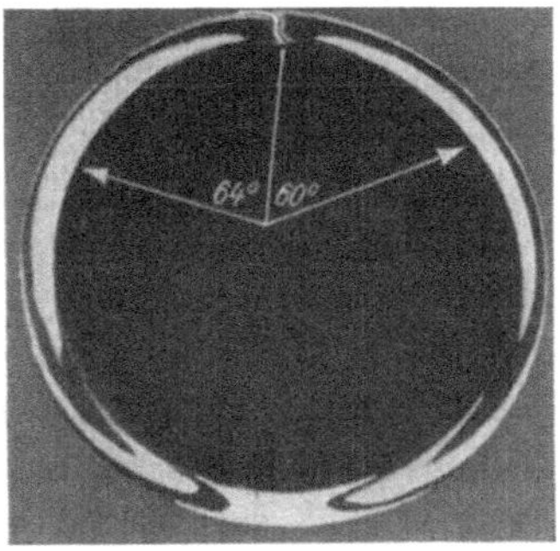

Fig. 349

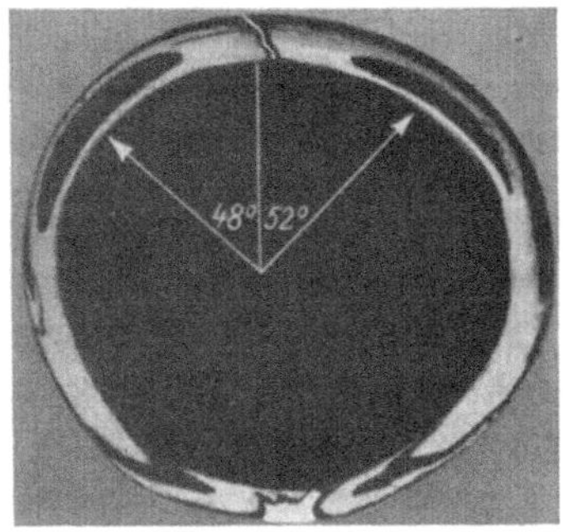

Fig. 350

Fig. 349 and 350. Cable sheathing produced in a model experiment made with plasticine of different colours. The section in Fig. 350 (right) is 2.9 m behind the section in Fig. 349. (According to v. GÖLER and SCHMID [397]). ×2

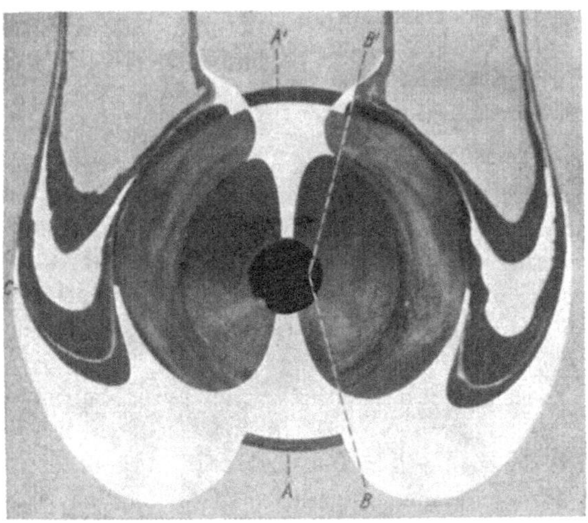

Fig. 351. Extrusion residue of the model cable sheathing of Fig. 349/350. Section in the plane of separation perpendicular to the direction of the cable. (According to v. GÖLER and SCHMID).

As the question of the welded joints has long been under consideration, many suggestions have been made for overcoming the defects indicated, sometimes with success. Extrusion with omission of the upper weld ([1120], SANDELIN [1045]) has already been mentioned. Fig. 352 represents the lower joint of a commercial lead-sheath from such a press. The welding is unexceptionable. Naturally it is important that the sheathing lead should contain no oxide inclusions, and further that the grades of lead used should be little prone to drossing. This question was thoroughly discussed on p. 72 and on p. 327. The damage of drossing can be reduced by suitable melting and casting (p. 321). The lead should be tapped from the melting pot as deeply as possible. Discharge through pipes is preferable to the use of troughs. Rapid lowering of the ram after charging is of importance. Installations have been described where the lead is passed in a closed, heated pipe system from a central kettle to the containers of several presses and returned again from there (SIMONDS [1125]). In other installations the closed melting space and the container filled with CO_2 and the lead first comes into contact with air as the finished cable-sheath [281]. The GLOVER-tray system consists of a funnelshaped receptacle at the upper end of the container, which is constantly filled with molten lead up to a minimum height. Below the surface of the melt are 4 to 6 slots for delivery of the molten lead into the container. The lead oxide floating on the surface therefore never has an opportunity of reaching the container. The ram is only about half as long as the container. Therefore when the ram descends, the lower half of the container remains filled with lead. This lead billet serves to cool the new charge; thus the billet is warmed and retains the temperature necessary for extrusion. The lower half of the container is therefore not cooled from outside, Fig. 353 (GIBLIN [376a]).

Fig. 352. Commercial lead cable sheath with only one welded joint. ×10

Further measures (abandoning complete prevention of formation of oxide) are aimed at removing the dross formed, which is less expensive. Skimming the drossed, still molten surface of the container charge in the so-called lead separating apparatus [291] has proved particularly successful, as also the filling attachment according to Fig. 354 (WIEGHARDT [1270]). The drossed surface of the lead billet is driven out in the molten or solid state by lowering the ram (which is of conical shape in front). The melt runs into the annular channel, divided by six ribs, from which it can easily be removed after solidifying. Lead which is already solid is extruded upwards as a sort of tube, which for better removal, is slit through the three bars touching the ram, on the inner wall of the apparatus.

It might be thought that the problem of dross-formation could also be solved by extruding clean lead billets inserted in the container. At the same time the period required for solidification of the lead melt in the

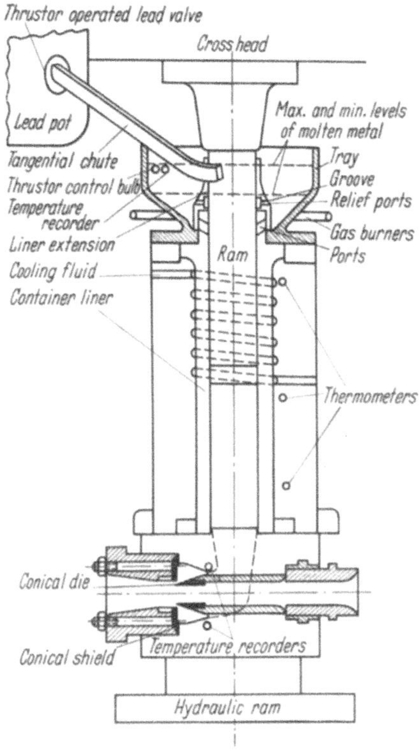

Fig. 353. The Glover Tray System.

Fig. 354. Filling attachment for lead cable sheathing presses. (German Patent Specification. D.R.P. 610,936).

container would be saved. Laboratory experiments (Meyer and Hanemann [847]), showed that the welding of different billets is made difficult by the oxide skin always present even on clean surfaces. Pearson [940] confirmed this finding from practical operation. He mentions the use

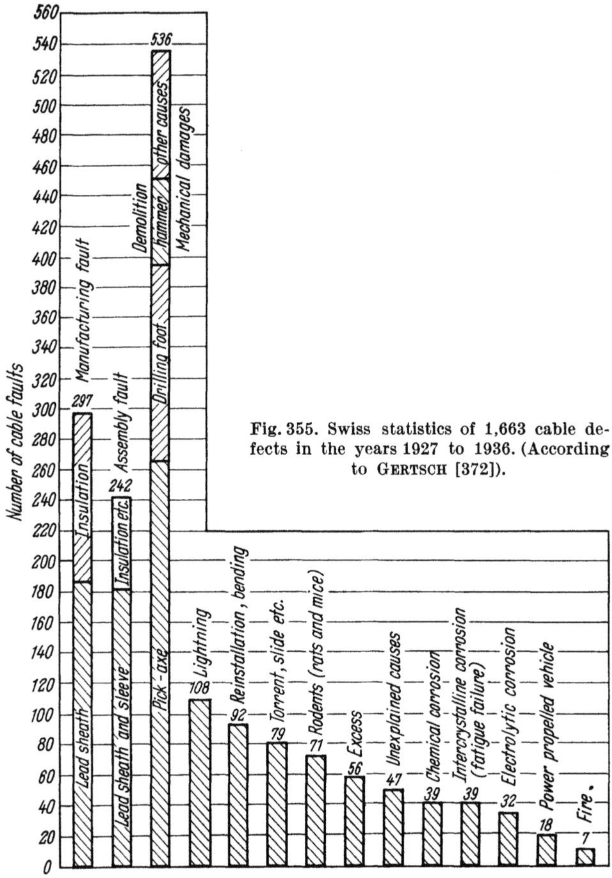

Fig. 355. Swiss statistics of 1,663 cable defects in the years 1927 to 1936. (According to Gertsch [372]).

of a press, which works with a solid charge, for jacketing rubber with lead. Here the lead sheath only acts as an aid in vulcanisation, and is subsequently removed.

Other types of cable sheathing defects are described in the literature. On interruption after extrusion of a container charge, a so-called "bamboo ring" is formed. At such places, undesired changes of cross-section exist; they do not occur in continuous presses. During the fabrication of cable sheaths, the alloying elements segregate somewhat in all types of presses. This segregation later promotes fatigue fracture, as shown by

ZAUSZNICA [1300b] who controlled the circular distribution of alloying elements spectrographically and made the zones of enrichment disappear by a suitable temperature distribution within the extruder. The general temperature level in the press is kept as low as possible.

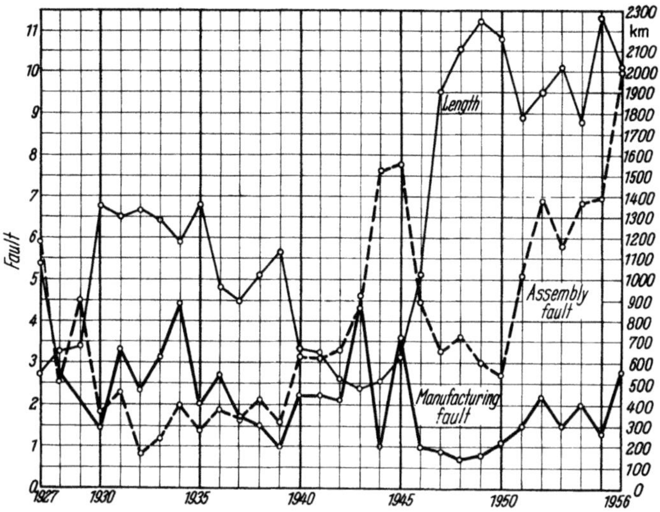

Fig. 356. Manufacturing defects and installation defects per 100 km of cable laid in the same year (Swiss PTT statistics).
Right: length in km of cable laid in the same year; *left:* defects per 100 km in cable laid in the same year. (According to HADORN and HAINFELD [462]).

Attack by animals, corrosion and fatigue fracture have been exhaustively treated elsewhere. Data on the avoidance of corrosion damage and the regular inspection of cables are given in the report of an American committee [1174].

The Swiss Postal, Telegraph and Telephone Administration PTT has, for decades, recorded all cable defects and classified them by the cause of the damage. The result of these statistical investigations has been indicated in several reports: 10 (Fig. 355), 20, 30 years of cable defect statistics (GERTSCH [371], GERTSCH and KOELLIKER [373], GERTSCH [372], HADORN and HAINFELD [462]). From the last report [462] some pertinent facts may be extracted. The annual growth of cable length since 1936 amounted to an average of 6%, while, on the other hand, the number of cable defects rose by an average of about 16% annually. Moreover the number of manufacturing defects, based on the annual length delivered, decreased over the period in question, an indication of improved production technique. On the contrary, defects, arising in the installation of cable, rose per 100 km of installed length, as shown in Fig. 356. The mechanical damage per km of the total cable network laid, also showed a great increase in recent years, which is mainly attributed to increased building activities. The number of cases of corrosion damage, and of fatigue fractures, likewise showed a tendency to increase. Considering the increasing average age of the cable network, both seem plausible. Among the cases of corrosion, in addition to electrolytic corrosion, the so-called phenol corrosion was prominent (p. 293). The significance of cathodic protection for cable-sheathing has been indicated elsewhere (p. 316).

3. Pipes, Wires, Bends (Traps)

a) Methods of Production. In the main, lead pipe is produced in vertical presses, either by the direct or the inverted method (p. 391). The lead is cast round the mandrel forming the axis of the container, and extruded between the mandrel and the die as a pipe. In presses which operate by the direct method, the mandrel is either firmly joined to the ram or it is fixed and moves in the ram which is provided with a borehole. The press illustrated in Fig. 357 is designed for an extrusion pressure of 300 metric tons and operates by the inverted method. In addition to the actual press, the installation also includes a coal-fired melting furnace, the reel with two interchangeable winding drums and the pumps for generating the hydraulic pressure. There is no hydraulic accumulator. The press is designed on the frame system, that is, the press cylinder situated under the working stage, the base plate, columns and presshead consist of a single steel casting. The hollow ram is fixed in the presshead. The heatable container, underneath, is screwed (not too rigidly) on to the press table which is carried by the press (hydraulic) cylinder. If the press cylinder is moved upward by hydraulic pressure, the container moves upwards over the ram. Bronze slides on the lower outer wall of the ram, which project a few millimetres, provide the necessary guidance.

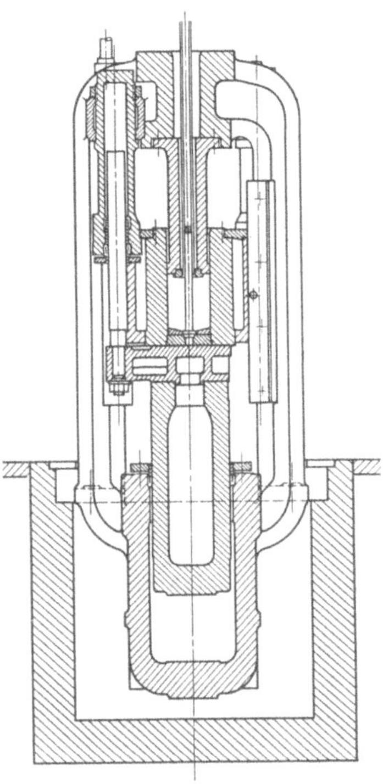

Fig. 357. Lead pipe press in extrusion position (Krupp-Gruson-Werk).

The die is fixed at the bottom of the ram. The mandrel is screwed into the so-called mandrel plate. It is set on the press table, the connection between the press table and the container being released and the latter moved up solely by means of the auxiliary cylinders situated laterally on the presshead. When the container is lowered, the mandrel plate with the mandrel moves into it and closes it at the bottom. The molten lead is top-poured into the container, which must be preheated. In order that the centring of the mandrel should not be disturbed, it is held firm by a centring piece. The centring piece is situated in the filling disc which is put on the container before pouring. The filling disc works like a sieve, and only allows the lead to fall vertically downwards. Thus, one-sided heating and bending of the mandrel is prevented. When the lead solidifies in the inlet trough of the filling disc, the disc is removed and the container moved up until the lead and the ram touch, so that the complete solidification of the lead takes place under pressure and piping is prevented. The lead content of the container is almost entirely extruded. The extrusion residue and the crust, that is, the cast skin of the billet, remain suspended on the ram when the container is lowered for recharging. Only the crust is removed, so that the extrusion residue welds up with the next charge. The purpose of this is not the production of continuous pipes as in extrusion of cable sheathing, but merely a shortening of the working process. The continuous pipe is cut at the junction of two successive charges in order to remove inferior parts.

Pearson [940] gives the cross-sectional illustration of a press (Fig. 358) which also works by the inverted process and allows the oxide-rich extrusion residue to be completely removed. The mandrel holder 3, of rigid design, passes through the entire length of the container and can be moved hydraulically up and down through the hole in the bottom of the container. The mandrel holder carries at its end the actual mandrel 2, which in this design can have a diameter as small as desired, without risk of distortion. At the end of extrusion, the impure extrusion residue stays on the tapering mandrel and can be removed from it on its downward movement following lowering of the container (Fig. 358).

Lead pipes are extruded in sizes ranging from the very thin up to internal diameters of 300 mm. Thin pipes are coiled on drums, thick ones are delivered in straight lengths. The strains arising from coiling can result in changes of structure, which will be discussed below. In very thin pipes, a mandrel of the length of the container (in the case of the design in Fig. 357) would not have the necessary rigidity. In this case, bridge mandrels (bridge cores) are used; here the lead is divided by the bridge and welded again beyond it. This principle is the basis of a design of extrusion presses for tubular tin solder. The "bridges" are formed here as a solid ring, which is provided with openings for the passage of the lead strands. Molten resin is fed into the tube through an aperture in the mandrel. Other presses for tubular tin solder are of design analogous to lead cable-sheathing extrusion presses with a vertical ram (Fig. 359) (MÜLLER [876]).

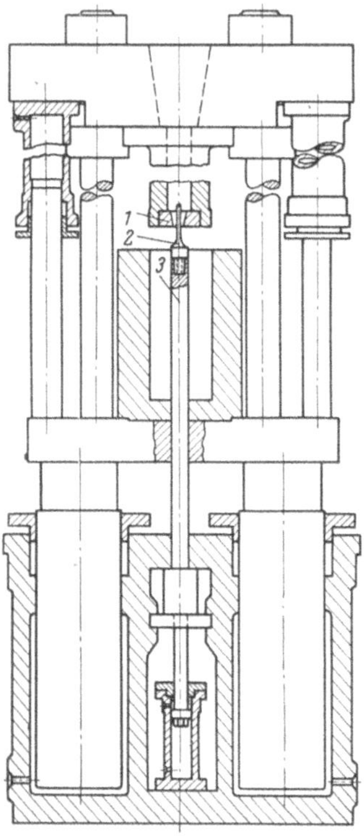

Fig. 358. Vertical lead pipe press. (According to PEARSON). *1* Die; *2* Tapering mandrel; *3* Mandrel holder.

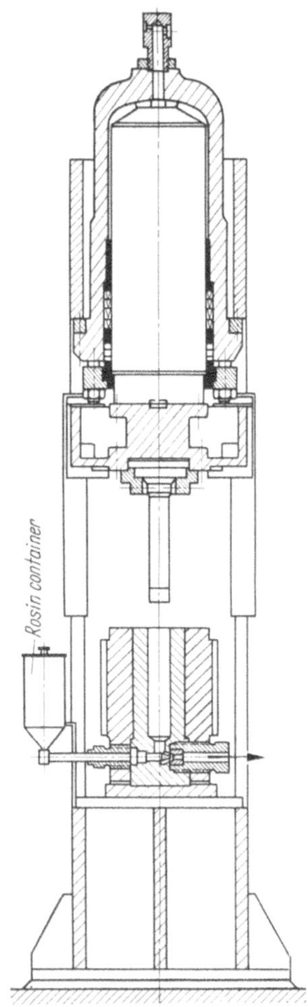

Fig. 359. Vertical press for the production of tin solder wire with resin core (Hydraulik, Duisburg). (According to MÜLLER [876]).

Coated pipes with tin linings, which are specially made for very soft, lead-solvent water are produced by inserting a central shell in the container, lead being cast round the vessel. After removal of the shell, tin is cast into the internal space. Extrusion must be carried out more slowly than in the case of lead pipe, as there is otherwise the risk of tearing the tin lining. A longitudinal metallographic section through the tin-lead transition zone in a pipe with a tin lining shows complete welding of the two metals (Fig. 360). Sulphur-lined pipes (p. 306) are produced by introducing sulphur into the front end of the pipe as soon as this has passed forward over the mandrel. An internal lining of lead sulphide and sulphur is formed. Lead-trap presses operate with two containers. The two lead strands emerging weld up to a single tube at opposite places on the mandrel. These are the same principles as in HUBER's older cable-sheathing presses (p. 406). The *S*-shape or other shapes (for preventing smells) arise from the fact that the two containers are operated at different extrusion pressures (Fig. 361). Experienced operators control the presses by visual observation alone.

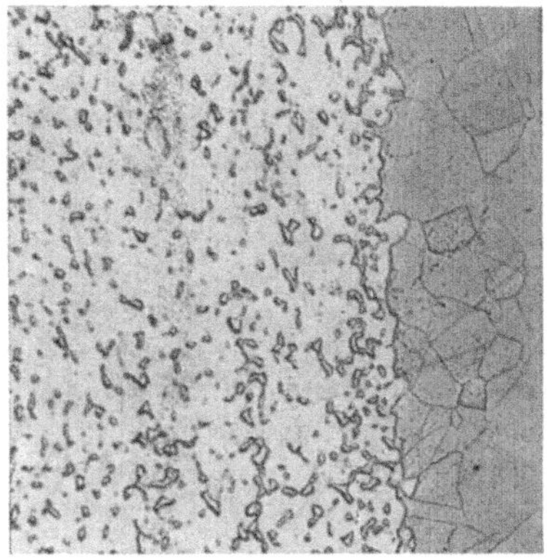

Fig. 360. Coated pipe with tin lining. Cross-section. Near-eutectic composition of the tin-rich layer (*left*). $\times 150$

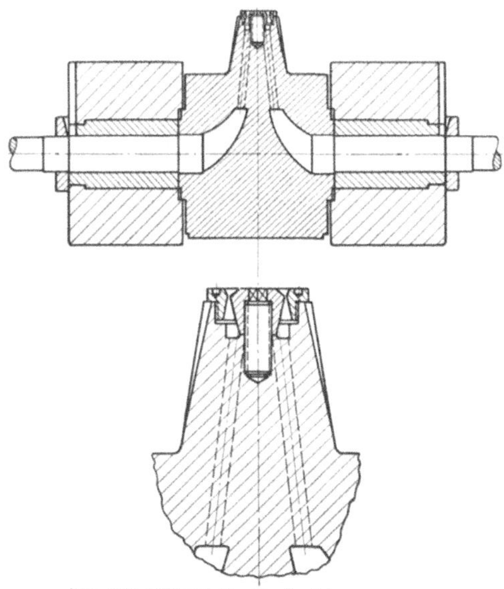

Fig. 361. Fittings for the lead trap press. (According to MÜLLER [876]).

Extrusion of wire, as well as of strip and sections, can be carried out in all lead pipe presses, if the proper dies are inserted. Dies with several apertures (see p. 394) are often used in wire extrusion. Lead wool, which is used for sealing purposes, can be produced by extrusion, as well as by planing (HEUSER [516]).

The use of lead string, so-called Riffelblei, instead of lead wool, offers certain advantages. According to United States Specifications (Lead [735]) sealing lead should be at least 99.73%, and should contain the minimum of hardening constituents (As, Sb, Sn).

b) Structure. Special Phenomena and Defects. Transverse metallographic sections through extruded wires and pipes give a good general view of their structure, and, in particular, give prominence to zones of differing grain size. Longitudinal sections are confined to specific places in the circumference, but in individual cases sometimes provide more information than transverse ones, as they usually reveal the flow process. Thus the section of an antimonial lead wire (Fig. 362) shows a linear structure, in which the lines originate from the lead-antimony eutectic which solidifies last. Linear structures can also often be observed in commercial lead pipes, as small amounts of other metals can bring about visible coring.

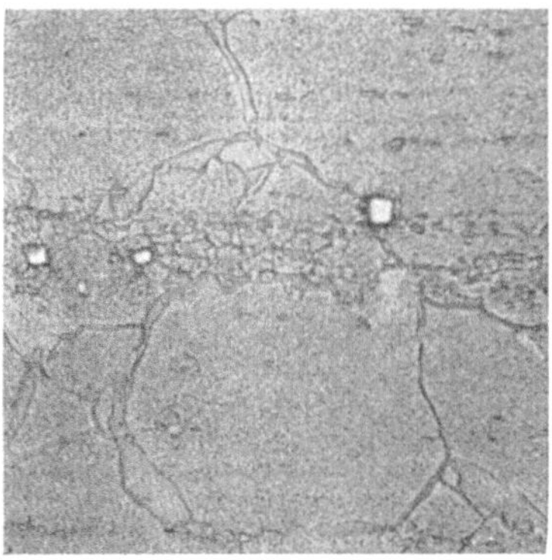

Fig. 362. 2.5% Sb. Wire in longitudinal section. Linear structure. Broadened grain boundaries similar to Fig. 273. ×500

As BUTLER [162] showed in a systematic series of experiments, the grain size of extruded lead products depends in the first place on the extrusion temperature. For example, the average grain size of Tadanac lead (99.999%) increased from about 0.5 to 4 mm if the extrusion temperature was raised from 150 to 300 °C and the extruded rods were allowed to cool in air. The extrusion ratio was 50:1. An increase of the degree of strain with the same original diameter of the billet caused a moderate reduction in grain size. This is attributed less to the strain as such than to the more rapid cooling of the thinner wire. The grain size of extruded pipes was found to be fairly independent of the mode of extrusion, direct or inverted. On the other hand, a much finer grain was obtained if the hot wires were water-cooled, and the more so the more quickly quenching follows extrusion, that is, the less time there is available for grain growth in the hot state, following recrystallisation (Fig. 363). Therefore, the effect of extrusion temperature on grain size is not so great in quenched as in air-cooled specimens. A decrease in the speed of extrusion in the region of 9 to 27 m/min resulted in hardly any change of grain size in air-cooled wires. However in quenched wires a small increase of grain size was observed on reduction of the speed of extrusion. Maintenance of a fixed ram speed of the press, with quenching of the extruded semi-fabricated product at a pre-arranged time was regarded as a suitable process for obtaining a specific grain size.

With regard to the degree of creep strength, and the desired low tendency to recrystallisation, medium grain size should be the aim. Moreover, as explained below, a grain as uniform as possible is advantageous. BUTLER [162] examined the stability of the structure by annealing the air-cooled or quenched extruded wires at 100 °C, and determining the time to the commencement of grain alteration. It can be seen in Fig. 364 that the stability of the structure increases with the initial grain size; and further the great effect of copper additions is to be noted. Similar results were obtained with lead pipe of commerce of purity 99.947 and 99.925% at annealing temperatures between 100 and 200 °C (BUTLER [163]).

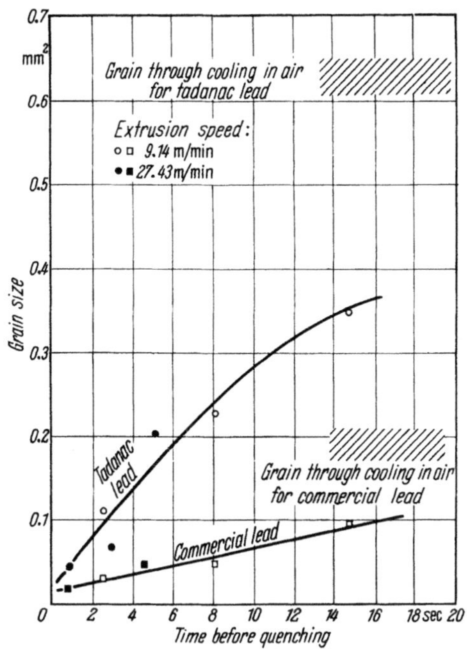

Fig. 363. Effect of the time delay before quenching, on the grain size of extruded lead pipe. Extrusion temperature 200 °C. (According to BUTLER).

According to BUTLER [163], regions of variable grain size can be due either to recrystallisation through slight strain, connected with coiling of the extruded pipe in the warm, or to recrystallisation owing to cold work in manufacture or installation. In the first case, coarse-grained zones are formed, which are opposite one another, and correspond to regions of tension or compression in bending (Fig. 365). The coarse-grained zones were produced, in the above-mentioned lead grades of 99.947 and 99.925% purity by, for instance, bending the pipes round a former immediately after extrusion at 180 °C. Coarse grain occurred here in the regions which exhibited a critical degree of strain of 2.3%. Consideration might be given to preventing the zone-formation due to hot bending, by coiling the pipes after previously quenching in water, rather than in the hot state. In this connection, the prerequisite for the absence of the coarse-grained zones is the non-occurrence of coarse-grained recrystallisation on storage after working at room temperature. Further the tendency to coarse-grained recrystallisation is greatest if the original grain is fine, that is, the recrystallisation threshold is low. In experiments at room-temperature on pipes of the purest lead, and on the two grades of lead already mentioned, zone-formation occurred only in the first material. For example, with an initial grain size of 0.077 mm² (quenched

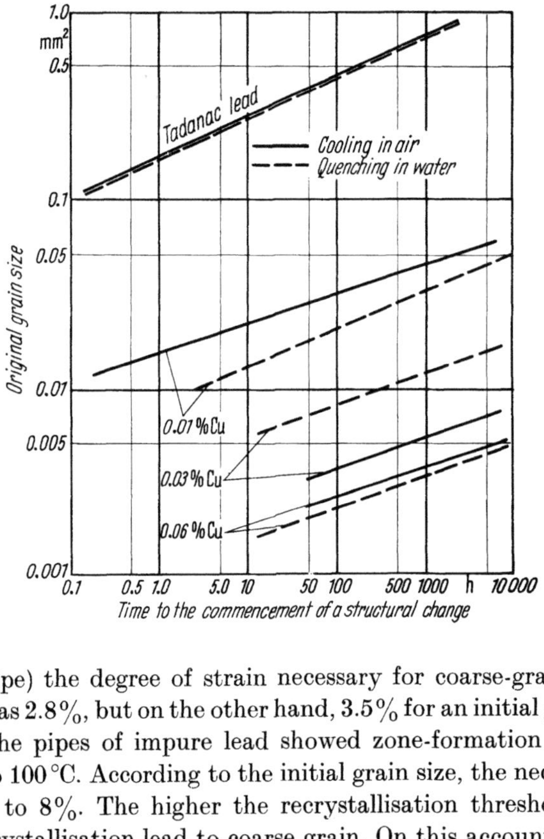

Fig. 364. Effect of grain size and of copper content on the structural stability at 100 °C of air- and water-cooled lead pipe. (According to BUTLER).

pipe) the degree of strain necessary for coarse-grained recrystallisation was 2.8%, but on the other hand, 3.5% for an initial grain size of 0.83 mm². The pipes of impure lead showed zone-formation only after annealing to 100 °C. According to the initial grain size, the necessary cold work was 2 to 8%. The higher the recrystallisation threshold, the less does recrystallisation lead to coarse grain. On this account, in quenching pipes, the moment of quenching should be as late as possible, thus (as mentioned above) ensuring a certain growth of the original grain. The experiments were continued on lead with small contents of alloying elements and with a standard lead of commerce B. S. 602, having 0.015% Sb, 0.006% Bi, 0.009% Cu and 0.001% Ag. Surprisingly, pipes of this lead required about 7% of strain in order to recrystallise in one hour at 100 °C. The small silver content of 0.001% is held responsible for this (see p. 191). The alloy with 0.06% Cu + 0.04% Te proved especially resistant to grain-coarsening after working and annealing. Here the tellurium content chiefly effects a delay in recrystallisation (p. 103) while copper mainly prevents grain-growth. For example, on annealing at 150 °C, specimens strained 6% required a period of 42 days for recrystallisation. The average grain diameter then amounted to only 0.12 mm. The alloy is therefore to be rated particularly good in respect of zone-formation. Alloys with 0.005% Ag + 0.005% Cu also behaved favourably.

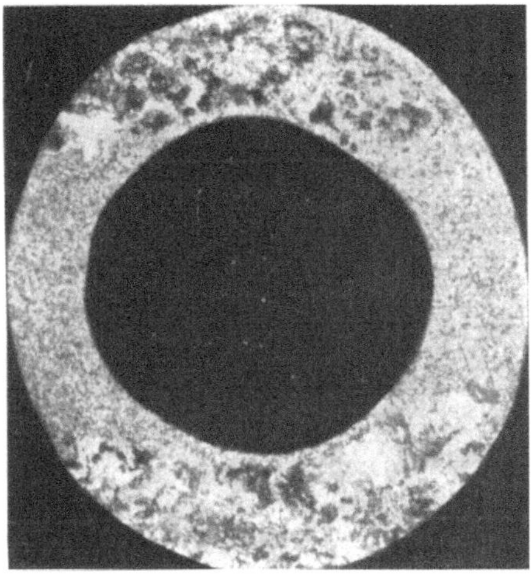

Fig. 365. Lead pipe exhibiting variable swelling due to non-uniform grain size. (According to JONES [633]).

In the creep tests under internal pressure, pipes with zones of coarse and of fine structure show non-uniform expansion. Here, as expected, the rate of creep is higher in the fine-grained zone. In the fatigue test, there is a tendency for cracks to begin at the boundary of the fine- and coarse-grained regions; BUTLER [164] attributes this to a stress concentration at the transition zone, on the basis of differing resistances of the two zones. The fatigue strength of pipes of varying grain size is therefore lower than that of pipes of uniform structure, whether this be coarse- or fine-grained.

The fatigue fracture of lead pipes generally occurs in the form of transverse cracks. Water pipes joined to branches of iron pipes, in domestic installations, are said to be particularly in hazard, especially if buried too deeply and the lead pipe without support. The structure (not reproduced) of the pipe with lateral cracking, illustrated in Fig. 366, exhibited remarkably large grain. At the grain boundaries cracks were seen, which in part formed a connected system, such as is characteristic of the fatigue fracture of lead (Fig. 268).

Cracks at grain boundaries, and transverse fractures originating from them, can occur, as experience has shown, over many years, even without marked alternating stress, particularly in pipes embedded in a solid medium, such as soil or cement. The diagrammatic representation of a flat longitudinal section through a pipe having lateral cracks, exhibited zones of very different grain size (Fig. 367). An inner fine-

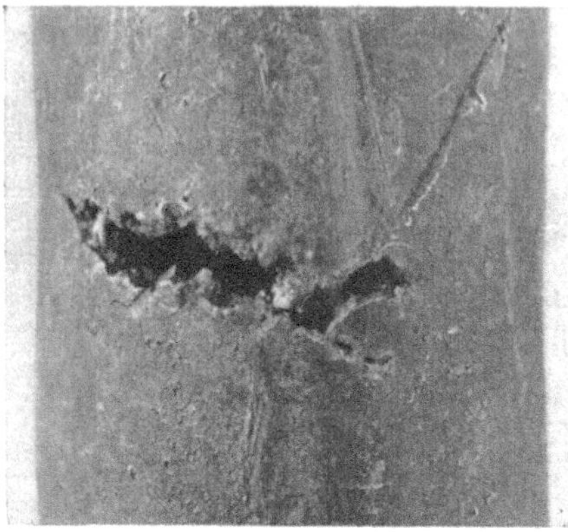

Fig. 366. Lead pipe with transverse crack due to fatigue. ×1.6

grained streak is enclosed by two regions of coarser grain. The edge zone on both sides is also (for reasons not explained in detail) particularly fine-grained. The occurrence of lateral cracks at the grain boundaries

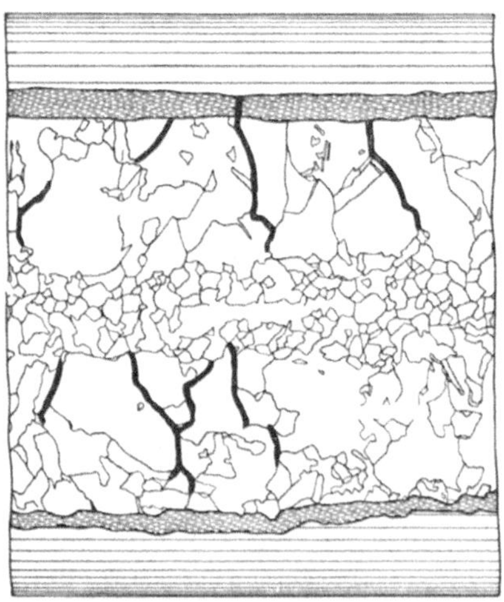

Fig. 367. Lead pipe with variable grain size in the longitudinal section. Transverse cracks in the coarse-grained zones. (According to JONES [633]).

of large crystals is characteristic (see also Fig. 370) and in agreement with similar observations by BUTLER [164]. The pipe in question had been embedded in cement. Corrosion cannot have led to the formation of grain boundary cracks, as these in part originated in the inside and have not penetrated to the surface. The cracks were explained by the different creep rates of coarse- and fine-grained zones (p. 211). Coarse-grained lead after long creep tends to brittle fracture (p. 231). Apart from internal pressure, creep can be brought about by expansion and con-

Fig. 368. Burst lead pipe. ×1.2

traction due to temperature fluctuations or by repeated, feeble bending to and fro. That longitudinal stresses, not hoop-stresses, are concerned in the formation of the cracks, is shown by their occurrence, transverse to the axis of the pipe.

Expansion and longitudinal cracks occur in lead pipe as a result of internal pressure too great for the wall-thickness. When this arises from the freezing of the contents of the pipe, the material offers hardly any protection. Lead-tellurium alloys have been proposed, which are said to behave more favourably than commercial lead or antimonial lead in the case of repeated frost with periods of thaw (SINGLETON and JONES [1126]), but sufficient practical experience is not available (p. 104). A typical longitudinal crack, arising from insufficient wall-thickness of the pipe, can be recognised in Fig. 368. The longitudinal grooves probably originate from extrusion grooves, possibly also from the linear structure, as the lead contained 0.4% Sb (cf. Fig. 362). Another pipe with expansions and longitudinal cracks, on investigation of its structure, showed particularly instructive phenomena. Thus, in the cross-section through the neighbourhood of a longitudinal crack (Fig. 369), it can be seen that from the base of the rounded cavity, where the stresses are locally increased, numerous further cracks penetrate into the inside. It is possible that here is a contribution from corrosive effects. The cross-section of a place in an expanded wall (Fig. 370) shows a circular zone of particularly coarse grain. Numerous inter-granular cracks, which stop at the boundary of the fine-grained structure, are present in this zone. This local limitation of the cracks can be explained as in Fig. 367.

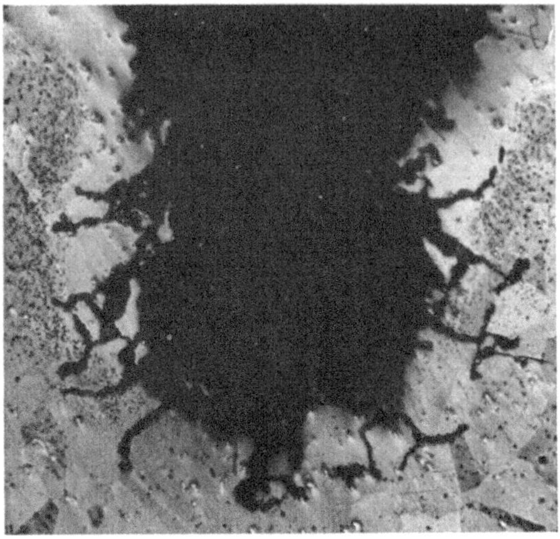

Fig. 369. Cross-section through the longitudinal crack of a pipe. Finer cracks emerging from the wall of the crack. × 150.

In contrast with Fig. 367, these cracks extend radially, and can therefore be attributed to hoop stresses due to internal pressure.

The circular zone with coarse grain arose not from the extrusion process but from subsequent working and recrystallisation, not ex-

Fig. 370. Lead pipe. Grain-boundary cracks in the coarse-grained zone. ×20

plicable in detail. Photographs at higher magnification showed directional precipitates in the crystals of the fine-grained zone—similar to Fig. 273—which are characteristic of age-hardened alloys which have been stored for a long time. These crystals are therefore the original ones formed during extrusion. On the other hand, the coarse crystals contained no little rod-shaped precipitates, but larger, round antimony particles, similar to Fig. 376.

Dross inclusions can be detected in some lead pipes. In the cross-section of a pipe (Fig. 371) a heavily-drossed zone with cavities can be recognised near the inner wall. The grain here is finer than in the outer parts of the pipe. Some of the dross obviously settled on the mandrel and from here reached the inner wall of the pipe. The blistered surface of the pipe illustrated in Fig. 372, no doubt originates from gas inclusions. They were compressed in the press, and expanded when the pipe came out of the die.

Fig. 371. 0.4% Sb. Cross-section of a lead pipe near the inner wall. Cavities and dross. ×500

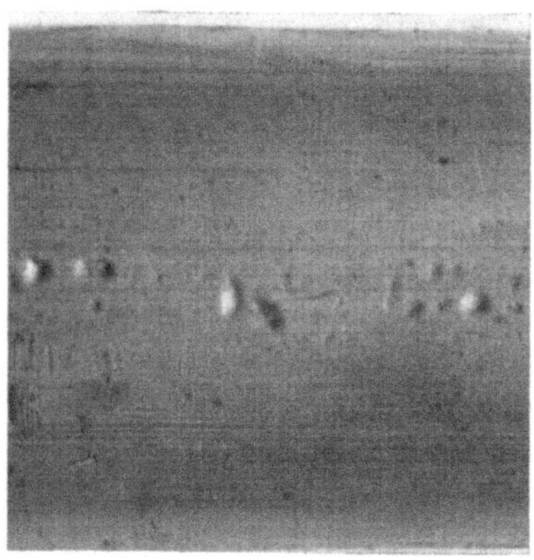

Fig. 372. Blistered lead pipe. ×1.5

c) Wall-Thickness and Creep Resistance of Commercial Lead and Antimonial Lead Pipe. For decades, efforts have been made in Germany and in other countries to utilise the higher strength of alloyed lead to reduce the wall-thickness of water-pipe (Hanemann [476], Stockmeyer and Hanemann [1147]). If it is here a question of security against sudden great increases of pressure by water-hammer or other causes, then it is the tensile strength of the alloy concerned which is the prime consideration. In reality however, it is not only the resistance to waterhammer which is of importance in water-pipes, but equally the resistance of the pipe wall against the continuous operating pressure; its creep resistance is therefore of great significance. Though in German standard specifications DIN 1261 and 1262 (p. 75), antimonial lead with 1 %Sb is permitted in addition to commercial lead for water pipe, and the wall-thickness of the antimonial lead pipe reduced by one-third as compared with commercial lead, the basis of this specification is that the creep resistance of the alloy is at least one-third higher than that of commercial lead.

As was shown in the discussion on creep strength (p. 236), this assumption is only valid if antimonial lead in the fully age-hardened condition is considered (HOFMANN and HANEMANN [563]). For this purpose it is first necessary that the extrusion of the antimonial lead pipe should take place at a temperature which is not too low, and secondly that the alloy contains a few multiples of 0.01% As (p. 87). In this case the hardness of the alloy reaches a value close to 10 Brinell units. Antimonial lead of this hardness was called antimonial lead *A* (Hartblei *A*) (HANEMANN and HOFMANN [478]). As shown in Fig. 247, it prossesses at high and at low stresses a creep strength which is far above one-third more than that of commercial lead, and therefore, with regard to the specified reduction of wall-thickness, still has a certain reserve of creep resistance. This reserve is important, as the most favourable condition cannot always be maintained in the working of the alloy. Furthermore, in age-hardened antimonial lead, there is the risk of a partial softening after storage for years (HOFMANN and HANEMANN [562]), particularly if a certain amount of cold work has taken place in uncoiling or fixing the pipe. Creep tests on a pipe of antimonial lead *A* with 1.15% Sb and 0.02% As, which had lain for seventeen years, still showed practically no loss of creep strength. Also the Brinell hardness still had almost its original value of 9.3 kg/mm² (HOFMANN and ENGEL [559]). A disadvantage of antimonial lead *A*, its greater rigidity, must be taken into account, If, as a result of absence of arsenic addition, or extrusion temperature which is too low, antimonial lead *A* age-hardens incompletely or not at all—antimonial lead *B* (Hartblei *B*)—then it still has an essentially higher tensile strength in comparison with commercial lead. Moreover the creep strength is even improved in the range of higher stresses (p. 238). On the other hand, in some circumstances antimonial lead of this kind creeps under low stresses just as quickly as commercial lead, and, with extremely fine grain, even more quickly. If now in such pipe the wall-thickness is also reduced, as compared with commercial lead pipe, the occurrence in service of swellings and longitudinal cracks should not cause surprise. In the investigation of a large number of such damaged antimonial lead pipes during the last war, it was found that in all cases the hardness was only 5.6 to 7 Brinell units and that there was no arsenic addition. Moreover, in many cases the antimony content was substantially below the specified 1%.

The superiority in creep strength of antimonial lead *A*, as compared with commercial lead, does not exist at higher temperatures (HOFMANN [556]) (Fig. 248). A reduction of the wall-thickness of lead pipe used for chemical plant—perhaps at temperatures of 80 °C—is, according to the results of HILLEN and HOFMANN [528], only wise if commercial lead is replaced by antimonial lead with at least 4% Sb.

In England there has been exhaustive study of the increase of the creep strength of lead pipe by alloying and by the form of working. In particular, examination has been made of the effect of small additions, for instance of copper and silver (p. 241).

4. The Rolling of Lead

The theory of the rolling process as a general method of shaping has been usefully elaborated in recent decades; this can be concluded from the comprehensive presentation by FORD [332] and the list of references which he gives. Here it will only be mentioned that the first fundamental formulations go back to v. KÁRMÁN [645] and OROWAN [924]. There the rolling process is treated as a problem of plane working, that is, the occurrence of spreading is disregarded. The hypothesis of the energy of change of shape due to v. MISES [858] is introduced as a condition of flow. LIPPMANN and MAHRENHOLTZ [754a] give a survey of the present position of the mathematical treatment of the rolling process.

Various fundamental investigations of the cold rolling of lead and its alloys have been carried out. As far as these were directed mainly to the elucidation of phenomena in the hot-rolling of other metals, only a brief report will be made here.

Compression tests on lead bodies of rectangular cross-section were undertaken by SIEBEL and OSENBERG [1117] with respect to the similar flow processes in rolling; here the parallelepipeds preferably deformed in the direction of the shorter edge, so that the rectangular cross-section gradually became a circle. This behaviour is conditioned by the friction on the compressing surfaces, and therefore the more marked at higher values of the latter.

Investigations on the spreading of lead in rolling, and its dependence on the experimental condition, gave the following results, (BENAD [75], EMICKE and BENAD [277]):

The spreading Δb is best represented by the formula for spreading due to SEDLACZEK-SIEBEL:

$$\Delta b = C \cdot \frac{1}{\left(\frac{b}{h}\right)^{1/2} + \left(\frac{h}{b}\right)^{3/2}} \sqrt{\frac{D}{h}} \cdot \Delta h$$

where b and h represent the width and the height of the specimen to be rolled and D the diameter of the rolls. For smooth condition of the roll surfaces, the constant C can be estimated at 0.19 in the case of lead and lead-antimony alloys. The equation shows that spreading increases with increase in the diameter of the rolls. This can also be understood from the compression tests described above, if it is borne in mind that on increase of the diameter of the rolls, the contact surface of material and roll is lengthened in the direction of rolling. This result is also confirmed by measurements in other laboratories (TAFEL and ANKE [1160]). This formula further signifies that the spreading depends on the ratio of the sides b/h. The function

$$\frac{1}{\left(\frac{b}{h}\right)^{1/2} + \left(\frac{h}{b}\right)^{3/2}}$$

and therefore the spreading, has a maximum when the value of b/h is 1.73 and falls steeply for lower values, but on the other hand, more slowly for higher values (SIEBEL [1114], TAFEL and KNOLL [1161]). In practice, spreading plays no essential part, as in general the material to be rolled has a high initial width, i.e., a high value of b/h. Within wide limits, the speed of rolling only affects the spreading slightly (TAFEL and ANKE [1160]). The condition of the surface of the rolls has hardly any effect when their diameter is no greater than 10 times the thickness of the material being rolled. On the other hand, for a greater value of the ratio D/h, the spreading with rough rolls increases more steeply than with smooth rolls, and is no longer proportional to the reduction of thickness Δh (Fig. 373). If, contrary to most formulae for spreading, TAFEL and ANKE [1160] found an increase of spreading with the power 1.5 of the reduction of thickness, it may consequently be suspected that the cause was a rough surface of the rolls. In a curve presented in a diagram, the ratio D/h usually was 22:1 for lead (TAFEL and ANKE [1160]).

The forward slip, that is, the stretching of the rolled material from the place of equal peripheral speed of the rolls and of the specimen surfaces in contact with them until exit from the rolls, depends very strongly on differences in friction for lead and for other metals. The forward slip increases with growing friction, so that the neutral point is displaced in the direction of the entry to the rolls (SIEBEL and OSENBERG [1117]).

Thorough investigations of the power requirements in the rolling of commercial lead and antimonial lead, in association with the measurements of spreading already mentioned, were carried out at the Bergakademie Freiberg (BENAD [75], EMICKE and BENAD [277]). The initial material was in part cast with the cross-section 80 · 80 mm; when it had already been rolled once, it was annealed at 100 °C. For determination of the power requirements, or the work of deformation, measurements of the rolling load were first taken. From the rolling load P_ω in kilograms the yield strength may be obtained as

$$k_f = \frac{P_w}{l_d \cdot b_m},$$

where l_d and b_m are respectively the compressed length and the average width of the material rolled. From the rolling load, the work of rolling follows as

$$A = P_w \cdot l_d \frac{l_1}{r\sqrt{\lambda}} \text{ cmkg},$$

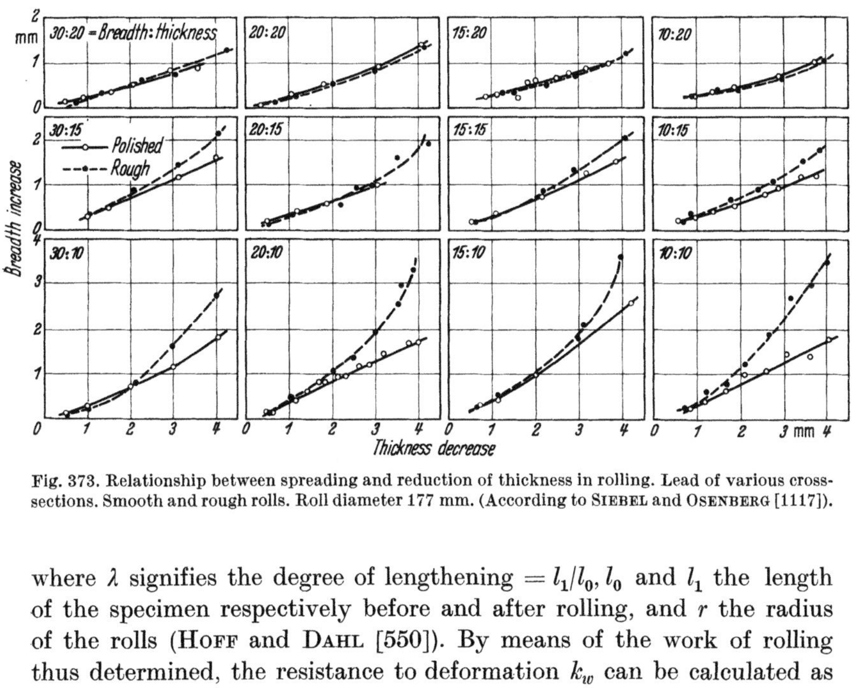

Fig. 373. Relationship between spreading and reduction of thickness in rolling. Lead of various cross-sections. Smooth and rough rolls. Roll diameter 177 mm. (According to SIEBEL and OSENBERG [1117]).

where λ signifies the degree of lengthening $= l_1/l_0$, l_0 and l_1 the length of the specimen respectively before and after rolling, and r the radius of the rolls (HOFF and DAHL [550]). By means of the work of rolling thus determined, the resistance to deformation k_w can be calculated as $\frac{\text{force}}{\text{surface}}$ or $\frac{\text{work}}{\text{volume}}$ from the equation $A = k_w \cdot V \cdot \ln \frac{l_1}{l_0}$ (FINK [316]). Here V is the volume of the material rolled.

Apart from the rolling load, the torsional moment M_d transferred to the coupling spindle was measured oscillographically (EMICKE, ALLHAUSEN and MAUKSCH [276]). While the equations hitherto have only contained static magnitudes, the measurement and basic application of the torsional moment M_d indicate the dynamic nature of the rolling process. If ω signifies the angular velocity of the rolls and t the duration of pass, then the work transferred to the coupling spindle is

$$A_{\text{mech}} = \int_{t_0}^{t_1} M_d \cdot \omega \cdot dt.$$

By subtraction of the frictional work A_r in the roll-bearings, the pure work of rolling results as $A_w = A_{\text{mech}} - A_r$. From A_w there follows, from the above-mentioned equation by FINK [316], the resistance to deformation $k_{w,\alpha}$. This is a sure basis for rolling mill calculations. A formula for ascertaining $k_{w\alpha}$ is given for the case in which only k_f can be determined.

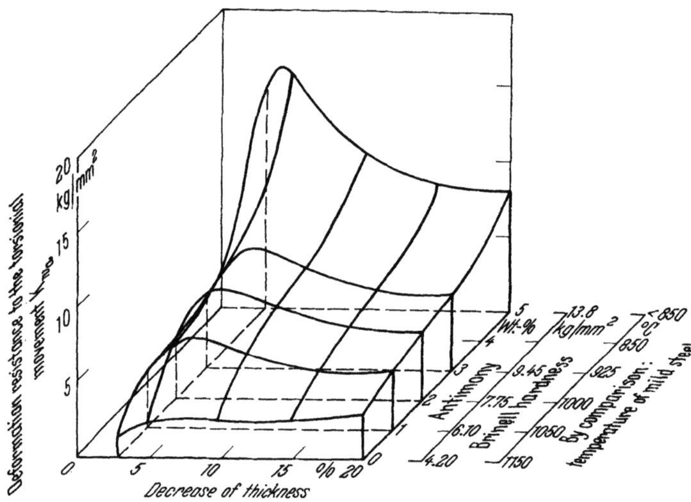

Fig. 374. Resistance to deformation, due to torsional moment, for various reductions and antimony contents at room temperature. (According to EMICKE and BENAD [277]).

The yield strength k_f, the resistance to deformation k_w and $k_{w\alpha}$ respectively, as well as the Brinell hardness, were measured in relation to the degree of deformation, the antimony content of the alloy, the thickness ratio h/D = thickness of material/diameter of roll and the speed of the rolls. All the magnitudes mentioned have the same trend, even if the absolute values differ more or less. The dependence of the resistance to deformation from the torsional moment on the degree of rolling and the antimony content, according to the adjacent diagram (Fig. 374), is particularly instructive. While the resistance to deformation of commercial lead constantly increases with the amount of rolling, a maximum of $k_{w\alpha}$ at a reduction of 5 to 7% becomes more and more marked with increasing antimony content. This maximum may be explained by the hardening of antimonial lead, in which perhaps a certain degree of age-hardening is indicated. The hardening is relieved at greater reductions by simultaneous crystal recovery and recrystallisation. The most favourable region of deformation is thus at quite small reductions of thickness and those of 15 to 20%. In respect of energy, the most favourable ratio thickness of material/roll diameter, is about 0.10. By increasing the speed of rolling, the resistance to deformation (as in extrusion) is only moderately increased, a gratifying fact for works' practice. From the difference in the work of rolling, as calculated from the torsional moment and the rolling load, the coefficient of friction for rolls and material is determined. The values for μ are between 0.01 and 0.20. In cone-compression tests similar values for lead were determined: less than 0.1; 0.3; about 0.4, according to the surface condition (SIEBEL and POMP [1118]). Still higher values of the coefficient of friction can occur on the slow extrusion of lead between oxidised steel compression plates (p. 226).

BENAD [75] carried out further rolling tests on composite specimens. These consisted of a concentric arrangement of three layers of differing antimony content, which in one series of experiments (A) increased towards the inner, and in the other (B) towards the outer side. Cases occurring in the hot-rolling steel in practice were imitated by the experiments. Here series A corresponds to the case of a billet which has not been warmed through, series B to a billet cooled on the outside by too slow working. Measurements of the power-requirement of layered metals showed that the hard external zone had a greater effect on the power-requirement than the hard core.

Lead rolling mills are built as reversing two-high rolling trains with roll-widths of below 1 to 4 m and corresponding powers between 40 and 150 horse-power. For the production of rolled lead, the cast billets with dimensions of, say $200 \cdot 170 \cdot 15$ cm, are first broken down hot (that is, at temperatures between 180 and 100 °C) with a reduction per pass of 3 to 5 mm. After the breaking-down, the thickness amounts to 25 to 30 mm. The slabs are now cut up into suitable pieces and finish-rolled. The reduction per pass in finish-rolling is below 1 mm since folds are formed with higher pressures. By contrast with breaking down, the ends usually do not go through the rolls. The rolling speed amounts to about 0.5 m/second. In order to achieve uniform thickness, the finished sheets are rerolled without pressure.

While the rolling of commercial lead offers no difficulties, antimonial lead is less easy to roll as, with the hardness the brittleness also increases and there is a risk of cracking, especially at the edges. Antimonial lead is therefore rolled with smaller reductions per pass. This however results in greater cooling of the material during rolling, which again gives rise to an increase in brittleness, resulting in greater waste on cutting. For this reason, rolling experiments on antimonial lead clad with commercial lead were carried out by the last-named author. The thickness of the antimonial lead plates was 77 mm, and that of the two commercial lead sheets which were applied, 3 mm each. The experiment showed a more favourable behaviour of the clad material as compared with the unclad, both in respect of possible reductions per pass as well as of cutting losses and of power requirements. Moreover, for some purposes, such a duplex material should also be expected to have an advantage in regard to corrosion behaviour. Fig. 375 represents a photomacrograph of an antimonial lead sheet which broke in rolling. The structure possibly points to the injurious effect of considerable piping of the cast billet. The next illustration, Fig. 376, shows special structural phenomena in an antimonial lead sheet with 6% Sb.

Lead foils are produced in thicknesses of 0.005 mm upwards. Foils of about 0.5 mm thickness are used for measuring explosion pressures (HOFMANN and MEIER [571]). The foils, fixed in a pressure cell,

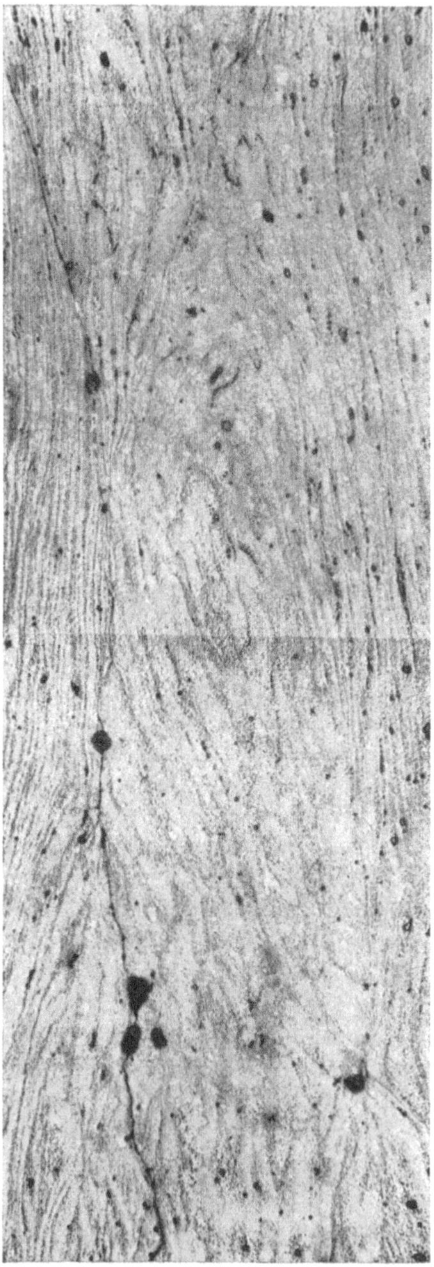

Fig. 375. Longitudinal section of antimonial lead with about 2% Sb, torn in rolling. Flow shown by lines of eutectic, very disturbed in places. Residues of incompletely welded pipes. Overlaps. ×40

bulge under one-sided pressure like a cup. The depth of the cup is a measure of the magnitude of the explosion pressure. In calibration tests on lead foils of different compositions, the accuracy of the method was measured, particularly the effect of time and of temperature. For measuring purposes, fine-grained antimonial lead is preferable to pure lead. Aluminium foils exhibit less time-effect than lead foils. Thicker foils, of about 1 mm

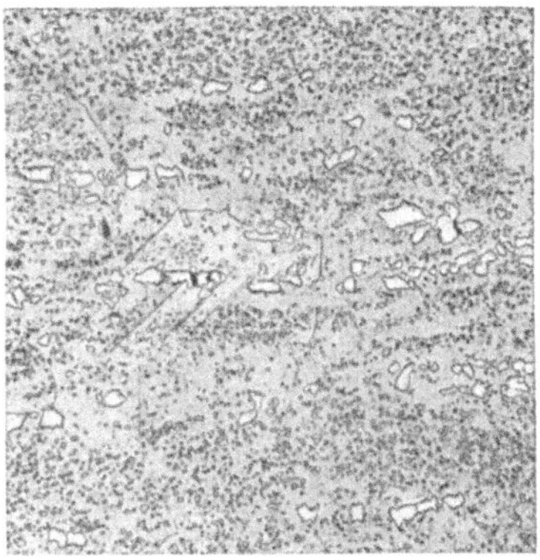

Fig. 376. Antimonial lead with about 6% Sb. Longitudinal section. Antimony inclusions with transverse cracks. Lead twins due to recrystallisation after rolling. ×500

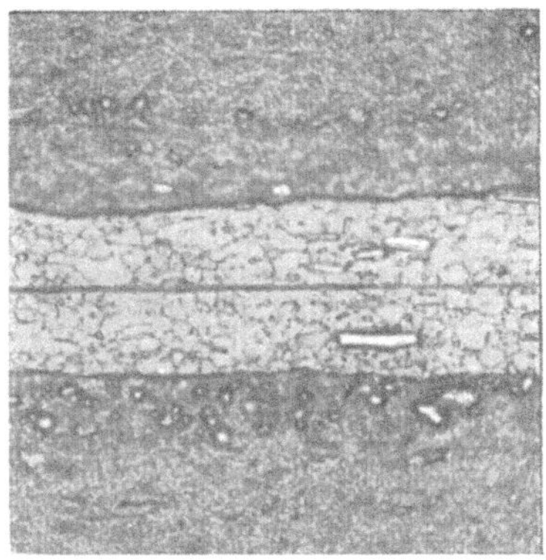

Fig. 377. Reduced lead-tin foil 0.1 mm thick. Tin layers of 2 specimens lying together, welded to lead. ×500

thickness, are used for lead stamping by the ALBERT-FISCHER method (p. 358). Foils for lead stamping must have faultless surface quality and require special care in their preparation (BAKER [35]). Foils of thickness a few tenths of a millimetre have been introduced for wrapping cables.

Tin-clad lead sheets are rolled to foil. These foils find application, for example, in the packing industry and as tinsel for Christmas trees. Cladding takes place in breaking-down. The plates are at first only warm to the touch; by using large passes a rise of temperature is obtained which favours welding (Fig. 377). Lead-tin foils can also be produced by casting molten tin over lead and then rolling.

5. Impact Extrusion and Deep Drawing

The production of tubes from soft metals, such as lead and tin, by impact extrusion is already over 100 years old (PEARSON [940]). Application of the process to aluminium has only been possible since 1920. Since the last war, impact extrusion has also been introduced for the production of thick-walled hollow steel bodies. The greatest known dimensions of lead tubes are 75 mm diameter and 300 mm length, the wall-thickness is usually in the region of 0.1 to 0.2 mm. In order to increase the stiffness of the tubes, low-antimonial lead is generally used instead of commercial lead; a content of, for example, 0.6% Sb is customary, such as is also established for cable-sheathing. For good appearance, as well as to avoid problems of health, tubes are often produced with a tin coating of at least 2.5% of the wall thickness on both sides.

The production of tubes takes place according to the diagram given in Fig. 378. Blanks are made from strip. In one mode of operation the round discs are perforated and worked. A lubricant is used to reduce

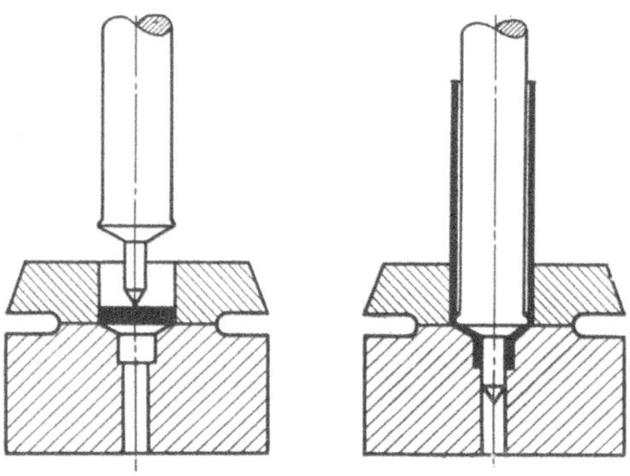

Fig. 378. Design and mode of operation of an impact extrusion press tool (Herlan and Co., Karlsruhe).

the friction between the tool and the workpiece. The production of a thread for closing the tube can be carried out in the working process and in that case the removal of the tube from the tool requires a rotary motion. Usually the thread is cut; in that case both ends of the tube can be trimmed in the same operation. Apart from the round cross-sections in Fig. 378, other cross-sections can be impact-extruded (Gohn [401]).

The extruded tubes, trimmed and provided with threads, are finally lacquered and printed. In some cases, to prevent corrosion, the tubes are sprayed internally with wax.

Bottle-caps are produced by the deep-drawing process on automatic machines. Use is made of tin-clad commercial lead, and for this reason these caps are known as 'tin foil caps'. Schikorr and Bergner [1063] tested the corrosion behaviour of tin-foil caps—with different thicknesses of the tin layer—for wine bottles. On the basis of their results they recommend a minimum of 3 μm for the thickness of the tin layer.

6. Lead in Powder Metallurgy

The production of lead powder for the purpose of storage battery technology has been mentioned elsewhere (see p. 342). Lead powder for powder metallurgy, for paint technology etc., is produced by atomising molten lead with compressed air. By this method, the lead is not so strongly oxidised as when production is carried out by grinding in a ball-mill. A powder thus obtained exhibits, for example, the following grain size distribution[1]:

Grain size (mm)	>0.075	>0.06	>0.04	<0.04
Fraction (%)	Trace	1	5	94

Bulk density 4.6 to 5.0 g/cm³

In a similar way, leaded bronze powder can also be produced. In this case the alloy melt is atomised by means of water under pressure instead of compressed air. By means of the more rapid cooling, segregation of the alloy droplets on solidification is prevented (compare SUGIYAMA [1154]). FAZEKAS [306a] described the production of lead powder by electrolysis. APPELT [22a] gives data on lattice disturbances in lead powder and their relationship with the absorption properties. Lead powder is winning increasing significance as a constituent of rust-preventive paints for steel. The average particle diameter of the lead pigment is 0.005 mm. Steel is passivated by the reaction product resulting from the lead; apart from this, lead in this fine metallic subdivision is said to behave anodically towards steel (RUDRAM [1036c]). The classical process of powder metallurgy can also be applied to lead. It is carried out by compressing the powder (as free as possible from oxide) and subsequent heating with exclusion of air (GOETZEL [400], KIEFFER and HOTOP [662]). According to the work of THÜMMLER [1195] lead, in its sintering tendency, stands between the metals zinc, cadmium and tin, which sinter badly, and, for example, iron, cobalt, nickel, copper and silver, which sinter well. CARLSON and WESTERMANN [180a] compressed antimonial lead spheres of 2.4 mm diameter at constant pressure and four temperatures above room temperature, as well as at a constant temperature of 205 °C and four different pressures. The degree of compaction during compression was measured and brought into relation with the initial porosity, the pressure and the time. Compression and heating can also be combined in one process: hot pressing. The

[1] Kindly communicated by Norddeutsche Affinerie, Hamburg.

application of heat is unnecessary in the case of lead. Thus, for instance, by rolling lead powder at room temperature it is possible to obtain a sintered strip which exhibits the density of solid lead, which means that it is completely free from pores (STORCHHEIM [1151]). The tensile strength of such a material amounted to 6 kg/mm², which is treble the usual value of the tensile strength. LENEL and BAUM [742] found a comparatively low elongation at fracture in extrusions of a lead/lead oxide-based material.

The increase of strength is attributed by LENEL and ANSELL [743] to the larger number of crystal structure defects in the materials worked by powder metallurgy. A further increase of strength results from the oxide present. The degree of subdivision of the included second phase is decisive for the behaviour of a sintered alloy of heterogeneous structure. According to the above-mentioned authors the yield point is inversely proportional to the square root of the average distance between the particles.

Just as lead oxide can be finely distributed in lead by powder metallurgy, this technique is also possible with other additions insoluble in lead. Such investigations were, at the instigation of the International Lead Zinc Research Organization, Inc. recently carried out by LENEL at the Rensselaer Polytechnic Institute, Troy, N.Y. LENEL [741a] produced, among others, alloys with respectively 0.5 to 8.1% Cu or 0.55% Al. For this purpose he started with alloy powder obtained by atomising and air-quenching the molten alloy heated to 850 to 1,000 °C. By compressing powder of grain size <44 μm, at pressure between 2,400 and 2,800 atmospheres (at), and subsequent extrusion at 240 °C, LENEL obtained wires of theoretical density. The tensile strength rises from 4.1 kg/mm² at a copper content of 0.5%, to 5.1 kg/mm² at 8.1% Cu. The elongation at fracture varies between 3 and 29%. In this case the dispersion hardening is to be attributed, beside the action of copper, to oxide formed in atomisation. In order to separate the two effects, LENEL reduced the powder (prior to compression) for several hours in hydrogen at 300 °C. Thus the oxygen content fell from the original 0.43 to 0.84% down to 0.03 to 0.07%. In these wires the tensile strength amounts, according to the copper content, to only just 2.6 to 4.4 kg/mm² with elongation values between 37 and 46%. Sintered bodies from coarse-grained powder showed essentially lower strengths than specimens pressed from fine powder. From metallographic sections, LENEL determined the distance apart of the copper particles as 1.17 to 1.50 μm from which a particle diameter of 0.33 to 0.69 μm could be calculated. The alloy with 0.55% Al gives a tensile strength of 3.8 kg/mm² and an elongation at fracture of 20%. The aluminium is present in the form of its extremely finely-divided oxide, which is formed during atomisation.

In respect of creep behaviour, the lead-aluminium alloy is far superior to all other known lead alloys. The lead-5%copper alloy investigated also shows favourable creep properties. The following table is a survey of the most important results and a comparison with conventional lead alloys.

	Testing temp. °C	Testing stress kg/cm²	Duration of test hrs.	Elongation in %/year extrapolated
Lead with 5% Cu, about 0.5% O, extruded at 240 °C, average	25	105	1,000	0.53
particle distance 1.85 μm	25	140	1,000	1.47
Lead with 5% Cu, about 0.5%O, extruded at 240 °C, average	53	27.4	200	0.92
particle distance 1.25 μm		35.8	400	1.47
Lead with 0.55% Al (Mostly as oxide) ex-	25	175	1,000	0.60
truded at 240 °C	100	84	333	0.59
Lead with 6% Sb, rolled	30	60.5	—	100
Lead with 6% Sb, cast according to Fig. 308	20	60	—	0.5 to 2
Commercial lead according to Fig. 246	20	20	—	0.5 to 30

The results obtained by LENEL on the lead-5%copper alloy agree with the theory of ANSELL and WEERTMAN. The latter two assumed that at low stresses the dislocations could not pass between the particles of the dispersed phase, but must individually climb over the particles. In this case the rate of creep is proportional to the applied stress. If a higher stress is applied, then the dislocation lines move between the particles, whereby constricted loops of dislocations remain behind. These, on their part, climb on to the obstacle and destroy themselves above. Thereby new loops can constantly form in the original slip plane. In this process the rate of creep is proportional to the fourth power of the applied stress.

In this connection mention may be made of the observation that lead-copper powders altered much less in air than pure lead powder. The latter oxidised noticeably on storage, passing through the colour-scale grey, black, brown to yellow. On the other hand, the alloy powder remained grey. ROBERTS and co-workers [1012a] report on further tests on sintered lead-lead oxide bodies. They confirm many of the results obtained by LENEL and co-workers. In their tests they establish a linear dependence of the tensile strength on the oxide content. If it is assumed that the oxide particles always have the same size, then this result indicates that the tensile strength is inversely proportional to the distance between the particles. Among the possible applications, storage battery plates, cable-sheathings and apparatus for radiation-protection are discussed.

Recently LUND [783b] has reported on the properties of sintered, and subsequently extruded, Pb/PbO and confirms the relationship between resistance and the reciprocal of the distances between the particles. Specimens from water blended powder give higher resistances than these from heated powder, though they tend to develop gassing when extruded at higher temperatures.

Possible applications of dispersion hardened lead, from lead/lead oxide, were proposed by GARZARD [360a] and LLOYD [755a]. According to LLOYD, sheets and pipes from this material should possess sufficient corrosion resistance against a large variety of chemicals, up to 120 to 150 °C. Components of DSL can be readily joined with a solder of lead plus 6% Sb.

In a process of a new kind, lead components can be produced by compression of lead powder in a container under the unilateral action of an explosion pressure (CROSS [224]). The use of the compression die which would otherwise be necessary, is obviated. Explosion pressures between 700 atmospheres and 3,500 atmospheres were used in the experiment. The initial particle size was between 5 μm and 131 μm. The measured compressive strength of the test pieces increased with increasing explosive pressure, and showed a maximum, depending on the particle size. Data on the creep behaviour of such bodies are not known.

Great use is made in powder metallurgy of lead as an addition to other metal powders. For example, sintered bodies of copper are produced with additions of 5 to 15% of lead and other metals, of graphite and oxides (HOFMANN and PIEPER [574]). Such friction materials have been introduced as brake- and clutch-linings in motor vehicles. Lead is said to favour the running-in process. In addition, however, so-called self-regulation is reported. If the temperature in the rubbing surfaces rises to the melting point of lead, then the latter forms a thin liquid film, which reduces the coefficient of friction. Consequently, the temperature falls and the lead again solidifies (GOETZEL [400]). The combination of plastics with finely-divided lead could also give rise to technically interesting properties, for instance, high density, sound damping (p. 446) or increased thermal conductivity (ZIEGFELD [1304]).

III. Metallic Coatings. Use of Sheet Lead. Jointing Methods

1. Coating with Lead

a) General. Lead coatings as a protection against corrosion are considered in the first place for steel, and for copper and copper alloys, zinc and aluminium. The action of a lead coating of steel is different from, for example, that of galvanising. In the case of galvanising, as the coating metal is less noble than the basis metal, there is a protection against corrosion by remote action, even in uncoated places. On the other hand, in solutions occurring in practice, lead behaves as electropositive towards iron and can only protect it completely as long as it is a sound coating. If the electrolyte penetrates to the steel through pores into the coating, then, in general, the anodic attack on the steel at these places is actually increased (see below). In the case of slight corrosive conditions, such as the action of an industrial atmosphere, the corrosion products formed by the local attack can lead to closure of the pores and thereby prevent further destruction of the basis metal (KNIGHT [681]). HUDSON [599] gives data on the durability of steel electroplated with lead, and of hot-dip and spray lead-coated steel, under different kinds of climatic and atmospheric conditions. Lead coatings can be applied by various methods. Lead coating by vacuum evaporation (vacuum metallising) is in course of development.

b) Hot Lead Coating (Dip Coating). Steel work-pieces which have been pickled in hydrochloric or sulphuric acid, and rinsed with water, are protected from oxidation until they are brought into the molten metal bath by wetting with zinc ammonium chloride solution. This bath is kept at a temperature of about 350 °C under a cover of zinc chloride. The pieces are immersed, receiving a coating thickness of 0.03 to 0.04 mm on both sides. Strip for so-called lead folded tubes, which are used in electrical installations, is continuously lead-coated. The strip runs through a degreasing trough, the pickling bath and the soldering flux trough (GÜNTHER [448]) into the lead bath covered with zinc chloride. Subsequently residues of dross, flux and melt which may remain adherent, are stripped off both sides. In special machines, the strip is bent together to a tube and folded, a tarred paper tube being simultaneously inserted. Large work-pieces, which cannot be immersed in the bath, are (after cleaning) heated zone-by-zone with a burner. In this case the molten lead is applied by wiping on with asbestos wool.

In general, alloyed lead is used to increase the hardness of the coating. Additions of antimony or of tin of about 5% are usual. As a hardener alloy the German producers favour antimonial lead 12 to 13, containing at least 0.05% As. Alloys with 2.5% Sb and Sn each, having in some circumstances 0.25 to 0.5% Ag or 1% Zn, are also used (KNIGHT [681]). It is assumed that the alloying elements also improve the bond between coating and basis metal (p. 57). Furthermore, with increasing proportion of alloying elements, particularly tin, the coatings obtain a lustrous appearance. For so-called terne plate [493], which is particularly common in America and England, baths with tin-contents up to 25% are used. This plate is very suitable for deep-drawing operations. In the first World War, alloys with about 1% Sn and 1% As, due to SCHLÖTTER (MAASS [787]), gave good service.

In order to increase the bond-strength and freedom from pores of the coating, and therewith its protective action, pre-tinning of low-carbon steel sheet and strip has been proposed. The thickness of the electroplated tin film is 0.1 to 0.5 μm. By annealing at about 350 °C in a reducing atmosphere, the tin layer is converted to an iron-tin alloy, on which the lead alloy, subsequently applied, adheres better (NEISH [892]).

Lead coatings also play a part in the plastic working of steel wire and rod, as they reduce the friction between material and tool and thus serve as a lubricant (BOWDEN and TABOR [120]). The use of such coatings, applied by dip-coating of lead, in the drawing of austenitic chromium nickel steel has been described [1280].

In the usual thicknesses, hot-dip lead coating does not produce completely sound coatings. It has behaved well in cases where the demands on corrosion-protection are not too great, for example, where it is a question of protection against condensed water, weathering effects, feebly acid gases and "other" things. A large field of application is the production of petrol cans, oil-storage tanks, cans for the transport industry and fire-extinguishing appliances. On the other hand, hot-dip lead coating is not recommended for the protection of steel against soil corrosion, as in this case it favours pitting (LOGAN and EWING [762]). The electric spot- and seam-welding of hot-dip lead coated steel sheet (terne plate) is treated in detail by GREER and BEGEMANN [436].

c) Homogeneous Lead Lining. Homogeneous Lead Lining or fusion lead coating is the perfect method of lead coating for the production of tanks and apparatus in the chemical industry, as—in addition to protection against corrosion—good thermal conductivity of the combined material is obtained. Of late homogeneously lead-lined steel, with lead thicknesses of 25 to 300 mm, has also achieved importance for radiation-protection in reactor technology. The process has often been described in detail in the literature (BILART [82], GREGER [438]). After careful cleaning, the steel surface is treated with soldering flux. A layer of tin solder is then generally applied. It is rubbed with asbestos wool, while heated from the outside with a hydrogen flame. A layer of lead is now melted on, so that beads of length 30 to 40 cm and breadth 3 to 4 cm are applied one by one. The free edge is sharply delimited by a square steel bar. On plane surfaces it is possible to work more quickly without laying down beads but rather by pouring lead on to rectangular areas bounded by square steel bars. According to the thickness desired, further layers of lead are applied on the first layer. The use of several layers offers the advantage that the tin in the intermediate coating does not reach the surface in the amount which would be the case if the lead were applied in one thick layer. The part of the surface to be coated

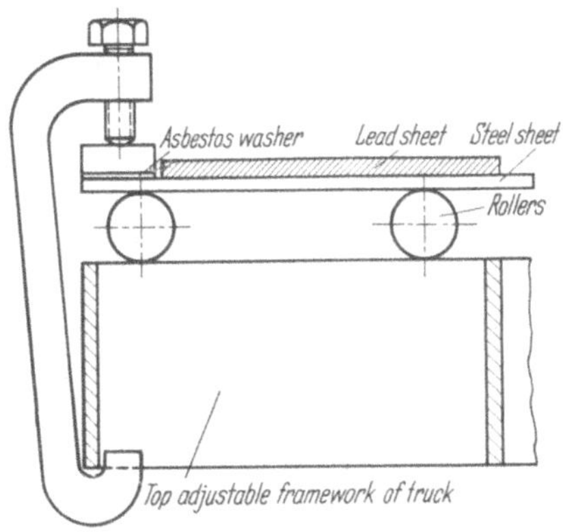

Fig. 379. Arrangement of lead sheet and steel sheet in mechanical homogeneous lead lining. The Port Pirie Smelter process.

with lead must always be set horizontal. In larger work-pieces, this necessitates special appliances for holding and adjusting. Tubes can be homogeneously lead-lined internally by the centrifugal process. At the Port Pirie Smelter of Broken Hill Associated Smelters, Krysko[1], Cunningham and Neve developed a process for mechanical homogeneous lead lining of steel sheets of width up to about 60 cm. With a group of 5 to 6 men, the process gives an output of about 7 square metres an hour. The dimensions of the steel sheet can be, for example, 2,489 · 635 mm, its thickness 6 mm and that of the lead coating 4.2 mm, corresponding to a weight of lead of 48 kg/m². The steel sheet is first prepared in a similar way to that described above for hot-dip lead coating, and tinned on both sides to a thickness of 0.01 mm by immersion in tin solder 60. On one side the solder serves as the basis for a later coating of paint. Subsequently the tinned steel sheet is laid horizontally on a transporting stand and tightly fixed by a frame round it (Fig. 379). Over it, fitting exactly into the steel frame, is laid the lead sheet—measuring 2,438 × 610 mm, with the above-mentioned thickness—which

[1] Communicated by W. W. Krysko, with the kind permission of the Board of Broken Hill Associated Smelters Ltd., Port Pirie.

has been cleaned with steel brushes. The transporting stand is moved to the lead coating machine. There the lead and the tin coating on the steel sheet are melted by means of the oxyacetylene flame, and bonding thus effected. The tubular burner is 610 mm long and equipped with 190 nozzles. Under the steel sheet, parallel with the burner, powerful water cooling is provided. This ensures that the lead solidifies at a distance of 10 cm from the burner. The homogeneously lead-lined plates can be bent into tubes or folded. Nikitin [898a] reports on a similar process.

According to a more recent process, the internal homogeneous lead-lining of tubes is achieved by setting them vertically (after first cleaning) and inserting a mandrel. Above the latter is lead which is kept molten by external heating with burners, and penetrates into the circular gap between mandrel and tube. Corresponding to the rate at which the lead solidifies there, by cooling from the outside, the mandrel with the burner are withdrawn (Fig. 380). According to Krysko [714c], short steel tubes of 50 cm diameter can be homogeneously lead-lined internally by casting. For this purpose, a hollow cylindrical core of thin sheet steel (provided externally with small pins of the desired thickness of the lead coating) is inserted in the vertical tube. Within the cylindrical core is another core tapering upwards. The molten lead is poured into the circular cavity between the two cores, so that the space between the tube and the outer core is filled with lead from the bottom. The solidification of the lead layer takes place from below upwards. On the other hand, molten lead can be sucked upwards from the supply of melt in the circular cavity. Steel surfaces can also be homogeneously lead-lined by casting a layer without an intermediate tin coating. Omission of the application of tin is said to offer the advantage that, as a result of the absence of the lead-tin eutectic melting at 183 °C, such lead coatings are mechanically stable up to temperatures above 200 °C. An international consortium produces under the designation *Insmetals*, homogeneously lead-lined steel, stainless steel, nickel, copper and aluminium [604a] without tin. In all these operations attention must be given to especially good ventilation of the working areas. In homogeneous lead lining for the chemical industry, layer thicknesses of 5 to 10 mm are preferably produced. The final surface is scraped to remove oxide, smoothed by hammering and tested for soundness and quality of bond.

For investigation of a homogeneous lead lining of 4.5 mm thickness under the microscope, the present author completely dissolved away the steel basis electrolytically. The structure is eutectical (Fig. 381). The microsection examination indicated a considerable content of tin in the lead: according to an earlier view this reduces the corrosion-resistance, but the opposite opinion can now be accepted without hesitation (p. 283). Fig. 381 allows the various applied layers to be clearly recognised. The topmost layer shows columnar crystals, and was therefore obviously not hammered.

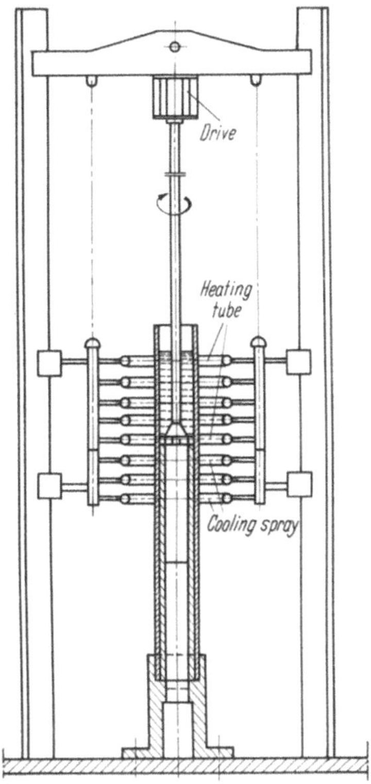

Fig. 380. Schematic illustration of internal homogeneous lead lining of tubes. (Works drawing from Schnakenberg & Co.).

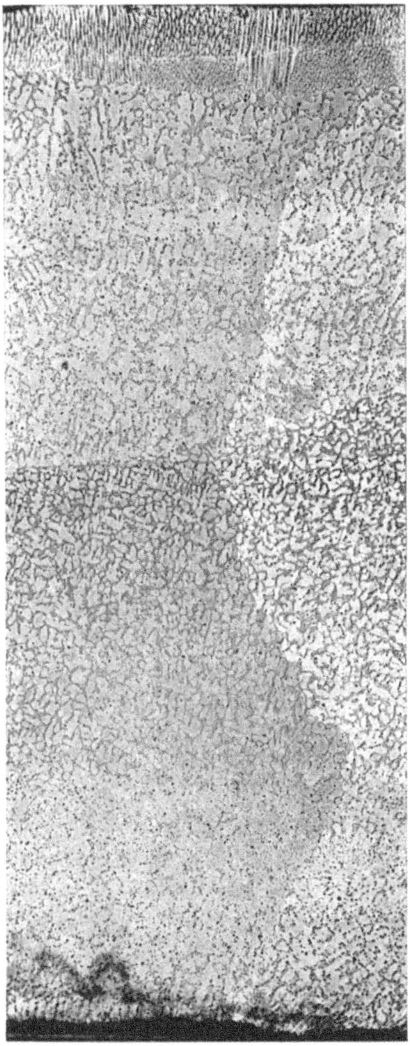

Fig. 381. Cross-section through a homogeneous lead lining 4.5 mm thick after electrolytic dissolution of the steel basis. Eutectic structure of tin-containing lead. ×30

In addition to the processes described, others are also in use, which however only in part deserve the designation of homogeneous lead lining. For example, molten tin solder is applied to the steel surface. When the solder has become pasty, a sheet of lead is placed on it and the lead thus soldered to the basis metal (THOMPSON [1191]). Thin intermediate layers of zinc or copper, which are formed by the use of corresponding fluxes, have been proposed in order to avoid tinning.

Homogeneous lead lining effects an adhesion of the coating on the basis metal by a purely metallic bond. The coating therefore cannot be loosened from the basis material even by bending the work-piece, by fluctuations of temperature and pressure, or by reduced pressure. Corrosion on bare spots is of purely local action. The heat transmission

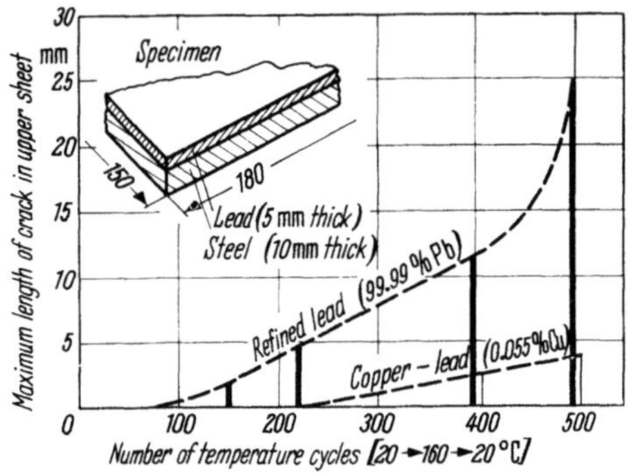

Fig. 382. Formation of surface cracks in homogeneous lead-lined steel sheets stressed by temperature changes. (According to HOFMANN and v. MALOTKI [569]).

is the best possible, which is, for example, particularly important in homogeneously lead-lined heating coils. Homogeneous lead linings can become cracked by very frequently repeated temperature changes. This form of stress was intensified in laboratory tests (HOFMANN and v. MALOTKI [569]). A furnace of particularly low heat capacity enabled the test-sheet to be heated to 160 °C and cooled to room temperature in regular repetitions of 75 minutes. Initiation and growth of crack formation were observed with a measuring microscope. In the representation of the result (Fig. 382) the lower cracking-tendency of homogeneous lead linings with lead containing copper can be recognised. In these experiments, loosening of the lead from the basis metal never occurred. In a continuation of this work, the widening of the cracks, passing from the lead surface to the interior, was followed. Simultaneously the increase of length and thickness of the lead coating was measured. The damage is due to the prevention of the free thermal expansion of the lead by the material below. The representation of the results in Fig. 382a therefore shows that the cracks appear the later, the smaller the temperature difference between the upper and lower limits of the cycle, and the smaller the difference in coefficient of thermal expansion of lead and the basis metal. This becomes particularly clear in the case of duralumin; here the deformation of the lead is least.

In a further series of experiments it was shown that frequent repetitions of temperature changes also reduced the strength of the lead/basis material bond. After 4,000 cycles of temperature two specimens were so ruined that, during the experiment, they fell apart. Although the cycle 30°/160°/30° chosen for this temperature is rarely attained in the practical use of homogeneous lead lining, the experimental results show that in the case of long and continuing extreme conditions of service, such damage must be taken into consideration. The experiments have again confirmed that unalloyed pure lead—owing to its great tendency to recrystalli-

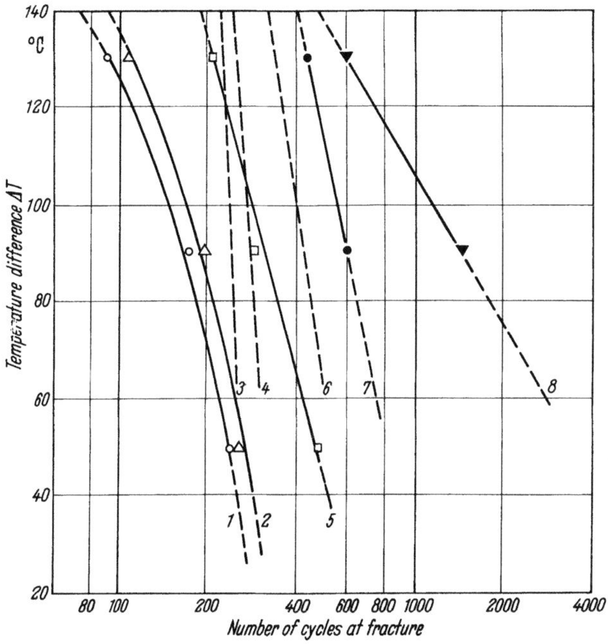

Fig. 382 a: Start of cracks in thermally cycled homogeneous lead lining on steel, copper and duralumin, amplitude of temperature cycles versus number of cycles.

1 6 mm lead (99.99) on steel
2 6 mm lead (0.025% Ni) on steel
3 10 mm lead (99.99) on steel
4 10 mm lead (0.025% Ni) on steel
5 6 mm lead (99.9% Pb, 0.05% Co) on steel
6 10 mm lead (99.9% Pb, 0.05% Co) on steel
7 6 mm lead (99.99) on copper
8 6 mm lead (99.99) on duralumin

sation under stress of temperature changes and the allied alternating strain—is damaged sooner than slightly alloyed lead.

Splits in homogeneous lead linings in a cracking plant in the petroleum industry, which occurred during the last war, led to the comparison of the endurance of various types of lead applied as homogeneous lining to steel (Hillen and Hofmann [528]). Formation of cracks by alternating bending was investigated by electrical resistance measurements. In hot tests (100 °C), lead with 0.05% Te behaved best, then followed lead with 0.06% Cu and lead with 0.02% Ni; in the lowest place was 99.99% lead.

The experiments were recently continued. The specimens were subjected to fluctuating stress. In addition, a heating element was built into the back of the specimens, so that raised temperatures could also be applied to them. The occurrence and propagation of cracks were determined and measured by a magnetoinductive method. The receiving head of the testing apparatus was, for this purpose, moved to and fro over the lead surface. The depth of crack indicated by the crack-testing apparatus was

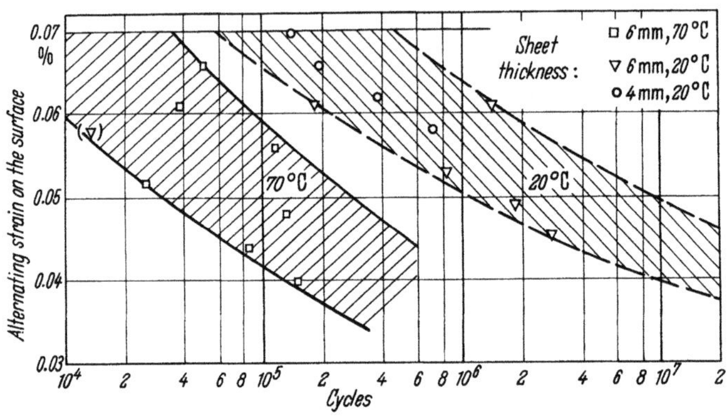

Fig. 382 b. Wöhler curves of lead-lined steel sheets.

furthermore registered by a recording instrument. The fatigue cracks originated predominantly on the surface of the lead coating. Fig. 382b shows Wöhler curves of a homogeneous lead lining with copper-containing lead on a steel sheet. As the crack development could be followed with the magnetoinductive method, in the maximum case to a depth of 1.8 mm, deeper cracks had to be measured by mechanical removal of the lead down to the bottom of the crack.

It was occasionally observed at very high amplitudes of strain that, simultaneously with the cracking of the lead coating, separation from the steel sheet along the soldered surface took place. At low alternating stress this type of damage only occurred in test-pieces with defective jointing between lead and steel. In these experiments the slightly alloyed type of lead again behaved more favourably than the unalloyed material (much coarser grain).

The advantages of homogeneous lead-lining justify the high production costs. It is carried out only with corrosion-resistant lead, thus, for example, very pure lead or lead with a few multiples of 0.01% Cu, Ni or Te. The use of zinc-containing lead, which has been suggested for the purpose of adhesion to iron without an intermediate layer, therefore seems questionable (Bray [127]). A change to antimonial lead can be made if a certain hardness of the coating is required. In addition to steel, copper, Monel, zinc, high-quality cast iron or cast steel can also be homogeneously lead-lined.

d) Spray-Coating with Lead. After an existence of about half a century, metal spraying has developed into an important method for applying metallic layers. Its main field of application is the repair of damaged machine-parts which are primarily stressed by pressure, for example, the spraying of steel and of non-ferrous metals on abraded axles, bearing bushes and pistons, as well as providing coatings for protection against corrosion, such as zinc, on steel. Information on the fundamentals and

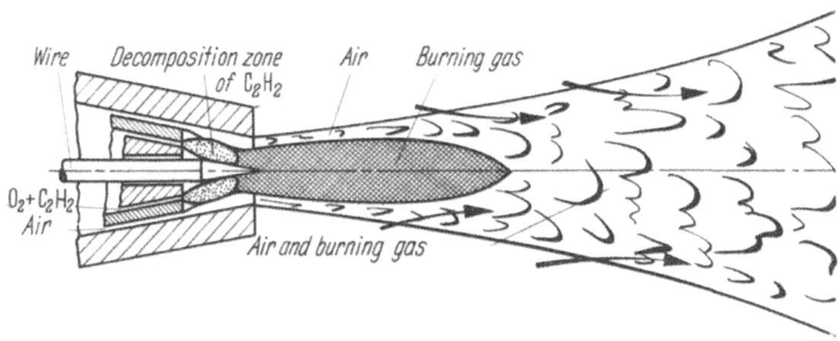

Fig. 383. Path of gas current in a metal spraying pistol. (According to MATTING and BECKER).

the application of the method is given in a series of comprehensive surveys and in trade literature from the appropriate companies (for instance, KREKELER and STEINEMER [703], REININGER [1003]). Compared with spraying by means of wire, metal spraying by means of lump metal which is melted in a pistol and the powder spraying method (ROLLASON [1030]) are receding in significance into the background. Most metal spraying devices require, for their operation, oxygen and a fuel gas such as hydrogen or acetylene, as well as compressed air. The metal to be sprayed, in the present case lead, is fed into the flame as wire of 2 to 3 mm thickness through a driving mechanism. The compressed air jet atomises the metal melted in the flame, and hurls it on to the surface, which has previously been cleaned and roughened by sand-blasting. By means of high-speed photographs, MATTING and BECKER [813] investigated the processes in the spraying of steel. Schlieren photographs reveal the zones shown diagrammatically in Fig. 383, namely a central stream (just at the mouth of the jet) formed by the combustion gases and compressed air, and this is succeeded, in the spraying jet, by a whirling region of compressed air and burning gas. The greatest velocity of the particles in the jet, depending on the operating conditions, was between 150 and 250 m/sec.

In metal spraying no metallic combination of the coating with the basis material occurs, connection therefore depending on adhesion and toothing. On impingement, the metal particles, already solidified to a pasty consistency, are pressed flat and matted together. The metallic continuity of the coating is always interrupted by small amounts of oxides and pores, as is shown in Fig. 384 by the metallographic section of a sprayed lead coating. For this reason, sprayed metal always has a smaller density and a lower elastic modulus than solid metal (v. HOFE, HOFMANN and SUCHAN [549]). In systematic experiments, EVERT [299] determined the operating conditions for obtaining, on the one hand the most, compact and, on the other, the most porous sprayed coatings. The compactness is, for instance, increased by a reduction in the spraying distance, further by a subsequent mechanical treatment, such as brushing or polishing, and by impregnation of the coating with a lacquer or a barium sulphate suspension.

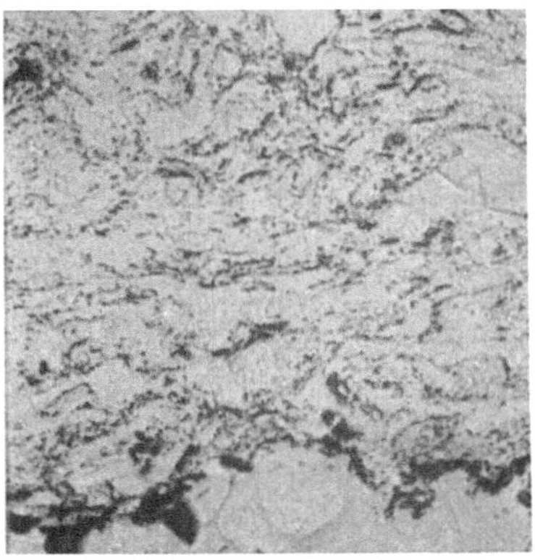

Fig. 384. Cross-section of a sprayed lead coating on lead. Defects principally on the surface of the foundation metal. ×250

EVERTS [299] further showed that the permeability to gas, which is always present, need not be combined with permeability to water. In spraying lead under the conditions which had been found most favourable, there was no longer any permeability to water at a layer thickness of 0.2 mm.

The appropriate operating pressure of the gases required for spraying is defined as follows by the makers of the apparatus: (In atmospheres excess pressure) acetylene 0.35 to 1.5, oxygen 0.6 to 2.5, compressed air 2.5 to 4.2. The speed of advance of the wire is between 6 and 11 m/minute, the spray-output between 12 and 38 kg/hour. For spraying lead and tin on a metallic basis material, the most favourable distance of the aperture of the apparatus from the work-piece is indicated as 150 to 300 mm. These distances correspond to values of about 30 to 55 mm for the diameter of the sprayed area. The application of sprayed lead coatings for protection against corrosion has repeatedly been described in the literature. Thus copper evaporating coils, destined for hot aluminium sulphate, were sprayed with lead in thickness up to 1.5 mm [839]. Latterly in Poland, elements (precipitating baths) of rayon spinning machines and electro filters in sulphuric acid works have been protected by spraying a lead coating of thickness of 0.3 to 1 mm (KOWALSKI [699]).

In spite of occasional applications for protection against corrosion, spray-coating with lead cannot compete with other methods of coating with lead, as while sprayed lead coatings can certainly be made impervious to liquids and gases by increasing the thickness, the corrosion-resistance is, after all, reduced by the oxide-content which is always present.

A possible region of application of spray-coating with lead might well be radiation-protection. Spraying of bearing metals containing lead on supporting shells must also be mentioned as a possibility.

e) Lead Electroplating. In the course of comprehensive surveys (SCHLÖTTER [1067], PFANHAUSER [956], GRAY [419], MACHU [790], the electroplating of lead is treated more thoroughly than is possible here. Deposition of lead from solutions of simple salts, such as lead nitrate or acetate, results in coatings which are technically useless. According to FISCHER [322] lead crystallises out here in the form of the so-called "Isolationstyp", that is, the crystals grow singly in needle shape; their orientation (without regard to the basis metal) is influenced solely by the electric field. Smooth and coherent coatings are only obtained from solutions of certain complex salts. Here lead is deposited in the form of the so-called "Reproduktionstyp", that is, the crystals grow oriented on those of the basis metal and show greater development in breadth. The solutions which have been most widely introduced are those of lead fluoborate with free hydrofluoboric acid, due to LEUCHS [744] (compare LEBRUN [739], GABRIELSON [354], GRAHAM [412a]), and of lead sulphamate with free sulphamic acid (GRAY [419]). In addition, mention must also be made of the baths of lead fluosilicate (BETTS [78]), lead perchlorate (MATHERS [804], SCHLÖTTER [1067]), lead phenolsulphonate (SCHLÖTTER [1066]), complex lead acetate (MACHU [790a]), as well as alkaline lead baths (MATHERS [805], MACHU [790]). In order to avoid spongy or coarsely-crystalline deposition, additions of organic substances (such as gelatin, peptone, cloveoil or tannic acid) must be made to the solutions mentioned. Electroplating of lead requires a pretreatment of the surfaces which are to be coated. They must be mechanically cleaned, in some circumstances ground and polished, and finally degreased and pickled. Electrodeposition takes place at current densities of, say, 2 A/dm^2 and voltage less than 2 volts. With correct ratio of lead to acid, the throwing power of the electrolytes is excellent. If, for example, in an experiment, an anode and a cathode (both in plate form) are placed opposite one another, the latter is also uniformly coated with lead on the back[1]. In addition to the usual method, in which the object to be electroplated with lead is immersed in the electrolyte, so-called dry lead coating has also been proposed. The hand anode is made in the form of an ebonite box, which contains lead over a perforated plate. The electrolyte, lead fluoborate, is applied by a brush over the object to be plated (FEDOTIEFF, ARTAMONOFF and RASMEROVA [308]). The electro-

[1] Kindly demonstrated by Professor M. SCHLÖTTER.

plating of lead on copper is possible without intermediate layers. On the other hand, the electroplating of lead on aluminium — which has been developed, for example, as bus bars of accumulator batteries — requires a previous nickel plating (SCARPA [1052]). Lead as pure as possible is used as the anode. Electrodeposited lead is very pure and quite sound as it has the density of cast lead.

As a substitute for the structure of a lead electroplate, the cross-section of a lead cathode employed in the electrolytic extraction of lead is shown in Fig. 385. The middle layer is the cathode sheet. The electrodeposited lead contains only small inclusions (not detectable under the microscope) of the colloids present in the bath and a little hydrogen. The hydrogen content is quoted as $0.8 \cdot 10^{-3}$ to $2 \cdot 10^{-3}\%$, that is 9 to 22 Nml hydrogen/100 g lead (KURREIN [723]). The tendency to recrystallisation or grain growth is said to be less than in other types of lead (FORSTNER [336]). HIRATA [530] established the growth texture of electrodeposited lead.

Fig. 385. Electrolytic lead. Section through a lead cathode. Individual zones parallel to the cathode sheet. Columnar crystals in the direction of the current. ×6

According to the American standard [27] the minimum layer thickness of lead coatings on steel must be 0.0064 to 0.025 mm. MACHU [790] quotes 0.075 mm as the thickness of coating necessary for freedom from porosity. There is practically no upper limit for the technically possible thickness. Electrodeposited lead has a very low hardness and hence has not been able to supplant hot-dip coatings, where a certain hardness of the coating is achieved. Lead electroplating serves, for example, as protection against rust for small mass-produced articles in electrotechnics and in chemical plant construction, and has been suggested for protection against corrosion for steel structures in railway stations, locomotive sheds and bridges (KURREIN [723], HAY [498], DUROSE [265]).

The possibility of electrodepositing lead alloys, as well as pure lead, is very important. The process is applied in practice for the production of lead-tin alloy coatings. They serve as lubricants in the plastic working of steel strip, and, apart from this, also form a subsequent protection against corrosion (Roehl [1018]). Cheffers gives details of the most favourable conditions for deposition from a fluoborate solution [192a]. According to Smart [1126b], the comparatively coarse distribution of the lead and tin phases in the layers points to high rate of migration of surface during deposition. White metal, that is, an alloy with 11% Sn, 7% Sb, remainder lead, could also be deposited from a fluoroborate solution of these metals as a running layer on a bearing with a silver bush. The advantage of such thin layers of bearing metal is mentioned elsewhere (pp. 370, 371).

f) Lead Cladding. The application of lead coatings by roll-cladding meets difficulties which arise from the great difference in hardness between lead and most technical metals. The method has therefore only been applied on the technical scale in the combinations of lead on zinc and tin on lead. In order to reduce the difference in hardness between zinc and lead, antimonial lead is used as coating instead of lead (Groove [440]). Lead-clad zinc is used for roofing and for cell-cups of Leclanché cells. Gumm [456] succeeded in rolling lead or lead-tin alloys on aluminium sheet also. When the aluminium, which had been roll-clad with tin solder, was heated to the melting point of the lead-tin eutectic, the molten solder wetted the aluminium surface. Thus, by heating sheets of lead and of tin solder-clad aluminium, soldered aluminium-lead sheet could be obtained.

g) Further Methods. Lund [783a] reports experiments on applying mixtures of lead and zinc powder to steel by hot or cold rolling.

h) Red Lead Priming. In the protection of steel against corrosion, lead in the form of red lead plays a most important part (Krenkler [707]). This subject is however outside the scope of this book and will therefore not be considered here.

2. Use of Sheet Lead in Chemical Technology and in Building

The lining of wood, steel or copper tanks with rolled lead is applied on a large scale in the chemical industry. Vats for storage of acids, alkalis and other liquids, for crystallising and settling, and for carrying out electrolytic processes, are "lead coated" in this way. Lead pipes for the transport of solutions or sea water are, similarly, drawn into steel pipes. In order that the sheet lead lining, especially on vertical walls, should have the necessary support, it is soldered or screwed in various places to the support, or spring clips of bars are used. Screw-heads and similar fixing devices must be covered with lead. Sheet lead lining in its simple state is unsuitable for operations in vacuo or at high pressures. Moreover, having regard to the creep of lead, it is not recommended if

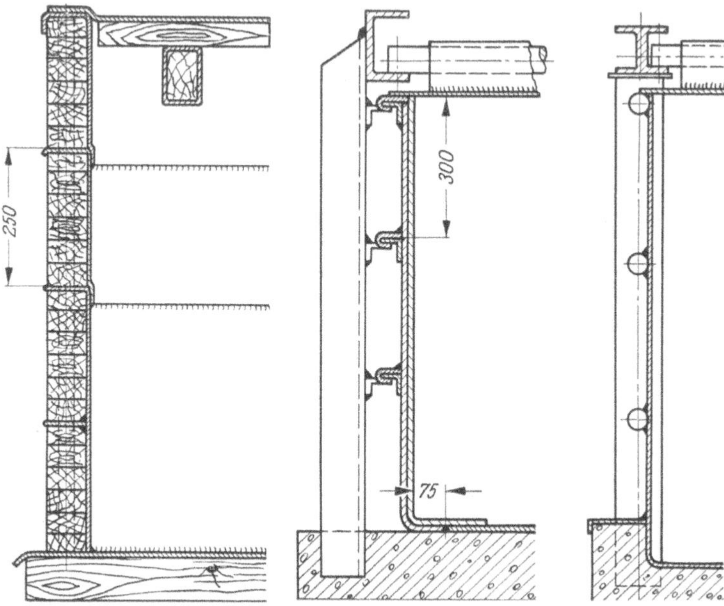

Fig. 386. Examples of designs of sheet lead linings [194]. Dimensions in mm.

temperatures above 50 °C are expected, where it is then usually replaced by homogeneous lead linings.

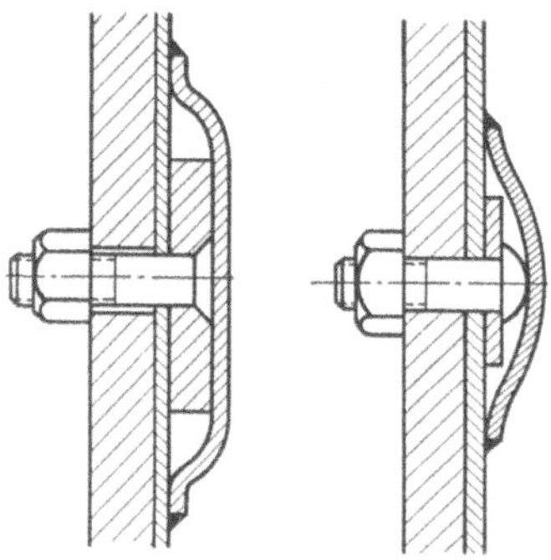

Fig. 387. Examples of the fixing of sheet lead linings by screws. (According to ROLL [1029]).

Fig. 386 [194] shows, as an example of the design of sheet lead linings, the possibilities of fixing a vertical lead wall to outer (steel or wood) supporting frames. In the case of fixing by screws, the protection of the screw-heads with lead can be carried out as in Fig. 387 (ROLL [1029]). The jointing of the lead sections to one another, in general, takes place in these constructions by welding.

For storage and transport containers, a large chemical company developed a combination of lead and glass fibre reinforced plastic. The lead cladding was first formed, the plastic laid on, and then joined. Lead sheet can also be used as a covering material on metal or wood containers to provide corrosion resistance (KLANT [668a]).

On the other hand, in lead roofing, jointing is carried out by folding. The necessary strength of the folded joints is generally achieved by means of copper clips (Fig. 388) [194]. HEUSER [516] describes the covering with lead of Cologne Cathedral and other important buildings. In recent years, lead has achieved increasing significance in building techniques as a material for damping sound and vibration (RADTKE [991 a]). For good sound damping, a material must fulfil the following conditions:

1. High specific gravity.
2. Low elasticity in bending.
3. Freedom from porosity.

The basic theory for the technique of sound attenuating lead laminates was recently developed by CREMER and v. MEIER [221 a] at the HEINRICH-HERTZ-Institute of the TU Berlin. From these investigations the results

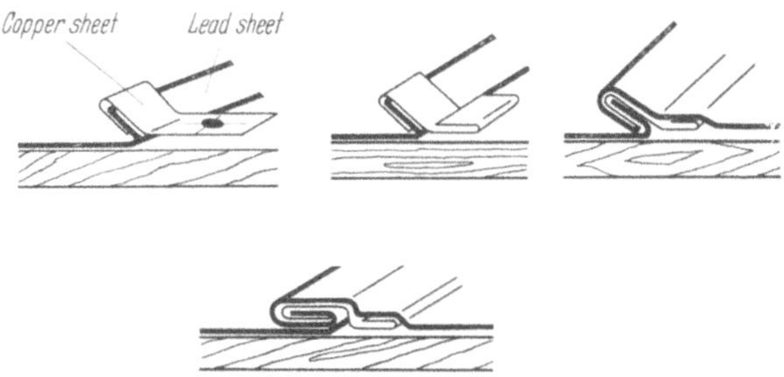

Fig. 388. Development of a folded joint for sheet lead roofing [194].

of two different rigid lead laminates are reproduced in Fig. 388a. Sound frequencies above about 3,000 Hertz are of little significance for sound attenuation. Further examples for uses of lead for sound attenuation are given by OSTERGAARD and GOFF [928a, 400a].

Lead fulfils these three important provisions almost completely. As thin layers of lead are already completely effective for sound prevention in architecture, lead will bear cost-comparison in any manner with other sound-resistant materials. In addition it has the advantage of permitting lighter and more space-saving construction. For the production of sound resistant walls, lead sheet 1.5 to 3 mm thick is combined with plywood, hard fibre, plaster of cement-asbestos sheets. Further, sound resistant walls are known consisting of plastic sheets containing lead in a finely-divided form. In cases where no solid walls are necessary or where the erection of such walls would be troublesome, sound-resistant textile mats, into which thin lead wires are woven, can be used.

For prevention of vibration, an American company produces laminated plates built up as follows: 3 mm sheet lead, 9.5 mm asbestos,

1 mm sheet steel, 9.5 mm asbestos, 3 mm sheet lead. The sheet lead not only forms the top and bottom layers, but encloses the whole in packet form. The plates are compressed under a pressure of 14 kg/cm² [736, 737]. Such lead-asbestos damping plates are used, for example, in building the foundations of skyscrapers in soil subject to vibration.

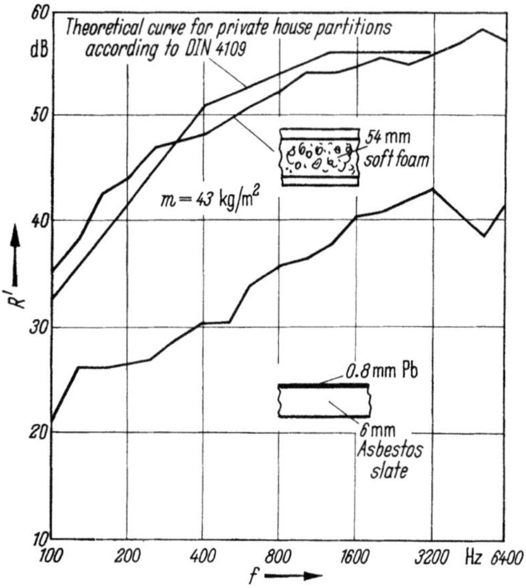

Fig. 388a. Sound absorption of a leaded laminate and of a wall constructed with such laminates on each face.

At this point, mention may also be made of lead linings for rooms exposed to radiation, and of lead sheet for sealing and for equalising pressure. The physical basis for protection by lead against radiation is treated on p. 21, and the mechanical behaviour of lead under the effect of pressure and friction in seals on p. 226.

3. Welding and Soldering of Lead

The jointing of lead pipe and sheet is particularly important for chemical plant construction, in, for example, the production of large pipeline systems in sulphuric acid works or in the construction of lead chambers. Having regard to the required corrosion-resistance, jointing takes place almost without exception by welding, which in the case of lead is, in general, erroneously designated soldering. In this matter, there is a most important investigation by the „Staatliches Sächsisches Hüttenwerk Freiberg“ (Brenthel [129]).

Thick sheets are fitted flush to one another, the edges to be joined being bevelled. Thin sheets are lap-welded. With vertical joints, it is preferable to work without a filler metal. The latter is applied only for horizontal joints. Thick-walled lead pipes can be butt welded, if high stress by pressure is not anticipated. If, on the other hand, they are to be stressed by pressure and bending, then socket pipe jointing is more suitable. A sleeve is flanged up on both sides and welded to the outsides of the two pipes. Thin-walled pipes are (after flanging up one of them) pushed into one another and the space between them filled with weld metal. If pieces of fixed pipe are to be replaced, then the jointing of the replacement piece to the existing pipe can be carried out in such a way that the lower part of the pipe is welded from inside after an opening has been cut in the upper half by bending back appropriate sections or taking out a piece. After finishing the lower half, the upper half of the pipe is welded from the outside, if necessary by incorporating a sheet lead half-round cover (Brenthel [129], Westbrook, Rodgers and Carter [1262]).

The places which are to be welded are cleaned by scraping off dirt and oxide. The thickness of the welding rod is governed by the sheet thickness. Fluxes are not used. Welding is carried out mainly with the oxyhydrogen flame. The flame should be regulated to neutral or with a slight excess of hydrogen so that the formation of oxide skins is avoided as far as possible. In general, the oxyacetylene flame is regarded as too hot. If in this case (as with the hydrogen flame) the work is not fast enough, defects can occur owing to melting, and the formation of holes. However, with sufficient skill, welding can also be carried out with the oxyacetylene flame with the consequent advantage of quicker and cheaper work (Schulze and Staebler [1089], Hunsicker [602]).

In welding sheet lead linings, the complete soundness of the weld is particularly important, as no liquid must penetrate through the lining to the material below it. For the purpose of testing soundness (Schulze and Staebler [1089]) the edges of the lining and of the outer cover are sealed with putty. Compressed air is passed between the outer cover and the lead coating by means of a connecting pipe. After spraying on a suitable liquid, for example "Neutrallösung" produced by Farbwerke Bayer, the welds are examined.

In addition to the jointing of lead pipe by welding, flanged joints are used to a small extent (Brenthel [129]). Pipes of small wall-thickness are turned over on to an iron flange. For thick-walled pipes, the iron flange is pushed over the pipe-end and a lead disc welded on to the latter. Cast antimonial lead flanges of high creep strength can be directly welded to pipe-ends.

Combinations of lead parts are also produced by actual soldering. Soldering can be done without hesitation if there is no risk of corrosion, for instance, in containers and piping for water [738]. It is usual for cable-sheathing [1176]. It is carried out by flame-soldering and tallow and colophony resin are used as fluxes. In horizontal pipes, the solder is applied to the highest point and spread in the pasty form to the sides and downwards by a greased rag (smear soldering). 33% tin solder ("plumber's solder") is preferably used, as it has a long solidification range (LEWIS [746], JOHNEN [622]).

Latterly, adhesive bonding has been considered for the jointing of lead to lead and to other materials (RADTKE [991b]).

4. Lead in Soft Solders

a) Solder Alloys and the Soldering Process. The classical soft solder — tin solder — is covered in Germany by the Specification DIN 1707, on lead and tin solders. It contains 10 lead-tin alloys with tin contents of 8, 25, 30, 33, 35, 40, 50, 60 and 90%; in addition it provides for a high lead solder with 98.5% lead. The most important additive is antimony. It derives from the so-called Mischzinn, a hardener alloy with about 54.5% Sn, 3.6% Sb and 41.9% Pb which is much used for the production of solder. Accordingly the maximum content of antimony is set at 0.5% in tin solder 8 and at 3.30% in tin solder 50. By and large, the British and American specifications provide for similar gradations of tin content; however, the antimony content of the alloys for certain applications, for instance, for soldered joints in electrotechnics, is more severely limited. Antimony is troublesome in the soldering of galvanized steel and of copper, as antimony-containing phases of higher melting points are formed [835a]. Apart from this, antimony certainly increases the strength of soldered joints, but in larger contents has an embrittling effect. Most lead-tin solders are hypoeutectic (Fig. 92); their structure consists therefore of lead-rich solid solutions and more or less extensive residual fields of eutectic. Larger antimony-contents make themselves known by inclusions of SbSn; as long as this material is not in equilibrium, it can be dissolved by homogenising at 150 °C (BAKER [37]). Copper, which can occur as an addition in the order of 0.1%, is present in the solder as the phase CuSn (BAKER [37]). According to experiments by KÜNZLER and BOHREN [720], melts of tin solder with copper additions up to 2% attack immersed copper foil less than do melts free from copper. The use of tin solders containing copper should therefore give rise to a longer life for copper soldering bits (LÜDER [782]). Aluminium and zinc impurities are detrimental and therefore not permissible even in the smallest quantities (of below 0.005%), while for iron maximum contents of a few multiples of 0.01% up to 0.1% have been set.

The difficulties of providing tin, during the last war, led, in a number of countries to efforts to develop tin-economy solders (LÜDER [781,

846]). For example, part of the tin was replaced by portions of other low-melting elements such as cadmium, bismuth, antimony, thallium or phosphorus ([845], Claus [200]); or attempts were made to introduce lead alloys free from tin, or lead itself, as solder (Lüder [779], Barham [44]). Some of these suggestions led to a result valid even in times of normal tin supply. In this respect, attention may be drawn to lead-silver alloys (p. 34) or to solders of high hot-strength. For soldering food cans, the contents of which are to be sterilised at a temperature of 120 °C, Bodymaker machines use, for instance, lead with only 1% tin (Panknin [930]). The high lead content of this solder is harmless in spite of the fact that the cans contain food, as we are dealing with external soldered joints and the small amounts of solder penetrating inwards are sealed off by a layer of lacquer. For internal soldered joints in contact with food, the current German law prescribes a tin solder with a maximum of 10% lead [376]. Such high-tin solders are less easy to manipulate — according to Lange and co-workers [730] — the surface tension of molten solder increasing with growing tin content, i. e., $\sigma_{Sn\,100} > \sigma_{Sn\,55} > \sigma_{Sn\,35}$. McKeown [818] has published a monograph on the investigations carried out by the British Non-Ferrous Metals Research Association during the war for the purpose of economising in tin. In a review of this book (Chadwick [187]) it is emphasised that in spite of the success then obtained, almost all tests reveal the advantage of high-tin solder in respect of speed of work and quality of the joints. Economic considerations however even now make the reduction of the tin content of solders interesting, especially in mass production. For example, in the soldering of automobile radiators a solder of low tin content could be adopted by changing over to dip soldering [735c].

Soldering presupposes the direct contact of the basis material and the solder without a separating oxide layer. The solution or coagulation of the oxide skin is achieved by a flux, for instance, by the much-used zinc ammonium chloride, by colophony resin or by other resins which work more mildly. McQuillan [829] and Künzler [719] give data on corrosive and non-corrosive fluxes (see p. 430). The flux promotes good wetting of the basis metal by the molten solder. Raman found the best wettability for copper by solders with 30 to 50% Pb. On both, copper and brass, zinc ammonium chloride provides better wetting than resin [991c]. In this connection there are some interesting experiments on applying solder electrolytically to steel or aluminium, thus facilitating and mechanising the soldering process (McConnell [815]). The occurrence of alloy-formation between solder-components and the basis material is not an unconditional assumption for the formation of a bond. For example, in soldering steel by means of pure silver (Fischer [324]), the formation of alloy layers cannot be assumed on the basis of the

Ag–Fe equilibrium diagram. It is further known, that metals, such as iron and lead (p. 57) — which do not dissolve in one another even in the molten state — can certainly be bonded to one another by cold pressure welding. In the case of soldering with lead-tin solders, alloys with tin have been detected on the edges of the basis metal (ROSTOSKY [1035]). In the soft-soldering of copper, CROW [225] observed the phase CuSn, MERZ and BRENNECKE [834] the phase Cu_3Sn. Alloy-formation takes place in this way, that, on the one hand, tin (or lead) migrates into the basis metal, while, on the other hand, the basis metal dissolves in the molten solder and diffuses away from the bounding surface. The process of solution is promoted by the diffusion away of the dissolved metal, as this forms the "slowest process" (MERZ and BRENNECKE [834]). In these experiments, MERZ allowed molten tin to act on solid lead and determined the diffusion coefficient of the dissolved lead in tin. Having regard to the effect of temperature, the values obtained at 260 and 310 °C — 1.24 and 1.59 cm^2/day respectively — agree well with older measurements by ROBERTS-AUSTEN [1013]. The nature of the alloy formed differs from case to case. While formation of solid solution is of the highest significance for hard soldering, in soft soldering (as shown by the examples given) it is more a question of the occurrence of intermetallic compound. In the soft soldering of steel, wetting by the solder and the formation of alloy is rendered more difficult; wetting is attained by "tinning" the steel surface prior to soldering. The soldering of aluminium by means of lead-tin solder is prevented by the adherent skin of aluminium oxide. This difficulty is avoided by the technique of friction-soldering or of ultra-sonic soldering.

MILNER [856] applied the theory of surface tension and that of liquid flow to the soldering process. When the melt and the solid metal make contact, then a definite angle of contact Θ is formed, which in the case of complete wetting is zero. For soft soldering, the condition $\Theta < 90°$ suffices. The rise of the melt in the gap as a static process can be assumed to be known. MILNER also treats this process dynamically. By experiments and by calculation he establishes that in the rise of the melt either laminar or tubulent flow is present. The rate of rise is in all cases so great, that there is no time for detrimental chemical reaction between the solder, the basis metal and the flux. For example, molten tin needed a time of 10^{-2} sec to rise 13 cm in a gap in solid tin. MILNER gives a possible interpretation of the action of the flux which completely diverges from existing views. In the case of the soldering of aluminium, the flux is said not to dissolve the oxide but to strip it off by means of an electrochemical process, by penetrating between the oxide and the metal. In this connection, experiments by HEUMANN [515a] are of interest. He observed a penetration by mercury, due to capillary forces, between metal and oxide skin.

b) Properties of Soft Solders and Soldered Joints. The mechanical properties of lead-tin solders are presented on the basis of an older investigation by GONSER and HEATH [406]. The tensile strength of lead-tin alloys increases with increasing tin content, falling only at very high tin content, which does not particularly concern solders (Fig. 389). The tensile strength is approximately parallel to the hardness (p. 204). The

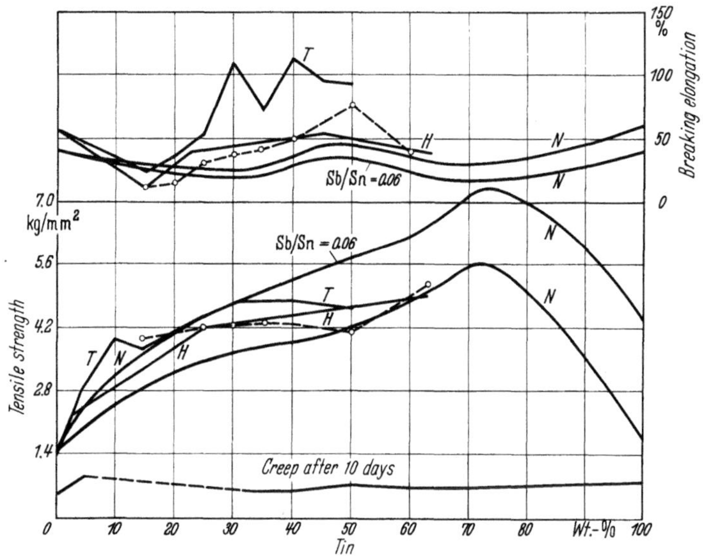

Fig. 389. Tensile strength, elongation and creep of lead-tin and antimonial lead-tin alloys. H. HIERS; N. NIGHTINGALE; T. THOMPSON. (According to GONSER and HEATH [406]).

elongation at fracture is very high over the whole range of concentration. A maximum is possibly present at middle tin contents. The minimum in the neighbourhood of the solubility limit on the lead side presumably disappears if the alloys soften on storage (p. 98). An antimony content of 5 to 6% of the tin content raises the strength of 10 to 20% (BAKER [37]). In the alloys which contain antimony, a transient age-hardening occurs before the softening. Measurements of the shear strength of cast specimens exhibited a steeper rise up to 30% Sn, then a more or less shallow rise up to 60% Sn (GONSER and HEATH [406]). The notch bar strength, measured on specimens with cast notches, had a maximum at tin contents of 35 to 40%.

Measurements of the creep strength (BAKER [37]) on alloys free from antimony, indicated the highest values at the eutectic composition. The creep strength is greatly increased by addition of antimony. Addition of 0.18% Cu caused a further rise. On the other hand, small contents of iron, copper and bismuth, as well as the process of softening, were without effect. The creep strength of solders with and without antimony fell at 80 °C to 20 to 30% of that at room temperature. Coarse-grained specimens showed intercrystalline fracture in the creep test (p. 231). In comparison with that of cast alloys, the creep strength of extruded eutectic lead-tin alloys is lower by an order of magnitude (cf. p. 236).

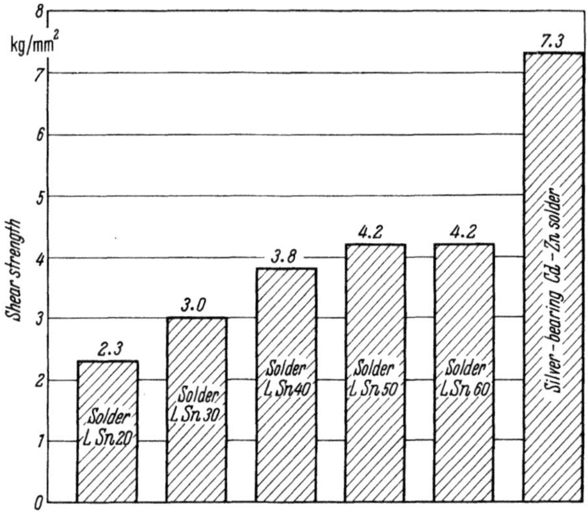

Fig. 390. Shear strength of soft-soldered joints in copper. (According to KOCH and WASSERBÄCH)
Experimental conditions:

Nature of soldered specimen	Plug joint i.e., end of a round rod soldered into hole, drilled in plate.
Soldered surface	140 mm²
Soldering gap	0.1 mm
Soldering temperature	80°C above liquidus of solder
Cooling	Air
Rate of loading	10 kg/sec
Duration of storage	1 day
Testing temperature	20 °C

Soft-soldered parts should as far as possible be stressed by shearing off, as this type of stressing corresponds best with the peculiarity of the joint. In this case the load-bearing capacity of soldered joints can be raised by increasing the length of overlap. Therefore the shear strengths of soldered parts are the most important criterion for the designer. The separating strength of solder is determined by the tensile fracture of flush-soldered rods. According to COLBUS [206] it is of the order of 7 kg/mm² for lead solder between copper rod, and of 17 kg/mm² for tin between copper rods, and the separating strength of hard soldered specimens increases with decreasing width of the soldering gap. Such joints can also be made by cold pressure welding or by cold pressure soldering (p. 199). After machining and tensile fracture of the "soldered" rods, the following values are obtained for the "separating strength" of soft solders [148]: Tin between copper rods, < 8.2 kg/mm²; lead between aluminium rods, 4.0 kg/mm²; eutectic lead-tin solder between Armco iron rods, 5.2 kg/mm². The values should only be regarded as guides, since no analysis of the state of stress in the soft metal has yet been carried out.

According to Koch and Wasserbäch [684, 1239] good values of the shear strength are obtained if the temperature of the solder is 60 to 90° above the liquidus temperature of the alloy. By and large, this determination agrees with a proposal by Mohler [861]. At the place of soldering, the work-piece must at least have reached the so-called working temperature [1095]. In the soft-soldering of steel and of copper, the highest values of the shear strength were obtained with a gap-width of 0.05 mm [1239]. According to Nightingale [898] and Leach [734], the soldering gap should be narrowed with increase of soldering temperature. The width of soldering gap should be above 0.03 mm. The quantitative relationship between soldering temperature and optimum width of gap put forward by both investigators has not proved reliable [1239]. The strength of soldered joints suffers a certain decrease on storage. This may be attributed to the formation of coarser precipitates from the supersaturated lead solid solution (p. 96). Thus Koch and Wasserbäch [684, 1239] found, for example in the soldering of copper, a decrease of shear strength from 3.6 to 2.8 kg/mm² after four months storage at room temperature. In order to obtain reproducible values of the strength of soldered joints, they propose the following conditions:

Width of soldering gap	0.1 mm
Soldering temperature	80 °C above the liquidus temperature of the solder
Duration of soldering	as short as possible
Rate of loading in strength testing	10 kg/sec
Temperature of test	18 to 24 °C

The specimen should be furnace-soldered, then slowly cooled in air, and tested for its strength about 24 hours after soldering.

In this way the values given in Fig. 390 and 391 were obtained for the strength in relation to the composition of solders.

In an older investigation by the „Staatliches Materialprüfungsamt“ in Berlin-Dahlem (Tonn and Günther [1204]), sheet zinc and tin-coated steel sheet (tinplate) were lap-soldered with pure lead-tin solders. While the strength of the soldered joints in sheet zinc increased with growing tin content, that of those in tinplate — except in the case of 10 and 15% solders — was only slightly influenced by the tin content. Determinations of the consumption of solder and of the duration of soldering showed

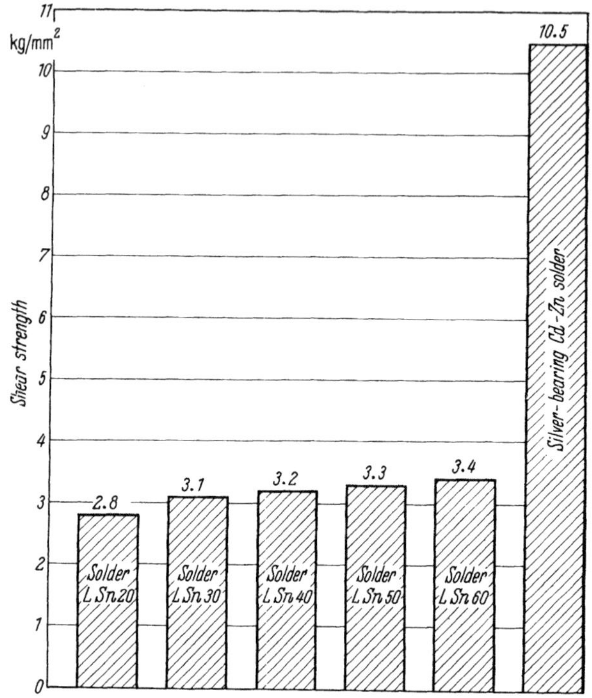

Fig. 391. Shear strength of soft-soldered joints in steel (St 37). (According to KOCH and WASSERBÄCH). Experimental conditions as in Fig. 390.

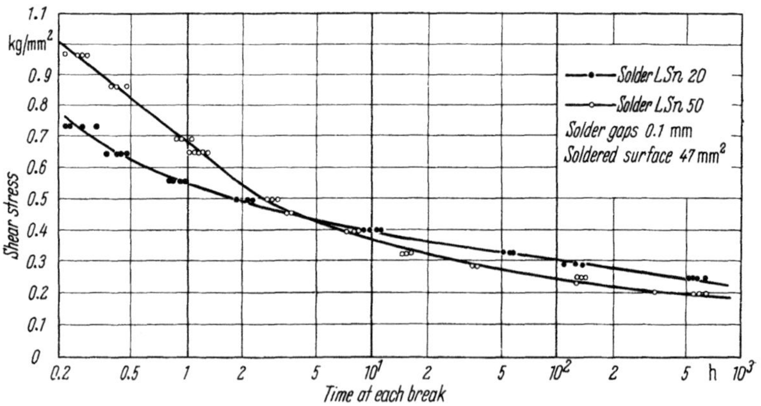

Fig. 392. Creep-to-rupture of soft-soldered joints [684, 1239].

Basis material copper (empty circles)
Soldering gap 0.1 mm
Soldered surface 141 mm²
Soldering temperature 290 °C

Basis material steel St 37 (full circles)
Duration of storage 1 day
Solder LSn 50

that both decrease with increasing tin content. From this follows that with the use of solder of low tin content, a part of the saving of tin is balanced by the increased consumption of solder. Soldered joints in sheet copper were tested for strength in another way (CHADWICK [186]). Each of two strips of sheet were bent to a right angle. The two strips were joined by soldering one leg of each together. Tensile testing was carried out along the two free legs. The strength was referred to the unit of length of the fractured joint and thus provides only a comparative value. It remains unchanged during the gradual splitting of the soldered joint. According to this investigation, it was not the properties of the solder which decided the strength of soldered joints in copper, but those of the intermediate copper-tin phases which were formed even with low tin solders. As the ε phase is said to have the lowest strength, fracture always takes place in the alloy veins formed by this phase. The strength of joint is therefore below that of the soldering alloy.

The strength properties of soldered joints are, as is to be expected, very dependent on the speed of testing, which should be borne in mind in comparing results from different laboratories (cf. KIES [664]). The results of creep tests provide a criterion of the permanent static load which can be expected in a soldered joint. According to Fig. 392, the creep-to-rupture of soldered joints if a life of one year is taken as a basis, amounts at the most to 0.3 kg/mm^2 and therewith to about 1/10 th of the value of the shear strength in the short-time test (cf. [415]). This relationship is of about the same order of magnitude as is applicable for lead alloys in the massive form. As usual, the creep-to-rupture decreases with increasing temperature of test (Figs. 393 and 394 [1239]). Soldered joints in copper and in brass proved superior in creep strength to those in steel. This is attributed to absorption of copper, or copper + tin, by the solder (BAKER [37]). In respect of creep-to-rupture of soldered steel, solders containing antimony excelled those free from antimony (BAKER [37]). KEIL [652] could not confirm the assumption that the creep strength of soldered joints could be increased by additions of silver to the lead-tin alloy. According to Fig. 395, solders of lower tin content give joints of higher hot-strength.

Fig. 396 [1239] provides a criterion for the endurance of soldered joints. As is usual for lead alloys (p. 248), the values for endurance lie above those for creep strength. The fatigue strength under fluctuating tension falls even in the range of high load cycles. This may be connected with a creep of the solder in this type of stress. According to investigations at the Battelle Memorial Institute, Columbus, Ohio, a higher fatigue strength is achieved in soft solders by the use of solders having more than 70% Sn and additions of 0.05 to 0.10% Ce [1201]. In order to combine the properties of lead-tin solders with those of steel, metal-reinforced

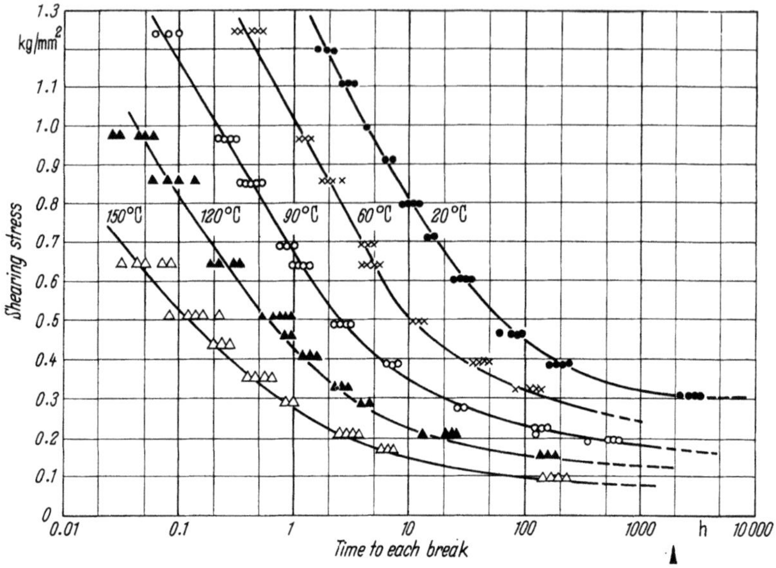

Fig. 393. Creep-to-rupture of soft-soldered joints at elevated temperatures [684, 1239].

Soldering gap 0.1 mm
Soldered surface 47 mm²
Soldering temperature 290 °C
Duration of storage 1 day
Basis material St 37
Solder LSn 50

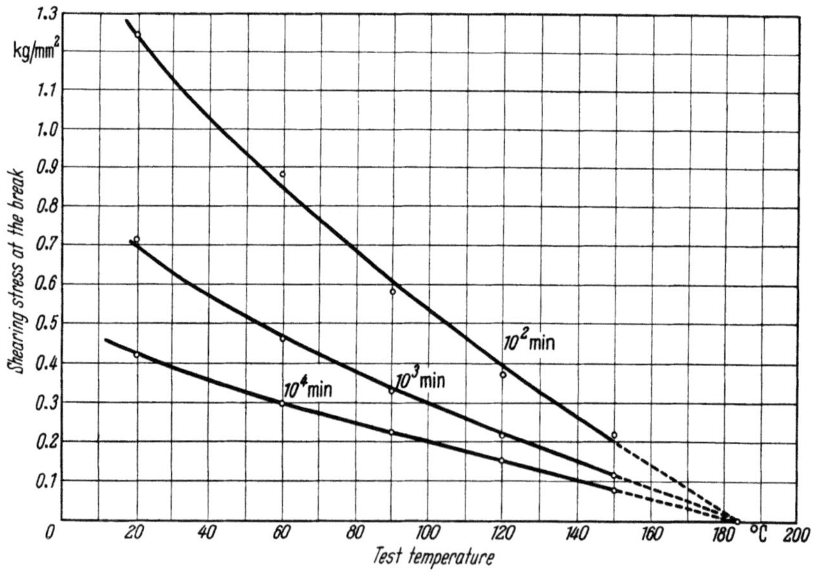

Fig. 394. Creep-to-rupture of soft-soldered joints in steel (St 37) at various temperatures, based on a life of 100, 1,000 and 10,000 minutes [684, 1239].

Soldering gap 0.1 mm Soldered surface 47 mm² Solder LSn 50

lead-tin alloys in the form of tapes were produced for experimental purposes. A felt of steel filaments was immersed in the molten alloy. The share of the volume occupied by the steel is, for instance, 10%. The

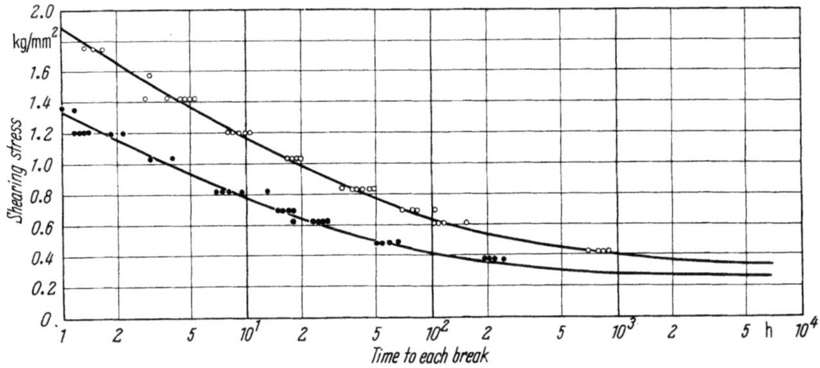

Fig. 395. Comparison of creep-to-rupture of soft-soldered joints. Use of two solders of different composition [684, 1239].

Soldering temperature 290 °C — Testing temperature 90 °C
Duration of storage 1 day — Basis material St 37

bodies were cold-rolled so that the thickness of the tapes amounted to 1.6 mm or 0.8 mm. By means of these tapes it is possible to produce soldered joints of higher strength. Attempts are also being made to press bearing bushings from the tapes [24 b].

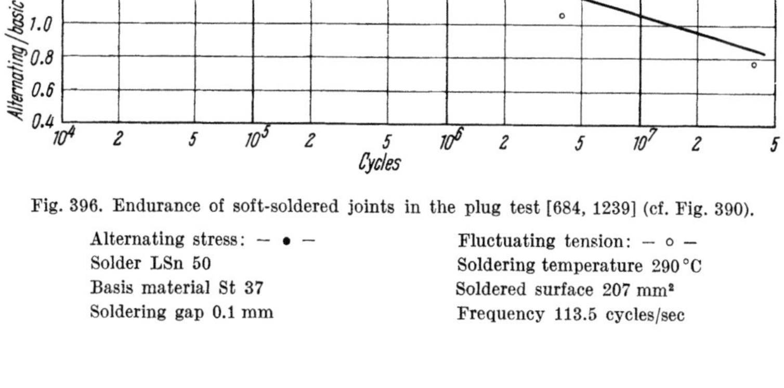

Fig. 396. Endurance of soft-soldered joints in the plug test [684, 1239] (cf. Fig. 390).

Alternating stress: – • – — Fluctuating tension: – ∘ –
Solder LSn 50 — Soldering temperature 290 °C
Basis material St 37 — Soldered surface 207 mm²
Soldering gap 0.1 mm — Frequency 113.5 cycles/sec

Bibliography

The numbers given in italics at the end of each bibliographic quotation, refer to the text of this book.

1. ABEL, E., and O. REDLICH: Z. anorg. allg. Chem. 161 (1927) 221. *126*
2. —, — and J. ADLER: Z. anorg. allg. Chem. 174 (1928) 269. *137*
3. —, — and F. SPAUSTA: Z. anorg. allg. Chem. 190 (1930) 79. *158, 159*

ACHATZ, R. V.: See [18].

4. ACKERMANN, K. L.: Metallwirtsch. 10 (1931) 593. *138*
5. —: Z. Metallkde. 24 (1932) 306. *128*
6. —: Metallwirtsch. 12 (1933) 618. *383*

ADAMS, L. H.: See [627].

ADLER, J.: See [2].

9. AGEEW, N. W.: Metal Ind. (London) 50 (1937) 3. *136*
10. — and N. Y. TALYZIN: Izvest. Sekt. Fiziko-Khimich. Analiza. 13 (1940) 251. *67*
11. AGRUSS, B., E. H. HERRMANN and F. B. FINAN: J. Electrochem. Soc. 104 (1957) 204. *353*

11a. AKUTAGAWA, T., and K. OGAWA: Nipp. Konzoku Gakkai-Si 24 (1960) 256. *41*

12. ALBANO, V. J.: Corrosion (Houston) 3 (1947) 488. *297*

ALCOCK, C. B.: See [197a].

14. ALDEN, T., D. A. STEVENSON and J. WULFF: Trans. AIME 212 (1958) 15. *41, 53, 69, 70, 354*

14a. ALDRED, A. T., and J. N. PRATT: Trans. Faraday Soc. 57 (1961) 611. *34*

ALLEMAN, N. I.: See [864].

15. ALLEMAND, D. A.: Rev. gén. Electr. 35 (1934) 886. *318*
16. ALLEN, N. P.: Proc. N. P. L. Symposium on Creep and Fracture of Metals at High Temperatures 1954. *233*

ALLHAUSEN, H.: See [276].

ALLISON, S. K.: See [210].

AL-SAMARRAI, S. K. H.: See [1125a].

AMBROJI, M. N.: See [1053].

17. American Smelting and Refining Comp.: F. Patent 835430, 1938. USA-P. Nr. 2142835, 3. 1. 1939. *354*
18. ANDEREGG, F. O., and R. V. ACHATZ: Bull. Engr. Exper. Station, Purdue Univ. 18 (1924); Teleph. Engr. 28 (1924) 34; J. Inst. Metals 33 (1925) 372. *310, 311*
19. ANDERSON, W. A., and R. F. MEHL: Inst. Met. Div. 181 (1945) 140. *186*
20. ANDRADE, E. N. DA C.: Nature 162 (1948) 410. *217*
21. ANGERSTEIN, H.: Bull. Acad. Pol. Sci. 3 (1955) 447: According to Met. Abstr. 23 (1956) 834. *316*

ANGLES, R. M.: See [1126b].

ANGSTADT, R. T.: See [1037].

ANKE, F.: See [1160].

Ansell, G. S.: See [743].
Aoki, O.: See [608].
22. Aoyama, S., and T. Fukuroi: Sci. Rep. Tohoku Imp. Univ. 29 (1941) 728. *99, 100*
22a. Appelt, K., M. Elbanowski and A. Nowacki: Electrochim. Acta 6 (1962) 207. *449*
23. Archbold, T. F., and R. E. Grace: Trans., AIME 212 (1958) 658. *327*
24. Archbutt, L.: Trans. Faraday Soc. 17 (1921/22) 22. *244*
24a. Armour Research Foundation Proj. 2745 (1962) 29, 84. *31, 74, 174, 191, 263*
24b. Armour Research Foundation, Illinois Inst. of Techn., New York, April 1963. *479*
Arndt, H.: See [58].
Arnold, S. M.: See [402].
Arnone, M.: See [179].
25. Arrowsmith, R.: J. Inst. Metals 55 (1934) 71. *385*
Artamonoff, B. P.: See [308].
26. Ascoli, A.: J. Inst. Metals 89 (1960/61) 218. *38*
26a. Ascoli, A., E. Germagnoli and L. Mongini: Nuovo Cimento 4 (1956) 123. *38*
Ashkroft, K.: See [1245a].
Ashley, R. W.: See [169, 172].
26b. Astakhov, I. I., E. S. Vaysberg and B. N. Kabanov: Doklady Akad. Nauk USSR. 154 (1964) 1414. *349*
27. ASTM Designation B200—45T: Electrodeposited Coatings of Lead on Steel. *464*
28. ASTM Standards: Part I Metals 789; Philadelphia, Pa: Amer. Soc. Test. Mater. 1936. *27, 28*
29. Atwater, H. A., and B. Chalmers: J. Appl. Physics 26 (1955) 918. *336*
29a. Aubrey, J. E.: Phil. Mag. 5 (1960) 1001. *20*
30. Aust, K. T., and J. W. Rutter: Transactions AIME 215 (1959) 119. *187*
30a. —, — Trans. AIME 215 (1959) 820. *187, 188*
31. —, —: Trans. AIME 218 (1960) 50. *188*
31a. —, and B. Chalmers: Proc. Roy. Soc. 201 (1950) 210; 204 (1951) 359. *17*
—: See [1039a, 1262a].
32. AWF: Der Spritzguß, 3. Aufl., Leipzig: Teubner 1937. *357*
33. Azzolino, J.: Plating 44 (1957) 1180. *387*

Badawy, M. L. A. E. S.: See [790a].
33a. Bagshaw, N. E., and T. A. Hughes: Preprint 4. Intern. Battery Symp. Brighton 1964. *356*
33b. Bailey, G. L.: Proc. First Intern. Conference on Lead, London 1962, Chap. 8, 93. *413*
33c. Bailey, J. A., and A. R. E. Singer: J. Inst. Metals 92 (1963/64) 404, 996. *206, 394*
Bainbridge, D. W.: See [414].
34. Baker, E. M., and P. J. Merkus: Electrochem. Soc. Preprint (1932) Apr. 23. *301*
35. Baker, H.: Proc. Graphical Arts Techn. Conference (The Amer. Soc. Mech. Engrs.) (1936) 19. *447*
36. Baker, J. B., B. B. Betty and H. F. Moore: Metals Techn. 5 (1938) No. 3. Trans. AIME, Techn. Publ. No. 906. *227*

37. Baker, W. A.: J. Inst. Metals 65 (1939) 277. *470, 473, 477*
38. —: J. Inst. Metals 65 (1939) 345. *71*
Baldwin, W. M.: See [1247].
39. Ballay, M.: C. R. Acad. Sci., Paris 202 (1936) 222. *203*
40. Balluffi, R. W., and L. L. Seigle: Acta Met. 5 (1957) 449. *212*
Bankes, P. E.: See [748].
41. Bardgett, W. E.: Iron and Steel 29 (1956) 392. *57*
42. Barducci, I.: Ricerca sci. 22 (1952) 1733. *100*
43. Bargilliat, A.: Chim. et Ind. 36 (1936) 3. *360*
44. Barham, B. S.: Metal Ind., London 52 (1938) 521. *471*
45. Barlow, W. E.: Z. anorg. allg. Chem. 70 (1911) 178. *129, 130*
45a Barnard, K. N., G. L. Christle and D. G. Gage: Corrosion (Houston) 15 (1959) 581. *301*
46. Barret, C. S.: Structure of Metals, New York: McGraw-Hill 1952. *235*
47. Barrs, C. E.: J. Soc. Chem. Ind. 38 (1919) T, S. 407, R, 452. *282*
48. Barteld, K.: Thesis T. H. Braunschweig 1955. *73, 74, 289*
49. —, and W. Hofmann: Erzmetall 5 (1952) 102. *72, 73*
49a. Barth, N.: Z. Physik 148 (1957) 646. *67*
49b. Barthel, J.: Freiberger Forschungshefte, R. B29, Berlin: Akademie-Verlag 1958. *332*
50. Barwell, F. T.: J. Appl. Phys. Suppl. 1 (1951) 32. *376*
Basil, I. L.: See [514].
51. Bassett, W. H.: Metals and Alloys 8 (1937) 185. *417*
52. —, and C. J. Snyder: Proc. ASTM 32 (1932) II 558. *26, 28*
53. —, —: Trans. AIME, Inst. Met. Div. 104 (1933) 254. *420, 421*
54. Bassow, A.: Werkst. u. Korrosion 7 (1956) 489. *338*
Bastien, P.: See [973].
55. Bastien, P. G., and M. Darmony: J. Met. 7 (1955) 1264. *95*
Basu, S. K.: See [653].
56. Bauer, H.: Aluminium-Archiv 24 (1939) 30. *121, 122*
57. Bauer, O.: Z. Metallkde. 16 (1924) 110. *334*
58. —, and H. Arndt: Z. Metallkde. 13 (1921) 497. *59*
59. —, and M. Hansen: Z. Metallkde. 21 (1929) 145. *154*
60. —, and G. Schikorr: Metallwirtsch. 14 (1935) 463. *307*
61. —, —: Mitt. dtsch. Mat. Prüf. Anst., Sonderheft 28 (1936) 67. *302, 303*
62. —, and H. Sieglerschmidt: Metallwirtsch. 15 (1936) 535. *79*
63. —, and W. Tonn: Z. Metallkde. 27 (1935) 183. *35, 37*
64. —, and O. Vollenbruck: Z. Metallkde. 22 (1930) 230. *317, 318*
65. —, and E. Wetzel: Mitt. Mat.-Prüf.-Amt Berlin-Lichterfelde-West 34 (1916) 346. *294, 295, 302, 306, 307*
—: See [520].
Baum, L. W.: See [742].
Baumann, M.: See [793].
67. Baxter, G. P., M. Guichard and R. Whytlaw-Gray: Thirteenth Report of the Committee on Atomic Weights of the International Union of Chemistry. J. Chem. Soc. (1947) 983. *9*
67a. Beck, A.: Z. Metallkde. 52 (1961) 13. *184, 186*
68. Beck, P. A.: Metals Techn. 6 (1939). AIME Techn. Publ. No. 1101. *103, 194*
69. —: Acta Met. 6 (1958) 287. *187*
—: See [960].
70. Beck, W.: Schweiz. Arch. 6 (1940) 201, 225. *317*

71. BECKER, E.: Chem. Apparatur (Werkst. u. Korrosion) 10 (1935) 1. *339*
BECKER, K.: See [813].
BECKER, M.: See [361].
72. BECKER, R.: Z. techn. Phys. 7 (1926) 547. *208*
73. BECKINSALE, S., and H. WATERHOUSE: J. Inst. Met. (London) 39 (1928) 375. *244, 258, 259, 293*
BECKMANN, J.: See [278].
BEDDOW, J. K.: See [334b].
BEDWORTH, R. E.: See [963].
BEGEMAN, M. L.: See [436].
74. Bell Telephone Laboratories: USP 2170650 of 2. 9. 1936. *354*
75. BENAD, H.: Thesis Freiberg 1939. *441, 442, 445*
—: See [277].
BENNEWITZ, R.: See [914].
BEREGEKOFF, D.: See [1222].
BERGNER, K. G.: See [1063].
BERL, E.: See [784].
76. BERNHARDT, E. O., and H. HANEMANN: Z. Metallkde. 30 (1938) 401. *231*
77. BETTERTON, J. O., and Y. E. LEBEDEFF: Trans. AIME 121 (1936) 205. *175*
78. BETTS, A. G.: D. R. P. Kl. 40c, No. 198288 2. 4. 1902 (13. 5. 1908). *463*
BETTY, B. B.: See [36, 865, 866].
79. BEUERLEIN, P.: Schmiertechnik, in: Kröners Taschenbuch der Maschinentechnik, Stuttgart 1955. *365*
BEYER, B.: See [729a].
80. BIDWELL, C. C.: Phys. Rev. 58 (1940) 561. *19*
81. —, and E. J. LEWIS: Phys. Rev. 33 (1929) 249. *19*
82. BILART, A.: Métaux, 11 (1936) 31. *454*
83. BIRCH, F.: Phys. Rev. 71 (1947) 809. *12*
84. BIRCUMSHAW, L. L.: Phil. Mag. 7 (1926) 341. *17*
85. —: Phil. Mag. 17 (1934) 181. *95*
86. —, and G. D. PRESTON: Phil. Mag. 25 (1938) 769. *326*
87. BISH, R. E.: J. Metals 4 (1952) 81. *54*
88. BISHOP, J. F. W.: Metallurgical Reviews 2 (1957) 361. *393, 400*
BISSETT, C. C.: See [883].
89. BLANDERER, J.: Metall 8 (1954) 288. *380*
BLANK, F.: See [202].
90. BLOCH, M. R., D. KAPLAN and J. SCHNERB: J. Appl. Chem. 8 (1958) 171. *297*
91. BLUMENTHAL, B.: Metals and Alloys 3 (1932) 181. *151*
92. —: Metals Techn. 10 (1943) AIME Techn. Publ. No. 1634 11. *113*
93. —: Trans. AIME 156 (1944) 240; Discussion 250. *76*
94. BLUTH, M., and H. HANEMANN: Z. Metallkde. 29 (1937) 48. *86, 87*
BOAS, W.: See [1070].
95. —, and R. W. K. HONEYCOMBE: J. Inst. Metals 73 (1947) 433. *374, 375*
96. BOCHVAR, A. A., and V. I. DOBATKIN: Tsvet. Metally (Non Ferrous Metals) (1940) 86. *97*
97. BODE, H., and E. VOSS: Z. Elektrochem. 60 (1956) 1053. *349*
98. BÖRGER, H.: Schriften d. Gesellschaft Deutscher Metallhütten- u. Bergleute e. V. (1960) No. 7, 10. *340*
99. BÖRSIG, F.: Masch.-Schad. 17 (1940) 61. *317*
100. BOESCHE, H.: DRP (1941) No. 709009. *416*
101. BOESONO: Physica 24 (1958) 71. *184*

102. Bogitsch, M. B.: C. R. Acad. Sci., Paris 159 (1914) 178. *116*
Bohren, H.: See [719, 720].
102a. Bolling, G. F.: Trans. AIME. 224 (1962) 635. *188*
102b. —: Trans. AIME 224 (1962) 1075. *188, 191*
102c. —, L. E. Hays and H. W. Wiedersich: Acta Met. 10 (1962) 185. *181*
102d. —, J. J. Kramer and W. A. Tiller: Trans. AIME 227 (1963) 47. *336*
103. —, T. B. Massalski and C. J. McHargue: Phil. Mag. 6 (1961) 491. *181*
104. —, and W. C. Winegard: Acta Met. 6 (1958) 283. *9, 187*
105. —, —: Acta Met. 6 (1958) 288. *38, 191*
106. —, —: J. Inst. Metals 86 (1958) 492. *17*
107. Bollinger, J.: Schweiz. Arch. 18 (1952) 321. *292*
108. Bolognesi, G.: Metallurgia Ital. 46 (1954) 9. *291*
108a. Bol'shanina, M. A., T. F. Elsukova, P. A. Kondrat'ev and M. A. Pomina: USSR Abstr. of Met. (1959) 63. *262, 418*
—: See [1216a].
108b. Bonde, G., and B. Lunn: Ingen. Intern. Edit. 2 (1958) 103. *293*
108c. Borchardt, H., and W. Thele: Siemens Z. 34 (1960) 794. *409*
109. Borchers, H., and K. Bunnenberg: Metall 15 (1961) 765. *93, 266*
109a. —, and M. Kainz: Metall 19 (1965) 103. *264*
109b —, M. Kainz and S. Pixner: Z. Metallkde. 52 (1961) 276. *93, 264*
109c. —, and M. M. Tenti: Metall 17 (1963) 784. *19*
110. Borel, I.: Bull. Ass. Suisse Electr. 28 (1937) 54. *315*
111. Borelius, G.: Arkiv Mat., Astron. Fysik 32 (1945) 9. *94*
112. —, F. Larris and E. Ohlsson: Arkiv. Mat., Astron. Fysik 31 (1944) 19. *96*
113. —, and L. E. Larsson: Inst. Metals: Symposium on Mechanism of Phase Transformations in Metals (1955) 67. *96*
114. —, and K. M. Säfsten: Arkiv Mat., Astron. Fysik 36 (1948) 5. *95*
115 Bornemann, K., and K. Wagenmann: Ferrum 11 (1913/14) 291, 309. *54, 55*
116. Bosch, C.: Z. Elektrochem. 24 (1918) 361. *296*
117. Botschwar, A. A., and K. W. Gorew: Z. anorg. allg. Chem. 210 (1933) 171. *140*
118. Bouton. G. M., and G. S. Phipps: ASTM Proc. 51 (1951) 761. *393, 419*
—: See [402, 1090].
119. Bowden, F. P., and D. Tabor: The Friction and Lubrication of Solids, Oxford: University Press 1950. *373, 374, 375*
120. — —: Reibung u. Schmierung fester Körper, Berlin/Göttingen/Heidelberg: Springer 1959. *454*
121. Bradhurst, D. H., and A. S. Buchanan: Australien J. Chem. 12 (1959) 523. *73*
Bradley, W. W.: See [212].
122. Brady, E. L.: J. Electrochem. Soc. 101 (1954) 466. *101*
123. Brady, F. L.: Metal Ind., London 40 (1932) 293. *307*
124. —: Met. Abstr. 3 (1936) 481. *293*
125. Brame, I. S. S.: J. Soc. chem. Ind. 37 (1918) 39. *294*
Brand, H.: See [938],
126. Brandt, A.: Brit. Non-Ferr. Met. Res. Assoc. Ser. 414 (1936); s. U. R. Evans [297]. *312*
De Brouckère, L.: See [748].
Brawley, D. J.: See [223].
127. Bray, J. L.: Trans. AIME 124 (1937) 199. *107, 460*
—: See [871].

128. BRČIĆ, B. S., and J. ŠIFTAR: Corrosion et Anticorrosion 6 (1958) 342. *294*
BRENNECKE, E.: See [834].
129. BRENTHEL, FR.: Chemiker-Ztg. 52 (1928) 169, 190, 250. *468, 469*
130. —: Z. Metallkde. 22 (1930) 23. *289*
132. BRETAG, W.: Z. VDI 96 (1954) 397. *357*
133. BRIESEMEISTER, S.: Z. Metallkde. 23 (1931) 225. *54, 151*
134. BRIGGS, G. W., D., and W. F. K. WYNNE-JONES: J. Chem. Soc. (1956) 2966. *296*
134a. Brit. Report No. A 1385, Brit. Non Ferrous Met. Res. Ass. *311*
135. BROCK, P.: J. Inst. Metals 83 (1954/55) 191. *217*
135a. BROCK, P., and J. MCKEOWN: Proc. Inst. Elect. Eng. 110 (1963) 174. *417*
BROOMFIELD, G. H.: See [980].
BRUNOLD, U. A.: See [1227c].
136. DE BRUYNE, N. A., and R. HOUWINK: Adhesion and Adhesives, Amsterdam: Elsevier 1951. *60*
BUCHANAN, A. S.: See [121].
137. BUDA, I. V.: Rev. chim. (Bucharest) 8 (1957) 166. *107*
138. BÜCKLE, H.: Z. Metallkde. 43 (1952) 82. *45*
139. —, and H. HANEMANN: Z. Metallkde. 32 (1940) 120. *349*
140. BÜHLER, H.: Stahl und Eisen 75 (1955) 420. *19*
141. BUEREN, H. G. VAN: Imperfections in Crystals, Amsterdam: North Holland 1960. *179*
BUKOWIECKI, A.: See [1065].
143. BUMM, H.: Thesis T. H. Berlin 1932. *124*
143a. BUNCE, R. G.: Proc. First Intern. Conf. on Lead, London 1962, Chap. 5, 57. *409*
144. BUNGARDT, W.: in: R. KÜHNEL: Werkstoffe für Gleitlager, Berlin: Springer 1939, 322. *384*
145. —: in: R. KÜHNEL: Werkstoffe für Gleitlager, 2nd Ed., Berlin/Göttingen/Heidelberg: Springer 1952. *380*
BUNGARDT, W.: See [290].
146. BUNNENBERG, K.: Thesis T. H., München 1960, See [109]. *93, 264*
147. U. S. Bur. Stand. Circular, Nr. 382 (1930), Bismuth. *44*, 364
148. BURAT, F., and W. HOFMANN: Feinwerktechnik 63 (1959) 82. *57, 475*
149. BURBANK, J.: J. Electrochem. Soc. 103 (1956) 87. *348*
150. —: J. Electrochem. Soc. 104 (1957) 693. *349*
150a. —: Electrochem. Soc. (USA) 111 (1964) 765. *349*
151. BURBANK, J. B., and A. C. SIMON: J. Electrochem. Soc. 100 (1953) 11. *288, 349*
152. BURGAN, B. R., R. C. HALL and R. F. HEHEMANN: J. Inst. Met. 80 (1951/52) 413. *59*
153. BURGERS, W. G.: Z. Metallkde. 41 (1950) 2. *184, 185*
153a. BURGERS, W. G.: Z. Metallkde. 52 (1961) 19. *185*
—: See [920, 1197].
BURKE, A. E.: See [767].
154. BURKHARDT, A.: Metallwirtsch. 10 (1931) 181. *275, 280*
155. —: Blei und seine Legierungen, Berlin: NEM-Verlag 1935. *62, 68, 193, 199, 200, 258, 261, 330, 332*
BURNET, G.: See [1142a].
156. BURNS, R. M.: Bell Syst. Techn. J. 15 (1936) 603. *310, 314*
157. —, and W. J. SALLEY: Industr. Engng. Chem. 22 (1930) 293. *312*
158. BUSKE, A.: Stahl u. Eisen 71 (1951) 1420. *366, 368*

159. Buss, G., and U. Meyer: Europ. Fernsprechdienst 50 (1938) 308. *416*
160, —, and H. K. Müller: Elektrizitätswirtsch. 49 (1950) 302. *316*
160a. Butcher, B. M., and A. L. Ruoff: J. Appl. Physics 32 (1961) 2036. *214*
161. Butcher, W. T.: Metal Ind., London 50 (1937) 105. *44*
162. Butler, J. M.: J. Inst. Metals 86 (1957/58) 145. *393, 432*
163. —: J. Inst. Metals 86 (1957/58) 155. *197, 433*
164. —: J. Inst. Metals 86 (1957/58) 161. *435, 436*
165. Butuzov, V. P., and M. G. Gonikberg: Met. Abstr. 21 (1953/54) 854. *14*
Butzik, M. G.: See [374].

166. Cahn, J. W., and H. N. Treaftis: Trans. AIME 218 (1960) 376. *95*
Cadoff, I.: See [855, 1101a].
Cadoff, I. B.: See [789a].
167. Calcott, W. S., J. C. Whetzel and H. F. Whittaker: Corrosion Tests at Materials of Construction, New York: Van Nostrand 1923. *290, 293*
Cale, G. N.: See [1156].
168. Cambi, L., and R. Piontelli: Chimica e Industria 20 (1938) 649. *298*
169. Campbell, A. N., and R. W. Ashley: Canad. J. of Research 18 (1940) 281. *34, 35*
170. —, and R. Kartzmark: Canad. J. Chem. *34* (1956) 1428. *123*
171. —, R. M. Screaton, T. P. Schaefer and C. M. Hovey: Canad. J. Chem. 33 (1955) 511. *156*
172. —, L. Yaffe, W. G. Wallace and R. W. Ashley: Canad. J. Research 19 (1941) 212. *109*
Campbell, H. S.: See [181].
173. Campbell, W.: Proc. Amer. Soc. Test. Mater. 30 I (1930) 380 and 384. *34*
Campney, K. N.: See [183].
174. Canad. Patent 503663, 1954. *355*
175. Cap, F.: Physik und Technik der Atomreaktoren, Wien: Springer 1957. *24*
176. Capua, Cl. di: Gazz. chim. ital. 55 (1925) 582. *131, 141*
178. —: Gazz. chim. ital. 55 (1925) 594. *134*
179. —, and M. Arnone: Atti Acad. Lincei 33 (1924) 28. *42*
180. Carius, C., and E. H. Schulz: In O. Bauer, O. Kröhnke, G. Masing: Die Korrosion metallischer Werkstoffe, Bd. 1, Leipzig: Hirzel 1936, 127 *304*
180a. Carlson, R. G., and F. E. Westermann: Planseeber. Pulvermet. 10 (1962) 15. *449*
181. Carter, V. E., and H. S. Campbell: Metal Finishing J. 8 (1962) 103. *301*
Carter, W. H.: See [1262].
182. Cartland, J.: J. Inst. Metals 56 (1935) 235. *336, 361*
—: See [883].
Casey, D. F.: See [1096].
183. Casey, E. J., and K. N. Campney: J. Electrochem. Soc. 102 (1955) 219. *273, 279*
184. Cathcart, J. V., and W. D. Manly: Corrosion (Houston) 12 (1956) 43. *41*
184a. Cavanaugh, C. R., and J. F. Elliott: Trans. AIME. 230 (1964) 633. *69*
Ceccarelli, O.: See [745].
185. Centre Belge d'étude de la Corrosion, „Cebelcor", Belg. P. 506131 16. 1. 1952. *295*
186. Chadwick, R.: J. Inst. Metals 62 (1938) 277. *477*
Chaikovsky, E. F.: See [660].
187. Chalmers, B.: Proc. Roy. Soc. 196 [A] (1949) 64. *471*

188. CHALMERS, B.: Proc. phys. Soc. London 47 (1935) 352. *13, 197*
—: See [29, 31a, 764b, 1187, 1236, 1279].
CHAND, R.: See [886].
189. CHARPY, M. G.: C. R. Acad. Sci., Paris 126 (1898) 1569 and 1645. *132, 133*
190. CHASTON, J. C.: Elektr. Nachr.-Wes. 13 (1934) 31. *315*
191. —: J. Inst. Metals 56 (1935) 87. *246*
192. —: J. Inst. Metals 57 (1935) 109. *190*
192a. CHEFFERS, R. R.: Electroplating 15 (1962) 274. *465*
193. Chemikerausschuß der Gesellschaft Deutscher Metallhütten- und Bergleute e. V.: Analyse der Metalle Vol. I, 2nd Ed., 1949; Vol. II, 1953; Vol. III, 1956; Berlin/Göttingen/Heidelberg: Springer. *9, 31, 80*
194. Chem. Engng. 63 (1956) 228. *466 467*
195. Chem. metall. Engng. 31 (1924) 74. *291, 292, 293, 294, 295, 296*
196. Chem. metall. Engng. 43 (1936) 526. *273, 280, 283, 292*
197. Chem. Process Engng. 37 (1956) 131. *292*
197a. CHENG, L. L., and C. B. ALCOCK: Trans. AIME. 221 (1961) 295. *75*
CHRISTLE, G. L.: See [45a].
CHUDOBIAK, J. M.: See [461c].
CLABAUGH, W. J.: See [623, 624].
198. CLAREBROUGH, L. M., M. E. HARGREAVES and G. W. WEST: Proc. Roy. Soc. 232 (1955) 252. *182*
199. CLARKSON, F., and H. C. HETHERINGTON: Chem. metall. Engng. 32 (1925) 811. *292*
200. CLAUS, W.: Gießerei-Ztg. 23 (1926) 463. *471*
201. —: Metallwirtsch. 13 (1934) 226. 54
202. —, and F. BLANK: Metallwirtsch. 18 (1939) 938. *95*
203. —, and I. HERMANN: Metallwirtsch. 18 (1939) 957. *121*
CLAYTON, C. Y.: See [486].
204. CLAYTON, D.: Brit. J. Appl. Phys. Suppl. 1 (1951) 25. *375*
COAKHAM, J. W.: See [1125b].
COCHARDT, A. W.: See [1006].
COE, H. C.: See [382].
205. COHEN, E.: Z. phys. Chem. 63 (1908) 631. *97*
COHEUR, P.: See [1005].
206. COLBUS, J.: Schweißen u. Schneiden 6 (1954) 287, 140 (S). *474*
COLE, J. H.: See [427, 428, 429].
207. COLES, E. L., and R. L. DAVIES: Chem. and Ind. (1956) 1030. *293*
208. —, J. G. GIBSON and R. M. HINDE: J. Appl. Chem. 8 (1958) 341. *292*
COLLINSWORTH, E. T.: See [977].
210. COMPTON, A. H., and S. K. ALLISON: X-Rays in Theory and Experiment, Toronto/New York/London: Van Nostrand 1949. *21*
211. COMPTON, K. G.: Corrosion (Houston) 12 (1956) 37. *316*
212. COMPTON, K. G., A. MENDIZZA and W. W. BRADLEY: Corrosion (Houston) 11 (1955) 383/92t. *319*
213. CONINX: Elektrizitätswirtsch. 28 (1929) 357. *402*
214. COOK, M.: J. Inst. Metals 31 (1924) 297. *142*
COPSON, R. L.: See [495].
CORDES, H.: See [252a].
214a —, and G. DOEGE: Z. f. Naturforschung 18a (1963) 835. *80*
215. Corrosion [Houston] 12 (1956) 355. *316*
215a COSTA, J. M., and T. P. HOAR: Corrosion Science (1962) 269. *300*
216. COTTON, W. J.: Corrosion (Houston) 11 (1955) 23. *289, 290*

217. COTTRELL, A. H.: Dislocations and Plastic Flow in Crystals, Oxford: Clarendon Press 1953. *179, 208, 234*
218. —, and M. A. JASWON: Proc. Roy. Soc. 199 (1949) 104. *216*
219. COURNOT, J.: C. R. Acad. Sci., Paris 186 (1928) 867. *53*
220. —: Techn. mod. 27 (1935) 635. *44, 389*
COUSLAND, MCK. S.: See [1284a].
221. COWAN, W. A., L. D. SIMPKINS and G. O. HIERS: Chem. metall. Engng. 25 (1921) 1181. *128, 129*
COX, G. L.: See [1020].
CRAIG, D. N.: See [1224].
221a. CREMER, L., and A. V. MEIER: Second Intern. Conf. on Lead Arnhem 1965 No. 3. *467*
222. CRENNELL, J. T., and A. G. MILLIGAN: Trans. Faraday Soc. 27 (1931) 103. *353*
CROCKER, A. J.: See [932].
223. CROCKFORD, H. D., and D. J. BRAWLEY: J. Amer. chem. Soc. 56 (1934) 2600. *273*
CRONE, W.: See [1164].
224. CROSS, A.: Iron Age 24 (1959) 48. *452*
225. CROW, T. B.: J. Inst. Metals 35 (1926) 55. *472*
226. CRUSSARD, C., and J. FRIEDEL: In „Creep and Fracture of Metals at High Temperatures", Proc. of a Symposium. Held at the National Physical Laboratory 1954, London: Her Majesty's Stationery Office 1956. *216*
227. CUTHBERTSON, J. W.: J. Inst. Metals 64 (1939) 485. *369*
228. CZOCHRALSKI, J., and E. RASSOW: Z. Metallkde. 19 (1927) 111. *61, 63*

DAHL, O.: See [1165].
Dahl, TH.: See [550].
DANILOV, V. I.: See [229].
229. DANILOVA, A. I., V. I. DANILOV and E. Z. SPEKTOR: Structure of Molten Tin, Bismuth and Lead. Doklady Akad. Nauk. SSSR 82 (4) (1952) 561. Met. Abstr. 22 (1954/55) 317. *10*
230. DANNAT, C. W., and F. D. RICHARDSON: Metal. Ind. 83 (1953) 63. *73*
DANNECKER, C.: See [685].
DANNÖHL, W.: See [294].
DARAGAN-SUSHCHOV, I. I.: See [800].
231. DARDEL, Y.: Light Metals 9 (1946) 220. *34*
231a. DARKEN, L. S., and R. W. GURRY: New York/London/Toronto: Mc Graw-Hill 1953. 31
DARMONY, M.: See [55].
232. DASSOJAN, M. A.: Dokl. Akad. Nauk. USSR 107 (1956) 683. *76*
233. —: Nachr. Elektro-Ind. (Russ.) 2 (1957). *352, 355*
234. DAUBENSPECK, G. W., and C. H. MCCLURE: Illinois State Dept. Labour, Div. Factory Inspection, Techn. Paper 1 (1941) 20. *356*
DAUNT, J. G.: See [595].
235. DAVEY, T. R. A.: J. Metals 8 (1956) 341. *8, 331*
235a. —: Bull. Inst. Min. Met. 72 (1963) 553. *124, 125, 148, 149, 150, 153, 176*
DAVIE, T. A. S.: See [768].
DAVIES, J. B.: See (636].
236. DAVIES, M. H.: J. Inst. Metals 81 (1952/53) 415. *122*
DAVIES, R. L.: See [207].

236a. DAVINI, G.: Proc. First Intern. Conf. on Lead, London, 1962, Chap. 7, 85. *409*
DAY, J H : See [982].
237. DAY, A. L., R. B. SOSMAN and J. C. HOSTETTER: N. Jb. Min. etc. Beil. Vol. 40 (1915) 119. *11*
238. DEAN, R. S., and J. E. RYJORD: Metals and Alloys 1 (1930) 410. *49, 259*
239. —, L. ZICKRICK and F. C. NIX: Trans. AIME 73 (1926) 505. *14, 79, 81, 82, 84, 88, 90*
240. DEHLINGER, U.: Chemische Physik der Metalle und Legierungen, Physik und Chemie, Vol. III, Leipzig: Akad. Verlagsges. 1939. *30*
241. —: Z. Metallkde. 31 (1939) 187. *229*
242. —: Handbuch der Physik, Vol. VII/2, Berlin/Göttingen/Heidelberg: Springer 1958. *263*
243. DELAHAY, P., M. POURBAIX und P. VAN RYSSELBERGHE: J. Electrochem. Soc. 98 (1951) 57. *271*
244. DEMING, A. F.: Proc. Jowa Acad. Sci. 45 (1938) 207. *177*
245. DENISON, I. A.: Trans. Electrochem. Soc. 81 (1942) 435. *308*
246. —, and M. ROMANOFF: J. Research. Nat. Bur. Standards 44 (1950) 259. *308, 309*
DERGE, G.: See [412].
247. DERYAGUIN, B. V.: Wear 1 (1957/58) 277. *372*
DETERT, K.: See [777].
248. —: Z. Metallkde. 45 (1954) 547. *96*
DEVER, J. L.: See [256].
248a. D'HEURLE, F. M., R. FEDER and A. S. NOWICK: J. Phys. Soc. Japan 18 (1963) Suppl. II, 184. *11*
DIETERLE, R.: See (686].
DIETRICH, A.: See [444].
249. DIETRICH, K. R.: Korrosion u. Metallsch. 5 (1929) 110. *298*
250. DILLER, L. W., R. L. ORR and R. HULTGREN: J. Phys. Chem. 64 (1960) 1736.
DISAM, A.: See [993a]. *76*
251. DITZ, H.: Z. angew. Chem. 13 (1913) 596. *294*
DOBATKIN, V. I.: See [96].
DOBBINS, A. G.: See [1091].
252. DODERO, M.: Métaux et Corrosion 24 (1949) 50. *295*
252a. DOEGE, G.: Thesis T. H. Braunschweig 1961. *80*
—: See [214a].
253. DOLLINS, C. W.: Univ. Illinois Bull. 45 (1948) 87. *240, 241, 242*
254. —: Univ. Illinois Bull. 48 (1950) 60. *230, 231, 238, 239, 240*
—: See [865, 866].
DOLGOPOLOVA, A. M.: See [307b].
DONAVAN, J. A.: See [927a].
254a. DORIN, V. A., and G. M. FILARETOVA: Fiz. Metal metalloved 3 (1960) 195. *75*
255. DORN, I. E.: J. Mech. Phys. Solids 3 (1955) 85; "Creep and Fracture of Metals at High Temperatures" Symposium, National Physical Laboratory 1954 London: Her Majesty's Stationery Office 1956. *213, 214*
—: See [1281].
DORNBLATT, A. J.: See [317].
256. DOUGLAS, T. B., and J. L. DEVER: U. S. Atomic Energy Commission Publ. (NBS-2544) (1953). *15, 40*
257. DOYLE, E. J.: Corrosion (Houston) 11 (1955) 17. *316*

258. DRAHT, G., and F. SAUERWALD: Z. anorg. allg. Chemie 162 (1927) 301. *17*
DRAPER, A.: See [1042].
258a. DREWS, TH., and J. GERLACH: Metall 17 (1963) 804. *331*
DREYER, K. L.: See [1166, 1167, 1168].
259. DROTSCHMANN, C.: Bleiakkumulatoren, Weinheim: Verlag Chemie 1951. *342, 344, 347, 348, 351, 356*
DUCKWORTH, W. E., and G. H. WALTER: See FORRESTER (Ref.) [334a].
260. —: Symposium on Powder Metallurgy (Iron Steel Inst.) Preprint, Group III (1954) 65. *387*
—: See [335].
DUGI, Z.: See [798].
261. VAN DUIJN, C.: Polytechn. Tijdschr. 6 (1951) 644a. *319*
262. DUNHAM, K. C.: International Geological Congress Report of the Eighteenth Session Great Britain 1948. Part VII, 11, London 1950. *2, 3*
DUNKERLEY, F. J.: See (642, 1034].
263. DUNSHEATH, P.: Nature (London) 139 (1937) 755. *420*
264. —, and H. A. TUNSTALL: J. Inst. Electr. Engers. 66 (1928) 280. *422*
265. DUROSE, A. H.: ASTM Symposium on Electrodeposited Metallic Coatings (1956) 97. *464*

266. EARLE, L. G.: J. Inst. Metals *72* (1946) 403. *114, 115*
267. EBORALL, R.: "Creep and Fracture of Metals at High Temperatures" Symposium, National Physical Laboratory 1954, London: Her Majesty's Stationery Office 1956. *217*
268. ECKEL, J. F.: ASTM Proc. 51 (1951) 745. *249, 253*
269. EDA, W.: Sci. Rep. Tohoku Univ. 20 (1931) 715. *10*
270. EGER, G.: Z. Metallkde. 23 (1931) 90. *298*
271. EHLERS, W., and H. LAU: Kabelherstellung, Berlin/Göttingen/Heidelberg: Springer 1956. *400, 403, 404*
EHRLINGER, H. P.: See [1162].
272. EISBEIN, W., and G. SACHS: Mitt. dtsch. Mat.-Prüf.-Anst., Sonderheft 16 (Spanlose Formung) (1931) 67. *392, 395, 396*
272a. EISENKOLB, F., E. STROBEL and W. HESSEL: Lead Abstr. 4 (1964) 74. *387*
273. EISENSTECKEN, F., and H. ROTERS: Öl und Kohle 15 (1939) 129. *298*
ELBANOWSKI, M.: See [22a].
ELDRIDGE, C. H.: See [318].
ELLIS, W. C.: See [403, 404].
ELLIOTT, J. F.: See [184a, 1254a, 1271a].
ELSUKOVA, T. F.: See [108a, 1216a].
274. EMANUELI, L.: Trans. Amer. Inst. Electr. Engrs. 47 (1928) 118. *402*
275. EMICKE, O.: Metall 4 (1950) 1, 48. *37, 284, 416*
276. —, H. ALLHAUSEN and W. MAUKSCH: Siemens-Z. 12 (1932) 341. *443*
277. —, and H. BENAD: Arch. Eisenhüttenw. 12 (1938/39) 365. *441, 442, 444*
278. EMMERICH, E., and J. BECKAMNN: Felten- und Guillaume-Rdsch. 33 (1951) 294. *43, 195, 196, 413*
279. ENDO, H.: Sci. Rep. Tohoku Univ. 13 (1924) 193. *12, 334*
ENGEL, M.: See [994].
ENGEL, R.: See [559, 560].
ENGELL, H.-J.: See [577a].
280. Engng. 139 (1935) 656, Ref. Metallwirtsch. 14 (1935) 839. *295*
281. Engng. 145 (1938) 595. *424*

282. Engineer, London 164 (1957) 609, 667. *390*
ENSSLIN, F.: See [365].
283. EPSTEIN, L. F.: Int. Conf. Peaceful Uses of Atomic Energy P/119 (1955) 22. *326*
284. EPSTEIN, S.: Graphic Arts Res. Bur. Preprint 1936; Bull. Brit. Non-Ferr. Met. Res. Ass. 89 (1936) 9. *363, 364*
284a. ERBEN, E. and R. LESSER: Metall 15 (1961) 678. *326*
285. ERDMANN-JESNITZER, F.: Metallwirtsch. 19 (1940) 627. *203*
286. —, and H. HANEMANN: Z. Metallkde. 32 (1940) 118. *197, 420*
EREMENKO, O. M.: See [287].
287. EREMENKO, V. N., and O. M. EREMENKO: Ukrain. Khim. Zhur. 18 (1952) 232. Met. Abstr. 22 (1954/55) 111. *76*
288. ERMANN, G. A., and W. SPRING: Pogg. Ann. 9 (1827) 564. *133*
289. ERMISCH, H.: Dipl.-Arbeit Bergakademie Clausthal 1948. *332*
290. ESSER, H., F. GREIS and W. BURGARDT: Arch. Eisenhüttenw. 7 (1933/34) 385. *18*
291. ETZ 55 (1934) 196. *424*
292. ETZ 59 (1938) 96. *404*
293. EUCKEN, A.: Metallwirtsch. 15 (1936) 27, 63. *15*
294, —, and W. DANNÖHL: Z. Elektrochem. 40 (1934) 814. *11*
295. —, and H. SCHÜRENBERG: Ann. Phys. 33 (1938) 1. *19, 64*
EVANS, B. S.: See [526].
296. EVANS, C. D.: J. Metals 6 (1953) 655. *14*
EVANS, E. LL.: See [715].
297. EVANS, U. R.: An Introduction to Metallic Corrosion, 2nd Ed., London: Arnold 1963. *269, 304, 305, 314, 315*
298. EVERS, D.: Metall 7 (1953) 881. *345, 347*
299. EVERTS, TH.: Z. Metallkde. 28 (1936) 143. *462*
300. EWALD, P. P., and C. HERMANN: Strukturbericht d. Z. Kristallogr. 1913 to 1928, Leipzig: Akad. Verlagsges. 1931. *10*
EWING, S. P.: See [762, 763].

301. Fachabt. Druckguß: VDI 2501, VDI-Richtlinien-Druckgußlegierungen, Gestaltungsrichtlinien, Wirtschaftlichkeit: Düsseldorf: Dtsch. Ing. Verlag 1950. *357*
302. Fachgruppe Metallgießerei: Handbuch über Nichteisen-Schwermetall-Guß, Berlin: NEM-Verlag 1937. *339*
FAIVRE, R.: See [813a].
303. FALK, L.: Chemiker-Ztg. 34 (1910) I 567. *268*
304. FALKENHAGEN, G., and W. HOFMANN: Z. Metallkde. 43 (1952) 69 *45, 101*
FANGMEIER, E: See [1116].
305 FANO, E DA: Telegr. u. Fernsprechtechn. 21 (1932) 267. *293*
FARMER, M. H.: See [826].
FARNHAM, G. S.: See [921].
FARNOW, H.: See [794].
306. FAUST, O., and G. TAMMANN: Z. phys. Chem. 75 (1911) 108. *197*
306a. FAZEKAS, S., and A. VERES: Am. Soc. Metals Rev. Lit. Abstr. 20 (1964) 21. *449*
FEATHERLY, R. L.: See [1015].
FEDER, R.: See [248a].
307. —, and A. S. NOWICK: Phys. Rev. 109 (1958) 1959. *11*
307a. FEDOROV, P. I., and V. I. SHACHNEV: Zhur. Neorg. Khim. 5 (1962) 66. *176*
307b. —., V. I. SHACHNER and A. M. DOLGOPOLOVA: Tsvetnaya Met. 5 (1962) 58. *176*

308. FEDOTIEFF, N. P., B. P. ARTAMONOFF and N. J. RASMEROVA: According to Met. Abstr. 4 (1937) 448. *463*
FEES, G.: See [1022a].
309. FEISER, J.: Blei, in: Ullmanns Enzyklopädie der technischen Chemie, Vol. IV, 3. Aufl. München/Berlin: Urban & Schwarzenberg 1953. *5, 7, 37*
310. FEITKNECHT, W.: Elektrochem. 62 (1958) 795. *349*
311. —, and A. GAUMANN: J. Chim. Phys. 49 (1952) C 135. *349*
FELGER, H.: See [1023].
312. FELTHAM, P.: Proc. Phys. Soc. (B) 69 (1956) 1173. *213*
313. —, and J. D. MEAKIN: Acta Met. 5 (1957) 555. *177, 179*
314. FENG, J.-M.: Stahl u. Eisen 75 (1955) 789. *375*
FENTON, R. G.: See [714d].
FIEK, G.: See [1043, 1119].
FILAVETONA, G. M.: See [254a].
FILAVETOVA, G. M.: See [254a].
FINAN, F. B.: See [11].
315. FINCH, G. I.: J. Appl. Phys. (1951) Suppl. 1, 34. *376*
316. FINK, C.: Z. Berg-, Hütt.- u. Salinenw. 22 (1874) 200. *443,444*
317. FINK, C. G., and A. J. DORNBLATT: Trans. Electrochem. Soc. 79 (1941) 269. *355*
318. —, and C. H. ELDRIDGE: Trans. Amer. Electrochem. Soc. 40 (1921) 51. *300*
319. FINKE, H.: Metallbörse 17 (1927) 2581. *135*
320. FINKELDEY, W. H.: ASTM, Committee B-3, Report of Sub-Committee (1944). *268*
321. FINNIE, I., and W. R. HELLER: Creep of Engineering Materials, New York: McGraw-Hill 1959. *222*
322. FISCHER, H.: Elektrolytische Abscheidung u. Elektrokristallisation von Metallen, Berlin/Göttingen/Heidelberg: Springer 1954. *463*
323. FISCHER, J.: Z. anorg. allg. Chem. 219 (1934) 1, 367. *14*
324. FISCHER, O.: Thesis T. H. Berlin 1938. *471*
325. FISHER, H. J., and A. PHILLIPS: J. Metals 6 (1954) 1060. *95, 96*
326. —, —: J. Metals 7 (1955) 1264. *95, 96*
327. FLEISCHER, R. L.: Metallurgy Reports 11 (1960) No. 5. Massachusetts Institute of Technology. *177, 179*
328. FLEISCHMANN, R.: Z. Phys. 41 (1927) 8. *133*
329. FÖRSTER, F., and W. KÖSTER: Z. Metallkde. 29 (1937) 116. *13*
330. —, and G. TSCHENTKE: Z. Metallkde. 32 (1940) 191. *15 20*
331. FOLBERTH, O. G., and A. KOCHENDÖRFER: Naturw. 39 (1952) 187. *229*
FORCH, K.: See [515a].
332. FORD, H.: Metallurgical Reviews 2 (1957) 1. *441*
333. FORETAY, E.: Bull. Assoc. Suisse Elect. 41 (1950) 433. *316*
334. FORRESTER, P. G.: J. Inst. Metals 73 (1947) 573, Met. Reviews 5 (1960) 507. *376*
334a. FORRESTER, P. G.: Metallurgical Reviews 5 (1960) 507; 521; 522. *370, 382*
334b. —, and J. K. BEDDOW: Powder Met. 5 (1960) 149. *387*
335. —, and W. E. DUCKWORTH: Symposium on Powder Metallurgy (Iron Steel Inst.) Preprint, Group III (1954) 71. *387*
—: See [767].
FORSTER, F. T.: See [1186].
336. FORSTNER, H. M.: Oberflächentechn. 11 (1934) 165. *464*

336a. Forsyth, P. J. E.: Acta Met. 11 (1963) 703. *247*
Forsythe, W. R.: See [830].
337. Frank, F. C.: Symposium 1954, Creep and Fracture of Metals at high Temperatures, 281. *216*
337a. Franklin, W. M., and J. B. Wagner: J. Appl. Phys. (USA) 34 (1963) 3121. *74*
Franz, H.: See [831, 832].
338. French, S. J.: Industr. Engng. Chem. 27 (1935) 1464, Industr. Engng. Chem. 28 (1936) 111. *388*
Frey, D. R.: See [1207].
339. Fricke, R.: Z. Elektrochem. 52 (1948) 72. *16*
Fried, L.: See [624].
340. Friedburg, H. R.: Plating 46 (1959) 834. *301*
341. Friedel, J.: Les Dislocations, Paris: Gauthier Villars 1956. *179*
—: See [226].
342. Friedensburg, F.: Erzmetall 11 (1958) 515. *1*
343. Friedli, J.: Schweiz. Arch. 15 (1949) 261. *295*
344. Friedrich, K., and A. Leroux: Metallurgie 4 (1907) 293. *111*
345. —, and M. Waehlert: Metall u. Erz 10 (1913) 578. *54*
346. Friend, J. N.: J. Inst. Metals 39 (1928) 111. *307*
347. —, and M. S. Tidmus: J. Inst. Metals 31 (1924) 177. *296*
348. Friend, W. Z., and H. O. Teeple: I. Oil and Gas 44 (1946) 87. *291*
349. Frisch, J., and E. G. Thomsen: Trans. Amer. Soc. Mech. Eng. 76 (1954) 599. *395*
349a. Fritze, H. W.: Erzmetall 16 (1963) 290. *334*
Frohberg, M. G.: See [1058].
350. Frommer, L.: Handbuch der Spritzgußtechnik, Berlin: Springer 1933. *357*
351. Fürth, R., S. Ornstein and W. Milatz: Proc. Nederl. Akad. Wetensch. 42 (1939) 107, Gießerei, techn.-wiss. Beih. (1958) No. 19, 1019. *12*
351a. Fujiki, Y., R. Suganuma and T. Yoshida: J. Phys. Soc. Japan 13 (1958) 969. *38*
Fujita, E.: See [890].
352. Fuke, Y., and A. Kondo: Nipp. Kinzoku Gakkai-Si 15 (1951) 67. *284*
353. —, —: Nipp. Kinzoku Gakkai-Si 16 (1952) 611. *33*
Fukuroi, T.: See [22].
Fukushima, S.: See [1242].
Furthmann, E.: See [1073].
Furukawa, T.: See [1154].

Gabriel, G.: See [863].
354. Gabrielson, G.: J. Appl. Chem. 8 (1958) 748. *353, 463*
Gage, D. G.: See [45a].
355. Gangler, J. J.: J. Amer. Ceram. Soc. 37 (1954) 312. *41*
Gardam, G. E.: See [597].
Garner, F. H.: See [1276, 1277].
356. Garre, B., and A. Müller: Z. anorg. allg. Chem. 190 (1930) 120. Auszug Z. Metallkde. 23 (1931) 236. *194*
357. —, —: Z. anorg. allg. Chem. 198 (1931) 297. *138, 199, 200, 206*
358. —, and F. Vollmert: Z. anorg. allg. Chem. 210 (1933) 77. *115*
359. Garten, W.: Der Blei-Akkumulator, München: Oldenbourg 1958. *342*
360. Gas- und Wasserfach 75 (1932) 943. *307*
Gaumann, A.: See [311].

360a. GAZZARD, S. T., D. H. ROBERTS, J. C. MOORE and N. A. RATCLIFF: Second Intern. Conf. on Lead, Arnhem 1965, No. 2. *452*
361. GEBHARDT, E., M. BECKER and E. TRÄGNER: Z. Metallkde. 46 (1955) 90. *63*
362. —, and K. KÖSTLIN: Z. Metallkde. 48 (1957) 636. *80, 95, 96*
363. —, —: Z. Metallkde. 49 (1958) 605. *18*
364. —, and W. OBROWSKI: Z. Metallkde. 45 (1954) 332. *144, 146*
364a. GEGUZIN, I. Y., and V. I. KUDRIK: Phys. Metals Metallography 7 (1959) 71. *229*
365. GEORGE, W., and F. ENSSLIN: Metall und Erz 37 (1940) 6, 109. *44*
366. GERING, K., and F. SAUERWALD: Z. anorg. allg. Chem. 223 (1935) 204. *18*
366a. GERLACH, J., and J. GRIGEL: Metall 17 (1963) 802. *75*
366b. —, and G. HERRMANN: Erzmetall 15 (1962) 132. *332*
366c. —, and E. A. MÜLLER: Metall 18 (1964) 807. *331*
—: See [258a].
GERMAGNOLI, E.: See [26a].
367. GERMAN, S.: Z. Metallkde. 45 (1954) 484. *16*
368. GERRARD, J., and J. R. WALTERS: According to Werkst. u. Korrosion 8 (1957) 499. *316*
369. GERSHMAN, R. B.: Zhur. Fiz. Khim. 31 (1957) 1573. *40*
370. —: Zhur. Fiz. Khim. 32 (1958) 12, 258. *133*
371. GERTSCH, R.: Techn. Mitt. PTT (1937) No. 6, 201. *427*
372. —: Ann. Post. Télégr. Téléph. 27 (1938) 880. *426, 427*
373. —, and H. KÖLLIKER: Techn. Mitt. PTT (1950) No. 1, 8. *427*
374. GERTSRIKEN, S. D., M. G. BUTZIK and Z. P. GOLUBENKO: Mem. Phys. Kiev 8 (1939) 55, Brit. Chem. Abs. 1940 (AI) 67, Met. Abstr. 1940, 434. *58*
375. —, and B. F. SLYUSAR: Physics of Metals and Metallography 6 (1958) 103. *11*
376. Gesetz betreffend den Verkehr mit blei- und zinkhaltigen Gegenständen vom 25. 6. 1887 (RGBLp. 273). *471*
GLAUQUE, W. F.: See [830].
376a. GIBLIN, J. F.: Proc. First Intern. Conf. on Lead, London 1962, Chap. 1, 1. *416, 424*
377. —, and W. T. KING: Proc. Inst. Elect. Eng. 101 (1954) 123. *318*
GIBSON, J. G.: See [208].
378. GIFKINS, R. C.: Trans. AIME 215 (1959) 1015. *189, 212*
379. —: J. Inst. Metals 87 (1958/59) 255. *217*
380. —: J. Inst. Metals 79 (1951) 233, J. Inst. Metals 81 (1952/53) 417. *106*
381. —: Bull. Inst. Metals 4 (1958) 117. *106, 244*
381a. —: Trans. AIME 218 (1960) 1128. *216*
381b. —: J. Austr. Inst. Met. 8 (1963) 418. *211*
382. —, and H. C. COE: Metallurgia 43 (1951) 47. *220*
382a. —, and J. M. NICHOLLS: J. Inst. Met. 88 (1959/60) 96. *29*
—: See [1085a].
383. GILETTE, H. C.: Trans. Amer. Electrochem. Soc. 41 (1922) 217. *353*
GILHAUS, F. J.: See [1085a].
384. GINDIN, L. G., and M. V. PAVLOVA: Zhur, Priklad, Khim. 24 (1951) 1026: According to Met. Abstr. 22 (1954) 129. *298*
385. GIOLITTI, F., and M. MARANTONIO: Gazz. chim. ital. 40 I (1910) 51. *151*
386. GLANDER, F., and W. GLANDER: Z. Metallkde. 44 (1953) 97. *315, 317*
387. —, —: Z. Metallkde. 46 (1955) 552. *151, 416, 419*
GLANDER, W.: See [386, 387].
GMACHL-PAMMER, J.: See [704].

388. GMELIN, and KRAUT: Handbuch der anorg. Chemie, 7th Ed., Vol. IV, Abt. 2, Heidelberg: Winter 1924. *302, 303*
389. GOEBEL, J.: Z. anorg. allg. Chem. 106 (1919) 209. *155*
390. —: Z. VDI 63 (1919) 424. *58, 155*
391. —: Z. Metallkde. 14 (1922) 357, 388, 425, 449. *41, 52, 65, 68, 81, 83*
392. GÖLER, F. K. Frhr. v.: Metallwirtsch. 16 (1937) 797. *238*
393. —: Gießerei 25 (1938) 242. *45, 48, 63, 79, 322*
394. —, and G. A. GREFF: Metallwirtsch. 18 (1839) 945. *404, 407, 417*
395. —, and H. PFISTER: Metallwirtsch. 15 (1936) 342. *383*
396. —, and F. SCHEUER: Z. Metallkde. 28 (1936) 121, 176. *128, 383, 384*
397. —, and E. SCHMID: Z. Metallkde. 31 (1939) 61. *406, 422, 423*
398. —, and N. WEBER: in: R. KÜHNEL: Werkstoffe f. Gleitlager, Berlin: Springer 1939, 367. *47, 61, 67, 386*
399. GOENS, E.: Ann. Phys. 38 (1940) 456. *12*
400. GOETZEL, C. G.: Treatise on Powder Metallurgy, III, New York: Interscience Publ. 1952. *35, 449, 452*
400a. GOFF, L. I., A. L. J. OAKLEY, L. N. ROWE and R. D. SEMMENS: Second Intern. Conf. on Lead, Arnhem 1965, No. 8. *467*
401. GOHN, G. R.: Bell. Labor. Rec. 17 (1938) 277. *448*
402. —, S. M. ARNOLD and G. M. BOUTON: Proc. ASTM 46 (1946) 990. *230, 231, 237, 239, 240, 242*
403. —, and W. C. ELLIS: ASTM Proceedings 48 (1948) 801. *230, 231, 242*
404. —, —: ASTM Proc. 51 (1951) 721. *248, 251, 253, 254, 255, 256*
405. GOLDSCHMIDT, V. M.: Geochemische Verteilungsgesetze der Elemente, Vol. IX. Schriften der Norwegischen Wiss. Akad. Oslo, I. Math.-Naturwiss. Kl. 1938. *3*
GOLUBENKO, Z. P.: See [374].
GONIKBERG, M. G.: See [165].
406. GONSER, B. W., and C. M. HEATH: AIME Inst. Metals Div., Techn. Publ. Nr. 727 (1936). *473*
407. GONSER, U.: Z. physikal. Chem. (Frankfurt) 1 (1954) 1. *41*
408. GORDON, R. B.: Acta Met. 7 (1959) 1. *13*
GOREW, K. W.: See [117].
409. GORMAN, J. W., and G. W. PRECKSHOT: Trans. AIME 212 (1958) 367. *55*
410. GOSDEN, J. H.: Chem. and Ind. (1956) 1069. *295*, 315
GOTTSCHOL, H.-J.: See [1019].
411. GOUGH, H. J., and W. G. SOPWITH: J. Inst. Metals 56 (1935) 55. *245, 258, 349*
411a. —: J. Inst. Metals 49 (1932) 17. *284*
GRACE, R. E.: See [23].
412. GRACE, R. E., and G. DERGE: J. Metals 7 (1955) 839. *41*
412a. GRAHAM, A. K., and H. L. PINKERTON: Proc. Amer. Electroplaters Soc. 50 (1963) 139. *463*
GRAHAM, J.: See [867, 997].
GRANT, N. J.: See [922].
413. GRANT, L. E.: Metals and Alloys 5 (1934) 161. *59, 63, 69, 129*
414. GRASSI, R. C., D. W. BAINBRIDGE and J. W. HARMAN: U. S. Atomic Energy Comm. Publ. AECU 2201 (1952) 72. *41*
415. GRASSMANN, P.: VDI-Z. 98 (1956) 1929. *477*
416. GRATSIANSKY, N. N., and M. L. KAPLAN: Zhur. Fiz. Khim. 30 (1956) 651. *60*
417. —, and A. K. RYABOR: Zhur. Fiz. Khim. 33 (1959) 1253. *60*

418. GRAVEMANN, H., and H. J. WALLBAUM: Z. Metallkde. 47 (1956) 433. *153*
419. GRAY, A. G.: Modern Electroplating, New York: Wiley 1953. *463*
GREENALL, C. H.: See [1205].
420. GREENAWAY, H. T.: J. Inst. Metals 74 (1948) 133. *17, 81*
421. GREENWOOD, H.: Intern. Tin Res. Develop. Council, Techn. Publ. Ser. No. A. 58 (1937). *104, 385*
422. GREENWOOD, J. N.: Chem. Engrs. Min. Rev. 28 (1936) 384. *34*
423. —: Proc. Amer. Soc. Test. Mat. 49 (1949) 834. *232, 241*
424. —: J. Iron and Steel Inst. (1952) 380. *216*
425. —: Proc. Amer. Soc. Test. Mat. 56 (1956) 859. *243*
426. —: J. Inst. Metals 87 (1958/59) 91. *58*
426a. —: Metallurg. Rev. 6 (1961) 279. *9*
427. —, and J. H. COLE: Metallurgia 36 (1947) 233. *241*
428. —, —: Metallurgia 39 (1949) 121. *232*
429. —, —: Metallurgia 39 (1949) 241. *239, 240*
430. —, and C. W. ORR: Proc. Austral. Inst. Min. Met. 109 (1938) 1. *55*
431. —, —: Proc. Ausrtal. Inst. Min. Met. 113 (1939) 1. *65*
432. —, and H. K. WORNER: Proc. Austral. Inst. Min. Met. 101 (1936) 57. *43, 104, 219, 220, 231, 242*
433. —, —: Proc. Ausrtal. Inst. Min. Met. 104 (1936) 385. *34*
434. —, —: J. Inst. Metals 64 (1939) 135. *109, 231, 237, 239, 241, 242*
435. —, —: J. Inst. Metals 65 (1939) 435. *75, 94, 101*
—: See [1132].
436. GREER, H. F., and M. L. BEGEMAN: Welding Journal, Research Suppl. 39 (1960) 247. *454*
GREFF, G. A.: See [394].
437. —, and K. LÖHBERG: Fernmeldetechn. Z. 3 (1950) 122. *317*
438. GREGER, E.: Apparatebau 48 (1936) 24. *454*
439. GREMPE, P. M.: Öl u. Kohle 11 (1935) 118. *297*
GRIGEL, J.: See [366a].
440. GROOVE, E.: French P. 765040 of 1933, Engl. P. 398318 of 1932. *465*
441. GROSHEIM-KRISKO, K. W.: Z. Metallkde. 34 (1942) 102. *80*
442. —, H. HANEMANN and W. HOFMANN: Z. Metallkde. 34 (1942) 97. *351, 352*
443. —, W. HOFMANN and H. HANEMANN: Z. Metallkde. 36 (1944) 85, 88. *44, 351, 352*
444. GRUBE, G., and A. DIETRICH: Z. Elektrochem. 44 (1938) 755. *38, 39*
445. —, and H. KLAIBER: Z. Elektrochem. 40 (1934) 745. *62*
GRUBE, K. R.: See [1273a].
GRUHL, U.: See [1241].
446. GRUHL, W.: Z. Metallkde. 40 (1949) 225. *326, 327, 328, 329, 332*
447. GRYMKO, S. M., and R. I. JAFFEE: Materials and Methods 31 (1950) 59. *60, 158*
GSCHNEIDER, K.: See [1233a].
448. GÜNTHER: Kalt-Walz-Welt 4 (1935) 30. *453*
GÜNTHER, H.: See [1204].
448a. GÜNZEL, E., and K. SCHUBERT: Z. Metallkde. 49 (1958) 124. *149*
449. GUERTLER, W.: Metall u. Erz 23 (1926) 468. *55*
450. —: Metallwirtsch. 19 (1940) 435. *381*
451. — at W. CLAUS: Z. Metallkde. 28 (1936) 86. *55, 75*
452. —, and G. LANDAU: Z. anorg. allg. Chem. 218 (1934). *147*
453. —, and F. MENZEL: Z. Metallkde. 15 (1923) 223. *144*
454. GURTLER, G., and E. SCHMID: Z. VDI 83 (1939) 749. *242, 248*
GUICHARD, M.: See [67].

455. GUITTON, L.: Bull. Soc. chim. Fr. 4 (1937) 570. *291*
456. GUMM, P.: Diplom-Arbeit T. H. Braunschweig 1960. *465*
457. Gummi-Ztg. 43 (1928/29) 1866. *296*
GURINSKY, D. H.: See [643].
GURRY, R. W. See [231a].
458. —: Met. Abstr. 23 (1956) 1012. *326*
459. GYGAX, A. F.: Moderne Chemigraphie in Theorie und Praxis, Frankfurt/M.: Polygraph-Verlag 1957. *358*

460. HAASE, L. W.: Gas- u. Wasserfach 75 (1932) 373. *306, 307*
461. —: Werkst. u. Korrosion 2 (1951) 90. *311*
HAASEN, P.: See [1100].
461a. HAASEN, P.: Metallkde. 55 (1964) 55. *262*
461b. —: Lecture at the AIME Seminar on Concent. Solid Solutions. Cleveland 1963. *247*
461c. HADDOW, J. B., and J. M. CHUDOBIAK: Canad. Met. Quart. 2 (1963) 325. *397*
—: See p. 91.
462. HADORN, E., and R. HAINFELD: Techn. Mitt. PTT 5 (1958) 177. *427*
463. HAEHNEL, O.: Z. Fernmeldetechn. 4 (1923) 49. *218, 313*
464. —: Elektr. Nachr.-Techn. 2 (1925) 230. *244, 307, 315*
465. —: Elektr. Nachr.-Techn. 4 (1927) 106. *315*
466. —: Elektr. Nachr.-Techn. 5 (1928) 171. *311*
467. —: ETZ 60 (1939) 713. *307, 311, 313, 314, 315, 317*
468. —: Korrosion u. Metallsch. 18 (1942) 297. *311*
HAHN, K. F.: See [1233].
469. HAIGH, B. P.: Trans. Faraday Soc. 24 (1928) 125. *249*
470. —, and B. JONES: J. Inst. Metals 43 (1930) 271. *245, 246*
471. —, —: J. Inst. Metals 51 (1933) 49. *201, 202*
472. HAINES, G. S.: Ind. Engng. Chem. 41 (1949) 2792. *297*
HAINFELD, R.: See [462].
HALL, L. D.: See [1036].
473. HALLA, F., and R. HERDY: Z. Elektrochem. 56 (1952) 213. *156*
474. HALLIDAY, W. M.: Precision Metal Moulding 11 (1953) 42. *357*
HAMMES, H.: See [618].
475. Handbook of Chemistry and Physics, 37. Aufl., Cleveland, Ohio: Chemical Rubber Publ. 1955. *19*
HANDLER, G. S.: See [888].
476. HANEMANN, H.: Metallwirtsch. 7 (1935) 133. *439*
477. —, K. v. HANFFSTENGEL and W. HOFMANN: Metallwirtsch. 16 (1937) 951. *237, 238*
478. —, and W. HOFMANN: Metallwirtsch. 14 (1935) 915. *237, 238, 440*
479. —, and A. SCHRADER: Atlas Metallographicus, Vol. III, Aluminium, Part 1, Berlin: Borntraeger 1941. *45*
480. —, —: Atlas Metallographicus, Vol. III, Ternäre Leg. d. Alum., 2. Teil, Düsseldorf: Verlag Stahleisen 1952, 122. *121*
—: See [*76, 94, 139, 286, 442, 443, 482, 483, 484, 485, 501, 502, 535, 548, 561, 562, 563, 564, 577, 747, 847, 992, 1081, 1082, 1147*].
HANFFSTENGEL, K. v.: See [*477*].
481. —: Dissertation T. H. Berlin 1937. *211*
482. —, and H. HANEMANN: Z. Metallkde. 29 (1937) 50. *209, 219, 220*
483. —, —: Z. Metallkde. 30 (1938) 41. *211, 214, 218, 222, 229, 231, 237*

484. —, —: Z. Metallkde. 30 (1938) 50. *63, 272*

485. —, —: Z..Metallkde. 30 (1938) 51. *257, 258, 259, 262*

486. HANLEY, H. R., C. Y. CLAYTON and D. F. WALSH: Trans. AIME, Yearbook 91 (1930) 275; item, General Vol. 96 (1931) 142. *298*

487. HANSEN, M.: Der Aufbau der Zweistofflegierungen, Berlin: Springer 1936. *58, 59, 76, 99, 106, 145*

488. HANSEN, M.: Constitution of Binary Alloys, New York/Toronto/London: McGraw-Hill 1958. *29, 34, 38, 53, 61, 71, 94, 105, 106, 107, 114, 135, 139*

HANSEN, M.: See [59].

489. HANSON, D., and M. A. WHEELER: J. Inst. Metals (London) (1931) 229. *216,*

489a. HANSSON, B. O. N.: Proc. First Intern. Conf. on Lead, London, 1962, Chap. 6, 73. *409, 419*

490. HARGREAVES, F.: J. Inst. Metals 44 (1930) 149. *10*

HARGREAVES, M. E.: See [198].

HARING, H. E.: See [1186].

491. HARING, H. E., and U. B. THOMAS: Trans. Electrochem. Soc. 68 (1935) 293. *353, 354*

492. HARING, M. M., M. R. HATFIELD and P. P. ZAPPONI: Electrochem. Soc. Preprint 1939 (April) 169. *58*

HARINGHUIZEN, D. J.: See [923].

HARMAN, J. W.: See [414].

493. HARN, O. C.: Lead, the precious Metal, New York and London: The Century Co. 1924. *454*

494. HARRIS, I. E.: Amer. Soc. Steel Treat. 17 (1930) 282. *200*

495. HARTFORD, CH. E., and R. L. COPSON: Ind. Eng. Chem. 31 (1939) 1123. *292*

496. HARTMANN, H., W. HOFMANN and W. STAHL: Erzmetall 11 (1958) 151. *9, 17, 326, 327, 328, 332, 363, 364*

496a. —, —, —: Z. Metallkde. 44 (1953) 123. *331, 332*

496b. HARTMANN, R. J.: Z. Metallkde. 55 (1964) 221. *247*

HARTNETT, J. P.: See [623, 624].

496c. HARVEY, D. J.: Trans. AIME. 221 (1961) 266. *36, 101*

HATFIELD, M. R.: See [491].

496d. HAUFFE, K.: Oxydation von Metallen und Metallegierungen, Berlin/Göttingen/Heidelberg: Springer 1956, 74. *327, 333*

497. HAUG, H.: In H. KRAUTMACHER and W. PÜNGEL, Stahl u. Eisen 77 (1957) 845. *323*

HAUPT, G.: See [970].

498. HAY, F. W.: Metallurgist (Suppl. to Engineer) 163 (1937) 23. *464*

HAYS, L. E.: See [102a].

499. HAYAKAWA, J. Y., H. KIDO and K. KURODA: J. Electrochem. Soc. Japan 21 (1953) 425: According to C. A. 48 (1954) 5059c. *295*

500. HEAP, J. H.: J. Soc. Chem. Ind. 32 (1913) 771. *304*

HEATH, C. M.: See [406].

501. HECKLER, O., and H. HANEMANN: Z. Metallkde. 30 (1938) 410. *272, 276, 277, 278, 303, 307*

502. —, W. HOFMANN and H. HANEMANN: Z. Metallkde. 30 (1938) 419. *47, 239, 263*

HEHEMANN, R. F.: See [152].

503. HEIDENREICH, R. D.: Acta Met. 3 (1955) 79. *33, 191*

504. HEIKE, W.: Int. Z. Metallogr. 6 (1914) 49. *35*

HEINRICHS, G.: See [917].

504a. HEINE, W., K. SAGEL and U. ZWICKER: Bleilegierungen als Abschirmmaterialien. Z. Kerntechnik. *24, 129*
505. HELD, CH., and O. GASSER: Siemens-Z. 19 (1939) 197. *402*
506. HELLER, H.: Z. phys. Chem. 89 (1915) 761. *10*
HELLER, W. R.: See [321].
HELMOLD, G.: See [1103].
508. HENGEMÜHLE, W.: Handbuch der Werkstoffprüfung, edited by E. SIEBEL, Vol. II, Berlin: Springer 1939, 352. *205*
509. HENGLEIN, E., and W. KÖSTER: Z. Metallkde. 39 (1948) 391. *117, 118, 154*
510. HENKY, H.: Z. angew. Math. Mech. 3 (1923) 241. *399*
511. —: Z. angew. Math. Mech. 4 (1924) 323. *399*
512. HENLEY, W. T.: Engng. 140 (1935) 412, compare Metal. Ind., London 47 (1935) 210. *408*
512a. HENSLER, J. H., and R. C. GIFKINS: J. Inst. Metals 92 (1964) 340. *211*
HERBIET, H.: See [1005].
HERDY, R.: See [473].
HERMANN, I.: See [203].
HERRMANN, E. H.: See [11].
HERRMANN, G.: See [366b].
513. HERRMANN, W., and G. PRÖPSTL: Z. Elektrochem. 61 (1957) 1154. *353*
513a. HERRNKIND, W.: Thesis T. H. Braunschweig 1967. *323*
514. HERSCHMANN, H. K., and I. L. BASIL: Proc. ASTM 32 (1932) 536. *383*
515. HESS, W.: Werkst. u. Korrosion 7 (1956) 649. *293*
HESSEL, W.: See [272a].
HETHERINGTON, H. C.: See [199].
515a. HEUMANN, T., and K. FORCH: Z. Metallkde. 53 (1962) 122. *472*
516. HEUSER, G.: Metall 10 (1956) 759. *431, 467*
517. HEVESY, G. v., W. SEITH and A. KEIL: Z. Physik 79 (1932) 197. *16*
518. HEWLETT, W. L.: J. Soc. Chem. Ind. Trans. 54 (1935) 1094. *295*
519. HEYER, G. O.: Seifensieder-Ztg. 60 (1933) 131, 165. *282, 293*
HEYER, R. H.: See [537].
520. HEYN, E., and O. BAUER: Untersuchungen über Lagermetalle, Antimon-Blei-Zinn-Legierungen, Berlin 1914. Beiheft der Verh. Ver. zur Beförderung des Gewerbefleißes, Febr. 1914. *100, 167, 170, 205, 383, 384*
521. HIDNERT, P.: J. Res. Nat. Bur. Stand. 17 (1936) 697. *81*
522. —, and W. T. SWEENEY: J. Res. Nat. Bur. Stand. 9 (1932) 703. *11*
HIERS, G. O.: See [221].
524. —: in Metals Handbook (Amer. Soc. for Metals) 1948 Edition. *295*
525. —, and G. A. STEERS: Metal Ind., London 54 (1939) 143. *104, 273*
526. HIGGS, D. G., and B. S. EVANS: Analyst 72 (1947) 105. *31*
526a. HIKI, Y.: J. Phys. Soc. Japan 14 (1959) 590. *13*
HILGERS, W.: See [1019].
527. HILL, R.: The Mathematical Theory of Plasticity, Oxford/London: Clarendon Press 1950. *399, 400*
528. HILLEN, G., and W. HOFMANN: Z. Metallkde. 44 (1953) 129. *244, 441, 459*
HINDE, R. M.: See [208].
529. HIRAMA, F., and N. WATANABE: Nipp. Kinzoku Gakkai-Si 22 (1958) 161, 345. *281*
530. HIRATA, H., Y. TANAKA and H. KOMATSUBARA: According to Met. Abstr. 2 (1935) 512. *464*
HIRSCH, P. B.: See [1194].
531. HIRST, H.: J. Inst. Metals 66 (1940) 39. *183*

532. HIRST, H.: Proc. Austral. Inst. Min. Met. 120 (1940) 777. *228*
533. —: Proc. Austral. Inst. Min. Met. 121 (1941) 11. *228*
534. —: Proc. Austral. Inst. Min. Met. 121 (1941) 29. *228*
534a. HISCOCK, S. A.: Ernest Benn, Ltd., London, by arrangement with Enfield-Standard Power Cables Ltd. 1961. *404, 416*
535. HO, T.-H., W. HOFMANN and H. HANEMANN: Z. Metallkde. 44 (1953) 127. *40, 43, 129, 132*
HOAR, T. P.: See [215a, 833].
536. —, and D. A. MELFORD: Trans. Faraday Soc. 53 (1957) 315. *95*
537. HODGE, J. M., and R. H. HEYER: Metals and Alloys 5 (1931) 297, 313. *107*
538. HODGMAN, CH. D.: Handbook of Chemistry and Physics, 37th Ed., Cleveland, Ohio: Chemical Rubber Publ. 1955. *268*
539. HOEHNE, E.: Chem. Fabrik 9 (1936) 259. *356*
540. —: Z. Metallkde. 30 (1938) 52. *343, 354*
541. —: Metallwirtsch. 23 (1944) 60. *352*
542. HOEHNE, E.: Arch. Metallkde. 2 (1948) 311. *354*
543. —: Arch. Metallkde. 3 (1949) 185. *353*
544. —, and H. D. GRAF V. SCHWEINITZ: Metallwirtsch. 21 (1942) 218. *351, 355*
545. HÖLL, K.: Gesundh.-Ing. 58 (1935) 323. *306*
546. HÖLTJE, H.: Elektrowärme 6 (1936) 336. *321, 322*
547. —: Techn. Zbl. prakt. Metallbearb. 48 (1938) Nr. 23/24. *321*
HÖRNLE, R.: See [716].
548. HOFE, H. v., and H. HANEMANN: Z. Metallkde. 32 (1940) 112. *41, 131*
549. —, W. HOFMANN and G. SUCHAN: Abhdlg. d. Brschweig. Wiss. Ges. 2 (1950) 175. *462*
550. HOFF, H., and TH. DAHL: Stahl u. Eisen 54 (1934) 655. *189, 443*
HOFFMAN, R. E.: See [598a].
551. HOFFMANN, O., and G. SACHS: Introduction to the Theory of Plasticity for Engineers, New York: McGraw-Hill 1953, 115. *399*
552. HOFMANN, K. B.: Das Blei bei den Völkern des Altertums, Berlin 1885. *1*
HOFMANN, W.: See [49, 148, 304, 442, 443, 477, 478, 496, 502, 528, 535, 549].
553. —: Z. Kristallographie 92 (1935) 161. *161*
554. —: Naturwiss. 24 (1936) 507. *182*
555. —: Z. Metallkde. 29 (1937) 266. *182, 192*
556. —: Chem. Apparatur 25 (1938) 92 u. 108. *238, 440*
558. —: Z. Metallkde. 33 (1941) 61. *149*
559. —, and R. ENGEL: Z. Metallkde. 44 (1953) 132. *76, 440*
560. —, —: Freiberger Forschungshefte, B 38 (1960) 45. *236*
561. —, and H. HANEMANN: Z. Metallkde. 30 (1938) 47. *33, 62, 192*
562. —, —: Z. Metallkde. 30 (1938) 416. *263, 267, 354, 440*
563. —, —: Z. Metallkde. 32 (1940) 109. *207, 237, 440*
564. —, —: Z. Metallkde. 33 (1941) 62. *103*
565. —, and J. KOHLMEYER: Z. Metallkde. 45 (1954) 339. *144, 146*
566. —, and T. B. LEY: Z. Metallkde. 34 (1942) 107. *177, 227, 228*
566a. —, and A. LUIS: Z. Metallkde. 55 (1964) 101. *108*
567. —, and J. MAATSCH: Z. Metallkde. 47 (1956) 89. *58*
568. —, and K. H. MAHLICH: Werkst. u. Korrosion 2 (1951) 55. *268*
569. —, and H. v. MALOTKI: Metall 13 (1959) 747. *458*
569a. —, —: Z. Metallkde. 55 (1964) 135. *34, 216, 223, 228*
569b. —, —: Erzmetall 17 (1964) 70. *225, 226*
570. —, —, and P. WEHR: Metall 14 (1960) 763. 416, 418

571. HOFMANN, W. and G. MEIER: Z. Metallkde. 45 (1954) 508. *446*
572. —, and R. MÜLLER: Erzmetall 7 (1954) 247. *248, 257, 418*
573. —, and H.-G. PETRI: Werkst. u. Korrosion 1 (1950) 232. *352*
574. —, and G. PIEPER: Z. Metallkde. 49 (1958) 80. *35, 452*
575. —, and H. J. SCHNEIDER: Gießerei, techn.-wiss. Beih. (1960) 28, 1567. *147*
576. —, —: Erzmetall 13 (1960) 484. *147*
577. —, A. SCHRADER and H. HANEMANN: Z. Metallkde. 29 (1937) 39. *77, 78, 84, 90, 191, 192, 203*
577a. —, P. WEHR and H.-J. ENGELL: Werkst. u. Korrosion 17 (1966) 277. *246*
HOFMEIER, F.: See [705].
578. HOGNESS, T. R.: J. Amer. Chem. Soc. 43 (1921) 1621. *17*
579. HOHLSTEIN, G., and E. PELZEL: Metall 14 (1960) 765. *277, 278*
579a. —, and E. PELZEL: Metall 16 (1962) 764. *276, 280*
HOHMANN, E.: See [671].
581. HOLLIGAN, P. T.: Metal Ind. (London) 70 (1947) 375, 402 and 419. *372*
582. HOLLOMON, J. H., and D. TURNBULL: J. Metals 3 (1951) 803. *96*
583. HOLM, R., and W. MEISSNER: Z. Phys. 74 (1932) 736. *203*
583a. HOLMES, J. F.: Metal Ind. (London) 101 (1962) 42. *23*
584. HONDA, K.: Ann. Phys., (Leipzig) 32 (1910) 1027. *20*
585. HONDA, K., and R. KIKUCHI: Sci. Rep. Tohoku Univ. 21 (1932) 575. *334*
HONEYCOMBE, R. W. K.: See [95].
586. HOOKER, E. J.: J. Inst. Metals 86 (1957/58) 98. *91*
587. —: J. Inst. Metals 88 (1959/60) 44. *11*
HOPKIN, L. M. T.: See [821, 822].
588. —, and C. J. THWAITES: J. Inst. Metals 81 (1952/53) 255. *36, 87, 90, 91*
589. —, —: J. Inst. Metals 82 (1953/54) 181. *221, 233, 234, 235, 236, 238, 241, 249, 256*
590. —, —: Met. Abstr. 21 (1954) 426. *208*
HORI, S.: See [857].
591. HORN, W.: Arch. Post. Telegr. 7 (1933) 165. *317*
592. —: Arbeiten über physiologische und angewandte Entomologie aus Berlin-Dahlem, I (1934) 4, 291. *317*
593. —: Arbeiten über physiologische und angewandte Entomologie aus Berlin-Dahlem, IV (1937) 4, 265. *317*
VAN HORN, K. R.: See [654].
594. HORNUNG, R.: Techn. Mitt. PTT 31 (1953) 265, 318. *316*
595. HOROWITZ, M., A. A. SILVIDI, S. F. MALAKER and J. G. DAUNT: Phys. Rev. 88 (1952) 1182. *15*
596. HORSLEY, G. W., and J. T. MASKREY: J. Inst. Metals 86 (1957/58) 446. *63*
HOSTETTER, J. C.: See [237].
597. HOTHERSALL, A. W., and G. E. GARDAM: Electrodepositors Techn. Soc., Advance Copy 27 (1951) 15. *301*
HOTOP, W.: See [662].
HOUCK, J. A.: See [1274].
598. HOUDREMONT, E.: Handbuch der Sonderstahlkunde, 3rd Ed., Berlin/Göttingen/Heidelberg: Springer 1956, 1423. *57*
HOUWINK, R.: See [136].
HOVEY, C. M.: See [171].
HOWE, H. E.: See [1128].
HOYT, E. W.: See [1273].
HUDRLIK, R. E.: See [977a].
598a. HUDSON, J. B., and R. E. HOFFMAN: Trans. AIME 221 (1961) 761. *16*

599. Hudson, J. C., and J. F. Stanners: Stahl und Eisen 74 (1954) 422. *453*
Hughes, J. E.: See [1012a].
Hughes, T. A.: See [33b].
Hult, J.: See [910a].
Hultgren, R.: See [250].
600. Hume-Rothery, W.: J. Inst. Metals 81 (1952/53) 259: See [588]. *88*
600a. Hume-Rothery, W.: Canad. J. Phys. 37 (1959) 1565. *10*
601. Humfrey, J. C. W.: Proc. Roy. Soc., London 70 (1902) 462. *10*
602. Hunsicker, L.: Das Schweißen von Blei und seine Anwendung, Halle/Saale: Marhold 1952. *469*
Hysel, V. B.: See [1031].

Igaki, K.: See [907].
602a. —, and S. Tanimoto: Nipp. Kinzoku Gakkai-Si 24 (1960) 121. *9*
Imoto, S.: See [857].
603. Ind. Chemist. Chem. Manufacturer 2 (1933) 110, 159. *301*
604. Industrial Heating 22 (1955) 1186. *345*
Intonti, R.: See [1137].
604a. Iron Age 186 (1960) 10, Chem. Engng. 67 (1960) 252. *456*
605. Isaac, E., and G. Tammann: Z. anorg. allg. Chem. 55 (1907) 58. *56*
606. Isihara, T.: Sci. Rep. Tohoku Univ. 18 (1929) 715. *134*
607. Ito, K.: Sci. Rep. Tohoku Univ. 12 (1923) 137. *203*
608. Iwasé, K., and O. Aoki: Kinzoku no Kenkyu 8 (1931) 253. *163, 165*

Jackson, K. A.: See [1197].
609. Jacoby, L., and J. A. Verö: Met. Abstr. 16 (1949) 484. *337*
610. Jacquet, P. A.: Bull. Soc. chim. Fr. 3 (1936) 705. *27*
611. —: Metall 8 (1954) 449. *27*
611a. Jaeger, Th.: Grundzüge der Strahlenschutztechnik, Berlin/Göttingen/Heidelberg: Springer 1960, 69, 145. *22, 24*
612. Jänecke, E.: Kurzgefaßtes Handbuch aller Legierungen, Leipzig: Spamer 1937; Nachtrag: Berlin: Kiepert 1940. *111, 112, 152, 153, 160*
613. —: Z. Metallkde. 26 (1934) 153. *141*
614. —: Z. Elektrochem. 42 (1936) 375. *147*
615. —: Z. Metallkde. 29 (1937) 367. *135, 176, 388*
616. —: Z. Metallkde. 30 (1938) 390. *137*
Jaffee, R. I.: See [447, 1273a, 1274].
617. —, and S. M. Weiss: Materials and Methods 36 (1952) 113. *60*
Jahn, E.: See [1057].
Jaswon, M. A.: See [218].
618. Jenckel, E., and H. Hammes: Z. Metallkde. 29 (1937) 89. *61, 190, 195, 337*
619. —, and Ch. Thierer: Z. Metallkde. 29 (1937) 21. *195, 337*
620. Jenge. W.: Z. anorg. allg. Chem. 118 (1921) 119. *65*
Jiřiště, J.: See [985].
Johannsen, F.: See [915].
621. —, and K.-H. Lange-Eichholz: Erzmetall 3 (1950) 397. *119, 120*
John, M. E.: See [1140].
Johnen, H.: See [1104, 1105].
622. —: Schweißen u. Schneiden 10 (1958) 281. *470*
Johnson, A. E.: See [1173].
623. Johnson, H. A., J. P. Hartnett and W. J. Clabaugh: Trans. Amer. Soc. Mech. Eng. 76 (1954) 513. *41*
624. —, —, —, and L. Fried: Trans. Amer. Soc. Mech. Eng. 79 (1957) 1079. *41*

625. JOHNSON, W.: J. Mech. Phys. Solids 4 (1956) 269. *391*
626. JOHNSON, W. A., and R. F. MEHL: Trans. AIME 135 (1939) 416. *185*
627. JOHNSTON, S., and L. H. ADAMS: Z. anorg. Chem. 72 (1911) 11. *14*
628. JOLLIVET, L.: Compt. Rend. 224 (1947) 1822. *120*
629. —: Compt. Rend. 226 (1948) 2076. *155*
630. —: Compt. Rend. 228 (1949) 1128. *175*
JONES, B.: See [470, 471, 1126].
631. —: J. Inst. Metals 52 (1933) 73. *25*
632. —: Metal. Ind. (London) 50 (1937) 335. *190, 193, 197*
633. —: Engineering 145 (1938) 285. *258, 260, 435, 436*
JONES, E. L.: See [112a, 1123, 1124].
634. JONES, D. W.: J. Soc. Chem. Ind. 47 (1928) 161 T. *287, 288, 290*
635. JONES, G. B.: Chem. Age 4 (1921) 494. *291*
636. JONES, W. R. D., and J. B. DAVIES: J. Inst. Metals 86 (1957/58) 164. *95*
637. JORDAN, L.: Min. u. Metall 17 (1936) 538. *66*
638. JUNGE, C. H.: ASTM-Bull. 198 (1954) 64. *387*
638a. JUNG-KÖNIG, W.: Metall 15 (1961) 1180. *418*
639. JUSCHMANOW, J. W., T. G. SHURAWLEWA and L. A. ORLOWA: Chem. Zbl. 1 (1939) 3622. *288*

KABANOV, B. N.: See [26a, 708].
640. Kabel- u. Metallwerke Neumeyer A. G., and E. Körner, Nürnberg, Engl. Pat. 449372 (1935). *103*
KADO, S.: See [900, 901].
KAHN, F.: See [958].
KAINZ, M.: See [109, 109a].
641. KAJA, P.: Z. angew. Chem. 40 (1927) 1167. *312*
642. KALISH, H. S., and F. J. DUNKERLEY: AIME Techn. Publication No. 2442 (Sept. 1948) 20. *99*
643. KAMMERER, O. F., J. R. WEEKS, J. SADOFSKY, W. E. MILLER and D. H. GURINSKY: Trans. AIME 212 (1958) 20. *41*
KAPLAN, D.: See [90].
KAPLAN, M. L.: See [416].
645. KÁRMÁN, T. v.: Z. angew. Math. u. Mech. 5 (1925) 142. *441*
KARTZMARK, R.: See [170].
646. KATZ, T.: Ann. chim. (Paris) 5 (1950) 5. *70, 71, 327*
647. KATZ, W.: Metalloberfläche [A] 7 (1953) 161. *283, 291, 294, 296*
KAUFMANN, S. J.: See [769].
648. KAUZMANN, W.: Trans. AIME 143 (1941) 57. *212*
649. KAWABA, E.: Jap. Patent 3459 (1954). *355*
KEIL, A.: See [517, 1106].
650. —: Metall 8 (1954) 515. *60, 134, 388, 389, 390*
651. —: Metall 9 (1955) 689. *116, 388*
652. —: Metall 11 (1957) 740. *477*
653. KELLOG, H. H., and S. K. BASU: Trans. AIME 218 (1960) 70. *175*
654. KEMPF, L. W., and K. R. VAN HORN: Metals Technology 5 (1938) AIME Techn. Publ. No. 990, 12. *121*
655. KENNEDY, A. J.: Proc. Phys. Soc. 68 [B] (1955) 257. *186, 218*
656. —: Nature (London) 178 (1956) 810. *232*
657. KENNEFORD, A. S., and H. O'NEILL: J. Inst. Metals 55 (1934) 51. *203*

657a. Kerr, H. W., A. Plumtree and W. C. Winegard: J. Inst. Metals 93 (1964/65) 63. *139*
657b. Kettler, L.: AZI-LIA Expand. Research Prog. New York 1962. *338*
658. Ketzer, P.: Erzmetall 6 (1953) 149. *1*
659. —: Erzmetall 11 (1958) 396. *2*
660. Khotkevich, V. I., E. F. Chaikovsky and V. V. Zashkvara: Doklady Akad. Nauk. USSR 96 (1954) 483. *182*
661. Kick, F.: Mechanische Technologie, 2nd Ed., Leipzig 1908. *392*
Kido, H.: See [499].
662. Kieffer, R., and W. Hotop: Sintereisen u. Sinterstahl, Wien: Springer 1948. *449*
663. Kienzle, O., and K. Lange: Werkstattstechnik u. Maschinenbau 42 (1952) 464. *74*
664. Kies, J. A., and W. F. Roeser: ASTM Preprint Nr. 32 (1944) 16. *477*
Kikuchi, R.: See [585].
665. Kikuta, T.: Sci. Rep. Tohoku Univ. 10 (1921) 139. *12, 13*
King, W. T.: See [377].
Kirkov, P.: See [799].
666. Kir'yakov, G. Z., and I. A. Korchmarek: Zhur. Priklad. Khim. 26 (1953) 921, Met. Abstr. 22 (1954) 792. *300*
667. —, and V. V. Stender: Zhur. Priklad. Khim. 24 (1951) 1263: According to Met. Abstr. 21 (1953) 803. *299*
668. —, —: J. Appl. Chem. USSR 25 (1952) 25, 33: According to Met. Abstr. 21 (1953) 805. *279, 300*
Klaiber, H.: See [445].
668a. Klant, H.: Second Intern. Conf. on Lead, Arnhem 1965, No. 5. *466*
Klein, H. A.: See [671].
669. Klein, M.: Kabeltechnik, Berlin: Springer 1929. *390, 400*
670. Kleinheisterkamp, H.: Erzmetall 1 (1948) 65. *124*
670a. Klement, W.: Calif. Inst. of Techn. Rep. 1962, TID-17334,9. *10*
Klemm, L.: See [671].
671. Klemm, W., L. Klemm, H. Hohmann, E. Volk, E. Orlamünder and H. A. Klein: Z. anorga. Chem. 256 (1948) 239. *158*
672. —, and H. Volk: Z. anorg. Chemie 256 (1947) 246. *59*
673. Kleppa, O. J.: J. Amer. Chem. Soc. 71 (1949) 3275. *38*
674. —: J. Amer. Chem. Soc. 74 (1952) 6047. *135*
675. Kleppa, O. J.: J. Phys. Chem. 59 (1955) 354. *41*
675a. —: J. Phys. Chem. 60 (1956) 446. *38*
676. —: Acta Met. 6 (1958) 233. 51
677. —, and J. A. Weil: J. Amer. Chem. Soc. 73 (1951) 4848. *55*
678. Klug, H. P.: J. Amer. Chem. Soc. 68 (1946) 1493. *10, 11*
679. Klut, H.: Korrosion und Metallsch. 1 (1925) 232. *304, 306*
680. Knappwost, A., and H. Restle: Z. Elektrochem. 58 (1954) 112. *11*
Knehans, K.: See [1051].
681. Knight, H. A.: Metals and Alloys 20 (1944) 1296. *454*
Knoll, W.: See [1161].
682. Knolle, K., and K. Löhberg: Z. Metallkde. 51 (1960) 350. *76*
683. Knuth-Winterfeldt, E.: Mikroskopie 5 (1950) 184. *27*
684. Koch, H., and W. Wasserbäch: Schweißen u. Schneiden 12 (1960) 2. *475, 476, 478, 479*
685. Koch, K. R., and C. Dannecker: Ann. Phys. (Leipzig) 47 (1915) 217. *12, 13*

686. Koch, K. R., and R. Dieterle: Ann. Phys. (Leipzig) 68 (1922) 441. *12, 13*
Kochendörfer, A.: See [331].
Koehler, J. S.: See [894, 895].
687. Köhler, W.: Werkst. u. Korrosion 6 (1955) 478. *297*
Koelliker, H.: See [373].
688. Kölsch, E.: Diplom-Arbeit T. H. Braunschweig 1957. *199*
690. Körber, F., and A. Krisch: Handbuch der Werkstoffprüfung, Editor: E. Siebel, Vol. II, Berlin: Springer 1939, 40, 95. *170, 202*
691. Köster, W.: Z. Metallkde. 39 (1948) 1. *13*
—: See [329, 509].
Köstlin, K.: See [362, 363].
692. Kogan, V. A., and A. A. Semionov: Zh. Fiz. Khim. (USSR) 37 (1963) 802. *165*
692a. —, and A. A. Semionov: Russian Castings Prod., 9 (1963) 424. *335*
692b. Kogan, V. S., and B. Y. Pines: Doklady Akad. Nauk SSSR 87 (1952) 771: According to Met. Abstr. 21 (1953/54) 246. *59*
Kohlmeyer, J.: See [565].
693. Kohlmeyer, E. J.: Chem. Z. 105 (1912) 993. *295*
694. —: Metall u. Erz 29 (1932) 108. *74*
694a. Kohlmeyer, J.: Dissertation T. H. Braunschweig 1955, 63. *146*
695. Kohlrausch, F.: Praktische Physik, II, Stuttgart: Teubner 1956. *12, 13, 19, 21, 202*
Komarek, K.: See [885, 1101a].
Komatsubara, H.: See [530].
Komnik, Y. F.: See [929a].
Kondic, V.: See [1297].
Kondo, A.: See [352, 353].
Kondraschow, J. D.: See [1047].
Kondrat'ev, P. A.: See [108a, 1216a].
696. Konno, S.: Phil. Mag. 40 (1920) 542. *19*
697. Konozenko, I. D., and V. I. Ust'yanov: Zhur. Tekhn. Fiziki 27 (1957) 1686. Met. Abstr. 25 (1958) 772. *81*
Korchmarek, I. A.: See [666].
698. Kornfeld, M. O.: Physik. Z. Sowjetunion 7 (1935) 432. *185, 186*
699. Kowalski, Z.: Zentralinstitut f. Schweißtechnik (ZIS): 1. Intern. Metallspritzenkonferenz in Halle/Saale 1956. *462*
Kramer, J. J.: See [102b].
700. Krause, H.: Metallfärbung, Berlin: Springer 1937. *338*
701. —: In O. Bauer, O. Kröhnke and G. Masing: Die Korrosion metallischer Werkstoffe, Vol. III, Leipzig: Hirzel 1940, 392. *338*
702. Krautmacher, H., and W. Püngel: Stahl u. Eisen 77 (1957) 837. *322*
703. Krekeler, K., and K. Steinemer: Metallspritzen, Werkstattbücher No. 93, Berlin/Göttingen/Heidelberg: Springer 1952. *461*
704. Kremann, R., and J. Gmachl-Pammer: Z. Metallkde. 12 (1920) 358. *65*
705. —, and F. Hofmeier: Sitzungsber. Akad. Wiss. Wien 297 (1911) 70. *116, 117, 120*
706. —, and A. Langbauer: Z. anorg. allg. Chem. 127 (1923) 239. *130*
707. Krenkler, K.: Werkst. u. Korrosion 10 (1959) 1. *465*
Krishnan, A. A.: See [991b].
708. Krivolapova, E. V., and B. N. Kabanov: Met. Abstr. 23 (1955/56) 557. *355*
Krisch, A.: See [690, 970].
Kröner, E.: See [640].

709. KRÖNER E.: Elektr. Nachr.-Techn. 12 (1935) 113. *34, 104, 195*
710. KROLL, W.: Metall u. Erz, 19 (1922) 13, 317. *175*
711. —: Techn. Zbl. prakt. Metallbearb. 47 (1937) 180. *66*
712. —: Z. Metallkde. 30 (1938) 373. *48*
713. KRONBERG, M. L., and F. H. WILSON: J. Metals 1 (1949) 501. *188*
KRUMBHAAR, W.: See [1121].
714. KRYSKO, W. W.: Erzmetall 6 (1953) 442. *334*
714a. —: Trans. AIME 244 (1962) 819. *71*
714b. —: Erzmetall 15 (1962) 545. *338*
714c. —: Werkst. u. Korrosion 15 (1964) 631; 726; 991. *456*
714e. —, and J. M. NEWBURN: Acta Tech. Acad. Sci. Hung. 54 (1966) 125. *203*
715. KUBASCHEWSKI, O., and E. Ll. EVANS: Metallurgical Thermochemistry, London: Pergamon Press 1956. *15*
716. —, and R. HÖRNLE: Z. Metallkde. 42 (1951) 129. *65*
—: See [1253].
KUDRIK, V. I.: See [364a].
717. KÜHNEL, R.: Gießerei 15 (1928) 441. *384*
718. —, and A. PUSCH: Hütte (Taschenbuch d. Stoffkunde). 2nd Ed. Berlin: Ernst & Sohn 1937, 396. *338*
719. KÜNZLER, H., and H. BOHREN: Techn. Mitt. PTT No. 9 (1954). *471*
720. —, —: Techn. Mitt. PTT 32 (1954) 329. *470*
KÜMMERLE, R.: See [939].
KURMANGALIYEV, M. K.: See [970a].
721. KURNAKOV, N. S., S. A. POGODIN and T. A. VIDUSOVA: Ann. Inst Anal. phys.-chim. (Leningrad) 6 (1933) 266. *65*
722. —, —, —: Izvest. Sekt. Fiziko-Khim. Anal. 15 (1947) 74. *64*
KURODA, K.: See [499].
723. KURREIN, H.: Z. VDI 76 (1932) 300. *464*
724. KURTEPOV, M. M.: Werkst. u. Korrosion 6 (1955) 40. *66*

725. LAIDLER, D. S.: Met. Ind. 83 (1953) 184. *274*
LANDAU, G.: See [452].
726. LANDER, J. J.: J. Electrochem. Soc. 99 (1952) 467. *350, 351*
727. —: J. Electrochem. Soc. 103 (1956) 1. *348*
728. —: J. Electrochem. Soc. 105 (1958) 289. *356*
729. LANDOLT-BÖRNSTEIN: Physikalisch-Chemische Tabellen and Ergänzungs-Bände, Berlin: Springer 1923, 1927, 1931, 1935, 1936. *11, 12, 14, 17, 269*
LANGBAUER, A.: See [706].
729a. LANGE, A., and B. BEYER: Neue Hütte 8 (1963) 585. *136*
730. LANGE, H., H. PARTHEY and I. N. STRANSKI: Grenzschichtprobleme beim Löten, Mitteil. d. Forschungsgesellschaft Blechverarbeitung e. V. No. 19. *471*
LANGE, K.: See [663].
LANGE-EICHHOLZ, K.-H.: See [621].
731. LAPKAMP, K., and D. MAGNUS: Nachrichtentechn. Z. 9 (1956) 415. *318*
LAPSLEY, J. T.: See [1192a].
732. LARIKOV, L. N.: Doklady Bolgar. Akad. Nauk 9 (1956) 53, 65. *99*
LARRIS, F.: See [112].
LARSON, A. C.: See [1233a].
LARSSON, L. E.: See [113].
733. LATIN, A.: Engineering 170 (1950) 121, J. Inst. Metals 81 (1952/53) 529. *411*
LAU, H.: See [271].
734. LEACH, R. H.: Met. Ind., London 49 (1936) 511. *475*

735. Lead, Nov. (1937) 6, 6. *431*
735a. Lead Abstr. 4 (1964) 174. *283*
735b. Lead. Proc. First Inter. Conf. on Lead, London 1962, Section 1. *404, 418*
735c. Lead Development Assoc. July 1964, 6. *471*
736. Lead Industries Association: Lead in Modern Industry, New York 1952, 92. *468*
737. Lead Industries Association: Lead 22 (1958) 2. *468*
738. Lead Ind. Development Council Lond. Lead Lined Tanks, Techn. Publ. Bull. No. 7. *469*
LEBEDEFF, Y. E.: See [77].
739. LEBRUN, C.: Galvano (Paris) 91 (1939) 11; Leuchs, DRP 38193, 1886. *463*
740. LEITGEBEL, W.: Z. anorg. allg. Chem. 202 (1931) 305. *14*
741. —, and E. MIKSCH: Metall u. Erz 31 (1934) 290. *75*
741a. LENEL, F. V.: Powder Met. 10 (1962) 119. *450*
742. —, and L. W. BAUM: Lead Powder Metallurgy, LP-8-59, Rensselaer Polytechnic Institute, Research Division, Troy, New York. *450*
743. —, and G. S. ANSELL: J. Metals 12 (1960) 376. *450*
LEROUX, A.: See [344].
LESSER, R.: See [284a].
744. LEUCHS, G.: D. R. P. 38193, 1886. *463*
745. LEVI-MALVANO, M., and O. CECCARELLI: Gazz. chim. ital. 41 II (1911) 269. *173*
745a. LEVIN, I. A., and E. M. SOLOMON: Khim. Prom. (USSR) 1963 69. *290*
LEWIS, A. M.: See [1152a].
LEWIS, E. J.: See [81].
746. LEWIS, W. R.: Int. Tin Res. Develop. Council, Publ. No. 93 (1939) 3. *470*
747. LEY, T. B., and H. HANEMANN: Z. Metallkde. 34 (1942) 105. *231*
—: See [566].
748. LIDDIARD, E. A. G., P. E. BANKES and L. DE BROUCKÈRE: J. Soc. Chem. Ind. 63 (1944) 39. *303*
749. LIEBSCHER, E.: Neue Hütte 4 (1959) 417. *347*
LIEPINA, L.: See [928].
750. LIESER, K. H., and H. WITTE: Z. Phys. Chem. 202 (1954) 321. *58*
751. LINDNER, E.: Dissertation T. H. Dresden 1949. *176*
752. LINGANE, J. I.: Amer. Chem. Soc. 60 (1938) 724. *269*
753. LINGNAU, E.: Werkst. u. Korrosion 7 (1956) 654. *297*
754. —: Werkst. u. Korrosion 8 (1957) 225. *291*
754a. LIPPMANN, H., and O. MAHRENHOLTZ: Feststellung des gegenwärtigen Standes der Plastizitätstheorie ..., Hannover: T. H. Hannover 1960. *441*
LISSNER, O.: See [1109].
754b. LIVINGSTON, J. D.: Acta Met. 11 (1963) 1371. *19, 78*
754c. —: J. Appl. Phys. (USA) 34 (1963) 3028. *52, 78, 97*
755. LLEWELYN, G.: Metallurgia 48 (1953) 215. *111*
755a. LLOYD, A., and E. R. NEWSON: Second Intern. Conf. on Lead, Arnhem 1965, No. 3. *452*
756. LOEBE: Metallurgie 8 (1911) 7, 33. *163*
LÖHBERG, K.: See [437, 682].
757. —: Z. Metallkde. 31 (1939) 279. *395*
758. —, and E. SCHULZ: Z. Metallkde. 43 (1952) 50. *75, 166, 365*
759. LOESCHMANN, A.: Erzmetall 5 (1952) 219. *415*

760. Loeschmann, A.: Erzmetall 13 (1960) 434. *411, 414, 418*
Löw, E.: See [879].
761. Logan, K. H.: J. Res. Nat. Bur. Stand. 17 (1936) 781, Res. Paper No. 945. *308, 309, 310*
762. —, and S. P. Ewing: J. Res. Nat. Bur. Stand. 18 (1937) 361. *308, 310, 313, 454*
763. —, —, and C. D. Yeomans: J. Res. Nat. Bur. Stand., Techn. Paper No. 368, 22 (1928) 447, also K. H. Logan: Bur. Stand. J. Res. 7 (1931) 585. *308, 310, 312*
764. —, and R. H. Taylor: J. Res. Nat. Bur. Stand. 12 (1934) 119. *308, 310*
764a. Loizu, N., and R. B. Sims: J. Mechanics Physics Solids 1 (1953) 234. *395*
764b. Lommel, J. M., and B. Chalmers: Trans. AIME. 215 (1959) 499. *95*
765. Loofs-Rassow, E.: Metallwirtsch. 10 (1931) 161. *190, 194, 195*
766. Lord, A. E., and N. A. Parlee: Trans. AIME 218 (1960) 644. *56*
766a. Lottermoser, W., and J. Meyer: Metall 16 (1962) 108. *20*
767. Love, P. P., P. G. Forrester and A. E. Burke: Proc. Inst. Mech. Eng. 168 (1953/54) 29. *370*
768. Loveless, A. H., T. A. S. Davie and W. Wright: J. Roy. Techn. Coll., Glasg. 3 (1933) 57, Metal Ind., (London) 42 (1933) 614. *289*
Lowrison, G. C.: See [971].
769. Lubarsky, B., and S. J. Kaufmann: (U. S.) Nat. Advis. Cttee. Aeronautics, Techn. Note 1955 (3336) 115. *41*
770. Lucas, F. F.: Proc. Inst. Metals Div., Amer. Inst. Min. Engrs. (1927) 481. *26*
771. Ludwig, N.: Metalloberfläche [A] 5 (1951) 38. *202*
772. Ludwik, P.: Z. Metallkde. 8 (1916) 53. *189*
773. —: Z. phys. Chem. 91 (1916) 237. *204*
774. —: Z. anorg. allg. Chem. 94 (1916) 161. *204*
775. Lücke, K.: Z. Metallkde. 41 (1950) 114. *185, 186*
776. —: Z. Metallkde. 44 (1953) 370, 418. *17*
776a. —: Z. Metallkde. 52 (1961) 1. *184*
—: See [802a].
777. —, and K. Detert: Acta Met. 5 (1957) 628. *187, 188, 196*
Lüder, E.: See [1035].
779. —: Z. VDI 79 (1935) 100. *471*
781. —: Masch.-Bau Betrieb 17 (1938) 77. *470*
782. —: Handbuch der Löttechnik, Berlin: Verlag Technik 1952. *470*
Luis, A.: See [566a].
Lukas, H. L.: See [1057a].
783. Lumsden, J.: Discussions Faraday Soc. 4 (1948) 60, Thermodynamics of Alloys, The Inst. of Metals, London 1952. *107*
783a. Lund, J. A.: Am. Zinc Inst.-Lead Ind. Assoc. LM-26-62 1962. *465*
783b. Lund, J. A., E. G. v. Tiesenhausen and D. Tromans: Second Intern. Conf. on Lead, Arnhem 1965, No. 1. *452*
783c. Lundquist, S., and B.-E. Lundin: Erzmetall 20 (1967) 117. *184*
785. Lunge, G., and E. Schmid: Z. angew. Chem. 5 (1892) 642; Z. anorg. Chem. 2 (1892) 451. *280, 287, 291*
Lunn, B.: See [108b].
786. —: Trans. Danish Acad. Techn. Sci. 2 (1952). *376*
Lyandres, A. Z.: See [801].

787. Maass, E.: Korrosion u. Metallsch. 1 (1925) 76. *454*
Maatsch, J.: See [567].

788. MAC CULLOUGH, G. H.: Trans. Amer. Soc. Mech. Engrs. 55 (1933) A. P. M. 55. *227*
789. MCGRAIL, F. I.: Foundry 52 (1935) 26, 80, 86, compare Gießerei 22 (1935) 598. *339*
789a. MACHONIS, A. A., and I. B. CADOFF: Trans. AIME. 230 (1964) 333. *101*
790. MACHU, W.: Moderne Galvanotechnik, Weinheim: Verlag Chemie 1954. *463, 464*
790a. MACHU, W., and M. L. A. E. S. BADAWAY: Werkst. u. Korrosion 14 (1963) 939. *463*
790b. MACK, D. J.: Am. Soc. Test. Mat. 45 (1945) 629. *284*
791. MAC VETTY, P. G.: Proc. ASTM 34 II (1934) 105. *219*
MAGEE, W. P.: See [958].
MAGNUS, D.: See [731].
792. MAHLICH, K. H.: Dissertation T. H. Braunschweig 1948. *328, 329*
—: See [568].
MAHRENHOLTZ, O.: See [754a].
MAJKA, S.: See [1279].
MALAKER, S. F.: See [595].
792a. MALMBERG, T.: J. Inst. Metals 89 (1960/61) 137. *56*
MALOTKI, H. v.: See [569, 570].
MANLY, W. D.: See [184].
793. MANNCHEN, W., and M. BAUMANN: Metall 9 (1955) 686. *58*
MARANTONIO, M.: See [385].
794. MARDER, M., and H. FARNOW: Dtsch. Kraftfahrforsch. 27 (1939): According to Chem. Zbl. II (1939) 4082. *297*
MARETTI, C.: See [935].
795. MARIN, J.: J. Appl. Mechanics 4 (1937) 21. *223*
796. MARKOVIĆ, T.: Werkst. u. Korrosion 5 (1954) 244. *311*
797. —: Werkst. u. Korrosion 6 (1955) 133. *302*
798. —, and Z. DUGI: Werkst. u. Korrosion 6 (1955) 294. *311*
799. —, and P. KIRKOV: Werkst. u. Korrosion 5 (1954) 241. *311*
799a. —, and LJ. RUBINIĆ: Werkst. u. Korrosion 12 (1961) 85. *311*
800. MASHOVETS, V. P., and I. I. DARAGAN-SUSHCHOV: Liteinoe Proizvodstvo 2 (1953) 18. *335*
801. —, and A. Z. LYANDRES: Zhur. Priklad. Khim. 21 (1948) 441. *352, 355*
802. MASING, G.: Lehrbuch der allgemeinen Metallkunde, Berlin/Göttingen/Heidelberg: Springer 1950, 293. *393*
—: See [778]
802a. —, K. LÜCKE and P. NÖLTING: Z. Metallkde. 47 (1956) 64. *187*
MASKREY, J. T.: See [596].
803. MASLANKA-ORMANOWA, Z.: Prace Inst. Minist. Hutn. 5 (1953) 89: According to Met. Abstr. 21 (1953/54) 716. *47*
803a. MASON, B.: California Inst. of Techn. TID-17334 1962. *3*
MASSALSKI, T. B.: See [103].
804. MATHERS, F. C.: Trans. Electrochem. Soc. 17 (1910) 261, USA P. 931944, 1909 *463*
805 —: Trans Electrochem. Soc. 38 (1930) 121. *463*
806. MATHEWSON, C. H., and W. M. SCOTT: Int. Z. Metallogr. 5 (1914) 1. *113*
807. MATIGNON, C.: C. R. Acad. Sci., Paris 154 (1912) 1609. *306*
808. MATSUYAMA, Y.: Sci. Rep. Tohoku Univ. 16 (1927) 447. *20, 41, 52, 95*
809. —: Sci. Rep. Tohoku Univ. 16 (1927) 555. *17*
810. —: Sci. Rep. Tohoku Univ. 17 (1928) 1. *362*

811. MATSUYAMA, Y.: Sci. Rep. Tohoku Univ. 18 (1929) 19. *11, 41, 52*
812. MATTHIESSEN, A.: Pogg. Ann. 110 (1860) 21, compare MAEY: Z. phys. Chem. 38 (1901) 295. *34*
813. MATTING, A., and K. BECKER: Schweißen u. Schneiden 6 (1954) 127. *461*
—: See [1233].
MAUKSCH, W.: See [276].
813a. MAYER, D., and R. FAIVRE: Compt. Rend. Paris 252 (1961) 1601. *190*
814. MAYER-WEGELIN, F.: Diplom-Arbeit T. H. Braunschweig 1957. *84, 85, 200*
815. McCONNELL, V. P.: Plating 41 (1954) 636, 645. *471*
McHARGUE, C. J.: See [103].
816. McKELLAR, W. M. G.: J. Soc. Chem. Ind. 40 (1921) 137. *289*
McKENZIE, J.: See [1245a].
McKEOWN, J.: See [135a].
817. —: J. Inst. Metals 60 (1937) 201. *230, 237*
818. —: British Non-Ferrous Metals Research Association, Research Monograph No. 5, London 1948. *260, 471*
820. —: Intern. Conf. on Fatigue of Metals, London 1956. *245, 249, 253*
821. —, and L. M. T. HOPKIN: Metallurgia 41 (1950) 135. *241, 242*
822. —, —: Metallurgia 41 (1950) 219. *242, 246*
823. McLEAN, D.: J. Inst. Metals 81 (1952/53) 293. *211*
824. —: Rev. Metallurg. 53 (1956) 139. *212*
825. —: Metal Treatment and Drop Forging 23 (1956) 3; 55; 91. *209*
826. —, and M. H. FARMER: J. Inst. Metals 85 (1956/57) 41. *210, 211*
827. —: Grain Bounderies in Metals, Oxford: Clarendon Press 1957. *210*
828. McMASTER, J. G.: Erzmetall 5 (1952) 363. *73*
829. McQUILLAN, W. P.: Amer. Machinist 100 (1956) 81. *471*
830. MEADS, P. F., W. R. FORSYTHE and W. F. GIAUQUE: According to Met. Abstr. 8 (1941) 319. *15*
MEAKIN, J. D.: See [313].
MEHL, R. F.: See [19, 626].
MEIER, A. V.: See [221a].
MEIER, G.: See [571].
MEISSNER, W.: See [583].
831. —, and H. FRANZ: Z. Phys. 65 (1930) 31. *19*
832. —, —, and H. WESTERHOFF: Ann. Phys., Leipzig 17 (1933) 593. *37*
MEISTER, H.: See [1227d].
MELFORD, D. A.: See [536].
833. —, and T. P. HOAR: J. Inst. Metals 85 (1956/57) 197. *17*
MENDIZZA, A.: See [212].
MENZEL, F.: See [453].
MERKUS, P. J.: See [34].
834. MERZ, A., and E. BRENNECKE: Z. Metallkde. 22 (1930) 185. *472*
835a. Met. Handbook 1961, 8th Edition, Vol. 1. The American Society for Metals, Cleveland 3/Ohio. *60, 369, 393, 470*
835b. Met. Handbook Edition 1955, The American Society for Metals, Cleveland 3/Ohio. *61, 128*
836. Metal Ind. 82 (1953) 360. *335*
837, Metal Ind. 83 (1953) 457, 477. *357*
838. Metal Ind., London 51 (1937) 185. *408*
839. Metal Ind., London 48 (1936) 579, 595. *462*
839a. Metall 17 (1963) 1240. *338*
841. Metallges. AG Frankfurt: DRP. 630666, Kl. 40b (1936). *63*

842. Metallges. AG Frankfurt: Statistische Zusammenstellungen 41. Jahrgang 1938—1952, Frankfurt/Main 1953. *2, 3*
844. Metallges. AG: Metallstatistik 1954—1963, 51. Jahrgang, Frankfurt/Main 1964. *1, 3*
845. Metallwirtsch. 16 (1937) 679. *471*
846. Metallwirtsch. 16 (1937) 862. *471*
847. MEYER, B., and H. HANEMANN: Z. Metallkde. 30 (1938) 422. *426*
848. MEYER, F. X.: Abh. Gesamtgebiet Hyg., No. 16 (1934). *305, 306*
849. MEYER, H.: Dissertation T. H. Braunschweig 1957. *208, 232*
849a, MEYER, H. W., and F. D. RICHARDSON: Bull. Inst. Min. Met. 71 (1962) 201. *176*
MEYER, J.: See [766a].
MEYER, U.: See [159].
850. MICHALKE: Elektr. Kraftbetr. u. Bahn 7 (1909) 226. *314, 315*
851. MIETHKE M., and G. WITT: Molkerei-Ztg. 48 (Hildesheim), (1934) 842: According to Chem. Abstr. 28 (1934) 6416 *296*
852. MIKRYUKOV, V. E., and N. A. TYAPUNINA: Fizika Metallov i Metallovedenie 3 (1956) 31. *41*
MIKSCH, E.: See [741].
MILATZ, W.: See [351].
853. MILAZZO, G.: Elektrochemie, Wien: Springer 1952. *269*
854. MILES, G.: J. Soc. Chem. Ind. 67 (1948) 10. *305, 306*
MILEY, H. A.: See [962].
MILLER, E.: See [1101a].
MILLER, W. E.: See [643].
855. MILLER, E., K. KOMAREK and I. CADOFF: Trans. AIME 218 (1960) 382. *101*
MILLIGAN, A. G.: See [222].
MILLS, G. J.: See [1034].
856. MILNER, D. R.: Brit. Weld. J. 5 (1958) 90. *472*
857. MIMA, G., S. HORI, T. MITSUI and S. IMOTO: Nipp. Kinzoku Gakkai-Si 15 (1951) 585. *189*
858. MISES, R. v.: Göttinger Nachr. Mathematik u. Physik (1913) 582. *399, 441*
MISETTI, R : See [935]
859. MITANI, Y.: Rep. Osaka Pref. Ind. Research Inst. 3 (1951) 36. *40*
MITSUI, T.: See [857].
860. MOHLER, J. B.: Metal Ind., London 86 (1955) 375. *387*
861. —: Designing for Soldered Joints Machine Design, June 14 (1956). *475*
862. MOLNAR, A.: C. R. Acad. Sci., Paris 190 (1930) 587. *189*
863. MONDAIN-MONVAL, P., and G. GABRIEL: Bull. Soc. chim. France, 7 (1940) 113. *173*
MONGINI, L.: See [26a].
MOORE, H. F.: See [36].
864. MOORE, H. F., and N. I. ALLEMAN: Univ. Illinois Bull. 29 (1932) No. 48. *238*
865. —, B. B. BETTY and C. W. DOLLINS: Univ. Illinois Bull. 32 (1935) No. 23. *210, 227, 231*
866. —, —, —: Univ. Illinois Bull. 35 (1938) No. 102. *209, 219, 230, 231, 232, 237, 239, 241, 248, 258, 259, 262*
867. MOORE, A., J. GRAHAM, G. K. WILLIAMSON and G. V. RAYNOR: Acta Met. 3 (1955) 579. *60*
MOORE, J. C.: See [360a].
868. MORGAN, J. H.: Corrosion Technol. 5 (1958) 347. *301*

869. MORGEN, R. A., L. G. SWENSON, F. C. NIX and E. H. ROBERTS: Trans. AIME Techn. Publ. 43 (1927). *131, 165, 168*
870. MORGENTHALER, R.: Erzmetall 7 (1954) 358. *68*
MORITZ, G.: See [1169].
871. MORRAL, F. R., and J. L. BRAY: Trans. Electrochem. Soc. 75 (1939) Preprint 6. *301*
872. MORRIS, W., W. A. TILLER, J. W. RUTTER and W. C. WINEGARD: Trans. Amer. Soc. Metals 47 (1955) 463. *96*
MORRIS-JONES, W.: See [1135].
873. MOTT, N. F.: Phil. Mag. Ser. VII 42 (1951). *214*
874. —: Phil. Mag. Ser. VII 44 (1953) 187. *213*
MOTZ, H.: See [1023].
MRDJENOVICH, R.: See [1197a].
MÜLLER, A.: See [356, 357].
MÜLLER, E. A.: See [366c].
876. MÜLLER, E.: Hydraulische Pressen u. Druckflüssigkeitsanlagen, Vol. III, Berlin/Göttingen/Heidelberg: Springer 1959. *430, 431*
877. MÜLLER, H.: Z. Metallkde. 21 (1929) 305. *382*
MÜLLER, H. K.: See [160].
878. MÜLLER, J.: Gas- u. Wasserfach 92 (1951) 39. *307*
MÜLLER, R.: See [572].
879. MÜLLER, W. J., and E. LÖW: Gas- u. Wasserfach 82 (1939) 82. *319*
880. —, and W. MACHU: Monatshefte Chem. 63 (1933) 347. *276*
881. —, —: Korrosion u. Metallsch. 16 (1940) 187. *276*
881a. MULHEARN, T. O., and D. TABOR: J. Inst. Metals 89 (1960) 7. *203*
882. MULLARKEY, E. J.: Corrosion (Houston) 15 (1959) 98. *295*
MUNAKATA, S.: See [1048].
883. MUNDEY, A. H., C. C. BISSETT and J. CARTLAND: J. Inst. Metals 28 (1922) 141. *365*
884. MURPHY, W. K., and R. A. ORIANI: Acta Met. 6 (1958) 556. *96*
MUSCOTT, A.: See [932].
885. MUYLDER, VAN J., and M. POURBAIX: Proc. 6th Meeting Internat. Cttee. Electrochem. Thermodynamics and Kinetics, Poitiers 1954, (1955) 342. *316*
886. MUZAFFAR, S. D., and R. CHAND: J. Amer. Chem. Soc. 66 (1944) 1374. *135*
MYGAGKOVA, A. R.: See [983a].

887. NABARRO, F. R. N.: Rep. Conf. on Strength of Solids Phys. Soc., London 1948, 75. *212*
888. NACHTRIEB, N. H., and G. S. HANDLER: J. Chem. Physics 23 (1955) 1569. *15*
889. NACHTIGALL, G.: Gas- u. Wasserfach 75 (1932) 941. *306*
890. NAGASAKI, S., and E. FUJITA: Nipp. Kinzoku Gakkai-Si 16 (1952) 313. *50*
NAPOLITAN, D. S.: See [1091].
891. NAUMANN, E.: Gas- u. Wasserfach 79 (1936) 214. *307*
892. NEISH, R. A.: USA P. 2643975, 1953: Werkst. u. Korrosion 5 (1954) 470. *454*
893. NEMILOV, V. A., and T. A. STRUNINA: Zhur. Priklad. Khim. 19 (1946) 449. *144*
NEUMAYR, S.: See [1306].
894. NEURATH, P. W., and J. S. KOEHLER: (U. S.) Nat. Advis. Cttee. Aeronautics Tech. Note (2322) 1951, 32. *208*

895. NEURATH, P. W. and J. S. KOEHLER: J. Appl. Physics 22 (1951) 621. *177*
NEWBURN, J. M.: See [714e].
NEWSOM, E. R.: See [755a].
NICHOLLS, J. M.: See [382a].
896. NICLASSEN, H.: Forschungsarbeiten zur Metallkde. No. 7., Berlin: Borntraeger 1922. *154*
897. NICOLAUS: Z. Metallkde. 17 (1925) 388. *361*
898. NIGHTINGALE, S. J.: Tin-Solders British Non-Ferr. Met. Res. Assoc., Research Monograph 1, London (1932). *475*
898a. NIKITIN, D. G., and V. Y. YURENKO: Byul. Tekhn. Ekon. Inform. Min. Morsk. Flota (USSR) (1963) No. 11; 21. *456*
899. NISHIHARA, K.: J. Elektrochem. Soc. Japan 17 (1949) 100, According to C. A. 48 (1954) 13479. *301*
900. NIWA, K., M. SHIMOJI, S. KADO and Y. WATANABE: Nipp. Kinzoku Gakkai-Si 18 (1954) 271. *134*
901. —, —, —, — and T. YOKOKAWA: Trans. AIME 209 (1957) 96. *41, 52, 80, 95*
NIX, F. C.: See [239, 869].
902. —, and E. SCHMID: Z. Metallkde. 21 (1929) 286. *336*
903. NODDACK, I., and W.: Z. angew. Chem. 49 (1936) 1. *3*
NÖLTING, P.: See [802a].
904. NORBURY, A. L.: Trans. Faraday Soc. 19 (1923) 140. *189*
905. Normblatt DIN 50910 April 1955. *315*
NOWACKI, A.: See [22a].
NOWICK, A. S.: See [248a, 307].
906. NOWOTNY, H.: Z. Metallkde. 33 (1941) 388. *62*
907. NOZATO, R., and K. IGAKI: Bull. Naniwa Univ. (Japan) A 3 (1955) 125. *94*

OAKLEY, A. L. J.: See [400a].
908. OBINATA, I., and E. SCHMID: Metallwirtsch. 12 (1933) 101. *78*
OBROWSKI, W.: See [364].
909. OBST, W.: Korrosion u. Metallsch. 7 (1931) 12. *295*
910a. ODQVIST, F. K. G., and J. HULT: Kriechfestigkeit metallischer Werkstoffe. Berlin/Göttingen/Heidelberg: Springer 1962. *222, 223*
911. Öl und Kohle 11 (1935) 259. *297*
912. ÖLANDER, A.: Z. Phys. Chem. A 168 (1934) 274. *105*
OELSEN, O.: See [916, 918].
OELSEN, W.: See [1170].
913. —: Arch. Eisenhüttenw. 26 (1955) 519. *50, 53*
914. —, and R. BENNEWITZ: Arch. Eisenhüttenw. 29 (1958) 663. *40*
915. —, F. JOHANNSEN and A. PODGONRIK: Erzmetall 9 (1956) 459. *76*
916. —, O. OELSEN and D. THIEL: Z. Metallkde. 46 (1955) 555. *15*
917. —, E. SCHÜRMANN and G. HEINRICHS: Arch. Eisenhüttenw. 30 (1959) 649. *56*
918. —, —, H. J. WEIGT and O. OELSEN: Arch. Eisenhüttenw. 27 (1956) 487. *50*
918a. OFTE, D., and L. J. WITTENBERG: Trans. AIME 227 (1963) 706. *18*
OGAWA, K.: See [11a].
OHLSSON, E.: See [112].
919. OKKERSE, B.: Acta Met. 2 (1954) 551. *16*
920. —, T. J. TIEDEMA and W. G. BURGERS: Acta Met. 3 (1955) 300. *16*
O'NEILL, H.: See (657].
921. —, and G. S. FARNHAM: J. Inst. Metals 57 (1935) 253. *326*
922. OPIE, W. R., and N. J. GRANT: Trans. AIME 191 (1951) 244. *58*
ORIANI, R. A.: See [884].
ORLAMÜNDER, E.: See [671].

923. Ornstein, L. S., D. J. Haringhuizen and D. A. Was: Met. Abstr. 5 (1938) 95. *298*
—: See [351].
924. Orowan, E.: Proc. Inst. Mech. Eng. 150 (1943) 140. *441*
925. —: West of Scotland, Iron and Steel Inst. 54 (1947) 45. *208*
Orr, C. W.: See [430, 431].
Orr, R. L.: See [250].
926. Osborg, H.: Metal Progr. 33 (1938) 43. *55*
927. —: USA P. 2140540, 1938. *63*
Osenberg, E.: See [1117].
927a. O'Shea, R. P., J. A. Donovan and E. A. Peretti: Trans. AIME 221 (1961) 1266. *134*
928. Ošis, Z., and L. Liepina: Latvijas PSR Zinātnu Akad. Vēstis (1954) 125: According to Met. Abstr. 23 (1955) 460. *302*
928a. Ostergaard, P. B.: Second Intern. Conf. on Lead, Arnhem 1965, No. 9. *467*
Ostrovskii, B. S.: See [750a].
929. Otten, F.: Siemens-Z. 16 (1936) 348. *404*

929a. Palatnik, L. S., and Y. F. Komnik: Phys. Met. Metallography 10 (1960) 141. *14*
930 Panknin, W : Verpackungs-Magazin (1959) 94. *471*
931. Parker, R.: Trans. Am. Soc. Metals 50 (1958) 52. *211*
Parkins, R. N.: See [940].
Parlee, N. A.: See [766].
932. Parr, N. L., A. Muscott and A. J. Crocker: J. Inst. Metals 87 (1958/59) 321. *349, 351*
933. Parravano, N.: Int. Z. Metallogr. 1 (1911) 89. *114*
934. —: Gazz. chim. ital. 44 (1914) 382. *143*
935. —, C. Maretti and R. Misetti: Gazz. chim. ital. 44 (1914) 478. *154*
936. —, and G. Sirovich: Gazz. chim. ital. 42 I (1912) 630. *176, 388*
Parthey, H.: See [730].
937. Pasternak, A.: Bull. intern. acad. polon. sci., Classe sci. math. nat., sér. A (1951), 177. *51*
938. Patterson, W., H. Brand and H. Trassl: Gießerei, techn.-wiss. Beih. (1960) No. 27, 1477. *321*
939. —, and R. Kümmerle: Gießerei 46 (1959) 897. *334*
Pauling, L.: See [1172].
Pavlova, M. V.: See [384].
Pawlek, F.: See p. 146.
940. Pearson, C. E.: The Extrusion of Metals, London: Chapman & Hall 1944; *390, 391, 393, 394, 397*
—, and R. N. Parkins: Second Revised Ed. 1960. *426, 429, 447*
941. —, and I. A. Smythe: J. Inst. Metals 45 (1931] 345. *391*
942. Pearson, W. B., and I. M. Templeton: Phys. Rev. 109 (1958) 1094. *19*
943. Pekrun, M.: Dissertation T. H. Braunschweig 1959. *365, 372, 376, 377, 378, 379*
944. Pelzel, E.: Metall 9 (1955) 692. *53, 69, 70*
945. —: Metall 10 (1956) 717. *66, 94*
946. —: Erzmetall 9 (1956) 17. *332*
947. —: Metall 11 (1957) 667. *144*
948. —: Z. Metallkde. 49 (1958) 102. *87*

949. PELZEL, E.: Z. Metallkde. 49 (1958) 236. *85, 205*
950. —: Metall 13 (1959) 552. *57*
951. —: Z. Metallkde. 50 (1959) 649. *329*
—: See [579, 579a].
951a. PEPLOW, D. B., and L. L. SHREIER: Corrosion Technol. 11 (1964) 16. *301*
PERETTI, E. A.: See [927a].
952. PERRY, R. I.: Corrosion (Houston) 12 (1956) 207. *295*
953. PERSHKE, V.: Chim. et Ind., 14 Congrès, Paris 1934 13. *288*
PETRI, H.-G.: See [573].
954. —: unpublished. *98*
955. PFANN, W. G.: Metallurgical Reviews 2 (1957) 29. *9*
956. PFANHAUSER, W.: Galvanotechnik, I u. II, Leipzig: Akad. Verlagsges. Becker u. Erler 1941. *463*
957. PFENDER, M., and R. SCHULZE: Z. Metallkde. 36 (1944) 94. *109, 251, 255, 417*
PFISTER, H.: See [395].
958. PHELPS, H. S., F. KAHN and W. P. MAGEE: Proc. ASTM 48 (1948) 815. *34*
PHILLIPS, A.: See [325, 326].
959. PHILLIPS, A. J.: Proc. ASTM 36 II (1936) 170. *220, 237*
960. —, A. A. SMITH and P. A. BECK: Proc. ASTM 41 (1941) 886. *383*
PHIPPS, G. S.: See [118].
PICHINOTY, F.: See [1108].
PIEPER, G.: See [574].
961. —: Thesis T. H. Braunschweig 1959. *35*
962. PIETENPOL, W. B., and H. A. MILEY: Phys. Rev. 34 (1929) 1588; Chem. Z. (1930) I 1601. *19*
963. PILLING, N. B., and R. E. BEDWORTH: J. Inst. Metals 29 (1923) 592. *327, 328*
PINES, B. Y.: See [692b].
PINKERTON, H. L.: See [412a].
PIONTELLI, R.: See [168].
964. —, and G. POLI: Z. Elektrochem. 62 (1958) 330. *269*
965. PIRELLI: In: W. EHLERS and H. LAU: Kabel-Herstellung, Berlin/Göttingen/Heidelberg: Springer 1956, 281. *408*
PIXNER, S.: See [109].
966. PLÜSS, M.: Z. anorg. allg. Chem. 93 (1915) 1. *80*
PLUMTREE, A.: See [657a].
967. PLYM, L. M.: Corrosion (Houston) 12 (1956) 331. *316*
PODGONRIK, A.: See [915].
968. POGODIN, S. A., and E. S. SHPICHINETSKY: Izvest. Sektera Fiz.-Khim. Anal. 15 (1947) 88. *62*
—: See [721, 722].
POLI, G.: See [964].
POMINA, M. A.: See [108a].
POMP, A.: See [1118].
969. —: Handbuch der Werkstoffprüfung, edited by E. SIEBEL, Vol. II, Berlin: Springer 1939, 232. *219*
970. —, A. KRISCH and G. HAUPT: Mitt. Kais.-Wilh.-Inst. Eisenforschg. 21 (1939) 219, 231. *199, 203, 206*
970a. PONOMAREV, V. D., and M. K. KURMANGALIYEV: USSR Abstr. of Metallurgy 1 Part A 9/10 (1959) 51. *172*
PONOMAREW, W. D.: See [1121a].

971. PORTER, J. J., and G. C. LOWRISON: Industrial Chemist 30 (1954) 369. *292*
972. PORTEVIN, A.: See KLEINHEISTERKAMP [670]. *124*
973. —, and P. BASTIEN: J. Inst. Metals 54 (1934) 45. *334*
974. —, and A. SANFOURCHE: Chim. et Ind., Sondernummer (1932) 360. *292*
POURBAIX, M.: See [243, 885].
975. PRANDTL, L.: Z. angew. Math. Mech. 3 (1923) 401. *399*
976. PRASAD, S. C., and W. A. WOOSTER: Acta Cryst. 9 (1956) 38. *12*
PRATT, J. N.: See [14b].
977. PRATT, W. E., and E. T. COLLINSWORTH: Corrosion (Houston) 5 (1949) 39. *319*
PRECKSHOT, G. W.: See [409].
978. PREDEL, B.: Z. Metallkde. 50 (1959) 663. *58*
979. —: Z. Metallkde. 51 (1960) 381. *95*
979a. PREDEL, B.: Z. Metallkde. 52 (1961) 507. *58*
979b. PREEKSHOT, G. W., and R. E. HUDRLIK: Trans. AIME 218 (1960) 516. *34*
PRESTON, G. D.: See [86].
980. PRESTON, C. T., and G. H. BROOMFIELD: J. Inst. Metals 87 (1958/59) 240. *41*
PRICE, E. G.: See [1002a]
PRINCE, M Y.: See [1233a].
PRÖPSTL, G.: See [513].
981. PRÜMM, W.: Felten u. Guilleaume Rdsch. 13/14 (1934) 20. *404, 406*
982. PRUTTON, C.F., and J.H. DAY: J. Physic Colloid Chem. 53 (1949) 1101. *298*
PRZYBYLA, E.: See [1127].
PÜNGEL, W.: See [702].
982a. PUGH, H. L. D.: J. Inst. Metals 90 (1961/62) 203. *391*
PUSCH, A.: See [718].
983. PUSCHIN, N. A., S. STEPANOVIĆ and V. STAJIĆ: Z. anorg. Chem. 209 (1932) 329. *57*
983a. PUTILOVA, I. N., and A. R. MYAGKOVA: Zhur. Fiz. Khim. 35 (1961) 1994. *293*
984. PUTNAM, G. L.: Indst. and Eng. Chem. 30 (1938) 1138. *61, 69*

985. QUADRAT, O., and J. JIŘIŠTĚ: Chim. and Ind. Paris (1934) 485, Chem. listy 29 (1935) 304. *76*
QUIMBY, S. L.: See [1255].

986. RABALD, E.: Chem. Fabrik 11 (1938) 293, Dechema Monogr. 11 (1939) 17. *291*
987. —: Werkst. u. Korrosion 7 (1956) 652. *281, 282*
989. RACHINGER, W. A.: J. Inst. Metals 81 (1952) 33. *211*
990. RADLEY, W. G.: Electr. Engng. 57 (1938) 168. *315*
991. RADTKE, S. F.: J. Metals 12 (1960) 486. *409*
991a. RADTKE, S.: Metall 16 (1962) 758. *467*
991b. —: ILZRO Research Digest No. 12, 7, October 1963. *470*
RAMACHANDRAN, V.: See [991b].
991c. RAMAN, K. S., A. A. KRISHNAN and V. RAMACHANDRAN: Ind. J. Technol. 1 (1963) 471. *471*
RAMSEY, W. J.: See [1046a].
992. RAPP, A., and H. HANEMANN: Z. Metallkde. 33 (1941) 64. *385*
RASMEROVA, N. J.: See [308].
RASSOW, E.: See [228].
RATCLIFF, N. A.: See [360a, 1012a].

993. RAUB, E.: Metalloberfl. Ausg. A 7 (1953) 17. *33*
993a. —, and A. DISAM: Metalloberfl. 16 (1962) 317. *38*
994. —, and M. ENGEL: Z. Elektrochem. 49 (1943) 89. *33*
995. RAYNOR, G. V.: Equilibrium Diagram of the System Lead-Tin, London: The Institute of Metals 1947. *95*
996. —: The Equilibrium Diagram of the System Antimony-Lead. London: The Institute of Metals 1951. *76*
997. —, and J. GRAHAM: Trans. Faraday Soc. 54 (1958) (2) 161, Met. Abstr. 26 (1958) 36. *175*
—: See [867, 1210].
998. READ, JR., W. T.: Dislocations in Crystals, New York: McGraw-Hill 1953. *179*
REDLICH, O.: See [1, 2, 3].
998a. REICHEL, C. J., and C. R. SIMCOE: Expand. Res. Progr. No. 11 (1963) 3. *63*
999. REICHENECKER, W. J.: Metal Progress 68 (1955) 110. *383*
1000. REINACHER, G.: Z. Metallkde. 47 (1956) 607. *26*
1001. REINER, S.: Z. Metallkde. 30 (1938) 277. *315*
1002. REINGLASS, P.: Chem. Techn. d. Leg. 1. Teil, Leipzig: Spamer 1919. *98*
1003. REININGER, H.: Gespritzte Metallüberzüge, München: Hanser 1952. *461*
1004. REINITZ, B. B.: Corrosion (Houston) 9 (1953) 425. *311*
RESTLE, H.:.See [680].
1004a. REUTEL, C.: Proc. First Intern. Conf. on Lead, London 1962, Chap. 11, 113. *342*
1005. REY, M., P. COHEUR and H. HERBIET: Trans. Electrochem. Soc. 73 (1938) 315. *299*
1006. RHINES, F. N., and A. W. COCHARDT: (U. S.) Nat. Advis. Cttee. Aeronautics Techn. Note 2746 (1952) 40. *210*
1007. RHINES, F. N., and W. TIMPE: Z. Metallkde. 48 (1957) 109. *33*
RHUDE, H. V.: See [1273].
1007a. RHYS, D. W., and E. G. PRICE: Metal Ind. 105 (1964) 243. *326*
1008. RICHARDS, TH. W., and CH. WADSWORTH: J. Amer. Chem. Soc. 38 (1916) 221, 1658. *10*
RICHARDSON, F. D.: See [230, 849a].
1009. RICHARDSON, W. H., V. WHITE and B. A. WEAVER: Phys. Rev. 61 (1942) 728. *99*
1010. RICHTER, F.: Metall 7 (1953) 520. *357*
1011. RICHTER, K.: Zink, Zinn u. Blei. Wien and Leipzig 1927. *81*
1011a. RIECHE, H.: Dissertation T. H. Braunschweig 1967. *223*
1011b. RIEMANN, H.: Techn. Mitt. Krupp-Werksberichte 20 (1962) 125. *322*
RITCHIE, L. M.: See [1225].
1012. RITTER, F.: Korrosionstabellen metallischer Werkstoffe. Wien: Springer 1952. *298*
1012a. ROBERTS, D. H., N. A. RATCLIFF and J. E. HUGHES: Powder Met. (1962) 132. *452*
—: See [360a].
ROBERTS, E. H.: See [869].
1013. ROBERTS-AUSTEN, W. C.: Phil. Trans. A 187 (1896) 383; Proc. Roy. Soc. 67 (1900) 101. *472*
1014. —: Phil. Trans. Roy. Soc. 187 (1896) 404. *37*
ROBINS, D. A.: See [1126b].
1015. ROBINSON, H. A., and R. L. FEATHERLY: Corrosion (Houston) 3 (1947) 349. *272, 291, 316*

1016. Robson, W. W., and A. R. Taylor: Chem. and Ind. (1956) 1111. *312*
1017. Rodyakin, V. V.: Tsvetnye Metally 31 (1958) 43. *47*
1018. Roehl, E. J.: Iron Age 173 (1954) 140. *465*
1018a. Roemer, E.: Metall 16 (1962) 126, 766. *370*
1019. Röntgen, P., W. Hilgers and H.-J. Gottschol: Erzmetall 6 (1953) 220.
Röschenbleck, B.: See [1068].
Roeser, W. F.: See [664].
1020. Roetheli, B. E., and G. L. Cox: Ind. Engng. Chem. 23 (1931) 1084. *297*
Rogers, F. E.: See [1262].
1021. Roggendorf, W.: Metalloberfläche 9 (1955) 76. *387*
1022. Rohrmann, F. A.: Chem. Metall. Engng. 42 (1935) 368. *292*
1022a. Roll, A., and G. Fees: Z. Metallkde. 51 (1960) 540. *95*
1023. —, H. Felger and H. Motz: Z. Metallkde. 47 (1956) 707. *20*
1024. —, and E. Uhl: Z. Metallkde. 50 (1959) 159. *34, 38*
1025. Roll, K. H.: Chem. Engng. 55 (1948) 231. *291*
1026. —: Chem. Engng. 56 (1949) 227. *291*
1027. —: Chem. Engng. 56 (1949) 201. *292, 297*
1028. —: Corrosion (Houston) 7 (1951) 454. *273, 274, 282*
1029. —: Chem. Engng. 59 (1952) 281. *466*
1030. Rollason, E. C.: J. Inst. Metals 60 (1937) 35. *461*
1031. —, and V. B. Hysel: Monthly J. Inst. Metals 5 (1938) 187. *51*
Romanoff, M.: See [246].
1032. Rose, K.: Materials and Methods 43 (1956) 104. *57*
1033. Rosenberg, A., and W. A. Tiller: Acta Met. 5 (1957) 565. *45, 336*
1034. Rosenthal, F. D., G. J. Mills and F. J. Dunkerley: Trans. AIME 212 (1958) 153. *107*
1035. Rostosky, L., and E. Lüder: Z. Metallkde. 21 (1929) 24. *472*
Roters, H.: See [273].
1036. Rothman, S. J., and L. D. Hall: Trans. AIME 206 (1956) 199, Short Report in: J. Metals 8 (1956) 199. *16, 41*
1036a. Rothwell, E.: J. Inst. Metals 90 (1961/62) 389. *18, 19*
Rowe, L. N.: See [400a].
1036b. Royuela, J. J., and M. Serra: Metales No Ferros 4 (1963) 135. *301*
1036c. Rudram, A. T. S., and L. M. Warner: Met. Rev., Inst. of Metals 9 (1964) 179. *449*
Rüdiger, H.: See [1171].
1037. Rüetschi, P., and R. T. Angstadt: J. Electrochem. Soc. 105 (1958) 555. *353*
Ruoff, A. L.: See [160a].
Russell, B.: See [132a, 132b].
Russell, R. P.: See [1264].
1038. Russell, R. S.: Proc. Australas. Inst. Min. Met. 95 (1934) 125. *9, 33*
1039. —: Proc. Australas. Inst. Min. Met. N. S. 101 (1936) 33. *34, 190*
Rutter, J. W.: See [30, 30a, 30b, 31, 872, 1198].
1039a, —, and K. T. Aust: Trans. AIME 218 (1960) 682. *188, 191*
Ryabor, A. K.: See [417].
Ryjord, J. E.: See [238].
Rysselberghe, P. van: See [243].

Sachs, G.: See [272, 551].
1040. —: Praktische Metallkde. 2, Berlin: Springer 1934, 182. *422*
1041. —: Abh. d. Braunschw. Wiss. Gesellschaft 9 (1957) 36. *207*

1042. Sachs, G., and A. Draper: Microtecnic (Suisse) 7 (1959) 22. *391, 396*
1043. —, and G. Fiek: Der Zugversuch, Leipzig: Akad. Verlagsgesellschaft 1926, 65. *200*
1044. Sacklowski, A.: Ann. Phys., Leipzig 77 (1925) 241. *65*
Sadofsky, J.: See [643].
Säfsten, K. M.: See [114].
Sagel, K.: See [504a].
Salkovitz, E. I.: See [1252].
Salley, W. J.: See [157].
Salmon, P.: See [1216].
1045. Sandelin, H. F.: Tekn. T. 67 (1937) 185. *423*
1046. Sandmeier, F.: Techn. Mitt. PTT 22 (1944) 187, 231. *293, 297, 317*
Sanfourche, A.: See [974].
Sargant, K. R.: See [1284a].
1047. Saslawski, A. J., J. D. Kondraschow and S. S. Tolkatschew: Ber. Akad. Wiss. UdSSR 75 (1950) 559. *349*
1048. Sato, T., and S Munakata: Bull. Research Inst. Mineral Dressing Met., Tohoku Univ. 10 (1954) 173. *95*
Sauerwald, F.: See [258, 366].
1050. Sauerwald, F.: Z. Metallkde. 41 (1950) 97. *12*
1051. —, and K. Knehans: Z. anorg. allg. Chem. 131 (1923) 57; 140 (1924) 227; Z. Metallkde. 16 (1924) 315. *205*
1051a. Sawyer, J. C.: Trans. AIME 227 (1963) 346. *74*
1052. Scarpa, O.: Alluminio 5 (1936) 1, 6 (1936) 85, Ref. Aluminium 18 (1936) 367. *464*
1053. Schachkeldian, A. B., and M. N. Ambroji: Abh. Staatsun. Saratow 1 (1936) 73: According to Korrosion u. Metallsch. 13 (1937) 405. *69*
1054. Schaefer, Cl.: Ann. Phys., Leipzig 4. Folge (1901) 220. *12, 13*
Schaefer, T. P.: See [171].
Schaller, I. E.: See [1064].
Schaub, C.: See [1109].
1055. Scheil, E.: Z. Metallkde. 29 (1937) 404. *337*
1056. —: Gießerei, techn.-wiss. Beih. (1958) No. 19, 1007. *12*
1057. —, and E. Jahn: Z. Metallkde. 40 (1949) 319. *35*
1057a. —, and H. L. Lukas: Z. Metallkde. 52 (1961) 417. *60, 105*
1058. —, and F. Wolf: Z. Metallkde. 50 (1959) 229. *64*
1058a. Schenck, H., and M. G. Frohberg: Erzmetall 15 (1962) 512. *53*
Scheuer, F.: See [396].
Schikorr, G.: See [60, 61].
1059. —: Z. angew. Chem. 44 (1931) 40. *305*
1060. —: Korrosion u. Metallsch. 16 (1940) 181. *271, 273, 291, 293, 294, 296, 303*
1061. —: Die Zersetzungserscheinungen der Metalle, Leipzig: Barth 1943. *271, 297*
1062. Schikorr, G.: Schweiz. Arch. angew. Wiss. Techn. 24 (1958) 33. *297*
1063. —, and K. G. Bergner: Werkst. u. Korrosion 8 (1957) 1. *448*
1064. —, and I. E. Schaller: Metallwirtsch. 20 (1941) 1135. *279, 281, 289*
1065. Schläpfer, P., and A. Bukowiecki: Schweiz. Arch. 14 (1948) 257. *298*
1066. Schlötter, M.: Korrosion u. Metallsch. 3 (1927) 30. *463*
1067. —: In O. Bauer, O. Kröhnke and G, Masing: Die Korrosion metallischer Werkstoffe 3, Leipzig: Hirzel 1940, 393. *463*
1068. Schmeling, E. L., and B. Röschenbleck: Werkst. u. Korrosion 9 (1958) 529, and 12 (1961) 215. *307*
Schmid, E.: See [397, 454, 785, 902, 908].

1069. SCHMID, E.: Z. Metallkde. 35 (1943) 85. *39, 65, 68, 386*
1070. —, and W. BOAS: Kristallplastizität, Berlin: Springer 1935. *177*
1071. —, and R. WEBER: Gleitlager, Berlin/Göttingen/Heidelberg: Springer 1953. *372*
1072. SCHMIDT, H.: Oesterr. Z. Telegr.-, Teleph.-, Funk- u. Fernsehtechn. 4 (1950) 109. *316*
1073. —, and E. FURTHMANN: Mitt. Kais.-Wilh.-Inst. Eisenforschg. 10 (1928) 225. *21*
1074. SCHMIDT, R.: Stahl u. Eisen 56 (1936) 228. *66*
1075. SCHNEEBERGER, P. E.: Bull. Ass. Suisse Electr. 29 (1938) 213. *390*
1076. SCHNEIDER, H. J.: Thesis T. H. Braunschweig, 1959. *145, 147*
—: See [575, 576].
SCHNERB, J.: See [90].
1077. SCHOPPER, H.: Z. Physik 143 (1955) 93. *38*
1078. SCHOPPER, W.: Blei, in: A. E. VAN ARKEL: Reine Metalle, Berlin: Springer 1939, 501. *9, 290*
1079. SCHRADER, A.: Ätzheft, Berlin: Borntraeger 1939. *25*
1080. —: Ätzheft, 4th Ed., Berlin: Borntraeger 1957. *27*
1081. —, and H. HANEMANN: Z. Metallkde. 29 (1937) 37. *25*
1082. —, —: Z. Metallkde. 33 (1941) 49. *149*
SCHRADER, A.: See [479, 480, 577].
SCHRAMM, G. N.: See [1226].
1082a. SCHRAMM, K. H.: Z. Metallkde. 53 (1962) 317. *12*
SCHUBERT, K.: See [448a].
1084. SCHÜNEMANN, A.: Korrosion u. Metallsch. 9 (1933) 325. *288*
SCHÜRENBERG, H.: See [295].
1085. SCHÜRMANN, E.: Arch. Eisenhüttenw. 30 (1959) 41. *50*
—: See [917, 918].
1085a. —, and F. J. GILHAUS: Arch. Eisenhüttenw. 32 (1961) 867. *95*
1086. SCHULZ, E.: Metallwirtsch. 20 (1941) 418. *26, 67*
—: See [758].
SCHULZ, E. H.: See [180].
1087. SCHULZE, B.: Dtschl. Buchdr., Jg. 46 (1934) No. 90, 880. *293*
1088. SCHULZE, F. A.: Drudes Ann. 314 (1902) 555. *41*
1089. SCHULZE, F., and J. STAEBLER: Autog. Metallbearb. 28 (1935) 68. *469*
SCHULZE, R.: See [957].
1090. SCHUMACHER, E. E., and G. M. BOUTON: Metals and Alloys 1 (1930) 405. *44, 48, 248, 258, 259*
1091. SCHUSSLER, M., A. G. DOBBINS and D. S. NAPOLITAN: U. S. Atomic Energy Commission Publ. K-1018 (1953). *291*
1092. SCHWARZ, M. Frhr. v.: Z. Metallkde. 28 (1936) 128. *128*
1093. —: Gießerei 27 (1940) 137. *128*
1094. SCHWARZ, R H, and J. H. WINKLER: Proc. Graphical Arts, Techn. Conference (Amer. Soc. Mech. Engrs.) (1936) 51. *361*
1094a. SCHWARZ, W.: Metall 10 (1956) 513, 921. *271*
SCHWEINITZ, H. D. v.: See [544].
1095. Schweißen u. Schneiden 10 (1958) 27. *475*
SCOTT, W. M.: See [806].
SCREATON, R. M.: See [171].
1096. SEBAN, R. A., and D. F. CASEY: Trans. Amer. Soc. Mech. Eng. 79 (1957) 1514. *41*

1097. Seeds, O. J.: Materials and Methods 32 (1950) 64. *388, 389*
1098. Seeger, A.: Report of a Conference on Defects in Crystalline Solids, Physical Society, London (1955) 328, in: Dislocations and Mechanical Properties of Crystals, London: Chapman & Hall 1957, 328. *10*
1099. —: In Handbuch der Physik, Vol. VII/1, Kristallphysik I, Berlin/Göttingen/Heidelberg: Springer 1955. *179*
1100. —: In Handbuch der Physik, Vol. VII/2, Kristallphysik II, Berlin/Göttingen/Heidelberg: Springer 1958. *177, 179, 184, 211, 214*
1101. —: Z. Metallkde. 45 (1954) 521. *212*
1101a. Seidman, D., I. Cadoff, I. Komarek and E. Miller: Trans. AIME 221 (1961) 1269. *94*
Seigle, L. L.: See [40].
1102. Seith, W.: Diffusion in Metallen, Berlin: Springer 1939. *34*
1103. —, and G. Helmold: Z. Metallkde. 42 (1951) 137. *118*
1104. —, and H. Johnen: Z. Elektrochem. 56 (1952) 140. *107*
1105. —, — and J. Wagner: Z. Metallkde. 46 (1955) 773. *55, 107, 135*
1106. —, and A. Keil: Z. Metallkde. 25 (1933) 104. *268*
—: See [517].
1107. Seljesater, K. S.: Trans. AIME 75 (1929) 573. *86*
Semionov, A. A.: See [692, 692a].
Semkova, E. N.: See [1201a].
Semmens, R. D.: See [400a].
1108. Senez, J. C., and F. Pichinoty: Corrosion et Anticorrosion 5 (1957) 203. *293*
Serra, M.: See [1036b].
Shachnev, V. I.: See [307a, 307b].
1109. Shanley, F. R., C. Schaub and O. Lissner: Proc. Colloquium on Fatigue, Stockholm 1955. Cf. Met. Abstr. 24 (1957) 641. *245*
Sherby, O. D.: See [1281].
1110. Sherman, W. L.: Wire and W. Prod. 7 (1932) 179, 204. *404, 423*
1110a. Shimaoka, G.: Trans. Nat. Res. Inst. Metals 1 (1959) 78. *327*
Shimoji, M.: See [900, 901].
1111. Shishokin, V. P.: In C. E. Pearson: The Extrusion of Metals, London: Chapman & Hall 1944. *393*
1112. Shooter, K. V.: J. Appl. Phys. Suppl. 1 (1951) 49. *376*
Shpichinetsky, E. S.: See [968].
Shreir, L. L.: See [951a].
Shurawlewa, T. G.: See [639].
1113. Siebel, E.: Die Formgebung im bildsamen Zustande, Düsseldorf: Verlag Stahleisen 1932. *392*
1114. —: Stahl u. Eisen 57 (1937) 413. *442*
1116. —, and E. Fangmeier: Mitt. Kais.-Wilh.-Inst. Eisenforschg. 13 (1931) 29. *391, 392, 393*
1117. —, and E. Osenberg: Mitt. Kais.-Wilh.-Inst. Eisenforschg. 16 (1934) 33. *441, 442, 443*
1118. —, and A. Pomp: Mitt. Kais.-Wilh.-Inst. Eisenforschg. 9 (1927) 157. *445*
1119. Sieglerschmidt, H., and G. Fiek: Z. Metallkde. 27 (1935) 38. *199*
—: See [62].
1120. Siemens-Schuckert-Werke: DRP 633767 (1932). *423*
1121. Sieverts, A., and W. Krumbhaar: Ber. Dtsch. chem. Ges. 43 (1910) 896. *58*
Siftar, J.: See [128].

1121a. Silina, Yu. K., and W. D. Ponomarev: Hochschulnachricht. Nichteisen-metallurg. 3 (1960) 91. *175*

Silvidi, A. A.: See [595].

Simcoe, C. R.: See [999a].

Simerska, M.: See [1203a].

1122. Simon, A. C.: Fourth Intern. Battery Symposium, Brighton 1964. *349*

1122a. —, and E. L. Jones: J. Electrochem. Soc. 100 (1953) 1. *77, 346*

1123. —, —: J. Electrochem. Soc. 101 (1954) 536. *346*

1124. —, —: J. Electrochem. Soc. 104 (1957) 133. *77*

—: See [151].

1125. Simonds, H. R.: Iron Age 134 (1934) 18. *424*

Simpkins, L. D.: See [221].

1125a. Singer, A. R. E., and S. H. K. Al-Samarrai: J. Inst. Metals 89 (1960) 225. *394, 396*

1125b. —, and J. W. Coakham: J. Inst. Metals 89 (1960) 177. *394*

—: See [33a].

1126. Singleton, W., and B. Jones: J. Inst. Metals 51 (1933) 71. *101, 102, 103, 104, 261, 416, 437*

Sirovich, G.: See [936].

1126a. Sistiaga, J. M., and J. L. Limpo: Bol. Inf. Técnica Depart. Metales, Madrid 1 (1960) 43. *244*

Slyusar, B. F.: See [375].

1126b. Smart, R. F., R. M. Angles and D. A. Robins: J. Inst. Metals 89 (1960) 349. *465*

1127. Smialowski, M., and E. Przybyla: Prace Glównego Inst. Met. 3 (1951) 501. *299*

1128. Smith, A. A., and H. E. Howe: Trans. AIME 161 (1945) 472. *243*

—: See [960].

1129. Smith, W. C.: Metals and Alloys 2 (1931) 236. *44*

Smith, F. D.: See [1284].

Smythe, J. A.: See [941].

1130. Snelling, R. J.: Metallwaren-Ind. Galvano-Techn. 44 (1953) 385, Werkst. u. Korrosion 6 (1955) 269. *61*

1131. Snelling, R., and E. R. Thews: Dtsch. Goldschmiede-Ztg. 41 (1938) 413. *338*

1131a. Snowden, K. U.: Phil. Mag. Ser. VIII, 6 (1961) 321. *246*

1131b. —: Acta Met. 12 (1964) 295. *246*

1132. —, and J. N. Greenwood: Trans. AIME 212 (1958) 626. *246*

1133. Snyder, C. J.: See Gohn and Ellis [404]. *249*

—: See [52, 53].

Snyder, C. L.: See [1224].

1134. Soderberg, C. R.: Trans. AIME 58 (1936) 733. *223*

1135. Solomon, D., and W. Morris-Jones: Phil. Mag. 11 (1931) 1090. *40*

Solomon, E. M.: See [745a].

Sopwith, W. G.: See [411].

1136. Sorbo, W. de, and D. Turnbull: Acta Met. 4 (1956) 495. *96*

1137. Sorrentino, E., and R. Intonti: Ann. chim. appl. 26 (1936) 385. *307*

Sosman, R. B.: See [237].

1138. Southerden, F.: J. Soc. Chem. Ind. 37 (1918) 85. *293*

1139. Späth, W.: Physik d. mech. Werkstoffprüfung, Berlin: Springer 1938. *229*

Spausta, F.: See [3].

Spektor, E. Z.: See [229].

1140. Spencer, J. F., and M. E. John: Proc. Roy. Soc. London 116 (A) (1927) 61. *34*
1141. Spengler, H.: Metall 9 (1955) 682. *390*
1141a. —: Metall 14 (1960) 789. *38, 176*
1142. Sperry, E. S.: J. Soc. Chem. Ind. 18 (1899) 113. *100*
Spring, W.: See [288].
1142a. Stachura, S. J., and G. Burnet: U. S. Atomic Energy Comm. Rp. IS-316 (1961) 32. *41*
Staebler, Johs.: See [1089].
1143. Stahl, W.: Dissertation T. H. Braunschweig 1957. *17, 268, 326, 327, 328, 332, 363, 364*
—: See [496].
Stajić, V.: See [983].
Stanners, J. F.: See [599].
1143a. Steeb, S.: Z. Metallkde. 52 (1961) 422. *10*
Steers, G. A.: See [525].
1144. Steidlitz, M. E.: Trans. AIME 212 (1958) 676. *175*
Steinemer, K.: See [703].
Steinitz, R.: See [1293a].
Stender, V. V.: See [667, 668].
1145. Stenquist, D.: Mitt. in Bayr. Industrie- u. Gewerbeblatt vom 22. Jan. 1921: Z. Metallkde. 13 (1921) 245. *66*
Stepanović, S.: See [983].
1145a. Stevenson, D. A., and J. Wulff: Trans. AIME 221 (1961) 271. *53, 56*
1145b. —, —: Trans. AIME 221 (1961) 279. *56, 69, 144*
Stevenson, D. A.: See [14].
1146. Stockdale, D.: J. Inst. Metals 49 (1932) 267. *96*
1147. Stockmeyer, W., and H. Hanemann: Z. Metallkde. 33 (1941) 67. *337, 439*
1148. Stoertz, H.: Brit. Patent 717796 Dez. 1951. *355*
1149. Stoffel, A.: Z. anorg. allg. Chem. 53 (1907) 137. *139, 140*
1150. Stokes, A. R., and A. J. C. Wilson: Proc. Phys. Soc. 53 (1941) 658. *11*
1151. Storchheim, S.: J. Metals 12 (1960) 225. *450*
Stranski, I. N.: See [730].
1152. Stribeck, R.: Z. VDI 46 (1902) 1341, 1432, 1463. *367*
Strobel, F. E.: See [272a].
Strong, R. L.: See [1187a].
Strunina, F. A.: See [893].
1152a. Strutt, P. R., A. M. Lewis and R. C. Gifkins: J. Inst. Metals 93 (1964) 71. *210*
Suchan, G.: See [549].
Suganuma, R.: See [351a].
1154. Sugiyama, M., H. Suzuki and T. Furukawa: Nipp. Kinzoku Gakkai-Si 22 (1958) 165. *449*
1155. Sully, A. H.: Metallic Creep and Creep-Resistant Alloys, London: Butterworths 1949. *207*
1156. —, G. N. Cale and G. Willoughby: Nature (London) 162 (1948) 411. *208*
1156a. Sulonen, M.: Fourth Intern. Sympos. on React. of Solids, Amsterdam 1960 (1961) 257. *97*
Suzuki, H.: See [1154].
Sweeney, W. T.: See [522].
Swenson, L. G.: See [869].
1157. Swift, I. H., and E. P. T. Tyndall: Phys. Rev. 61 (1942) 359. *12*

TABOR, D.: See [119, 120, 881a].
1158. TAFEL, V.: Lehrbuch der Metallhüttenkunde, Vol. II, 2nd Ed., Blei, Antimon, Zinn, Zink, Kadmium, Leipzig: Hirzel 1953. *5*
1159. —: Lehrbuch der Metallhüttenkunde, Vol. I u. Vol. II, Leipzig: Hirzel 1927 1929. *300*
1160. TAFEL, W., and F. ANKE: Z. Metallkde. 9 (1927) 225. *442*
1161. —, and W. KNOLL: Metallwirtsch. 10 (1931) 799. *442*
1162. TAINTON, U. C., A. G. TAYLOR and H. P. EHRLINGER: Trans. AIME 77 (1929) 192. *298*
1163. TAKASE, T.: According to Met. Abstr. 6 (1939) 97. *41*
TALYZIN, N. Y.: See [10].
1164. TAMMANN, G., and W. CRONE: Z. anorg. allg. Chem. 187 (1930) 289. *194*
1165. —, and O. DAHL: Z. anorg. allg. Chem. 144 (1925) 1. *171*
1166. —, and K. L. DREYER: Z. anorg. allg. Chem. 191 (1930) 65. *194*
1167. —, —: Z. Metallkde. 25 (1933) 64. *339*
1168. —, —: Ann. Phys. Leipzig 19 (1934) 680. *189*
1169. —, and G. MORITZ: Z. anorg. allg. Chem. 214 (1933) 414. *140*
1170. —, and W. OELSEN: Z. anorg. allg. Chem. 186 (1930) 257. *57, 69*
1171. —, and H. RÜDIGER: Z. anorg. allg. Chem. 192 (1930) 1. *61*
—: See [306, 605].
TANAKA, Y.: See [530].
1172. TANG, Y.-C., and L. PAULING: Acta Cryst. 5 (1952) 39. *105*
TANIMOTO, S.: See [602a].
1173. TAPSELL, H. J., and A. E. JOHNSON: J. Inst. Metals 57 (1935) 121 *226*
TAYLOR, A. G.: See [1162].
TAYLOR, A. R.: See [1016].
TAYLOR, J. W.: See [1237].
TAYLOR, R. H.: See [764].
1174. Technical Unit Committee T 4 B: Corrosion (Houston) 10 (1954) 445. *427*
1175. Technical Unit Committee T 4 B: Corrosion (Houston) 12 (1956) 257. *294*
TEEPLE, H. O.: See [348].
1176. Telegraphenbauordnung, part II, Kabel, Berlin: Reichsdruckerei 1933. *400, 470*
1176a. TELLIER, R.: Proc. First Intern. Conf. on Lead, London 1962, Chap. 2, 13. *416*
TEMPLETON, I. M.: See [942].
TENTI, M. M.: See [109c].
1177. TERASAWA, M.: Nippon Kinzoku Gakkai-Si 22 (1958) 217, 283, 347. *94*
1177a, —, and H. YOSHIDA: Nipp. Kinzoku, Gakkai-Si 24 (1960) 487. *94*
THALL, B. M.: See [1279].
1178. THAU, A.: Z. VDI, Beiheft Verfahrenstechnik 4 (1937) 105. *296*
THELE, W.: See [108c].
1179. THEWS, E. R.: Metallurgische Verarbeitung von Altmetallen und Rückständen, Vol. I: Altweißmetalle, München: Hanser 1948. *5*
1180. —: Metalloberfl. 5 (1953) 134. *60*
—: See [1131].
1181. THIEL, A.: Ber. Dtsch. Chem. Ges. 53 B (1920) 1052. *10*
THIEL, D.: See [916].
1182. THIEL, V.: Felten u. Guilleaume-Rdsch. 23/24 (1938) 21. *404*
THIERER, CH.: See [619].
1183. THIRSK, H. R., and W. F. K. WYNNE-JONES: J. Chim. Phys. 49 (1952) C 131. *341*

1184. THOMAS, U. B. jr.: Bell. Lab. Rec. 16 (1937) 12. *353*
1185. THOMAS, U. B.: Trans. Amer. Electrochem. Soc. 94 (1948) 42. *341*
1186. —, F. T. FORSTER and H. E. HARING: Trans. Amer. Electrochem. Soc. 92 (1947) 313. *355*
—: See [492].
1187. THOMAS, W. R., and B. CHALMERS: Acta Met. 3 (1955) 17. *187*
1187a. THOMPSON, B. A., and R. L. STRONG: J. Phys. Chem. 67 (1963) 594. *327*
1188. THOMPSON, C. H.: Corrosion (Houston) 5 (1949) 151. *312*
1189. THOMPSON, H. G.: U. S. Bur. Stand. J. Res. 5 (1930) 1085. *42, 43, 44*
THOMPSON, C.: See [1290].
1190. THOMPSON, P. F.: Proc. Austral. Inst. Min. Met. 87 (1932) 193. *281*
1191. THOMPSON, W.: Chem. Age 34 (1936) 428. *457*
1192. THOMSEN, E. G.: Trans. Amer. Soc. Mech. Eng. 77 (1955) 515. *400*
1192a. —, and J. T. LAPSLEY: Proc. Soc. Exper. Stress Analysis 11 (1953/54) 59. *396*
1193. THOMPSON, N., and N. J. WADSWORTH: Advances in Physics 7 (1958) 72. *247*
THOMSEN, E. G.: See [349, 1296].
1194. THORNTON, P. R., and P. B. HIRSCH: Phil. Mag. Ser. VIII, 3 (1958) 738. *10*
1195. THÜMMLER, F.: Stahl u. Eisen 78 (1958) 1134. *449*
THWAITES, C. J.: See [334b (Ref. Forrester), 588, 589, 590].
1196. TICE, E. A.: Corrosion (Houston) 7 (1951) 128. *298*
TIDMUS, J. S.: See [347].
1197. TIEDEMA, T. J., and W. G. BURGERS: Appl. Sci. Research 4 (1954) 243. *96*
—: See [920].
TIESENHAUSEN, E. G. v.: See [783b].
1197a. TILLER, W. A., and R. MRDJENOVICH: J. Appl. Physics 34 (1963) 3639. *96*
1198. —, and J. W. RUTTER: Canad. J. Phys. 34 (1956) 96. *9, 32*
—: See [102b, 872, 1033].
1199. TIMMERHOFF, W.: Z. Metallkde. 34 (1942) 102. *10, 322, 323, 363*
TIMPE, W.: See [1007].
1200. Tin and its Uses 32 (1955) 10. *361*
1201. Tin and its Uses 42 (1958) 14. *477*
TITOV, P. S.: See [1300a].
1201a. TITOV, V. A., and E. N. SEMKOVA: Zhur. Priklad. Khim. 32 (1959) 2492. *311*
1202. TOBER, H.: Metalen 8 (1953) 431. *96*
1203. TÖDT, F.: Korrosion u. Korrosionsschutz, Berlin: de Gruyter 1955. *273*
TOLKATSCHEW, S. S.: See [1047].
1203a. TOMAN, K., and M. SIMERSKA: Czechoslov. J. Physics 8 (1958) 94, 101, 233. *183*
1204. TONN, W., and H. GÜNTHER: Z. Metallkde. 28 (1936) 237. *475*
—: See [63].
1205. TOWNSEND, I. R., and C. H. GREENALL: Proc. ASTM 30 (1930) 395. *258, 259*
TRASSL, H.: See [938].
TRÄGNER, E.: See [361].
TREAFTIS, H. N.: See [166, 1208].
TROMANS, D.: See [783b].
TSCHENTKE, G.: See [330].
1205a. TSUKI, M., and Y. UENO: Gifu Daigaku Kogakubu Kenkyu Hohkoku 8 (1958) 41. *300*

1206. Tufts, B. I.: Analyt. Chemistry 31 (1959) 238. *340*
Tunstall, H. A.: See [264].
1207. Turnbull, D., and D. R. Frey: J. Physic. Coll. Chem. 51 (1947) 681. *298*
1208. —, and H. N. Treaftis: Acta Met. 3 (1955) 43; Trans. AIME 112 (1958) 33. *96*
—: See [247a, 582, 1136].
1208a. Turner, A., J. R. Wellington and L. Williams: Canad. J. Chem. Ing. 37 (1959) 55. *283*
Tyapunina, N. A.: See [852].
Tyndall, E. P. T.: See [1157].
1209. „Type F — 3 Alloy Sheath", Anaconda Wire and Cable Company, 27. *414*
1210. Tyzack, G., and G. V. Raynor: Acta Cryst. 7 (1954) 505. *31, 58*

Ueno, Y.: See [1205a].
Uhl, E.: See [1024].
1211. Uhlig, H. H.: The Corrosion Handbook, 5th Ed., New York/London 1955. *282, 294, 297, 298, 308, 312, 316*
1212. Ullmann, F.: Enzyklopädie der technischen Chemie, Vol. IX, 2nd Ed., Berlin/Wien 1932, 232. *292*
1213. Unckel, H.: Z. Metallkde. 32 (1940) 1. *98*
Ust'yanov, V. I.: See [697].

1214. Väth, A.: Metallwirtsch. 17 (1938) 879. *126, 128*
1215. Valentiner, S.: Z. Metallkde. 49 (1958) 375. *59*
1216. Valland, A., and P. Salmon: Corrosion et Anticorrosion 4 (1956) 135. *318*
1216a. Vasil'ev, L. I., T. F. Elsukova, M. A. Bol'shanina and P. A. Kondrat'ev: USSR Abstracts of Metallurgy A, 8 (1959) 62. *232*
Vaysberg, E. S.: See [26a].
1217. VDE-Norm 0510/60: Bestimmungen für Akkumulatoren u. Akkumulatoren-Anlagen. *343, 356*
1218. Vegesack, A. v.: Z. anorg. allg. Chem. 54 (1907) 367. *159, 160*
1219. Vernon, W. H. J.: Trans. Faraday Soc. 23 (1927) 113. *268*
Veres, A.: See [306a].
Verö, J. A.: See [609].
1220. Veszelka, J.: Banyaszati es Kohaszati Lapok 65 (1932) 212, 237. *151*
Vidusova, T. A.: See [721, 722].
1221. Vilella, J. R.: Metals Handbook, American Society for Metals, Cleveland 1948, 955. *29*
1222. —, and D Beregekoff: Industr. Engng. Chem. 19 (1927) 1049. *26*
1223. Vinal, G. W.: Storage Batteries, 4th Ed., New York 1955. *353, 355*
1224. —, C. N. Craig and C. L. Snyder: U. S. Bur. Stand. J. Res. 10 (1933) 795, Res. Paper No. 567, World Power 20 (1933) 318. *53*
1225. —, and L. M. Ritchie: Chem. Metall. Engng. 27 (1922) 1116. *353*
1226. —, and G. N. Schramm: J. Amer. Inst. Electr. Engng. 44 (1925) 128. *353*
1227. Vitovec, F.: Proc. ASTM 57 (1957) 977. *233*
1227a. Vögtli, K.: Techn. Mitt. PTT 37 (1959) 430. *293*
1227b. —: Techn. Mitt. PTT 35 (1957) 489. *297*
1227c. —, and A. Brunold: Techn. Mitt. PTT 34 (1956) 286. *297*
1227d. —, and H. Meister: Techn. Mitt. PTT 36 (1958) 109. *310*
1228. Vogel, R.: Z. anorg. allg. Chem. 126 (1923) 1. *182*
1229. —: Z. Metallkde. 44 (1953) 133. *112*

1230. VOGEL, R.: and A. ZASTERA: Z. Metallkde. 41 (1950) 14. *161*
1231. VOGELPOHL, G.: Z. VDI 96 (1954) 261. *367*
1232. —: Betriebssichere Gleitlager, Berlin/Göttingen/Heidelberg: Springer 1958. *372*
1232a. —: Forsch. Ing. Wes. 26 (1960) 108. *378*
1233. —, A. MATTING and K. F. HAHN: Aluminium 31 (1955) 544. *367*
VOLK, H.: See [671].
—: See [672].
VOLLENBRUCK, O.: See [64].
VOLLMERT, F.: See [358].
VOSS, E.: See [97].

1233a. **WABER**, J. T., K. GSCHNEIDER, JR., A. C. LARSON and M. Y. PRINCE: Trans. AIME 227 (1963) 717. *31*
WADSWORTH, CH.: See [1008].
WADSWORTH, N. J.: See [1193].
1234. WAEHLERT, M.: Z. VDI 62 (1918) 518. *388*
—: See [345].
WAGENMANN, K.: See [115].
1234a. WAGNER, C.: Z. phys. Chemie (B) 21 (1933) 25. *327, 333*
WAGNER, J.: See [1105].
WAGNER, J. B.: See [337a].
WALLACE, W. G.: See [172].
WALLBAUM, H. J.: See [418].
1235. WALSH, S.: Brit. P. 712718, (1954). *354*
WALSH, D. F.: See [486].
WALTER, G. H.: See Ref. FORRESTER [334b].
WALTERS, J. R.: See [368].
1236. WALTON, D., and B. CHALMERS: Trans. AIME 215 (1959) 447. *336*
1237. WARD, A. G., and J. W. TAYLOR: J. Inst. Metals 85 (1956) 145. *56*
WARNER, L. M.: See [1036c].
1238. WARTENBERG, H. v.: Verh. dtsch. phys. Ges 12 (1910) 105 *21*
WAS, D. A.: See [923].
1239. WASSERBÄCH, W.: Dissertation T. H. Hannover 1959. *475, 476, 477, 478, 479*
—: See [684].
1240. WASSERMANN, G.: Texturen metallischer Werkstoffe, Berlin: Springer 1939. *182*
1241. —, and U. GRUHL: Metall 6 (1952) 769. *388*
1242. WATANABE, M., and S. FUKUSHIMA: Sci. Rep. Research Inst. Tohoku Univ. 8 (A) (1956) 142: According to Met. Abstr. 24 (1956) 186. *299*
WATANABE, J.: See [1294].
WATANABE, N.: See [529].
WATANABE, Y.: See [900, 901].
1243. WATERHOUSE, H.: Brit. N. F. Met. Res. Ass. Res. Rep. Nr. 440 (1937) London: According to Met. Abstr. 4 (1937) 425. *258, 259*
1244. —, and R. WILLOWS: J. Inst. Metals 46 (1931) 139. *138, 141*
—: See [73]
1245. WATKINS, M. T.: J. Inst. Metals 90 (1961/62) 209. *394*
1245a. —, K. ASHKROFT and J. MCKENZIE: Mech. Eng. Research Lab. Plasticity Div. Rep. 102 (1955). *394*

1246. WEAVER, F. D.: J. Inst. Metals 56 (1935) 209. *163, 167*
WEAVER, B. A.: See [1009].
1247. WEBER, E., and W. M. BALDRIN: J. Metals 4 (1952) 854. *268, 326, 327*
WEBER, H.: See [398].
1248. WEBER, R.: Metall 12 (1958) 96. *369*
—: See [1071].
1249. WEBER, W. C.: Chem. metall. Engng. 39 (1932) 542. *292*
WEEKS, J. R.: See [643].
1250. WEERTMAN, J.: J. Appl. Phys. 26 (1955) 1213. *214*
1251. —: Trans. AIME 218 (1960) 207. *215*
1252. —, and E. I. SALKOVITZ: Acta Met. 3 (1955) 1. *13*
1252a. WEHR, P.: Thesis T. H. Braunschweig 1965. *283, 284 289*
WEHR, P.: See [570, 577a].
1253. WEIBKE, F., and O. KUBASCHEWSKI: Thermochemie der Legierungen, Berlin: Springer 1943. *95*
1254. WEIGERT, K. M.: Metalloberfl. 6 B (1954) 77. *116*
WEIGT, H. J.: See [918].
WEIL, J. A.: See [677].
1254a. WEINSTEIN, M., and J. F. ELLIOTT: J. Electrochem. Soc. 110 (1963) 792. *175*
WEISS, S. M.: See [617].
1254b. WEISSBACH, H.: Werkst. u. Korrosion 15 (1964) 555. *280, 283*
1255. WELBER, B., and S. L. QUIMBY: Acta Met. 6 (1958) 351. *13*
1256. WELLINGER, K.: Mitteilungen d. V. G. B. 15 (1951) 1. *220*
1256a. WELSHER, R. A. G.: Proc. First. Intern. Conf. in Lead, London 1962, Chap. 17, 141. *37*
WELLINGTON, J. R.: See [1208a].
1257. WENGER, E.: Stahl u. Eisen 74 (1954) 1202. *367*
1258. —: Diplom-Arbeit T. H. Braunschweig 1956. *377*
1259. WERNER. M.: Z. Metallkde. 22 (1930) 342. *289*
1260. —: Z. Metallkde. 24 (1932) 85. *196*
1261. —: In O. BAUER, O. KRÖHNKE and G. MASING: Die Korrosion metallischer Werkstoffe 2, Leipzig: Hirzel 1938. *82, 85, 282, 289, 301*
1261a. WESSON, H. C.: Corrosion Prevention and Control 6 (1959) 9, 12. *290, 304*
WEST, G. W.: See [198].
1262. WESTBROOK, F. A., F. E. ROGERS and W. H. CARTER: Metal. Ind., London 49 (1936) 560; Metal. Ind. N. Y. 29 (1931) 152. *469*
1262a. WESTBROOK, J. H., and K. T. AUST.: Acta. Met. 11 (1963) 1151. *188*
WESTERHOFF, H.: See [832].
WESTERMANN, F. E.: See [180a].
WETZEL, E.: See [65].
WHEELER, M. A.: See [489].
WHETZEL, J. C.: See [167].
WHITE, F. T. M.: See [132a, 132b].
WHITE, H. V.: See [1009].
1263. —: Bull. Virginia Polytech. Inst. Engrs., Exper. Sta. Series Bull. 13 (1933) 1: Chem. Abstr. 27 (1933) 3692. *339, 364*
1264. WHITMAN, W. G., and R. P. RUSSELL: Korrosion u. Metallsch. 1 (1925) 43; Ind. Engng. Chem. 17 (1925) 348. *291, 293*
WHITTAKER, H. F.: See [167].
WHYTLAW-GRAY, R.: See [67].
1265. WICKERT, K.: Korrosion u. Metallsch. 19 (1943) 120. *290, 296*

1266. Wickert, K.: Korrosion u. Metallsch. 20 (1944) 147. *280, 282, 288*
1267. —, and B. Zenker: Chem. Fabrik 14 (1941) 311. *288*
1268. Wiedemann, E.: Ann. Phys. Leipzig 20 (1883) 228. *41*
1269. Wiederholt, W.: Z. VDI, Beiheft Verfahrenstechnik (1939) 196. *293*
Wiedersich, H. W.: See [102a].
1270. Wieghardt, P.: ETZ 55 (1934) 339. *424*
1271. Wiemer, H.: Sintermetalle in: Werkstoffe für Gleitlager, 2nd Ed., edited by R. Kühnel, Berlin/Göttingen/Heidelberg: Springer 1952. *367*
1271a. Wilder, T. C., and J. F. Elliott: J. Electrochem. Soc. USA 111 (1964) 352. *34, 120*
1272. Wilkenson, W. D.: U. S. Atomic Energy Comm. ANL-5262 (1954) 73. *134*
1273. —, E. W. Hoyt and H. V. Rhude: U. S. Atomic Energy Comm. Pub. ANL-5449 (1955). *41*
1273a. Williams, D. N., K. R. Grube and R. I. Jaffee: Trans. Amer. Soc. Metals 54 (1961) 96. *264*
1274. Williams, D. N., J. A. Houck and R. I. Jaffee: Metal Progress 77 (1960) 79. *264*
1274a. Williams, G. K.: Trans. AIME 121 (1936) 226. *332*
Williams, L.: See [1208a].
1274b. Williams, R. O.: Trans. AIME 224 (1962) 719. *182, 264*
1274c. Williams, R. S.: Z. anorg. Chemie 55 (1907) 31. *66*
Williamson, G. K.: See [867].
1275. Willihnganz, E.: Iron Age 163 (1949) 62. *346*
Willoughby, G.: See [1156].
Willows, R.: See [1244].
Wilson, A. J. C.: See [1150].
1276. Wilson, B. S., and F. H. Garner: J. Inst. Petrol 37 (1951) 225. *298*
1277. Wilson, F. H.: See [713]. *371*
1278. Wilson, T. C.: J. Chem. Phys. 8 (1940) 13. *62*
1278a. Winegard, W. C.: Metallurgical Rev. 6 (1961) 57. *31*
1279. Winegard, W. C., S. Majka, B. M. Thall and B. Chalmers: Canad. J. Chem. 29 (1951) 320. *96*
—: See [104, 105, 106, 657a, 872].
Winkler, J. H.: See [1094].
1280. Wire Industry 21 (1954) 523. *454*
1281. Wiseman, C. D., O. D. Sherby and J. E. Dorn: Trans. AIME 209 (1957) 57. *213*
1282. Witt, D.: Gas- u. Wasserfach 83 (1940) 341. *297*
Witt, G.: See [851].
Witte, H.: See [750].
Wittenberg, J.: See [918a].
Wolf, F.: See [1058].
1283. Wolzogen-Kühr, C. v.: Water 40 (1956) 281. *312*
1284. Wood, A. B., and F. D. Smith: Proc. Phys. Soc., London 47 (1935) 149, 369. *20*
Wood, D. H.: See [1046a].
1284a. Wood, W. A., S. McCousland and K. R. Sargant: Acta. Met. 11 (1963) 643. *247*
Wooster, W. A.:See [976].
1285. Worner, H. K.: Proc. Austral. Inst. Min. Met. 117 (1940) 29. *203*
—: See [432, 433, 434, 435, 1287].

1286. WORNER, H. W.: J. Inst. Metals 66 (1940) 131. *71*
1287. WORNER, H. W. and H. K. WORNER: J. Inst. Metals 66 (1940) 45. *29*
1288. WRIGHT, C. R. A.: Proc. Roy. Soc., London 50 (1891/92) 372. *173*
1289. —: Proc. Roy. Soc., London 52 (1892/93) 530. *171*
1290. —, and C. THOMPSON: Proc. Roy. Soc., London 48 (1890) 25. *116*
1291. —: Proc. Roy. Soc., London 49 [1891) 174. *116*
WRIGHT, W.: See [768].
1292. WÜST, F.: Metallurgie 6 (1909) 769. *79, 97*
WULFF, J.: See [14, 1145a, 1145b].
1293. WUNDER, A.: In [1261]. *82*
1293a. WURMS, C., and R. STEINITZ: Trans. AIME 224 (1962) 1283. *326*
WYNNE-JONES, W. F. K.: See [134, 1183].

YAFFEE, L.: See [172].
1294. YAMAMOTO, M., and J. WATANABE: Nipp. Kinzoku Gakkai-Si 21 (1957) 732. *10*
1295. YANAGIHARA, T.: Nipp. Kinzoku Gakkai-Si 15 (1951) 74, 122. *97*
1296. YANG, C. T., and E. G. THOMSEN: Trans. Amer. Soc. Mech. Eng. 75 (1953) 575. *396*
1297. YAO, T. P., and V. KONDIC: J. Inst. Metals 81 (1952) 17. *18*
YEOMANS, C. D.: See [763].
YOKOKAWA, T.: See [901].
YOSHIDA, H.: See [1177].
YOSHIDA, T.: See [351a].
1298. YURKOV, V. A.: Doklady Akad. Nauk USSR 91 (1953) 891: According to Met. Abstr. 21 (1953/54) 608. *81*
YUVENKO, V. Y.: See [898a].

1299. ZAHN, H. E.: Precision Metal Moulding 12 (1954) 68. *346*
1299a. —: Chem. Abstr. 55 (1961) 1397a. *356*
1300. —: Iron Age 174 (1954) 174. *346*
1300a. ZAITSEV, Y. V., and P. S. TITOV: Tsvetnaya Met. 5 (1962) 136. *60*
ZAPPONI, P. P.: See [491].
ZASHKVARA, V. V.: See [660].
ZASTERA, A.: See [1230].
1300b. ZAUSZNICA, A.: Second. Intern. Conf. on Lead, Arnhem 1965, No. 31. *427*
1300c. Z. Metallkde 24 (1932) 306. *383*
1301. ZENER, C.: Univ. of Chicago Press (1948) 147. *217*
1302. —: Reference in Acta Met. 6 (1958) 287. *196*
ZENKER, B.: See [1267].
1303. ZICKRICK, L.: Trans. AIME, Inst. Met. Div. 99 (1932) 345 *405, 419*
—: [239].
1304. ZIEGFELD, R. L.: Pres. at the fifteenth annual Meeting of the Metal Powder Industries Federation, 1959, Detroit/Michigan, USA. *452*
1304a. —: Proc. First Intern. Conf. on Lead 1962, Chap. 12, 121. *356*
1305. ZINK, J.: Z. anal. Chem. 91 (1933) 246. *302, 303*
1306. ZINTL, E., and S. NEUMAYR: Z. Elektrochem. 39 (1933) 86. *46, 100*
1308. ZUNKER, P.: Metallwirtsch. 19 (1940) 223, 247. *173, 381*
ZWICKER, U.: See [504a].

Subject Index

Acetic acid, corrosive properties 293, 314
Acid lead, composition 6
Acid solutions, corrosive properties 291, 319
Acids, corrosive properties 371
Acoustic properties 20
Activation, thermal 208
Activation energy 11, 15, 214
  calculation 213
  deformation effects 186
  gold additions effects 191
  oxidation 327
  silver additions effects 191
Activation stress 208
Additions
  effects
    on age hardening 86
    on corrosion resistance 271, 273, 276
    on drossing 328
    on grain growth 194
  to bearing metals 370, 383
  to soft solders 470
  to storage battery grids 352—353
  to type metals 360
Aeration cells 269, 305, 312
Age-hardening 30, 39, 42, 47—50, 62, 64, 67—69, 86, 169, 192—193, 235, 238, 261, 263—264, 266—268, 358, 386
  additions effects 86
  arsenic effects 86, 440
  at elevated temperatures 89
  deformation effects 93, 192
  effects
    on creep 235, 238
    on recrystallization 192
  grain boundary effects 266, 414
  microscopic precipitation relationship 64
  precipitates 266
  segregation 93
Ageing, cable sheathing 411
Albert Fischer printing block moulding method 358
Alcohols, corrosive properties 298
Alkalis, corrosive properties 294—295, 312
Alloy concentration, effects on fatigue strength 250
Aluminium
  addition
    to Bahnmetall 386
    to type metals 364
    to white metal 383
  cable sheathing 401
  effects on drossing 328
  in die casting alloys 357
  in type metals 363
  in soft solders 470
  lead coating 464
Aluminium sulphate, corrosive properties 296
Ammonia, corrosive properties 295
Ammonium carbonate, corrosive properties 295
Ammonium hydrogen fluoride solution, corrosive properties 297
Ammonium phosphate, corrosive properties 292
Ammonium sulphate, corrosive properties 296
Andrade's creep law 208
Anodes 298—301
  cathodic protection 291, 301, 316
  copper plating 300—301
  electroplating 299—301
  lead-silver 299
    casting 338
  lead-thallium 299
  magnesium-aluminium-zinc 316
  oxygen overpotential 299
  sulphate solution electrolysis 299—300
  zinc electrolysis 298—299
Antimonial lead, *see* Lead-antimony alloys

Antimony
  addition
    for corrosion resistance 276
    for inhibition of recrystallization 190
    to lead coating bath 453
    to soft solders 470
  effects on grain refinement 196
  in storage battery grid pastes 353
Arsenic
  addition
    to bearing metals 383
    to lead-antimony alloys 86, 355, 440
    to storage battery grid plates 352
    to white metal 383
  effects
    on age-hardening 86, 440
    on storage battery grid plates 355
  in bearing metals 383
  in type metals 364
  on remelting 35
Art casting, *see* Casting
Asphalt, cable sheathing 315
Atomic radii 29
Atomic weight 9, 11

Babbitt metals 380
Backing metals 359
Bahnmetall 34, 63 (*see also* Bearing metals)
  age hardening 69, 386
  aluminium additions 386
Basic lead carbonate 302—303
Bearing metals 60, 63, 111, 365—390 (*see also* Bearings; White metal)
  additions 370, 383
  arsenic additions 383
  Bahnmetall 34, 63
  casting 384
  chill casting 384
  compressive strength 384
  copper content 382
  corrosion 371—372
  embedding of hard constituents 112, 370
  fatigue strength 370—371
  friction 376
  grain refinement effects 383
  hardness 112, 369, 385
  hot hardness 112, 369
  Lagerhartblei 150
  lead-bismuth-antimony-tin alloys 389
  Lichtenberg metal 388
  Lipowitz metal 388
  Matrix lead-bismuth-antimony-tin alloy 389
  mixed friction 367, 372
  Newton metal 388
  nickel content 382
  pitting by acids 371
  powder metallurgy 387
  properties 369—372
  Rose's metal 133, 389
  running behaviour 372—380
    supporting crystal theory 372
  standard specifications 381, 387
  structure 382
  ternary eutectic 165—166, 382
  testing machine 378
  thermal conductivity 384
  thermal expansion 375, 390, 428
  toothing theory 375
  Vogelpohl band 377—378
  wear 365—369, 372, 375, 379
Bearings
  friction 372
  hydrodynamic lubrication 365
  journal 366—367
  leaded bronze 371, 374
  linings, fatigue strength 370
  lubrication, lead coatings 374
  plain 365—369
  running properties 367—372
  running surfaces, wetting by lubricant 379
  three-layer 371
Bending slip 182
Bending values 85, 262
Benzene, corrosive properties 298
Betatron radiation 21
Binary alloys 29—109
Bismuth
  addition
    for corrosion resistance 276
    to soft solders 470
    to storage battery plates 352
    to type metals 364
  extraction from lead 8
Blistering, pipes and tubes 439
Boiling point 14
Bond strength 454, 458
Bottle caps, deep drawing 448
Boundary lubrication 375

Brasses
  lead coating 339
  leaded 154—155
Brines, corrosive properties 296
Bromine, corrosive properties of vapour 297
Bronzes
  cast lead coating 339
  leaded 152, 387
    bearings 371, 374
    casting 387
    composition 380—387
    lead-indium plating 60
    properties 380—387
    tin-leaded, casting 387
Building, sheet lead 467
Burgers vector 178

Cable sheathing 63, 195, 411—417 (*see also* Cables; Corrosion; Extrusion; Extrusion presses)
  age hardening 414
  ageing 411
  alloys 410—419
  alternating impact strength 419
  aluminium 401
  asphalt 315
  bending value 262
  cathodic protection 316
  compressive strength 418
  copper 401
  corrosion resistance 307—308, 311—314, 404, 414
  creep rate 412
  creep strength 219, 412
  damage
    by animals 317
    by bacteria 311
    by insects 317—318
  defects 420—428
  electrolytic corrosion 428
  elongation at fracture 411
  extrusion 109, 151, 407—408
  fatigue fracture 244, 413, 427
  fatigue strength 251, 255, 414, 418, 428
  flat 403
  grain growth 411
  grain refinement 195—196, 416
  grain size 411—412
  intercrystalline fracture 412
  lead-oxygen system 424
  linear structure 414
  mechanical properties, grain size effects 411
  mine winding gear cables 404
  overhead cables 244, 403—404, 414
  "phenol-corrosion" 293, 428
  pitting
    by soils 309—313
    by stray currents 314—317
  production 404—410
  sockets, casting 338
  standard specifications 414, 416
  steel 401
  tensile strength 417
  tongues, model experiments 422
  welding 406, 409, 420, 423
Cables 400—404
  cathodic protection 314, 316
  communication 413
  corrosion
    by sea water 307—308, 404, 416
    by soils 314, 404
  damage 311, 317—318
  design 400—404
  diaphragm 402
  ducted, corrosion 314, 404
  Höchstädter 401
  overhead 244, 318, 403—404, 414
  power 400—401, 412—414
  sockets 338
  underground 308, 314, 404
Cadmium
  addition
    for corrosion resistance 276
    to soft solders 470
  impurity in white metal 383
  solid solution in lead, oriented precipitates 51—52
Calcium, effects on grain refinement 196
Calcium carbonate, as protective film 294, 305
Carbonic acid, corrosive properties 304
Carbon dioxide, corrosion inhibition properties 302
Cast alloys
  creep strength, grain growth effects 236
  crystal growth 337
  crystal structure, coring 414
  grain size 337
Castability, *see* Casting properties

Casting 334—340
  alloys, age hardening 265
  bearing metals 384
  cable sockets 338
  centrifugal, leaded bronzes 387
  chill 334—340, 344, 384
    mould temperature 334, 384
    shrinkage 334
  fluidity effects 335
  for art purposes 338
  health hazards 339—340
  hollow 338
  moulds, spiral-chill 334, 361, 363
  pipe formation
    storage battery plates 345
    type metal 363
  properties 97, 334, 361
  sand 337, 339
  shape 337, 344
  shrinkage 36, 79, 97, 334—335, 362
    temperature effects 79
  temperature, effects on shrinkage 79
  with eutectics 335
Castings, lead-alloy, texture 336
Cathodic protection
  anodes 291, 300—301, 316
  cable sheathing 316
  cables 314, 316
  lead 316
Caustic soda solution, corrosive properties 294
Cement, corrosive properties 294, 313
Chemical lead 9
Chemical plant
  homogeneous lead coatings 454
  lead-antimony alloys applications 339
  lead-copper alloys applications 56
  sheet lead applications 465, 468, 476
Cinders, corrosive properties 311
Chromic acid, corrosive properties 301
Cladding 465
Clay, corrosive properties 311
Coalescence with zinc 107
Coarse grain 194 (*see also* Grain growth)
  cable sheathing 411
  effects on creep strength 231
  pipes and tubes 435, 437
Coatings 453—479 (*see also* Cladding; Electroplating; Homogeneous lead coatings; Hot dip coatings; Sprayed lead coatings)
  asphalt, on cable sheathing 315
  cast lead 339
  lead
    bearing lubrication 374
    bond strength 454
    crystallization 463
    on aluminium 464
    on copper 461
    porosity 453
    vacuum evaporation 453
  lead-tin, on copper 308
  tin
    casting 431
    for homogeneous lead coatings 456
    for hot dip lead coatings 454
    on pipes and tubes 431
    on sheet lead 447
    on steels 456
Cobalt
  addition to storage battery plates 352
  solubility in lead melts 53
Cobalt sulphate, addition to storage battery acids 355
Cohesive strength 199
Coke, corrosive properties 311
Commercial lead
  bending value 262
  casting, shrinkage 335
  creep rate 237, 439, 451
    stress effects 231
  fatigue strength 262
  hardness 203
  melts, corrosive properties 322
  rolling 445
  sintered powders 450—452
  tin-clad, applications to bottle caps 448
Compressive modulus 12
Compressive strength 37, 100, 170—171, 361, 384
  temperature effects 205
  tests 205—206
Concentration cells 312
Concrete, corrosive properties 294, 295, 311
Conductivity
  electrical 11, 13
  thermal 19—20
    bearing metals 384
    lead-antimony alloys 81
    lead-bismuth alloys 41
Consumption 1, 3, 5
Cooling brines, corrosive properties 296

Co-ordination number 10
Copper
  addition
    for corrosion resistance 273, 276, 288
    to soft solders 470
    to storage battery grid plates 352
    to type metals 360
  cable sheathing 401
  effects
    on cable sheathing extrusion pressure 151
    on grain refinement 197
    on homogeneous lead coatings 459
  homogeneous lead coating 456
  in bearing metals 382
  in homogeneous lead coatings 458
  in soft solders 470, 473
  in type metals 360
  in white metal 383
  lead coating 461
  lead-tin coating 308
  pitting of lead 319
  soldering 470, 472, 474, 477
Copper sulphate, corrosive properties 296
Coring, *see* Crystal structure; Segregation
Corrosion 9, 268, 319—320
  anodic 69, 341, 348, 352, 356, 453
    eutectic structure effects 289, 349
  atmospheric 268, 297
  by commercial lead melts 322
  by lead-antimony melts 322
  by lead-bismuth melts 41
  by type-metal melts 363
  contact 319—320
  copper additions effects 273, 276, 288
  electrochemical theory 65, 268, 279, 305, 312
  electrolytic 314, 428
  electroplated lead coatings 319
  eutectic structure effects 289, 349
  fluid flow effects 282
  hydrogen evolution effects 271
  intercrystalline 65, 193, 271—272
    anodes 298
    melting pots 323
  lead-indium alloys 60
  mechanical stress effects 289
  oxygen-absorption, *see* Oxidation
  "phenol" 293, 428
  protective films 268, 271—273, 291—292, 294, 296—297, 302—305, 307, 310
  rate, measurement 272
  storage battery positive plates 348, 352
  stray-current 314, 317
    detection 315
    electrical drainage 316
  stress 284—287, 289
  sulphite pulp equipment 292
  underground 308—317, 454
    concentration cells 312
  underwater 307
Corrosion products 276, 302—304, 307, 311, 315—317, 431
Corrosion resistance 268—320, 371—372
  additions 271, 273, 276, 288
  grain size effects 289
  to acetic acid 293, 314
  to acid solutions 291, 319
  to alcohols 298
  to alkalis 294—295
  to ammonia 295
  to ammonium carbonate 295
  to ammonium hydrogen fluoride solution 297
  to ammonium phosphate 292
  to ammonium sulphate 296
  to benzene 298
  to brines 296
  to bromine vapour 297
  to carbolic acid 304
  to caustic soda solution 294
  to cement 294, 313
  to chromic acid 301
  to concrete 294—295, 311
  to copper sulphate 296
  to distilled water 294, 296, 301—303
  to fatty acids 293
  to fluorine 297
  to fluosilicic acid 291
  to gases 297
  to humus 311
  to hydrobromic acid 271
  to hydrochloric acid 271, 290
  to hydrofluoric acid 291
  to hydrogen sulphide 297, 307
  to illuminating gas 297
  to inorganic acids 290—292
  to insulating oils 298
  to light oils 298

to limestone 311, 313
to liquid fuels 297
to lubricating oils 60
to magnesium chloride 296
to marl 311
to mortar 294, 311
to nitrates 311
to nitric acid 287, 291
to nitrosylsulphuric acid 287—289
to oils 60, 298
to organic acids 60, 292—293, 298, 304, 318
to perchloric acid 341
to phosphates 295
to phosphoric acid 292
to potassium chloride 296
to potassium hydrogen fluoride 297
to potassium sulphate 296
to quicklime 294
to salt solutions 295—297
to sandy soil 311
to sea water 304, 307—308, 404, 416
to sodium bicarbonate 295
to sodium carbonate 295
to soils 294, 308—317, 454
to sulphate solutions 296
to sulphur 297
to sulphuric acid 44, 69, 74, 272—290, 319—341
effects of impurities 290
lead monoxide effects 289
to sulphurous acid 292
to tap water 304—307
to town gas 297
to wood 293
Crack formation 257, 262, 435, 437
Creep 203
age hardening effects 235, 238
Andrade's law 208
diffusion effects 212, 214
effects on embrittlement 200, 216, 231, 412
grain boundary effects 209—216
instantaneous 207, 218
intercrystalline fracture 200, 216, 231
multi-axial stress effects 222—227
secondary 208
skin effect 220
steady-state 209—216
stress effects 212—216
temperature effects 212—216
tertiary 208, 216—217
testing 217—220
time to rupture 230—231, 476
true creep limit 229—230
Creep rate 231, 235, 245, 439, 451
embrittlement effects 200, 216, 231, 412
grain boundary effects 211
grain size effects 412, 435
initial strain 219
limiting creep rate 219
logarithmic creep law 208
measurement 217
polygonization effects 209, 222
pre-strain effects 220—222
recrystallization effects 217
softening effects 213
standard specifications 218
strain hardening effects 213
stress effects 231
temperature effects 209, 212—216, 242, 477
Creep strength 34, 207—244, 354—355, 412, 417, 433, 436, 439, 450—451, 474
age hardening effects 235, 238
coarse grain effects 231
dynamic 231—233
embrittlement effects 216, 231
grain growth effects 210, 233, 236, 412, 433, 436
measurement 217
model tests 207
reference stress/reference elongation ratio 224
structure effects 233—244
Creep stress
limiting creep stress 219
single crystals 227, 242
temperature effects 241
Creosote, corrosive properties 314
Critical dispersion 266
Cross slip, *see* Crystal structure
Crucible tests 334
Crushing pressure, *see* Deformation
Cryogenics 19
Crystal coring, *see* Crystal structure
Crystal growth 31, 337
Crystal structure 10—11, 33, 40, 44, 58, 64, 67, 75, 94, 100, 150
break planes 209
co-ordination number 10
coring 32, 45, 77, 102, 105, 414
cross slip 180, 211

cubic close packing 10, 30, 40, 177, 181 (*see also* Solidification)
defects 11, 16, 177
edge dislocation 16, 178
effects on creep strength 233–244
Frank-Reed dislocation lines 178
linear structure 414, 432
polygonization 184, 209
slip 177, 182, 209, 228
stacking fault energy 10, 181
supporting crystal theory 372
toothing theory 375
twinning 10, 18, 181
vacancies 11, 211–212
welding theory 372
Crystallization 97, 337, 463

Damage
by animals 317
by bacteria 311
by insects 317–318
De-coppering 72, 150, 155
Deep drawing 85, 448, 454
Deformability 369–370
Deformation 177–182
critical 194, 433
crushing pressure 419
easy glide region 179
edge dislocation 16, 178
effects
on activation energy 186
on age hardening 93, 192
on recrystallization 292
grain fragmentation 180
plastic yielding 398–400
resistance 74
rolling 444
slip line field 399
toothing theory 375
welding theory 372
work load 170, 393, 443
work of deformation, rolling 442–446
Density 11
De-silvering 8, 116, 120
Parkes process 8, 119–120, 155, 194
De-zincification, vacuum method 8
Die casting alloys 356
aluminium as impurity 357
Diffusion 16, 329, 333
effects
on creep 212, 214
on drossing 327
on melting 327, 329
Dispersion hardening 264, 452
Distilled water
carbon dioxide content 302
corrosive properties 294, 296, 301–303
lead hydride content 302
solubility for lead salts 303
Drossing 53, 326–334, 364, 420, 423 (*see also* Oxidation)
crucible test 334
diffusion effects 327
lead monoxide 326
parabolic time law of thickness increase 327

Easy glide region, *see* Deformation
Edelblei 69, 275
Edge dislocation, *see* Crystal structure; Deformation
Elastic properties 12–13, 20, 227 (*see also* Yield point)
Electro-deposition, *see* Electroplating
Electrolysis
sulphate solutions 299–300
zinc, anodes 298–299
Electrolytic lead 299
casting 329
composition 6
recrystallization diagrams 195
Electroplating 463–465
brass, on lead 338
copper, anodes 300–301
copper, on lead 338
gold, on copper plating 338
lead, 453, 463
anodes 299
corrosion 319
crystallization 463
on aluminium 464
on cast iron 461
on cast steel 461
on zinc 461
pitting of steel 319
lead-antimony-tin, on silver 465
lead-indium, on leaded bronzes 60
lead-silver, supersaturated solid solutions 33
lead-tin, on steels 464
white metal, on silver 465
Electrotyping 358
Elongation at fracture 84, 200 (*see also* Fracture)
cable sheathing 411

necking 202
rate of test effects 197, 200
reference elongation 224
temperature effects 205
Embrittlement, creep effects 216, 231, 412
Endurance, *see* Fatigue strength
Equilibrium segregation, *see* Recrystallization
Erosion, by sea water 308
Etching 25, 28, 129
effect on fatigue strength 246
pipes and tubes 27
type metal 29
zinc printing plates 361
Eutectic structure
effect on corrosion 289, 340
Evaporation, latent heat 15
Expansion, thermal 11—12, 375, 390, 428
Extrusion 390—400
explosive-pressure 452
frictional forces 397
ideal plastic isotropic material 397
impact 447—448
indirect 390, 428
law of similitude 392
oxides in residue 420
plane strain 398, 400
pressure 109, 151, 392—396, 407—408, 428—429, 431
rate 395, 407, 432
recrystallization effects 292
residues 420
temperature 266, 393, 407, 414, 420, 432
Extrusion presses 390, 405, 408—410, 428—431
die boxes 395, 407, 422, 431
filling attachment 424
hydraulic 408, 428
inverted 390, 428
screw-operated 408—410
twin-ram 406
Extrusions, solid solution breakdown 266

Fatigue fracture 244, 257, 262, 370, 413, 427, 435
Fatigue strength 57, 244—262, 370—371, 414, 417—418, 428, 459, 477, 479 (*see also* Compressive strength)
alloy concentration effects 250, 252
environmental effects 245
etching effects 246
grain size effects 250—251, 435, 437
Lüders bands 246
notch effect 246
rate of test effects 84, 112
recrystallization effects 250
slip bands 247
stress frequency effects 248
temperature effects 253
testing machine 418
Fatty acids, corrosive properties 293
Faure storage battery plates 341—356
Feinblei, composition 6
Flash test 281
Flow resistance 189, 197, 369, 392, 419, 443
Fluid flow, effects on corrosion 282
Fluidity 18, 335—336
Fluorine, corrosive properties 297
Fluosilicic acid, corrosive properties 291
Fluxes
soft soldering 390, 453, 471
type metal re-melting 363
Foil
lead-tin 447
printing block moulds 358
production 447
statistics 5
texture 182
Food cans, solders 471
Foundry alloys 321—390 (*see also* Bearing metals; Casting; Fusible alloys; Melting; Pressure die casting; Printing; Storage batteries)
drossing 364
Fracture, *see also* Elongation at fracture
ductile 206
intercrystalline 200, 216, 231
Frank-Reed dislocation lines 178
Frary metal 128—129
Free machining alloys 34, 57, 121
Friction
boundary lubrication 375
cone compression test 445
extrusion 297
internal 13, 18
materials 34, 452
mixed 367, 372
pressure-equalizing pads 224
rolling of lead 442, 445

soldering 472
"stick-slip" effect 376
toothing theory 375
Vogelpohl band 377—378
welding theory 372
Frost risk 104, 202, 307
Fuels, liquid, corrosive properties 297
Furnaces, melting 321, 327, 347
Fusible alloys 387—390

Galena 2
Gaseous inclusions 439, 464
Gases, corrosive properties 297
Geochemistry 2—3
Glass, wetting 60
Gold
additions, effects on grain growth 38
electroplating, on copper 338
Grain boundaries 42, 161, 187
displacement 210, 212
effects
on age hardening 266, 414
on creep 209—216
energy 16
large angle 184, 186, 211, 212
migration 185—189
slip 227
small angle 16
Grain growth 10, 194—197, 411
effects on creep strength 210, 233, 236, 412, 433, 436
gold additions effects 38
rate 194—197
silver additions effects 38
Grain refinement 383, 416
additions effects 196—197
Grain size 190, 194—197, 233, 337, 411—412
deformation effects 196
effects
on corrosion resistance 289
on creep rate 210, 233, 236, 412, 435
on fatigue strength 250—251, 435, 437
on mechanical properties 199
on recrystallization 190
extrusion temperature effects 432
non-uniform 226, 432, 435
Graphite, as friction material 35, 452
Guinier-Preston zones 263
Gunmetal, cast lead coatings 339

Halides, anodic polarization of lead 296
Hall effect, lead-antimony alloys 81
Hansson-Kobertson-Press 410
Hardener alloys 46, 76, 103, 322, 470
Hardening 262—267
dispersion 264, 452
Hardness 36, 84, 112, 166—168, 202—205, 262, 369, 385, 444
Brinell 202—204
diamond cone 204
dynamic 205
high-purity lead 203
hot 112, 369
lead-antimony alloys 84
lead-silver-copper alloys 112
in rolling 444
Mohs 202
temperature effects 204
tensile strength relationship 204
Harris lead 6, 8
Hartblei 440
Heat content 15—16 (*see also* Latent heat)
Heat of mixing 50, 61, 95
Heat transfer 19—20, 41, 458
Heat treatment 48, 128, 262—267
leads baths 321
Henley extrusion press 408
High-purity lead 69, 275
hardness 203
History 1
Höchstädter cables 401
Homogeneous lead coatings 454—461
bond strength 458
casting 456
copper content 458
corrosion resistance to soils 454
cracking, additions effects 459
fatigue strength 57, 459
heat transmission 458
internal 456
lead-copper alloy 458
on copper 456
on iron 57
on pipes and tubes 456
on steel, omission of tin coating 456
radiation shielding 23, 454
temperature variation effects 458—459
underground corrosion 454
uses in chemical plant 454
Homogenizing 36, 48
Honeycomb structure 45, 51, 77

Hot dip coatings 453—454
lead-antimony alloys 453
lead-arsenic-tin alloys 454
lead-tin 454
on folded pipes 453
wire drawing lubrication 454
Hot hardness, *see* Hardness
Huber extrusion press 406
Humus, corrosive properties 311
Hydraulic Press 410
Hydrobromic acid, corrosive properties 271
Hydrochloric acid, anodic polarization of lead 296
corrosive properties 271, 290
Hydrogen
dissolution in lead 58
over-potential 269, 301
Hydrofluoric acid, corrosive properties 291
Hydrogen sulphide, corrosive properties 297, 307

Illuminating gas, corrosive properties 297
Impact strength 171, 205—206
alternating 419
notch test 205
oxygen effects 74
temperature effects 205—206
tests 205—206
Ingots 335
Inorganic acids, corrosive properties 290—292
Insmetals homogeneous lead coating process 457
Insulating oils, corrosive properties 298
Interfacial tension, *see* Surface energy
Internal friction, *see* Friction
Intertype line setting and casting machines 351
Ions, complex-formation 271
Iron
addition for corrosion resistance 276
cast lead coating 339, 461
corrosion resistance
to lead melts 322
to type metal melts 363
homogeneous lead coating 57

Jointing 468 (*see also* Soldered joints; Soldering; Welding)
cable sheathing, lead-oxygen system 420
flanged joints 469
folded joints, roofing 467
socket pipe 469

Kick law of similitude 392
Kroll-Betterton bismuth extraction process 8

Lagerhartblei 150
Latent heat 15—16
of evaporation 15
of fusion 15
of sublimation 16
Lead, *see also* Chemical lead; Electrolytic lead; High-purity lead; Parkes lead; Pattinson lead; Rolled lead; Secondary lead; Sheet lead; Sintered lead alloys; Smelter lead; Soft lead
acoustic properties 20
anodic polarization 296
atomic radii 29
atomic weight 9
crystal structure 10
drossing, inhibition by additions 328
types 5—9
Lead anodes 298
Lead baths, heat treatment 44, 321
Lead bullion 8
Lead bricks 8
Lead carbonate
as corrosion product 302—304
as protective film 292, 294, 302, 305
Lead chamber process 287
Lead chloride
as corrosion product 315
as protective film 307
in corrosion products 315
Lead chlorocarbonate, as protective film 307
Lead dioxide 71
as corrosion product 317
in storage battery plates 341, 352
Lead fluoride, as protective film 297
Lead hydrides, in distilled water 302
Lead hydroxide 302—303
Lead monoxide 326 (*see also* Drossing; Lead-oxygen systems; Litharge)
as protective film 294
corrosion resistance to sulphuric acid 289
formation 294
Lead oxychloride, as protective film 307

Lead patenting, pans 322
Lead salts
solubility 268—270
in distilled water 303
in organic acids 292
Lead scrap, *see also* Secondary lead
solubility for oxygen 72
Lead sheet, *see* Sheet lead
Lead string 431
Lead sulphate
as corrosion product 276
as protective film 272—273, 307
solubility in sulphuric acid 273
in storage batteries 343, 347—348
Lead sulphide 74—75
as corrosion product 307, 311, 431
as protective film 307
Lead wool 5, 431
Lead-alkali metal systems 175, 322
Lead-alkaline earth metal systems 44, 175, 322
Lead-aluminium alloys 34—35
free-machining properties 34
sintered powders 450—452
wear 35
Lead-aluminium-bismuth alloys 120
Lead-aluminium-copper alloys 121
Lead-aluminium-magnesium alloys 121—123
Lead-aluminium-tin alloys 122—123
Lead-antimony alloys 28—29, 75—94
age hardening 83, 86—94, 266—267, 412, 414
arsenic effects 86, 440
at elevated temperatures 89
segregation 93
silver effects 88
arsenic additions 86, 355, 440
as hardener alloys 76
bending value 85, 262
cable sheathing 411—416
casting
for art purposes 238
properties 76, 79—82, 335
compressive strength 170
contact corrosion 319
corrosion resistance 276
to soils 310
to sulphuric acid 275, 282—284
to tap water 307
creep 231, 233, 238, 242—243, 354, 439, 451
crystal structure, coring 77, 414
deep drawing 85
drossing 328, 331
fatigue strength 250, 256, 260
rate of test effects 84, 190, 197, 200
grain growth 196
grain refinement 196
Hall effect 81
Hartblei 440
honeycomb structure 77
hot dip coatings 453
hot hardness 84
in chemical plant 339
master alloys 76
mechanical properties 82—94
melts
corrosive properties 322
fluidity 336
physical properties 79—82
pipes and tubes 448
pitting by chromium electrolyte 319
pressure die casting 356
production 76—79
recrystallization 89—92, 192
rolling, work of deformation 442—446
shear modulus 83
shot 339
sintered powders 451—452
softening 36, 83, 90—91
solid solutions
breakdown 84, 89
precipitates 77, 79
standard specifications 75
storage batteries 347—356
plates 345, 347, 355
strain rate 85
structure 76—79
sulphur additions 77
thermal conductivity 81
workability 84
Lead-antimony-tellurium alloys, corrosion resistance to sulphuric acid 283
Lead-antimony-tin alloys 28, 163—171, 358 (*see also* Type metals; White metal)
age hardening 169, 358
casting properties 335
compressive strength 170—171, 361, 384
electroplating on silver 465
fatigue strength 256
hardness 166—168

impact strength 171
mechanical properties 166—171
melts, fluidity 335
Mischzinn 470
pressure die casting 356
rolling, model tests 445
tensile strength 168—170
Lead-antimony-zinc alloys 171—173
Lead-arsenic alloys 29, 35—37
cable sheathing 417
casting, shrinkage 36
compressive strength 37
creep strength 240
grain size, deformation effects 196
hardness 36
homogenizing 36
shot 37
softening 36
superconductivity transition point 37
Lead-arsenic-antimony alloys 126—128
age hardening 86
creep strength 235
fatigue strength 261
shot 339
Lead-arsenic-copper alloys 123—124
Lead-arsenic-iron alloys 124—125
Lead-arsenic-nickel alloys 125
Lead-arsenic-sodium alloys 175
Lead-arsenic-tin alloys, hot dip coating 454
Lead-arsenic-zinc alloys 175
Lead-barium alloys 29, 38—40
age hardening 39
Lead-barium-calcium alloys 128—129
Frary alloys 128
heat treatment 128
radiation shielding 129
Lead-bismuth alloys 8, 28—29, 40—44
age hardening 42
cable sheathing 195
corrosion resistance 276
to sulphuric acid 44, 275, 278
crystal structure 40
fatigue strength 43
grain boundaries 42
mechanical properties 43
melts, corrosive properties 41
nuclear reactors 41
solders 44
storage battery plates 352
thermal conductivity 41
Lead-bismuth-antimony alloys 131—132
low-melting solders 131
Lead-bismuth-antimony-tin alloys, bearing metals 389
Lead-bismuth-cadmium alloys 129—131
Lead-bismuth-cadmium-tin alloys 129—131
bearing metals 388
Wood's metal 388
Lead-bismuth-cadmium-zinc alloys 176
Lead-bismuth-calcium alloys, heat treatment 48
Lead-bismuth-lithium alloys, cable sheathing 63
Lead-bismuth-tellurium alloys 134
Lead-bismuth-tin alloys 132—135
low-melting solders 134
melts, fluidity 335
Rose's metal 133, 389
wetting properties 389—390
Lead-bismuth-tin-zinc alloys 176
Lead-bismuth-zinc alloys 135—136
Lead-cadmium alloys 28—29, 50—53
corrosion resistance 276
creep strength, grain size effect 233
equilibrium segregation 234
fatigue strength 259
grain size, effect on creep strength 233
heat of mixing 50
honeycomb structure 51
solid solution
breakdown 51
precipitates 79
Lead-cadmium-antimony alloys 137—139
fatigue strength 261
storage battery plates 355
Lead-cadmium-magnesium alloys 137
Lead-cadmium-thallium alloys 141—142
Lead-cadmium-tin alloys 139—141
fatigue strength 261
low-melting solders 140
soft solders 140—141, 158
ternary eutectic 387—388
Lead-cadmium-tin-zinc alloys 176
Lead-cadmium-zinc alloys 142—143
Lead-calcium alloys 28—29, 44—50, 69—70
age hardening 47, 192, 267
as hardener alloys 46
cable sheathing 417
corrosion resistance to distilled water 303

creep 232, 235, 239—240, 354—355, 417
crystal structure, coring 44—45
drossing 329
fatigue strength 259, 262
grain size 195
hardness 36
heat treatment 48
homogenizing 36
honeycomb structure 45
melting 329
recrystallization 192
solid solution
precipitates 45—46
supersaturation 45, 69, 101
storage battery plates 354
Lead-calcium-copper alloys
cable sheathing 417
creep 417
fatigue strength 417
mechanical properties 417
Lead-calcium-lithium alloys
anodes 298
intercrystalline corrosion 272
Lead-calcium-sodium alloys 136—137
Lead-chromium alloys 29, 53
Lead-cobalt alloys 29, 53
hardening 264
Lead-copper alloys 8, 29, 54—56, 221, 233, 235, 240—241, 451
age hardening 264
cable sheathing 411, 416
corrosion resistance 273, 276
to sulphuric acid 275, 288
creep 220, 231, 235, 242—244
fatigue strength 257, 260
grain growth 194, 196
homogeneous coatings 458—459
in chemical plant 56
recrystallization 194, 196
sintered powders 450—452
stresses, temperature effects 458
structure, stability 56
Lead-copper-antimony alloys 149—151
crystal structure 150
de-coppering 150
Lagerhartblei 150
Lead-copper-nickel alloys 143—144
Lead-copper-oxygen system 144—147
Lead-copper-sulphur alloys 147—149
Lead-copper-tellurium alloys 153—154
Lead-copper-tin alloys 151—153
Lead-copper-zinc alloys 154—155
de-coppering 155
Lead-gallium alloys 29, 58
Lead-germanium alloys 29
Lead-gold alloys 29, 37—38
activation energy, grain growth 191
Lead-hydrogen system 58
Lead-indium alloys 29, 59—61
corrosion resistance 60
creep 233
electroplating on leaded bronzes 60
equilibrium segregation 234
solders 60
wetting of glass 60
Lead-indium-thallium alloys 175
Lead-indium-tin alloys 156—158
soft solders 158
Lead-iron alloys 56—57
corrosion resistance 276
free-machining properties 57
welding, cold pressure 57
Lead-lanthanum alloys 29
Lead-lithium alloys 29, 61—63
age hardening 62
heat of mixing 61
Lead-magnesium alloys 29, 63—66
age hardening 64
crystal structure 64
intercrystalline corrosion 65
shot 66
Lead-magnesium-antimony alloys 158—159
Lead-magnesium-tin alloys 159—160
Lead-manganese alloys 29, 66—67
Lead-mercury alloys 29, 58—59
corrosion resistance 276
creep rate 245
crystal structure 58
low -melting solders 59
Lead-mercury-sodium alloys 155—156
Lead-monel metal alloys, corrosion resistance to sulphuric acid 319
Lead-nickel alloys 69—70
anodic corrosion 69
corrosion resistance 273—276
to sulphuric acid 69, 275—276
homogeneous coatings 459
Lead-oxygen system 70—74
cable sheathing 424
welding 420
melting 326—334, 424
storage batteries 342—343

Lead-oxygen-sulphur system 175
Lead-palladium alloys 29
Lead-potassium alloys 29, 61
Lead-selenium alloys 29, 94
  crystal structure 94
Lead-silicon alloys 29, 94
  corrosion resistance to sulphuric acid 280
Lead-silver alloys 29, 32—34
  activation energy 191
  anodes 299
    casting 338
  corrosion resistance 273, 276
  creep 34, 242
  crystal structure 33
    coring 32
  electroplating 33
Lead-silver-aluminium alloys 109—110
Lead-silver-antimony alloys 113—114
Lead-silver-arsenic alloys, Tainton lead 298
Lead-silver-calcium alloys, anodes 298
Lead-silver-copper alloys 110—112
  creep 242
  embedding of hard constituents 112
  fatigue strength, rate of test effects 112
  hot hardness 112
  strain rate 112
Lead-silver-sulphur alloys 112—113
Lead-silver-tin alloys 114—116
  soft solders 141
Lead-silver-zinc alloys 116—120
  de-silverizing 116, 120
Lead-sodium alloys 29, 67—69
  age hardening 67
  as corrosion product 316
  crystal structure 67
  solid solutions, supersaturation 69
Lead-sodium-tin alloys 161
Lead-strontium alloys 29, 100
  crystal structure 100
Lead-sulphur alloys 29, 74—75
  crystal structure 75
Lead-sulphur-tin alloys 161—163
Lead-tellurium alloys 28—29, 101—104
  age hardening 103
  as hardener alloys 103
  cable sheathing 411, 417
  corrosion resistance 273, 276
    to chromic acid 301
    to phosphoric acid 292
    to sulphuric acid 274, 276, 290
  creep 231, 242—244
  crystal structure 102
  drossing 329
  fatigue strength 261—262, 459
  hardness 262
  homogeneous coatings 459
  recrystallization 103—104
  semiconducting properties 101
  solid solutions, supersaturation 101
  storage battery plates 352
  strain hardening 102
  thermal activation 180, 280
Lead-thallium alloys 29, 105—106
  anodes 299
  corrosion resistance 276
  crystal structure, coring 105
Lead-tin alloys 28—29, 95—100 (*see also* Soldered joints; Soldering; Solders)
  cable sheathing 412, 414
  casting, shrinkage 97
  coatings on copper 308
  compressive strength 100
  contact corrosion 319
  corrosion resistance
    to soils 310
    to sulphuric acid 276
  creep 219—233, 236, 239, 243, 474
  crystallization 97
  electroplating on steels 465
  fatigue strength 260, 414
  heat of mixing 95
  hot dip coatings 454
  melts 95
  organ pipes 100
  sheet, casting 338
  softening 98, 189
  solid solutions, precipitates 79, 96
Lead-tin-cadmium-antimony alloys, fatigue strength 261
Lead-tin-zinc alloys 173
Lead-titanium alloys 105
Lead-uranium alloys 106
Lead-zinc alloys 29, 107—109
  corrosion resistance 276
Lichtenberg bearing metal 388
Light oils, corrosive properties 298
Limestone, corrosive properties 311, 313
Linotype alloys, casting 361—362
Linotype line setting and casting machines 357
Lipowitz bearing metal 388

Liquid fuels, corrosive properties 297
Lubricating oils, corrosive properties 60
Lubrication
  boundary 375
  hydrodynamic 365
  lead coatings 374, 454
Lüders bands 246

Magnesium, pitting due to lead content 319
Magnesium chloride, corrosive properties 296
Magnetic properties 20
Marl, corrosive properties 311
Master alloys 76
Matrix bearing metal 389
Mechanical properties 47—50, 82—94, 166—171, 174—175, 197—206, 470—479
  grain size effects 199, 411
Melting 321—334
  crucible test 334
  diffusion effects 327, 329
  fluxes 363
  furnaces 321, 347
  lead-oxygen system 326—334, 424
  losses by burning off 322, 386
  pots, intercrystalline corrosion 323
  zone 9, 17
Melting point 13—14, 16
Melts 321 (*see also* Drossing; Melting; Oxidation; Re-melting)
  aluminium, lead additions 35
  commercial lead, corrosive properties 322
  lead 321—334
    corrosive attack 322—326
    crystal structure 10
    density 10
    solubility for cobalt 53
    solubility for oxygen 4, 71—73
    viscosity 18—19
    volume increase 12
  lead-antimony 322—336
  lead-antimony-tin, fluidity 335
  lead-bismuth, corrosive properties 41
  lead-bismuth-tin, fluidity 335
  lead-magnesium, electrochemical potential 65
  lead-tin, properties 95
  lead-zinc, corrosive attack on steels 323
  supercooling 76—77
  type metal
    oxidation 331, 364
    corrosive attack 363
Mercury
  addition
    for corrosion resistance 276
    to storage battery plates 352
Microsections, preparation 25—29
Mine winding gear cables 404
Mischzinn hardener alloy 470
Miscibility gap 29—30
Molten lead, *see* Melts, lead
Monotype letter setting and casting machines 357
Mortar, corrosive properties 294, 311
Moulds, spiral-chill 334, 361—363
Multi-alloy systems 109, 175

Necking 198
Neutron absorption 24, 41
Newton bearing metal 388
Nickel
  addition
    for corrosion resistance 273, 276
    to storage battery plates 352
    to type metals 360
    to white metal 383
  effects
    on grain refinement 197
    on homogeneous lead coatings 459
  in bearing metals 382
  pitting of lead 319
Nitrates, corrosive properties 310
Nitric acid, corrosive properties 287, 291
Nitrosylsulphuric acid, corrosive properties 74, 284, 287—289
Notch effect 246
Notch-impact test 205—206
Nuclear reactors, lead-bismuth alloys 41

Oils, corrosive properties 60, 298
Optical properties 21
Ore deposits 2—3
Organ pipes 97, 100, 339
Organic acids
  corrosive properties 60, 292—293, 298, 304, 318
  solubility for lead salts 292
Oxidation 268—271, 331, 364, *see also* Drossing
  activation energy 327

detection 327–329, 334
parabolic time law of thickness increase 327
rate 8, 268, 328–329, 364
scaling constant 328
Oxide
as cathode 277
as protective film 268, 306
Oxygen
effects on impact strength 74
solubility
in lead melts 41, 71–73
in lead scrap 72
in smelter lead 72
overpotential 299

Parkes lead, recrystallization diagrams 195
Parkes de-silverizing process 8, 116, 119–120, 155, 194
Passivation 276–279, 319
Pattinson lead 194–195
Pekrun bearing testing machine 378
Perchloric acid, corrosive properties 341
"Phenol" corrosion 273, 428
Phosphates, corrosive properties 295
Phosphoric acid, corrosive properties 292
Phosphorus, addition to soft solders 470
Physical properties 9–24, 79–82
Pickling, steel 453, 463
Pipes and tubes 5, 428–441 (*see also* Corrosion, underground; Water pipes)
alloys, standard specifications 439
blistering 439
coarse grain effects 435, 437
corrosion resistance 294, 296, 307, 311–313, 318
crack formation 257, 435, 437
creep 35, 412, 433, 436, 439–441
critical deformation 433
deep drawing 448
defects 431–439
etching 27
expansion 411, 437
extrusion 390–400, 428–432, 447–448
fatigue fracture 435
flanged joints 469
folded joints 453
frost risk 104, 437
gaseous inclusions 439
grain size 432, 435
homogeneous lead coating 456
hot dip lead coating 453
impact extrusion 447–448
internal pressure 411
jointing 469
linear structure 432
"phenol" corrosion 428
pitting 307–317
recrystallization 433–435
soldering 476
structure 431–439
tin coating 307, 431
underground corrosion 312
wall thickness 439–441
welding 476
Pirelli extrusion press 408
Pitting
by acids 371
by bi-metallic contact 319
by chromium electrolyte 319
by copper contact 319
by nickel contact 319
by nitric acid 288
by sea water 309
by soils 309–313
by stray currents 314–317
by sulphuric acid 288
by tap water 307
Planté storage battery plates 341–356
Plastic working 390–452 (*see also* Cable sheathing; Deep drawing; Extrusion; Pipes and tubes; Powder metallurgy; Wire drawing)
Plastic yielding, *see* Deformation
Plasticity 177, 369
Poisoning, *see* Toxicity
Polygonization 184, 209, 222
Porosity in lead coatings 453
Potassium chloride, corrosive properties 296
Potassium hydrogen fluoride, corrosive properties 297
Potassium sulphate, corrosive properties 296
Potential, electrochemical 65, 269, 271, 277, 316
Powder metallurgy 342, 387, 449–452
Precipitates
age hardening 266
oriented 51–52, 266

solid solution, *see* Solid solutions, precipitates
Pressure die casting 356—357
Pressure equalizing pads 224
Pre-strain, effect on creep rate 220
Printing 357—365 (*see* also Electrotyping; Linotype; Monotype; Sterotype alloys; Type metals)
backing metals 359
block moulds 358
matrices 358, 361
plates 358—361
Production statistics 1, 4—5
Protective films, *see* Corrosion
Purity, degree of 8

Quaternary systems 129—131, 176, 261, 388—389
Quicklime, corrosive properties 294

Radiation shielding 21—24, 129
homogeneous lead coatings 23, 454
neutron absorption 24
Rate of test
effects
on elongation 197, 200
on fatigue strength 84, 190, 197, 200
Rats, damage to lead products 317
Recovery of lead 103, 180, 184—186, 192, 222
Recrystallization 10, 25, 26, 184—197 (*see also* Grain boundaries)
age hardening effects 192
deformation effects 292
diagrams 195
grain size effects 190
effects
on creep rate 217
on extrusion 292
on fatigue strength 250
equilibrium segregation 187, 234
grain boundary migration 185—187
inhibition 89—92, 103—104, 190—197, 235, 435
nucleus formation 184—185
oriented relationships 188
polygonization 184, 209, 222
rate 187, 194—197
secondary 185
texture 183
threshold 190—194
Red lead 5, 71, 326, 342, 465
Refining 8, 71, 72
Relaxation tests 207, 220
Remelted lead, *see* Secondary lead
Remelting 363
Resistance, electrical 11, 19
Resistance to flow, *see* Flow resistance
Riffelblei lead string 431
Rolled lead and lead alloys, *see* also Sheet lead
age hardening 266
grain growth 197
solid solutions, breakdown 266
texture 182—183, 192
Rolling 441, 447
forward slip 442
friction 442, 445
cone compression tests 445
hardness 444
plane strain 441
resistance to deformation 444
speed, effects on spreading 442—445
thickness ratio 444—445
work of deformation 442—446
yield strength 444
Rolling mills 445
Röntgen unit of ionic damage 22
Roofing 467
Rose's metal 133, 389

Salt solutions, corrosive properties 295—297
Scaling constant 328
Scrap, *see* Lead scrap
Sea water, corrosive properties 304, 307—308, 404, 416
Sealing with lead 431, 468
Secondary lead, *see also* Lead scrap
composition 6
Segregation, single-phase 93
Self-diffusion 15—16
Semiconducting properties 101
Separation strength 199, 344, 474
Shear strength, soldered joints 12—13, 83, 474—475
Sheet lead 465—468 (*see also* Rolled lead)
building 465—468
cladding 465
folded joints 467
radiation shielding 468
roofing 467

soldering 476
tin-clad 447
uses in chemical plant 465, 468, 476
uses in sealing 468
Sheet lead-antimony alloy, pitting by chromium electrolyte 319
Sheet lead-tin alloy, casting 338
Sheet steel, pitting by lead plating 319
Sheet zinc, soldered joints 475—476
Shipbuilding 307
Shot 37, 66, 339
Shrinkage, casting 79, 334, 362
Silicates, corrosive properties 310
Silver
addition
for corrosion resistance 273, 276
for recrystallization inhibition 191
to lead-antimony alloys 77, 355
to storage battery acids 356
to storage battery plates 352
to type metals 364
white metal electroplating 465
Single crystals 15 (*see also* Crystal structure)
creep stress 227, 242
grain boundary slip 227
migration 268
plasticity 177
Sintered powders 450—452
Slip, *see* Crystal structure
Slip line field 399
Smelter lead
production 5—9
solubility for oxygen 72
Sockets, cable 338
Sodium bicarbonate, corrosive properties 295
Sodium carbonate, corrosive properties 295
Soft lead, composition 6, 36, 83, 90—91, 98
Softening
effects on creep rate 213
torsion tests 189
Soils
corrosive properties 294, 308—317, 454
sandy, corrosive properties 311
Soldered joints 473—479
creep 476—479
fatigue strength 477, 479
shear strength 474—475
tensile strength 477
Soldering 470—472
alloy formation 471
copper 470, 472, 474, 477
electrolytic 471
flame 470
fluxes 390, 453, 471
friction 472
gap 472, 474—475
hard 472
soft, fluxes 390, 453, 471
ultrasonic 472
working temperature 475
Solders
contact corrosion 319
food cans 471
glass-to-metal 44
high-lead 470—471
lead, separation strength 474
lead-bismuth 44
lead-bismuth-antimony 131
lead-bismuth-tin 134, 389—390
lead-cadmium-tin 140—141, 158
lead-indium 60
lead-indium-tin 158
lead-mercury 59
lead-silver-tin 116
lead-tin 319, 470—479
low-melting 59—60, 131, 134, 140
separation strength 474
soft 60, 116, 141, 158, 470—479
aluminium as impurity 470
additions 470
creep strength 475
mechanical properties 470—479
separation strength 475
standard specifications 470
toxicity 471
wetting properties 389—390, 465, 471—472
tin, extrusion 429
tin-economy 470—471, 477
wettability 471
Solid solutions
extrusions, breakdown 266
precipitates 45—46, 51—52, 77, 79, 92, 96, 263, 266
supersaturation 45, 69, 101
Solidification 29—32, 79, 334—337, 362—363
Sommerfeld number 368, 377
Specific heat 15, 335

Sprayed lead coatings 461–463
Spreading, rolling 441–445
Stacking fault energy 10, 181
Standard specifications
  bearing metals 381, 387
  cable sheathing alloys 414, 416
  corrosion rate, measurement 272
  creep rate 218
  lead-antimony alloys 75
  pipes and tubes, alloys 439
  soft solders 470
  storage batteries 343
  type metal 358, 360, 363
Starter batteries, *see* Storage batteries
Statistics 3–5
  cable sheathing defects 426–427
Steels
  anodic corrosion 453
  cable sheathing 401
  cast, lead coatings 461
  cast lead coating 339
  cathodic protection, anodes 291, 301
  corrosion resistance 322, 366
  free-machining properties 57
  homogeneous lead coating 456
  lead-tin electroplating 465
  pitting by lead plating 319
  tin coating 456
Stereotype alloys 358, 360
Storage batteries 341–356
  acids
    additions 355–356
    concentration 347
  capacity 343, 347
  charging 354
  classification 343
  defects 350–351
  discharge 343, 355
  grid pastes 341–343
    antimony additions 353
    lead powders 342
    separation strength 344
  grids 341–356
    additions 352–353
    corrosion resistance 348, 352, 355
    creep 354
    lead-oxygen system 342–343
  plates 37, 53, 75, 341–356
    anodic corrosion 341, 348
    arsenic additions 355
    bismuth additions 352
    box type 341
    casting 344–345
    cobalt additions 352
    cores 344
    corrosion resistance 341, 348, 352, 356
    durability 347–356
    frame 341
    grain growth 350
    large area 341, 344
    lead dioxide formation 341, 352
    production 344–347
    properties 344–347
    reinforcement 341, 344
    tubular 341
  production statistics 5
  self-discharge 352–353
  standard specifications 343
  starter 343–344, 353, 356
  toxicity 357
Strain hardening 102, 179–180, 189–190, 197
  dynamic recovery 180
  effects on creep rate 213
  slip bands 247
  torsion test 189
Strain rate 85
  temperature effects 244
Stray currents, *see also* Corrosion
  pitting effects 314–317
Stress and strain, *see also* Corrosion
  alternating strain 252
  effects
    on corrosion 289
    on creep 212–216, 222–229, 231
  frequency, effects on fatigue strength 248
  multi-axial stress 202, 222
  plane strain 398, 400, 441
  reference stress/reference elongation ratio 224
  slip line field 399
  strain rate 85, 112, 190, 197, 200
  stress-strain diagrams 201–202
  triaxial stress 200, 202
Sulphate solutions
  corrosive properties 296
  electrolysis 299–300
Sulphite pulp equipment, corrosion 292
Sulphur
  addition to lead-antimony alloys 77
  corrosive properties 297
  trace in type metals 75, 166–167, 365

Sulphuric acid, corrosive properties 44, 69, 74, 272—290, 319, 341
additions effects 273—284
corrosion fatigue 284—287
stress corrosion 284—287
Sulphurous acids, corrosive properties 292
Superconductivity 13, 19, 37
Supersaturation, *see* Solid solutions
Superstructures 33, 265
Surface energy 16
Surface tension 16—17, 333, 472

Tainton lead 298
Tantalum, corrosion resistance to lead melts 326
Tap water
corrosive properties 304—307
hardness 305
Tar and tar products, corrosive properties 314
Tellurium
addition
for corrosion resistance 373, 376
for recrystallization inhibition 192, 435
to storage battery plates 352
Temperature
effects
on casting shrinkage 79
on creep rate 209, 212—216, 241—242
on creep stress 241
on compressive strength 205
on elastic modulus 13
on elongation at fracture 205
on fatigue strength 253
on grain refinement 197
on hardness 204
on homogeneous lead coatings 458—459
on impact strength 205—206
on strain rate 244
on tensile strength 241
on vapour pressure 14
variation, effects on stress and strain 458
Tensile strength 168—170, 197—201, 447, 477
hardness relationship 204
temperature effects 241
testing 197
Tensile-compression testing machine 418
Ternary alloys 109—176
Ternary eutectic
bearing metals 165—166, 382
lead-cadmium-tin alloys 387—388
type metals 358
Terne plate 454—455
Tetraethyl lead 6, 68
Texture 182—186
castings 336
foil 182
recrystallization 183
rolled lead 182—183, 192
wire 182
Thallium
addition
for corrosion resistance 276
to soft solders 470
Thermal expansion 11—12, 375
Thermoelectric power 81
Tin
addition
for recrystallization inhibition 192
to lead coating bath 453
to storage battery plates 352
effects
on drossing 328
on grain refinement 196
Tin pest 97
Tinsel 447
Toothing theory 375
Torsion tests 189
Town gas, corrosive properties 297
Toxicity 14, 304, 339—340, 357, 471
Traps 431
Tubes, *see* Pipes and tubes
Tungsten, corrosion resistance to lead melts 326
Type line setting and casting machines 357
Type metals 357—365 (*see also* Lead-antimony-tin alloys)
additions 360, 364
age hardening 361
casting 357, 361—365
composition 6
compressive strength 361
etching 29
hand setting 360
matrices 361
melts, corrosive properties 363

monotype alloys 360, 364
oxidation 331, 364
remelting 363
solidification 362—363
standard specifications 358, 360, 363
sulphur traces 75, 166—167, 365
ternary eutectic 358
Type setting machines 357
Types, cast, defects 363
Typograph line setting and casting machines 357
Types of lead 5—6

Ulco metal 128—129

Vacancies 11
Vacuum evaporated lead coatings 453
Vaporization, *see* Evaporation
Vapour pressure 14
Vibration
damping 13, 249, 468
Viscosity 18—19, 80, 321
Vogelpohlband 377—378

Water 301—308 (*see also* Brines; Corrosion, underwater; Distilled water; Sea water; Tap water)
solubility for lead salts 270
Water pipes 37, 75, 219, 301—308, 428—441 (*see also* Pipes and tubes)
Wear 365—369, 372, 375, 379
Welding 468—470
cable sheathing 406, 409, 420, 423
cold pressure 57, 199
pipes and tubes 476
seams, contact corrosion 320
sheet lead 476
testing 469
Wetting
bearing running surfaces 379
by solders 389—390, 465, 471—472
cold pressure welding 57
glass 60
White metal 29, 371, 374, 380—387 (*see also* Bearing metals; Lead-antimony alloys)
additions 383
casting 384
composition 380—387
drossing 383
electroplating on silver 465
fatigue strength 370
hardness 383
properties 380—387
structure 380—387
Wire, drawing 431, 454
texture 182
Wood, corrosive properties 293
Wöhler curves 245, 248, 252—253, 255
Wood's metal 388
Workability 84, 200
Wrought alloys, creep strength 236

X-ray examination 10—11, 77, 183, 235

Yield point 13, 197, 245, 450

Zinc
addition
for corrosion resistance 276
to storage battery plates 352
to type metals 364
to white metal 383
coalescence with lead 107
de-silvering 8, 119—120, 155, 194
electrolysis, anodes 298—299
extraction from lead 8
lead coatings 461
lead-clad 465
pitting by lead contact 319
soldered joints 475—476
Zone melting, 9, 17

GPSR Compliance

The European Union's (EU) General Product Safety Regulation (GPSR) is a set of rules that requires consumer products to be safe and our obligations to ensure this.

If you have any concerns about our products, you can contact us on ProductSafety@springernature.com

In case Publisher is established outside the EU, the EU authorized representative is:

Springer Nature Customer Service Center GmbH
Europaplatz 3
69115 Heidelberg, Germany

Batch number: 10704711

Printed by Printforce, the Netherlands